Waves in Complex Media

This book offers a clear and interdisciplinary introduction to the structural and scattering properties of complex photonic media, focusing on deterministic aperiodic structures and their conceptual roots in geometry and number theory. It integrates important results and recent developments into a coherent and physically consistent story, balanced between mathematical designs, scattering and optical theories, and engineering device applications. The book includes discussions of emerging device applications in metamaterials and nano-optics technology. Both academia and industry will find the book of interest as it develops the underlying physical and mathematical background in partnership with engineering applications, providing a perspective on both fundamental optical sciences and photonic device technology. Emphasizing the comprehension of physical concepts and their engineering implications over the more formal developments, this is an essential introduction to the stimulating and fast-growing field of aperiodic optics and complex photonics.

Luca Dal Negro is Full Professor in the Department of Electrical and Computer Engineering, a member of the Boston University Photonics Center, and he holds joint appointments in the Department of Physics and the Division of Materials Science at Boston University. He received his Ph.D. in Physics from the University of Trento, Italy. Professor Dal Negro is a fellow of the Optical Society of America (OSA). His research interests concern the behavior of electromagnetic fields in nanostructured materials and the engineering of complex photonic media.

Waves in Complex Media

LUCA DAL NEGRO

Boston University

CAMBRIDGE
UNIVERSITY PRESS

University Printing House, Cambridge CB2 8BS, United Kingdom

One Liberty Plaza, 20th Floor, New York, NY 10006, USA

477 Williamstown Road, Port Melbourne, VIC 3207, Australia

314–321, 3rd Floor, Plot 3, Splendor Forum, Jasola District Centre,
New Delhi – 110025, India

103 Penang Road, #05–06/07, Visioncrest Commercial, Singapore 238467

Cambridge University Press is part of the University of Cambridge.

It furthers the University's mission by disseminating knowledge in the pursuit of
education, learning, and research at the highest international levels of excellence.

www.cambridge.org
Information on this title: www.cambridge.org/9781107037502
DOI: 10.1017/9781139775328

First published 2022

Printed in the United Kingdom by TJ Books Limited, Padstow Cornwall

A catalogue record for this publication is available from the British Library.

Library of Congress Cataloging-in-Publication Data
Names: Dal Negro, Luca, author.
Title: Waves in complex media : fundamentals and device applications /
    Luca Dal Negro.
Description: New York : Cambridge university Press, 2021. |
    Includes bibliographical references and index.
Identifiers: LCCN 2021026903 (print) | LCCN 2021026904 (ebook) |
    ISBN 9781107037502 (hardback) | ISBN 9781139775328 (epub)
Subjects: LCSH: Electromagnetic waves. | Electromagnetic waves–Scattering. |
    BISAC: TECHNOLOGY & ENGINEERING / Electronics / Optoelectronics
Classification: LCC QC670 .D34 2021 (print) | LCC QC670 (ebook) |
    DDC 530.14/1–dc23
LC record available at https://lccn.loc.gov/2021026903
LC ebook record available at https://lccn.loc.gov/2021026904

ISBN 978-1-107-03750-2 Hardback

Dedicated to my parents, Roberto and Paola,
Wonder begins with your Love.

Contents

Appendix E   The Multifractal Formalism   631

Appendix F   Aperiodic Arrays and Spectral Graph Theory   634

Appendix G   Essentials of Fractional Calculus   639
           G.1   Fractional Differential Equations   642
           G.2   Fundamental Solutions of Fractional Diffusion   644

Appendix H   Mie–Lorenz Field Components   648

           References   650
           Index   690

Preface

Why should we learn about the behavior of waves in optical media with irregular, nonperiodic, and even disordered structures? First of all, because they can be found everywhere around us, from complex functional materials to the arrangement of leaves on plant stems and even at the inner core of number theory! Second, because the scattering behavior of waves in complex media surprises us with emergent phenomena driven by interference effects. Third, because waves in complex media unveil profound analogies between the classical and quantum transport regimes beyond standard diffusion theory, for example, Anderson light localization. However, while waves in periodic structures have been deeply investigated for more than a century and are discussed by the majority of graduate-level textbooks on optics and photonics, the behavior of waves in more general aperiodic environments is almost exclusively addressed in the specialized literature and consequently has limited impact on graduate curricula. It is my goal to bridge this gap by offering a comprehensive and interdisciplinary textbook that systematically addresses both the conceptual foundation and the scattering properties of optical waves in complex media. This task is made even more compelling by the readily available fabrication techniques that currently enable the precise realization of optical structures using a range of materials (metallic, dielectric, linear, and nonlinear) with nanoscale features and almost arbitrary morphology. In this context, the robust understanding of waves in complex media can enable not only the advancement of optical science but also the engineering of more efficient optical devices for sensing, imaging, energy conversion, spectroscopy, optical communications, and signal processing.

This book proposes a comprehensive introduction to the structural and scattering properties of complex photonic media with particular focus on deterministic aperiodic structures and their conceptual roots in geometry and number theory. Today, the concept of aperiodic order plays an ever-increasing role in several fields of pure and applied sciences that include mathematics, physics, chemistry, biology, economics, finance, and engineering. Structurally complex nonperiodic media can be efficiently generated using precise algorithmic rules that interpolate in a tunable fashion between random and periodic behavior, offering unique opportunities to engineer optical media with tailored light-matter interaction, electromagnetic transport, and novel light-scattering phenomena. The structural complexity of such media profoundly influences their linear and nonlinear optical properties, giving rise to functionalities uniquely enabled by the engineering of aperiodic order. An impressive wealth of research activities has been recently carried out under the broad umbrella of complex optics and photonics, making it impossible to cover all aspects. As a result, this book represents a special viewpoint that focuses on the fundamental aspects of the single and multiple scattering regimes of two-dimensional and three-dimensional optical media.

This textbook is intended to teach students at the graduate level or advanced undergraduate level about the fundamental principles and engineering device applications of wave diffraction and scattering in structurally complex optical environments. The book is of interest to both academia and industry as it develops the underlying physical and mathematical background in partnership engineering device applications, providing a perspective on both fundamental optical sciences and photonic device technology. A very rich bibliography is also provided, enabling further readings and more in-depth investigations. This book emphasizes the comprehension of physical concepts and their engineering implications over the more formal developments, thus aiming to attract graduate students and researchers with varying backgrounds to the stimulating and fast-growing field of aperiodic optics and complex photonics.

Acknowledgments

I wish to express my thanks for the input received from various colleagues, researchers, and graduate students. In particular, students in my classes and in my research group at Boston University have helped me in the past few years with their challenging questions, careful readings of parts of this book, suggestions, and concrete inputs that have improved the clarity of my explanations and eliminated mistakes. I am sincerely indebted to Carlo Forestiere, Antonio Capretti, Fabrizio Sgrignuoli, Ren Wang, Yuyao Chen, Sean Gorsky, and Wesley Britton. It was also a great pleasure to discuss various topics of the book with Professor Felipe Pinheiro, Professor Salvatore Torquato, Professor Nader Engheta, Professor Vito Mocella, Professor Jacopo Bertolotti, Professor Marcel Filoche, Professor Diederik Wiersma, Professor Enrique Maciá-Barber, and Professor Uwe Grimm. Additionally, I am grateful to a large number of colleagues at Boston University and elsewhere, whose names cannot all be possibly listed here, with whom I shared in the past few years my enthusiasm for optics and complex photonic structures. A special thank you goes to my wife, Laura. The blessings of her unwavering support cannot possibly be overstated. Finally, I would like to thank the team at Cambridge University Press, particularly Julie Lancashire, Nicholas Gibbons, Samuel Fearnley and Julia Ford, whose invaluable assistance, patience, and support during this long and challenging project have been truly exceptional.

1 Introduction

Everything that is not done with utter devotion falls into oblivion and, in fact, does not deserve to be remembered.

Physics and Beyond, Werner Heisenberg

In his delightful small book *An Introduction to Mathematics*, published in 1911, the English mathematician and philosopher Alfred North Whitehead reflected on the role of periodicity in Nature and remarked [1], "The whole life of Nature is dominated by the existence of periodic events, that is, by the existence of successive events so analogous to each other that, without any straining of language, they may be termed recurrences of the same event." He further elaborates, "The presupposition of periodicity is indeed fundamental to our very conception of life. We cannot imagine a course of nature in which, as events progressed, we should be unable to say: 'This has happened before'." It is indeed the case that many natural phenomena manifest *periodic order*. It is enough to consider the rotation of the Earth, the yearly recurrence of seasons, the phases of the Moon, and even the cycles of our bodily life, such as the beatings of our heart or the recurrence of our breathing, as Whitehead observed in his book.

Most likely providing an evolutionary advantage, humans developed the remarkable ability to easily recognize periodic patterns even in the presence of substantial noise and perturbations. In contrast, we perform very poorly at perceiving randomness or at recognizing different degrees of nonperiodic behavior. For instance, when people are asked to assess by inspection the randomness of different binary sequences, it is found that their perceptions simply reflect the inability to efficiently encode them [2]. Experimental psychology has shown that the binary sequences recognized as random invariably contain more alternations between zeros and ones than what are expected for random processes where long runs of zeros or ones that counter the common intuition are occasionally exhibited. Moreover, studies performed by asking people to select integers at random in a given interval demonstrate a systematic bias in favor of smaller numbers, a fact known in cognitive psychology as the "mental compression of large numbers" [3]. Given how deeply rooted in our cognitive structures the idea of periodic order appears to be, it comes as no surprise that such periodic behavior is considered emblematic of structural order.

However, the idea of order without periodicity, or *aperiodic order*, recently began to be fully appreciated thanks to important discoveries in the mathematical and physical sciences. In particular, during the 1920s Harald Bohr developed a general theory of almost periodic functions that describes the generic quasiperiodic behavior manifested, for instance, by the Lissajous figures for certain choices of parameters. These curves can exhibit intricate patterns arising from the composition of two orthogonal harmonic vibrations, represented as follows:

$$x = A \sin(\omega_1 t + \phi_1) \tag{1.1}$$

$$y = B \sin(\omega_2 t + \phi_2). \tag{1.2}$$

When time progresses, the point P with coordinates (x, y) traces a curve whose shape depends on the choice of the six parameters introduced above. For example, when $\omega_1 = \omega_2$ and the phase difference $\phi_2 - \phi_1 = \pi/2$, the point P traces an ellipse with axes along the x- and y-directions. For an arbitrary phase difference, the ellipse will be tilted and its shape and orientation will vary as a function of the phase difference, switching from a circle (in the case $A = B$) to the straight lines with equations $y = \pm x$, and vice versa. A more complex behavior is obtained when the angular frequencies are different ($\omega_1 \neq \omega_2$). However, as long as their ratio ω_1/ω_2 can be expressed using rational numbers, the corresponding curves eventually repeat themselves, giving rise to an elaborate but periodic motion. Examples are shown in Figure 1.1a–c where we consider frequency ratios of 1, 5/3, and 13/8, respectively. In this example, we also set equal amplitudes ($A = B$) and $\pi/2$ phase difference between the two components.

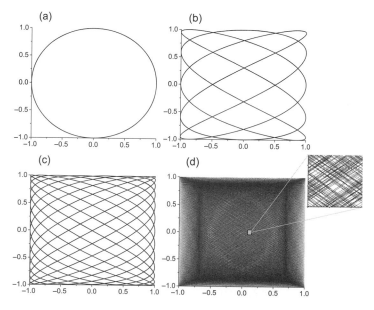

Figure 1.1 Lissajous figures with $A = B$, $\phi_2 - \phi_1 = \pi/2$, and a frequency ratio equal to (a) 1, (b) 5/3, (c) 13/8, and (d) the golden mean τ. The inset in (d) shows a close-up region around the origin of the unit square, highlighting the complex threading of the aperiodic trajectory of point P.

However, the situation changes dramatically when ω_1/ω_2 is irrational, i.e., for incommensurate frequencies. In this case, as shown in Figure 1.1d for ω_1/ω_2 approximately equal to the golden mean $\tau = (1 + \sqrt{5})/2 = 1.618034\ldots$, the trajectory of point P never retraces itself and densely fills the unit square while coming arbitrarily close to every point in a nonperiodic fashion. Since there are many more irrational numbers than rational ones, a random choice for the ratio ω_1/ω_2 will most likely produce nonperiodic trajectories, though with a perfectly deterministic behavior. The simple situation that we have described is typical of the more general quasiperiodic dynamics manifested by conservative systems characterized by a finite number of incommensurate frequencies in their spectra. For instance, as originally recognized by Poincaré at the end of the nineteenth century, the general orbits of the three-body problem of celestial mechanics in which three bodies attract each other gravitationally as in the Earth–Moon–Sun system, are indeed quasiperiodic [4]. Moreover, a fundamental result of dynamics known as the Kolmogorov–Arnold–Moser (KAM) theorem states that such quasiperiodic motions are also stable under small perturbations and therefore represent the norm rather than the exceptional cases. Clearly, this is a remarkable shift from the "presupposition of periodicity." In addition, important milestones in the study of aperiodic order originated from the geometrical theory of tilings or tessellations, which are sets of planar figures that perfectly fill the plane without leaving any gaps. In particular, the mathematician Roger Penrose[1] discovered in 1974 the existence of two simple polygonal shapes that tile, following deterministic prescriptions, the infinite Euclidean plane without any spatial periodicity. Shortly thereafter, generalizations to the three-dimensional (3D) space were provided by Robert Ammann, who produced a pair of rhombohedra that perfectly fill the 3D space without periodicity.

The mathematical models of aperiodic tilings and the theory of quasiperiodic functions found physical applications in the early 1980s in the study of quasicrystals, which are the prototypical examples of perfectly ordered structures without periodicity. Dan Shechtman was awarded the 2011 Nobel Prize in Chemistry for his pioneering work on the electron diffraction of aluminum–manganese alloys that provided the first experimental demonstration of synthetic materials with quasicrystal structures. In 2009, the first known naturally occurring quasicrystal mineral, called icosahedrite, was discovered after a 10-year-long search by an international team of scientists led by Luca Bindi and Paul J. Steinhardt [5].

Today, the phenomenon of aperiodic order plays an important role in several fields of mathematics, biology, chemistry, physics, economics, finance, and engineering where deterministic chaotic systems with long-term unpredictable, though not random, behavior are widespread. For instance, the connection between aperiodic order and unpredictability, which is central to number theory, resulted in numerous engineering applications to cryptography and coding theory. In this context, cryptographically secure deterministic pseudorandom generators (DPRG) have been

---

[1] Roger Penrose gave numerous pioneering contributions to mathematical physics, particularly in the areas of general relativity and cosmology, and was awarded the Nobel Prize in Physics in 2020 "for the discovery that black hole formation is a robust prediction of the general theory of relativity."

developed that produce numerical sequences featuring statistical randomness (no recognizable patterns or regularities) and exhibit spectral properties that are virtually indistinguishable from those of uncorrelated random noise. Moreover, generating deterministic structures such as automatic sequences by aperiodic substitutions or exploiting number-theoretic concepts such as Galois fields, algebraic rings, or elliptic curves offers novel opportunities for research areas ranging from acoustics diffusers to radar imaging (stealthy surfaces), remote sensing, and photonics technology.

It should be clear from this introductory discussion that a sharp dichotomy between periodic order and random behavior is untenable in view of the recent developments in the field of deterministic aperiodic order leading to the demonstration of novel structures that interpolate in a tunable fashion between periodic crystals and disordered media. Moreover, such a divide has limited our ability to explore and engineer the novel behavior of waves in complex aperiodic media with varying degrees of structural organization, locally defined symmetries, and distinctively rich spatial correlation properties that result in novel wave transport and localization phenomena. Examples discussed in this book include amorphous photonic and hyperuniform media [6], quasicrystals, fractal systems, structures based on automatic sequences [7], and number-theoretic concepts that display inherent chaotic and "random-like" Poisson behavior [8–10].

1.1   Aperiodically Modulated Oscillator Chains

It is remarkable that many important features of the complex behavior of waves in aperiodic systems can already be understood quite generally by considering the vibration modes of one-dimensional systems of harmonic oscillators. Although such "toy models" do not describe actual systems, they quantitatively illustrate the fundamental physical differences in the wave dynamics between periodic and aperiodic structures for both classical and quantum particles.

The simplest dynamical model is the *modulated spring model* consisting of a linear chain of identical particles with nearest-neighbor harmonic interactions described by spring constants that vary along the chain. The spring constants effectively represent atomic interactions in a general material. The relevant equations of motions are given by

$$\frac{d^2 u_n}{dt^2} = -\kappa_{n+1}(u_n - u_{n+1}) - \kappa_n(u_n - u_{n-1}), \tag{1.3}$$

where u_n represents the small displacement of a unit mass at position n from its equilibrium. We first discuss the case where spring constants vary according to the following modulation function:

$$\kappa_n = 1 - \epsilon \cos\left(\frac{2\pi a}{b}n + \phi\right) = 1 - \epsilon \cos\left(\frac{2\pi}{N}n + \phi\right). \tag{1.4}$$

The period of the modulation function is $N = b/a$, where a and b are integers, ϕ is the phase shift of the modulation with respect to the chain, and ϵ is the modulation amplitude. The spectrum of the normal modes of vibration is simply obtained by considering the frequency-domain solutions $u_n \sim \tilde{u}_n \exp(-j\omega t)$, where $\tilde{u}_n$ is the spatial envelope of the mode and j denotes the imaginary unit. Substituting into (1.3), we obtain the following eigenvalue problem:

$$\omega^2 \begin{bmatrix} \tilde{u}_1 \\ \tilde{u}_2 \\ \tilde{u}_3 \\ \vdots \\ \tilde{u}_n \end{bmatrix} = \begin{bmatrix} \kappa_1 + \kappa_2 & -\kappa_2 & & & \\ -\kappa_2 & \kappa_2 + \kappa_3 & -\kappa_3 & & \\ & -\kappa_3 & \ddots & \ddots & \\ & & \ddots & \ddots & -\kappa_{n-1} \\ & & & -\kappa_{n-1} & \kappa_{n-1} + \kappa_n \end{bmatrix} \begin{bmatrix} \tilde{u}_1 \\ \tilde{u}_2 \\ \tilde{u}_3 \\ \vdots \\ \tilde{u}_n \end{bmatrix} \quad (1.5)$$

This problem is defined by a tridiagonal matrix and therefore can be efficiently solved numerically to obtain the spectrum of frequency eigenvalues and the eigenvectors corresponding to the spatial envelopes of the modes at each frequency.

A complete equivalence exists between the modulated spring model and the one in which the masses of the particles are modulated instead, providing a link with the traditional approach to lattice waves in periodic lattices [11, 12]. In particular, when the unit cell of the chain contains N different spring constants or masses ($b = Na$), the modulation function takes on N distinct values, i.e., $\kappa_{n+N} = \kappa_n$ and the solutions of the system (1.5) consist in N frequency branches and $N - 1$ spectral gaps that are opened by the periodic modulation. For example, when $b = 2a$ there will be one gap in the spectrum analogously to the case of a diatomic chain with two distinct masses in the unit cell. Furthermore, when $b = a$ we recover the spectrum of the monoatomic chain, in which there are no gaps. In addition, for a periodic modulation with $b = Na$ the spectral gaps will have maximal width $\Delta\omega \sim \epsilon$ for $\phi = 0$ and will be minimal for $\phi = \pi/N$. In the special case $b = 2a$, the gap will vanish for $\phi = \pi/2$. Therefore, the number of spectral gaps is determined by the factor $2\pi/N$ that appears in the argument of the modulation function. When $N = b/a$ is not an integer, the number of gaps can be found by expressing it as a rational fraction with the smallest denominator. This gives rise to a very large number of gaps when $b \approx a$. Indeed, the number of gaps fluctuates wildly with the values of N for which the oscillation spectrum is computed. Importantly, when $N = b/a$ cannot be expressed as the ratio of two integers, there will be no lattice periodicity in the chain because the spring constants have a period that is incommensurate with the one of the chain. In this case, the equations (1.3) form an infinite set of coupled equations. Since any irrational number can be approximated by rational numbers arbitrarily well, the gap structure can be understood by considering the limit behavior of the spectra for successive rational approximations of N in continued fractions.

A striking consequence of the approximation theory of numbers in rational fractions is that the spectral gaps of the harmonic chain arrange in complex hierarchical structures when we sweep the values $1/N$. As an example of this general phenomenon,

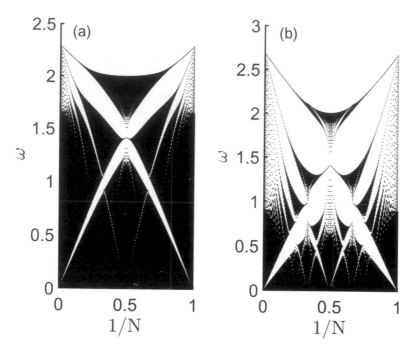

Figure 1.2 Spectrum of the modulated spring model as a function of the modulation period (a) at small modulation amplitude $\epsilon = 0.3$ and (b) at larger modulation amplitude $\epsilon = 0.8$. We considered $\phi = 0$ and 1,000 unit masses.

we show in Figure 1.2 the calculated frequency spectra of the modulated spring model as a function of the inverse period for two different values of the modulation amplitude. The white areas denote the spectral gap regions, while the black ones correspond to transmission states. We recognize that the distribution of largest gap regions is very smooth as a function of $1/N$. In addition, narrower gap regions form visible "branches" that diagonally cross and merge in very complex patterns. Note in particular how the two main "diagonal branches" cross at $1/N = 0.5$, indicating the presence of only one spectral gap, corresponding to the diatomic chain. Additional small gaps can be observed branching off the main diagonal regions with a characteristic self-similar organization that becomes more evident by increasing the modulation parameter ϵ, as shown in Figure 1.2b. Here the same "butterfly-like pattern" appears at multiple scales all over the diagram. Interestingly, this striking recursive structure is also exhibited by the Hofstadter model for the dynamics of a Bloch electron in a crystal under an external magnetic field [13], which in fact has the same mathematical structure of the modulated spring model introduced here [12]. In both cases, when N tends to infinity the total length of the bands goes to zero, producing a "pure-point" spectrum of discrete points.

The transport and spatial localization properties of the different modes depend dramatically on their locations in the spectrum. For instance, transmission eigenstates bounded by large spectral gaps correspond to the extended modes of the system.

In contrast, the very narrow transmission bands embedded in the fragmented spectral regions that are visible in the upper part of the frequency range contain peculiar eigenstates with wildly fluctuating envelopes and an increased localization character. These modes exemplify the characteristic behavior of the *critical modes* that appear in structures with quasiperiodic spring–mass modulations, i.e., obtained when a and b are incommensurate. This scenario corresponds to the Aubry–André or Harper tight-binding model that describes the lattice dynamics of one-dimensional systems in the presence of a gauge field and supports a universal critical behavior characterized by a fractal spectrum with critically localized modes [13–15].

In general, the frequency spectra of incommensurate and quasiperiodic structures are completely determined by the continued fraction expansions of their irrational periods, which produce a self-similar arrangement of spectral gaps forming a Cantor set fractal. A striking example of this intriguing behavior is provided by one-dimensional structures with quasiperiodic modulations that follow the binary Fibonacci sequence. In the context of the modulated spring model, we consider the *Fibonacci chain* defined by the modulation function:

$$
\kappa_n = \mathrm{sgn}\left[\cos\left(\frac{2\pi n}{\tau} + \phi\right) - \cos\left(\frac{\pi}{\tau}\right)\right].
\tag{1.6}
$$

In the preceding expression, τ is the golden mean and sgn denotes the sign opera-tor that makes the modulation binary with $\kappa_n = \pm 1$ values of the spring constant distributed according to the binary Fibonacci sequence. We also remark that by vary-ing the phase angle ϕ we obtain equivalent representations of the Fibonacci chain. Interestingly, the resulting structures flip abruptly when continuously varying ϕ due to the effect of the sgn operation, creating *phason flips*. Moreover, this simple model manifests a complex scenario with interesting topological properties that have been recently discussed in a number of papers [16–18].

In Figure 1.3, we show the integrated density of states (IDOS) of the Fibonacci chain, which is defined as follows:

$$
\Omega(\omega) = \int_0^\omega \rho(\omega')d\omega',
\tag{1.7}
$$

where we introduced the *vibrational density of states* that is obtained from the distribution of eigenfrequencies ω_n, as follows:

$$
\rho(\omega) = \sum_n \delta(\omega - \omega_n).
\tag{1.8}
$$

The plateau regions in the normalized IDOS correspond to the spectral gaps of the Fibonacci chain that are distributed in frequency according to a universal fractal staircase function derived from a Cantor set. Remarkably, the self-similarity of the Fibonacci spectrum is largely independent of the specific physical nature of the

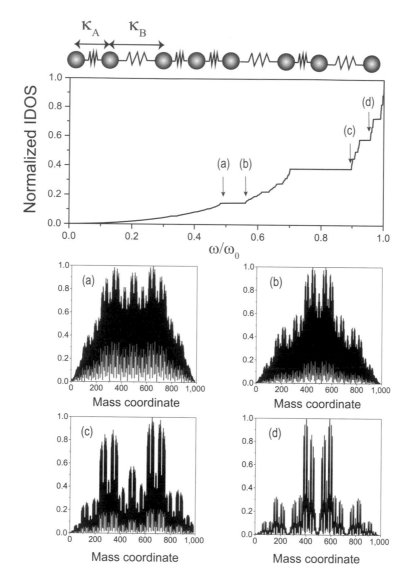

Figure 1.3 Top: integrated density of states of a Fibonacci harmonic chain of 978 masses. Panels (a–d) show the spatial profiles of the selected vibration modes at the frequency positions indicated by the corresponding arrows in the top panel.

system and only reflects the self-similar process of continued fraction expansion of the incommensurate modulation period of the chain:

$$\tau = 1 + \cfrac{1}{1 + \cfrac{1}{1 + \cfrac{1}{1 + \ddots}}} \tag{1.9}$$

In particular, each level of fractional approximation of the golden mean, i.e., each approximant value, is responsible for the splitting of one frequency band into two sub-bands with a width proportional to the corresponding approximant. However, the actual width of each fractal sub-gap will also depend on the actual modulation amplitude. We note that in the low-frequency regime, the normalized IDOS is almost identical to the one of a periodic chain and the corresponding modes are extended. On the other hand, in the high-frequency regime, the IDOS becomes self-similar with a dense set of gaps and critical modes that exhibit increasing spatial localization. In Figure 1.3, we plot four representative (normalized) vibration modes obtained by the direct diagonalization of the harmonic spring model with Fibonacci modulation. These eigenvectors are critical band-edge modes characterized by distinctive envelope fluctuations with multifractal scaling properties, as demonstrated using wavelet analysis in the next section. The fractal dimensions of the critical modes and of the frequency spectra of 1D quasiperiodic structures determine the asymptotic scaling of the width of a spreading wave packet, resulting in anomalous diffusion properties [19–22].

The peculiar hierarchical structure of critical modes can be quite generally understood on the basis of the *Conway theorem* [23]. This result states that every finite region of characteristic length L in a quasicrystal admits at least one identical replica within a distance $\sim L$. As a result, the frequent and regular reappearances of identical local environments along the Fibonacci chain give rise to tunneling of exponentially localized wave packets from one cluster to its nearby copies in the structure, forming the characteristic critical mode envelopes displayed in Figure 1.3. The distinctive spatial decay of the critical modes in quasicrystal of size L is characterized by considering the scaling with respect to L of the number $N(\epsilon)$ of sites in the structure where $|u_n|^2$ exceeds a small threshold ϵ [11]. We note that for periodic structures $N(\epsilon) \sim L$ because they support extended modes with roughly equal intensity at all sites. On the other hand, $N(\epsilon)$ is bounded by a constant in random structures supporting localized states, since their intensity will vanish almost everywhere except for a few sites. In contrast, for critical states we have that $N(\epsilon) \sim L^\alpha$ with $0 < \alpha < 1$. According to this criterion, we say that critical modes display a power-law spatial decay that is intermediate between extended and exponentially localized states.

1.2 A Brief History of Aperiodic Optics

The optical properties of periodic and disordered random media have been deeply investigated for centuries, leading to a number of spectacular physical effects that unveil deep analogies between the behavior of quantum and electromagnetic waves [24–27]. The first scientific analysis of waves on a periodic lattice of one-dimensional oscillators, i.e., the spring–mass model, can be traced back to Newton in his attempt to derive a mathematical expression for the velocity of sound. A general theory of wave propagation in periodic structures can already be found in the work of Brillouin [28]. On the other hand, the radiative transfer equation (RTE) for the propagation of light in randomly structured (turbid) media consisting of isotropically scattering centers was first introduced in 1887 by Eugen von Lommel [29], and subsequently generalized to include arbitrarily polarized waves by Chandrasekhar [30]. However, the radiometry of randomly structured media has a long and fascinating history that dates back to the pioneering works of Pierre Bouguer and Johann Lambert around the middle of the eighteenth century [29]. In contrast, the propagation of optical waves in strongly scattering aperiodic media is a rather recent field of study. In fact, it was not until the pioneering work of Anderson on quantum and classical wave scattering in disordered media that profound questions on the nature of wave transport in aperiodic media began to be directly addressed [31, 32]. One of the first phenomena to be investigated in this context was the coherent backscattering of light [33–35], or *weak localization*, which is a direct manifestation of wave interference in photon transport. During the 1980s, intense research activities focused on the search for Anderson light localization [32, 36], where wave propagation in a random medium is dominated by interference effects and light transport can be completely suppressed.

In 1984, Dan Shechtman et al. [37] demonstrated experimentally the existence of physical structures with noncrystallographic rotational symmetries. In their study of the electron diffraction spectra from certain metallic alloys (Al_6Mn), they discovered sharp diffraction peaks arranged according to the icosahedral point group symmetry. The sharpness of the measured diffraction peaks, which is a measure of the coherence of the spatial interference, turned out to be entirely comparable with what typically observed in ordinary periodic crystals. Intrigued by these findings, Dov Levine and Paul Steinhardt promptly formulated the notion of quasicrystal in a seminal paper titled [38]: "Quasicrystals: a new class of ordered structures." It then became apparent that icosahedral structures can be obtained by projecting periodic crystals from a higher-dimensional (six-dimensional in this case) superspace, which became a central idea in geometrical diffraction theory and quasicrystallography [11, 39, 40].

1.2.1   One-Dimensional Aperiodic Structures

The first study of electron transport across a one-dimensional (1D) quasiperiodic Fibonacci structure was performed by Merlin et al. [41] in 1985. This work stimulated a flurry of experimental and theoretical work on the propagation of electron waves through quasiperiodic media. Largely stimulated by these results, the interest in quasiperiodic photonic structures began in 1987 with the first investigation of 1D dielectric multilayers arranged in a Fibonacci sequence [42]. Such structures display a very complex transmission spectrum characterized by a hierarchical organization of pseudogap regions and support resonant states with fractal scaling properties that are similar to what were previously discussed for the modulated spring model. In fact, photonic quasicrystals of the Fibonacci type exhibit an energy spectrum that consists of a self-similar Cantor set with zero Lebesgue measure, implying the presence of narrow transmission pseudogaps with vanishingly small widths in the limit of infinite-size systems.

The transmission spectrum of transverse electric (TE) waves through a photonic multilayer medium constructed by arranging two types of layers according to a prescribed deterministic sequence can be conveniently calculated using the standard transfer matrix technique [43]. In this method, the amplitudes of the electric and magnetic field vectors on the two sides of a dielectric layer with thickness d_i and refractive index n_i are related through the transfer matrix:

$$\begin{bmatrix} E_\ell \\ B_\ell \end{bmatrix} = \mathbf{M}(d_i) \begin{bmatrix} E_r \\ B_r \end{bmatrix} = \begin{bmatrix} \cos \delta_i & -jp_i^{-1} \sin \delta_i \\ -jp_i \sin \delta_i & \cos \delta_i \end{bmatrix} \begin{bmatrix} E_r \\ B_r \end{bmatrix}, \tag{1.10}$$

where the indices ℓ and r refer to the left and right interfaces of the layer, $\delta_i = k_0 n_i d_i \cos \theta_i$ is the phase change of layer i when the light makes an angle θ_i with the normal to the interface, and $p_i = n_i/c_0 \cos \theta_i$ (k_0 is the wave vector and c_0 the speed of light in vacuum). By imposing the continuity conditions of the electric and magnetic fields, we can construct the transfer matrix $\mathbf{M}_N$ for a system with N layers by the matrix multiplication:

$$\mathbf{M}_N = \prod_{i=1}^{N} \mathbf{M}(d_i) = \begin{bmatrix} M_{11} & M_{12} \\ M_{21} & M_{22} \end{bmatrix}. \tag{1.11}$$

If p_0 and p_m are the values of p_j for the incident and transmitted waves, respectively, then the transmissivity T can be expressed as [43]:

$$T = \frac{4p_0 p_m}{p_0^2 M_{11}^2 + p_m^2 M_{22}^2 + (p_0 p_m)^2 |M_{12}|^2 + |M_{21}|^2 + 2p_0 p_m}. \tag{1.12}$$

For a 1D multilayer system, it is also possible to compute the electric field distribution inside the structure. In order to achieve this, we first need to calculate the reflectivity coefficient of the whole structure, after which the incident and the total reflected light can be used as boundary conditions to obtain the field amplitude of all layers inside the structure. The electric field on the kth interface can be expressed as follows [44]:

$$E(z^{(k)}, \omega) = [1 + r(\omega)]m_{22}^{(k)} - p_0[1 - r(\omega)]m_{12}^{(k)}, \tag{1.13}$$

where $m_{ij}^{(k)}$ are the elements of the transfer matrix from the first to the k-th interface.

It can be shown that the recursive structure of the Fibonacci numbers given by $F_{j+1} = F_j + F_{j-1}$ for $j \geq 1$ with $F_0 = F_1 = 1$ gives rise to a corresponding recursion of the transfer matrices:

$$\mathbf{M}_j = \mathbf{M}_{j-2}\mathbf{M}_{j-1}, \tag{1.14}$$

where $\mathbf{M}_j$ is the transfer matrix of a Fibonacci structure with F_j layers [42]. Equation (1.14) has been shown to be equivalent to the *dynamical map*:

$$x_{j+1} = 2x_j x_{j-1} - x_{j-2}, \tag{1.15}$$

where $x_j = (\mathrm{Tr}\,\mathbf{M}_j)/2$ denotes half the trace of the transfer matrix and the allowed frequencies in the spectrum generally satisfy the constraint $-1 \leq x_j \leq 1$ [45, 46]. The dynamical map has a constant of motion that quantifies the strength of the

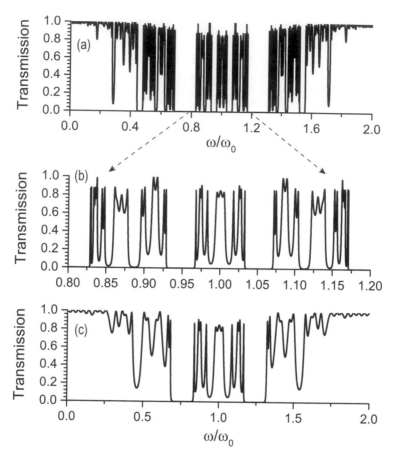

Figure 1.4 Scaling of Fibonacci spectra. (a) Optical transmission spectrum of a Fibonacci multilayer structure with 233 layers as a function of the normalized frequency where $\omega_0 = 2\pi c_0/\lambda_0$; and (b) enlarged view of the spectrum to around the scaling region $\omega/\omega_0 \approx 1$. (c) Optical transmission spectrum of a Fibonacci multilayer structure with 55 layers. The A and B symbols denote two dielectric layers with refractive indices $n_A = 2$ and $n_B = 3$ and optical thickness determined by the $\lambda/4$ Bragg condition.

effect of quasiperiodicity and is maximized at the quarter-wavelength layer condition $(n_i d_i = \lambda_0/4)$ [42, 45]. Moreover, under such condition the cyclic character of the Fibonacci trace map implies that the transmissivity $T(S_j)$ of the Fibonacci structure S_j repeats every three generations of the sequence, i.e., $T(S_{j+3}) = T(S_j)$, and exhibits a self-similar behavior around $\delta = \pi/2$, as shown in Figure 1.4 [42, 47]. The self-similarity of the transmission spectrum of Fibonacci dielectric multilayers was first demonstrated by Gellerman et al. [47]. The fractal nature of the energy and of the transmission spectra of Fibonacci photonic quasicrystals drastically modifies the dynamics of optical wave packets, resulting in largely tunable anomalous transport properties [19, 20, 22]. Moreover, the critical modes of photonic quasicrystals exhibit multiscale intensity fluctuations described by multifractal geometry, similarly to the

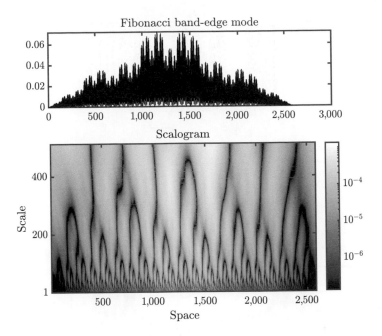

Figure 1.5 Wavelet-based scalogram of the first band-edge mode (shown in the top panel) of an optical Fibonacci multilayer (1,597 layers). The self-similar nature of the critical mode oscillations is displayed clearly by the scalogram (shown in the bottom panel).

wave functions of random media at the Anderson localization transition [48–50]. This is visibly demonstrated by the self-similar structure of the wavelet scalogram of the first band-edge optical mode in a Fibonacci quasicrystal, as shown in Figure 1.5. Wavelet-based techniques for self-similarity detection and multifractal analysis are introduced in Chapter 9. Here we remark the impressive similarity between the optical mode shown in the top panel of Figure 1.5, which is the mode of an optical system, and the one shown in Figure 1.3, which is a vibrational band-edge mode of the harmonic Fibonacci chain. Despite the very different physical nature of the two systems, their critical mode patterns are virtually identical, demonstrating the primary role played by the Fibonacci quasiperiodic modulation. This fact generalizes to other 1D aperiodic structures as well, such as, for instance, the ones with the binary Thue–Morse modulation defined recursively by the following:

$$\epsilon_0 = 0, \quad \epsilon_{2n} = \epsilon_n, \quad \epsilon_{2n+1} = 1 - \epsilon_n. \tag{1.16}$$

The preceding equations generate an infinite binary string of digits that never repeat. The same string can alternatively be generated by the simple substitution rule: $0 \rightarrow 01$, $1 \rightarrow 10$, and several other characterizations exist as well [51]. Interestingly, the Thue–Morse sequence remains invariant upon the decimation of every other digits, and therefore it is not only aperiodic but also self-similar.

After the breakthrough discovery of quasicrystals and the fabrication of Fibonacci as well as Thue–Morse semiconductor heterostructures [41, 52], it became clear that

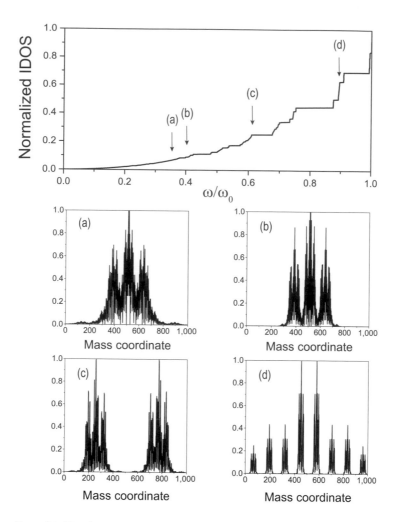

Figure 1.6 Top: integrated density of states of a Thue–Morse harmonic chain of 1,024 masses. Panels (a–d) show the spatial profiles of the selected vibration modes at the frequency positions indicated by the corresponding arrows in the top panel.

physical systems could be engineered with singular-continuous energy spectra not previously encountered in any natural system. Figure 1.6 summarizes the relevant properties of the Thue–Morse harmonic chain. A fractal distribution of gaps similar to the case of the Fibonacci chain appears in the IDOS of the Thue–Morse structure and the critically localized modes shown in Figure 1.6a–d demonstrate a characteristic triplication pattern. Similar results are obtained for optical multilayers as well [53–56].

1.2.2   Energy Spectra and Nonlinear Dynamics

Seminal work on the nature of the electronic and optical spectra of Fibonacci structures and the properties of the corresponding eigenmodes was performed by Kohmoto, who established an exact isomorphism between the 1D Schrödinger equation with aperiodic electron potentials and the Helmholtz equation for classical waves [42, 45, 57]. In these papers, a transfer matrix method was utilized that enabled the exact treatment of 1D scattering problems for electronic and optical excitations on the same footing. An alternative approach was also introduced by Kohmoto, Kadanoff, and Tang [58, 59] that exploits the recursion relations of the transfer matrices to define an iterative map for their traces and antitraces, leading to the efficient calculation of the energy spectra of large-scale 1D aperiodic structures. The resulting formalism, known as the *trace map* approach, describes the physical properties of deterministic aperiodic structures by analyzing the orbits of corresponding discrete dynamical systems [60]. The study of the trace map of aperiodic structures enables the computation of their spectrum without calculating the full product of matrices, thus significantly reducing the complexity of the problem. Considering the nonlinear nature of these dynamical systems, Kohmoto and Oono [61] were able to discover the Smale horseshoe mechanism, which is considered the hallmark of chaos in dynamical systems theory, in the trace map of Fibonacci multilayers. Wave propagation through Cantor-set fractal media and its relation to chaos theory is discussed in the recent paper by Esaki et. al. [62]. In particular, it was demonstrated that the phase-space trajectories of the trace map and their basins of attraction in the two-dimensional trace space (or a higher-dimensional) space provide universal information on the nature of resonant states and transmission spectra in quasiperiodic structures. This very elegant method has been also extended to more complex aperiodic systems, such as the Thue–Morse structure and its generalizations [51] as well as the generalized Fibonacci structures (GFS). The GFS are defined following the inflation scheme [63]:

$$S_{j+1} = S_j^m S_{j-1}^n, \qquad (1.17)$$

where, for instance, $S_0 = B$ and $S_1 = A$, m and n are integers, and S_j^m denotes m adjacent repetitions of the string S_j. This inflation scheme is equivalent to the substitution rule $A \rightarrow A^m B^n$, $B \rightarrow A$, where A^m is a string of m A's. The total number of A's and B's is the string S_j is equal to the *generalized Fibonacci number* F_j defined by the recurrence relation $F_j = m F_{j-1} + n F_{j-2}$ with $F_0 = F_1 = 1$. In these structures, the ratio of A's and B's for $j \rightarrow \infty$ is $\tau' = \sigma/n$, where

$$\sigma = \lim_{j \to \infty} \frac{F_j}{F_{j-1}} = \frac{1}{2}[m + \sqrt{m^2 + 4n}]. \qquad (1.18)$$

When $m = n = 1$, we obtain the standard Fibonacci sequence and τ' reduces to the golden mean τ. However, when $m = 2$ and $n = 1$, we obtain the so-called *silver mean sequence* with $\sigma = 1 + \sqrt{2}$; when $m = 3$ and $n = 1$, we obtain the *bronze mean sequence* with $\sigma = (3 + \sqrt{13})/2$; when $m = 1$ and $n = 2$, we have the *copper mean sequence* with $\sigma = 2$; and finally, when $m = 1$ and $n = 3$, we obtain the *nickel mean sequence* with $\sigma = (1 + \sqrt{13})/2$. Similarly, the Thue–Morse sequence can be generalized following the substitution rule $A \rightarrow A^m B^n$, $B \rightarrow B^n A^m$. In this case, the ratio of A and B letters in the sequence is equal to m/n. Differently from the Fibonacci case, these systems are characterized by two-dimensional, nonlinear, and area nonpreserving trace maps and exhibit a degree of aperiodicity intermediate between quasiperiodic and random systems.

In order to illustrate the main ideas behind the trace map method we consider the trace map of the standard Thue–Morse sequence [51]:

$$x_{j+1} = y_j \tag{1.19}$$

$$y_{j+1} = 4x_j^2(y_j - 1) + 1 \tag{1.20}$$

and the one of the copper mean sequence [46]:

$$x_{j+1} = y_j \tag{1.21}$$

$$y_{j+1} = (4x_j^2 - 2)y_j + \gamma, \tag{1.22}$$

where γ is a parameter. Depending on the regions of trace space, these maps may exhibit bounded or unbounded orbits, periodic points, and attractors with very complex and even fractal shapes. However, these maps provide access to properties of the corresponding systems that are completely independent of the nature of the transfer matrices associated to the binary blocks A and B in the structures. In fact, the energy spectra of any 1D physical system specified by unimodular transfer matrices, such as an electronic or an optical multilayer structure, can be accurately computed simply considering the dynamical properties of the maps. This is so because the allowed energy bands are always characterized by stable orbits and fixed points. Therefore, to find the spectra of a given model it will suffice to plot the curve generated by the initial values of the traces (x_1, y_1) as a function of energy (or optical phase β) and look for intersections with various regions of the *fractal basins of attraction* of the trace map. Given the chaotic nature of the fractal basins of attraction,[2] the transmission spectra of aperiodic 1D structures generally display very irregular distributions of peaks and gaps with remarkable scaling properties. Following this general method, one can predict exactly the transmission spectra without ever iterating the trace map. In Figure 1.7, we show the transmission spectra of dielectric multilayer structures corresponding to (a) the Thue–Morse sequence, (b) the copper mean sequence, and (c) the silver mean sequence. A complex structure of nested gaps and transmission states populate the frequency space of these remarkable structures.

We now illustrate how the trace map approach can be used the determine the transmission spectra of the Thue–Morse and the copper mean optical multilayer structures assuming for simplicity that $\beta_j = \beta$ for all j. In the case of the Thue–Morse multilayer, the initial conditions for the trace map can be readily obtained from the optical transfer matrices and the trajectory of the point $(x_1(\beta), y_1(\beta))$ in the (x, y) trace space defines a parabolic curve. The special values of the optical phase β that correspond to transmission peaks are then found by finding the intersections of such parabola of initial trace values with the subset of the basin of attraction of the point $(1, 1)$, for the given number of generations of the predecessors of the $(1, 1)$ point [51].

---

[2] The basin of attraction of the map for a given stable point is the set of points in trace space that are mapped to that point upon iteration.

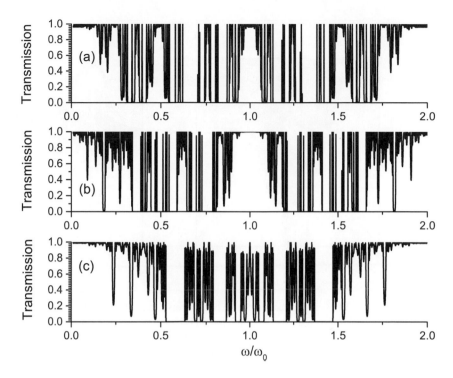

Figure 1.7 Transmission spectra of optical multilayer structures with aperiodic distribution of layers according to the (a) Thue–Morse sequence, (b) the copper mean sequence, and (c) the silver mean sequence. The A and B symbols denote two dielectric layers with refractive indices $n_A = 2$ and $n_B = 3$ and optical thickness determined by the $\lambda/4$ Bragg condition.

The case for 12 generations of the predecessors is shown in Figure 1.8a, where the invariant parabola is indicated in gray color. The outstanding complexity and chaotic behavior of the basin of attraction is evident in the figure. On the other hand, in order to find the transmission peaks of copper mean structures, we need to compute the basin of attraction of the $y_\ell = 0$ line in trace space for the corresponding trace map [46]. We then proceed as in the Thue–Morse case and look for the intersections of the basin of attraction with the invariant curve of initial conditions. In the case of the copper mean map, the invariant curve is a Lissajous figure with the parametric representation $\{\cos t, \cos(2t - \alpha)\}$ with $\cos \alpha = \gamma$ and $t \in (0, 2\pi)$. The chaotic attractor of the copper mean map is shown in Figure 1.8b for the first eight generations of predecessors of $y_\ell = 0$ and the invariant Lissajous curve is highlighted in gray.

The dynamical trace map approach was also applied to investigate the properties of arbitrary substitutional sequences by Kolář and Nori [64]. The rich physical behavior of critical states, including the presence of extended fractal wavefunctions at the band-edge energy of the Fibonacci spectrum, and their relation to optical transport are analytically studied by Kohmoto and Maciá-Barber using rigorous discrete tight binding method and a transfer matrix renormalization technique, respectively [59, 65]. The existence of critical modes in quasiperiodic Fibonacci systems was

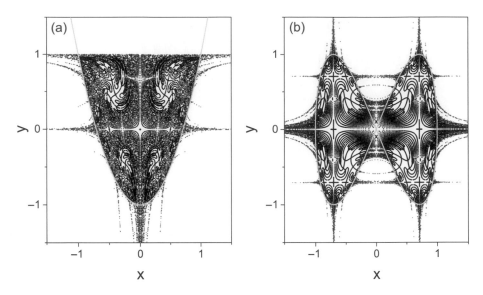

Figure 1.8 (a) Basin of attraction of the Thue–Morse map (black crosses). The first 12 generations of the predecessors of the point $(1, 1)$ and the invariant parabola (gray curve). (b) Basin of attraction of the copper-mean map with $\gamma = 0$ (black crosses). The first eight generations of the predecessors of the $y_\ell = 0$ line and (gray curve) the invariant Lissajous figure.

demonstrated experimentally by Desideri and coworkers [66] who investigated the propagation of Rayleigh surface acoustic waves on a quasiperiodically corrugated solid. Characteristic spatial patterns with remarkable scaling features were shown from optical diffraction experiments [67]. The complex photonic dispersion of optical waves transmitted through a Fibonacci multilayer structure was demonstrated experimentally by Hattori and coworkers [68] while Dal Negro and collaborators used ultrashort pulse interferometry to demonstrate strongly suppressed group velocity at the Fibonacci band-edge states [44]. Recently, the optical properties of multilayered structures based on the Thue–Morse sequence, characterized by a singular continuous Fourier spectrum, have also attracted considerable attention [51, 53, 54, 69]. These structures are characterized by a self-similar hierarchy of optical gaps that scale according to a triplication pattern of symmetry-induced perfect transmission states [53], as shown in Figure 1.6 for the harmonic Thue–Morse chain.

In applied optics and photonics, aperiodic optical media generated by deterministic mathematical rules attract significant interest due to their simplicity of design, fabrication, and full compatibility with current materials deposition and device technologies. In particular, deterministic aperiodic nanostructures (DANS) generated by substitution rules and number-theoretic functions have been recently demonstrated, and technological applications have been found in different areas of nanophotonics and plasmonics in relation to improved optical sensing, spectroscopy, broadband light sources, lasers, and nonlinear optical elements [70–74].

1.3   Scope of This Book

The study of waves in complex media deals with a broad spectrum of electromagnetic scattering phenomena that occur in structurally aperiodic environments composed of resonant optical materials, with applications to both photonic technology and fundamental optical science. Also known as *complex photonics*, this fascinating research field combines electromagnetic scattering and diffraction theory in dispersive materials with the mathematical characterization and design of structures that lack periodicity and instead manifest aperiodic spatial correlations [75]. The present book provides a multidisciplinary and comprehensive introduction to the single and multiple scattering properties of electromagnetic waves in complex media, with emphasis on the foundational aspects of aperiodic order that play an important role in applied and basic optical science. This graduate-level textbook introduces the wave physics and engineering of both disordered and deterministic aperiodic media that interpolate in between periodic crystals and random structures. Besides providing a self-consistent discussion of the analytical framework of single and multiple light scattering theory, this book attempts to integrate a vast number of recent results and developments into a coherent story that balances electromagnetic physics, engineering design concepts, mathematical ideas, and advanced device applications to metamaterial, photonic, and plasmonic technologies.

In order to achieve these goals, the book systematically introduces and develops all the concepts and techniques that are essential to understand the complex behavior of light waves in aperiodic and random media without requiring prior knowledge on the subjects beyond what is normally acquired in the undergraduate physics and engineering curricula. The wide spectrum of selected topics is developed to position graduate students in the conditions to deepen their understanding of physical phenomena in strong partnership with semi-analytical design and modeling techniques for waves in complex media. The combination of conceptual aspects and rigorous design of complex media will allow students to directly engage with the more specialized literature on the subject. Key concepts and results of single scattering and multiple scattering theories of scalar and vector waves, which are essential tools for every researcher, are introduced and developed at the graduate level. The emphasis is on analytical formulations and general results, since they provide physically transparent insights into the complexities of aperiodic optics and serve as rigorous starting points for many numerical techniques utilized to study complex optical media. Moreover, our interdisciplinary detours into several mathematical aspects of the subject in combination with extensive references to device applications should also appeal to practitioners in both industry and academia who may approach complex wave phenomena from different disciplines, including mathematics, physics, astronomy, engineering, and biology.

We begin by providing in Chapter 2 the background concepts of classical electromagnetic field theory. In particular, we start from Maxwell's equations in free space and discuss fundamental theoretical concepts, such as the radiation potentials and the connection between the Green function and the density of states in optical systems. We then introduce macroscopic fields and the macroscopic Maxwell's equations needed for the classical description of the optical behavior of materials. We then proceed to discuss different field representations and wave equations in materials. The dyadic formalism is introduced in the context of a general radiation and scattering problem whose solution is provided by volume integral equations in terms of the dyadic Green function. The resonant response of dispersive materials is often a key element in the description of strongly scattering media, requiring the formulation of dispersion models introduced in Chapter 3. Here we rigorously analyze the behavior of waves in dispersive media and introduce polariton waves, which play a central role in current nano-optical technology. This leads us to discuss the effective optical response of heterogeneous complex media made of small scattering inclusions and introduce the idea of electromagnetic metamaterials. In Chapter 4, we study the propagation of optical wave packets through dispersive media and introduce the basic equations of pulse dynamics along with a powerful solution approach based on neural networks. We then proceed with a discussion of scalar diffraction theory including the modern formulation based on the angular spectrum representation of waves. This chapter also introduces an alternative formulation of paraxial light diffraction based on the fractional Fourier transform and concludes with a brief account of the Stratton–Chu integral formula in the context of vector diffraction theory. Design concepts and advanced engineering applications of optical diffraction are discussed in Chapter 5 within the elegant framework of linear systems theory. In particular, complex diffractive optical elements are presented with application to both space and time domains. Structural complexity in diffraction theory is addressed in the last two sections, where we introduce the general framework of catastrophe optics and provide several examples of engineered caustics. Finally, optical rogue waves in aperiodic diffractive media are introduced and the fascinating concepts of singular optics and meta-optics are briefly introduced. Chapter 6 introduces the general formalism of wave scattering theory for both perfectly and partially coherent fields. The basic concepts of the unified theory of coherence and polarization are summarized and their applications to the scattering of partially coherent radiation by deterministic and random media are provided.

The fundamental connection between wave diffraction and the kinematic scattering theory is addressed in Chapter 7, where we establish the physical and geometrical meaning of the structure factor for arbitrary arrays of scattering particles and provide numerous examples to radiation engineering using deterministic aperiodic structures. This chapter introduces the main spectral types of deterministic aperiodic structures in one spatial dimension and addresses their diffraction properties. Chapter 7 closes with a discussion of the resonant behavior of periodic arrays of nanoholes and nanoparticles that found with numerous device applications in relation to the extraordinary light transmission properties of periodically nanostructured metallic surfaces. In Chapter 8, we address the important topic of tailored disorder and introduce the statistical description of random media. The effects of long- and short-range spatial correlations are discussed starting from a brief introduction to optical speckles and correlated random walks. This chapter introduces the concepts of hyperuniform random and deterministic media based on engineered subrandomness and discusses their application to the design of weakly scattering and stealthy media.

The following three chapters provide an interdisciplinary journey through number theory and diffraction geometry to explore randomness and aperiodic order at its conceptual roots. In Chapter 9, we introduce arithmetic functions and analyze their distinctively multiscale aperiodic behavior using multifractal scaling analysis. We then focus on the Riemann zeta function and the aperiodic distribution of prime numbers, as well as surprising connections with chaotic systems and the theory of random matrices. Finally, we introduce the relevant notions of continued fractions and Diophantine approximation theory that often appear in the rigorous analysis of aperiodic order. Chapter 10 deals with quasiperiodicity and the Poisson behavior in number theory, and it exploits "arithmetic randomness" for the design of novel aperiodic point patterns. Aperiodic arrays based on prime numbers are then introduced along with the relevant concepts from algebraic number theory. The aperiodic structures of Galois fields and elliptic curves (over finite fields) are presented together with deterministic methods that efficiently generate weak and strong pseudorandomness. Chapter 10 ends with a broader discussion of the concept of randomness in the contexts of algorithmic information theory and statistical analysis. The Fourier spectral properties of aperiodic point patterns are discussed in Chapter 11 in the general context of diffraction geometry, and the main classes of aperiodic structures are discussed together with their mathematical diffraction properties. Quasicrystals and the approach of higher-dimensional quasicrystallography are introduced as well as additional types of aperiodic structures from phyllotaxis and number theory.

In Chapter 12, we return to wave scattering physics and introduce the powerful method of Green's function perturbation theory. Within this general framework, we derive the multiple scattering equations for scalar and vector electromagnetic waves and introduce the coupled dipole numerical technique for the design of scattering arrays of electric and magnetic dipoles. The dyadic Green's matrix method is introduced and applications to the multiple scattering physics of different types of deterministic aperiodic media are provided. In particular, we show how the analysis of the spectral properties of the dyadic Green's matrix provide invaluable information on the wave transport and localization properties of large aperiodic scattering arrays. The resonant scattering properties of particles beyond the dipole approximation are presented in Chapter 13 within the general approach of the Mie–Lorenz theory. Localized surface plasmon and surface phonon resonances in small nanoparticles are discussed and applications to plasmonics technology are highlighted. We also discuss light scattering from magnetic spheres and the application of the Mie–Lorenz theory to the design of three-dimensional metamaterials. Generalizations of the Mie–Lorenz scattering theory to two-dimensional and three-dimensional arrays are also addressed and examples provided in the context of aperiodic photonic bandgap structures and plasmonic molecules. The chapter ends with a brief account of the more general T-matrix null-field method for nonspherical particles.

The textbook concludes with Chapter 14, where we address the light transport and localization properties of continuous and discrete random media. The so-called microscopic or "first-principle" multiple scattering theory of random media is explained and the Dyson and Bethe–Salpeter transport equations derived under the assumption of Gaussian disorder. The effective medium solution is obtained and its applications to the design of metamaterials are addressed. This final chapter also deals with the detailed solutions for the intensity transport in a slab geometry and introduces key ideas behind mesoscopic scattering phenomena in strongly scattering random media. The final part of the chapter discusses the applications of random scattering media to laser devices, i.e., random lasers, with uncorrelated and correlated disorder within the framework of fractional diffusion.

2   Electromagnetics Background

There can be little doubt that the most significant event of the 19th century will be judged as Maxwell's discovery of the laws of electrodynamics.

Lectures on Physics, Richard P. Feynman

The study of electromagnetic waves and their interaction with materials, which paved the way to Einstein's relativity theory [76] and quantum mechanics [77], also underpins the many spectacular achievements of the electrical and optical technologies that drive the current information age [78, 79]. Maxwell's equations, together with the Lorentz force law, are the foundation of classical electrodynamics and describe how electric and magnetic fields are generated and modified by charges, currents, and by each other. They are named after James Clerk Maxwell, who published them between 1861 and 1862 [80]. Maxwell's theory not only unified the seemingly distinct realms of electricity and magnetism with optics, but it also established the primacy of the field concept in physics, originally recognized by Michael Faraday in his pioneering experimental research [81].

Starting from Maxwell's equations, this Chapter summarizes the fundamentals of classical electromagnetic theory that forms the necessary conceptual basis for this book. Our discussion is limited to the basic aspects of electromagnetic fields in materials and radiation phenomena. For more advanced treatments of these extensive topics, the readers are referred to standard textbooks on electromagnetism such as the books by Jackson [82] and Garg [83].

2.1   Maxwell's Equations

Maxwell's equations, together with the Lorentz force law, are the foundation of classical electrodynamics and describe how electric and magnetic fields are generated and modified by charges, currents, and by each other.

The Maxwell-Lorentz equations describe the *joint dynamics* of the electromagnetic field and of a set of charged particles in vacuum. Maxwell's equations relate the electric field $E(\mathbf{r},t)$ [V/m][1] and the magnetic flux density or magnetic induction

---

[1] The symbols in square brackets indicate the units in which the preceding quantities are measured.

$\boldsymbol{B}(\mathbf{r},t)$ [Wb/m^2] to the **total** charge density[2] $\rho_{tot}(\mathbf{r},t)$ [C/m^3] and current density $\boldsymbol{J}_{tot}(\mathbf{r},t)$ [A/m^2]. In international system (SI) units, Maxwell's equations are as follows:

$$\nabla \cdot \boldsymbol{E}(\mathbf{r},t) = \frac{\rho_{tot}(\mathbf{r},t)}{\epsilon_0} \tag{2.1}$$

$$\nabla \cdot \boldsymbol{B}(\mathbf{r},t) = 0 \tag{2.2}$$

$$\nabla \times \boldsymbol{E}(\mathbf{r},t) = -\frac{\partial \boldsymbol{B}(\mathbf{r},t)}{\partial t} \tag{2.3}$$

$$\nabla \times \boldsymbol{B}(\mathbf{r},t) = \epsilon_0 \mu_0 \frac{\partial \boldsymbol{E}(\mathbf{r},t)}{\partial t} + \mu_0 \boldsymbol{J}_{tot}(\mathbf{r},t) \tag{2.4}$$

In the preceding equations, $\epsilon_0 = 8.854 \times 10^{-12}$ [F/m] is the free-space electric permittivity, $\mu_0 = 4\pi \times 10^{-7}$ [H/m] is the free-space magnetic permeability, and the two constants are related to the speed of light in free space c_0 by the relation $\epsilon_0 \mu_0 = 1/c_0^2$.

2.2   What Do They Mean?

The divergence of a vector field at a given point measures the total outgoing flux (per unit volume) in the neighborhood of that point whereas the curl measures the total field rotation (per unit surface) around that point. Therefore, according to equation (2.1), the positive (negative) electric charge density at a given point acts as the source (sink) of the electric field. On the other hand, equation (2.2) tells us that there are no sources (or sinks) for the magnetic flux density. Equations (2.1) and (2.2) are known as the Gauss's electric and magnetic laws, respectively. Equation (2.3) is Faraday's induction law, which establishes how a time-varying magnetic flux generates an electric field with rotation. Similarly, equation (2.4), which is known as Ampère–Maxwell's law, specifies how a *time-varying electric field or a current generate a magnetic flux with rotation*. On a purely theoretical ground,[3] Maxwell modified the original Ampère's law by the addition in equation (2.4) of the term known as the *displacement current*. This term is responsible for the coupling of the magnetic flux density to the time-varying electric field, and it enables the wave propagation of the electromagnetic fields away from their sources.

---

[2] Remember that 1 Wb/m^2 = 1 Tesla = 10^4 Gauss. As a reference, the strength of the Earth's magnetic induction is about 0.5 Gauss.

[3] Maxwell's insight was to add an extra term to Ampère's law in order to make it consistent with the charge continuity equation in the dynamic regime.

By integrating Maxwell's equations over the volume V and using the divergence theorems of vector calculus, we can express them in integral form:

$$\int_{\partial V} \mathbf{E}(\mathbf{r},t) \cdot \hat{n} dA = \frac{1}{\epsilon_0} \int_V \rho_{tot}(\mathbf{r},t) d^3\mathbf{r} = Q_{tot} \tag{2.5}$$

$$\int_{\partial V} \mathbf{B}(\mathbf{r},t) \cdot \hat{n} dA = 0 \tag{2.6}$$

$$\int_{\partial S} \mathbf{E}(\mathbf{r},t) \cdot dl = -\int_{\partial V} \frac{\partial \mathbf{B}(\mathbf{r},t)}{\partial t} \cdot \hat{n} dA = -\frac{d}{dt} \Phi_B(t) \tag{2.7}$$

$$\int_{\partial S} \mathbf{B}(\mathbf{r},t) \cdot dl = \int_{\partial V} \left[\mathbf{J}_{tot} + \frac{\partial \mathbf{E}(\mathbf{r},t)}{\partial t} \right] \cdot \hat{n} dA = I_{tot} + \frac{d}{dt} \Phi_E(t) \tag{2.8}$$

In the preceding equations, dA denotes a surface element, $\hat{n}$ the unit vector normal to the surface (pointing outward), ∂V denotes the surface S of the volume V, and ∂S is the border of the surface S. We indicate by $\Phi_B(t)$ and $\Phi_E(t)$ the time-varying magnetic and electric fluxes across the surface S, respectively. The integral form of Gauss's law (2.5) connects the total charge Q_{tot} contained within volume V to the flux of the electric field through the closed surface ∂V that bounds the volume. Its magnetic counterpart, equation (2.6), shows that the magnetic flux across any closed surface is always zero. As a result, the magnetic flux density $\mathbf{B}(\mathbf{r},t)$ is a solenoidal, or divergence-free, vector field whose lines form *close loops*. Equation (2.7) expresses *Faraday's induction law* and connects the line integral of E around a closed loop, known as the *electromotive force*, to the variation of the flux of B through the area encircled by the loop. Similarly, Ampère–Maxwell's law (2.8) states that the integral of B around a loop is contributed by the total current passing through the loop, which is the original Ampère's law, plus the rate of variation of the flux of E through the area encircled by the loop.

2.3 The Lorentz Equation

Maxwell's equations, as stated previously, can only be solved provided the total charges $\rho_{tot}(\mathbf{r},t)$ and currents $\mathbf{J}_{tot}(\mathbf{r},t)$ in the system are exactly known. This unfortunately cannot be achieved inside materials, requiring a macroscopic formulation complemented by suitably chosen *constitutive relations* that model the coupling of the sources with the fields, as we will discuss in Section 2.4. However, Maxwell's equations (2.1)–(2.4) or (2.5)–(2.8) describe the behavior of the microscopic world made of electrons and nuclei, considered as point systems over length scales that are much larger than the actual nuclear dimensions of $\sim 10^{-14}$ m. Under these assumptions, $\rho_{tot}(\mathbf{r},t)$ and $\mathbf{J}_{tot}(\mathbf{r},t)$ provide microscopic (i.e., fast-varying) charge and current densities, while E and B become *microscopic fields*, often denoted using the lowercase letters $e(\mathbf{r},t)$ and $b(\mathbf{r},t)$. For these reasons, Maxwell's equations as introduced earlier are referred to as *microscopic Maxwell's equations*. They are complemented by the *Lorentz equation of motion*, which describes the dynamics of a system of charged particles with masses m_i, electric charges q_i, position vectors $\mathbf{r}_i(t)$, and velocities $\mathbf{v}_i(t)$ moving under electric and magnetic forces exerted by the fields:[4]

$$m_i \frac{d^2}{dt^2} \mathbf{r}_i(t) = q_i \left[\mathbf{E}(\mathbf{r}_i(t)) + \mathbf{v}_i(t) \times \mathbf{B}(\mathbf{r}_i(t)) \right] \qquad (2.9)$$

The right-hand side in equation (2.9) defines the *Lorentz force*. In such systems, the total charge and current densities are expressed as a function of the variables $\mathbf{r}_i(t)$ and $\mathbf{v}_i(t)$, as follows:

$$\rho_{tot}(\mathbf{r},t) = \sum_i q_i \delta[\mathbf{r} - \mathbf{r}_i(t)] \qquad (2.10)$$

$$\mathbf{J}_{tot}(\mathbf{r},t) = \sum_i q_i \mathbf{v}_i \delta[\mathbf{r} - \mathbf{r}_i(t)]. \qquad (2.11)$$

In general situations, Maxwell's equations and the Lorentz force equation are coupled, in the sense that the evolution of fields depends on the particles through $\rho_{tot}(\mathbf{r},t)$ and $\mathbf{J}_{tot}(\mathbf{r},t)$, while the motion of the particles depends on the values of the fields $\mathbf{E}(\mathbf{r},t)$ and $\mathbf{B}(\mathbf{r},t)$. As a result, the state of the coupled system composed of the fields plus the particles is determined by specifying at some instant of time t_0 the values of the fields $\mathbf{E}(\mathbf{r},t_0)$ and $\mathbf{B}(\mathbf{r},t_0)$ *at all points in space* as well as the positions $\mathbf{r}_i(t_0)$ and velocities $\mathbf{v}_i(t_0)$ of each individual particle in the system.

2.3.1   Electromagnetic Potentials

Maxwell's equations (2.2) and (2.3) are automatically satisfied if we introduce an auxiliary vector field $A(\mathbf{r},t)$ and an auxiliary scalar field $\phi(\mathbf{r},t)$ defined by the following:[5]

$$\mathbf{B}(\mathbf{r},t) = \nabla \times \mathbf{A}(\mathbf{r},t) \qquad (2.12)$$

$$\mathbf{E}(\mathbf{r},t) = -\frac{\partial \mathbf{A}(\mathbf{r},t)}{\partial t} - \nabla \phi(\mathbf{r},t). \qquad (2.13)$$

The vector field $A(\mathbf{r},t)$ is called the *vector potential* and the scalar field $\phi(\mathbf{r},t)$ the *scalar potential* of the electromagnetic field. Inserting the preceding definitions into the two remaining Maxwell's equations (2.1) and (2.4), and using the well-known vector identity $\nabla \times \nabla \times \mathbf{A} = -\nabla^2 \mathbf{A} + \nabla(\nabla \cdot \mathbf{A})$, we obtain the following:

---

[4] This equation is valid only in the nonrelativistic limit, i.e., for small velocities compared to the speed of light. Moreover, by using either the Lagrangian or the Hamiltonian formalism, the Lorentz equation can be rigorously derived as the equation of motion for a charged particle interacting with the electromagnetic field.

[5] This fact follows from the well-known vector identity $\nabla \cdot \nabla \times \mathbf{A} = 0$ valid for any vector field $A(\mathbf{r},t)$ and $\nabla \times \nabla \phi = 0$ for any scalar field ϕ ($\mathbf{r},t$).

$$\nabla^2\phi + \frac{\partial}{\partial t}(\nabla \cdot \boldsymbol{A}) = -\rho/\epsilon_0 \tag{2.14}$$

$$\nabla^2\boldsymbol{A} - \frac{1}{c_0^2}\frac{\partial^2 \boldsymbol{A}}{\partial t^2} - \nabla\left(\nabla \cdot \boldsymbol{A} + \frac{1}{c_0^2}\frac{\partial\phi}{\partial t}\right) = -\mu_0 \boldsymbol{J}, \tag{2.15}$$

where we have written $\boldsymbol{J}_{tot} \equiv \boldsymbol{J}$ to simplify the notation (and similarly for the charge density). The two preceding equations form a system of second-order coupled partial differential equations that is entirely equivalent to Maxwell's equations. Furthermore, we can uncouple these equations by exploiting the arbitrariness involved in the definitions of the potentials. To achieve this goal, we recognize that the fields $\boldsymbol{E}$ and $\boldsymbol{B}$ are invariant under the following *gauge transformation* of the potentials:

$$\boldsymbol{A} \to \boldsymbol{A}' = \boldsymbol{A} + \nabla\chi \tag{2.16}$$

$$\phi \to \phi' = \phi - \frac{\partial\chi}{\partial t}, \tag{2.17}$$

where $\chi(\mathbf{r}, t)$ is a scalar function, known as the *gauge function*. This invariance gives us the freedom to choose a *gauge condition* that fixes the value of $\nabla \cdot \boldsymbol{A}$ by an appropriate choice of χ (note that $\nabla \times \boldsymbol{A}$ is already determined by (2.12)). The two most commonly used gauges for the potentials are the Lorenz gauge and the Coulomb gauge, also known as the radiation or transverse gauge.[6]

The *Lorenz gauge condition* is defined by choosing the electromagnetic potentials $(\boldsymbol{A}, \phi)$ to satisfy the following:

$$\nabla \cdot \boldsymbol{A} + \frac{1}{c_0^2}\frac{\partial\phi}{\partial t} = 0. \tag{2.18}$$

This gauge decouples equations (2.14) and (2.15) giving rise to the following *inhomogeneous wave equations* for the potentials:

$$\nabla^2\phi - \frac{1}{c_0^2}\frac{\partial^2\phi}{\partial t^2} = -\rho/\epsilon_0 \tag{2.19}$$

$$\nabla^2\boldsymbol{A} - \frac{1}{c_0^2}\frac{\partial^2\boldsymbol{A}}{\partial t^2} = -\mu_0\boldsymbol{J} \tag{2.20}$$

Therefore, in the Lorenz gauge both potentials are treated on an equal footing, and this condition is *relativistically covariant*, meaning that the potentials preserve their form upon a Lorentz transformation connecting two inertial reference frames.

The *Coulomb gauge condition*, or the *radiation gauge*, which is particularly useful in the quantization of the electromagnetic free field [85], is defined by the following condition:

$$\nabla \cdot \boldsymbol{A} = 0. \tag{2.21}$$

---

[6] Additional gauges can be introduced in order to simplify the solutions of specific problems. An excellent review of the gauge-fixing problem can be found in [84].

In the Coulomb gauge, the inhomogeneous potential equations (2.14) and (2.15) become the following equations:

$$\nabla^2 \phi = -\rho/\epsilon_0 \tag{2.22}$$

$$\nabla^2 \boldsymbol{A} - \frac{1}{c_0^2}\frac{\partial^2 \boldsymbol{A}}{\partial t^2} = -\mu_0 \boldsymbol{J} + \frac{1}{c_0^2}\nabla\left(\frac{\partial \phi}{\partial t}\right) \tag{2.23}$$

We recognize in equation (2.22) the well-known *Poisson's equation* with the following general solution:

$$\phi(\boldsymbol{r},t) = \frac{1}{4\pi\epsilon_0}\int \frac{\rho(\boldsymbol{r}',t)}{|\boldsymbol{r}-\boldsymbol{r}'|}d^3\boldsymbol{r}'. \tag{2.24}$$

This solution yields the instantaneous Coulomb potential at any position $\boldsymbol{r}$ and at time t, as the superposition of contributions due to the charge density values at distant points $\boldsymbol{r}'$ is calculated at the same time t, and it thus appears to violate causality. However, a detailed analysis can show how the fields $\boldsymbol{E}$ and $\boldsymbol{B}$, which are obtained by differentiating the potentials, always remain causal [86].

When no sources are present in the region of interest, we have $\phi(\boldsymbol{r},t) = 0$. In this case, the electromagnetic field far from its sources is simply described by the following equations:

$$\boldsymbol{E} = -\frac{\partial \boldsymbol{A}}{\partial t} \tag{2.25}$$

$$\boldsymbol{B} = \nabla \times \boldsymbol{A} \tag{2.26}$$

The Coulomb gauge becomes particulary useful in the quantum theory of the electromagnetic field confined inside cavities where in fact only the vector potential is subjected to the quantization procedure [87, 88].

2.3.2   Green Function and Radiation Potentials

The inhomogeneous wave equations (2.19) and (2.20) share the same mathematical form:

$$\nabla^2 \psi - \frac{1}{c_0^2}\frac{\partial^2 \psi}{\partial t^2} = -S(\boldsymbol{r},t) \tag{2.27}$$

where ψ represents any scalar component of the potential and $S(\boldsymbol{r},t)$ is a known source term. Equation (2.27) can be solved by using the *Green function method*, which we will now briefly review. We first eliminate the explicit time dependence from the inhomogeneous wave equation by introducing the *Fourier representations*:

$$\psi(\boldsymbol{r},t) = \frac{1}{2\pi}\int_{-\infty}^{\infty} \psi(\boldsymbol{r},\omega)e^{-j\omega t}d\omega \tag{2.28}$$

$$S(\mathbf{r}, t) = \frac{1}{2\pi} \int_{-\infty}^{\infty} S(\mathbf{r}, \omega) e^{-j\omega t} d\omega. \tag{2.29}$$

Inserting into (2.27), we deduce that each Fourier component $\psi(\mathbf{r}, \omega)$ obeys the *inhomogeneous Helmholtz wave equation*:

$$\nabla^2 \psi(\mathbf{r}, \omega) + k_0^2 \psi(\mathbf{r}, \omega) = (\nabla^2 + k_0^2)\psi(\mathbf{r}, \omega) = -S(\mathbf{r}, \omega). \tag{2.30}$$

Here we have introduced the wavenumber $k_0 = \omega/c_0$ associated with the frequency ω. The resulting *Helmholtz equation* (2.30) is an elliptical equation similar to the Poisson equation, which in fact can be recovered when $k_0 = 0$.

We now define the *scalar Green function* appropriate to solve the scalar Helmholtz equation as the function $G(\mathbf{r}, \mathbf{r}')$ that satisfies the following:[7]

$$\boxed{(\nabla^2 + k_0^2)G(\mathbf{r}, \mathbf{r}') = -\delta(\mathbf{r} - \mathbf{r}')} \tag{2.31}$$

where $\delta(\mathbf{r} - \mathbf{r}')$ is the Dirac delta function representing an *impulsive source* located at position $\mathbf{r} = \mathbf{r}'$. Notice that when there are no boundary surfaces to perturb the problem, as it is the case for homogeneous media, the Green function only depends on the relative distance from the source, i.e., $\mathbf{R} = \mathbf{r} - \mathbf{r}'$. Moreover, if the medium is *isotropic* (e.g., free space), the corresponding Green function will only depend on the amplitude $R = |\mathbf{R}|$.

Let us now find an explicit expression for the Green function of an *homogeneous and isotropic medium*. Under these assumptions, we can use the expression of the Laplacian operator in spherical coordinates. For a spherically symmetric medium, the Helmholtz equation simplifies to the following ordinary differential equation, valid everywhere in space except at $R = 0$:

$$\frac{d^2}{dR^2}(RG) + k_0^2(RG) = 0 \tag{2.32}$$

with well-known solutions:

$$G_0^\pm(R) = \frac{e^{\pm jk_0 R}}{4\pi R}. \tag{2.33}$$

These are the *free-space Green functions in frequency domain*.[8] The *retarded Green function* G_0^+ represents a diverging spherical wave that propagates away from the origin, while the *advanced Green function* G_0^- represents a spherical wave converging toward the origin.

Next, we show that the solution of a nonhomogeneous (i.e., a driven) linear partial differential equation can be expressed in terms of the corresponding Green function.

---

[7] Note that some authors use a positive Dirac delta function as the inhomogeneous term in equation (2.31). In this case, their Green function will have to be negative in order to maintain consistency with equation (2.27), which determines the correct sign of the Green function.

[8] The factor $1/4\pi$ is introduced for consistency with the Poisson solution that must be obtained in the limit of $k_0 = 0$.

For this purpose, we consider a general nonhomogeneous linear differential equation in the following form:

$$L\psi(\mathbf{r}) = -S(\mathbf{r}) \tag{2.34}$$

where L denotes a linear operator acting on the unknown scalar function ψ and S is the known source term. We define the Green function G *associated to the linear operator* L as:

$$LG(\mathbf{r}, \mathbf{r}') = -\delta(\mathbf{r} - \mathbf{r}') \tag{2.35}$$

Postmultiplying the last equation by $S(\mathbf{r}')$ on both sides and integrating over the volume V in which $S \neq 0$, we obtain the following:

$$\int_V LG(\mathbf{r}, \mathbf{r}') S(\mathbf{r}') d^3\mathbf{r}' = -\int_V \delta(\mathbf{r} - \mathbf{r}') S(\mathbf{r}') d^3\mathbf{r}' = -S(\mathbf{r}). \tag{2.36}$$

By comparing with equation (2.34), we immediately obtain the following:

$$\int_V LG(\mathbf{r}, \mathbf{r}') S(\mathbf{r}') d^3\mathbf{r}' = L\psi. \tag{2.37}$$

The operator on the left-hand side of (2.37) can be taken out of the integral since it is linear and acts on the variable $\mathbf{r}$ alone.[9] Therefore, the solution of the equation (2.34) has been expressed in the form:

$$\psi(\mathbf{r}) = \int_V G(\mathbf{r}, \mathbf{r}') S(\mathbf{r}') d^3\mathbf{r}' \tag{2.38}$$

For this reason, the Green function is sometimes called the *fundamental solution* associated to the operator L. However, not every operator L admits a Green function. Moreover, in many situations of interest the Green functions are not even standard functions but are generalized functions, or distributions. Regardless, for most systems described by linear partial differential equations with constant coefficients, the Green functions can be readily calculated.[10]

Let us now apply the Green's function approach to the Helmholtz operator (i.e., $L \equiv \nabla^2 + k_0^2$) for the electromagnetic potentials in Cartesian orthogonal coordinates. Considering the retarded Green function only, the electromagnetic potentials in the Lorenz gauge can be readily obtained:

$$\mathbf{A}(\mathbf{r}) = \mu_0 \int_V \mathbf{J}(\mathbf{r}') G_0^+(\mathbf{r}, \mathbf{r}') d^3\mathbf{r}' = \frac{\mu_0}{4\pi} \int_V \mathbf{J}(\mathbf{r}') \frac{e^{jk_0 R}}{R} d^3\mathbf{r}' \tag{2.39}$$

---

[9] However, particular care must be exercised if L is an integral operator and the Green function has a singularity at $\mathbf{r} = \mathbf{r}'$. In this case, an infinitesimal exclusion volume surrounding the point $\mathbf{r} = \mathbf{r}'$ must be introduced, as we will discuss in Section 2.9.1.

[10] Sturm–Liouville systems with nonconstant coefficients also admit a unique Green function, but its construction is not always straightforward [89].

$$\phi(\mathbf{r}) = \frac{1}{\epsilon_0} \int_V \rho(\mathbf{r}') G_0^+(\mathbf{r}, \mathbf{r}') d^3\mathbf{r}' = \frac{1}{4\pi\epsilon_0} \int_V \rho(\mathbf{r}') \frac{e^{jk_0 R}}{R} d^3\mathbf{r}', \qquad (2.40)$$

where $\mathbf{A}(\mathbf{r})$ and $\rho(\mathbf{r})$ are the Fourier amplitudes of the corresponding time-domain quantities, obtained by Fourier transformation.

An analogous approach can be utilized to solve the wave equation directly in the *time domain*. In this case, the free-space Green functions are defined by the following equation:

$$\left(\nabla^2 - \frac{1}{c_0^2} \frac{\partial^2}{\partial t^2} \right) G_0^\pm(\mathbf{r}, t; \mathbf{r}', t') = -\delta(\mathbf{r} - \mathbf{r}')\delta(t - t') \qquad (2.41)$$

The time-domain Green functions can be obtained by Fourier antitransformation of their frequency domain counterparts:

$$G_0^\pm(R, \tau) = \frac{1}{2\pi} \int_{-\infty}^{\infty} \frac{e^{\pm jk_0 R}}{4\pi R} e^{-j\omega\tau} d\omega, \qquad (2.42)$$

where $\tau = t - t'$ and $R = |\mathbf{r} - \mathbf{r}'|$. By remembering that $k_0 = \omega/c_0$ as well as the following Fourier transform pair:

$$\delta(\tau - R/c_0) \leftrightarrow e^{-j\left(\frac{R}{c_0}\right)\omega}, \qquad (2.43)$$

we can easily derive the explicit expressions of time-dependent Green functions:

$$G_0^\pm(R, \tau) = \frac{1}{4\pi R} \delta\left(\tau \mp \frac{R}{c_0} \right) \qquad (2.44)$$

or, equivalently:

$$G_0^\pm(\mathbf{r}, t; \mathbf{r}', t') = \frac{\delta\left(t' - \left[t \mp \frac{|\mathbf{r} - \mathbf{r}'|}{c_0} \right] \right)}{4\pi|\mathbf{r} - \mathbf{r}'|} \qquad (2.45)$$

The preceding expressions are the *retarded* (G_0^+) and the *advanced* (G_0^-) Green functions of nondispersive and isotropic media expressed in the time domain [82]. The argument of G_0^+ shows that any effect observed at point $\mathbf{r}$ and at time t is determined by the action of sources (i.e., charges and currents) that are located a distance R away *at an earlier time* $t' = t - R/c_0$. The time difference $\tau = t - t' = R/c_0$ is the time of propagation of the signal, and it reflects the causal behavior of the retarded Green function. In the case of G_0^-, the potentials at time t are determined by the behavior of the sources at later times, which is generally dismissed as unphysical. However, as we will fully appreciate only in Chapter 14, both the advanced and the retarded Green functions are necessary to compute physically acceptable solutions in situations involving the intensity propagation through complex scattering media.

The time-dependent Green functions allows us to express the general solutions of the inhomogeneous time-dependent wave equation (2.27) as the *space-time superposition integral*:

$$\psi^{\pm}(\mathbf{r},t) = \psi_0(\mathbf{r},t) + \int\int G_0^{\pm}(\mathbf{r},t;\mathbf{r}',t')S(\mathbf{r}',t')d^3\mathbf{r}'dt', \qquad (2.46)$$

where $\psi_0(\mathbf{r},t)$ is a suitable solution of the homogeneous equation,[11] which is dictated by the specific nature of the problem (e.g., a plane wave excitation in scattering problems). The integral is computed over the volume of the source $S(\mathbf{r},t)$ and, when using the retarded Green function, at all prior times extending from $t = -\infty$ up to the observation time t.

In classical and quantum mechanical scattering theory [90], the Green function G_0^+ is interpreted as a *propagator* since it describes the effects of a source originally at $\mathbf{r}',t'$ when it reaches the point $\mathbf{r}$ at the later time t. Applying the Green function method and assuming $\psi_0 = 0$, we can directly write the causal solutions of the wave equations for the electromagnetic potentials in the Lorenz gauge as follows:

$$\phi(\mathbf{r},t) = \frac{1}{4\pi\epsilon_0}\int\frac{\rho\left(\mathbf{r}',t - \frac{|\mathbf{r}-\mathbf{r}'|}{c_0}\right)}{|\mathbf{r}-\mathbf{r}'|}d^3\mathbf{r}' \qquad (2.47)$$

$$A(\mathbf{r},t) = \frac{\mu_0}{4\pi}\int\frac{J\left(\mathbf{r}',t - \frac{|\mathbf{r}-\mathbf{r}'|}{c_0}\right)}{|\mathbf{r}-\mathbf{r}'|}d^3\mathbf{r}'. \qquad (2.48)$$

The quantities appearing in the preceding equations are known as *retarded potentials* because they are formally identical to the potentials generated by static charge and current distributions evaluated at the retarded time $t' = t - |\mathbf{r}-\mathbf{r}'|/c_0$, fully manifesting their causal nature. The expressions (2.47) and (2.48) for the retarded potentials formally solve the general problem of determining the fields produced by arbitrary charge and current distributions.

2.3.3   Green Function and Density of States

In the previous section, we have seen how the Green function is equivalent to the full solution of the wave equation problem. Here we show that the Green function carries additional information on wave problems through its fundamental connection with the *density of states (DOS)*:

$$\boxed{\rho(\omega) = \sum_n \delta(\omega - \omega_n)} \qquad (2.49)$$

The preceding expression, assumed normalized per unit volume, quantifies the density of oscillation modes per frequency interval in a generic wave system and plays a fundamental role in the semiclassical theory of light–matter interaction, where it

---

[11] This is the solution of the equation in the absence of any forcing terms in its right-hand side.

determines the likelihood for radiative transitions [91]. In order to establish the important connection between the DOS and the Green function, we need to consider the spatial Fourier transform of the free-space Green function in equation (2.33), which is [24]:

$$G_0^\pm(k, \omega) = \frac{1}{k^2 - k_0^2 \pm j\epsilon}, \quad (2.50)$$

where k_0 and ω are related by the free-space dispersion relation and the small imaginary quantity ϵ is added to ensure convergence when transforming back to the spatial domain. In addition, we will make use of the well-known identity:

$$\frac{1}{x + j\epsilon} = P.V.\frac{1}{x} - j\pi\delta(x), \quad (2.51)$$

where $P.V.$ denotes the principal value integral. Due to the linear dispersion $\omega = k_0 c$, we can express the density of states as follows:

$$\rho(\omega) = \frac{1}{c} \sum_k \delta(k - k_0), \quad (2.52)$$

where $k = |\mathbf{k}|$. The imaginary part of the retarded Green function in (2.50) can be written as follows:

$$\Im\{G_0^+(k, \omega)\} = \frac{\pi}{2k_0}[\delta(k + k_0) + \delta(k - k_0)]. \quad (2.53)$$

We can now use the useful identity (2.51) and obtain the following:

$$\boxed{\rho(\omega) = \frac{2k_0}{\pi c} \sum_k \Im\{G_0^+(\mathbf{k}, \omega)\}} \quad (2.54)$$

This important result provides the density of states per unit volume in terms of the Green function. Since the density of states does not depend on the specific representation, i.e., it is the same in the momentum or in the position representation, we can also express the previous result in the equivalent form:

$$\rho(\omega) = \frac{2k_0}{\pi c} \int d\mathbf{r}\Im\{G_0^+(\mathbf{r}, \mathbf{r}, \omega)\}. \quad (2.55)$$

The integrand in the preceding expression is the *local density of states* (LDOS):

$$\boxed{\rho(\omega, \mathbf{r}) = \frac{2k_0}{\pi c}\Im\{\text{Tr}\, G_0^+\}} \quad (2.56)$$

where Tr denotes the trace. The LDOS provides the total number of electromagnetic modes per unit volume and per unit frequency at a given location of the medium and plays an important role in the description of light–matter interactions in nonhomogeneous media. A related concept is the *partial local density of states* $\rho_p(\omega, \mathbf{r})$ that enables the calculation of the spontaneous decay rate of an emitting dipole embedded in a nonhomogeneous system. A clear derivation of this quantity is provided in [92] and will be discussed in Chapter 12 in relation to the radiation of light from arbitrary systems of vector dipoles.

2.4   Maxwell's Equations in Materials

The direct application of the microscopic Maxwell's equations inside materials presents unsurmountable difficulties since the total charge and current densities include the detailed behavior of matter at the atomic scale. Inside material bodies, the microscopic electromagnetic fields produced by point charges vary quite rapidly in space and time, with fluctuations occurring over distances of the order of the atomic Bohr radius $\sim 10^{-10}$ m or less, and temporal fluctuations with periods as small as 10^{-17} s, which approximately corresponds to the electronic orbital motion. Therefore, in order to smooth out the fast-varying microscopic field fluctuations that exist at the molecular/atomic structure levels, suitable macroscopic average field quantities must be introduced. This is accomplished by the so-called *macroscopic Maxwell's equations*, which deal with continuous and macroscopic vector quantities and with *auxiliary fields* describing the large-scale electromagnetic behavior of matter without having to consider explicitly the atomistic details. The resulting equations provide the general framework of *macroscopic electrodynamics*, which is concerned with the study of electromagnetic fields inside materials [93, 94]. Within this approach, the complex response of materials can be modeled by suitable macroscopic parameters and *constitutive relations* that couple matter and fields in various types of media.

2.4.1   Macroscopic Field Averages

In order to operate with continuous fields inside optical materials, it is customary to introduce *macroscopic field averages* taken over spatial regions that are much larger than the typical interatomic spacings $d \approx 10^{-10}$ m, yet small enough compared to the scale of the macroscopic structures under consideration. In most practical situations, a value of $L \cong 100d$ suffices to wash out the microscopic field fluctuations. In addition, since the time fluctuations of microscopic fields over distances of the order of L are typically uncorrelated, only the spatial averages need to be considered when dealing with macroscopic field quantities.

The spatial average of a microscopic field quantity $f(\mathbf{r},t)$ is defined with respect to a specified *window function* $W(\mathbf{r})$ according to the following:

$$F(\mathbf{r},t) \equiv \langle f(\mathbf{r},t) \rangle = \int_V W(\mathbf{r}')f(\mathbf{r} - \mathbf{r}',t)d^3\mathbf{r}'. \tag{2.57}$$

The window function must be real, nonzero over the length scale L (typically assumed centered around $\mathbf{r} = 0$), and normalized to unity over all space. It turns out that *the actual choice of the averaging window is unimportant* as long as it is smooth and approximately constant for $r < L$ while vanishing for $r \gg L$. Usually, isotropic averaging functions over a volume of radius R are utilized, either in the form of a spherical volume region or of a Gaussian window [82, 93].

Using the specified averaging procedure, it is possible to introduce the macroscopic fields $E(\mathbf{r},t)$ in terms of the spatial averages of their microscopic counterparts $e(\mathbf{r},t)$ and $b(\mathbf{r},t)$, as follows:

$$E(\mathbf{r},t) \equiv \langle e(\mathbf{r},t) \rangle \tag{2.58}$$

$$B(\mathbf{r},t) \equiv \langle b(\mathbf{r},t) \rangle. \tag{2.59}$$

Since the spatial averaging operation defined in equation (2.57) commutes with the space and time-differentiation operations, the macroscopic Maxwell's equations can be readily obtained starting from their microscopic counterparts.

2.4.2 Polarization and Magnetization Response

The macroscopic Maxwell's equations can be solved only after an appropriate treatment of their source terms. In order to achieve this, it is convenient to split the total charge and current density into *bound* and *free* contributions:

$$\rho_{tot} = \langle \varrho(\mathbf{r},t) \rangle = \rho_{boud} + \rho_{free} \tag{2.60}$$

$$\mathbf{J}_{tot} = \langle \mathbf{j}(\mathbf{r},t) \rangle = \mathbf{J}_{bound} + \mathbf{J}_{free}, \tag{2.61}$$

where the symbols inside the angle brackets denote microscopic quantities. Bound charges and currents are localized around specific atomic or molecular positions while free ones are delocalized inside materials and "free to move," such as charges that flow in wires or in electrolyte solutions. More generally, free charges include the ones over which we have direct control in a given situation, such as source charges that are external to the system and are also called impressed or external sources. On the other hand, bound charges automatically ensue from the (generally unknown) internal polarization and magnetization response mechanisms of materials. The specification of macroscopic sources requires the knowledge of the internal response mechanisms of materials, which is provided by the macroscopic polarization $P(\mathbf{r},t)$ [C/m^2] and magnetization density $M(\mathbf{r},t)$ [A/m] vectors. These two vector fields are defined respectively as macroscopically averaged electric and magnetic dipole moments induced by an electric field and a magnetic flux density inside a material medium.[12] These fields vanish outside the medium and can be directly related to the polarization and magnetization vectors, as shown in Appendix A.

---

[12] It is worth noting that the contribution of quadrupolar and higher-order multipolar terms (electric and magnetic) are neglected in these basic definitions on the basis of their very small effects in standard materials. However, they can become relevant in special situations.

2.4.3   The Auxiliary Macroscopic Fields

In addition to the polarization and magnetization vectors discussed, it is convenient to introduce two auxiliary macroscopic fields, the *displacement field*:

$$D = \epsilon_0 E + P, \tag{2.62}$$

measured in C/m^2, and the *magnetic field*:

$$H = B/\mu_0 - M, \tag{2.63}$$

measured in A/m. These quantities are introduced in order to account for the internal response of materials that is due to bound charges and currents. Representing the total charge density as the sum of free and bound charges:

$$\rho_{tot} = \rho_{pol} + \rho_{free}, \tag{2.64}$$

and using the expression $\rho_{pol} = -\nabla \cdot P$ derived in Appendix A, we can write the divergence of the displacement field as follows:

$$\boxed{\nabla \cdot D = \rho_{free}} \tag{2.65}$$

This equation, which is Gauss's law for the displacement field in matter, involves only the density of free charges. However, since in a polarizable material $P \neq 0$, we have that $\nabla \times D \neq 0$ as well. Therefore, the free charges are not the only sources of the vector D[13].

Analogously, we decompose the total current density as follows:

$$J_{tot} = J_{free} + J_{pol} + J_{mag}, \tag{2.66}$$

where the polarization and magnetization current densities J_{pol} and J_{mag} are derived in Appendix A. It is also customary to split the free current density as follows:

$$J_{free} = J_s + J_c, \tag{2.67}$$

where J_s is the contribution due to the external (source or impressed) current density, and J_c is the one due to the conduction current, which is determined by strength of the electric field. For fields that are not too strong, the conduction current density depends linearly on the electric field and can be written as $J_c = \sigma E$, where σ is the *conductivity* of the material. However, in many applications related to the optical behavior of conductive media the conductivity term is absorbed in the definition of the frequency-dependent effective dielectric function, as we will later discuss in Section 3.1. Perhaps a physically more transparent approach splits the total charges and currents into external and internal contributions only, the former being the ones that are completely under

---

[13] Recall that any vector field that decays to zero at infinity is uniquely specified if and only if its divergence and its curl are simultaneously specified at every point in space. This important result is known as the Helmholtz decomposition theorem. See Appendix B.

our control while the latter are the ones that originate from the generally unknown response of the material media to an external excitation. Combining Maxwell's equation (2.4) with the curl of the magnetic field in (2.63) and using the expressions derived for $\boldsymbol{J}_{pol}$ and $\boldsymbol{J}_{mag}$, we obtain the following:

$$\nabla \times \boldsymbol{H} = \boldsymbol{J}_{free} + \frac{\partial \boldsymbol{D}}{\partial t}. \tag{2.68}$$

This equation is the expression of Ampère–Maxwell's law that is valid inside materials. However, since even in the static case when $\partial \boldsymbol{D}/\partial t = 0$ $\nabla \cdot \boldsymbol{H} \neq 0$ when $\boldsymbol{M} \neq 0$, we should not think that the magnetic field is determined completely by the free current density.

We can now summarize the macroscopic Maxwell's equations that are valid inside materials:

$$\nabla \cdot \boldsymbol{D}(\mathbf{r}, t) = \rho_{free}(\mathbf{r}, t) \tag{2.69}$$

$$\nabla \cdot \boldsymbol{B}(\mathbf{r}, t) = 0 \tag{2.70}$$

$$\nabla \times \boldsymbol{E}(\mathbf{r}, t) = -\frac{\partial \boldsymbol{B}(\mathbf{r}, t)}{\partial t} \tag{2.71}$$

$$\nabla \times \boldsymbol{H}(\mathbf{r}, t) = \frac{\partial \boldsymbol{D}(\mathbf{r}, t)}{\partial t} + \boldsymbol{J}_{free}(\mathbf{r}, t) \tag{2.72}$$

This system of equations clearly contains more unknown quantities than equations and therefore cannot be solved unless we postulate some relation that links the fields $\boldsymbol{D}$ and $\boldsymbol{H}$ with $\boldsymbol{E}$ and $\boldsymbol{B}$, respectively. In particular, we need to know how the polarization and magnetization densities $\boldsymbol{P}$ and $\boldsymbol{M}$ are generated by the fields $\boldsymbol{E}$ and $\boldsymbol{B}$. This can be achieved by introducing appropriate *constitutive relations* for the media at hand.

Inserting the expressions derived in Appendix A for the induced charges and currents, we obtain the alternative form of the macroscopic Maxwell's equations inside materials where only the electric field and the magnetic flux density appear:

$$\nabla \cdot \boldsymbol{E} = \frac{\rho_{free} - \nabla \cdot \boldsymbol{P}}{\epsilon_0} \tag{2.73}$$

$$\nabla \cdot \boldsymbol{B} = 0 \tag{2.74}$$

$$\nabla \times \boldsymbol{E} = -\frac{\partial \boldsymbol{B}}{\partial t} \tag{2.75}$$

$$\nabla \times \boldsymbol{B} = \epsilon_0 \mu_0 \frac{\partial \boldsymbol{E}}{\partial t} + \mu_0 \left(\boldsymbol{J}_{free} + \frac{\partial \boldsymbol{P}}{\partial t} + \nabla \times \boldsymbol{M} \right) \tag{2.76}$$

The macroscopic Maxwell's equations in the preceding form make it very clear that in order to obtain a solution subject to prescribed boundary conditions at the interface between different materials, we need to know the polarization and magnetization vectors, i.e., the charges and currents induced inside materials.

2.4.4   Constitutive Relations

The derivation of the polarization and magnetization vectors in terms of microscopic bound charges and currents is an extremely difficult task in general, which is addressed by solid-state physics. However, at the macroscopic level, appropriate *constitutive relations* are employed in order to connect the E and B fields with the polarization and magnetization densities in materials, respectively. The introduction of suitable constitutive relations allows us to conveniently describe certain *response mechanisms* such as polarization, magnetization, and electrical conduction. However, although constitutive relations provide only approximate descriptions of the complex response of materials, they can be used as convenient means to broadly classify materials into various categories. For example, materials that are *linear, homogeneous, isotropic, and nondispersive*, such as gases, liquids, amorphous or cubic crystalline materials at frequencies far from resonant optical transitions, can be well described by the following constitutive relations:

$$P = \epsilon_0 \chi_e E \tag{2.77}$$

$$M = \chi_m H \tag{2.78}$$

$$J_c = \sigma E \tag{2.79}$$

where the dimensionless proportionality constants χ_e and χ_m are the *electric and magnetic susceptibilities* and σ is the *conductivity* (measured in Siemens/m). The preceding relations can be generalized to describe *inhomogeneous media* when their optical parameters do not vary significantly over the length scale of the wavelength. In such cases, the material parameters χ_e, χ_m and σ simply become functions of the position vector **r**.

However, when the applied electric fields are very large, e.g., in excess of the typical atomic value of $\sim 10^8 \, V/m$, they induce an anharmonic electron displacement inside materials that results in a nonlinear polarization response. This behavior, which is studied by *nonlinear optics* [95], gives rise to many fascinating optical effects, such as optical harmonic generation and frequency mixing.

Using the linear constitutive relations (2.77) and (2.78) into the definitions of the auxiliary fields, we obtain the following simple expressions:

$$D = \epsilon_0 (1 + \chi_e) E = \epsilon E \tag{2.80}$$

$$B = \mu_0 (1 + \chi_m) H = \mu H, \tag{2.81}$$

where the *electric permittivity* ϵ and the *magnetic permeability* μ of the medium have been defined. In the case of *anisotropic media*, i.e., media whose optical properties depend on the orientation of the applied fields, the permittivity and permeability become second-rank tensors. The *relative permittivity and permeability* of the medium are defined by the following:

$$\epsilon_r = \epsilon/\epsilon_0 = 1 + \chi_e \tag{2.82}$$

$$\mu_r = \mu/\mu_0 = 1 + \chi_m \tag{2.83}$$

In the dynamic regime, which is specifically addressed in the next section, the relative electric permittivity is also known as the *dielectric function* of the medium. The aforementioned quantities are generally referred to as the *constitutive parameters* of the medium.

2.4.5   Optical Dispersion and Nonlocality

Optical dispersion effects are always associated to the presence of inherent time or spatial scales that are characteristic of the medium. In particular, temporal dispersion is associated to intrinsic time scales of materials, such as the ones that correspond to the resonant frequency of their atomic or molecular components. On the other hand, spatial dispersion arises from fundamental length scales such as, in the case of a crystal structure, its lattice periodicity. Temporal and spatial dispersion give rise to *nonlocal relations* between physical variables such as the electric displacement D and the electric field E. Such dependencies are usually expressed in convolution integral form. Temporal dispersion results in the frequency dependence of constitutive parameters such as their susceptibilities $\chi_e(\omega)$ and $\chi_m(\omega)$ or the conductivity $\sigma(\omega)$. A general discussion of classical models for the frequency dependence of materials parameters will be discussed in detail in Section 3.1.

When an optical signal that contains multiple frequencies propagates through a temporally dispersive medium, its shape will be generally distorted because different frequency components propagate at different speeds. Due to the nonlocal nature of the optical dispersion effects, the constitutive relations can no longer be written as in equations (2.80) and (2.81). In particular, for linear and homogeneous media they are expressed in *convolution form* as follows:

$$D = \epsilon_0 E + \epsilon_0 \int_{-\infty}^{t} \tilde{\chi}_e(t - t')E(t')dt' = \epsilon_0 E + \epsilon_0 \tilde{\chi}_e * E \qquad (2.84)$$

$$B = \mu_0 H + \mu_0 \int_{-\infty}^{t} \tilde{\chi}_m(t - t')H(t')dt' = \mu_0 H + \mu_0 \tilde{\chi}_m * H \qquad (2.85)$$

where $\tilde{\chi}_e$ and $\tilde{\chi}_m$ are the electric and magnetic susceptibilities of the medium in the time domain and the symbol $*$ denotes the temporal convolution operator.

The physical meaning of such convolution-type responses is that, due to the presence of intrinsic time scales, the medium cannot polarize or magnetize instantaneously in response to an applied electric or magnetic field. By taking the Fourier transform of equations (2.84) and (2.85) and by using the convolution property of the Fourier transform, we can express the constitutive relations of a temporally dispersive medium in the *frequency domain*:

$$\mathbf{D}(\omega) = \epsilon_0[1 + \chi_e(\omega)]\mathbf{E} = \epsilon(\omega)\mathbf{E}(\omega) \qquad (2.86)$$

$$\mathbf{B}(\omega) = \mu_0[1 + \chi_m(\omega)]\mathbf{H} = \mu(\omega)\mathbf{H}(\omega), \qquad (2.87)$$

where the frequency-dependent susceptibilities $\chi_e(\omega)$ and $\chi_m(\omega)$ are obtained by Fourier transforming the corresponding time-dependent quantities $\tilde{\chi}_e$ and $\tilde{\chi}_m$ and the fields **D**, **E**, **H**, and **B** denote the frequency-domain complex Fourier amplitudes of their time-dependent (real) counterparts.[14]

Additionally, if the material is also *spatially dispersive*, the constitutive parameters ϵ and μ acquire a dependence on the spatial wavevector **k**, which varies at different length scales. In these situations, the amplitude of the induced displacement field **D** can be written as follows:

$$\mathbf{D}(\mathbf{k}, \omega) = \epsilon(\mathbf{k}, \omega)\mathbf{E}(\mathbf{k}, \omega), \tag{2.88}$$

and analogously for the magnetic flux density. Here, $\mathbf{E}(\mathbf{k}, \omega)$ denotes the complex-valued Fourier amplitude component of the real-valued electric field $\mathbf{E}(\mathbf{r}, t)$. By taking the inverse space-time Fourier transform of equation (2.88), we obtain the following:

$$\mathbf{D}(\mathbf{r}, t) = \epsilon_0 \int_{-\infty}^{t} \int_{-\infty}^{\mathbf{r}} \tilde{\epsilon}(\mathbf{r} - \mathbf{r}', t - t')\mathbf{E}(\mathbf{r}', t')d^3\mathbf{r}'dt', \tag{2.89}$$

where $\tilde{\epsilon}$ is the space-time *impulse response function* of the linear, shift-invariant (uniform in space), and time-invariant (uniform in time) medium. Equation (2.89) indicates that the displacement field **D** at time t depends on the electric field at all times $t' < t$, which is the essence of temporal dispersion, i.e., *nonlocality in time*. In addition, due to spatial dispersion the displacement field **D** at position **r** depends on the values of the electric field at all the neighboring points $\mathbf{r}'$, which is the manifestation of *nonlocality in space*.

The wavevector dependence of the dielectric function can be neglected for bulk matter, but it becomes important when dealing with sub-wavelength nano-structures. In particular, as the dimensions of metallic structures become much smaller than the optical wavelength, the electrons begin to feel the influence of their surfaces. This happens for very small objects whose sizes are comparable with the collision mean free path of the electrons in materials.

2.4.6   Boundary Conditions

At the boundaries separating different media, the materials properties exhibit jump discontinuities and Maxwell's equations in differential form cannot be applied directly.[15] However, Maxwell's equations are valid in their integral form even across boundaries, which allows one to derive specific boundary conditions for the normal and the tangential components of fields.

---

[14] In order to distinguish the complex fields from their real counterparts, we introduced the bold uppercase notation.

[15] This statement is not entirely true. In fact, as brilliantly shown by Feynman in his lectures [96], a general methodology based on balancing field discontinues at the interfaces of different media can be devised, but this implicitly makes use of generalized (delta) functions.

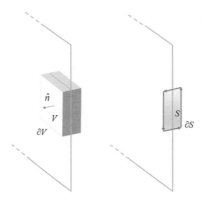

Figure 2.1 Integration paths for the derivation of the boundary conditions.

The macroscopic Maxwell's equations in integral form are written as follows:

$$\int_{\partial V} \boldsymbol{D}(\mathbf{r},t) \cdot \hat{n} dA = \int_V \rho_{free}(\mathbf{r},t) d^3\mathbf{r} = Q_{free} \tag{2.90}$$

$$\int_{\partial V} \boldsymbol{B}(\mathbf{r},t) \cdot \hat{n} dA = 0 \tag{2.91}$$

$$\int_{\partial S} \boldsymbol{E}(\mathbf{r},t) \cdot d\boldsymbol{l} = -\int_S \frac{\partial \boldsymbol{B}(\mathbf{r},t)}{\partial t} \cdot \hat{n} dA \tag{2.92}$$

$$\int_{\partial S} \boldsymbol{H}(\mathbf{r},t) \cdot d\boldsymbol{l} = \int_S \left[\boldsymbol{J}_{free} + \frac{\partial \boldsymbol{D}(\mathbf{r},t)}{\partial t} \right] \cdot \hat{n} dA. \tag{2.93}$$

Equations (2.90) and (2.91) are applied to a small box of volume V that straddles the boundary surface separating two media with different electromagnetic properties (see Figure 2.1). The boundary of volume V is the surface denoted by ∂V. If the volume V of the box is sufficiently small, the boundary surface can be considered locally flat and the fields as homogeneous on both sides, which simplifies the calculation of their fluxes across the surface S enclosing V. In the limit of a very small volume V, only the largest faces of the box contribute to the flux of $\boldsymbol{D}$ and $\boldsymbol{B}$ across ∂V, so that the boundary conditions for the normal components of $\boldsymbol{D}$ and $\boldsymbol{B}$ immediately follow:

$$(\boldsymbol{D}_2 - \boldsymbol{D}_1) \cdot \hat{n} = \sigma_{free} \tag{2.94}$$

$$(\boldsymbol{B}_2 - \boldsymbol{B}_1) \cdot \hat{n} = 0, \tag{2.95}$$

where σ_{free} is the free surface charge density [C/m^2] distributed on the boundary and $\boldsymbol{D}_1$ and $\boldsymbol{D}_2$ ($\boldsymbol{B}_1$ and $\boldsymbol{B}_2$) denote the fields on both sides of the boundary, infinitesimally close to it.

The boundary condition in equation (2.94) reflects the fact that, when crossing the boundary between two different materials, the normal component of the $\boldsymbol{D}$ field is discontinuous, and its discontinuity is equal to the surface charge density. On the other hand, equation (2.95) shows that the normal component of the $\boldsymbol{B}$ field is always continuous. Using the definition of the displacement vector D, we can also write the boundary condition as follows:

$$\boxed{\epsilon_0(E_{1n} - E_{2n}) = \sigma_{free} - (P_1 - P_2) \cdot \hat{n} = \sigma_{tot}} \tag{2.96}$$

This equation explicitly accounts for the bound charges that accumulate on the surface at the boundary between two dielectric materials. Note that the negative sign in front of the polarization in medium 1 originates because the surface normal $\hat{n}$ points outward from the dielectric. This boundary condition shows clearly that polarization bound charges accumulating at a dielectric interface give rise to a discontinuity in the normal component of the electric field even in the absence of free charges, where we have the following:

$$D_{1n} - D_{2n} = \epsilon_{r1} E_{1n} - \epsilon_{r2} E_{2n} = 0. \tag{2.97}$$

The last expression shows that, when an external electromagnetic field is incident on a dielectric interface with large permittivity, the induced surface polarization charges can drive a larger electric field intensity at the interface. This effect is maximized for p-polarized waves[16] at the critical angle for total internal reflection [92]. An even stronger enhancement of the surface electric field occurs due to the excitation of internal polariton resonances of materials, such as the surface plasmon polaritons at metal-dielectric interfaces, which we will discuss in Chapter 3.

The boundary conditions for the tangential fields are obtained by applying equations (2.92) and (2.93) to a small rectangular loop that crosses the boundary between the media (see Figure 2.1). The boundary of the surface S is the contour line denoted by ∂S. If the area S enclosed by the loop is sufficiently small, the electric and magnetic fluxes through S vanish. However, this argument does not apply to the free current density term of equation (2.93) because free current could flow directly on the boundary with a density given by K [A/m]. Therefore, the boundary conditions for the tangential field components are as follows:

$$\boxed{\begin{aligned} \hat{n} \times (E_2 - E_1) &= 0 \\ \hat{n} \times (H_2 - H_1) &= K \end{aligned}} \tag{2.98} \tag{2.99}$$

These equations show that the tangential component of E across an interface is continuous while the tangential component of H is discontinuous. Note that the (vector) discontinuity is equal in magnitude to the surface current density $|K|$ and points along the direction given by $K \times \hat{n}$. This can be immediately appreciated because the vector K is purely tangential and any vector V can be decomposed as the sum of a component tangential to the surface V_t and a component V_n perpendicular to it, i.e., $V = V_t + V_n$. In particular, we can represent the electric field using the following vector identity:

---

[16] When unpolarized light is incident on a plane boundary, we must distinguish between two independent polarization states: one with the electric field vector parallel (p-polarization) and one perpendicular (s-polarization) to the plane of incidence, defined as the plane that contains the direction of propagation and the normal to the boundary.

$$E = E_t + E_n = \hat{n} \times (E \times \hat{n}) + \hat{n}(\hat{n} \cdot E) \tag{2.100}$$

and rewrite the boundary conditions in the alternative vector form:

$$\hat{n} \times (E_1 \times \hat{n}) - \hat{n} \times (E_2 \times \hat{n}) = 0 \tag{2.101}$$

$$\hat{n} \times (H_1 \times \hat{n}) - \hat{n} \times (H_2 \times \hat{n}) = K \times \hat{n}. \tag{2.102}$$

Finally, since the fields on both sides of the boundary are related by Maxwell's equations, the boundary conditions for the normal and the tangential components are not independent of each other. On the contrary, the boundary conditions for the normal components are automatically satisfied if the boundary conditions for the tangential components hold everywhere at the boundary.

2.5 Spectral Representation of Time-Dependent Fields

We now discuss in more detail the complex field representation of Maxwell's equations and consider arbitrary time-dependent fields and for harmonic fields. These representations are very useful since they allow one to lower the dimensionality of Maxwell's equations and greatly simplify the solutions of *linear problems* with the aid of Fourier transform techniques.

The Fourier transform of an arbitrary time-dependent real field $E(r, t)$ is defined at position r by the integral:

$$\boxed{E(\mathbf{r}, \omega) = \int_{-\infty}^{\infty} E(\mathbf{r}, t) e^{j\omega t} dt} \tag{2.103}$$

where $j = \sqrt{-1}$ and ω [rad/s] is the angular frequency related to the time frequency ν [1/s or Hertz] by the usual relation $\omega = 2\pi\nu$. This integral provides the *complex Fourier amplitude* of the field and resolves the original time-dependent signal $E(r, t)$ into its frequency or *spectral components*. We remark that since the field $E(r, t)$ is a real-valued function, its spectral components obey the symmetry condition $E(\mathbf{r}, -\omega) = E^*(\mathbf{r}, \omega)$. Direct Fourier transformation of the materials' Maxwell's equations (2.69)–(2.72) produces the corresponding equations for their complex Fourier amplitudes:

$$\nabla \cdot \mathbf{D}(\mathbf{r}, \omega) = \varrho_{free}(\mathbf{r}, \omega) \tag{2.104}$$

$$\nabla \cdot \mathbf{B}(\mathbf{r}, \omega) = 0 \tag{2.105}$$

$$\nabla \times \mathbf{E}(\mathbf{r}, \omega) = j\omega \mathbf{B}(\mathbf{r}, \omega) \tag{2.106}$$

$$\nabla \times \mathbf{H}(\mathbf{r}, \omega) = -j\omega \mathbf{D}(\mathbf{r}, \omega) + \mathbf{J}_{free}(\mathbf{r}, \omega). \tag{2.107}$$

Note that in these equations the time derivatives do not appear anymore due to the fact that $\partial/\partial t \rightarrow -j\omega t$ in the transformed field quantities. Therefore, Fourier transforming the derivative of a field quantity is equivalent to multiplying it by a constant. Finally, once the Fourier amplitude $E(r, \omega)$ has been determined for the problem at hand, the corresponding time-dependent field can be recovered by the inverse Fourier transformation as follows:

$$E(\mathbf{r},t) = \frac{1}{2\pi} \int_{-\infty}^{\infty} \mathbf{E}(\mathbf{r},\omega)e^{-j\omega t}d\omega \tag{2.108}$$

This integral can be regarded as the linear superposition of complex fields that harmonically oscillate at different angular frequencies ω and with complex amplitudes $\mathbf{E}(\mathbf{r},\omega)$, that is, the Fourier transform, with respect to time, of the real field quantity $E(\mathbf{r},t)$.

2.5.1   Time-Harmonic Fields

Many problems of practical interest deal directly with *monochromatic* or time-harmonic fields, which are fields that oscillate at a single frequency. Since general time-dependent fields can be decomposed into a superposition of time-harmonic fields with different frequencies, time-harmonic fields play a pivotal role in the study of general time-varying problems.

A time-harmonic wave field is expressed as follows:

$$E(\mathbf{r},t) = E_0(\mathbf{r})\cos[\varphi(\mathbf{r}) - \omega t], \tag{2.109}$$

where E_0 denotes the (real or complex) field amplitude, $\varphi(\mathbf{r})$ the phase and ω the angular frequency. However, it is often more convenient to express a monochromatic real field as the real part of a corresponding *complex wave function*:

$$E(\mathbf{r},t) = \Re\{\mathbf{E}(\mathbf{r},\omega)e^{-j\omega t}\} = \frac{1}{2}\left[\mathbf{E}(\mathbf{r},\omega)e^{-j\omega t} + \mathbf{E}^*(\mathbf{r},\omega)e^{j\omega t}\right] \tag{2.110}$$

where $\Re\{\cdot\}$ and $\Im\{\cdot\}$ denote the real and imaginary parts of the quantity in brackets. The *complex amplitude* of the field is the space-dependent part of its complex wave function and can be written as follows:

$$\mathbf{E}(\mathbf{r},\omega) \equiv \mathbf{E}(\mathbf{r}) = |\mathbf{E}(\mathbf{r})|e^{j\varphi(\mathbf{r})}. \tag{2.111}$$

The amplitude of the real field and its phase can be obtained by the complex amplitude as follows:

$$E_0(\mathbf{r}) = |\mathbf{E}(\mathbf{r})| \tag{2.112}$$

$$\varphi(\mathbf{r}) = \arctan[\Im\{\mathbf{E}(\mathbf{r})\}/\Re\{\mathbf{E}(\mathbf{r})\}], \tag{2.113}$$

where we omitted the explicit reference to ω in the complex amplitude. The real harmonic wave field can be directly written in terms of its complex amplitude as follows:

$$\boxed{\boldsymbol{E}(\mathbf{r},t) = \Re\{\mathbf{E}(\mathbf{r})\}\cos(\omega t) + \Im\{\mathbf{E}(\mathbf{r})\}\sin(\omega t)}$$ (2.114)

where the in-phase component of the wave field $\boldsymbol{E}(\mathbf{r},t)$ is associated to the real part of its complex amplitude and its quadrature component dependents on the imaginary part of $\mathbf{E}(\mathbf{r})$.

For time-harmonic fields, $\partial/\partial t \rightarrow -j\omega t$, thus eliminating the time variable and reducing Maxwell's equations to the following simpler form:

$$\nabla \cdot \mathbf{D}(\mathbf{r}) = \varrho_{free}(\mathbf{r})$$ (2.115)

$$\nabla \cdot \mathbf{B}(\mathbf{r}) = 0$$ (2.116)

$$\nabla \times \mathbf{E}(\mathbf{r}) = j\omega\mathbf{B}(\mathbf{r})$$ (2.117)

$$\nabla \times \mathbf{H}(\mathbf{r}) = -j\omega\mathbf{D}(\mathbf{r}) + \mathbf{J}_{free}(\mathbf{r}).$$ (2.118)

All the quantities appearing in these equations are complex vectors or scalar functions that oscillate at the angular frequency ω, and the harmonic time dependence $\exp(-j\omega t)$ is therefore considered as implicit. The equations coincide with what previously derived using the spectral Fourier components of the field quantities in the time domain, with the difference that we dropped the explicit reference to the frequency ω.

2.6   Maxwell's Equations in K-Space

Using Fourier analysis in the space domain, we can obtain the so-called plane wave representation, or k-space representation, of general space-dependent fields, which is particularly useful in the context of scattering problems that involve layered media. At a given time t, we define the *Fourier spatial component*, or the (complex) spatial spectrum, of an arbitrary time- and space-dependent (real) field $\boldsymbol{E}(\mathbf{r},t)$ as follows:

$$\boxed{\mathbf{E}(\mathbf{k},t) = \int_{-\infty}^{\infty} \boldsymbol{E}(\mathbf{r},t)e^{-j\mathbf{k}\cdot\mathbf{r}}d^3\mathbf{r}}$$ (2.119)

The vector $\mathbf{k} = (\omega/c)\hat{\mathbf{n}}$ (where c is the speed of light and $\hat{\mathbf{n}}$ is a unit vector along the propagation direction of the wave) is the wavevector, and its amplitude $k = (\omega/c) = 2\pi/\lambda$ (where λ is the wavelength) is called the wavenumber. Note that the phase of the exponential term in equation (2.119) has an opposite sign convention compared to equation (2.103). This choice is physically more convenient because it results in forward-propagating plane wave components with positive phase velocity.

The inverse Fourier transform returns back the real wave field $\boldsymbol{E}(\mathbf{r},t)$ as a superposition of its *Fourier spatial harmonics*:

$$\boxed{\boldsymbol{E}(\mathbf{r},t) = \frac{1}{(2\pi)^3} \int_{-\infty}^{\infty} \mathbf{E}(\mathbf{k},t)e^{j\mathbf{k}\cdot\mathbf{r}}d^3\mathbf{k}}$$ (2.120)

We remark that since the wave field $E(\mathbf{r},t)$ is real, the following symmetry property (Hermitian symmetry) of its spatial spectrum must hold:

$$\mathbf{E}^*(\mathbf{k},t) = \mathbf{E}(-\mathbf{k},t). \tag{2.121}$$

The materials' Maxwell's equations in k-space can be immediately written down by spatial Fourier transformation:

$$j\mathbf{k} \cdot \mathbf{D}(\mathbf{k},t) = \varrho_{free}(\mathbf{k},t) \tag{2.122}$$

$$j\mathbf{k} \cdot \mathbf{B}(\mathbf{k},t) = 0 \tag{2.123}$$

$$j\mathbf{k} \times \mathbf{E}(\mathbf{k},t) = -\frac{\partial \mathbf{B}(\mathbf{k},t)}{\partial t} \tag{2.124}$$

$$j\mathbf{k} \times \mathbf{H}(\mathbf{k},t) = \frac{\partial \mathbf{D}(\mathbf{k},t)}{\partial t} + \mathbf{J}_{free}(\mathbf{k},t). \tag{2.125}$$

Note that similarly to the previously discussed case of time harmonic fields, when the gradient operator ∇ is applied to a spatial Fourier harmonic it produces the scalar or dot-multiplication by the vector $j\mathbf{k}$.

The time derivatives appearing in the preceding equations only depend on the values of the fields at the same point $\mathbf{k}$. Therefore, Maxwell's equations in k-space are always local. Moreover, we can deduce that for a linear, homogeneous, isotropic, and source-free medium, the electric and magnetic fields are orthogonal to each other and to the propagation direction $\mathbf{k}$, thus forming a right-handed triad at any given time t. In addition, when the fields are also monochromatic at frequency ω, their complex wave functions are homogeneous plane waves:

$$\mathbf{E}(\mathbf{r},t) = \mathbf{E}_0 e^{j(\mathbf{k}\cdot\mathbf{r}-\omega t)} \tag{2.126}$$

$$\mathbf{H}(\mathbf{r},t) = \mathbf{H}_0 e^{j(\mathbf{k}\cdot\mathbf{r}-\omega t)}. \tag{2.127}$$

In this case, Maxwell's equations in k-space simply reduce to the following:

$$\mathbf{k} \times \mathbf{E}_0 = \omega\mu\mathbf{H}_0 \tag{2.128}$$

$$\mathbf{k} \times \mathbf{H}_0 = -\omega\mu\mathbf{E}_0, \tag{2.129}$$

where $\mathbf{E}_0$ and $\mathbf{H}_0$ are constant vectors called complex envelopes. The other two Maxwell's equations (2.122) and (2.123) are identically satisfied in this case because the divergence of a uniform plane wave is always zero. Since the E and H fields are at all times mutually orthogonal and perpendicular to $\mathbf{k}$, the solutions of equations (2.128) and (2.129) are called *transverse electromagnetic* (TEM) waves. From equations (2.128) and (2.129), it follows that the ratio between the amplitudes of the electric and the magnetic field is as follows:

$$\eta = \frac{|\mathbf{E}_0|}{|\mathbf{H}_0|} = \sqrt{\frac{\mu}{\epsilon}}, \tag{2.130}$$

which is known as the *impedance* of the medium. For free space, $\epsilon_r = \mu_r = 1$, and the impedance reduces to $\eta_0 \approx 377\,\Omega$.

We conclude this section with a derivation of the Green functions of a homogeneous medium in the k-space representation. This task can easily be accomplished by first representing the real-space Green functions $G_0^\pm(R,\omega)$ and $\delta(\mathbf{r} - \mathbf{r}')$ in plane wave components:

$$G_0^\pm(R,\omega) = \int_{-\infty}^{\infty} \frac{d^3k}{(2\pi)^3} e^{jk\cdot(\mathbf{r}-\mathbf{r}')} G_0^\pm(k,\omega) \tag{2.131}$$

$$\delta(R) = \int_{-\infty}^{\infty} \frac{d^3k}{(2\pi)^3} e^{jk\cdot(\mathbf{r}-\mathbf{r}')}, \tag{2.132}$$

where $G_0^\pm(k,\omega)$ are the *k-space Green's functions*. Note that the wavenumber k is the Fourier conjugate variable of the relative displacement $R = |\mathbf{r}-\mathbf{r}'|$ between the source and the observation points, which reflects the homogeneity of the medium. We then insert the preceding expressions into the defining equation for the real-space Green functions:

$$\nabla_r^2 G_0^\pm(R,\omega) + k_0^2 G_0^\pm(R,\omega) = -\delta(R) \tag{2.133}$$

and note that the Laplacian ∇_r^2 acts only on the variable r, which allows us to obtain the k-space Green functions:

$$\boxed{G_0^\pm(k,\omega) = \frac{1}{k^2 - k_0^2 \pm j\eta}} \tag{2.134}$$

The small imaginary part $\pm j\eta$ has been added in the denominator of the last equation in order regularize the Green functions and avoid the divergence when the k-vector components of the source coincide with the dispersion relation of the medium $k^2 = k_0^2(\omega)$. This happens when the system is excited by any of its *natural oscillation modes*, similarly to the case of a forced harmonic oscillator. The regularization procedure permits us to perform the inverse transformation back to the spatial domain in all cases.

Finally, it is also important to realize that the expression (2.134) for the k-space Green functions is independent of the spatial dimensionality d of the system. This is to be contrasted with the real-space Green functions, whose expressions sensibly depend on the space dimensionality. For example, if we consider a homogeneous medium in $d = 2$ dimensions, we have the following:

$$G_0^\pm(R,\omega) = \pm\frac{j}{4} H_0^{(1,2)}(k_0|\rho - \rho'|), \tag{2.135}$$

where $H_0^{(1,2)}$ are zero-order Hankel functions of the first (1) and second (2) kind, physically representing outgoing or incoming cylindrical waves. On the other hand, for a homogeneous medium in dimension $d = 1$, the real-space Green functions are simple plane waves:

$$G_0^{\pm}(R,\omega) = \pm \frac{j}{2k_0} e^{jk_0|x-x'|}. \tag{2.136}$$

The coefficients in the expressions (2.135) and (2.136) are chosen such that the equations are consistent with the Poisson problems ($k_0 = 0$) for the appropriate dimensionality of the space.

2.7   General Wave Equations

We now discuss the general form of the electromagnetic field equations in materials. In order to achieve this goal, we first recall the definitions of the auxiliary fields D and H:

$$D(\mathbf{r},t) = \epsilon_0 E(\mathbf{r},t) + P(\mathbf{r},t) \tag{2.137}$$

$$H(\mathbf{r},t) = B(\mathbf{r},t)/\mu_0 - M(\mathbf{r},t). \tag{2.138}$$

If we substitute these definitions into Maxwell's curl equations (2.71) and (2.72), take the curl on both sides, and combine the two resulting expressions, we obtain the generally valid inhomogeneous wave equations:

$$\nabla \times \nabla \times E + \frac{1}{c_0^2} \frac{\partial^2 E}{\partial t^2} = -\mu_0 \frac{\partial}{\partial t} \left(J_{free} + \frac{\partial P}{\partial t} + \nabla \times M \right) \tag{2.139}$$

$$\nabla \times \nabla \times H + \frac{1}{c_0^2} \frac{\partial^2 H}{\partial t^2} = \nabla \times J_{free} + \nabla \times \frac{\partial P}{\partial t} - \frac{1}{c_0^2} \frac{\partial^2 M}{\partial t^2} \tag{2.140}$$

where $c_0^2 = 1/\epsilon_0 \mu_0$ and c_0 is the speed of light in free space (i.e., in vacuum). Note that the source terms in the inhomogeneous wave equations include the contributions of the polarization and magnetization currents. The equations (2.139) and (2.140) are derived directly from Maxwell's equations without any assumptions and therefore are *generally valid in any medium*.

2.7.1   The Complex Refractive Index

We look at how the general wave equations can be reduced to a simpler form in a few particular cases of interest. Let us first consider the case of a nonmagnetic ($M = 0$), dielectric ($\sigma = 0$), and homogeneous medium, such as a semiconductor material in a source-free volume region. Using the well-known vector identity $\nabla \times \nabla \times E = \nabla(\nabla \cdot E) - \nabla^2 E$ along with the first Maxwell's equation, we obtain the wave equation, for the electric field:

$$\nabla^2 E - \frac{1}{c_0^2} \frac{\partial^2 E}{\partial t^2} = \mu_0 \frac{\partial^2 P}{\partial t^2} \tag{2.141}$$

Equation (2.141), which describes the behavior of doped semiconductors from the infrared (IR) to optical frequencies, is known as the *semiconductor wave equation* (SVE). This equation is often used as the starting point for the theory of wave propagation in nonlinear media [95]. In the particular case of *linear, isotropic, and nondispersive media*, the polarization vector is linearly related to the field (see equation (2.77)) and the SVE reduces to the following:

$$\nabla^2 E - \epsilon_0 \mu_0 \epsilon_r \frac{\partial^2 E}{\partial t^2} = 0, \tag{2.142}$$

where $\epsilon_r = 1 + \chi_e$. If we include the magnetic properties in our description, the previous wave equation becomes:

$$\nabla^2 E - \epsilon_0 \epsilon_r \mu_0 \mu_r \frac{\partial^2 E}{\partial t^2} = 0. \tag{2.143}$$

Equation (2.143) allows us to define the speed of light in a material as c_0/n where n is the *refractive index* of the medium given by the following:

$$n = \pm \sqrt{\epsilon_r \mu_r}, \tag{2.144}$$

where the negative sign is chosen if ϵ_r and μ_r are both negative, as it is the case for *doubly negative metamaterials*. The general solutions of equation (2.143) are all waves of the form:

$$E(\mathbf{r}, t) = E_0 f(\mathbf{k} \cdot \mathbf{r} - \omega t), \tag{2.145}$$

where E_0 is a constant vector and f is an arbitrary function whose second derivative exists. Inserting the ansatz (2.145) into equation (2.143), we find the *dispersion relation* that connects the angular frequency and the wavenumber with the speed of light in the medium:

$$\frac{\omega}{k} = \frac{c_0}{n}, \tag{2.146}$$

where $k = |\mathbf{k}| = 2\pi/\lambda$ and $\lambda = \lambda_0/n$ are the wavenumber and the wavelength in the medium, respectively. In a temporally dispersive medium, the relative permittivity and permeability become *frequency dependent quantities*. In this case, the constitutive relations (2.77)–(2.79) connect the *complex amplitudes* of the respective field quantities (i.e., their Fourier transforms), and the causality principle requires the relative permittivity and permeability to become complex-valued functions of frequency. This leads to define the *complex refractive index*:

$$\tilde{n} = \sqrt{\epsilon_r(\omega)\mu_r(\omega)} = n + j\kappa, \tag{2.147}$$

where n is the *real refractive index* and κ is called the *extinction coefficient*. The wave equation for the complex field amplitude $\mathbf{E}(\mathbf{r}, \omega)$ becomes the Helmholtz equation:

$$\boxed{\nabla^2 \mathbf{E}(\mathbf{r}, \omega) + \omega^2 \epsilon_0 \mu_0 \epsilon_r(\omega) \mu_r(\omega) \mathbf{E}(\mathbf{r}, \omega) = 0} \tag{2.148}$$

where

$$\omega^2 \epsilon_0 \mu_0 \epsilon_r(\omega)\mu_r(\omega) = \left(\frac{\omega}{c_0}\right)^2 \epsilon_r(\omega)\mu_r(\omega) = k_0^2 \tilde{n}^2 = \tilde{k}^2. \tag{2.149}$$

Therefore, wave propagation through a dispersive medium is described by a Helmholtz wave equation with a complex wavenumber $\tilde{k} = k_0 \tilde{n}$ (and a complex wavevector $\tilde{\mathbf{k}} = \mathbf{k}_0 \tilde{n}$). It is therefore clear that a progressive plane wave propagating through a medium with a complex wavenumber $\tilde{k}$ will experience attenuation. In fact, for a plane-wave propagating along the z-direction, we can write the following:

$$\mathbf{E} = \mathbf{E}_0 e^{j(\tilde{n}k_0 z - \omega t)} = \mathbf{E}_0 e^{-\frac{2\pi\kappa z}{\lambda}} e^{j\left(\frac{2\pi n z}{\lambda} - \omega t\right)}. \tag{2.150}$$

Since the wave intensity is proportional to the square of its complex function, equation (2.150) implies that the intensity decays as follows:

$$I = I_0 e^{-\alpha z}. \tag{2.151}$$

The $1/e$ decay length of the intensity is called the *absorption coefficient*. We see from equation (2.150) that the absorption coefficient is related to the imaginary part of the complex refractive index:

$$\boxed{\alpha = \frac{4\pi\kappa}{\lambda}} \tag{2.152}$$

In the case of a nonmagnetic material, the real and imaginary parts of the complex refractive index are simply connected to the real and imaginary parts of the complex permittivity $\epsilon_r = \epsilon_r' + j\epsilon_r''$. By squaring equation (2.147) and equating the coefficients of the real and imaginary parts, we can express the real and imaginary parts of the permittivity in terms of the refractive index and the extinction coefficient:

$$\epsilon_r' = n^2 - \kappa^2 \tag{2.153}$$
$$\epsilon_r'' = 2n\kappa. \tag{2.154}$$

By solving for n and κ, we can invert these equations and express n and κ as a function of ϵ_r' and ϵ_r'':

$$\boxed{n = \sqrt{\frac{\sqrt{\epsilon_r'^2 + \epsilon_r''^2} + \epsilon_r'}{2}}} \tag{2.155}$$

$$\boxed{\kappa = \sqrt{\frac{\sqrt{\epsilon_r'^2 + \epsilon_r''^2} - \epsilon_r'}{2}}} \tag{2.156}$$

We finally consider the case of wave propagation in a medium that is *inhomogeneous* with a permittivity that depends on the position vector $\mathbf{r}$ (but not on time). If the medium is additionally *linear, isotropic, and nondispersive*, it can be obtained from

Maxwell's equations that the real-valued electric field obeys the following *inhomogeneous wave equation:*[17]

$$\nabla^2 \mathbf{E} + 2\nabla(\mathbf{E} \cdot \nabla \ln n) - \frac{1}{c^2}\frac{\partial^2 \mathbf{E}}{\partial t^2} = 0. \tag{2.157}$$

It is evident that in the case of an homogeneous medium, equation (2.157) reduces to (2.143). For a monochromatic complex field $\mathbf{E}(\mathbf{r}, \omega)$ and a dielectric medium with dielectric function $\epsilon(\mathbf{r}, \omega)$, we can express equation (2.157) as follows:

$$\nabla^2 \mathbf{E}(\mathbf{r}, \omega) + k_0^2 \epsilon(\mathbf{r}, \omega)\mathbf{E}(\mathbf{r}, \omega) + \nabla\left[\mathbf{E}(\mathbf{r}, \omega) \cdot \nabla \ln \epsilon(\mathbf{r}, \omega)\right] = 0. \tag{2.158}$$

For the magnetic field, it can easily be shown that the following equation holds:

$$\nabla^2 \mathbf{H}(\mathbf{r}, \omega) + k_0^2 \epsilon(\mathbf{r}, \omega)\mathbf{H}(\mathbf{r}, \omega) + \nabla \ln \epsilon(\mathbf{r}, \omega) \times (\nabla \times \mathbf{H}(\mathbf{r}, \omega)) = 0. \tag{2.159}$$

The preceding two vector equations are quite complicated to solve, but for most problems of light propagation in inhomogeneous media, they can be replaced by simpler scalar Helmholtz wave equations valid for each component of the electric and magnetic field vectors. It can be simply shown by estimating the order of magnitude of the third terms in equations (2.158) and (2.159) that their contributions to the wave equation can be neglected when the relative change of the dielectric constant $\epsilon = n^2$ over a distance of the order of the wavelength is much smaller than unity, or $\Delta n^2/n^2 \ll 1$. Under these condition, which defines the regime of *weak inhomogeneity*, the scalar Helmholtz equation will be approximately satisfied by each field component, generically denoted by the complex amplitude ψ. This implies that the electromagnetic field components become uncoupled and that the polarization of the field is preserved through weakly inhomogeneous media, or *GRIN media*.

On the other hand, when the refractive index n (or the permittivity ϵ) vary strongly with position, the inhomogeneous gradient terms in the wave equations will become significant and will introduce *coupling between the different components of the electric field*. In this case, the full-vector wave equations (2.158) and (2.159) must be considered, and the simpler scalar approximation will fail. This situation regularly takes place when a wave that propagates in a homogeneous medium encounters a *sharp interface* with a different homogeneous medium, such that the field boundary conditions must be applied. However, these effects and the deviations from the scalar optical theory are generally small if the boundary conditions affect only a small region compared to the one where the wave field is propagating. This is the case, for instance, in light diffraction from a large aperture compared to the wavelength, since the fields are modified by the edges of the aperture over a region that extends only few wavelengths beyond the aperture. Unless we are interested in describing the field in very close proximity to the diffraction apertures, we can neglect the full-vector coupling effect and use scalar wave optics. However, vector wave coupling becomes essential

---

[17] The gradient term can be neglected when the optical constants of the inhomogeneous medium vary only slightly over distances compared to the wavelength. In such cases, the inhomogeneous nature of the medium can be well approximated by simply considering position-dependent optical constants.

when light is diffracted by optically thick inhomogeneous and dispersive media such as grating structures (dielectric or metallic) due to predominant edge effects or in the study of subwavelength apertures or scattering structures frequently encountered in *plasmonics technology* [71, 97–99].

2.8   Optical Radiation

In this section, we apply the Green function approach to the analysis of optical radiation. After introducing the important concept of the dyadic functions, we will discuss general source–field relations and specialize them to handle vector scattering problems reformulated in terms of equivalent radiation problems. A derivation of the volume integral equations of vector scattering will be provided and an application to the far-field scattering solution by small objects will be addressed.

2.8.1   The Dyadic Green Function

In this section, we will extend the Green function approach in order to directly connect the electromagnetic field with its sources in order to obtain a consistent description of vector radiation problems. This will be achieved by introducing the all-important dyadic notation and the dyadic Green function that we will later utilize to reformulate general scattering problems in terms of integral equations.

Let us start by restating the field equations (2.14) and (2.15) in the frequency domain representation where the complex-valued electromagnetic potentials obey the following equations:

$$\mathbf{B}(\mathbf{r}) = \nabla \times \mathbf{A}(\mathbf{r}) \tag{2.160}$$

$$\mathbf{E}(\mathbf{r}) = j\omega\mathbf{A}(\mathbf{r}) - \nabla\phi(\mathbf{r}). \tag{2.161}$$

Here $\mathbf{A}(\mathbf{r})$ and $\phi(\mathbf{r})$ are complex vectors. These equations are complemented by the frequency domain counterpart of the Lorenz gauge condition (2.18):

$$\nabla \cdot \mathbf{A}(\mathbf{r}) - j\omega\epsilon\mu\phi = 0. \tag{2.162}$$

Considering a homogeneous and linear medium characterized by ϵ and μ and assuming the Lorenz gauge, the potentials $\mathbf{A}(\mathbf{r})$ and $\phi(\mathbf{r})$ satisfy the Helmholtz equations:

$$(\nabla^2 + k^2)\mathbf{A}(\mathbf{r}) = -\mu\mathbf{J}(\mathbf{r}) \tag{2.163}$$

$$(\nabla^2 + k^2)\phi(\mathbf{r}) = -\frac{\varrho}{\epsilon} \tag{2.164}$$

obtained by the Fourier transform of the corresponding time-domain equations (2.19) and (2.20). As discussed in Section 2.3.2, the vector potential $\mathbf{A}(\mathbf{r})$ can be written in terms of the frequency-domain (retarded) Green function as follows:

$$A(\mathbf{r}) = \mu \int_V \mathbf{J}(\mathbf{r}') G_0^+(\mathbf{r},\mathbf{r}') d^3\mathbf{r}' = \frac{\mu}{4\pi} \int_V \mathbf{J}(\mathbf{r}') \frac{e^{jkR}}{R} d^3\mathbf{r}', \tag{2.165}$$

where $R = |\mathbf{r} - \mathbf{r}'|$, $\mu = \mu_0 \mu_r$ and k refers to the wavenumber in the medium. The current density $\mathbf{J}$ contains all the sources of the fields. The solution for the scalar electric potential, which we already found in Section 2.3.2, is given by the following:

$$\phi(\mathbf{r}) = \frac{1}{\epsilon_0} \int_V \rho(\mathbf{r}') G_0^+(\mathbf{r},\mathbf{r}') d^3\mathbf{r}' = \frac{1}{4\pi\epsilon_0} \int_V \rho(\mathbf{r}') \frac{e^{jk_0 R}}{R} d^3\mathbf{r}'. \tag{2.166}$$

Once the vector potential $\mathbf{A}$ has been determined, we can obtain the electromagnetic field produced by the source current density $\mathbf{J}$:

$$\mathbf{E} = j\omega\mathbf{A} - \frac{1}{j\omega\epsilon\mu}\nabla(\nabla \cdot \mathbf{A}) \tag{2.167}$$

$$\mathbf{H} = \frac{1}{\mu}\nabla \times \mathbf{A}, \tag{2.168}$$

where in equation (2.167) the scalar potential ϕ has been eliminated using the Lorenz condition (2.162). It is clear from equation (2.165) that the vector potential $\mathbf{A}$ is parallel to the current density $\mathbf{J}$ that generates it. However, the electric and magnetic fields as obtained through the vector operations in (2.167) and (2.168) depend in a more complex way on the orientation of the source current density. In particular, a current density aligned along a coordinate axis (e.g., the x-axis) will create an electric and magnetic field with nonzero x-, y-, and z-components. It is then clear that the relation between a unit current source and the resulting electric and magnetic fields is expressed by a more general Green function with a tensor nature. This more general type of Green function is the *dyadic Green function* [100] that provides the solution of the following *vector wave equation*:[18]

$$\nabla \times \nabla \times \mathbf{E}(\mathbf{r}) - k^2 \mathbf{E}(\mathbf{r}) = j\omega\mu\mathbf{J}(\mathbf{r}). \tag{2.169}$$

The dyadic free-space Green function $\overleftrightarrow{\mathbf{G}_0}(\mathbf{r},\mathbf{r}')$ associated to the vector wave equation is the vector propagator or impulse response that satisfies the following:[19]

$$\nabla \times \nabla \times \overleftrightarrow{\mathbf{G}_0}(\mathbf{r},\mathbf{r}') - k^2 \overleftrightarrow{\mathbf{G}_0}(\mathbf{r},\mathbf{r}') = \overleftrightarrow{\mathbf{I}}\,\delta(\mathbf{r} - \mathbf{r}'), \tag{2.170}$$

where $\overleftrightarrow{\mathbf{I}} = \hat{\mathbf{x}}\hat{\mathbf{x}} + \hat{\mathbf{y}}\hat{\mathbf{y}} + \hat{\mathbf{z}}\hat{\mathbf{z}}$ is the dyadic representation of the identity operator (i.e., the unit dyadic). The dyadic Green's function gives the vector electric field at position $\mathbf{r}$ due to a point source excitation located at $\mathbf{r}'$, *for any possible orientation of the source*. In Cartesian coordinates, $\overleftrightarrow{\mathbf{G}_0}$ is a second-rank tensor describing the general linear relation between the fields at $\mathbf{r}$ and an arbitrarily oriented point current source located at $\mathbf{r}'$. In particular, the first column of the 3×3 matrix representation of $\overleftrightarrow{\mathbf{G}_0}$ is

---

[18] The vector wave equation can immediately be obtained by combining via a curl operation the two dynamic Maxwell's equations in the frequency domain.

[19] Note here the positive sign in front of the delta excitation. This is to maintain consistency with the positive source term in equation (2.169).

the vector field due to a point source aligned along the x-direction, the second column is the field due to a point source in the y-direction, and the third column is the field due to a point source in the z-direction. As a result, we should consider equation (2.170) a compact notation for the three vector Green functions that specify the field response to the three principal orientations of a vector point current source.

2.8.2   Dyads and Dyadics

Before proceeding with the derivation of the general link between the current sources and the resulting fields, we will review basic concepts related to dyads and dyadics that are used extensively in the analysis of electromagnetic problems [100].

A dyad $\overleftrightarrow{\mathbf{D}}$ is an algebraic object formed by placing two vectors $\mathbf{A}$ and $\mathbf{B}$ side by side:[20]

$$\overleftrightarrow{\mathbf{D}} \equiv \mathbf{AB}. \tag{2.171}$$

Obviously, the dyad $\overleftrightarrow{\mathbf{D}} = \mathbf{AB}$ contains six independent scalar quantities. Dyads are defined by the way they act on vectors. It is convenient to introduce two scalar and vector products for dyads. The *anterior scalar product* of the dyad $\overleftrightarrow{\mathbf{D}}$ with the vector $\mathbf{C}$ is as follows:

$$\mathbf{C} \cdot \overleftrightarrow{\mathbf{D}} = (\mathbf{C} \cdot \mathbf{A})\,\mathbf{B}, \tag{2.172}$$

where the dot in parentheses is the familial scalar product between vectors. Note that the anterior scalar product with $\mathbf{C}$ is a vector parallel to $\mathbf{B}$. The *posterior scalar product* is defined by the following:

$$\overleftrightarrow{\mathbf{D}} \cdot \mathbf{C} = \mathbf{A}\,(\mathbf{B} \cdot \mathbf{C}), \tag{2.173}$$

which is a vector parallel to $\mathbf{A}$, and therefore $\mathbf{C} \cdot \overleftrightarrow{\mathbf{D}} \neq \overleftrightarrow{\mathbf{D}} \cdot \mathbf{C}$. It is also useful to define two vector products, anterior and posterior. In particular, the *anterior vector product* is as follows:

$$\mathbf{C} \times \overleftrightarrow{\mathbf{D}} = (\mathbf{C} \times \mathbf{A})\,\mathbf{B}, \tag{2.174}$$

where the cross in parentheses is the familial vector product between vectors, which generates another vector. As a result, $\mathbf{C} \times \overleftrightarrow{\mathbf{D}}$ is a dyad, not a vector, since it is obtained by placing the two vectors $\mathbf{C} \times \mathbf{A}$ and $\mathbf{B}$ side by side. Analogously, the *posterior vector product* is as follows:

$$\overleftrightarrow{\mathbf{D}} \times \mathbf{C} = \mathbf{A}\,(\mathbf{B} \times \mathbf{C}), \tag{2.175}$$

which is a different dyad compared to the anterior product. A dyad also has its curl and divergence, and a dyad is formed by applying a gradient to a vector.

---

[20] The word *dyad* means pair. This notation was originally established by the American physicist and mathematician Josiah Willard Gibbs (1839–1903), who greatly contributed to the development of modern vector analysis.

A *dyadic* is a linear combination of dyads. If we select a Cartesian reference system with orthogonal unit vectors denoted by $\hat{\mathbf{x}}$, $\hat{\mathbf{y}}$, and $\hat{\mathbf{z}}$, we can write a *general dyadic* as:

$$\overleftrightarrow{\mathbf{D}} = \mathbf{D}_x\hat{\mathbf{x}} + \mathbf{D}_y\hat{\mathbf{y}} + \mathbf{D}_z\hat{\mathbf{z}}, \tag{2.176}$$

where $\mathbf{D}_x = (D_{xx}, D_{yx}, D_{zx})$, $\mathbf{D}_y = (D_{xy}, D_{yy}, D_{zy})$ and $\mathbf{D}_z = (D_{xz}, D_{yz}, D_{zz})$ are vectors. Therefore, we can expand the general dyadic (2.176) in Cartesian components as follows:

$$\overleftrightarrow{\mathbf{D}} = D_{xx}\hat{\mathbf{x}}\hat{\mathbf{x}} + D_{yx}\hat{\mathbf{y}}\hat{\mathbf{x}} + D_{zx}\hat{\mathbf{z}}\hat{\mathbf{x}} + D_{xy}\hat{\mathbf{x}}\hat{\mathbf{y}} + D_{yy}\hat{\mathbf{y}}\hat{\mathbf{y}} + D_{zy}\hat{\mathbf{z}}\hat{\mathbf{y}}$$
$$+ D_{xz}\hat{\mathbf{x}}\hat{\mathbf{z}} + D_{yz}\hat{\mathbf{y}}\hat{\mathbf{z}} + D_{zz}\hat{\mathbf{z}}\hat{\mathbf{z}}. \tag{2.177}$$

The expansion (2.177) shows that a general dyadic has nine components, similarly to a second-rank tensor. In fact, we can regard a dyadic as a second-order tensor written in a special notation, formed by juxtaposing pairs of vectors, along with rules for manipulating such expressions analogous to the rules for matrix algebra. However, while tensor components are usually defined with respect to transformations of different reference systems, the emphasis with dyadics is on their algebraic properties within a given reference frame. This is consistent with the fact that when a dyadic is applied to a vector (i.e., its scalar product with a vector) it linearly changes both the magnitude and the direction of that vector, exactly like tensors do.

A special dyadic is the unit dyadic, defined in Cartesian coordinates by the following:[21]

$$\overleftrightarrow{\mathbf{I}} = \hat{\mathbf{x}}\hat{\mathbf{x}} + \hat{\mathbf{y}}\hat{\mathbf{y}} + \hat{\mathbf{z}}\hat{\mathbf{z}}. \tag{2.178}$$

It is evident by the previous definitions that $\mathbf{C} \cdot \overleftrightarrow{\mathbf{I}} = \overleftrightarrow{\mathbf{I}} \cdot \mathbf{C} = \mathbf{C}$, as expected for the unit operator. For additional properties and theorems on dyads and dyadics in electromagnetic theory, we refer to the specialized literature on the subject, such as the excellent book by Tai [100].

2.8.3 Dyadic Source-Field Relations

In the previous section, we expressed the electromagnetic field $(\mathbf{E}, \mathbf{H})$ in terms of the vector potential, which in turn is related to the current sources. In this formulation, the vector potential plays the role of an intermediate function, with no direct physical meaning. Here we will show how the fields can be obtained directly in terms of their sources without resorting to potentials. This goal is achieved by the dyadic Green function that enables the solution of inhomogeneous vector equations with arbitrary vector sources. For this purpose, let us consider the time-harmonic Maxwell's equations in a uniform and linear medium:

$$\nabla \cdot \mathbf{E} = \varrho/\epsilon \tag{2.179}$$

---

[21] In other orthogonal coordinate systems the unit dyadic is defined analogously. For instance, in spherical polar coordinates, it is given by $\overleftrightarrow{\mathbf{I}} = \hat{\mathbf{r}}\hat{\mathbf{r}} + \hat{\boldsymbol{\theta}}\hat{\boldsymbol{\theta}} + \hat{\boldsymbol{\varphi}}\hat{\boldsymbol{\varphi}}$, where $\hat{\mathbf{r}}$, $\hat{\boldsymbol{\theta}}$, and $\hat{\boldsymbol{\varphi}}$ are the corresponding unit vectors.

$$\nabla \cdot \mathbf{H} = 0 \tag{2.180}$$

$$\nabla \times \mathbf{E} = j\omega\mu\mathbf{H} \tag{2.181}$$

$$\nabla \times \mathbf{H} = -j\omega\epsilon\mathbf{E} + \mathbf{J}, \tag{2.182}$$

where we have indicated by $(\varrho, \mathbf{J})$ the (free) sources of the fields, which are assumed to be perfectly specified for radiation problems. The more difficult case of scattering problems, where sources are not known a priori, will be addressed in the next section.

If we take the curl of equation (2.181) and substitute from equation (2.182), we get the vector wave equation:

$$\nabla \times \nabla \times \mathbf{E} - k^2\mathbf{E} = j\omega\mu\mathbf{J}, \tag{2.183}$$

where $k^2 = \omega^2\epsilon\mu$. Using the identity $\nabla \times \nabla \times \mathbf{E} = -\nabla^2\mathbf{E} + \nabla(\nabla \cdot \mathbf{E})$, the equation (2.179), and the continuity equation in order to eliminate the charge density $\varrho/\epsilon = \nabla \cdot \mathbf{J}/j\omega\epsilon$, we can recast the vector wave equation in the following form:

$$\nabla^2\mathbf{E} + k^2\mathbf{E} = -j\omega\mu\left[\mathbf{J} + \frac{\nabla(\nabla \cdot \mathbf{J})}{k^2}\right]. \tag{2.184}$$

The solution of the equation (2.184) at points outside the source volume V can be found using the general Green function method, i.e., by integrating the Green function of the unbounded homogeneous medium over the source term. This will provide the *particular solution* for the problem. By then adding a homogeneous field solution $\mathbf{E}_0$, we obtain the *general solution* as follows:

$$\mathbf{E}(\mathbf{r}) = \mathbf{E}_0(\mathbf{r}) + \frac{j\omega\mu}{4\pi}\int_V \left\{\mathbf{J}(\mathbf{r}') + \frac{\nabla'[\nabla' \cdot \mathbf{J}(\mathbf{r}')]}{k^2}\right\}\frac{e^{jk|\mathbf{r}-\mathbf{r}'|}}{|\mathbf{r} - \mathbf{r}'|}d^3\mathbf{r}', \tag{2.185}$$

where ∇' indicates that the derivatives are taken with respect to the variable $\mathbf{r}'$ and the integral extends over the volume V, where the sources are different from zero. Similarly, for the magnetic field $\mathbf{H}$, we obtain the following:

$$\mathbf{H}(\mathbf{r}) = \mathbf{H}_0(\mathbf{r}) + \frac{1}{4\pi}\int_V \{\nabla' \times \mathbf{J}(\mathbf{r}')\}\frac{e^{jk|\mathbf{r}-\mathbf{r}'|}}{|\mathbf{r} - \mathbf{r}'|}d^3\mathbf{r}'. \tag{2.186}$$

An alternative form of the field equations (2.185) and (2.186) can be found by substituting the expressions for the electromagnetic potentials (2.165) and (2.166) into $\mathbf{E}(\mathbf{r}) = j\omega\mathbf{A}(\mathbf{r}) - \nabla\phi(\mathbf{r})$, allowing us to express the electric field generated by the current source $\mathbf{J}$ as follows:

$$\mathbf{E}(\mathbf{r}) = j\omega\mu\int_V G_0^+(\mathbf{r}, \mathbf{r}')\mathbf{J}(\mathbf{r}')d^3\mathbf{r}' - \frac{\nabla}{j\omega\epsilon}\int_V G_0^+(\mathbf{r}, \mathbf{r}')\nabla' \cdot \mathbf{J}(\mathbf{r}')d^3\mathbf{r}', \tag{2.187}$$

where we have also used the continuity equation $\nabla \cdot \mathbf{J}(\mathbf{r}) = j\omega\rho$. Integrating equation (2.187) by parts, remembering that $\mathbf{J}$ is supported only within the volume V, and using the identity $\nabla G_0^+(\mathbf{r},\mathbf{r}') = -\nabla' G_0^+(\mathbf{r},\mathbf{r}')$, we obtain the following:

$$\mathbf{E}(\mathbf{r}) = j\omega\mu \int_V G_0^+(\mathbf{r},\mathbf{r}')\mathbf{J}(\mathbf{r}')d^3\mathbf{r}' - \frac{\nabla\nabla\cdot}{j\omega\epsilon}\int_V G_0^+(\mathbf{r},\mathbf{r}')\mathbf{J}(\mathbf{r}')d^3\mathbf{r}' \qquad (2.188)$$

This expression can be written more compactly using dyadic notation,[22] which allows us to express the electric and the magnetic fields directly in terms of their vector sources:

$$\mathbf{E}(\mathbf{r}) = \mathbf{E}_0(\mathbf{r}) + j\omega\mu \int_V \overset{\leftrightarrow}{\mathbf{G}}_0(\mathbf{r},\mathbf{r}') \cdot \mathbf{J}(\mathbf{r}')d^3\mathbf{r}' \qquad (2.189)$$

$$\mathbf{H}(\mathbf{r}) = \mathbf{H}_0(\mathbf{r}) + \int_V \left[\nabla \times \overset{\leftrightarrow}{\mathbf{G}}_0(\mathbf{r},\mathbf{r}') \right] \cdot \mathbf{J}(\mathbf{r}')d^3\mathbf{r}' \qquad (2.190)$$

where we have introduced the free-space *dyadic Green function* $\overset{\leftrightarrow}{\mathbf{G}}$ as:

$$\overset{\leftrightarrow}{\mathbf{G}}_0(\mathbf{r},\mathbf{r}') = \left[\overset{\leftrightarrow}{\mathbf{I}} + \frac{\nabla\nabla}{k^2} \right] G_0^+(\mathbf{r},\mathbf{r}') \qquad (2.191)$$

and G_0^+ is the scalar (retarded) Green function that satisfies the following equation:

$$(\nabla^2 + k^2)G_0^+(\mathbf{r},\mathbf{r}') = -\delta(\mathbf{r} - \mathbf{r}'). \qquad (2.192)$$

Note that equation (2.191) solves the defining equation (2.170), as it can be verified by direct substitution using the identity $\nabla \times \nabla \times (\overset{\leftrightarrow}{\mathbf{I}} G_0^+) = \nabla\nabla G_0^+ - \nabla \cdot (\nabla G_0^+)\overset{\leftrightarrow}{\mathbf{I}}$.

As a consequence of the *reciprocity theorem* [101], it can also be shown that the dyadic Green function must obey the following symmetry relation:

$$\overset{\leftrightarrow}{\mathbf{G}}_0^T(\mathbf{r}',\mathbf{r}) = \overset{\leftrightarrow}{\mathbf{G}}_0(\mathbf{r},\mathbf{r}'), \qquad (2.193)$$

where $\overset{\leftrightarrow}{\mathbf{G}}^T$ denotes the transpose. Taking the transpose of equation (2.189) we can alternatively express the source–field relation as follows:

$$\mathbf{E}(\mathbf{r}) = \mathbf{E}_0(\mathbf{r}) + j\omega\mu \int_V \mathbf{J}(\mathbf{r}') \cdot \overset{\leftrightarrow}{\mathbf{G}}_0(\mathbf{r}',\mathbf{r})d^3\mathbf{r}'. \qquad (2.194)$$

As we already discussed in the previous chapter, a source current along the u-direction ($u = x,y,z$) leads to an electric/magnetic field with x, y, and z components. The dyadic Green function relates all the components of the source to all the components of the fields. For each u-component of $\mathbf{J}$, there is a corresponding vector Green function $\mathbf{G}_{0,u}$ that allows us to write the electric field generated at $\mathbf{r}$ by an infinitesimal, u-directed current element at $\mathbf{r}'$ in the form $j\omega\mu\mathbf{G}_{0,u}(\mathbf{r},\mathbf{r}')$. On the other

---

[22] Notice that the Green function is singular at $r = r'$ and an infinitesimal exclusion volume surrounding r' has to be introduced. As we will show in the next section, the shape-dependent depolarization of the principal volume must be considered. However, as long as we consider field points outside of the source volume V, there is no need to expand on this issue now.

hand, the dyadic Green function $\overleftrightarrow{\mathbf{G}}_0$ accounts simultaneously for all the orientations of the current element and can be represented in the following matrix form:

$$\overleftrightarrow{\mathbf{G}}_0 = \begin{bmatrix} \mathbf{G}_{0,x}^{(1)} & \mathbf{G}_{0,y}^{(1)} & \mathbf{G}_{0,z}^{(1)} \\ \mathbf{G}_{0,x}^{(2)} & \mathbf{G}_{0,y}^{(2)} & \mathbf{G}_{0,z}^{(2)} \\ \mathbf{G}_{0,x}^{(3)} & \mathbf{G}_{0,y}^{(3)} & \mathbf{G}_{0,z}^{(3)} \end{bmatrix} \tag{2.195}$$

As evident in this representation, the first, second, and third columns of the matrix yield the field components due to a point current source oriented along the x-, y-, or z-direction, respectively.

The compact operator form of the dyadic Green function (2.191) is generally valid and can be used to obtain explicit representations in any desired coordinate system. For example, in the Cartesian coordinate system, it is represented in matrix form as follows:

$$\overleftrightarrow{\mathbf{G}}_0 = \begin{bmatrix} k^2 + \frac{\partial^2}{\partial x^2} & \frac{\partial^2}{\partial x \partial y} & \frac{\partial^2}{\partial x \partial z} \\ \frac{\partial^2}{\partial y \partial x} & k^2 + \frac{\partial^2}{\partial y^2} & \frac{\partial^2}{\partial y \partial z} \\ \frac{\partial^2}{\partial z \partial x} & \frac{\partial^2}{\partial z \partial y} & k^2 + \frac{\partial^2}{\partial z^2} \end{bmatrix} \frac{e^{jk|\mathbf{r}-\mathbf{r}'|}}{4\pi k^2 |\mathbf{r}-\mathbf{r}'|}, \tag{2.196}$$

which shows very clearly the symmetry property of the dyadic Green function. Finally, by setting $\mathbf{R} = \mathbf{r} - \mathbf{r}'$ and after some vector manipulations, we can find another convenient form of the dyadic Green function that is very useful for practical scattering calculations:

$$\boxed{\overleftrightarrow{\mathbf{G}}_0(\mathbf{r},\mathbf{r}') = \left[\left(\frac{3}{k^2 R^2} - \frac{3j}{kR} - 1 \right) \hat{\mathbf{R}}\hat{\mathbf{R}} + \left(1 + \frac{j}{kR} - \frac{1}{k^2 R^2} \right) \overleftrightarrow{\mathbf{I}} \right] G_0^+(R)}$$

$$\tag{2.197}$$

where $R = |\mathbf{r} - \mathbf{r}'|$ and $\hat{\mathbf{R}}$ is the unit vector in the direction of $\mathbf{R} = \mathbf{r} - \mathbf{r}'$. However, when dealing with the computation of the field within the source region, the preceding expressions become singular. This singularity problem can be solved by decomposing the dyadic Green function into a singular part and a principal value ($P.V.$) part [101]:

$$\overleftrightarrow{\mathbf{G}}_0(\mathbf{r} - \mathbf{r}') = P.V.\overleftrightarrow{\mathbf{G}}_0(\mathbf{r} - \mathbf{r}') - \frac{\overleftrightarrow{\mathbf{L}} \delta(\mathbf{r} - \mathbf{r}')}{k_0^2}, \tag{2.198}$$

where $\overleftrightarrow{\mathbf{L}}$ is a constant dyad that depends on the shape of the exclusion domain chosen to define the $P.V.$ As we will make clear in the next section, a principal value integral is an improper integral defined everywhere except within a small exclusion volume surrounding the observation vector $\mathbf{r}$. For a spherical exclusion volume, we have $\overleftrightarrow{\mathbf{L}} = \overleftrightarrow{\mathbf{I}}/3$. The dyadic Green function defined as in (2.198) is valid for all observation points $\mathbf{r}$ in space. Rigorous derivations of the Green dyadic decomposition can be found in [100, 101]. The specific application to a general light-scattering problem is discussed in the next section.

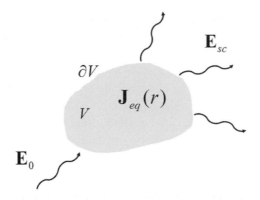

Figure 2.2 Schematics of the scattering geometry for the derivation of volume integral equations.

2.9 Light Scattering as a Radiation Problem

In many problems of practical interest, the sources of the radiated fields are not explicitly known but depend on the solution of the problem. This is always the case in electromagnetic scattering problems, where an external incident wave induces an a priori unknown current density inside a volume V occupied by an arbitrary scattering particle and on its surface ∂V. In such situations, the dyadic field equations (2.189) and (2.190) become integral equations for the induced polarization current density that radiates the scattered field, as schematically illustrated in Figure 2.2. According to this picture, the lightscattering problem of a body illuminated by a field generated by known external currents becomes entirely equivalent to the radiation of an equivalent current source distributed within the volume of the body and on its surface. This *equivalent problem* can be treated rigorously using the Green function integral method.

2.9.1 Light Scattering and Volume Integral Equations

In general, integral equation methods are very accurate frequency domain techniques for lightscattering simulations and, when compared to alternative approaches based on difference equations, such as the finite difference time domain (FDTD) method, feature uniquely attracting characteristics. In particular, (i) only the volumes of the scattering materials need to be discretized, (ii) the regularity of the fields at infinity (i.e., the Sommerfeld radiation condition) is automatically satisfied, and (iii) the physics behind complex multiple scattering processes can be made manifest (as it will become evident in subsequent chapters). Moreover, Green function–based integral methods provide the general framework of multiple scattering theory in which the dyadic Green's function acts as a classical propagator. Several equivalent formulations of the integral equation methods have been developed, broadly differing in the types of vector variables being considered (e.g., equivalent currents or electric/magnetic fields). Historically, volume integral methods in electromagnetic scattering became practical only after the work of Van Bladel [102] on Green function singularity extraction and the introduction of the *method of moments* (MoM) for the numerical solution of integral equations [103].

We now utilize the dyadic Green function method to establish the classical electric field integral equation (EFIE) model of electromagnetic scattering originally introduced in [104]. In particular, we will consider a penetrable particle of arbitrary shape and volume V bounded by a closed surface ∂V and illuminated by the known incident field $\mathbf{E}_0$, as illustrated in Figure 2.2. The scattering particle is described by a local permittivity $\epsilon(\mathbf{r})$ and conductivity $\sigma(\mathbf{r})$. The current induced in the particle radiates the scattering field $\mathbf{E}_{sc}$. This field can be obtained by replacing the scattering particle with the equivalent free-space current density $\mathbf{J}_{eq}$, given by the following:

$$\mathbf{J}_{eq}(\mathbf{r}) = [\sigma(\mathbf{r}) - j\omega\epsilon_0(\epsilon_r - 1)]\mathbf{E}(\mathbf{r}) = \tau(\mathbf{r})\mathbf{E}(\mathbf{r}), \qquad (2.199)$$

where $\mathbf{E} = \mathbf{E}_0 + \mathbf{E}_{sc}$ is the *total electric field* inside the scatterer, and the two terms in equation (2.199) represent the induced conduction and the polarization current density $\mathbf{J}_b = \partial\mathbf{P}/\partial t$, respectively.

The field inside the particle can be expressed in terms of the equivalent current density $\mathbf{J}_{eq}$ using the general source–field relation (2.189). However, when the observation field points are inside the body of the scattering particle the dyadic Green function $\overleftrightarrow{\mathbf{G}}_0(\mathbf{r}, \mathbf{r}')$ has a singularity and must be evaluated with care in order to obtain a unique solution. In fact, due to the presence of the double differentiation operator, the dyadic Green function is *hypersingular*. A rigorous technique, known as the principal volume method, has been developed by Van Bladel [102, 105, 106] to treat this type of singularity, allowing to express the scattered electric field at an arbitrary point inside the particle as follows:

$$\mathbf{E}_{sc}(\mathbf{r}) = j\omega\mu \int_V \left[P.V. \overleftrightarrow{\mathbf{G}}_0(\mathbf{r}, \mathbf{r}') - \frac{\overleftrightarrow{\mathbf{L}} \, \delta(\mathbf{r} - \mathbf{r}')}{k_0^2} \right] \cdot \mathbf{J}_{eq}(\mathbf{r}') d^3\mathbf{r}'$$

$$= j\omega\mu P.V. \int_V \overleftrightarrow{\mathbf{G}}_0(\mathbf{r}, \mathbf{r}') \cdot \mathbf{J}_{eq}(\mathbf{r}') + \frac{\overleftrightarrow{\mathbf{L}} \cdot \mathbf{J}_{eq}(\mathbf{r})}{j\omega\epsilon_0} \qquad (2.200)$$

where $P.V.$ denotes the principal value integral obtained by excluding from the full integration range the small singular volume δV (i.e., the exclusion volume), and taking the following limit:

$$P.V. \left[\int_V \overleftrightarrow{\mathbf{G}}_0(\mathbf{r}, \mathbf{r}') \cdot \mathbf{J}_{eq}(\mathbf{r}') d^3\mathbf{r}' \right] = \lim_{\delta V \to 0} \int_{V - \delta V} \overleftrightarrow{\mathbf{G}}_0(\mathbf{r}, \mathbf{r}') \cdot \mathbf{J}_{eq}(\mathbf{r}') d^3\mathbf{r}'. \quad (2.201)$$

Note that the principal value integral converges to a value that depends on the shape of the exclusion volume δV. This makes the source–field relation as defined by equation (2.189) *nonunique*. In order to obtain a converging and unique expression for the electric field, equation (2.200) has to be augmented by the additional term containing the dyad $\overleftrightarrow{\mathbf{L}}$, which also depends on the shape of δV. Physically, this term removes the contribution of the surface charges that accumulate around the exclusion volume. In this way, the dyadic Green function is uniquely defined. When considering isotropic exclusion volumes (i.e., small spheres or cubes), the exclusion dyad acquires the particularly simple form $\overleftrightarrow{\mathbf{L}} = \overleftrightarrow{\mathbf{I}}/3$. A detailed derivation of equation (2.200) can be found in [101], along with explicit forms of $\overleftrightarrow{\mathbf{L}}$ for exclusion volumes of different shapes.

The volume integral equation for the total electric field can now be established by writing the *total electric field inside the particle* $\mathbf{E}(\mathbf{r})$ as the sum of the scattered field $\mathbf{E}_{sc}(\mathbf{r})$ and the incident electric field $\mathbf{E}_0(\mathbf{r})$. Considering a spherical exclusion volume and rearranging terms, we get the following:

$$\boxed{\left[1 - \frac{\tau(\mathbf{r})}{3j\omega\epsilon_0}\right]\mathbf{E}(\mathbf{r}) - j\omega\mu P.V. \int_V \tau(\mathbf{r}')\overleftrightarrow{\mathbf{G}}_0(\mathbf{r},\mathbf{r}') \cdot \mathbf{E}(\mathbf{r},\mathbf{r}')d^3\mathbf{r}' = \mathbf{E}_0(\mathbf{r})} \quad (2.202)$$

The only unknown in this equation is the total electric field inside the particle, or the *internal field*. This integral equation can be solved using the method of moments. However, one of the major challenges related to moments method solution of (2.202) is the so-called *low-frequency breakdown problem*, which consists in the strong ill-conditioning of the relevant stiffness matrix due to the very different frequency scaling exhibited by the solenoidal and non-solenoidal components of the unknown fields. This problem is particularly severe in the modeling of surface plasmon oscillations of metallic particles that are small compared to the wavelength, necessitating proper scaling, at a given frequency, of the solenoidal with respect to the nonsolenoidal components [107, 108].

In the context of strongly dispersive optical structures, often encountered in nano-optics and plasmonics technologies, an alternative and more efficient integral equation model based on the radiation potentials has been recently developed [107]. An important feature of this formulation is that it only uses the scalar free-space Green function, as opposed to the hypersingular dyadic Green function discussed earlier. This choice, together with the representation of the unknown equivalent current density as the sum of a solenoidal and nonsolenoidal components, leads to a broadband-efficient numerical scheme, whose implementation is provided in [107]. The derivation of this alternative integral formulation of the scattering problem is provided in the following series of equations considering a scattering particle with a frequency-dependent complex permittivity $\epsilon(\omega)$ that includes both polarization and conduction mechanisms.[23] The induced polarization inside the volume of the material is given by the following:

$$\mathbf{P}(\mathbf{r}) = \epsilon_0(\epsilon_r - 1)\mathbf{E}(\mathbf{r}). \quad (2.203)$$

This gives rise to a polarization current density:

$$\mathbf{J}(\mathbf{r}) = \frac{\partial \mathbf{P}}{\partial t} = -j\omega\mathbf{P} = -j\omega\epsilon_0(\epsilon_r - 1)\mathbf{E}(\mathbf{r}). \quad (2.204)$$

---

[23] Equivalently, it is possible to separate the contribution due to the conductivity of free electrons by introducing the effective relative permittivity: $\epsilon_{eff} = \epsilon_d + j\sigma/\omega$, where σ is the conductivity.

If the material is homogeneous inside the volume V, bound charges only accumulate on the surface of the scattering particle, giving rise to a surface density equal to the following:

$$\sigma_S = \frac{1}{-j\omega} \mathbf{J} \cdot \hat{\mathbf{n}}, \tag{2.205}$$

where $\cdot \hat{\mathbf{n}}$ denotes the outward normal on ∂V. Additionally, we have the following:

$$\int_{\partial V} \sigma_S(\mathbf{r}') d^2 \mathbf{r}' = 0 \tag{2.206}$$

A full-wave integral equation model for the scattering problem can now be obtained by rewriting the total electric field in terms of its potentials:

$$\mathbf{E}(\mathbf{r}) = j\omega \mathbf{A}(\mathbf{r}) - \nabla \phi(\mathbf{r}) + \mathbf{E}_0. \tag{2.207}$$

Inserting the general relations (2.39) and (2.40) that link, within the Lorentz gauge, the electromagnetic potentials and the induced charges into the last equation, we obtain the integral equation satisfied inside the volume V:

$$\frac{\mathbf{J}(\mathbf{r})}{j\omega(\epsilon_0 - \epsilon)} + j\omega\mu_0 \int_V \mathbf{J}(\mathbf{r}') G_0^+(\mathbf{r}, \mathbf{r}') d^3 \mathbf{r}' \tag{2.208}$$

$$+ \frac{1}{j\omega\epsilon_0} \nabla \int_{\partial V} \mathbf{J} \cdot \hat{\mathbf{n}}(\mathbf{r}') G_0^+(\mathbf{r}, \mathbf{r}') d^2 \mathbf{r}' + \mathbf{E}_0(\mathbf{r}) = 0 \tag{2.209}$$

This integral equation model has been derived in [107] and it has been successfully utilized to describe plasmon oscillations in metallic nanoparticles of arbitrary shapes [108–110].

2.9.2   Far-Field Scattering by Small Objects

We now illustrate a simple application to the problem of *far-field light scattering* by a particle of arbitrary shape with dielectric constant ϵ embedded in free space and illuminated by an external monochromatic wave $\mathbf{E}_0$. The external excitation gives rise to an induced polarization current density $\mathbf{J}_b = -j\omega(\epsilon - \epsilon_0)\mathbf{E}$, where $\mathbf{E}$ is the total electric field inside the particle. The induced current density in turn radiates the scattered field. The radiated electric and magnetic fields can be expressed directly in terms of the vector potential from Maxwell's equations:

$$\mathbf{H} = \frac{1}{\mu} \nabla \times \mathbf{A} \tag{2.210}$$

$$\mathbf{E} = \frac{j}{\omega\epsilon\mu} \nabla \times \nabla \times \mathbf{A}. \tag{2.211}$$

The magnetic vector potential can be formally obtained by integrating the Green function over the induced current density. In many practical applications, it is important

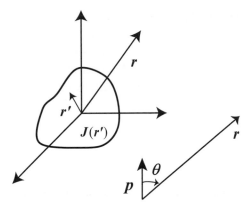

Figure 2.3 Schematics of the radiation geometry for the scattering by a small dielectric particle.

to obtain an expression for the scattered field in the far-field zone, several wavelengths away from the object. This can be accomplished by exploiting the far-field approximation for the magnetic vector potential:

$$A = \mu \int J_b(\mathbf{r}')G_0(\mathbf{r},\mathbf{r}')d^3\mathbf{r}' \approx \mu \frac{e^{jkr}}{r} \int J_b(\mathbf{r}')e^{-jk\hat{s}\cdot\mathbf{r}'}d^3\mathbf{r}', \tag{2.212}$$

where $\hat{s} = \mathbf{r}/|\mathbf{r}|$. Substituting the preceding expression into equation (2.211) and using the identity $\nabla \times \nabla \times [JG_0(\mathbf{r}),\mathbf{r}'] = \nabla \times (\nabla G_0 \times J)^{24}$, we obtain the far-field scattered electric field:

$$E_{sc} = -j\omega\mu\frac{e^{jkr}}{4\pi r}\mathbf{s} \times \mathbf{s} \times \int_V J_b(\mathbf{r}')e^{jk\mathbf{s}\cdot\mathbf{r}'}d^3\mathbf{r}', \tag{2.213}$$

where $\mathbf{s} \equiv \hat{\mathbf{r}} = \mathbf{r}/|\mathbf{r}|$ is the unit radial vector. As we discussed in the previous section, the induced current density is not known a priori since it depends on the *internal field* inside the particle $E = E_0 + E_{sc}$. In general, this problem is solved by the integral equation approach explained in the previous section. However, in the case of a particle that is small compared to the wavelength, we can immediately find the solution. In fact, for a small particle the phase in the exponential term under the integral sign of equation (2.213) is approximately constant and

$$\int_V J_b(\mathbf{r}')d^3\mathbf{r}' = -j\omega(\epsilon - \epsilon_0)VE \equiv -j\omega\mathbf{p}, \tag{2.214}$$

where we have defined the vector dipole moment $\mathbf{p}$ of the scatterer. Therefore, we derive a simple expression for the far-field scattered by a small (i.e., dipolar) particle as follows:

$$\boxed{E_{sc} = -\omega^2\mu\frac{e^{jkr}}{4\pi r}\hat{\mathbf{r}} \times \hat{\mathbf{r}} \times \mathbf{p}} \tag{2.215}$$

---

[24] Realize that the operator $\nabla \times (\cdot)$ does not act on $J(\mathbf{r}')$.

This equation provides the far-field radiated by an elementary dipole $\mathbf{p}$ measured in the direction $\hat{\mathbf{r}}$, as illustrated in Figure 2.3. Note that the term $\hat{\mathbf{r}} \times \hat{\mathbf{r}} \times \mathbf{p}$ extracts the component of the dipole moment vector $\mathbf{p}$ that is perpendicular to the scattering direction $\hat{\mathbf{r}}$, as expected for a transverse far field. Moreover, the angular dependence of scattered is described by $\sin\theta$, where θ is the angle between the dipole moment vector and the scattering direction.

3   Optical Dispersion Models

We can sum up the preceding considerations by saying that an electron does not emit
energy so long as it has a uniform rectilinear motion, but that it does as soon as its
velocity changes either in magnitude or in direction.

The Theory of Electrons, Hendrik Antoon Lorentz

3.1   Classical Dispersion Models

In this Chapter, we introduce the most general classical model that accounts for the
polarization response of materials and gives rise to their frequency dependent, or
dispersive, optical behavior. This simple yet powerful model, which was originally
developed by Lorentz, consists of a collection of identical, independent and isotropic
harmonic oscillators that are subject to the driving force of local electromagnetic
fields. Based on the Lorentz oscillator model, we then discuss wave propagation
through dispersive media and introduce the idea of *polaritons* as hybrid electromagnetic and polarization waves in materials. Motivated by the current advances
of *polaritonics* and *plasmonics* technologies, we address the behavior of surface-
localized waves in dispersive media with a focus on surface-plasmon polaritons. The
theory of electromagnetic waves in resonant media is relevant for the understanding of
resonance scattering and field localization phenomena that will be discussed in later
chapters. Finally, we consider classical dispersion models of structurally complex
heterogeneous materials and demonstrate the application of basic electromagnetic
mixing rules in the context of current metamaterials technology.

3.1.1   The Lorentz Oscillator Model

The Lorentz oscillator model, originally developed by Hendrik Antoon Lorentz in a
series of lectures delivered at Columbia University in 1906 [111], is the most successful and ubiquitous classical description of the interaction of an electromagnetic
field with electrons and ions inside materials. The dynamic behavior of an ensemble
of independent harmonic oscillators interacting with an electromagnetic field also
provides the most general classical description of light–matter interaction phenomena,
and can easily be generalized to account for more specialized scenarios that involve
optical nonlinearities or anisotropy (see Figure 3.1). Even though rigorous treatments

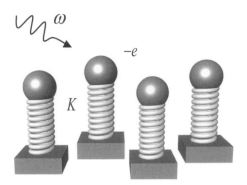

Figure 3.1 Ensemble of Lorentz harmonic oscillators that are used to classically model the optical properties of dielectrics. In such media, electrons are bound to fixed positive nuclei. The resonant oscillation frequency of each electron is related to its mass m and the restoring spring constant K according to: $\omega_0 = \sqrt{K/m}$.

of these problems require a quantum mechanical theory of the properties of atoms and solids and of their interaction with light, the basic features of light–matter coupling can already be clearly understood within the *effective physical picture* introduced by Lorentz. In fact, the more advanced quantum mechanical results turn out to be formally identical to the classical Lorentz's predictions. By properly reinterpreting its parameters, the classical Lorentz model can even be used to quantitatively describe experimental results and reliably extract optical properties of materials.

We now introduce the Lorentz microscopic model of polarizable matter using the words of ITS creator, Hendrik Antoon Lorentz himself [111, p. 99]:

Let each atom (or molecule) contain one single electron, having a definite position of equilibrium, towards which it is drawn back by an elastic force, as we shall call it, as soon as it has been displaced by one cause or another. Let us further suppose this elastic force, which must be considered to be exerted by the other particles in the atom, but about whose nature we are very much in the dark, to be proportional to the displacement.

As clearly stated, *classical oscillators* are associated to each moving electron or electrically charged ion in a solid, and these oscillators are located at fixed equilibrium positions determined by internal forces, which turned out to be quantum mechanical in nature. However, the picture introduced by Lorentz only requires these forces to be elastic, irrespective of their detailed physical nature. According to the Lorentz classical model, when a light wave propagates through the solid, its electric field exerts a force on the charged particles that displaces them from their equilibrium positions. Displaced particles (i.e., electrons or ions) are also subjected to the interaction with surrounding atoms, resulting in an elastic restoring force, proportional to their displacement x, which tends to drive them back into their original equilibrium positions. In a first approximation, which is valid for small values of the local field intensity (i.e., linear optics), the charged particles perform harmonic oscillations around their equilibrium positions at the *natural oscillation frequency* ω_0. These oscillations will be generally damped by a phenomenological friction force that effectively accounts for the incoherent scattering of the oscillating particles with their environments.

The dynamics of a generic Lorentz electron of charge $-e$ bound by an harmonic force and subjected to the driving action of the *local electromagnetic field* $\mathbf{E}_{loc}(\mathbf{r}, t)$ is described by the classical equation of motion:

$$m\ddot{\mathbf{r}} + m\gamma\dot{\mathbf{r}} + m\omega_0^2\mathbf{r} = -e\mathbf{E}_{loc}(\mathbf{r}, t) \tag{3.1}$$

where γ is a phenomenological damping force accounting for the scattering losses of the electron. Here we neglected radiation damping and the contribution of magnetic forces since they are usually much smaller than electrical ones. Note also that the local field $\mathbf{E}_{loc}(\mathbf{r}, t)$ is the field "perceived" by the charge of a single oscillator at its position $\mathbf{r}$ plus the contribution of the field scattered by all the other oscillators in the material. Therefore, the local field $\mathbf{E}_{loc}$ is different in general from both the externally applied field $\mathbf{E}_{ext}$ and the total macroscopic electric field $\mathbf{E}(\mathbf{r}, t)$, which is the field averaged over a region containing many oscillators. In what follows, we assume a low density of oscillators (i.e., diluted media approximation) so that $\mathbf{E} \approx \mathbf{E}_{ext} \approx \mathbf{E}_{loc}$. The more general relation between $\mathbf{E}_{loc}$, $\mathbf{E}_{ext}$, and $\mathbf{E}$ can only obtained within the framework of *multiple scattering theory*, established in Chapter 12 or, in a more simplified manner, using the effective medium approach described in Section 3.4.

Taking the Fourier transform of equation (3.1), we immediately obtain the frequency-domain (i.e., steady-state) solution for the (complex) Fourier amplitude of the dipole moment contributed by a single Lorentz electron:

$$\mathbf{p} = -e\mathbf{r} = \frac{e^2}{m}(\omega_0^2 - \omega^2 - j\omega\gamma)^{-1}\mathbf{E}. \tag{3.2}$$

If there are N bound electrons per unit volume, then the macroscopic polarization vector $\mathbf{P}$ of the medium is given by the following:

$$\mathbf{P} = -N e\mathbf{r} = \frac{Ne^2}{m}(\omega_0^2 - \omega^2 - j\omega\gamma)^{-1}\mathbf{E} = \epsilon_0\chi_e\mathbf{E}. \tag{3.3}$$

At this point, we can immediately write the corresponding expressions for the electric susceptibility χ_e and the relative permittivity ϵ_r. In particular, we obtain the following:

$$\chi_e(\omega) = \frac{\omega_p^2}{(\omega_0^2 - \omega^2 - j\omega\gamma)} \tag{3.4}$$

and

$$\epsilon_r = 1 + \chi_e(\omega) = 1 + \frac{\omega_p^2}{(\omega_0^2 - \omega^2 - j\omega\gamma)}, \tag{3.5}$$

where we have defined the *plasma frequency* of the material as follows:

$$\omega_p^2 = \frac{Ne^2}{\epsilon_0 m}. \tag{3.6}$$

The quantum mechanical nature of the atomic potential and the band structure of the crystal are not explicitly taken into account by the classical Lorentz model. However, some aspects of them are reflected in the values of *effective optical mass m* of the electron, which does not coincide in general with the mass of a free electron.

Equations (3.4) and (3.5) are frequency-dependent *response (or transfer) functions* that connect the Fourier amplitudes of **P** and **D** with the ones of the electric field **E** at the electron's position.

The real and imaginary parts of the complex susceptibility function $\chi_e(\omega) = \chi_e'(\omega) + j\chi_e''(\omega)$ can be obtained as follows:

$$\chi_e'(\omega) = \frac{\omega_p^2(\omega_0^2 - \omega^2)}{(\omega_0^2 - \omega^2)^2 + (\gamma\omega)^2} \tag{3.7}$$

$$\chi_e''(\omega) = \frac{\omega_p^2\gamma\omega}{(\omega_0^2 - \omega^2)^2 + (\gamma\omega)^2} \tag{3.8}$$

We note that in the limit of ω approaching infinity $\chi_e(\infty) = 0$, the polarization of the medium cannot respond fast enough to the time-varying electric field and the oscillator's displacement is oppositely phased (i.e., π-phase shift) with respect to the exciting field. On the other hand, in the limit of very low frequency, the susceptibility tends to its DC value $\chi_e(0) = \omega_p^2/\omega_0^2$ and the Lorentz electron responds in-phase (i.e., 0-phase shift) to the exciting field.

In Figure 3.2, we show the real and imaginary components of the complex susceptibility and of the complex refractive index (in normalized frequency units) for a generic Lorentz medium (i.e., a medium composed of Lorentz electrons). On both

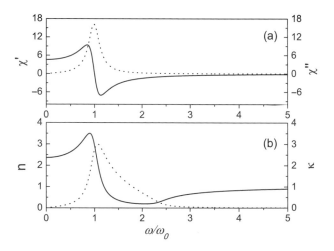

Figure 3.2 (a) Real part (solid line) and imaginary part (dotted line) of the electric susceptibility and (b) refractive index (solid line) and extinction coefficient (dotted line) of a Lorentz medium. The parameters used in the graphs' calculations are as follows: $\omega_p = 15 \times 10^{15}$ Hz, $\omega_0 = 8 \times 10^{15}$ Hz, $\gamma = 9 \times 10^{14}$ Hz.

sides of the resonance region, the refractive index n increases with frequency, giving rise to the *normal dispersion* behavior. On the other hand, in the immediate vicinity of the resonance frequency ω_0 the refractive index decreases with frequency, which is characteristic of the so-called *anomalous dispersion* region. The anomalous dispersion behavior of the Lorentz medium occurs in regions of large absorption, as shown in Figure 3.2, where the extinction coefficient features a prominent peak around ω_0. It is also interesting to observe that the refractive index is less than one in the entire range $\omega > \omega_0$, implying directly that the phase velocity of light v in the Lorentz medium is *superluminal* ($v > c_0$) in this range. Fortunately, this intriguing fact leads to no contradiction with the causal restrictions imposed by relativity theory, because no physical signal can be transported by the phase velocity of a plane wave due to the fact that a plane wave is present in all space at all times.[1]

A necessary and sufficient condition for the signal velocity in a medium to be less than c_0 is that the real and imaginary parts of its optical constants (i.e., complex refractive index or dielectric function) satisfy the Kramers–Kronig causality relations that follow directly from the Hilbert transform. In Figure 3.3, we plot the real and imaginary components of the dielectric function of the Lorentz medium. A region of negative dielectric function appears when $n < \kappa$, which corresponds to the (usually narrow) frequency region between the oscillation resonance ω_0 and the frequency where $\epsilon'(\omega) = 0$, known as the *longitudinal frequency* ω_L. Note that ω_L is also the frequency where the complex dielectric function nearly vanishes due to the typically small contribution of the imaginary part in this frequency range. Therefore, we can write $\epsilon(\omega_L) \approx 0$. *At this frequency, the Lorentz medium supports longitudinal wave solutions.* These are wave solutions of Maxwell's equations that are irrotational and propagate with a wavevector $\mathbf{k}$ parallel to the electric field vector $\mathbf{E}$. These intriguing solutions can occur in the Lorentz medium since $\nabla \cdot \mathbf{D} = \epsilon \nabla \cdot \mathbf{E} = 0$ is automatically satisfied at $\omega = \omega_L$ without requiring the transversally of the wave solution. In turn, if the medium is source-free and nonmagnetic ($\mathbf{M} = 0$), then from Maxwell's equations we deduce that also $\nabla \times \mathbf{B} = 0$, which implies $\mathbf{B} = 0$ and consequently $\nabla \times \mathbf{E} = 0$, showing that the electric field is *purely longitudinal*. Moreover, since at $\omega = \omega_L$ we have that $\mathbf{D} = 0$, the medium can support *longitudinal polarization waves* with $\mathbf{P} = -\epsilon_0 \mathbf{E}$.

The engineering of longitudinal polarization waves in materials provides opportunity to enhance light–matter interactions due to their spatial localization that can be tuned deep into the subwavelength regime. This is possible in many dielectric materials, such as SiC, in the frequency region $\omega_0 < \omega < \omega_L$. This region typically occurs in the infrared spectrum and is called the *Reststrahlen band*. These intriguing aspects of the optical response of polarizable materials are intensively investigated in order to resonantly enhance the electric and magnetic field strengths within small nanostructures in *polaritonic technology*. The connection between the negative

---

[1] Note to this regard that, if originating from a coherent source, a plane wave would require a certain finite time to be established.

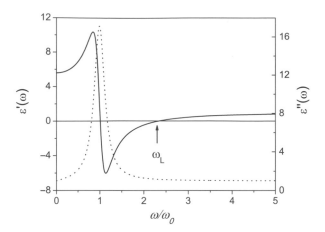

Figure 3.3 Real and imaginary parts of the complex dielectric function of a Lorentz medium with parameters: $\omega_p = 15 \times 10^{15}$ Hz, $\omega_0 = 8 \times 10^{15}$ Hz, $\gamma = 9 \times 10^{14}$ Hz.

permittivity of materials in certain frequency bands and the resulting resonant behavior can be simply understood based on the linear constitutive relation $\mathbf{P} = \epsilon_0(\epsilon_r - 1)\mathbf{E}$. In fact, when $\epsilon_r < 0$ in some frequency range, the depolarization field $\mathbf{E}_d$ created in the material will reinforce the externally applied field $\mathbf{E}_0$, leading to a strong enhancement of the total field $\mathbf{E}_{in} = \mathbf{E}_0 + \mathbf{E}_d$ inside the material. This effect will be particularly strong in small nanostructures with typical dimensions of the order of the metal skin depth (i.e., few tens of nanometers). The polaritonic resonant behavior of small dielectric and metallic spheres is discussed in greater detail in Chapter 13.

Let us now discuss the dielectric function of materials in more detail. First of all, at frequencies $\omega \gg \omega_p$, a residual polarization contribution due to the positive background produced by the ion cores needs to be included in the total polarization (3.3), resulting in the more general dielectric function:

$$\epsilon_r = \epsilon_b + \frac{\omega_p^2}{\left(\omega_t^2 - \omega^2 - j\omega\gamma\right)} \tag{3.9}$$

where the background contribution to the polarization density has been written as[2]: $\mathbf{P}_b = \epsilon_0\chi_b = \epsilon_0(\epsilon_b - 1)\mathbf{E}$, where ϵ_b is called the *background dielectric constant*. This modification reflects the fact that Lorentz electrons are not embedded in free space, but in a complex environment that supports, in general, multiple *polarization channels* beyond the specific contribution of Lorentz electrons. The background permittivity is particulary important when multiple and well-separated resonances are simultaneously present in a material. This is the case in semiconductors that typically feature not only one type of oscillator (with one resonance frequency ω_0) but many of them (at frequencies ω_{0j}) corresponding to the different polarization channels. Examples of polarization channels are the atomic lattice vibrations (i.e., phonons), or the bound

---

[2] Some authors prefer to indicate the background permittivity by ϵ_∞ instead of ϵ_b.

electron-hole pair states (i.e., excitons) that form in polar semiconductors. Moreover, in metals, different types of electron wavefunctions (i.e., conduction s-electrons or valence d-electrons) can participate in the optical transitions and will contribute with different polarization terms. Therefore, in the most general case, the total dielectric function of a Lorentz medium is expressed as a sum over various types of oscillators:[3]

$$\epsilon_r = 1 + \sum_i \frac{\omega_{pi}^2}{\left(\omega_{0i}^2 - \omega^2 - j\omega\gamma_i\right)}. \tag{3.10}$$

However, when analyzing the optical dispersion around a specified resonant frequency ω_0', we can neglect the contributions from all lower-lying resonances ($\omega_{0i} \ll \omega_0'$) (since their high-frequency polarization contribution will vanish close to the resonance of interest) and the constant contributions of all higher-lying resonances ($\omega_{0i} \gg \omega_0'$) will be incorporated in the background dielectric constant ϵ_b. Clearly $\epsilon_b = 1$ for the highest resonance present in the system, which in many cases resides in the X-ray spectral regime and is associated to the K-absorption edge of materials.

3.1.2   The Drude–Sommerfeld Model

The Lorentz harmonic oscillator model allows to accurately understand the optical properties of dielectric materials (i.e., insulators and semiconductors containing prevalently bound electrons) over a wide frequency range. However, the same model can be easily specialized to describe the optical properties of conductive materials such as metals, which can be regarded in a first approximation as a gas of free electrons that move against a fixed background of positive ion cores. In this model, electrons oscillate in response to the driving electromagnetic field and experience a damping force due to collisions (with lattice ions, other electrons or with defects in the crystal) that is described by the collision frequency $\gamma = 1/\tau$, where τ is the *relaxation time* of the free electron gas. At room temperature, τ is principally determined by electron–phonon scattering and has a typical value on the order of 10^{-14} s, which corresponds to $\gamma = 100$ THz. This is also the fundamental physical reason why many metals present significant optical losses in the optical regime.

The classical *plasma model* of metals, also known as the *Drude–Sommerfeld model*, follows directly from the Lorentz oscillator picture by setting the restoring harmonic force to zero, or equivalently $\omega_0 = 0$. Under this condition, we can directly obtain the dielectric function for the plasma model of free electrons from equation (3.5), which reduces to the following:

$$\epsilon_r = 1 - \frac{\omega_p^2}{\left(\omega^2 + j\omega\gamma\right)} \tag{3.11}$$

---

[3] However, notice that if all the oscillators are included in the summation, then $\epsilon_b = \epsilon(\infty) = 1$.

where ω_p is now the plasma frequency of the free electron gas. We note that in the limit of zero electron damping ($\gamma = 0$), we obtain immediately the *real* dielectric function of the *lossless electron plasma*:

$$\epsilon_r = 1 - \frac{\omega_p^2}{\omega^2}. \tag{3.12}$$

Alternatively, we can derive the Drude–Sommerfeld dielectric function by starting from the equation of motion of a generic electron in the plasma under a local field $\mathbf{E}_{loc}$:

$$m\ddot{\mathbf{r}} + m\gamma\dot{\mathbf{r}} = -e\mathbf{E}_{loc}(\mathbf{r},t). \tag{3.13}$$

Assuming again that $\mathbf{E} \approx \mathbf{E}_{loc}$, the frequency domain solution of equation (3.13) is obtained again by Fourier transformation, yielding the following:

$$\mathbf{r} = \frac{e}{m\left(\omega^2 + j\omega\gamma\right)}\mathbf{E}. \tag{3.14}$$

The expression for the polarization follows immediately:

$$\mathbf{P} = -\frac{Ne^2}{m\left(\omega^2 + j\omega\gamma\right)}\mathbf{E}. \tag{3.15}$$

Inserting the polarization (3.15) into the auxiliary field definition $\mathbf{D} = \epsilon_0\mathbf{E} + \mathbf{P}$, we can immediately write down the dielectric function of the plasma model, which is the same as in equation (3.11).

The dielectric function of free electrons is directly linked to the classical Drude model for the AC conductivity $\sigma(\omega)$ of metals. This can be easily accomplished by rewriting (3.13) as an equation governing the electron's velocity $\mathbf{v}$:

$$m\dot{\mathbf{v}} + m\gamma\mathbf{v} = -e\mathbf{E}(\mathbf{r},t). \tag{3.16}$$

In the presence of a time-harmonic electromagnetic field, the solution for the velocity can be easily obtained by Fourier transformation allowing us to express the current density $\mathbf{J}$ due to the flow of N electrons per unit volume as follows:

$$\mathbf{J} = -Ne\mathbf{v} = \frac{Ne^2}{m\left(\gamma - j\omega\right)}\mathbf{E} = \sigma(\omega)\mathbf{E}, \tag{3.17}$$

from which we obtain the *Drude AC condictivity*:

$$\boxed{\sigma(\omega) = \frac{Ne^2}{m\left(\gamma - j\omega\right)} = \frac{\sigma_0}{(1 - j\omega\tau)}} \tag{3.18}$$

where the DC (constant) conductivity $\sigma_0 = Ne^2\tau/m$ has been defined. Finally, by comparing the expressions in equations (3.11) and (3.18), we obtain an alternative form for the dielectric function of free electrons:

$$\boxed{\epsilon_r = 1 + \frac{j\sigma(\omega)}{\epsilon_0\omega}} \tag{3.19}$$

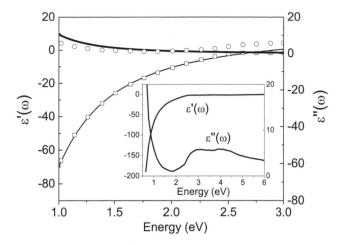

Figure 3.4 Real and imaginary parts of the dielectric function of Au according to the Drude–Sommerfeld model ($\epsilon_b = 9.5$, $\omega_p = 1.366 \times 10^{16}$ Hz, $\gamma = 2.0 \times 10^{14}$ Hz). Experimental data of the real (open squares) and imaginary (open circles) parts. Inset: measured imaginary part of the dielectric function of Au over an extended energy range. Experimental data extracted from [112].

The simple dielectric function (3.11) of free electrons can be directly utilized to describe the optical properties of alkali metals over a broad spectral range, which extends up to the ultraviolet. However, in the case of noble metals, most notably Au, Ag, and Cu, *interband transitions* of bound d-electrons occur in the visible range and severely limit the applicability of the free-electron Drude–Sommerfeld model, as shown for the case of Au in Figure 3.4. In this case, the simple plasma model needs to be modified by adding the contributions of interband electron transitions that are described, at visible wavelengths, by the previously discussed Lorentz model of bound electrons, yielding a more general *Lorentz–Drude model* dielectric function:

$$\epsilon_r = 1 - \frac{\omega_{pe}^2}{(\omega^2 + j\omega\gamma_e)} + \sum_i \frac{\omega_{pi}^2}{(\omega_{0i}^2 - \omega^2 - j\omega\gamma_i)} \qquad (3.20)$$

where the subscript e is appended to the free-electron parameters. The expression (3.20) clearly shows that in order to accurately describe the optical properties of a material *with both polarization and conductivity properties*,[4] we must express the dielectric function (and the susceptibility) as the sum of two terms, one describing bound polarized charges (i.e., ϵ_d) and the other unbound (i.e., free) conduction charges. Such a generalized permittivity, which is sometimes called *effective permittivity* of the medium, can written in the form:

---

[4] This is the case of noble metals at optical and infrared frequencies, which are relevant in plasmonics technology.

$$\epsilon(\omega) = \epsilon_d(\omega) + \frac{j\sigma(\omega)}{\omega} \qquad (3.21)$$

Note that this more general permittivity treats the free (conduction) charges on the same footing as the polarization bound charges, thus eliminating de facto the somewhat arbitrary distinction between free and bound charges. This is particularly pleasing since the traditional distinction between free and bound charges is quite artificial. In fact, there are many nonequivalent ways to define the dielectric constant, polarization, and conductivity of a medium. For instance, depending on the materials and the frequency range of interest, it may be convenient to group valence electrons into a constant polarization background while considering the other electrons as free. This ambiguity is eliminated by simply distinguishing between charges and currents that are external to the system (ρ_{ext}, $\mathbf{J}_{ext}$) and charges and currents that are *induced inside materials* (ρ_{in}, $\mathbf{J}_{in}$), so that

$$\rho_{tot} = \rho_{ext} + \rho_{in} \qquad (3.22a)$$

$$\mathbf{J}_{tot} = \mathbf{J}_{ext} + \mathbf{J}_{in}. \qquad (3.22b)$$

According to this interpretation, external charges and currents are the ones that drive the system, while the internal ones, regardless if they originate from polarization or conduction mechanisms, respond to the external driving fields.

3.2   Waves in Dispersive Media

Now that we gained a better understanding of the physical origin of optical dispersion, we can address the more complex problem of wave propagation through dispersive media by coupling Maxwell's equations to the response of materials. In particular, the coupling of electromagnetic waves to the mechanical degrees of freedom of materials such as oscillating electrons, lattice vibrations, etc., leads us to introduce the important concept of *polariton wave excitations*, which play an important role in current nano-optics technology [92].

3.2.1   Waves in Plasmas

Following reference [153], let us consider first the propagation of a plane, linearly polarized transverse electromagnetic (TEM) wave in a homogeneous dispersive medium composed of a plasma of free electrons described by a lossless Drude model. An electromagnetic wave propagating in a polarizable medium obeys the dynamic Maxwell's equations:

$$\nabla \times \boldsymbol{E} = -\mu_0 \frac{\partial \boldsymbol{H}}{\partial t} \qquad (3.23a)$$

$$\nabla \times \boldsymbol{H} = \epsilon_0 \frac{\partial \boldsymbol{E}}{\partial t} + \frac{\partial \boldsymbol{P}}{\partial t}, \qquad (3.23b)$$

where P is the polarization of the medium. Considering a y-polarized TEM wave propagating along the x-direction, which is described by an electric field vector $\mathbf{E} = [0, E(x,t), 0]$ and by a magnetic field vector $\mathbf{H} = [0, 0, H(x,t)]$, the preceding general system reduces to the following:

$$\frac{\partial E}{\partial x} = -\mu_0 \frac{\partial H}{\partial t} \tag{3.24a}$$

$$\frac{\partial H}{\partial x} = -\frac{\partial}{\partial t}(\epsilon_0 E + P). \tag{3.24b}$$

The Lorentz force exerted on the moving electrons is as follows:

$$\mathbf{F} = -e\mathbf{E} - e(\mathbf{v} \times \mathbf{B}). \tag{3.25}$$

Moreover, since we will discuss nonrelativistic electrons for which $v \ll c$, the Lorentz force simplifies to $\mathbf{F} = -e\mathbf{E}$.

In the case of a lossless Drude medium, the dynamics of oscillating electrons and the polarization of the medium follows from the equations:

$$m\frac{d^2 y}{dt^2} = -eE(x,t) \tag{3.26a}$$

$$P = -eNy, \tag{3.26b}$$

where m is the electron mass and N is the number density of electrons in the medium. Combining the last two equations, we immediately obtain an equation for the dynamic polarization:

$$\frac{d^2 P}{dt^2} = \frac{e^2 N}{m} E. \tag{3.27}$$

Therefore, the propagating electromagnetic field couples to the mechanical polarization of the medium according to the following system of equations:

$$\frac{\partial E}{\partial x} = -\mu_0 \frac{\partial H}{\partial t} \tag{3.28}$$

$$\frac{\partial H}{\partial x} = -\frac{\partial}{\partial t}(\epsilon_0 E + P) \tag{3.29}$$

$$\frac{\partial^2 P}{\partial t^2} = \frac{e^2 N}{m} E \tag{3.30}$$

The preceding system can be solved for the electric field, resulting in the following wave equation:

$$\frac{\partial^2 E}{\partial t^2} - c_0^2 \frac{\partial^2 E}{\partial x^2} + \omega_p^2 E = 0 \tag{3.31}$$

where $\omega_p^2 = e^2 N/\epsilon_0 m$ is the plasma frequency, as already encountered in Section 3.1.1. The modified wave equation that we have obtained for the electromagnetic waves in a plasma has the form of the relativistic *Klein–Gordon equation* that appears in the context of scalar (spinless) quantum field theory [113]. The characteristic dispersion relation for waves in plasmas can be obtained by substituting the harmonic ansatz $E \sim \exp[j(kx - \omega t)]$, from which we obtain the following:

$$\omega^2 = \omega_p^2 + c_0^2 k^2$$

(3.32)

We notice that this is a nonlinear (quadratic) dispersion relation that yields propagating wave solutions only when $\omega > \omega_p$, such that the wavenumbers are purely real. On the other hand, in the interval $0 < \omega < \omega_p$ there exist solutions with imaginary wavenumbers, which do not correspond to propagating waves. Our example of waves in a lossless electron plasma has taught us a general lesson: the coupling of electromagnetic waves with (dynamic) polarization degrees of freedom, i.e., electron oscillations in this simple case, gives rise to a new coupled wave solution with profoundly modified wave dispersion properties compared to the straight line $\omega = \pm c_0 k$ observed in a nondispersive medium. These hybrid wave solutions consisting of electromagnetic waves coupled to electron oscillations in the medium are an example of so-called *polariton waves*. The dispersion relation for TEM waves in a plasma is illustrated by the black line in Figure 3.5. We observe that in the neighborhood of $k = 0$, the polariton wave dispersion is horizontal. This regime is called "particle-like" because it describes purely electronic mechanical oscillations in the medium. On the other hand, the dispersion asymptotically approaches the light line $\omega = c_0 k$ when $k \to \infty$, which describes the "wave-like" regime of free photons in a uniform nondispersive medium.

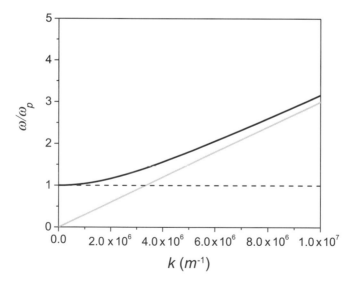

Figure 3.5 Black line: dispersion relation of TEM waves in an electron plasma. Gray line: dispersion relation of a TEM wave in free space. Dashed line: dispersion relation of longitudinal electron oscillations at the plasma frequency $\omega = \omega_p$. For frequencies in the interval $0 < \omega < \omega_p$, there are no propagating wave solutions in the plasma. $\omega_p = 10^{15}$ rad/s has been used in this example.

Interestingly, the dispersion bends quadratically upward at intermediate frequencies, where it describes mixed polariton waves sharing both particle-like and wave-like properties in a way that is uniquely associated to their dispersion relation. In analogy with the electronic band structure of solids, in this regime we can also regard polaritons as the propagation of "heavy photons" with a positive effective mass in a uniform medium. Finally, at $\omega = \omega_p$ the electrons undergo longitudinal bulk oscillations enabled by the vanishing of the electric permittivity of the medium.

3.2.2 Waves in Dielectrics

In this section, we address the propagation of classical electromagnetic waves in the more general homogeneous dielectric medium described by the Lorentz oscillator model, here referred to as a *Lorentz medium*. We recall that such a medium classically describes materials with bound electrons such as transparent dielectrics or semiconductors whose bandgap frequency coincides with the transverse oscillation frequency ω_0 of the Lorentz oscillator model. Considering again a y-polarized TEM wave that propagates along the x-direction in the Lorentz medium, we write down a system of equations for the electromagnetic field coupled to the polarization of the medium by simply replacing equation (3.30) in the previous system with the one derived from the Lorentz model:

$$\frac{\partial^2 P}{\partial t^2} + \gamma \frac{\partial P}{\partial t} + \omega_0^2 P = \frac{e^2 N}{m} E_{loc}, \tag{3.33}$$

where γ is the phenomenological damping force of the Lorentz model. Combining equations (3.28) and (3.29), we easily obtain the self-consistent equations for the electromagnetic waves interacting with a system of Lorentz oscillators:

$$\frac{\partial^2 E}{\partial t^2} - c_0^2 \frac{\partial^2 E}{\partial x^2} = -\frac{1}{\epsilon_0} \frac{\partial^2 P}{\partial t^2} \tag{3.34}$$

$$\frac{\partial^2 P}{\partial t^2} + \gamma \frac{\partial P}{\partial t} + \omega_0^2 P = \frac{e^2 N}{m} E_{loc} \tag{3.35}$$

Substituting the harmonic ansatz $E \sim \exp[j(kx - \omega t)]$ in the preceding system, we obtain after few steps the dispersion relation of the polariton for the Lorentz model:

$$(\omega^2 - c_0^2 k^2)(\omega_0^2 - \omega_p^2/3 - \omega^2 - j\omega\gamma) + \omega^2 \omega_p^2 = 0, \tag{3.36}$$

where we used the expression $E_{loc} = E + P/3\epsilon_0$ for the local electric field (see discussion in Section 3.4). We note that since the local field E_{loc} is different from the average field E in media with sufficiently large density, their resonance frequency is red-shifted by the contribution of the local field via the term $\omega_p^2/3$. This effect is generally referred to as *local field correction*, and it is relevant only for high-density media. Moreover, the ω_p^2 term acts as a coupling term for the aforementioned polariton dispersion. In fact, when $\omega_p = 0$ this dispersion degenerates into two independent branches that reduce, when $\gamma = 0$, to the light line $\omega^2 = c_0^2 k^2$ of free photons and the horizontal dispersion $\omega^2 = \omega_0^2$ describing a mechanical system of uncoupled identical harmonic oscillators.

The dispersion relation of the Lorentz medium can be solved for the wavenumber k in the medium and expressed in the equivalent form:

$$c_0^2 k^2 = \omega^2 \left[1 + \frac{\omega_p^2}{(\omega_0^2 - \omega_p^2/3) - \omega^2 - j\gamma\omega} \right] \equiv \omega^2 \epsilon(\omega) \qquad (3.37)$$

where the dielectric function $\epsilon(\omega)$ defined earlier is the general Lorentz model dielectric function that includes local field corrections to the oscillation frequency. On the other hand, in low-density media where $\omega_p^2 \ll \omega_0^2$ the local field and the average field are approximately equal and we can neglect the $\omega_p^2/3$ frequency shift in equation (3.37), which reduces it to the standard Lorentz formula derived in Section 3.1.1. We refer to equation (3.37) as to the *bulk polariton dispersion* for the Lorentz medium. The bulk polariton dispersion relation is computed based on equation (3.37) and shown in Figure 3.6, where the gap region has been highlighted. The dispersion consists of two branches: (i) the upper polariton branch (UPB) that begins at the longitudinal frequency ω_L and increases quadratically, and (ii) the lower polariton branch (LPB) that starts from zero frequency and plateaus horizontally at a frequency slightly below the transverse oscillation frequency ω_0 due to dissipation ($\gamma \neq 0$) and local field corrections. As a special case, let us now consider the nondissipative

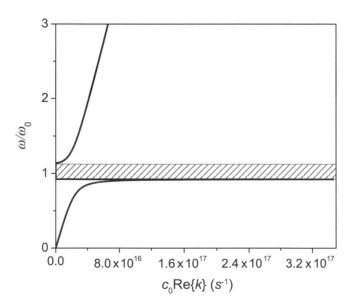

Figure 3.6 Black line: dispersion relation of TEM waves in a nondissipative ($\gamma = 0$) Lorentz medium. The real part of $c_0 k$ in units of inverse seconds. The gap region where there are no propagating wave solutions is highlighted. $\omega_p = 15 \times 10^{15}$ rad/s and $\omega_0 = 1.5 \times \omega_p$ have been used in this example.

polariton dispersion ($\gamma = 0$) in the low-density limit ($\omega_p^2 \ll \omega_0^2$), where $E_{loc} \simeq E$ and local field corrections can be neglected. In this case, the dispersion then simplifies to the following:

$$c_0 k = \omega n = \pm\omega \left(\frac{\omega_L^2 - \omega^2}{\omega_0^2 - \omega^2} \right)^{1/2}, \qquad (3.38)$$

where we can simply express the *longitudinal frequency* as $\omega_L^2 = \omega_0^2 + \omega_p^2$. We should now clearly appreciate that for frequencies inside the interval (ω_0, ω_L), the wavenumber becomes purely imaginary and therefore no propagating wave solution exist. Therefore, the interval (ω_0, ω_L) determines a characteristic *frequency gap* in the polariton dispersion. This gap coincides in fact with the Reststrahlen band previously introduced in Section 3.1.1.

The effect of nonzero dissipation on bulk polaritons is illustrated in Figure 3.7, where we plot the dispersion relations corresponding to three different values of the parameter γ, as indicated in the legend. The finite dissipation is reflected in the presence of a maximum wavenumber $k < \infty$ in the polariton dispersion with an amplitude that decreases by increasing the value of γ. This limits the localization length of the polariton in proximity of ω_0 compared to the non-dissipative case. Moreover, transmission states appear inside the Reststrahlen gap region in the presence of dissipation, with increasing propagation losses associated to the imaginary part of the wavenumber inside this gap.

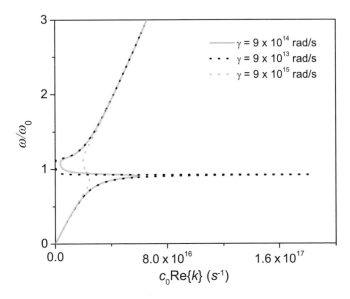

Figure 3.7 Black line: dispersion relation of TEM waves in a dissipative Lorentz medium. The real part of $c_0 k$ in units of inverse seconds. The values of the dissipation parameters γ are reported in the legend. $\omega_p = 15 \times 10^{15}$ rad/s and $\omega_0 = 1.5 \times \omega_p$ have been used in this example.

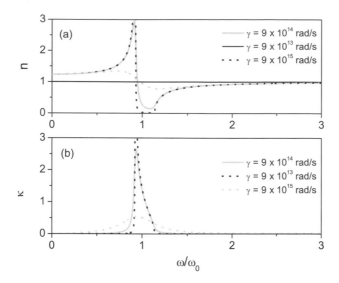

Figure 3.8 (a) Refractive indices of a dissipative Lorentz medium and (b) corresponding extinction coefficients. In this example, we used $\omega_p = 15 \times 10^{15}$ rad/s and $\omega_0 = 1.5 \times \omega_p$. The values of the dissipation parameters are reported in the legend.

The complex refractive index associated to the general dispersion relation (3.37) of bulk polaritons in the Lorentz dielectric media is shown in Figure 3.8. The real part of the complex refractive index n decreases with frequency within the Reststrahlen band, where the polariton dispersion becomes anomalous. In this region, the corresponding extinction coefficient κ increases dramatically, giving rise to large losses where the refractive index n drops below unity.

3.2.3   Polaritons as Hybrid Light–Matter Excitations

In the last sections, we discussed wave propagation in conductive and dielectric media and derived the characteristic dispersion relation of the Lorentz medium from which all the relevant wave properties can be systematically derived. We now address the more general situation of electromagnetic wave propagation in a dispersive medium with optical response described by a dielectric function $\epsilon(\omega)$ and a relative magnetic permeability $\mu(\omega)$. In analogy with equation (3.37), we can express its general dispersion relation in the following form:

$$\boxed{\frac{c_0^2 k^2}{\omega^2} = \epsilon(\omega)\mu(\omega)} \tag{3.39}$$

This expression is the *classical polariton dispersion*. It is satisfied by new wave solutions that correspond to electromagnetic waves coupled to polarization and magnetization waves in materials. These "mixed states" are generally called *polaritons*. They are the most general solutions of wave propagation problems in solids and can be naturally understood in quantum mechanics as "hybrid states" where the quanta of the free electromagnetic field, i.e., photons, couple to the quanta of the polarization or magnetization fields in materials.

Elementary excitations in solids, such as lattice vibrations or charge density waves, are classified in condensed matter theory either as quasiparticles or collective wave-like oscillations described by plane-wave parameters (ω, k). These concepts are introduced in order to reduce the outstanding complexity of many-body systems with strongly interacting components, such as the electron motion in solids, to the simpler behavior of weakly interacting quasi particles that propagate in free space with effective or renormalized parameters, typically an effective mass m^* and a finite lifetime τ_p. For example, individual electrons or electron–hole pairs bound by the Coulomb interaction (i.e., excitons) in semiconductor crystals are quasiparticles with an effective mass that accounts for their complicated quantum mechanical interactions with the atoms in the crystal and with the other electrons. Another typical example is the polaron quasiparticle, which was first proposed by Lev Landau in 1933 [114] to describe the motion of an electron in a dielectric crystal where atoms move from their equilibrium positions and effectively screen the charge of an electron, effectively creating a "phonon cloud" around the electron. Therefore, polarons reflect the electron interaction with the polarization of its surrounding ions, and are experimentally observed in organic semiconductors.

More generally, quasiparticles are weakly interacting *dressed particles* composed of a "bare" particle core that is clothed, or dressed, by the interactions with the other surrounding particles in their local environments. The energy difference between the renormalized quasiparticle and the bare particle is called the *self-energy* of the quasiparticle, and it measures the interaction strength in the many-body system [115].

On the other hand, collective excitations arise from the aggregate or collective behavior of the entire system, and there is no individual real particle forming its core. A standard example of a collective excitation in a solid is the phonon, which corresponds to the vibrational motion of every ion in the crystal lattice. In most circumstances, quasiparticles behave as fermions while collective excitation as bosons. Their quantum mechanical description can be obtained in terms of a stream of particle-like excitations or quanta described by well-defined energy and momentum variables (E, p), in complete analogy to the quantization of the free electromagnetic field. The connection between the classical and the quantum description is captured by the energy and momentum dispersion relations $E = \hbar\omega$ and $p = \hbar k$, respectively. The most frequently encountered elementary excitations in solids are listed in Table 3.1

Table 3.1 Examples of elementary excitations in solids.

Type of excitation	Physical nature	Polariton quasiparticle
Phonon	Lattice vibration	Phonon-polariton
Plasmon	Charge density wave	Plasmon-polariton
Magnon	Magnetic spin wave	Magnon-polariton
Exciton	Electron–hole bound states	Exciton-polariton

along with the corresponding polariton type. In the quantum mechanical picture, a polariton is defined as the quantum of a quantized electromagnetic wave coupled with a quantized materials' excitations. For example, a phonon-polariton is the quantum of the electromagnetic field coupled to lattice oscillations, and a plasmon-polariton the quantum of the electromagnetic field coupled to electron density oscillations.

It is possible to show that a full quantum mechanical formulation, which is beyond the scope of our discussion, leads to a bulk polariton dispersion relation that is formally identical to the one obtained using classical physics:

$$\boxed{\frac{c_0^2 k^2}{\omega^2} = \epsilon_b + \frac{f}{\omega_0^2 - \omega^2 - j\gamma\omega}} \tag{3.40}$$

where ϵ_b is the background permittivity due to the ion cores and

$$f = \frac{2N\omega_0}{\epsilon_0 \hbar} |\langle \psi_j | H^{(D)} | \psi_i \rangle|^2 \tag{3.41}$$

is the *oscillator strength* for the dipole transition. Here the spatial dispersion that arises from the coupling of the different oscillators has been neglected. We refer the interested reader to the references [116–118] for a systematic treatment of the quantum theory of polaritons.

3.3   Surface Polariton Waves

Surface polaritons (SPs) are wave solutions of the homogeneous Maxwell's equations (i.e., eigenmodes) that propagate along the interface between two media with different dielectric constants. Surface waves decay exponentially in the direction vertical to the interface, to which are bound by evanescent fields. Surface waves appear in many fields of research, such as seismology, where several types of elastic waves can travel along the surface of solids, or in fluid dynamics, where surface gravity waves propagate along the surface of liquids as in the familiar case of the ocean waves. In the context of polaritons, surface waves can appear for a general Lorentz medium inside the Reststrahlen band at frequencies $\omega_0 < \omega < \omega_L$, where there is no propagation of bulk waves. For standard semiconductor materials, this condition is satisfied in the far-infrared spectral range, where the permittivity has a negative real part. The most interesting case is the one of surface plasmon polaritons (SPPs) that occur at the interface between a metal and a transparent dielectric medium. Since in a metal we have that $\omega_0 = 0$ and $\omega_L = \omega_p$, SPP waves can exist across a very large spectrum of frequencies that extends for DC ($\omega = 0$) up to the plasma frequency of the metal. Therefore, in contrast to SPs at the interface between two dielectric media, SPPs in metals are extremely broadband phenomena and can be excited from the visible spectrum to the infrared range. However, due to the intrinsically lossy character of the metallic dispersion, they do not propagate over large distances in the visible and near-infrared range, as we will now discuss in more detail.

3.3.1   Dispersion of Surface Plasmon Polariton Waves

Let us now specifically address the homogeneous solutions of Maxwell's equations that describe localized waves at the interface between two media with dielectric functions ϵ_1 and ϵ_2. Following the derivation in [92], we will assume ϵ_1 to be real while no constraints will be imposed on ϵ_2. Our goal is to derive the conditions on the dielectric functions under which a surface wave can propagate along the interface (i.e., along the x-axis) while confined in the z-direction normal to the interface.

We begin by considering the Maxwell's vector wave equation for time-harmonic fields that is generally valid in source-free (not necessarily homogeneous) dielectric media:

$$\nabla \times \nabla \times \mathbf{E}(\mathbf{r}, \omega) - \frac{\omega^2}{c_0^2} \epsilon(\mathbf{r}, \omega) \mathbf{E}(\mathbf{r}, \omega) = 0 \tag{3.42}$$

where $\epsilon(\mathbf{r}, \omega) = \epsilon_1(\omega)$ for $z > 0$ and $\epsilon(\mathbf{r}, \omega) = \epsilon_2(\omega)$ for $z < 0$. Moreover, we will consider only p-polarized waves with (x, z) nonzero electric field components because no surface bound modes exist for the s-polarization for nonmagnetic media, as it can be easily verified. Therefore, we represent the electric field amplitude of the surface wave in the two half-spaces as the following:

$$\mathbf{E}_i = \begin{bmatrix} E_{x,i} \\ 0 \\ E_{z,i} \end{bmatrix} e^{jk_x x - j\omega t} e^{jk_{z,i}}, \tag{3.43}$$

where $i = 1, 2$ identifies the half-spaces $z > 0$ and $z < 0$, respectively. Notice that the transverse wavevector component k_x is conserved (i.e., it is the same in both media) due to the homogeneity of the planar slab geometry and therefore is not labeled by index i, and k-conservation can be expressed as follows:

$$k_x^2 + k_{z,i}^2 = \epsilon_i k_0^2. \tag{3.44}$$

In both half-spaces, Gauss's law $\nabla \cdot \mathbf{D} = 0$ implies the two equations

$$k_x E_{x,i} + k_{z,i} E_{z,i} = 0 \tag{3.45}$$

for $i = 1, 2$. Imposing the electromagnetic boundary conditions at the interface between the two media, we write the following immediately:

$$E_{x,1} - E_{x,2} = 0 \tag{3.46a}$$

$$\epsilon_1 E_{z,1} - \epsilon_2 E_{z,2} = 0 \tag{3.46b}$$

Equations (3.45) and (3.46a) form a homogeneous linear system that admits a non-zero solution when the determinant of its matrix form vanishes, which yields the important relation connecting the transverse wave-vector components:

$$\frac{k_{z,2}}{\epsilon_2} = \frac{k_{z,1}}{\epsilon_1} \tag{3.47}$$

Combining the last equation with the k-conservation relation in (3.44), we can obtain the dispersion relation:

$$\beta^2 = \frac{\omega^2}{c_0^2} \frac{\epsilon_1 \epsilon_2}{\epsilon_1 + \epsilon_2} \tag{3.48}$$

$$k_{z,i}^2 = \frac{\omega^2}{c_0^2} \frac{\epsilon_i^2}{\epsilon_1 + \epsilon_2} \tag{3.49}$$

where we define the propagation parameter $\beta \equiv k_x$. The preceding dispersion relations are generally valid and allow us to identify the conditions for an interface mode to exist. In fact, the surface wave must propagate along the x-axis and therefore its propagation parameter β must be real. Therefore, from equation (3.48) we deduce that the product and the sum of the dielectric functions must be both positive or both negative. On the other hand, for the propagating wave to be evanescently bound to the interface, its electric field must decay exponentially along the z-axis, which implies from equation (3.49) that $\epsilon_1 + \epsilon_2 < 0$.

Therefore, surface wave solutions exist when the following conditions are simultaneously satisfied:

$$\epsilon_1(\omega)\epsilon_2(\omega) < 0 \tag{3.50}$$
$$\epsilon_1(\omega) + \epsilon_2(\omega) < 0 \tag{3.51}$$

These conditions imply that, if both the dielectric functions are real, they must have opposite signs and the absolute value of ϵ_2 must exceed the positive value of ϵ_1 such that their overall sum remains negative. When ϵ_2 is complex, the same argument applies limited to its real part. It should now be clear that the frequency interval where materials exhibit negative permittivity plays a fundamental role for the engineering of surface polariton waves. As we already remarked, in the case of metals this frequency interval extends to $0 < \omega < \omega_p$.

Let us now choose to consider medium 2 specifically as a metallic material (e.g., Al or Ag to a good approximation in the visible spectrum) described by the lossless Drude model permittivity $\epsilon_2(\omega) = \epsilon_D(\omega)$, where $\epsilon_D(\omega)$ is explicitly shown in equation (3.12). According to the general relation in equation (3.47) the transverse field amplitude of the SPP wave will symmetrically decay in both media when $k_{z,2} = -k_{z,1}$, which generally occurs at the *surface plasmon frequency* ω_{sp} defined by the condition:

$$\epsilon_2(\omega_{sp}) = -\epsilon_1. \tag{3.52}$$

When $\epsilon_2(\omega) \equiv \epsilon_D(\omega)$, the surface plasmon frequency of the lossless Drude model can be simply determined:

$$\omega_{sp} = \frac{\omega_p}{\sqrt{1 + \epsilon_1}} \tag{3.53}$$

The surface plasmon frequency ω_{sp} is the frequency of maximum transverse localization of the SPP wave in the dielectric medium. This frequency can be significantly lowered by depositing metallic layers atop dielectric materials with large refractive index, i.e., large ϵ_1. In particular, the sensitivity of the SPP waves to their dielectric environments motivates their applications to optical biochemical sensing, which are driven by the large enhancement of the p-polarized SPP electric field amplitude that occurs at the metal–dielectric interface [119, 120]. Noble metals such as Au and Ag are characterized by dielectric functions with small imaginary parts. Also in this case we can use the general relations (3.48) and (3.49) by substituting the complex permittivity $\epsilon_2 \rightarrow \epsilon_2 = \epsilon_2' + j\epsilon_2''$, which renders the propagation constant β and the transverse wavevector components $k_{z,i}$ complex quantities. In particular, the imaginary part β'' of the SPP propagation constant accounts for the damping of the wave as it propagates along the interface, and allows us to define the SPP propagation length as follows:

$$L = \frac{1}{2\beta''}. \tag{3.54}$$

This quantity L provides the $1/e$ decay length of the SPP field intensity. On the other hand, the real part β' of the propagation constant determines the SPP wavelength according to the following:

$$\lambda_{SSP} = \frac{2\pi}{\beta'}. \tag{3.55}$$

A comparison of the SPP characteristics of different noble metal films is shown in Figure 3.9. At near-infrared frequencies the typical field penetration depth of SPP waves bound at the interface between air and noble metals ranges in the 20–30 nm inside the metal (comparable to the skin depth) and is of order 300–500 nm in air. The corresponding SPP propagation length ranges between 10 μm and few hundreds μm,

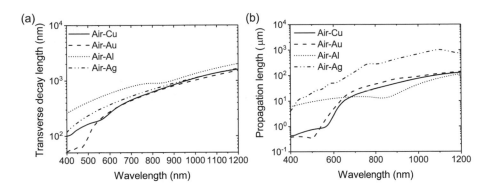

Figure 3.9 Comparison of the SPP properties of different materials. (a) Transverse decay length in the dielectric medium of surface plasmon polaritons (SPPs) computed as $1/\Im(k_{z,1})$, where the complex wavevector $k_{z,1}$ is given by the general relation (3.49). (b) SPP propagation length computed from the general expression (3.54). Materials dispersion parameters have been obtained from tabulated data in references [112, 121, 122].

depending on the materials of choice and on the operation wavelength. For noble metals such as Au and Ag, we can expand in Taylor series the complex propagation constant with resect to the small parameter $|\epsilon_2''|/|\epsilon_2'| \ll 1$. Using equation (3.48) and separating real and imaginary parts, it is possible to obtain first-order expressions for the real and imaginary parts of β and of the transverse wave-vector components [92].

3.4   Polarizability and Local Field Corrections

Our discussion of the Lorentz model so far only considers independent oscillators excited by a local field that can be approximated by the externally applied electric field. This model is quite accurate for low-density media. However, in more general situations, the oscillators respond to local fields that are influenced by nearby oscillators as well. Moreover, retardation effects between the different oscillators turn this apparently simple problem into a very complex one, where each oscillator can in principle couple with any other oscillator in the system via radiation fields. This general case can only be accurately described by the Foldy–Lax self-consistent multiple scattering theory that will be introduced in Chapter 12.

In this section, we will consider only low-frequency (i.e., $\omega \simeq 0$) local field corrections, and we will discuss, neglecting any retardation effect, the fundamental *Lorentz–Lorenz formula* that connects the microscopic polarizability α with the macroscopic dielectric function of materials. When a local field $\mathbf{E}_{loc}$ acts upon a single oscillator, it polarizes it, giving rise to a microscopic dipole moment $\mathbf{p} = \alpha \mathbf{E}_{loc}$. The local electric field, or the locally exciting field, has been derived by Lorentz within the electrostatic approximation that is valid for particles much smaller than the wavelength. This classical results reads as follows:

$$\mathbf{E}_{loc} = \mathbf{E} + \frac{\mathbf{P}}{3\epsilon_0}$$

(3.56)

where 1/3 is the so-called *depolarization factor* of the sphere [101].

Lorentz's approach in the derivation of the preceding local field expression considers a composite medium as a mixture of small dielectric inclusions that radiate as dipolar oscillators, all spherical in shape, immersed in an external average field. These inclusions can be thought of as elementary dipoles (i.e., atoms or molecules) that make up the material, but any subwavelength scattering (radiating) particle in an artificial array will play essentially the same role in the Lorentz theory. To calculate the local field that excites a given radiating oscillator or inclusion Lorentz, assumed that the overall effect of all the other nearby inclusions in the medium is simply to generate an average, or macroscopic, polarization field $\mathbf{P}$. This polarization spreads uniformly across the medium except within a fictitious spherical cavity that surrounds a given inclusion. This cavity is called the Lorentz exclusion volume. The local electric field $\mathbf{E}_{loc}$ is defined within the volume of this cavity, and its amplitude is increased with respect to the external average $\mathbf{E}$ field due to the effect of induced polarization charges that are created on the surface of the cavity region. As we show in Appendix A, the surface polarization charges are linked to the normal component of the polarization field. Therefore, by applying Gauss's law on the surface of a suitably oriented pillbox (cubic) volume that crosses the small cavity region, we can immediately obtain equation (3.56). However, in general the calculation of the Lorentz local field is not so trivial because its value is affected by the shape and orientation of the exclusion cavity around the inclusion. Note that this problem is exactly the same that we already encountered in Section 2.9.1 in relation to the regularization of the dyadic Green function.

It should also be clear that in equation (3.56) the field $\mathbf{E}$ is the total macroscopic field in the medium, while $\mathbf{E}_{loc}$ is the field at the inclusion's position due to everything else except the radiation of the particular oscillator (i.e., atom, molecule, or inclusion) under consideration. A similar logic will be applied in Chapter 12, where the concept of a "locally exciting field" will be introduced in the formulation of the multiple scattering Foldy–Lax problem. The local field polarizes the single oscillator, leading to the macroscopic polarization density $\mathbf{P} = N\alpha\mathbf{E}_{loc}$, where N is the number density of oscillators/inclusions in the material and α is the microscopic polarizability. Due to the complex interactions with all the neighboring particles and the randomness of the mixture, a simple relation between the dipole moment and the externally applied field on the single particle does not apply here, and we need to proceed as follows.

Eliminating $\mathbf{P}$ in equation (3.56), we obtain the following:

$$\mathbf{E}_{loc} = \frac{1}{1 - \frac{N\alpha}{3\epsilon_0}}\mathbf{E} = \phi_{loc}\mathbf{E}, \tag{3.57}$$

where the *local field factor* ϕ_{loc} describes the enhancement of the local field with respect to the macroscopic field in terms of the microscopic polarizability. We can also express ϕ_{loc} in terms of the experimentally available macroscopic dielectric function. Since for linear media, we can write $\epsilon_r = 1 + \mathbf{P}/\epsilon_0\mathbf{E}$, the validity of equations (3.56) and (3.57) implies the following:

$$\epsilon_r = 1 + \frac{\frac{N\alpha}{\epsilon_0}}{1 - \frac{N\alpha}{3\epsilon_0}}. \tag{3.58}$$

The last expression can be rearranged to obtain the following:

$$\boxed{\frac{N\alpha}{3\epsilon_0} = \frac{\epsilon_r - 1}{\epsilon_r + 2}} \tag{3.59}$$

Equation (3.59) is known as the *Clausius–Mossotti equation*. When the oscillators, or the dipoles that they effectively describe, are embedded in a host matrix with a permittivity ϵ_h, we must substitute $\epsilon_0 \rightarrow \epsilon_h$ in the Lorentz field expression (3.56), from which we can rewrite the effective permittivity ϵ of the oscillator medium in the more general form:

$$\epsilon = \epsilon_h + \frac{N\alpha}{1 - \frac{N\alpha}{3\epsilon_h}}. \tag{3.60}$$

When the density N is small, the Clausius–Mossotti formula can be approximated as follows:

$$\epsilon \approx \epsilon_h + N\alpha, \tag{3.61}$$

which directly connects the dielectric properties of a macroscopic medium with the electric polarizability of its microscopic constituents. Interestingly, when expressed in term of the optical refractive index,[5] the Clausius–Mossotti equation goes under the name of the *Lorentz–Lorenz formula*. Using the Clausius–Mossotti equation, we can finally express the local field factor as follows:

$$\boxed{\phi_{loc} = \frac{\epsilon_r + 2}{3}} \tag{3.62}$$

Notice that in a medium with negative permittivity such that $\epsilon_r = -2$, the total internal field can be resonantly enhanced, as previously discussed. As we will see in Chapter 13, this is the prevalent case in the optical response of metallic nanospheres across the visible spectral range. The preceding equations work very well for gases and surprisingly for solids as well. In fact, more advanced treatments that include dynamic retardation effects that are relevant when $\omega \neq 0$ turned out to confirm that the static result can indeed be applied also at finite frequencies. In particular, all the preceding equations remain valid, and one just needs to replace α by $\alpha(\omega)$ and ϵ_r by $\epsilon_r(\omega)$ so that the explicit consideration of retardation can be neglected in a first approximation.[6]

3.4.1   Electromagnetic Mixing Rules

Closed-form expressions for the effective permittivity of heterogeneous or composite media are referred to as *mixing rules*. Simple mixing formulas and approximations for the effective optical response of composite media are nicely reviewed in [123]. The electromagnetic properties of composite materials are the primary subject of the *effective medium theory* (EMT), which describes the macroscopic properties of multiply scattering composites such as porous media, aerosols, interstellar dusts, or more recently, man-made metamaterials. At their constituent level, heterogeneous composites are inhomogeneous media and a detailed (i.e., not averaged) description of their optical properties can only be obtained by solving the full-vector multiple scattering equations, which in general is a very difficult task. However, several analytical approaches have been developed that can provide useful approximations of heterogeneous random media in terms of few macroscopic parameters and properties such as the relative fractions of their components (i.e., the density of each phase) or their statistical properties as captured by pair-correlation functions.

---

[5] This is simply done by using the relation $\epsilon_r = n^2$ valid for dielectric materials.
[6] In the context of the more advanced effective medium theory, radiation corrections do play a role, and more complex expressions for the macroscopic parameters have been developed. See, for instance, [123].

The main goal of EMT is to identify the average or *coherent field*[7] that propagates in a composite random medium with the field that propagates within an equivalent homogeneous medium described by an effective permittivity.

The oldest example of an analytical EMT is the *Maxwell–Garnett (MG)* [124] treatment of the optical properties of dipolar spherical scatterers randomly dispersed in a host dielectric matrix of permittivity ϵ_h. In the case of a material that consists of a mixture of two or more species, we can write equation (3.59) as follows:

$$\sum_i \frac{N_i \alpha_i}{3\epsilon_h} = \frac{\bar{\epsilon} - \epsilon_h}{\bar{\epsilon} + 2\epsilon_h}, \tag{3.63}$$

where the contribution from each species is indexed by i in the preceding sum, and N_i indicates the number of particles per unit volume. Note that we denote by $\bar{\epsilon}$ the effective or average permittivity of the composite medium. The corresponding MG mixing rule follows directly by substituting inside the general Clausius–Mossotti relation for a multiphase material the well-known expression for the polarizability of a small sphere of radius a and dielectric constant ϵ_i, which is as follows:

$$\alpha = 4\pi a^3 \epsilon_h \frac{\epsilon_i - \epsilon_h}{\epsilon_i + 2\epsilon_h}. \tag{3.64}$$

This is the regime of the *quasistatic approximation* that is valid for particles with sizes much smaller than the wavelength so that the dynamic Helmholtz wave equation reduces to the static Laplace equation. Therefore, for a homogeneous composite medium made of quasistatic dielectric spherical inclusions with dielectric constant ϵ_i, we can write the following:

$$\boxed{\frac{\bar{\epsilon} - \epsilon_h}{\bar{\epsilon} + 2\epsilon_h} = f \frac{\epsilon_i - \epsilon_h}{\epsilon_i + 2\epsilon_h}} \tag{3.65}$$

where the dimensionless quantity f is the volume fraction of the inclusions in the mixture. Note that in equation (3.65), which is known as the *Rayleigh mixing formula*, only the volume fraction of the inclusions and their permittivities appear to be relevant. In particular, no assumptions on the size of the individual spheres are made provided that they are much smaller than the wavelength, and the detailed spatial distribution of the inclusions is not relevant as long as the particles' positions remain uncorrelated. This condition is expressed more precisely by the so-called quasistatic Rayleigh limit $ka \ll 1$, where k is the wavenumber in the host medium.

A useful mixing formula for the effective permittivity $\bar{\epsilon}$ can be readily derived from (3.65) rewritten as follows:

$$\boxed{\bar{\epsilon} = \epsilon_h + 3f\epsilon_h \frac{\epsilon_i - \epsilon_h}{\epsilon_i + 2\epsilon_h - f(\epsilon_i - \epsilon_h)}} \tag{3.66}$$

---

[7] The coherent field in a random medium is obtained by considering the average field over several realizations of the disorder, often modeled as a Gaussian random process.

This equation is the *Maxwell–Garnett (MG) mixing formula*. This simple formula depends only on two parameters, which are the permittivity contrast ϵ_i/ϵ_h and the volume fraction of the inclusions, and it has a surprisingly large range of applicability.

The classical MG formula is generally considered valid for situations where the volume fraction of the composite is small (typically it is assumed that $f < 0.1$). For larger volume fractions, the *Bruggeman mixing rule* is believed to have superior accuracy [125]:

$$f\frac{\epsilon_i - \epsilon_b}{\epsilon_i + 2\epsilon_b} + (1 - f)\frac{\epsilon_h - \epsilon_b}{\epsilon_h + 2\epsilon_b} = 0 \tag{3.67}$$

where ϵ_b is the Bruggeman effective permittivity. This expression applies for a two-phase mixture. Note also that the Bruggeman mixing rule treats the inclusions and the host environment symmetrically. In other words, a mixture and its complement have exactly the same effective permittivity. This is not the case for the MG approach.

3.4.2   Radiative Corrections

Radiation and scattering effects lead to corrections to the classical MG formula that become important at high frequency, such as in the visible spectrum for the case of metallic inclusions. Radiation effects are generally included through a size correction to the permittivity. Rigorous theoretical analysis shows that for sufficiently small particles that radiate like dipoles in free space, the *radiation-corrected MG mixing formula* can be expressed as follows [126]:

$$\bar{\epsilon} = 1 + 3f\frac{\beta}{1 - \beta f}\left[1 + \frac{2j}{3}(ka)^3\frac{\beta}{1 - \beta f}\right], \tag{3.68}$$

where the *spectral parameter* $\beta = (\epsilon_i - 1)/(\epsilon_i + 2)$ has been introduced. It is important to realize that for EMT that explicitly account for radiation or scattering effects at high frequency, the effective permittivity and the associated effective wavenumber $\bar{K}^2 = k_0^2\bar{\epsilon}$ become complex even when the composite random medium has no absorption losses. In these cases, the complex effective wavenumber results in an extinction coefficient that physically describes the exponential attenuation of the coherent field due to the multiple scattering of the radiation inside the random medium. In contrast, the classical MG formula derived within the quasistatic approximation does not contain an imaginary term and therefore it does not attenuate the wave field. However, this limitation can be remedied by introducing the simple expression for the radiation-corrected Clausius–Mossotti dynamic polarizability α_{rad} as [127]:

$$\alpha_{rad} = \alpha_0\left[1 - \frac{2j}{3}\beta(ka)^3\right]^{-1}, \tag{3.69}$$

where $\alpha_0 = 4\pi a^3\epsilon_h\beta$ is the static polarizability of a small sphere.

An example that illustrates the general contribution of radiation corrections to the classical Maxwell–Garnett dispersion is shown in Figure 3.10. We consider a

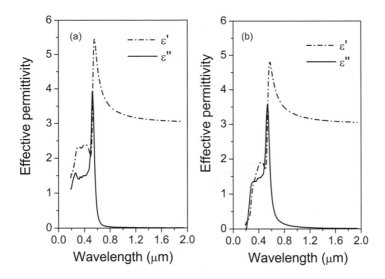

Figure 3.10 (a) Real and imaginary parts of the Maxwell–Garnett effective dielectric function for a uniform random mixture of Au nanospheres with radius $R = 40$ nm and volume filling fraction $f = 0.4$ without radiation corrections. (b) Real and imaginary parts of the Maxwell–Garnett effective dielectric function of the same mixture, including radiation corrections according to equation (3.68). In both cases, the Au dispersion data are obtained from the tabulated values in [112].

uniformly random mixture of Au nanospheres of radius $R = 40$ nm and compare the wavelength dependence of the real and the imaginary parts of the effective dielectric functions computed using the standard Maxwell–Garnett formula in Figure 3.10a or the radiation corrected formula (3.68) in Figure 3.10b. In both cases the sharp peak in the real part of the permittivity corresponds to the excitation of a dipole surface *plasmon mode of the mixture*. The broad shoulder at short wavelengths (around 400 nm) reflects the contribution of materials absorption losses in Figure 3.10a and scattering losses combined with materials losses in Figure 3.10b. We observe additionally that in Figure 3.10b the peak in the real part of the permittivity is significantly broadened and reduced in its intensity when radiation corrections are included.

3.4.3 Spatial Correlation Corrections

Numerical studies [126] have shown that the radiation-corrected MG mixing formula has a surprisingly wide range of applicability that can even be extended to large volume fractions $f \approx 0.4$ and strong dielectric contrast provided the positions of the inclusions remain *statistically uncorrelated*. More advanced theories [128, 129], such as the quasicrystalline approximation (QCA), the effective-field approximation (EFA), or the coherent-potential approximation (CPA) account for radiation effects and also for the statistical correlations in the positions of the particles in the random medium. These correlations are generally described through the *radial or pair distribution function g(r)*, which we discuss in great detail in Chapter 7. Here it suffices to realize that $g(r)$ measures the probability of finding a pair of particles separated by a distance r relative to the probability expected for a uniformly random distribution at the same density. Clearly, any deviation of $g(r)$ from a constant line is indicative of local (i.e., short-range) correlations among the positions of the particles in the system.

Theories that account for correlation effects are based on a renormalization procedure [130] that yields the effective permittivity of an unbounded random medium as the solution of a dispersion relation for its effective wavenumber $\bar{K}^2 = k_0^2 \bar{\epsilon}$. However, a very useful and simpler homogenization procedure has been recently established based on a perturbation of the classical MG formula [131]. Rigorous derivations of several dynamic formulas for finite-size aggregates of small (dipolar) particles that include radiation and correlation corrections are presented in [126, 131] and in the references therein. A key result derived in [131] for short-range spatial correlations permits us to express the radiation and correlation-corrected effective permittivity of a mixture as follows:

$$\bar{\epsilon} = 1 + 3f \frac{\beta}{1 - \beta f} \left[1 + \frac{2j}{3} (ka)^3 \frac{\beta'}{1 - \beta f} (1 + 3f M_2) \right.$$
$$\left. + \frac{11j}{10} (ka)^2 \frac{\beta''}{1 - \beta f} (1 + 2f M_1) \right], \tag{3.70}$$

where

$$M_n = \int_0^\infty [g(u) - 1] u^n \, du \tag{3.71}$$

and $g(u)$ is the suitably normalized radial distribution function [131]. The results in (3.70) coincide analytically with the low-frequency QCA approximation [132] for lossless particles ($\beta'' = 0$). Moreover, when spatial correlation effects can be neglected (i.e., $M_n = 0$), the formula (3.70) yields a more accurate approximation of radiation correction compared to (3.68) [131].

In Figure 3.11, we apply the high-frequency QCA formula (3.70) to short-range correlated random mixtures of Au particles with two different values of the volume filling fraction, which are reflected in the different $g(r)$ curves shown in Figure 3.11a. In Figure 3.11b, we show the computed real and imaginary parts of the effective QCA permittivity. The results illustrate vividly the significant effects of short-range spatial correlations that strongly modify, due to the enhanced scattering losses, the spectrum of the imaginary permittivity when increasing the filling fraction of the correlated mixture. On the other hand, the spectrum of the real part of the permittivity was not found to be significantly influenced by correlation.

While the incorporation of radiation corrections in the homogenized random medium invariably increases the losses in wave propagation, the role of arbitrary structural correlations is very complex and sensitive to the specific type of disorder model. Moreover, in the general case where the spatial correlations are not short-range, such as in structurally complex media of finite size, the effective parameters become position dependent, i.e., $\beta = \beta(\mathbf{r})$. In such situations, it is meaningful to

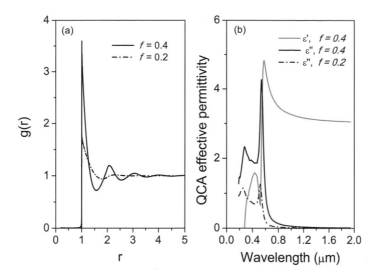

Figure 3.11 (a) Pair distribution function $g(r)$ (in units normalized to the diameter of the particles) for two different densities as specified in the legend. The pair distribution is computed as the solution of the Percus–Yevick equation for hard spheres, as discussed in Chapter 7. (b) Real and imaginary parts of the correlation- and radiation-corrected Maxwell–Garnett effective dielectric function using the QCA approximation in [131] for a nonuniform mixture of Au spheres with radius $R - 40$ nm and $f = 0.4$. The Au dispersion data are obtained from the tabulated values in [112].

consider a spatially averaged spectral parameter $\bar{\beta}$ over the test volume and then obtain the corresponding effective permittivity as follows:

$$\bar{\epsilon} = \frac{1 + 2\bar{\beta}}{1 - \bar{\beta}}. \tag{3.72}$$

However, to what extent EMT approaches can be applied to general inhomogeneous media remains a matter of current research [133–137]. Finally, EMTs and mixing formulas can also be obtained for nonspherical inclusions as well as for anisotropic and nonlinear mixtures within the quasistatic regime. A comprehensive survey on these topics can be found in [123].

3.5   What Are Metamaterials?

Metamaterials are artificially engineered media with constituents that are much smaller than the wavelength scale. They can be designed to display unusual electromagnetic properties that emerge from the resonant interaction of electromagnetic waves with their subwavelength components [138–140]. Metamaterials are typically made of two different building blocks that alternate periodically with dimensions and periods that are much smaller (typically by a factor of 10) than the optical wavelength of interest. Metamaterials can be engineered to exhibit both negative permittivity ϵ and permeability μ, near-zero complex permittivity/permeability, and zero-averaged refractive index [141, 142], resulting in *very unusual optical responses* such as backward propagating waves [138] and negative refraction [143], near-field amplification for superresolution imaging applications [144], and optical magnetism [139, 140], to name a few. The research on metamaterials was pioneered by the work of Smith et al. [145] stimulated by Veselago's seminal paper on negative refractive index [146]. However, the concept of a man-made metamaterial dates back to a paper by Bose published in 1898 and dealing with the effects of artificially twisted structures on the polarization of electromagnetic waves [147]. Due to the subwavelength nature of their components, metamaterials are often designed using EMT. However, it is important to realize that the EMT approach is correct only when considering local responses, and it may fail at shorter wavelengths due to multiple scattering and collective resonances in nonrandom geometries. This is because the periodic/aperiodic positional ordering of subwavelength building blocks invariably introduces nonlocal effects in the optical response of the resulting metamaterials, which are difficult to take into account using electromagnetic mixing rules.

There are currently several paths available for the design of optical metamaterials. The most direct one relies on the *numerical optimization* of specific metamaterials functions, such as broadband nanofocusing, negative refraction, reduced optical losses, etc. Genetic optimization algorithms have been successfully applied to the engineering of plasmonic nano-antennas [148] and negative-index metamaterials [149]. Alternative ways are provided by digital [150] and coding, or programmable, metamaterials [151]. According to this paradigm, it is possible to synthesize optical structures that perform desired functions by properly sculpting only two material "bits" into local functional units called *metamaterial bytes*. Differently from traditional structures, often limited to periodic assemblies, digital metamaterials possess well-defined structural units designed by simple mixing of two material bits at the nanoscale. Carefully designed local spatial ordering contributes to determine the effective (meta)properties of digital metamaterials, analogously to ordered sequences of zeros and ones that convey information by their local position in digital messages. In the related approach of coding metamaterials, the resulting bytes do not even need to be described by effective medium parameters. Such metamaterials only rely on the presence of two types of unit cells with 0 and π phase responses, which can be controlled by field-programmable gate arrays (FPGAs). The desired optical properties are achieved by "programming" two materials bits that display, when assembled into a specific pattern, the targeted responses.

4 Wave Propagation and Diffraction

Lumen propagatur seu diffunditur non solum Directe, Refracte, ac Reflexe, sed etiam alio quodam quarto modo, Diffracte. [Light is propagated or diffused not only directly, or via Refraction and Reflection, but also in another and fourth way, Diffraction.].

Physico-mathesis de lumine, coloribus, et iride, aliisque adnexis libri duo (1665), Francesco Maria Grimaldi

In this Chapter, we address the propagation of optical pulses through dispersive media and introduce the rigorous framework of scalar wave diffraction theory that will be applied to complex optical structures in later chapters. After introducing the concepts of group and energy velocities, we show how to derive transport equations that govern the pulse dynamics and how they can be solved using emerging neural networks approaches. The scalar wave diffraction theory is then presented following the historically motivated approach based on the Helmholtz and Kirchhoff integral theorem as well as the angular spectrum approach to optical waves that is used in advanced diffractive optical technology and Fourier optics. The full equivalence between these two formulations is also established. A self-contained introduction to vector wave diffraction based on the Stratton–Chu integral formula concludes the chapter.

4.1   Optical Wave Packets

Optical pulses or wave packets can be generally obtained by the coherent mixture (linear superposition) of waves with different frequencies, amplitudes, and phases according to the inverse Fourier transform:

$$E(z,t) = \frac{1}{2\pi} \int_{-\infty}^{\infty} e^{j(kz-\omega t)} \mathbf{E}(0,\omega)d\omega = \frac{1}{2\pi} \int_{-\infty}^{\infty} e^{-j\omega t} \mathbf{E}(z,\omega)d\omega, \qquad (4.1)$$

where $\mathbf{E}(0,\omega)$ is the Fourier transform of the initial field $\mathbf{E}(0,t)$ at $z = 0$ and

$$\mathbf{E}(z,\omega) = e^{jkz}\mathbf{E}(0,\omega) \qquad (4.2)$$

is the propagated temporal spectrum at an arbitrary distance z. A direct consequence of the scaling property of the Fourier transform[1] is that the duration of an pulse is inversely proportional to the range of frequencies involved in the mixture, or the frequency spread of the harmonic wave components. Moreover, the wavenumber is generally frequency dependent in a dispersive linear medium, e.g., $k(\omega) = \omega\sqrt{\epsilon(\omega)\mu_0}$, where we neglect the magnetic response.

As a result, a wave packet travels across a nondispersive medium with the same shape (no distortion). On the other hand, truly dispersive media characterized by a nonlinear dispersion relation $k = k(\omega)$ that will generally alter the shape of propagating wave packets because different frequency components in the spectrum of the initial pulse will travel at different speeds, thus "dispersing" its original shape.

4.1.1   Velocity of Wave Packets

In order to better understand the physical meaning of the propagation velocity of a general wave packet, we first consider the case of a simple packet created by the interference of two monochromatic plane waves. In particular, we study waves that propagate along the z-direction with equal amplitudes and slightly different frequencies ($\omega_1 \approx \omega_2 = \omega$) in a nondispersive medium. In this case, the total electric field of the superposition is as follows:

$$E_T(z,t) = E_0[\cos(k_1 z - \omega_1 t) + \cos(k_2 z - \omega_2 t)]. \tag{4.3}$$

Using the well-known trigonometric identity for the sum of two cosine functions, we can express the previous field in the following product form:

$$\boxed{E_T(z,t) = 2E_0 \cos(k_p z - \omega_p t)\cos(k_g x - \omega_g t)} \tag{4.4}$$

This equation describes a *beat phenomenon* and gives rise to a train of localized wave packets created by the product of a fast-oscillating carrier wave and a slow-oscillating envelope function. This scenario is characteristic of amplitude wave modulation problems, such as those encountered in the transmission of radio waves. The carrier wave amplitude is characterized by the "fast" parameters:

$$\omega_p = \frac{\omega_1 + \omega_2}{2} \tag{4.5}$$

$$k_p = \frac{k_1 + k_2}{2}, \tag{4.6}$$

while the slow amplitude variations of the envelope are described by the "slow" parameters:

$$\omega_g = \frac{\omega_1 - \omega_2}{2} \tag{4.7}$$

$$k_g = \frac{k_1 - k_2}{2}. \tag{4.8}$$

---

[1] If $f(x) \to f(ax)$, then in the Fourier domain $\tilde{f}(v) \to (1/|a|)\tilde{f}(v/a)$, where v is the temporal frequency.

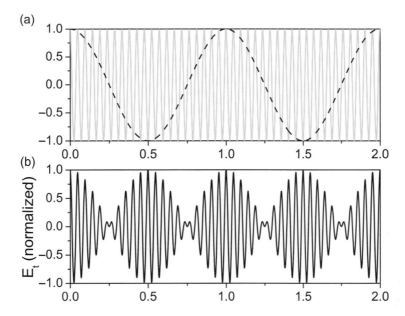

Figure 4.1 (a) Representative carrier and envelope (dashed line) waves. (b) Modulated wave form corresponding to equation (4.4) with the carrier and the envelope components displayed in panel (a). In this example, we considered $\omega_1/\omega_2 = 1.1$.

Carrier and envelope waves are shown in Figure 4.1a. The time (spatial) evolution of the modulated or total waveform at an arbitrary position (time) is illustrated in Figure 4.1b. The carrier wave propagates with the *phase velocity* $v_p = \omega_p/k_p \approx \omega/k$, where $\omega_1 \approx \omega_2 = \omega$ and $k_1 \approx k_2 = k$. The phase velocity is simply the ratio between the wavelength $\lambda = 2\pi/k$ and the optical period $T = 2\pi/\omega$ of a wave, and was already introduced by Newton in his *Principia*. On the other hand, the envelope of the wave packets propagates at the different velocity v_{gr}, called the *group velocity*, determined by the ratio:

$$v_g = \frac{\omega_2 - \omega_1}{k_2 - k_1} = \frac{\Delta\omega}{\Delta k} \tag{4.9}$$

The definition of the group velocity needs to be modified in the case of two infinitesimally close frequencies and wave numbers that fall within the continuum spectrum of a more general wave packet. In this situation, we express the group velocity as follows:

$$v_g = \frac{d\omega}{dk} = \omega'(k), \tag{4.10}$$

where the prime indicates the first derivative of the frequency with respect to the wavenumber. The group velocity as defined in (4.10) is the velocity of signal propagation, and it also provides the velocity of energy transport in a nondispersive (lossless) medium. However, the group velocity may differ significantly from the phase velocity for media that are dispersive, i.e., when the phase velocity is wavelength dependent. In fact, remembering that $v_p = \omega/k$, the group and the phase velocities are simply related as follows:

$$v_g = \frac{d}{dk}(kv_p) = v_p + k\left(\frac{dv_p}{dk}\right) = v_p - \lambda\left(\frac{dv_p}{d\lambda}\right) \qquad (4.11)$$

This expression is known as the *Rayleigh's formula*. From this formula, it is clear that the phase and group velocity are only equal when v_p does not depend on wavelength, i.e., in nondispersive media. In general, the group velocity may be either larger or smaller than the phase velocity depending on the sign of the derivative of the phase velocity with respect to the wavelength. Alternatively, since $v_p = c_0/n(\lambda)$, where c_0 is the speed of light in free space and $n(\lambda)$ is the wavelength-dependent refractive index in a dispersive medium, we can rewrite the Rayleigh's formula as follows:

$$v_g = v_p\left[1 + \frac{\lambda}{n}\left(\frac{dn}{d\lambda}\right)\right] \qquad (4.12)$$

In this expression, the dispersive nature of the medium is captured by the wavelength dependence of its refractive index. In particular, we say that the medium dispersion is normal when $dn/d\lambda < 0$ (the refractive index decreases with wavelength), which implies $v_g < v_p$. On the opposite, when $dn/d\lambda > 0$ the dispersion is called anomalous because in this regime v_p and v_g may have different signs. Anomalous dispersion is naturally encountered when electromagnetic waves propagate at frequencies that are near to resonant absorption lines of materials, as in the cases presented in Chapter 3. Finally, in media characterized by significant dispersion, a third velocity must be introduced, the so-called *energy velocity* v_e. This is defined as the velocity of the energy transport, which is given by the ratio between the amplitudes of the time-averaged Poynting vector $\bar{S}$ and of the time-averaged energy density $\bar{w}$:

$$v_e = \frac{\bar{S}}{\bar{w}} \qquad (4.13)$$

When the medium is nondispersive, we have that $v_e = v_g$ and the transport of the time-averaged energy obeys the advection equation:

$$\frac{\partial \bar{w}}{\partial t} + \frac{\partial(v_g\bar{w})}{\partial x} = 0. \qquad (4.14)$$

In particular, equation (4.14) implies that the energy of a propagating wave packet remains constant when transported.

4.1.2   Propagation of Wave Packets

We consider here in more detail the propagation of an optical wave packet in a medium characterized by the dispersion relation $\omega = f(k)$. In the simplest case of a homogeneous isotropic medium, we have the linear dispersion $\omega = ck$, where c is the constant phase velocity of the wave. However, when ω and k are nonlinearly related, then the phase velocity depends on frequency and the different spectral components of the signal propagate at different velocities. As a consequence, the pulse of light will disperse as it propagates. The dispersion equation can be written in the implicit form $\varphi(\omega, k) = 0$, and it may have several roots, or *dispersion branches*, corresponding to different normal modes. An example is the polariton dispersion in dielectric media or in a plasma, which we encountered in Section 3.1. We remark that a nonlinear dispersion relation $\omega = f(k)$ always arises due to inherent temporal of spatial scales in the medium, and it induces a nonlocal dependence between the physical variables describing the response of the system. Additionally, when a dispersive medium is subject to geometrical constraints, arising for instance from its shape or the presence of boundaries, we speak of *geometrical dispersion* in addition to *material dispersion*. Strong geometrical dispersion of electromagnetic waves arises, for instance, in waveguides or in optical fibers.

We now consider in more detail the propagation problem for an unbounded dispersive medium and represent a scalar pulse as the Fourier superposition integral of plane waves in the form:[2]

$$u(z,t) = \int_{-\infty}^{+\infty} A(k)e^{j[kz-\omega(k)t]}dk. \tag{4.15}$$

The initial wave packet is specified at $t = 0$ by a modulated wave with a *narrow spectrum* centered around $k = k_0$:

$$u_0(z) = A(z)e^{jk_0 z} \tag{4.16}$$

and with zero initial ($t = 0$) time derivative. It is also assumed that the function $A(z)$ is slowly varying compared to the carrier wave $\exp(jk_0 z)$ and that it smoothly decays to zero when $z \to \pm\infty$. We can then look for the solution of the propagation problem by rewriting equation (4.15) in the form of a space- and time-modulated wave:

$$u(z,t) = e^{j[k_0 z - \omega_0 t]} \int_{-\infty}^{+\infty} A(k)e^{j[(k-k_0)z-(\omega-\omega_0)t]}dk. \tag{4.17}$$

Since we are dealing with a narrow spectrum, the dispersion relation $\omega(k)$ can be approximated in the neighborhood of $k = k_0$ using the Taylor expansion:

$$\omega(k) = \omega_0 + \omega'(k_0)(k - k_0) + \frac{1}{2}\omega''(k_0)(k - k_0)^2 + \cdots, \tag{4.18}$$

where the prime denotes the derivative with respect to the argument and therefore $\omega'(k_0) = v_g$. Let us now consider the first-order approximation of dispersion theory,

---

[2] The wave packet can alternatively be expressed in the following form:
$u(z,t) = \int_{-\infty}^{\infty} A(\omega)\exp\{j[k(\omega)z - \omega t]\}d\omega$, in which the dispersion is provided by the $k = f^{-1}(\omega)$ inverse function.

which consists in neglecting all the nonlinear terms in the expansion in (4.18). Substituting the linearized Taylor expansion into the integral (4.17), and using the well-known shifting property theorem of the Fourier transform, we obtain the following:

$$u(z,t) = A(z - v_g t)e^{j[k_0 z - \omega_0 t]}. \tag{4.19}$$

This result shows that to first order in dispersion theory, the carrier wave propagates at the phase velocity ω_0/k_0 while the slowly varying pulse envelope propagates without distortion at the group velocity v_g and obeys the transport equation:

$$\boxed{\frac{\partial A}{\partial t} + v_g \frac{\partial A}{\partial z} = 0} \tag{4.20}$$

We now consider pulse propagation beyond the restriction of first-order group velocity by including the higher-order terms that appear in equation (4.18). In this case, the pulse envelope will no longer propagate unaltered across the medium, and its shape will change in space and in time, even though very slowly compared to the scales of variation of the carrier wave. We will represent the general wave packet by the modulated wavefunction:

$$u(z,t) = A(z,t)e^{j[k_0 z - \omega_0 t]}, \tag{4.21}$$

where we still require the envelope $A(z,t)$ to be a slowly varying compared to the wavelength and the frequency of the carrier wave, which is as follows:

$$\frac{\partial A}{\partial t} \ll A\omega_0 \tag{4.22}$$

$$\frac{\partial A}{\partial z} \ll Ak_0. \tag{4.23}$$

The equation describing the complex amplitude beyond the first-order group velocity approximation can be derived directly from the general dispersion relation (4.18) through the substitutions $\partial/\partial z \to j\Delta k$ and $\partial/\partial t \to -j\Delta\omega$. The correspondence between equations and polynomial dispersion relations is evident because monochromatic plane waves must satisfy the associated equations, yielding the same polynomial form in the transformed variables [152]. Implementing this approach and neglecting terms higher than quadratic, we obtain the following:[3]

$$\boxed{j\left(\frac{\partial A}{\partial t} + v_g \frac{\partial A}{\partial z}\right) + \beta \frac{\partial^2 A}{\partial z^2} = 0} \tag{4.24}$$

where $\beta = \omega''(k_0)/2$ is the dispersion parameter, called *group velocity dispersion* (GVD), which determines the broadening or the compression of the pulse upon

---

[3] Note that if we specified the dispersion relation as $k = k(\omega)$ and expanded it in Taylor series through powers of ω, we would have ended up with a second-order derivative in time instead of in space replacing the last term in equation (4.24).

propagation. The solutions of equation (4.24) can be found by rewriting it in a traveling reference system that moves along with the pulse. This amounts to considering the change of variables $t = t'$ and $z' = z - v_g t$, which transform the previous equation in the following simpler form:

$$j\frac{\partial A}{\partial t'} + \beta\frac{\partial^2 A}{\partial z'^2} = 0 \qquad (4.25)$$

We note that the last equation is formally identical to the quantum mechanical Schrödinger equation, and therefore can be solved with the same methods. It is customary to represent the complex envelope as follows:

$$A(z,t) = a(z,t)e^{j\varphi(z,t)}, \qquad (4.26)$$

where $a(z,t)$ and $\varphi(z,t)$ are real functions that describe the modulated amplitude and phase of the wave packet. Note that the partial derivatives $\partial\varphi/\partial t$ and $-\partial\varphi/\partial z$ represent the corrections to the frequency ω_0 and to the wavenumber k_0 of the carrier wave, respectively. An exact solution of equation (4.25) is obtained by considering as an initial condition the Gaussian envelope [153]:

$$A(z,0) = a_0 e^{-\gamma z^2} \qquad (4.27)$$

with $\gamma = \Delta_0^{-2} + j\delta$, where Δ_0 determines the initial width of the wave packet and 2δ is the constant rate of linearly modulated frequency or wavenumber (compare to the representation in equation (4.26)). The solution to this initial-value problem can be obtained using standard Fourier transform methods and can be expressed as follows [153]:

$$A(z,t) = \frac{a_0}{\sqrt{1 + j\alpha t}}\exp\left(\frac{-\gamma x^2}{1 + j\alpha t}\right), \qquad (4.28)$$

where $\alpha = 4\beta\gamma$. The width of the wave packet $\Delta(t)$ and the maximum value of the real amplitude $a(t)$ change with time according to [153]:

$$\Delta(t) = \Delta_0\sqrt{T(t)} \qquad (4.29)$$

$$a(t) = \frac{a_0}{\sqrt[4]{T(t)}}, \qquad (4.30)$$

where $T(t) = (1 - 4\beta\delta t)^2 + (4\beta\Delta_0^{-2}t)^2$. In particular, the solution shows that depending on the sign of the product of δ and the dispersion parameter β, the wave packet will compress ($\delta\beta > 0$) or stretch ($\delta\beta < 0$) as it propagates. In the case of compression, the wave packet will reach a minimum width and a maximum amplitude at a characteristic compression time after which the wave packet will start spreading [153]. On the other hand, in the absence of modulation ($\delta = 0$), the wave packet will only spread as it propagates and its width will increase by $\sqrt{2}$ times in a characteristic dispersion spreading time $t_{ds} = \Delta_0/(2\sqrt{\beta})$.

The analysis based on the pulse propagation equation (4.25) has shown that wave packet compression and stretching are phenomena that occur already in the second-order approximation of dispersion theory. However, in situations where the group velocity reaches an extremum ($\beta = 0$), we would need to consider the cubic term in the dispersion relation (4.18). Under these circumstances, it has been shown that the complex wavefunction obeys the linearized Korteweg–de Vries (L-KdV) equation:

$$\frac{\partial A}{\partial t'} - \beta_1 \frac{\partial^3 A}{\partial z'^3} = 0, \tag{4.31}$$

where $\beta_1 = \omega'''(k_0)/6$. In this case, an initial pulse changes its shape as it propagates in the medium, and the velocity of its center of mass is generally different from the group velocity of the wave due to the significant reshaping effects. For example, when the initial condition is a narrow pulse specified by a δ function, the envelope solution of the L-KdV equation is given in terms of the Airy function $Ai(z)$ [152]:

$$Ai(z) = \frac{1}{\pi} \int_0^\infty \cos(tz + t^3/3) dt. \tag{4.32}$$

This function gives rise to a fast-decreasing, high-frequency oscillating envelope behind the main head of the pulse and to an exponentially decreasing amplitude in front of the head of the pulse [153]. The full nonlinear KdV equation that involves an additional term proportional to $A \partial A / \partial z$ was originally introduced by Korteweg–de Vries in the modeling of waves on shallow water surfaces. This nonlinear dispersive equation admits solitary wave solutions, or *solitons*, that maintain their shape during propagation due to the balancing effects of nonlinearity and dispersion in the medium. Solitons display many fascinating properties, including the fact that they maintain their individuality, i.e., they do not change in shape when passing through each other. The KdV equation appears in electrical transmission lines, plasma waves, ion-acoustic wave propagation, and modified KdV-type equations with cubic nonlinearity. The equation also describes the propagation of electromagnetic waves in nonlinear optical fibers for high-speed communications [153].

4.1.3   Pulse Dynamics with Neural Networks

Solving the nonlinear equations associated to general wave transport problems can be extremely complicated, especially when we seek to determine some unknown model parameters, as in the case of inverse problems. Recently, the growing interest for machine learning algorithmic development in engineering led to successful applications to inverse problems in imaging and tomography [159, 160], potentially opening new territories with respect to medical imaging, microscopy, and remote sensing technologies. In this section, we discuss the novel technique of physics-informed neural networks (PINNs) that is a general framework for solving both forward and inverse problems of partial differential equations [161, 162] (see Figure 4.3).

Box 4.1 What Are Artificial Neural Networks?

Artificial neural networks (ANNs) are massively parallel and distributed information-processing systems made of a large number of nonlinear process-ing units, called artificial neurons, arranged in highly interconnected architec-tures loosely inspired by the human brain. Their strength relies on the ability to perform complex computations through a *learning process*, which makes them essential tools in the pervasive fields of machine learning [154] and deep learning [155] for applications that range from email spam filters, pattern recog-nition and data mining, and autonomous vehicles control to medical diagnostics. A model of a single neuron is shown in Figure 4.2a, consisting of a vector of inputs

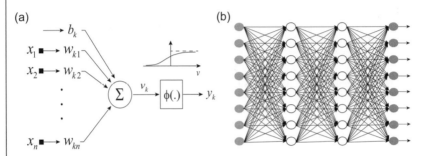

Figure 4.2 (a) Model of a single neuron showing the input vector x, the vector of synaptic weights w_k and the nonlinear activation function ϕ is also sketched. (b) Example of a feedforward ANN architecture with two hidden layers.

$\mathbf{x} = (x_1, x_2, \ldots, x_n)$, a vector of synaptic weights $\mathbf{w}_k = (w_{k1}, w_{k2}, \ldots, w_{kn})$, the bias value b_k, and a nonlinear *activation function* $\phi(\cdot)$ that is applied to the summation Σ of the products of the weights with the inputs values, resulting in the output value y_k of the kth neuron of the network. More precisely, the neuron output can be written as follows:

$$y_k = \phi(v_k) = \phi(\mathbf{w}_k \cdot \mathbf{x} + b_k), \tag{4.33}$$

where we defined the activation potential v_k. Typically, the activation function is a sigmoid that ranges from -1 to 1, similar to the function sketched in Figure 4.2a. An example of a feedforward ANN architecture with a total of four neuron layers, of which the two internal ones are called "hidden layers," is shown in Figure 4.2b. The hidden layers are not seen directly from either the input or the output layers of the network and intervene indirectly implementing a nonlinear input–output mapping capable of approximating any continuous function. This fundamental property of multilayered ANNs is established by the universal approximation theorem (UAT) [156, 157], which provides the rigorous underpinning behind their enormous versatility and success [158].

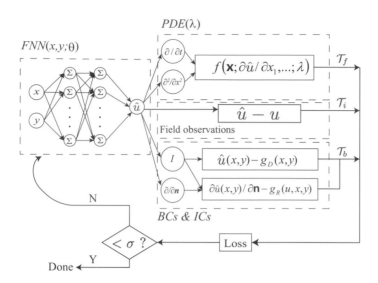

Figure 4.3 Schematic of a PINN for solving inverse problem in photonics based on partial differential equations. The left part neural network represents a surrogate model u of the PDE solution. The right part shows the loss function to restrict u to satisfy PDE in the domain Ω, boundary conditions (BC) $\hat{u}(x, y) = g_D(x, y)$ on $\Gamma_D \subset \partial\Omega$, and $\frac{\partial \hat{u}}{\partial \mathbf{n}}(x, y) = g_D(u, x, y)$ on $\Gamma_R \subset \partial\Omega$. The initial condition (IC) is treated as a special type of boundary condition. For inverse problem, we have also loss term from $\hat{u} - u$ on residual points. We optimize the neural network's weights and biases to obtain loss smaller than σ.

The approach leverages the capabilities of deep neural networks as universal function approximators. However, differently from standard deep learning approaches, PINNs restrict the space of admissible solutions by enforcing the validity of partial differential equation (PDE) models governing the actual physics of the problem. Moreover, PINNs use only one training dataset to achieve the desired solutions, thus relaxing the burdens often imposed by the massive training datasets necessary in alternative, i.e., non-physics-constrained, deep learning approaches. Importantly, PINNs solve highly nonlinear and dispersive inverse problems on the same footing as forward problems, by the simple addition of an extra loss term to the overall loss function for the minimization of the residuals of PDEs and their boundary conditions [161]. Therefore, PINNs are particularly effective in solving *ill-posed* inverse problems, which are often intractable with available mathematical formulations.

The main idea of PINNs is schematically illustrated in Figure 4.3. First, we construct a neural network with output $\hat{u}(\mathbf{x}; \boldsymbol{\theta})$ as a surrogate of the PDE solution $u(\mathbf{x})$. For practical purposes, it is often sufficient to consider a relatively simple feedforward neural network (FNN). Here $\boldsymbol{\theta}$ is a vector containing all weights and biases in the neural network that needs to be trained. The next step is to constrain the neural network's output $\hat{u}$ to satisfy the PDE as well as data observations u. This is achieved by constructing the loss function by considering terms corresponding to PDE, boundary conditions and initial conditions, and field observations in the case of inverse scattering problems.

Specifically, we consider the general PDE problem with unknown parameter λ for the solution $u(\mathbf{x})$ with $\mathbf{x} = x_1, \ldots, x_d$ defined on a domain $\Omega \subset \mathbb{R}^d$:

$$f\left(\mathbf{x}; \frac{\partial \hat{u}}{\partial x_1}, \ldots, \frac{\partial \hat{u}}{\partial x_d}; \frac{\partial^2 \hat{u}}{\partial x_1 \partial x_1}, \ldots, \frac{\partial^2 \hat{u}}{\partial x_1 \partial x_d}; \ldots; \lambda\right) = 0, \quad \mathbf{x} \in \Omega. \quad (4.34)$$

We denote $\mathcal{T}_i \subset \Omega$ to be the points where we have the PDE solution $u(\mathbf{x})$. The loss function is then defined as follows:

$$\mathcal{L}(\boldsymbol{\theta}, \lambda) = w_f \mathcal{L}_f(\boldsymbol{\theta}, \lambda; \mathcal{T}_f) + w_i \mathcal{L}_i(\boldsymbol{\theta}, \lambda; \mathcal{T}_i) + w_b \mathcal{L}_b(\boldsymbol{\theta}, \lambda; \mathcal{T}_b), \quad (4.35)$$

where

$$\mathcal{L}_f(\boldsymbol{\theta}, \lambda; \mathcal{T}_f) = \frac{1}{|\mathcal{T}_f|} \sum_{\mathbf{x} \in \mathcal{T}_f} \left\| f\left(\mathbf{x}; \frac{\partial \hat{u}}{\partial x_1}, \ldots, \frac{\partial \hat{u}}{\partial x_d}; \frac{\partial^2 \hat{u}}{\partial x_1 \partial x_1}, \ldots, \frac{\partial^2 \hat{u}}{\partial x_1 \partial x_d}; \ldots; \lambda\right) \right\|_2^2$$

$$(4.36)$$

$$\mathcal{L}_i(\boldsymbol{\theta}, \lambda; \mathcal{T}_i) = \frac{1}{|\mathcal{T}_i|} \sum_{\mathbf{x} \in \mathcal{T}_i} \|\hat{u}(\mathbf{x}) - u(\mathbf{x})\|_2^2 \quad (4.37)$$

$$\mathcal{L}_b(\boldsymbol{\theta}, \lambda; \mathcal{T}_b) = \frac{1}{|\mathcal{T}_b|} \sum_{\mathbf{x} \in \mathcal{T}_b} \|\mathcal{B}(\hat{u}, \mathbf{x})\|_2^2 \quad (4.38)$$

and w_f, w_b, and w_i are the weights of the network. $\mathcal{T}_f$, $\mathcal{T}_i$, $\mathcal{T}_b$ denote the residual points from partial differential equations, training dataset, and ICs and BCs, respectively. Here, $\mathcal{T}_f \subset \Omega$ is a set of predefined points that are used to measure the matching degree of the neural network output $\hat{u}$ to the PDE. $\mathcal{T}_f$ can be chosen as grid points or random points; see more details and discussion in reference [162]. In the last step, we train the neural network by minimizing the loss function $\mathcal{L}(\boldsymbol{\theta})$. Note that using PINNs, the only difference between a forward and an inverse problem is the addition of an extra loss term $\mathcal{L}_i$ to equation (4.35), which comes at an insignificant computational cost. As an example, we use DeepXDE [162], a user-friendly Python library, for the code implementation of the advection and KDV equations and show the results in Figure 4.4.

An application of the powerful PINN approach for the solution of representative inverse scattering problems in photonic metamaterials and nano-optics technologies has been recently demonstrated by Chen et al. [137].

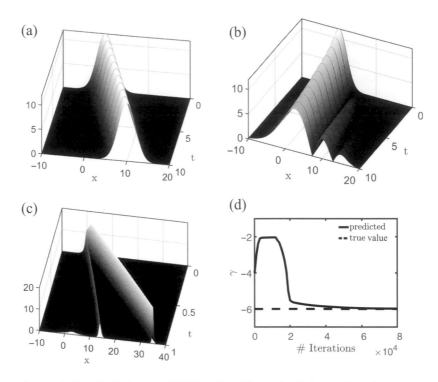

Figure 4.4 (a) and (b) show the PINN results of forward solution to advection equation with v_g equals to 1 and the L-KDV equation with $\beta_1 = 1$, respectively. (c) shows the forward solution for nonlinear KDV equation $\frac{\partial A}{\partial t} + \frac{\partial^3 A}{\partial x^3} - 6A\frac{\partial A}{\partial x} = 0$ with initial condition $A(x,0) = -6sech(x)^2$. Panel (d) shows the inverse parameter finding of C in the nonlinear KDV $\frac{\partial A}{\partial t} + \frac{\partial^3 A}{\partial x^3} + C\frac{\partial A}{\partial x} = 0$ using the PINN with the forward solution obtained from (c). For all the models implemented in the PINN, the neural networks composed of 3 hidden layers with 20 neurons each. We trained the neural networks using the Adam method with the learning rate equals to 0.001 and the iteration steps is 10^4.

4.2   Scalar Diffraction Theory

One of the most striking consequences of the propagation of waves is their ability to deviate from their rectilinear paths and spread around obstacles encountered along their trajectories. While characteristic of any type of waves, classical or quantum in nature, wave diffraction phenomena cannot be easily captured by simple definitions. An often quoted attempt by Arnold Sommerfeld refers to diffraction as "any deviation of light rays from rectilinear paths which cannot be interpreted as reflection or refraction" [163]. In essence, wave diffraction refers to the angular spreading of waves that are laterally confined by obstacles or obstructions impeding their wavefronts, for instance by optical components such as lenses or apertures. Consider, for example, the light focusing by a simple refractive lens. If the lens is free from aberrations, the smallest spot that can be achieved at the back focal plane will be limited in size by diffraction effects occurring at the edges of the lens, and for this reason iy is called diffraction limited. Therefore, a quantitative understanding of diffraction effects is paramount for the design of imaging systems and more general optical devices that laterally confine propagating waves. The resulting diffraction effects become most appreciable when the confinement occurs over length scales comparable with the wavelength. As we will discuss in Section 4.2.5, diffraction effects naturally originate from a fundamental type of uncertainty relation that finds its origin within the general Fourier theory of linear systems. However, before addressing the modern approach to wave propagation and diffraction in Section 4.2.5, we first introduce the traditional one that hinges on the direct application of a general integral representation of wave phenomena originally developed by Helmholtz and Kirchhoff. Together with Huygen's principle, this is the starting point for the more advanced and mathematically consistent formulations of scalar wave diffraction developed by Rayleigh and Sommerfeld. Extensions of diffraction theory to full-vector waves have been developed but are significantly more complex than their scalar counterparts. Beyond our introductory discussion in Section 4.3, interested readers can find detailed expositions of vector diffraction theory and its numerous applications to optical design in our cited references [164–169].

4.2.1 A Brief History of Wave Diffraction

Diffraction theory has a long and interesting history in optics and greatly contributed to establish the correct wave picture of light that is valid in classical physics. Pioneering experiments were already conducted by Antonio Grimaldi and published in 1665. Grimaldi observed the fine structure of the shadow cast by an aperture in an opaque screen illuminated by a small light source. The light intensity was recorded across a plane located at some distance behind the aperture screen. Instead of observing a shadow with sharp edges, as predicted by the then popular corpuscular theory of light that expected straight light propagation, Grimaldi reported a gradual transition from light to darkness extending even behind the object's geometrical shadow. However, due to the low spectral purity of the light source available at his time, Grimaldi could not observe the even more striking bright and dark interference fringes far into the geometrical shadow of the screen. Seminal contributions toward the correct wave theory of light were provided by Christian Huygens, who introduced the concept of wavefront and in 1678 proposed the idea of regarding each point on the wavefront of a propagating wave as a new source of "secondary" spherical waves. The wavefront at a later instant of time could then be found by constructing the envelope of the secondary spherical wavefronts. However, this simple picture does not allow one to correctly account for the quantitative aspects of waves that diffract at obstacles. An important step in this direction was taken by Thomas Young, who in 1804 performed his famous two-slit experiment and introduced the idea of interference in the wave theory of light. According to this theory, light could be added to light and produce either light or darkness depending on appropriate conditions. The first quantitative picture of light diffraction was developed by Augustin Jean Fresnel in his 1818 paper submitted to a prize committee of the French Academy of Sciences. In this work, Fresnel combined Young's idea of interference with the Huygen's picture of the propagating secondary waves. Specifically, he considered rather arbitrary assumptions about the amplitudes and phases of Huygens's secondary sources and allowed the various wavelets to interfere, which enabled the first quantitatively accurate predictions of the light intensity distribution past obstructions, i.e., diffraction patters. Fresnel's theory was heavily criticized by the famous French mathematician Poisson, who was chairing the prize committee. Poisson tried to discredit Fresnel's idea by showing that if Fresnel's theory was correct, then an obstacle in the shape of an opaque circular disc should produce a bright spot right at the center of its shadow. Arago, who was also a member of the committee, decided to perform an experiment and observed the predicted bright spot, since then known as the "Poisson–Arago spot." Fresnel won the prize and Arago's experimental observation contributed to confirm to a large extent the wave nature of light.

4.2.2   Helmholtz and Kirchhoff Integral Theorem

Light diffraction phenomena follow naturally from Maxwell's electromagnetic theory, which we consider with respect to the scalar approximation. The starting point for the mathematical formulation of the scalar diffraction theory is the integral field representation known as the *Helmholtz and Kirchhoff theorem*. This generally valid result allows one to express the field at any point in space in terms of its "boundary values" on any closed surface surrounding that point. More precisely, the value of a smooth scalar field at a point inside a volume V can be obtained as an integral of the field value and its derivative on the surface S that bounds the volume, provided the validity of the homogeneous scalar wave equation and the Sommerfeld radiation condition at infinity. The starting point to derive this important result is Green's identity of complex vector calculus. For any two smooth functions ψ and G continuous and with continuous first- and second-order partial derivatives within volume V, Green's identity can be written as follows:

$$\int_V (\psi \nabla^2 G - G \nabla^2 \psi) dv = \int_{\partial V} \left(\psi \frac{\partial G}{\partial n} - G \frac{\partial \psi}{\partial n} \right) ds \qquad (4.39)$$

where the normal derivative is considered in the outward direction at each point on the surface $S = \partial V$ that bounds the volume V. This expression is also known as *Green's theorem*. Let us now consider the observation point P_0 (with vector position $\mathbf{r}_0$ with respect to an arbitrarily selected reference) of the field function U, and let $S = \partial V$ be an arbitrary closed surface surrounding P_0. Kirchhoff's problem was to express the field function ψ at P_0 in terms of its values on the surface S. In order to solve this problem, Kirchhoff applied Green's theorem directly to ψ and to an auxiliary function G that he chose as a spherical wave expanding from P_0, i.e., the free-space scalar Green's function $G(P_1) = \exp(jkr_{01})/r_{01}$, where P_1 is an arbitrary point. The quantity r_{01} denotes the length of the vector $\mathbf{r}_{01}$ that points from P_0 to P_1, as

(a)   (b)

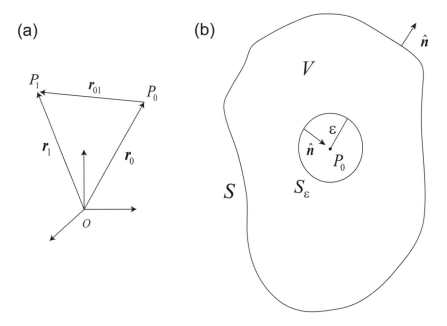

Figure 4.5 (a) Definition of the observation points and their vector positions with respect to the origin O of an arbitrary reference system. (b) Schematic representation of the surface of integration for the derivation of the integral theorem of Helmholtz and Kirchhoff. Note the spherical exclusion volume around P_0 and the unit normal vector that points radially inward on the surface of the exclusion sphere S_ϵ.

shown in Figure 4.5a. In order for this function and its derivatives to be continuous inside the volume V, we need to avoid the discontinuity of the function G at P_0 by removing from the integration volume V a spherical volume of small radius ϵ around P_0, as shown in Figure 4.5b. We can then apply Green's theorem over the remaining integration volume V' that lies in between the two surfaces S and S_ϵ. As a result, the corresponding integration surface is the surface $S' = S + S_\epsilon$. We should also realize that the normal unit vector $\hat{n}$ to this composite surface points outward on S but inward (toward P_0) on S_ϵ. Since both ψ and G obey the homogeneous Helmholtz equation in V', substituting into the left-hand side of Green's theorem we obtain the following:

$$\int_{S'} \left(\psi \frac{\partial G}{\partial n} - G \frac{\partial \psi}{\partial n} \right) ds = 0. \tag{4.40}$$

The preceding surface integral is contributed by the two surfaces S_ϵ and S that make up the compound surface S. Therefore, we can equivalently write the following:

$$\int_S \left(\psi \frac{\partial G}{\partial n} - G \frac{\partial \psi}{\partial n} \right) ds = -\int_{S_\epsilon} \left(\psi \frac{\partial G}{\partial n} - G \frac{\partial \psi}{\partial n} \right) ds. \tag{4.41}$$

The integral over the small exclusion sphere S_ϵ can be simply evaluated in the limit $\epsilon \to 0$. For this purpose, we note that, for a generic observation point P_1 on S_ϵ, we have the following:

$$\frac{\partial G(P_1)}{\partial n} = \cos(\hat{\mathbf{n}}, \hat{\mathbf{r}}_{01}) \cdot \nabla G(P_1) = -1\left(jk - \frac{1}{\epsilon}\right)\frac{e^{jk\epsilon}}{\epsilon}, \tag{4.42}$$

where the cosine refers to the angle between the outward normal $\hat{\mathbf{n}}$, which points radially toward P_0 on S_ϵ, and the vector $\mathbf{r}_{01}$ connecting P_0 to P_1. Therefore, due to its radial symmetry the $\epsilon \to 0$ limit of the surface integral over S_ϵ that appears in (4.41) can be simply evaluated as follows:

$$\lim_{\epsilon \to 0} 4\pi\epsilon^2 \left[\psi(P_0)\frac{e^{jk\epsilon}}{\epsilon}\left(jk - \frac{1}{\epsilon}\right) + \frac{\partial\psi(P_0)}{\partial n}\frac{e^{jk\epsilon}}{\epsilon}\right] = -4\pi\psi(P_0). \tag{4.43}$$

Substitution of this result in (4.41) yields the following:

$$\boxed{\psi(P_0) = \frac{1}{4\pi}\int_S\left[\frac{\partial\psi}{\partial n}\left(\frac{e^{jkr_{01}}}{r_{01}}\right) - \psi\frac{\partial}{\partial n}\left(\frac{e^{jkr_{01}}}{r_{01}}\right)\right]ds} \tag{4.44}$$

This expression is the *integral theorem of Helmholtz and Kirchhoff*, which provides the foundation of scalar diffraction. According to this theorem, the scalar field at any point P_0 within V can be expressed in terms of the boundary values of the field on any closed surface surrounding that point. A proper selection of this surface will provide the Kirchhoff solution of wave diffraction by a planar screen, as discussed in the following section.

4.2.3   The Fresnel–Kirchhoff Diffraction Formula

Based on the Helmholtz and Kirchhoff integral theorem, we want to find the solution of scalar wave diffraction by an aperture in an infinite opaque screen. The geometry of reference is illustrated in Figure 4.6, where an incoming wave from the left gives rise to a diffracted field at the observation point P_0 behind the aperture. Kirchhoff chose the surface of integration in Figure 4.6, as the total closed surface $S = S_1 + S_2$ composed of the plane surface S_1, lying directly behind the diffracting screen, and the large spherical cap S_2 of radius r that is centered at the observation point P_0. He then applied the integral theorem (4.44) on the total surface $S = S_1 + S_2$ and noticed that the Green function $G(P_1)$ evaluated on the surface S_2 of radius R can be simplified, for large R values, as follows:

$$\frac{\partial G}{\partial n} = \left(jk - \frac{1}{R}\right)\frac{r^{jkR}}{R} \approx jkG. \tag{4.45}$$

Therefore, the contribution to the Helmholtz and Kirchhoff integral theorem that originates from the surface S_2 at large R (compared to the wavelength) can be expressed as follows:

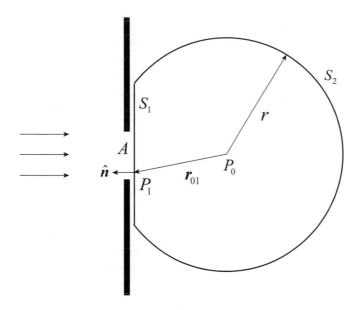

Figure 4.6 Geometry for the Kirchhoff formulation of the scalar wave diffraction by an aperture with surface A in an infinite opaque screen. The diffracted field behind the aperture is probed at point P_0. Adapted from reference [173].

$$\int_{S_2} \left(G \frac{\partial \psi}{\partial n} - \psi jkG \right) ds = \int_{\Omega} G \left(\frac{\partial \psi}{\partial n} - jk\psi \right) R^2 d\omega, \qquad (4.46)$$

where $d\omega = ds/R^2$ is a differential solid angle and Ω is the solid angle subtended by S_2 at P_0. Since the product RG is bounded on S_2, the preceding integral will vanish at large R provided:

$$\lim_{R \to \infty} R \left(\frac{\partial \psi}{\partial n} - jk\psi \right) = 0 \qquad (4.47)$$

This condition is known as the *Sommerfeld radiation condition* (SRC), which plays a crucial role in the scattering theory discussed in Chapter 6. Enforcing the SRC, we can then express the field at P_0 in terms of the integral on the surface S_1 only:

$$\psi(P_0) = \frac{1}{4\pi} \int_{S_1} \left(\frac{\partial \psi}{\partial n} G - \psi \frac{\partial G}{\partial n} \right) ds \qquad (4.48)$$

The preceding integral expression is generally valid for scalar waves provided ψ and G satisfy the homogeneous Helmholtz wave equation and the Sommerfeld radiation condition is satisfied.

Note that S_1 denotes the infinite planar surface immediately behind the diffracting screen, as shown in Figure 4.6. Since the screen is assumed to be perfectly opaque (absorbing), it is therefore reasonable to assume that the dominant contributions to the preceding integral will come from the points inside the aperture area A. The remaining task is now to determine the boundary conditions for the field at the aperture. This is a very complicated problem even for apertures of simple shapes in a perfectly conducting screen due to the coupling of the different vector field components at the edge of the aperture. In fact, the first rigorous full-vector solution of the diffraction by small circular holes in a perfectly conducting screen of zero thickness was provided by Hans Bethe only in 1944 [170].

In his 1883 seminal contribution [171], Kirchhoff simply assumed that (i) the wave incident on the aperture was not affected by the presence of the aperture and (ii) he neglected the field amplitude in the geometrical shadow of the screen. These simplifying assumptions turned the otherwise difficult integral equation (4.48) into a simple integration restricted over the surface area A of the aperture. It is worth mentioning at this point that Kirchhoff boundary conditions as stated previously cannot be physically correct and are also mathematically inconsistent. In fact, the very presence of the aperture indeed perturbs the incoming field profile over the surface A due to the vector coupling effects arising from the electromagnetic boundary conditions at the rim of the aperture. Moreover, the induced charges and currents at the edge of the aperture give rise to a nonzero field amplitude that extends even behind the opaque screen over distances of a few wavelengths. Finally, the theory of analytic complex functions tells us that if a field ψ along with normal derivative vanishes along a finite surface, then ψ must vanish over the entire space. As a result, the two Kirchhoff boundary conditions would result in a zero field everywhere, in stark contradiction with experimental situations.

Before discussing the more rigorous approach developed by Sommerfeld, let us consider the case of the aperture illumination by a single spherical wave radiating from a point P_2 at a distance r_{21} from P_1, as shown in Figure 4.7. Therefore, at point P_1 the wave disturbance will have an amplitude $\psi(P_1) = C \exp(jkr_{21})/r_{21}$, where C is an arbitrary constant. When the distance r_{01} from the aperture to the observation point P_0 is large compared to the wavelength, we have $k \gg 1/r_{01}$, which allows us to write the following:

$$\frac{\partial G(P_1)}{\partial n} = \cos(\hat{\mathbf{n}}, \hat{\mathbf{r}}_{01}) \left(jk - \frac{1}{r_{01}} \right) \frac{e^{jkr_{01}}}{r_{01}} \approx jk \cos(\hat{\mathbf{n}}, \hat{\mathbf{r}}_{01}) \frac{e^{jkr_{01}}}{r_{01}}. \tag{4.49}$$

An analogous approximation can be obtained for $\partial \psi(P_1)/\partial n$ when r_{21} is many optical wavelengths away from the screen. Substituting these approximations in the integral (4.48) restricted to the surface area A of the aperture, we obtain the *Fresnel–Kirchhoff diffraction formula*:

$$\psi(P_0) = \frac{C}{j\lambda} \int_A \frac{\exp[jk(r_{21} + r_{01})]}{r_{21}r_{01}} \left[\frac{\cos(\hat{\mathbf{n}}, \hat{\mathbf{r}}_{01}) - \cos(\hat{\mathbf{n}}, \hat{\mathbf{r}}_{21})}{2} \right] ds \tag{4.50}$$

Note the complete symmetry of the preceding expression with respect to the respective roles of the source point P_2 and the observation point P_0, which can be exchanged with each other without altering the resulting integral. The Fresnel–Kirchhoff diffraction

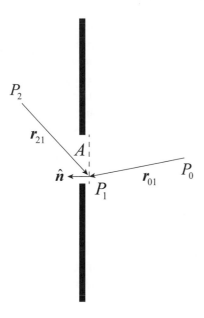

Figure 4.7 Spherical wave illumination of an aperture in an opaque diffracting screen used for the derivation of the Fresnel–Kirchhoff diffraction formula. Adapted from reference [173].

formula can be rewritten in a way that is compatible with the Huygens's principle, according to which the diffracted wave can be obtained by superimposing the contributions of secondary or "fictitious" point sources located within the aperture region:

$$\psi(P_0) = \int_A \psi'(P_1) \frac{\exp(jkr_{01})}{r_{01}} ds. \tag{4.51}$$

The complex amplitude of each secondary waves is given by the following:

$$\boxed{\psi'(P_1) = \frac{C}{j\lambda} \frac{\exp[jk(r_{21})]}{r_{21}} \left[\frac{\cos(\hat{n}, \hat{r}_{01}) - \cos(\hat{n}, \hat{r}_{21})}{2} \right] ds} \tag{4.52}$$

This depends on amplitude of the illuminating spherical wave as well as on the *obliquity factor* in square brackets that contains the angles of illumination and observation. Therefore, Kirchhoff's theory of diffraction mathematically determines the precise form of the complex amplitude of the fictitious secondary waves inside the aperture without resorting to ad hoc assumptions that are extrinsic to the theory (as done by Fresnel). In particular, we should easily recognize based on the preceding formula that the fictitious sources are not traditional spherical waves since they do not radiate in the backward direction and their amplitudes are maximized in the forward direction.

4.2.4 The Rayleigh–Sommerfeld Formulation of Diffraction

Despite the fact that Kirchhoff diffraction theory often produces accurate results when compared to experimental scenarios, it remains flawed by the mathematical inconsistencies that were mentioned in the previous section. These problems are related to the need to impose boundary conditions to the field amplitude and to its normal derivative at the same time, and were successfully removed by the formulation of diffraction introduced by Sommerfeld, known as the *Rayleigh–Sommerfeld theory*.

Sommerfeld important insight was to define an alternative Green function $G_1(P_1)$ that would vanish by construction on the aperture's plane A. This can be simply obtained by considering G_1 generated by two point sources, one located at P_0 and another at the point $\tilde{P}_0$, that is the mirror-symmetric image of P_0 with respect to the screen. Moreover, this symmetric pair source is assumed to oscillate with the opposite phase with respect to the one at P_0, so that the new Green function becomes the following:

$$G_1(P_1) = \frac{\exp(jkr_{01})}{r_{01}} - \frac{\exp(jk\tilde{r}_{01})}{\tilde{r}_{01}}, \tag{4.53}$$

where $\tilde{r}_{01}$ denotes the distance between $\tilde{P}_0$ and P_1. Since G_1 evidently vanishes identically on the plane of the aperture, Kirchhoff boundary conditions now apply to ψ only, without the need to separately enforce them also on its normal derivative. Therefore, using the new Green function in (4.48), we obtain the following:

$$\boxed{\psi(P_0) = -\frac{1}{4\pi} \int_A \psi \frac{\partial G_1}{\partial n} ds} \tag{4.54}$$

This result is known as the *first Rayleigh–Sommerfeld solution*.

The normal derivative of the Green function G_1 appearing in the expression (4.54) can be explicitly computed when realizing that for points P_1 on the surface of the screen, the mirror symmetry between P_0 and $\tilde{P}_0$ implies $r_{01} = \tilde{r}_{01}$. Moreover, the angles that the unit normal $\hat{\mathbf{n}}$ makes with the directions $\hat{\mathbf{r}}_{01}$ and $\tilde{\mathbf{r}}_{01}$ differ by an angle π and the corresponding cosines are equal and opposite. Therefore, substituting this information into the general expression for the normal derivative, we can obtain the following:

$$\frac{\partial G_1(P_1)}{\partial n} = 2\cos(\hat{\mathbf{n}}, \hat{\mathbf{r}}_{01}) \left(jk - \frac{1}{r_{01}} \right) \frac{\exp(jkr_{01})}{r_{01}}. \tag{4.55}$$

Substituting into the first Rayleigh–Sommerfeld solution and enforcing the Kirchhoff boundary conditions[4] we obtain the exact form of the first Rayleigh–Sommerfeld solution:

$$\boxed{\psi(P_0) = -\frac{1}{2\pi} \int_A \psi(P_1) \left(jk - \frac{1}{r_{01}} \right) \frac{\exp(jkr_{01})}{r_{01}} \cos(\hat{\mathbf{n}}, \hat{\mathbf{r}}_{01}) ds} \tag{4.56}$$

---

[4] We remind that this choice requires that the field vanishes on the surface S_1 right behind the screen while it is unperturbed by the screen inside the open aperture A.

This solution can be expressed in convolution form, motivating the more general linear systems approach to wave diffraction that we will later develop in Section 4.2.5. Choosing a reference "input plane" at the surface A of the plane aperture, where we denote the input field as $\psi(x, y; 0)$, we can express the diffracted field at the observation plane with $z \neq 0$ as follows:

$$\psi(x, y; z) = h(x, y; z) * \psi(x, y; 0), \tag{4.57}$$

where $*$ denotes the two-dimensional convolution and the *impulse-response function* for the Rayleigh–Sommerfeld diffraction is explicitly given as follows:

$$h(x, y; z) = \frac{1}{2\pi} \frac{z}{r} \left(\frac{1}{r} - jk \right) \frac{\exp(jkr)}{r} \tag{4.58}$$

where $r = \sqrt{x^2 + y^2 + z^2}$ and $\cos(\hat{n}, \hat{r}_{01}) = z/r$. Note that the preceding general impulse-response function reduces, for $r \gg \lambda$, to the simpler far-field approximation:

$$h(x, y; z) = \frac{z}{j\lambda} \frac{\exp(jkr)}{r^2} \tag{4.59}$$

However, using this form of the impulse-response close to the screen (compared to the wavelength) leads, to significant errors and the most general form in (4.58) must be employed instead.

In situations for which $r_{01} \gg \lambda$, we can simplify the expression (4.56) in the following form:

$$\psi(P_0) = \frac{1}{j\lambda} \int_A \psi(P_1) \frac{\exp(jkr_{01})}{r_{01}} \cos(\hat{n}, \hat{r}_{01}) ds. \tag{4.60}$$

The mathematical inconsistency of the Kirchhoff theory is now resolved since no boundary condition needs to be applied on the normal derivative of the field. Consistently, the Rayleigh–Sommerfeld formulation reproduces the imposed boundary values when the field point approaches the aperture's plane [172]. We notice in passing that an alternative formulation of the Rayleigh–Sommerfeld diffraction can be obtained using a Green function G_2 in the form of the superposition of two mirror-symmetric point sources that oscillate in phase. This leads to the so-called second Rayleigh–Sommerfeld solution where the normal derivative of the field appears explicitly and the corresponding boundary value problem of diffraction is of the Neumann type, i.e., requiring the enforcing of boundary conditions on the normal derivative of the field. We will not consider this equivalent formulation and refer the interested reader to the discussions in [43, 173].

The first Rayleigh–Sommerfeld integral (4.60) can be specialized to the common situation when the screen is illuminated by a diverging spherical wave from a point P_2 in front of the aperture, as schematized in Figure 4.7. In this situation, the incident wave on A is $\psi(P_1) = C \exp(jkr_{21})/r_{21}$, which, when substituted in the last integral, yields the following:

$$\psi(P_0) = \frac{C}{j\lambda} \int_A \frac{\exp[jk(r_{21} + r_{01})]}{r_{21}r_{01}} \cos(\hat{\mathbf{n}}, \hat{\mathbf{r}}_{01}) ds \qquad (4.61)$$

This integral is the analogue to the Sommerfeld theory of the Fresnel–Kirchhoff diffraction formula and is known as the *Rayleigh–Sommerfeld diffraction formula*. Notice that this formula differs from the Fresnel–Kirchhoff diffraction formula only in the expression of the obliquity factor, which is the directivity pattern describing the angular radiation properties of the secondary (fictitious) sources within the aperture region.[5] In particular, for an aperture illuminated by a normal incident plane wave, which corresponds to a point source at infinity, the obliquity factor from Kirchhoff theory reduces to $[1 + \cos \theta]/2$, and the one of the Rayleigh–Sommerfeld theory is simply $\cos \theta$, where θ is the angle between the normal vector $\hat{\mathbf{n}}$ and the direction $\hat{\mathbf{r}}_{01}$.

4.2.5   Angular Spectrum Representation of Fields

The modern approach to diffraction engineering is based on linear systems theory for the solution of the Helmholtz wave equation in homogeneous linear media. This formulation, known as the *angular spectrum representation*, connects the transverse spatial spectra (Fourier k-space) of three-dimensional field quantities evaluated at different reference planes. The fruitful combination of linear systems and wave optics gives rise to the powerful approach of Fourier optics that provides a quantitative analysis of diffractive and imaging systems in the language of spatial filters and transfer functions that behave as propagators for scalar and vector wave fields [173].

We consider the scalar Helmholtz equation for a generic time-harmonic field component $\psi(\mathbf{r}, \omega) \equiv \psi(\mathbf{r})$, which is as follows:

$$\nabla^2 \psi(\mathbf{r}) + k^2 \psi(\mathbf{r}) = 0. \qquad (4.62)$$

We have denoted by $\psi(\mathbf{r})$ any component of the complex field $\mathbf{E}(\mathbf{r}, \omega)$ at the angular frequency[6] ω. Without loss of generality, we assume that the wave propagates along the z-coordinate axis, and we can regard the quantity $\psi(\mathbf{r})$ as completely equivalent to the expression $\psi(x, y; z)$, where the dependence on the two-dimensional transverse plane coordinates (x, y) has been made explicit while the z-dependence along the propagation axis is considered separately as a parameter.[7] Therefore, the quantity $\psi(x, y; z)$ represents the two-dimensional spatial distribution (profile) of the wave specified at an arbitrary plane that is transverse to the propagation z-axis.

---

[5] Remember that the actual sources are located at the edge of the aperture, and there are no physical sources inside the aperture. This is why the secondary sources are called fictitious.

[6] Remember that in this book we chose to separate out the harmonic time dependence with the factor $\exp(-j\omega t)$ that implicitly multiplies any complex amplitude.

[7] This makes sense because of the privileged role played by the z-axis as the propagation direction in a diffraction problem.

This interpretation allows us to elegantly solve the general problem of wave propagation/diffraction by introducing the z-parameterized *transverse spatial Fourier transform pairs*:

$$\hat{\psi}(k_x,k_y;z) = \int\int \psi(x,y;z)e^{-j(k_x x+k_y y)}dxdy \qquad (4.63)$$

$$\psi(x,y;z) = \frac{1}{4\pi^2}\int\int \hat{\psi}(k_x,k_y;z)e^{j(k_x x+k_y y)}dk_x dk_y \qquad (4.64)$$

where the double integrals in this section are assumed to extend from $-\infty$ to $+\infty$. The quantity $\hat{\psi}(k_x,k_y;z)$ is also referred as the *spatial spectrum* at z of the field component. Taking the 2D transverse Fourier transform of the scalar Helmholtz equation, we obtain the simpler ordinary differential equation:

$$\frac{d^2}{dz^2}\hat{\psi}(k_x,k_y;z) + k^2\left(1 - \frac{k_x^2}{k^2} - \frac{k_y^2}{k^2}\right)\hat{\psi}(k_x,k_y;z) = 0, \qquad (4.65)$$

whose forward-propagating solution can be written as follows:

$$\hat{\psi}(k_x,k_y;z) = \hat{\psi}(k_x,k_y;0)e^{jkz\sqrt{1-\frac{k_x^2}{k^2}-\frac{k_y^2}{k^2}}}. \qquad (4.66)$$

The quantities k_x/k and k_y/k are the cosines of the angles that the propagation k-vector makes with respect to the corresponding coordinate axes. For this reason, the approach introduced here is also called the *angular spectrum approach*. The preceding solution defines the *k-space propagator* or the *transfer function* $H(k_x,k_y;z)$ for the propagation of the initial spectrum at $z = 0$ up to the generic distance $z \neq 0$:

$$H(k_x,k_y;z) \equiv \frac{\hat{\psi}(k_x,k_y;z)}{\hat{\psi}(k_x,k_y;0)} = e^{jkz\sqrt{1-\frac{k_x^2}{k^2}-\frac{k_y^2}{k^2}}} \qquad (4.67)$$

The propagator can give rise to oscillatory wave propagation or to exponential wave attenuation along the z-axis depending on the sign of the expression that appears under the square root in equation (4.67). According to this picture, a general field propagation problem can be regarded as a *filtering problem* in k-space such that the spectrum of the propagated field (output quantity) is obtained from the initial spectrum of the field (at $z = 0$) by multiplication with the transfer function as follows:

$$\hat{\psi}(k_x,k_y;z) = H(k_x,k_y;z)\hat{\psi}(k_x,k_y;0). \qquad (4.68)$$

The spatial field distribution in the transverse plane located at $z \neq 0$ can be retrieved by inverse Fourier transforming the equation (4.66), which produces the following:

$$\psi(x,y;z) = \frac{1}{4\pi^2}\int\int H(k_x,k_y;z)\hat{\psi}(k_x,k_y;0)e^{j(k_x x+k_y y)}dk_x dk_y \qquad (4.69)$$

A completely equivalent description can be formulated directly in the spatial domain by introducing the propagation *impulse response function* $h(x,y;z)$, or the Green function, which is the two-dimensional inverse Fourier transform of the transfer function:

$$h(x, y; z) = \frac{1}{4\pi^2} \int \int H(k_x, k_y; z) e^{j(k_x x + k_y y)} dk_x dk_y. \tag{4.70}$$

Using the impulse response function, we can express the propagated complex field solution as a two-dimensional spatial convolution:

$$\boxed{\psi(x, y; z) = \int \int \psi(x', y'; 0) h(x - x', y - y'; z) dx' dy'} \tag{4.71}$$

Equations (4.69) and (4.71) provide the solutions to a general propagation or diffraction problem. For instance, they allow one to express the complex amplitude of a field in the transverse plane at z in terms of the field distribution on a diffracting aperture located in the plane at $z = 0$. The same approach describes the propagation of a field in a homogeneous medium once the initial field distribution is known at a reference plane, e.g., the propagation of optical beams. The results discussed here are generally valid within the limits of scalar wave optics and in fact are completely equivalent to the (first) Rayleigh–Sommerfeld solution (4.58) of wave diffraction theory [173].

The equivalence between the angular spectrum and the scalar optics pictures of diffraction can be explicitly recognized by transforming the impulse response function (4.70) in cylindrical coordinates via the change of variables $x - x' = r_\perp \cos \theta$, $y - y' = r_\perp \sin \theta$ and $k_x = 2\pi \rho \cos \phi$, $k_y = 2\pi \rho \sin \phi$. The relevant quantities that appear in this derivation are indicated in the schematics of the diffraction geometry shown in Figure 4.8. The transformed integral can be evaluated analytically with the

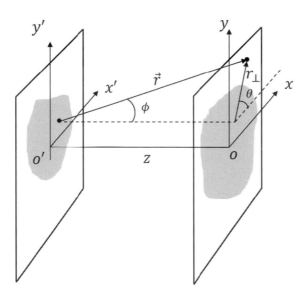

Figure 4.8 Schematics of the diffraction geometry used in the derivation of the Fresnel diffraction formula starting from the general Rayleigh–Sommerfeld impulse response function. Adapted from reference [173].

help of *Erdélyis's formula* yielding, after some manipulations, the following expression for the impulse response [174]:

$$h(r) = \frac{e^{jk\sqrt{r^2+z^2}}}{j\lambda\sqrt{r^2+z^2}} \frac{z}{\sqrt{r^2+z^2}} \left(1 - \frac{1}{jk\sqrt{r^2+z^2}}\right). \tag{4.72}$$

So far, no approximations have been made. In fact, we recognize that the preceding result coincides with the Rayleigh–Sommerfeld impulse response function (4.58) that was previously derived based on the approach of scalar diffraction theory. However, in many applications it is convenient to simplify this expression by considering the following circumstances:

- $z \gg \lambda$. In this case, the second term in the parentheses is $\ll 1$ and can be neglected.
- We observe that $z/\sqrt{r^2+z^2} = \cos\phi$, where ϕ is the angle between the positive z-axis and the line passing through the points $(x', y', 0)$ and (x, y, z). This angular term is called the *obliquity factor*. In many applications where the dimensions of the region of interest are small compared to z, we have $\cos\phi \simeq 1$. Replacing the obliquity factor by unity restricts the diffraction to occur in the *paraxial regime*.
- The quantity $\sqrt{r^2+z^2}$ is the distance between the points with coordinates $(x', y', 0)$ and (x, y, z). In the paraxial approximation, this quantity is replaced by z in the denominator of equation (4.72), while it is expanded to second order using the binomial expansion in the phase of the equation (4.72), where we will substitute $\sqrt{r^2+z^2} \simeq z + (x^2 + y^2)/2z$.

Implementing the series of approximations identified in this list yields the *paraxial approximation* of the impulse response function:

$$\boxed{h(x, y; z) = \frac{e^{jkz}}{j\lambda z} e^{jk(x^2+y^2)/2z}} \tag{4.73}$$

The corresponding paraxial transfer function is obtained by Fourier transforming this expression, which can be found analytically resulting in the following [173]:

$$H(k_x, k_y; z) = e^{jkz} \exp\left[-j\frac{(k_x^2 + k_y^2)z}{2k}\right]. \tag{4.74}$$

Note that the same expression can also be obtained by starting with the exact propagator in the expression (4.67) and directly using the binomial approximation $\sqrt{r^2+z^2} \simeq z + (x^2 + y^2)/2z$, which is valid for $k_x^2 + k_y^2 \ll k^2$, i.e., for small diffraction angles with respect to the propagation z-axis.

Therefore, within the paraxial approximation of the impulse response, we can express the propagated/diffracted complex field by the convolution integral:

$$\boxed{\psi(x, y; z) = \frac{e^{jkz}}{j\lambda z} \int \int \psi(x', y'; 0) \exp\left\{j\frac{k}{2z}[(x-x')^2 + (y-y')^2]\right\} dx' dy'}$$

$$\tag{4.75}$$

which is known as the *Fresnel diffraction formula*. It is a simple exercise to show that if the quadratic term $\exp\left[jk(x^2 + y^2)/2z\right]$ of the Fresnel diffraction formula is factored outside the integrals (expanding the squares that appear in the argument), we obtain the alternative form:

$$\psi(x, y; z) = \frac{e^{jkz}}{j\lambda z} e^{j\frac{k}{2z}(x^2+y^2)} \mathfrak{F}_{2D}\left\{\psi(x', y'; 0)e^{j\frac{k}{2z}[(x')^2+(y')^2]}\right\} \qquad (4.76)$$

In this formula, the symbol $\mathfrak{F}_{2D}$ denotes the transverse two-dimensional Fourier transform of its argument, which consists of the product of the initial wave amplitude multiplied by a quadratic phase modulation. It is also to be noted that the transverse Fourier transform is here a function of the wave vector components $k_x = xk/z$ and $k_y = yk/z$.

Box 4.2 Resolution Limit in Optical Diffraction

The linear systems or Fourier optics approach to wave diffraction introduced in this section allows one to understand the resolution limit of any focusing system as a manifestation of the general Fourier uncertainty principle. In fact, in free space, any diffracted field of wavevector $k = 2\pi/\lambda$ can be expanded in a Fourier integral with wavevector components k_x, k_y, k_z obeying the following:

$$k^2 = k_x^2 + k_y^2 + k_y^2. \qquad (4.77)$$

The maximal localization of the field in a transverse plane is achieved when $k_\perp^{max} \equiv k_x^2 + k_y^2 = k^2$, because for far-field propagating waves we have $k_z^2 \geq 0$. Therefore, we obtain the following:

$$\delta \sim \frac{2\pi}{k} = \lambda, \qquad (4.78)$$

assuming that no evanescent components are present in the accessible field scattered from the object. The evanescent component of the diffracted field decays exponentially with the distance z from the object according to the following:

$$e^{-|k_z|z} = \exp\left[-z\sqrt{k_x^2 + k_y^2 - \left(\frac{2\pi}{\lambda}\right)}\right], \qquad (4.79)$$

showing that spatial features on the wavelength scale cannot be resolved by any conventional lens because the evanescent field components do not propagate at distances greater than approximately λ from the object. The sub-wavelength evanescent field components that constitute the near-field of the object can only be detected by approaching it at distances shorter than λ, which is achieved using nanometer tips in near-field optical microscopy [92].

The Fresnel diffraction formula is only applicable when the truncation of the binomial expansion is accurate, for propagation distances not too close to the diffractive screen, typically when $z \geq 10\lambda$. However, at large distances from the initial (source) plane, the Fresnel diffraction formula simplifies significantly. In fact, if we restrict the attention to propagation distances that satisfies the following:

$$z \gg z_R = \frac{k[(x')^2 + (y')^2]_{max}}{2}, \tag{4.80}$$

where z_R is called the *Rayleigh range*, we can neglect the contribution of the quadratic term that appears inside the argument of the transverse Fourier transform in equation (4.76). This leads to the *Fraunhofer diffraction formula*:

$$\psi(x, y; z) = \frac{e^{jkz}}{j\lambda z} e^{j\frac{k}{2z}(x^2 + y^2)} \mathfrak{F}_{2D} \{\psi(x', y'; 0)\} \tag{4.81}$$

The Fraunhofer diffraction formula is a limiting case of the Fresnel formula and it is valid at large propagation distances satisfying $z \gg z_R$. For a typical visible wavelength of 600 nm and transverse dimensions of the input field of 1 mm, we observe that $z \gg 5$ m guarantees the accuracy of the Fraunhofer far-field regime. We remark that an alternative and less stringent far-field condition states that for an aperture of maximum linear dimension D, the Fraunhofer approximation can be stated as follows:

$$z > \frac{2D^2}{\lambda}. \tag{4.82}$$

This condition is known as the "antenna designer's formula" for the far-field regime due to its frequent use in antenna engineering [175].

4.2.6 The Paraxial Helmholtz Equation

In many applications of interest to optical technology, the spatial variation of the dielectric function $\epsilon(\mathbf{r})$ can be considered small over the distance of one wavelength, so that we can neglect the gradient term in the inhomogeneous wave equation (2.157). In the scalar approximation, this choice leads to the following wave equation:

$$\nabla^2 \psi - \epsilon(\mathbf{r})\mu \frac{\partial^2 \psi}{\partial t^2} = 0, \tag{4.83}$$

where $\epsilon(\mathbf{r}) = \epsilon(x, y, z)$ depends on position. The preceding equation can be transformed into a more convenient form if we represent the unknown complex field as follows:

$$\psi(\mathbf{r}, t) = \psi_e(\mathbf{r})e^{-j(\omega_0 t - k_0 z)}, \tag{4.84}$$

where $\psi_e(\mathbf{r})$ denotes a slowly varying position-dependent complex function known as the *envelope function* of the wave, and k_0 is a constant number called the *propagation constant*. Substituting the preceding solution ansatz into the scalar equation (4.83) leads to the following equation for the complex envelope:

$$\nabla_T^2 \psi_e + \frac{\partial^2 \psi_e}{\partial z^2} + 2jk_0 \frac{\partial \psi_e}{\partial z} - [k_0^2 - \epsilon(\mathbf{r})\mu\omega_0^2]\psi_e = 0, \tag{4.85}$$

where $\nabla_T^2 = \partial^2/\partial x^2 + \partial^2/\partial y^2$ is the transverse Laplacian. We will consider the situation where the spatial variations of the wave envelope and of its derivatives are slow within the distance of the wavelength, so that the wave representation in (4.84) approximately maintains its plane wave nature. This implies that, within a distance $\Delta z = \lambda$, the change of the envelope function is much smaller than the function itself, i.e., $\Delta\psi_e = (\partial\psi_e/\partial z)\lambda \ll \psi_e$. This condition implies the following restrictions:

$$\frac{\partial\psi_e}{\partial z} \ll k_0\psi_e \tag{4.86}$$

$$\frac{\partial^2\psi_e}{\partial z^2} \ll k_0\frac{\partial\psi_e}{\partial z}, \tag{4.87}$$

defining the so-called *slowly varying envelope approximation*. Applying this approximation to the equation for the complex envelope allows us to neglect the second-order partial derivative with respect to z and obtain the following:

$$\boxed{\nabla_T^2\psi_e + 2jk_0\frac{\partial\psi_e}{\partial z} - [k_0^2 - \epsilon(\mathbf{r})\mu\omega_0^2]\psi_e = 0} \tag{4.88}$$

which is the *paraxial Helmholtz equation* for a nonuniform medium. When the medium is uniform, i.e., $\epsilon(\mathbf{r})$ is a constant, it reduces to the standard-form paraxial Helmholtz equation:

$$\boxed{\nabla_T^2\psi_e + 2jk_0\frac{\partial\psi_e}{\partial z} = 0} \tag{4.89}$$

This equation for the evolution of the paraxial field envelope is a parabolic equation similar in structure to the heat diffusion equation $\partial u/\partial t = D\nabla^2 u$ for the scalar field u (e.g., the density of the diffusing substance), as we immediately recognize by the substitutions $z \to t$ and $j/2k_0 \to D$, where D is the constant diffusivity (or diffusion constant). However, the presence of an imaginary diffusivity introduces novel wave-like solutions and makes equation (4.89) formally identical, after the substitution $z \to t$, to the Schrödinger equation of quantum mechanics for two-dimensional problems. The formal "isomorphism" between paraxial wave propagation and two-dimensional quantum mechanics has been recently exploited in photonic lattices of evanescently coupled waveguides to demonstrate intriguing physical phenomena, such as one-way light transport and topological phases, originally predicted for strongly correlated electronic quantum waves [176, 177]. Among the simplest solutions of the paraxial Helmholtz equation are paraboloidal waves, which are the paraxial approximations of spherical waves, as well as Gaussian beams [78, 79]. We will see in the next section that a "complex diffusion equation" also describes the spreading of linear waves in dispersive media, unveiling a duality between paraxial wave diffraction in space and pulse dynamics in time [178]. According to this duality, a monochromatic paraxial wave "spreads" is the transverse (k_x, k_y)-space as a narrowband z-directed plane wave will do in angular frequency space when propagating in a medium with dispersion $\omega(k)$.

The paraxial Helmholtz equation for a homogeneous medium can be solved by considering its transverse Fourier transform counterpart:

$$\frac{d\hat{\psi}_e}{dz} = \frac{(k_x^2 + k_x^2)}{2jk_0}\hat{\psi}_e. \tag{4.90}$$

Given an initial envelope in k-vector space $\hat{\psi}_e(k_x, k_y; 0)$, we write the solution at any other $z \neq 0$ as follows:

$$\hat{\psi}_e(k_x, k_y; z) = \hat{\psi}_e(k_x, k_y; 0)H_e(k_x, k_y; z), \tag{4.91}$$

where

$$H_e(k_x, k_y; z) = \exp\left[-j\frac{(k_x^2 + k_y^2)z}{2k_0}\right]. \tag{4.92}$$

We recognize that the function $H_e(k_x, k_y; z)$, which propagates the initial Fourier spectrum of the envelope of the field, follows from the Fresnel propagator (4.74) of the field spectrum obtained in Section 4.2.5. Therefore, there is fundamentally no distinction between paraxial wave propagation and paraxial wave diffraction as described in the spectral (k-space) domain. The solution of the paraxial Helmholtz equation (4.88) for nonhomogeneous media can be efficiently obtained in the spectral domain via the split-step beam propagation method (BPM). Interested readers will find a clear derivation of this numerical method in [179], on which our discussion has mostly been based.

4.2.7 Fractional Fourier Transform in Wave Diffraction

We discuss in this section the very interesting connection between wave diffraction/propagation and the generalized form of Fourier transform called the *fractional Fourier transform*. In particular, this approach provides an equivalent formulation of the paraxial wave propagation and Fresnel scalar diffraction theory [180] according to which light propagation can be regarded as a process of continual fractional transformation of increasing order. Moreover, the fractional Fourier transform (FRFT) has successfully been applied to the study of quadratic phase systems, imaging systems, and diffraction problems in general [181, 182].

Linear fractional operators, which have a long history in pure mathematics dating back to Leibniz, are becoming more and more popular in the applied sciences and engineering thanks to their ability to efficiently model complex systems with long-range correlations [183]. Particularly, fractional diffusion and wave operators play a crucial role in the analysis of anomalous diffusion through fractals and multifractal structures where geometrical correlations significantly alter the classical diffusion picture that is valid for homogeneous random media. An introductory discussion of fractional calculus is presented in the Appendix G, and applications to random lasers are addressed in Chapter 14.

We introduce the FRFT by first recalling the following eigenequation:

$$\boxed{\mathfrak{F}\{\phi_n(x)\} = (-j)^n \phi_n(x)}$$ (4.93)

that is obeyed by the usual Fourier transform operator $\mathfrak{F}\{\cdot\}$. The eigenfunctions of the Fourier transform are the *Hermite-Gaussian* functions given by the following:

$$\phi_n(x) = \frac{2^{1/4}}{\sqrt{2^n n!}} e^{-\pi x^2} H_n(\sqrt{2\pi}x),$$ (4.94)

where

$$H_n(x) = (-1)^n e^{x^2} \frac{d^n}{dx^n} e^{-x^2}$$ (4.95)

are the *Hermite polynomials* that form an orthogonal basis of the Hilbert space of finite energy functions $f \in L^2(\mathbb{R})$, satisfying the following:

$$\int_{-\infty}^{+\infty} |f(x)|^2 w(x) dx < \infty,$$ (4.96)

where $w(x) = \exp(-x^2)$ is the Gaussian weight function. Due to the completeness of the Hermite–Gaussian basis, we can expand any finite-energy function $g(x)$ in the form:

$$g(x) = \sum_{n=0}^{\infty} a_n \phi_n(x)$$ (4.97)

with the expansion coefficients obtained as follows:

$$a_n = \int_{-\infty}^{+\infty} g(x) \phi_n(x) dx.$$ (4.98)

We can now utilize the preceding eigenequation to compute the Fourier transform of $g(x)$ as follows:

$$\mathfrak{F}\{g(x)\} \equiv G(v_x) = \sum_{n=0}^{\infty} a_n (-j)^n \phi_n(v_x) = \int_{-\infty}^{+\infty} g(x) \sum_{n=0}^{\infty} \phi_n(x) \phi_n(v)_x dx,$$ (4.99)

where we also inserted the definition (4.98) of the expansion coefficients.

The last expression allows one to introduce the fractional-order Fourier transform of order α as follows:

$$\boxed{\mathfrak{F}^\alpha\{g(x)\} = \sum_{n=0}^{\infty} a_n (-j)^{n\alpha} \phi_n(v_x)}$$ (4.100)

which satisfies the required additivity property:

$$\mathfrak{F}^\alpha \mathfrak{F}^\beta = \mathfrak{F}^{\alpha+\beta}. \tag{4.101}$$

Using the Mehler's identity for the Hermite functions, we can express the FRFT of order α in the explicit form:

$$\mathfrak{F}^\alpha \{g(x)\} = \int_{-\infty}^{+\infty} g(x) K(x, v_x, \alpha) dx, \tag{4.102}$$

where the *fractional kernel* is given by [180]:

$$K(x, v_x, \alpha) = A_\alpha \exp\left\{ j\pi \left[(x^2 + v_x^2) \cot\left(\frac{\alpha\pi}{2}\right) - 2x v_x \csc\left(\frac{\alpha\pi}{2}\right) \right] \right\}, \tag{4.103}$$

where $A_\alpha = \sqrt{1 - j \cot(\alpha\pi/2)}$. The αth- order FRFT defined in (4.102) coincides with the usual (i.e., integer order) Fourier transform when $\alpha = 1$ and with the identity operator when $\alpha = 0$. The FRFT maps a function to an intermediate domain between time and frequency and can be interpreted as a rotation in the time-frequency plane. This perspective is generalized by the linear canonical transformations, which will be addressed in the next section. Many additional properties satisfied by the FRFT are found in [180].

The interest of the FRFT in optical diffraction consists on the fact that the gradual evolution of the diffracted field from the object plane at $z = 0$ to the far-field can be efficiently computed by the FRFT with fractional orders continuously varying between $\alpha = 1$, corresponding to the object plane ($z = 0$), and $\alpha = 1$, corresponding to the Fraunhofer far-field.

In Figure 4.9, we show the computed amplitudes (normalized to one) of the FRFTs corresponding to different values of the fractional order α for a rectangular aperture with unit width. We can see that for $\alpha = 0$ the FRFT perfectly reproduces the rectangular aperture, while for $\alpha = 1$ it provides the Fraunhofer diffraction solution. We also can clearly appreciate how by continuously increasing the fractional order in the range $0 < \alpha < 1$ we can obtain the diffraction amplitudes at the intermediate distances between the aperture plane and the Fraunhofer far-field, simply as the result of fractional Fourier transformation. Therefore, while the usual, i.e., integer-order, Fourier transform provides the link between the field distribution in the object plane and in the Fraunhofer far field, the fractional-order Fourier transform extends this correspondence to any finite distance.

The FRFT method can be naturally generalized to two-dimensional functions where it provides an efficient solution to large-scale diffraction problems that are often prohibitively difficult to treat using traditional numerical methods such as finite difference or finite element solvers of Maxwell's equations. A recent application of this powerful approach to the study of the propagation the scattered field from large-scale structures with aperiodic Vogel spiral geometry is discussed in [184].

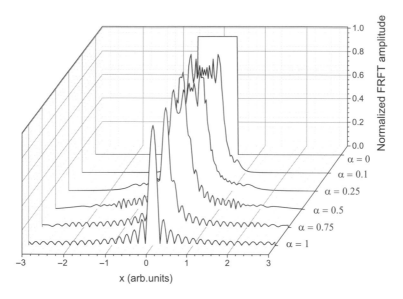

Figure 4.9 Normalized amplitudes of the fractional Fourier transforms of a rectangular function with unit width computed for the different values of the fractional order α indicated in the figure. We can clearly recognize in the evolution of the FRFT amplitude with respect to the order α the Fresnel diffraction solutions at increasing distances from the rectangular aperture under uniform illumination.

4.3 The Stratton–Chu Integral Representation

Vector diffraction theory is a difficult and specialized subject. Therefore, our goal here is limited only to introducing the key ideas while we refer the interested readers to the cited references for more details. Different types of vector diffraction theories can be derived from a vector analogue of Green's theorem, which, as we already have come to appreciate, is the mathematical underpinning of scalar diffraction theory [166]. In particular, the seminal work of Stratton and Chu [185] provided a rigorous approach to the solution of diffraction and focusing problems reformulated in terms of boundary value problems where the field values within a volume V are obtained based on the knowledge of the field values on its surface S. In particular, their work established an exact integral representation for general vector fields that is equivalent to the solution of Maxwell's equations, known as the *Stratton–Chu integral formula* [185]. This approach fully generalizes the Helmholtz and Kirchhoff integral theorem to vector waves that solve Maxwell's equations and is often utilized in conjunction with numerical techniques as an efficient transformation that propagates the computed field values from the near- or intermediate finite field region in proximity of devices to the far-field region [167].

The rigorous derivation of the Stratton-Chu integral formula is a somewhat delicate matter that has been recently addressed in great detail in [166, 167], to which we direct interested readers. The starting point in the derivation is to establish vector analogues of Green's first identity and of Green's theorem (or second Green's identity). In the original paper by Stratton and Chu [185], this was simply achieved by applying the divergence theorem to the vector quantity $\mathbf{P} \times \nabla \times \mathbf{Q}$, where $\mathbf{P}$ and $\mathbf{Q}$ are two smooth[8] vector fields within the closed volume V and on its surface S:

$$\int_V \nabla \cdot (\mathbf{P} \times \nabla \times \mathbf{Q})dv = \int_S (\mathbf{P} \times \nabla \times \mathbf{Q}) \cdot \hat{\mathbf{n}}ds, \qquad (4.104)$$

where $\hat{\mathbf{n}}$ is the unit normal vector directed outward from S. Remembering the general vector identity:

$$\nabla \cdot (\mathbf{a} \times \mathbf{b}) = \mathbf{b} \cdot (\nabla \times \mathbf{a}) - \mathbf{a} \cdot (\nabla \times \mathbf{b}), \qquad (4.105)$$

which is valid for any pair of vectors $\mathbf{a}$ and $\mathbf{b}$, we can expand the integrand in the preceding volume integral and obtain the vector analogue of Green's first theorem:

$$\int_V [(\nabla \times \mathbf{P}) \cdot (\nabla \times \mathbf{Q}) - \mathbf{P} \cdot \nabla \times (\nabla \times \mathbf{Q})]dv = \int_S (\mathbf{P} \times \nabla \times \mathbf{Q}) \cdot \hat{\mathbf{n}}ds \qquad (4.106)$$

Interchanging the roles of $\mathbf{P}$ and $\mathbf{Q}$ in the last equation and subtracting the resulting equation from (4.106), we obtain the vector analogue of Green's theorem:

$$\int_V [\mathbf{Q} \cdot \nabla \times (\nabla \times \mathbf{P}) - \mathbf{P} \cdot \nabla \times (\nabla \times \mathbf{Q})]dv$$

$$= \int_S [\mathbf{P} \times (\nabla \times \mathbf{Q}) - \mathbf{Q} \times (\nabla \times \mathbf{P})] \cdot \hat{\mathbf{n}}ds. \qquad (4.107)$$

The two vector Green's theorems (4.106) and (4.107) form the basis for the derivation of the Stratton–Chu integral formula that is obtained by setting $\mathbf{P} = \mathbf{E}$ and $\mathbf{Q} = G\mathbf{a}$, where $\mathbf{E}$ is the electric field and $G = \exp(jkr)/r$ is the scalar Green's function and $r = |\mathbf{r}_{01}|$, both required to satisfy the scalar wave equation inside V and on S, while $\mathbf{a}$ is a constant arbitrary unit vector. Implementing the preceding substitutions for $\mathbf{P}$ and $\mathbf{Q}$ in the vector Green's theorem (4.107) and using the divergence theorem again, it is possible to obtain a surface integral expression for the electric and magnetic fields [167]:

$$\boxed{\int_S [j\omega\mu(\hat{\mathbf{n}} \times \mathbf{H})G + (\hat{\mathbf{n}} \times \mathbf{E}) \times \nabla G + (\hat{\mathbf{n}} \cdot \mathbf{E})\nabla G]ds = 0} \qquad (4.108)$$

This result, also referred to as the *Stratton–Chu integral theorem* [166], made use of the the transversality of the electric field $\nabla \cdot \mathbf{E} = 0$ since the homogeneous and isotropic medium filling volume V is assumed to be source-free (no charges and currents), used the vector wave equation for $\mathbf{E}$, introduced the magnetic field via the time-harmonic Maxwell's equation $\nabla \times \mathbf{E} = j\omega\mu\mathbf{H}$, and leveraged the arbitrariness

---

[8] These vector functions of positions are assumed to have first and second derivatives that are continuous inside V and on its surface.

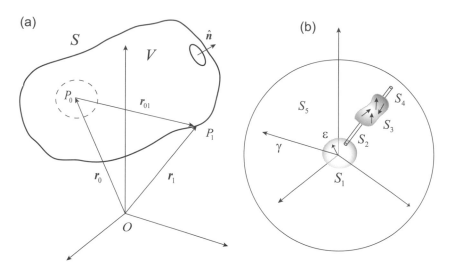

Figure 4.10 (a) Geometry for the Stratton–Chu problem illustrating the closed integration surface with the fixed observation point P_0 and the boundary (integration) point P_1 running on the surface S. The dashed circle around P_0 indicates the small spherical exclusion volume to avoid the singularity of the Green function when the observation and the boundary point coincide on the surface. (b) Closed integration surface (smooth, orientable, and simply connected) composed by the surfaces S_i, $i = 1, \ldots, 5$ used in the derivation of the Stratton–Chu formula for the electric and magnetic field at any exterior point of a closed surface bounding all sources and sinks. The small exclusion sphere S_1 is centered on the observation point P_0 and has radius ϵ while the large sphere S_5 has radius γ.

of vector **a**. See [166, 167] for more details on this derivation. The geometry used in the derivation of the Stratton–Chu formula is illustrated in Figure 4.10, where we denote $r = |\mathbf{r}_{01}| = |\mathbf{r}_1 - \mathbf{r}_0|$.

The delicate parts in the derivation have to do with the removal of the singularity of the Green function for $r = 0$ (i.e., observation and boundary point coincide on the surface S) as well as the exclusion of the radiation sources, i.e., currents and charges from the considered volume, since from the beginning we assumed the validity of the homogeneous Helmholtz equation for ψ and G. These goals can be achieved by a careful redefinition of the boundary surface S that is best achieved by first shifting the origin of the reference system to the fixed observation point P_0, as shown in Figure 4.10b. The main idea is to define a boundary surface where to integrate equation (4.108) without incurring into any singularity of the integrand [167]. To this purpose, a small exclusion sphere with surface S_1 and radius ϵ is considered wrapping around the new origin of the Cartesian system. In addition, all the field sources within V must be encircled by the volume with boundary surface S_3. We then construct a large spherical surface S_5 of radius γ also centered on P_0 and connect the surfaces $S_1, S_3,$ and S_5 using two small tubular volumes of surfaces S_2 and S_4, as schematized in Figure 4.10b. As a result, the original surface S is now replaced by the union of the elementary surfaces

S_i, with $i = 1, \ldots, 5$ that define a piecewise smooth, orientable, and simply connected boundary surface over which to evaluate (4.108).

In the implementation of equation (4.108), we should realize that, analogously to the situation previously encountered with respect to the Helmholtz and Kirchhoff integral theorem, the outward normal unit vector $\hat{\mathbf{n}}$ in Figure 4.10 will point inward over the surfaces $S_1, \ldots, S_4$ and point outward over S_5. Moreover, the integration over S_1 can easily be performed retracing the same steps that led to the Helmholtz and Kirchhoff integral theorem, leading to a contribution $4\pi\mathbf{E}$ when $\epsilon \to 0$. On the other hand, the contributions to the integral in (4.108) due to the surfaces S_2 and S_4 of the two tubes clearly vanish when the diameters of the tubes are infinitesimally small. The next contribution to be evaluated is the one of the large surface S_5, which is made to encompass the entire space, i.e., it is considered in the limit $\gamma \to \infty$. Since this surface is clearly a mathematical construction and not a physical surface, its contribution to the integral (4.108) must be made to vanish. The exact conditions for this to happen have been obtained in [167] and essentially reduce to enforce the vector analogue of the Sommerfeld radiation condition $\mathbf{r} \times (\nabla \times \mathbf{E}) + jkr\mathbf{E} \to 0$ in the limit $r \to \infty$. This condition is called the *vectorial radiation condition* [167]. In addition, the field $\mathbf{E}$ must decay to zero at infinity at least as $1/r$ so that the product $|r\mathbf{E}|$ remains bounded. Putting together all the correct expressions for the contributions to the surface integral in (4.108), it is possible to obtain an integral expression for the electric field that is given over the open exterior region of S_3:

$$\mathbf{E}(P_0) = \frac{1}{4\pi} \int_{S_3} [j\omega\mu(\hat{\mathbf{n}} \times \mathbf{H})G + (\hat{\mathbf{n}} \times \mathbf{E}) \times \nabla G + (\hat{\mathbf{n}} \cdot \mathbf{E})\nabla G]dS_3 \qquad (4.109)$$

The vector integral above is the *Stratton–Chu formula for the electric field* that enables its computation at a given point P_0 anywhere *outside* S_3 once its boundary values on S_3 are known. Due to the duality symmetry of Maxwell's equations, the corresponding Stratton–Chu formula for the magnetic field can be simply obtained based on the one for the electric field via the substitutions $\mathbf{E} \to \mu\mathbf{H}$ and $\mathbf{H} \to -\epsilon\mathbf{E}$. An alternative expression to (4.109), where the surface S_3 encompasses a source-free region of space can be derived as well. In this case, the electric field is given within the *closed interior* of S_3, i.e., within the source-free volume. As shown in [167], this leads to a simple sign difference between these two formulations in agreement with the expression published by Stratton and Chu [185]. In order to uniquely obtain the solution for a given diffraction problem, the Stratton–Chu formula (4.109) requires the knowledge of the tangential components of both $\mathbf{E}$ and $\mathbf{H}$ together with the normal components of $\mathbf{E}$, which comprises a total of five field components. As originally pointed out by Stratton and Chu [185], these components correspond to the fields that would be produced by a distribution of electric current over S with a density $\mathbf{K}_e = -\hat{\mathbf{n}} \times \mathbf{H}$, a distribution of magnetic current with density $\mathbf{K}_m = \hat{\mathbf{n}} \times \mathbf{E}$, and a surface electric charge of density $\rho_e = -\epsilon\hat{\mathbf{n}} \cdot \mathbf{E}$. It is also known from the general *uniqueness theorem* [186] that an electromagnetic field within a bounded domain is completely determined by the knowledge of just the tangential components of either

E or **H** on a closed surface.[9] Therefore, the fields appearing in the Stratton–Chu formula are overspecified, which is a consequence of having reduced all four Maxwell's equations into a single expression. In fact, it has been recently proven that the Stratton–Chu formula satisfies Maxwell's equations [166, 167]. Several examples of numerical implementations of the Stratton–Chu formula are discussed in [166–168] that we strongly recommend to interested readers.

---

[9] This is so because **E** and **H** are ultimately connected by Maxwell's equations and, within a source-free medium, the tangential components can be obtained from the normal components of the fields using the corresponding divergence-free equations or the transversality condition.

5 Diffraction in Optical Engineering

Having spoken of the rays of the sun which are the focus of all the heat and light that we enjoy, you will undoubtedly ask, what are these rays? This is, beyond question, one of the most important inquiries in physics, as from it an infinite number of phenomena is derived.

Letters to a German Princess, Leonhard Euler

The ability to propagate and control optical beams using the Fourier optics method introduced in the previous chapter is central to the engineering of optical devices, such as diffractive lenses and gratings, which exploit wave diffraction phenomena to achieve intended functionalities. In principle, by tailoring the spatial distribution of the field amplitude and optical phase across the input plane we could entirely control the diffraction and propagation properties of optical beams. This is indeed the approach leveraged by diffractive optical elements (DOEs) that are used in many research, military, and commercial applications to optical imaging, sensing, and spectroscopy.

In this Chapter, we discuss focusing and dispersive devices in the space and time domains. In addition, we address the natural focusing behavior of diffractive structures with complex phase distributions from the general viewpoint of catastrophe theory in optics. We close the chapter by introducing the emerging singular optics and meta-optics approaches that provide unprecedented opportunities to engineer focusing and diffraction phenomena using compact dielectric structures with subwavelength thickness.

5.1 Linear Optical Systems

The approach of Fourier optics, introduced in the previous chapter in relation to wave diffraction, fruitfully combines wave optics and linear systems theory. In this framework, any linear optical system is characterized by an *impulse response function* or, in the case of coherent illumination, by an amplitude point-spread function $h(\mathbf{r}, \mathbf{r}')$, which is the field response at $\mathbf{r}$ due to a point-source located at $\mathbf{r}'$. Here $\mathbf{r}$ and $\mathbf{r}'$

are the *transverse* spatial coordinates in the image and object planes, respectively. In particular, a linear imaging system is always described by the superposition integral:

$$U_2(\mathbf{r}) = \int_{-\infty}^{+\infty} h(\mathbf{r}, \mathbf{r}')U_1(\mathbf{r}')d\mathbf{r}' \tag{5.1}$$

where $U_1(\mathbf{r}')$ represents the object function, i.e., the field distribution (complex amplitude) on the object plane, and $U_2(\mathbf{r})$ is the image function, i.e., the field distribution (complex amplitude) on the image plane. The same expression can be applied to both coherent and incoherent illumination conditions. However, in the coherent case the system is linear in the complex amplitude while in the incoherent case it is linear in the intensity and the corresponding impulse response function is given by $|h(\mathbf{r}, \mathbf{r}')|^2$. We also recall that when $h(\mathbf{r}, \mathbf{r}') = h(\mathbf{r} - \mathbf{r}')$, the system is said to be *linear and shift-invariant* (LSI) (see Appendix C.4), and in this case, the convolution theorem can be used to relate the image and the object spectra as follows:

$$\tilde{U}_2(\mathbf{k}) = H(\mathbf{k})\tilde{U}_1(\mathbf{k}) \tag{5.2}$$

where $H(\mathbf{k})$, called the amplitude *transfer function*, is the Fourier transform of $h(\mathbf{r} - \mathbf{r}')$. Hence the image spectrum $\tilde{U}_2(\mathbf{k})$ is given by the product of the transfer function and the object spectrum $\tilde{U}_1(\mathbf{k})$. It follows that a linear imaging system can be viewed as a *filter* in the wave vector or (spatial) frequency domain that modifies the original object spectrum according to the properties of the transfer function.

Realistic optical systems such as telescopes or microscopes are complex assemblies of several discrete optical components that include lenses, mirrors, and irises. However, as long as image formation is the result of a linear process, all their complexity can be effectively reduced to an effective model where a "black box" containing a number of discrete optical components is placed in between two apertures, called the entrance and exit pupils. Moreover, it is assumed that light propagation between the entrance and exit pupils is adequately described by geometric optics [173]. These apertures limit the collection of light due to diffraction effects and set the performance limit of the imaging system. In fact, not all the propagating spatial frequencies[1] will be transmitted through the pupils, limiting the spatial resolution achievable by the optical system. It is shown in the theory of image formation that it is possible to associate all the limitations arising from diffraction effects with either of the two pupils, leading to two equivalent approaches to image formation. The view that diffraction effects result from the entrance pupil only leads to the Abbe theory of imaging while the equivalent approach, originally introduced by Lord Rayleigh and more frequently used in optical engineering, attributes all diffraction effects to the exit pupil [173].

---

[1] Remember that free-space propagation itself acts as a low-pass filter and the evanescent field components with wave numbers in excess of the free-space cutoff value k_0 are exponentially attenuated, i.e., do not propagate more than few wavelengths away from the object. This phenomenon sets the fundamental limit to any far-field imaging techniques based on the linearity principle.

Box 5.1 Diffraction-Limited Imaging Systems

An imaging system is considered *diffraction-limited* when the following conditions are satisfied:

- A diverging spherical wave emanating from a point-source object is mapped by the system into a spherical wave converging toward a point on the image plane.
- The location of the generic image point is related to the location of the corresponding object point via a scaling factor that is constant across all points of the image field.

Therefore, in a diffraction-limited imaging system a diverging spherical wave incident on the entrance pupil is converted into a converging spherical wave at the exit pupil. However, in any realistic imaging system the aforementioned requirements are only satisfied over finite regions of the object and image planes. Outside these ranges, the wavefronts leaving the exit pupils will significantly deviate from the ideal spherical shape, leading to an aberrated image. The amplitude point-spread function of a coherently illuminated diffraction-limited system is equal to the Fraunhofer diffraction patterns of its exit pupil, and its transfer function equals the rescaled (symmetric) pupil function $P(x, y)$ [173]:

$$H(k_x, k_y) = P(z_i k_x / k, (z_i k_y / k)), \tag{5.3}$$

where z_i denotes the distance (along the optical axis) from the exit pupil to the image plane. The case of coherent illumination also serves as the basis to understand the frequency response of an incoherently illuminated diffraction-limited system, characterized by the *optical transfer function* (OTF) denoted by $\mathcal{H}(k_x, k_y)$. In fact, it can be shown that the OTF is the normalized autocorrelation function of the amplitude transfer function $H(k_x, k_y)$ [173].

In the simple case of an imaging system consisting of a single lens, the entrance and exit pupils coincide and are equal to the physical size of the lens, whose imaging performances are in fact limited by the diffraction effects caused by its finite size. Note that the finite spatial resolution that is characteristic of any wave optical system implies that the transfer function has a finite support or, equivalently, the impulse response function of the system is *bandlimited*. Bandlimited functions are defined by the property that their Fourier transforms vanish outside a finite interval and play a crucial role in imaging theory due to their remarkable analytical properties [187, 188]. In particular, the Paley–Wiener theorem establishes a correspondence between bandlimited functions and entire functions,[2] which are analytic, i.e., continuous and differentiable any number of times everywhere in the complex plane. It follows that if an object is finite in size and is described by a space-limited function, then its image in Fourier space (its spectrum) is an entire function. Moreover, entire functions have

---

[2] The theorem is limited to entire functions $f(z)$ of exponential type, whose growth is bound by an exponential according to: $|f(z)| \leq M \exp(\alpha|z|)$, where M and α are constants.

the remarkable property that if known over a finite interval, they can be known every-where by analytic continuation. This mathematical property provides opportunities to reconstruct an object completely by extrapolating its observed spectrum. This idea is at the origin of the *bandwidth extrapolation technique* that is a computational approach to improve the spatial resolution of a linear optical system beyond the diffraction limit [189].

5.1.1   Fourier Properties of Paraxial Focusing

An optically thin, converging lens is modeled within the paraxial approximation by the following phase transformation [173]:

$$t_f(x, y) = \exp\left[-j\frac{k}{2f}(x^2 + y^2)\right],\tag{5.4}$$

where $f > 0$ is the focal length. A very remarkable property that follows from this phase transformation is that a thin lens naturally performs the two-dimensional spatial Fourier transform of an illuminated transparency and displays it on its back focal plane, i.e., a plane normal to the lens axis located a focal distance f behind the lens (along the light propagation direction). We can easily appreciate this behavior by placing an input transparency with complex amplitude transmittance $t_0(x, y)$ right against the lens. This transparency may describe an optically thin plate with prescribed amplitude and/or phase variations such as, for instance, a spatial light modulator (SLM) whose transmittance is externally controlled by an electrical signal. When the transparency-lens combination is normally illuminated by a unit amplitude plane wave, the field immediately behind it is represented by the product $t_0 t_f$. The paraxial field distribution at the back focal plane of the lens ($z = f$) is computed using the general Fresnel diffraction formula 4.76:

$$\psi(x, y; f) = \frac{e^{jkf}}{j\lambda f}e^{j\frac{k}{2f}(x^2+y^2)}\mathfrak{F}_{2D}\left\{\psi(x', y'; 0)e^{j\frac{k}{2f}[(x')^2+(y')^2]}\right\}.\tag{5.5}$$

Substituting $\psi(x', y'; 0) \rightarrow t_0(x', y')t_f(x', y')$ in the Fresnel formula, we find that the complex field in the focal plane is proportional to the Fourier transform of the input transparency:

$$\boxed{\psi(x, y; f) = \frac{e^{jkf}}{j\lambda f}e^{j\frac{k}{2f}(x^2+y^2)}\mathfrak{F}_{2D}\left\{t_0(x', y')\right\}}\tag{5.6}$$

where the Fourier transform is evaluated at coordinates $(k_x = k(x/f), k_y = k(y/f))$. The phase curvature factor in front of the Fourier transform can be eliminated by positioning the input transparency in the front focal plane of the lens, i.e., on a plane located at a distance $d = f$ in front of the lens. The input transparency can also be placed behind the lens at a distance d from its back focal plane, providing the additional opportunity to scale the Fourier transform by tuning the d/f ratio [173].

Note that in the special case when $t_0(x, y) = 1$ (uniform incident plane wave), the focal field $\psi(x, y; f) \propto \delta(x, y)$ and a perfectly sharp focal spot are achieved by a lens of infinite size. However, real lenses have a finite extension specified by the *pupil function* $P(x, y)$, typically a circular aperture in shape. The additional effect of the pupil function of the lens is naturally taken into account by placing it directly in front of an ideal (infinite-size) lens so that the field distribution in the back focal plane becomes

$$\psi(x, y; f) = \frac{e^{jkf}}{j\lambda f} e^{j\frac{k}{2f}(x^2 + y^2)} \mathcal{F}_{2D} \left\{ t_0(x', y') P(x', y') \right\}. \tag{5.7}$$

Using the convolution theorem, we can substitute the Fourier transform in the last expression with a convolution and obtain the following:

$$\boxed{\psi(x, y; f) \propto (T_0 * \tilde{P})} \tag{5.8}$$

where T_0 and $\tilde{P}$ are the Fourier transforms of the input transparency and of the pupil function, respectively. Therefore, due to the diffraction effects introduced by the finite pupil, any physical lens (finite-size) will broaden the focal spot compared to an ideal lens. We can also deduce from the last expression the general form of the point-spread function of a finite-size lens illuminated by a normally incident (unit-amplitude) plane wave. In fact, removing the input transparency is equivalent to setting $t_0(x, y) = 1$, and we immediately obtain that $\psi(x, y; f) \propto \tilde{P}$, since in this case $T_0 = \delta$. Therefore, we found that the *point-spread function* of a thin lens is proportional to the Fourier transform (or the Fraunhofer diffraction pattern) of its pupil function, provided the validity of the paraxial optics approximation.[3] For the usual case of a circular pupil, the intensity distribution in the focal plane of a lens with radius a is described by an Airy disk intensity pattern:

$$\boxed{I(r) = \left(\frac{A}{\lambda f}\right)^2 \left[2 \frac{J_1(kar/\lambda)}{kar/\lambda}\right]^2} \tag{5.9}$$

where r is the radial distance from the observation point to the optical axis, $A = \pi a^2$, and J_1 is the zeroth order Bessel function of the first kind. From this, it follows that the first dark ring in the diffraction pattern occurs when $\sin\theta = 0.61(\lambda/a)$, where θ is the angle at which the first minimum of intensity occurs, measured from the direction of incoming light.

The Airy disk pattern also enters the *Rayleigh criterion* for resolving two point objects, such as stars observed using a diffraction-limited system with a circular pupil, such as a telescope. According to this criterion, the two objects will be barely distinguishable when the center of the Airy disk generated by the first object occurs at the first minimum of the Airy disk of the second. In particular, the *lateral resolution* of

---

[3] Remember that we started from the Fresnel diffraction formula and considered the quadratic phase profile of a paraxial lens in expression (5.4). The general situation is more complex and can only be dealt with numerically, except for rare cases.

a system is measured by the minimum distance r at which the point sources can be distinguished. Using the Rayleigh criterion, the following can be readily shown:

$$r = \frac{0.61\lambda}{n \sin \theta} = \frac{0.61\lambda}{NA} \tag{5.10}$$

where θ is the the half angle subtended by the exit pupil of the system as seen from the image plane, n is the index of refraction of the medium surrounding the point sources, and $NA = n \sin \theta$ is the numerical aperture of the system. Expression (5.10) sets a fundamental limitation to the spatial resolution of any linear far-field imaging technique that does not use some prior information about the object. A broader discussion of optical resolution that surveys additional criteria based on more advanced approaches is provided in the paper by A. J. den Dekker and A. van den Bos [190].

5.1.2   Diffractive Optical Elements

One of the most basic operations in optics is light focusing. This effect is traditionally achieved using refractive lenses, which are typically introduced based on the laws of geometrical or ray optics instead of the more general diffractive theory. According to the ray optics picture, light travels in straight lines in a medium of constant refractive index, and the application of Snell's law at the interfaces with different media determines the direction of refracted light rays. A phase delay (with respect to propagation in free space) is imparted to the incident wave by the refractive elements whose shape and refractive index uniquely determine their imaging properties. As a result, traditional refractive optical components such as lenses, objectives, etc., consist of individual bulky units where most of the incident light undergoes refraction while only a small amount diffracts at the edges due to the finite size of the components. The presence of diffraction in optical components, which depends on their size with respect to the wavelength of the incident light, sets a fundamental limit to the smallest spot size obtained at the focal plane (diffraction-limited spot).

Unlike their refractive counterparts, diffractive optical elements achieve wavefront control by diffraction effects. These elements are composed of many different units that are comparable in size to the incident wavelength. The coherent superposition (interference) of the fields diffracted from each unit gives rise to the desired optical responses, i.e., a well-defined focal spot in the case of a diffractive lens. This idea was exploited by Augustin-Jean Fresnel, who replaced the heavy spherical lenses used in lighthouses by a thin device with different sections arranged in a plane, as shown in Figure 5.1a,b. The enabling idea is to realize that it is only the curvature of the lens material at the glass–air interface (together with the refractive index of the glass material) that determines the light-bending profile of incident optical rays. Consequently, we can remove the "inert layers" of glass material that are present in

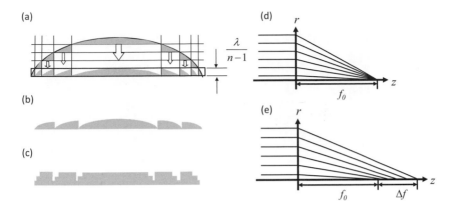

Figure 5.1 (a–b) Schematic processes showing the formation of a Fresnel lens starting from a conventional (refractive) lens in panel (a) and the creation of (c) a four-level discretized Fresnel lens device formed in a dielectric transparent medium. The thickness of the discretized layers is chosen to correspond to a phase difference 2π with respect to an incident wave in air. The focusing behavior of a traditional lens is displayed by the ray picture in panel (d). The distinctive focusing behavior of an axilens is illustrated in panel (e).

a conventional lens, i.e., all the layers with a thickness that corresponds to inessential 2π phase shifts, without altering the functionality of the lens.[4]

This is equivalent to consider a phase distribution modulo 2π, which we write for a focusing device as follows:

$$\phi(r)_{FL} = \left[\frac{2\pi}{\lambda}\left(\sqrt{f^2 + r^2} - f\right)\right]_{2\pi}, \tag{5.11}$$

where f is the focal length. The phase distribution in (5.11) describes the behavior of a *Fresnel lens*. A binary version of the Fresnel lens where the phase only assumes two values equal to 0 and π has been invented by Lord Rayleigh. It is called a *Fresnel zone plate* (FZP) and can achieve efficiency as high as 40% [191]. The modulo 2π operation is employed to convert the continuous phase profile of a refractive element, a Fresnel lens in this example, into discrete zones whose phase variation always lies in the interval $[0, 2\pi]$. Furthermore, the 2π-reduced phase distribution of the Fresnel lens can be discretized into a desired number n of layers each contributing $2\pi/n$ to the reduced phase in order to achieve similar performances with respect to the continuous phase profile, as illustrated in Figure 5.1c. Multilevel DOEs are designed to generate a phase distribution as close as possible to that of refractive optical elements and

---

[4] Fresnel's original design involved a flat lens whose different units were still larger in size compared to the incident wavelength. However, if the same elements are reduced to sizes comparable to the wavelength of light, then the resulting lens will be predominantly a diffractive optical lens.

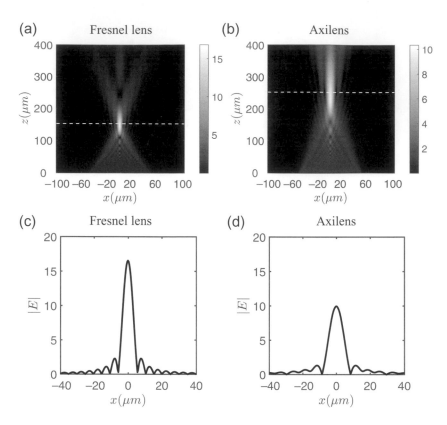

Figure 5.2 (a) Transverse view of the diffracted electric field from a Fresnel lens located at $z = 0$ with radius $R = 50$ μm, focal length $f_0 = 150$ μm. Light propagates toward the positive z-axis and is focused at the predicted focal length. (b) Transverse view of the diffracted electric field from a diffractive axilens (located at $z = 0$) with radius $R = 50$ μm and focal length $f_0 = 150$ μm. The designed focal depth is $\Delta f = 100$ μm. The white dashed lines in (a) and (b) indicate the positions for the one-dimensional field cuts shown in (c) and (d) for the Fresnel lens at $z = 150$ μm and for the diffractive axilens at $z = 250$ μm, respectively.

achieve high diffraction efficiency $>90\%$ since their efficiency increases rapidly with the number of phase discrimination levels. When lateral feature sizes are of the order of several wavelengths, scalar theory is enough to describe the behavior of DOEs. On the other hand, for subwavelength features, scalar optics becomes inaccurate and a rigorous vector theory is necessary to design DOEs and their intended effects on incident light beams.

In Figure 5.2, we provide some examples of the focusing behavior of multilevel Fresnel lens devices for light focusing. All the results discussed in this section have been obtained using the general first Rayleigh–Sommerfeld formulation of scalar diffraction, which can be expressed in convolution form as follows:

$$U_2 (x, y; z) = U_1 (x, y; 0) * h(x, y, z) \tag{5.12}$$

$$h(x, y, z) = \frac{1}{2\pi} \frac{z}{r} \left(\frac{1}{r} - jk \right) \frac{e^{(jkr)}}{r}, \tag{5.13}$$

where $*$ denotes convolution, U_1, U_2 are the transverse field distributions in the object and image planes, k is the incident wave number, and $r = \sqrt{x^2 + y^2 + z^2}$ and z is the distance between object and image plane. In Figure 5.2a,c, we show the spatial distribution of the electric field diffracted by a uniform plane wave at normal incidence to a Fresnel lens element located in the $z = 0$ plane. Figure 5.2b,d illustrates the distinctive focusing behavior of a DOE called *axilens*, which enables precise engineering of the focal depth and its location. Differently from a standard Fresnel lens, which has only one single focal length, the axilens is characterized by a radially dependent focal length:

$$f(r) = f_0 + \left(\frac{\Delta f}{R} \right) r \tag{5.14}$$

where f_0 if the focal length, Δf is the focal depth, and R is the radius of the element. See the illustration in Figure 5.2e. As a result, the 2π-reduced phase distribution of the axilens is given by the following:

$$\phi_{ax}(r) = \left[\frac{2\pi}{\lambda} \left(\sqrt{f(r)^2 + r^2} - f(r) \right) \right]_{2\pi}. \tag{5.15}$$

The increased depth of focus of the axilens compared to the Fresnel lens is clearly visible in Figure 5.2b,d, as well as its reduced intensity in the focal region. The results in the figure are obtained by approximating the continuous phase variations using four discretization levels. One of the key features of axilenses is that they support an achromatic plane where light is brought to focus over a broad spectral band [192–194].

5.1.3 Fresnel Zone Plates

In Figure 5.3, we illustrate two more examples that play an important role in scalar diffraction theory. In particular, we show in Figure 5.3a the binary amplitude mask corresponding to a simple circular disk region obstructing a normally incident plane wave at visible wavelength. As we already commented in our brief historical introduction in Section 4.2.1, this very simple setup led Arago to discover a small spot of light right within the geometrical shadow of the disk, which was arguably the first unambiguous confirmation of Fresnel's wave diffraction theory. In Figure 5.3c we show the radial intensity distribution of the diffracted radiation across a transverse plane at a distance $z = 5$ mm from the disk. A clear intensity peak can be observed in the middle, corresponding to the Arago spot. The width of the spot intensity peak depends on the distances between disk, source, and detection screen, as well as the source's wavelength and the diameter of the circular disk. When exciting with a plane wave source, the radial intensity distribution on the diffraction screen can be approximated

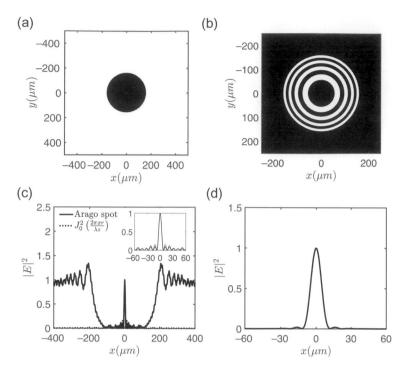

Figure 5.3 (a) Binary mask amplitude profile for the generation of the Arago spot. The blocked region in black is a disk with radius $r = 160$ μm. (b) Binary mask amplitude profile of a Fresnel zone plate (FZP). The FZP mask has a radius $r = 160$ μm. (c) Transverse intensity distribution of the diffraction by the disk in panel (a) showing the Arago spot at the distance $z = 5$ mm from the mask. In the inset, we also plot the analytical approximation to the field intensity distribution, demonstrating excellent agreement with the numerical data. (d) Intensity distribution at the focal plane of the Fresnel zone plate. The diffraction patterns have been obtained under normally incident plane wave excitation of wavelength $\lambda = 633$ nm.

in the proximity of the optical axis by a squared zeroth-order Bessel function of the first kind [195]:

$$\psi(r, z) \propto J_0^2 \left(\frac{\pi r D}{\lambda z} \right), \tag{5.16}$$

where D is the diameter of the disk and z is the distance between disk and screen. The accuracy of this approximation is demonstrated in the inset of Figure 5.3c where equation (5.16) is compared with the numerical data in the proximity of the optical axis.

A natural extension of the focusing ability of the Arago spot is provided by the Fresnel zone plate, which is the binary counterpart of the phase distribution of a standard Fresnel lens. In Figure 5.3b, we show the amplitude mask of a FZP. This consists of a set of radially symmetric rings, known as *Fresnel zones*, which alternate between opaque and transparent. The rings are spaced so that the light diffracted by the opaque zones constructively interferes at the desired focal spot. For this to happen, the radius of each zone must obey the equation:

$$r_n = \sqrt{\frac{n^2 \lambda^2}{4} + nf\lambda}, \tag{5.17}$$

where the label n is an integer, $f = 5$ mm is the principal focal length, and $\lambda = 633$ nm is the wavelength of the normally incident light. This simple diffractive focusing device is of great importance in contexts where no conventional materials can be utilized to fabricate refractive lenses, as is the case at X-ray frequencies. Unlike a standard lens, a binary FZP produces intensity maxima along the optical z-axis at odd fractions $f/3, f/5, f/7$, etc. These maxima are progressively sharper and contain less energy than the principal focus, although they possess the same maximum intensity. The focusing ability of a FZP is demonstrated in Figure 5.3d, where we show the transverse intensity distribution at it principal focus. Zone plates can also be constructed so that their transmission varies smoothly in space, for instance according to the function $[1 \pm \cos(ar^2)]/2$, where r is the radial coordinate from the center of the device and a is a scaling parameter. Wave diffraction by such a continuous zone plate gives rise to only a single focal point. A detailed engineering analysis of FZP and their modulated counterparts can be found, for instance, in the excellent book by Iizuka [196].

5.1.4 Optical Superoscillations

The far-field propagation of the subwavelength spatial features of wave fields that satisfy the Helmholtz wave equation in a homogeneous medium is limited by exponential attenuation, as previously discussed in Box 4.2. However, several approaches based on the mathematical properties of analytic functions are available to substantially improve field localization and optical resolution in imaging systems without violating the constraints imposed by the Fourier theory [189]. Due to its fundamental connection with aperiodic structures, we introduce here the one based on optical superoscillations [197–200].

It may come as a surprise to realize that *bandlimited* functions, which are functions whose Fourier transform vanishes outside some finite interval, can exhibit local oscillations at a rate that far exceeds their largest Fourier components. This mathematical property of bandlimited functions, called *superoscillation*, has been known for some time in various contexts, including quantum mechanics [201], signal processing [202, 203], and Fourier optics, where the local spatial frequencies $v_x \equiv (1/2\pi)\partial\phi(x)/\partial x$ of the function $f(x) = a(x)\exp[j\phi(x)]$ are generally distinct from the frequency components of its Fourier spectrum [173].

More generally, a complex signal of the form $\psi(\mathbf{r}) = \rho(\mathbf{r})\exp[j\phi(\mathbf{r})]$ is superoscillatory at points where the *local wavenumber* $k(\mathbf{r}) > k_0$, where

$$\boxed{k(\mathbf{r}) = |\nabla\phi(\mathbf{r})| = |\Im\{\nabla \log \phi(\mathbf{r})\}|} \tag{5.18}$$

where $\Im\{\}$ denotes the imaginary part. Many rigorous mathematical results on superoscillatory functions are presented by Ferreira and Kempf [203]. In particular, it is

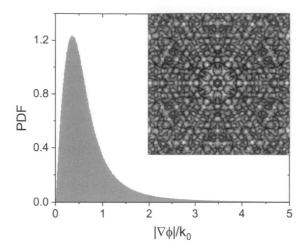

Figure 5.4 Probability density function (PDF) of the local wavenumbers in the optical intensity diffracted from an aperiodic Eisenstein prime array of subwavelength apertures. The diffraction plane is located at the distance $z = 15\ \mu m$ from the plane of the array. The inset shows the spatial distribution of the diffracted field intensity at $z = 15\ \mu m$. The considered nanohole array consists of 1,920 apertures with average separation $a = 1.2\mu m$ and constant $D = 200$ nm. The wavelength of the normally incident plane wave is $\lambda = 632.8$ nm and the calculation was performed using the Rayleigh -Sommerfeld (RS) diffraction theory.

established superoscillations can only occur in signals that possess amplitudes of widely different scales, which makes multifractal and more general aperiodic systems ideal candidates for the superoscillatory behavior.

Optical superoscillations are known to occur in random fields such as speckle patterns and in the neighborhood of optical vortices. In these situations, phase singularities develop where the field intensity vanishes but the phase is undefined and can vary arbitrarily fast, leading to the formation of subwavelength spatial structures without the involvement of evanescent field components [197]. The application of superoscillations to imaging beyond the diffraction limit, i.e., superresolution, was originally proposed by Berry and Popescu [201] that also considered the example of the following superoscillatory periodic function:

$$f(x) = (\cos x + ja \sin x)^N, \tag{5.19}$$

where the parameter $a > 1$ controls the degrees of superoscillation and $N \gg 1$ determines the extent of superoscillation. This function is bandlimited with maximum wavenumber N, but around $x = 0$ it oscillates as $\exp(jaNx)$. However, superoscillating functions attenuate exponentially in the superoscillatory region, requiring very large dynamical ranges that practically limit the recovery of sub-wavelength field information [203].

Optical superoscillations have been discovered in the field diffracted by a quasiperiodic array of nanoholes in a metal screen, showing that it can focus light into subwavelength spots in the far-field without contributions from evanescent fields [198].

More recently, a superresolution optical microscope with a resolution better than $\lambda/6$ was demonstrated based on engineered superoscillation binary masks, or "super-oscillatory lenses" [199]. A very beautiful review on the history and basic physics behind the phenomenon of superoscillation and its applications in optical imaging is provided in [204]. The design of superoscillation masks that enable subwavelength focus without evanescent fields is discussed in [200] along with applications to superresolution imaging.

An important question concerns the characterization of suitable bandlimited structures that exhibit superoscillating behavior. In fact, while it is known that superoscillations can only occur in structures that possess amplitudes of widely different scales [203], the connection between their multiscale spectral properties and superoscillations has not been established. As a result, optimization methods are often utilized to design and engineer suitable "superoscillation masks." In Figure 5.4, we demonstrate the formation of superoscillations in Eisenstein prime arrays (see Chapters 7 and 10 for a description of these aperiodic structures) by computing the probability density of the (normalized) local phase gradient associated to the diffracted light intensity shown in the inset. Superoscillation phenomena occur when $|\nabla\phi| > k_0$, reflecting the multiscale nature of the aperiodic structures with singular-continuous diffraction spectra [10, 205].

5.1.5 Phase-Modulated Axilenses

Recently, diffractive axilenses with spatially modulated phase variations have been introduced for the focusing of infrared radiation [192–194]. Spatially modulated axilens devices flexibly combine efficient point focusing and grating selectivity within a scalable top-down fabrication based on a four-level phase mask configuration. Engineering phase-modulation in achromatic diffractive lenses provides opportunities to achieve large focusing efficiency over a broad spectral range, polarization insensitivity, and adds spectroscopic capabilities to planar chips, enabling monolithic integration with infrared focal plane arrays for multi-spectral imaging, detection, and spectroscopy [192].

In Figure 5.5, we illustrate the effects of a periodic and chirped phase modulation in a four-level discretized axilens device. Phase-modulated axilenses are examples of multifunctional DOEs, which are used for beam shaping [206, 207]. In Figure 5.5a, c, we show the phase profiles of a modulated axilens summed (modulo 2π) with a two-dimensional (2D) periodic function and a $45°$ chirped function, respectively. The characteristic phase of the axilens ϕ_{ax} is modulated by adding the phase function ϕ_m according to the following expression:

$$\phi = [\phi_{ax} + \phi_m]|_{2\pi} \tag{5.20}$$

where the 2π subscript indicates that the phase is reduced by modulo 2π. The specific phase profiles shown in Figure 5.5a,c are characterized by $\phi_m = 2\pi(x' + y')/p$ and $\phi_m = \frac{\pi}{2}[\text{sign}(\cos(2\pi x'/p)) + \text{sign}(\cos(2\pi y'/p)] + \pi$, respectively, with $p = 15$ µm.

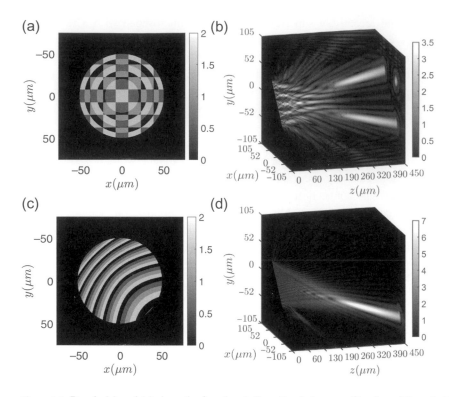

Figure 5.5 Panels (a) and (c) show the four-level discretized phase profiles for a 2D periodically modulated axilens and a 45° chirped axilens. The parameters are chosen as the diameter $d = 100 \ \mu m$, $f_0 = 250 \ \mu m$, $\Delta f = 120 \ \mu m$, and periodicity $p = 15 \ \mu m$. Panels (b,d) show the diffraction field 3D view for the phase profiles in (a) and (c), respectively. The wavelength of the incident radiation is $\lambda = 9 \ \mu m$. A host medium with a refractive index $n = 3.3$ was considered.

All the modulated axilenses have a diameter $D = 100 \ \mu m$, which is comparable to the typical size of alternative devices based on metalenses [208, 209]. Figure 5.5b,d illustrate, the three-dimensional distributions along a plane cut at 45° of the diffracted field amplitudes calculated using the Rayleigh–Sommerfeld formulation that are focused by the 2D modulated axilens and the chirped axilens shown in Figure 5.5a,c, respectively. The periodically modulated axilens simultaneously focuses incident radiation into four focal spots arranged along the main diagonal directions of the focal plane. This is a consequence of the angular dispersion introduced by the ±1 diffraction orders associated to the periodic phase modulation that we introduced along the orthogonal x- and y-directions, which splits an incident wave while simultaneously focusing it (with a larger depth of focusing) due to the underlying phase behavior of the axilens. In Figure 5.5b, we show only two focal spots along one diagonal direction. On the other hand, the incident beam does not split for the phase-chirped axilens, but it is focused and steered "as a whole" at different locations along the diagonal 45° line cutting through the focusing plane. Therefore, we show that the combination of simultaneous focusing and steering of radiation enabled by phase-modulated axilenses can be designed to achieve controllable focusing, on the same plane, of different wavelengths at different locations.

The wavelength-dependent shifts of the focused incident radiation across the achromatic pane of a diffractive axilens device enables the realization of compact single-lens microspectrometers with subwavelength thickness [210]. Therefore, these multi-functional diffractive devices add fundamental imaging and spectroscopic capabilities to traditional DOEs and provide compact solutions for emerging applications to spectral sorting, multiband imaging, photodetection, and spectroscopy.

5.1.6   Time-Domain Diffraction

In this section, we show how to utilize the Rayleigh–Sommerfeld first integral approach for the simulation of diffraction problems in the time domain. We consider an input Gaussian pulse represented in the frequency domain by the spectral density:

$$\hat{S}(v) = \frac{2\sqrt{\ln 2}}{\sqrt{\pi}\Delta v} \exp\left[-\frac{4\ln 2\,(v - v_0)^2}{\Delta, v^2}\right] \tag{5.21}$$

where Δv is the FWHM of the pulse in the frequency domain and v_0 is the center frequency of the pulse. The corresponding pulse in the time domain can be obtained by inverse Fourier transformation as follows:

$$S(t) = \int_0^\infty \hat{S}(v)\exp(-j2\pi vt)dv$$
$$= \exp\left[-\left(\frac{\pi\Delta vt}{2\sqrt{\ln 2}}\right)^2\right]\exp\left(-j2\pi v_0 t\right). \tag{5.22}$$

Therefore, by performing simulations at different frequencies of the diffracted field $U_2(x, y; z, v)$ from a normally incident plane-wave pulse modulated by a Gaussian distribution, we can construct the time-domain response of the diffraction using the Fourier representation:

$$U_2(x, y; z, t) = \int_{-\infty}^\infty \hat{S}(v)U_2(v, z; x, y)\exp(-j2\pi vt)dv \tag{5.23}$$

Note that it is essential to properly sample the input pulse in both frequency and time domains in order to achieve accurate results using this direct method, and that any arbitrary pulse shape can be utilized to compute the time-domain diffracted field. In Figure 5.6, we show two relevant application examples for the approach discussed: the space-time diffraction picture of a Gaussian pulse incident on a narrow circular aperture (a) and on a Fresnel lens with a larger aperture size (b). The time evolution of the spatial spreading of the diffracted field is clearly displayed in Figure 5.6a, while Figure 5.6b shows how the lens focusing effect unfolds in the space-time domain.

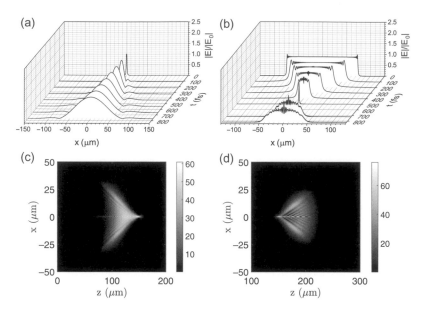

Figure 5.6 (a,b) Time-domain RS diffraction calculations showing the space-time evolution of the diffracted field amplitudes of a Gaussian pulse diffracted by a circular aperture and focused by a Fresnel lens, respectively. The Gaussian pulse has a frequency $\nu_0 = 429$ THz, and a width $\Delta \nu = 6$ THz. The aperture is located in the xy plane and the incident light propagates along the z-axis. The circular aperture diameter is $D = 2.8$ μm. The focal length of the Fresnel lens is $f_0 = 150$ μm designed for a central wavelength $\lambda_0 = 700$ nm and has a circular aperture radius $r = 75$ μm. Panels (c) and (d) show the field amplitude distribution for the pulse before and after the focal spot at times t $= 392$ fs and 658 fs, respectively.

We note that the linear cut of the electric field amplitude of the pulse immediately after the large aperture of the lens resembles the well-known Fresnel diffraction by the two edges of a rectangular aperture. At longer times, the field amplitude is focused by the lens and reaches the minimum transverse spread after approximately 500 fs, corresponding to the designed focal length $f_0 = 150$ μm. For even longer times, the diffracted field solution broadens and turns into a smoother Airy disk profile with high-frequency oscillations due to the interference of different spectral components. Figure 5.6c,d displays the xz plane side view of the diffracted fields at two time instants immediately before and after the focusing condition. Due to the nonparaxial nature of the focusing, the intensity distributions extend considerably across the z-axis and feature characteristic spatial structures of polychromatic refraction caustics [211].

5.1.7   Space-Time Duality

In Section 4.1.2, we analyzed the propagation of wave packets in dispersive media, and generally showed that pulse reshaping occurs upon propagation. We established our results by considering the Taylor expansion of the dispersion relation $\omega = \omega(k)$, which is justified for narrowband pulses. We now specify the dispersion relation of the medium in terms of a frequency-dependent propagation constant, i.e., $k = k(\omega)$, following the widespread engineering convention. Following the same steps that led to equation (4.24), we readily obtain the following for the slowly varying pulse envelope:

$$\left(\frac{\partial A(z,t)}{\partial z} + \frac{1}{v_g}\frac{\partial A}{\partial t}\right) = j\frac{k''}{2}\frac{\partial^2 A(z,t)}{\partial t^2} \qquad (5.24)$$

where the relevant group dispersion parameter is now given by $k'' = d^2 k/d\omega^2$.

This equation can be expressed in the traveling coordinate system defined by the change of variables: $\tau = t - z/v_g$ and $\xi = z$, where it takes the simpler form:

$$\frac{\partial A(\xi,\tau)}{\partial \xi} = j\frac{k''}{2}\frac{\partial^2 A(\xi,\tau)}{\partial \tau^2}. \qquad (5.25)$$

We can immediately realize that the last equation for the narrowband dispersion problem is also a complex diffusion equation structurally identical to the paraxial diffraction equation (4.89). In the dispersion case, the time coordinate τ is the analogue of the transverse x- and y-coordinates that appear in the paraxial diffraction equation. This observation constitutes the basis of *space-time duality* and leads to the interesting theory of *temporal imaging* [178].

Using the formal analogy between the heat diffusion equation, paraxial propagation, and the narrowband dispersion equation, we can easily deduce their general solutions using simple substitutions of variables. Let us provide an example by first recalling that the paraxial equation is readily solved in terms of the previously obtained field envelope propagator $H_e(k_x, k_y; z)$ as follows:[5]

$$\psi_e(x,y;z) = \frac{1}{4\pi^2}\int\int H_e(k_x,k_y;z)\hat{\psi}_e(k_x,k_y;0)e^{j(k_x x + k_y y)}dk_x dk_y. \qquad (5.26)$$

The general solution of the narrowband pulse dispersion problem governed by equation (5.25) is readily obtained by implementing the formal substitutions:

$$z \to \xi \qquad (5.27)$$

$$x \to \tau \qquad (5.28)$$

$$k_x \to \omega \qquad (5.29)$$

$$\frac{j}{2k_0} \to j\frac{k''}{2} \qquad (5.30)$$

in the one-dimensional version of (5.26). This leads to the following solution:

$$A(\xi,\tau) = \frac{1}{2\pi}\int_{-\infty}^{+\infty} A(0,\omega)e^{-j\frac{\xi}{2}k''\omega^2}e^{j\omega\tau}d\omega \qquad (5.31)$$

---

[5] This solution follows immediately from the general relation in (4.69). Note that the factor e^{jkz} that appears in the Fresnel propagator formula (4.74) is absent in the envelope propagator H_e, consistently with the pulse envelope representation in (4.84).

where the space variable ξ is placed in the first position inside the argument of the pulse envelope A to be consistent with our representation in (4.21). The presented analysis shows that the solutions of the narrowband pulse propagation can be obtained by multiplying the initial spectra by a phase filtering term that is quadratic in frequency and linear in the propagation variable. As a consequence, when pulses propagate the rapid oscillations of the quadratic phase term contribute to average out the contributions at the highest frequencies, leading to pulse spreading in space. However, since only the phase of the initial spectrum is modified, the amplitude of the propagating pulse always remains constant (in the absence of dissipation phenomena).

5.1.8   Time Lenses and Temporal Optics

In the previous section, we have shown how the space-time duality allows one to regard dispersion as the time-domain analog of paraxial diffraction, both special cases of a complex diffusion process. Building on this fundamental insight, it becomes possible to construct time-domain analogues of space diffraction phenomena. Of particular importance in dispersion engineering is the time-domain analogue of a traditional space lens, or a "time lens."

The concept of a time lens follows naturally when we realize that the transverse space coordinate x of diffraction theory is mapped into the time variable τ of dispersion theory. This can be immediately appreciated by recalling that a thin space lens is described in the paraxial approximation by the quadratic phase transformation [173]:

$$H_S(x, y) = \exp\left[-j\frac{k_0}{2f}(x^2 + y^2)\right],\qquad(5.32)$$

where f is the focal length and the constant phase associated to the total thickness of the lens has been neglected.[6] Therefore, the time-domain analogue of the space lens transformation is realized when considering a quadratic phase modulation in the time variable τ, according to the expression [178]:

$$H_T(\tau) = \exp\left[-j\frac{\omega_0\tau^2}{2f_T}\right],\qquad(5.33)$$

where the parameter f_T describes the "focal time" of the time lens. The explicit expression of f_T depends on the particular physical mechanism used for producing the quadratic phase modulation. Note that, in complete analogy with its spatial counterpart, the signal obtained in the focal plane of a time lens will be the Fourier transform of the input signal. A detailed analysis of time lens parameters, such as its f-number, magnification, and numerical aperture, considering electro-optic phase modulation is provided in [178]. In addition, using linear systems theory, Kolner [178] showed that a suitable combination of dispersion in the medium with the quadratic phase modulation

---

[6] Note that here for simplicity we consider an ideal lens with infinite aperture size, implying that its focal spot is point-like, i.e., described mathematically by a delta function. Instead, realistic lenses have a finite extension described by a pupil function, typically circular in shape, which gives rise to finite spatial resolution due to light diffraction with the aperture.

of a time lens enables the creation of the time-domain analog of an imaging system. For example, a scaled replica of the incident wavefront, magnified or compress in time, is created at the *temporal-imaging condition* [178]:

$$\left(\xi_1 \frac{d^2 k_1}{d\omega^2}\right)^{-1} + \left(\xi_2 \frac{d^2 k_2}{d\omega^2}\right)^{-1} = -\frac{\omega_0}{f_T} \qquad (5.34)$$

where $k_1 = k_1(\omega)$ and $k_2 = k_2(\omega)$ are the dispersion parameters of the two media placed before and after a time lens of focal time f_T. Expression (5.34) is the temporal analogue of the well-known Gauss's imaging condition of ray optics [78, 79]. Building on the fundamental equivalence between light diffraction in space and pulse dispersion in time, temporal magnifications, temporal Fourier transform, temporal signal processing, and even temporal depth imaging systems have been demonstrated by implementing optical schemes from space to time [212, 213]. A simple way to realize a time lens is to modulate the phase of the optical signal using a phase modulator. However, the time window over which the lens operates is typically limited because in a modulating voltage with a sinusoidal waveform is used in practice to provide only a locally quadratic phase. Alternatively, nonlinear optical processes can be applied to impart a quadratic phase to the input signal, but these often require high pump powers. Recently, the parametric four-wave mixing (FWM) process has been used to produce a time lens in a silicon nanowaveguide for ultrafast temporal processing [212].

The diffraction behavior of time-varying media has also attracted considerable attention for the engineering of novel optical effects and materials in the time domain [214–217]. Of particular interest in this context is the phenomenon of *time refraction* that occurs when an optical pulse approaches a moving temporal boundary across which the refractive index changes in time. The space-time duality in paraxial optical wave propagation implies that the pulse undergoes a temporal equivalent of reflection and refraction of optical beams at a traditional spatial boundary. However, in this case the role of the incident/refraction angles that appear in the usual Snell's law is played here by a change in the pulse frequency. Specifically, the change in the refractive index of the medium breaks its time translation symmetry, inducing a frequency shift while leaving the pulse wavevector unchanged. An analytic derivation of the expressions for these frequency shifts at temporal boundaries under different conditions can be found in [218]. Large frequency shifts up to $\approx 6.5\%$ of the carrier frequency have recently been demonstrated in indium tin oxide (ITO) thin films, which are highly nonlinear materials with a low-refractive index [217]. Frequency conversion by time modulation effects can pave the way to build more efficient on-chip frequency tuners and converters for coherent optical communication and quantum information technology. Spatiotemporal light control for on-chip wavefront modulation has also been recently demonstrated by combining a passive metasurface with a frequency comb source [216]. Laser frequency combs are broad spectra composed of equidistant narrow lines and revolutionized time and frequency metrology in the late 1990s by providing "frequency rules" that measure large optical frequency differences. More recently, frequency combs found many applications to attosecond science, astronomy, and broadband spectroscopy [219, 220]. In the time domain, they correspond to ultra-short optical pulses generated by the coherent interference of many phase-locked frequency lines, i.e., the resultant sum of time-domain phasors with a constant relative phase. A *frequency comb* pulse $\psi(t) = \psi_e(t) \exp(-j\omega_0 t)$ has an envelope composed of $2N + 1$ frequency lines:

$$\psi_e(t) = \sum_{n=-N}^{N} c_n e^{-jn\Delta\omega t} \tag{5.35}$$

where $\omega_n = \omega_0 + n\Delta\omega$. Using this ultrafast source to illuminate a metasurface establishes a linear frequency gradient across it that redirects the incident plane wave toward a different, fixed direction by tilting the incoming phase front. The resulting interference pattern will evolve in time, producing a characteristic dynamic beam steering effect with time-dependent steering angle and a steering rate that is directly proportional to the frequency gradient across the metasurface [216]. Angular scanning rates of approximately $2.5°/\text{ps}$ have been experimentally achieved following this approach, demonstrating a new strategy for nonmechanical, ultrafast dynamic beam steering on a silicon chip.

5.2   Natural Focusing and Diffraction Caustics

In the previous sections, we discussed light diffraction in situations where a uniform plane wave is incident on an aperture or obstruction in a flat screen that bounds the incoming planar wavefronts. However, very interesting wave diffraction phenomena arise even for unobstructed wavefronts when they are not initially flat. In this case, phase variations result from an uneven height distribution across the initial wavefront, producing very complex diffraction patterns when the waves propagate. This situation is typical of the wave transmission or reflection across phase screens with unintended height fluctuations such as rippled glass screens or irregularly shaped water droplets. Interestingly, wave propagation through transparent media with fluctuating phase profiles can give rise to *stable focusing effects*, i.e., focal points that are robust to the perturbations in the model's parameters. A familiar example is provided by the refraction of sunlight through the wavy water surface of a swimming pool, which produces a complex network of rapidly moving patches of concentrated light that are visible on the bottom of the pool. Similar effects can be observed when light is reflected by the rippled surface of the water on the side hulls of harbored ships or directly under bridges spanning across water canals. All these *natural focusing* phenomena correspond to optical caustics, typically discussed using the ray optics picture. Ray optics, also called geometrical optics, can be formulated in a way that is formally analogous to the the classical Newtonian mechanics of particles moving in a potential force field [221]. For optical rays, this potential depends on the refractive index distribution in the medium. Consistently, geometrical optics admits an elegant Lagrangian and Hamiltonian formulation that provided inspiration for the early development of

Schrödinger's wave mechanics as the natural generalization of Newtonian mechanics, to which in fact it reduces in the "geometric limit" of $\lambda \to 0$.

In the context of geometrical optics, caustics are defined as the *envelopes of light rays* reflected or refracted by extended surfaces with varying curvature.[7] Since a caustic is a curve or a surface to which each of the light rays is tangent, it defines a region of space where all the light rays are concentrated or focused. Therefore, the corresponding light intensity diverges near caustics, which are in fact the singularities of geometrical optics [211, 222]. A frequently encountered caustic's structure[8] consists of bright lines of nonparaxial rays, called the "fold," that meet at a sharp "cusp" point (i.e., the paraxial focus) and is easily observed when light rays reflect off a curved surface. This is in fact a well-known example of a spherical aberration, which was already investigated in Leonardo da Vinci's clear geometrical constructions shown on folio 84v of the *Codex Arundel* [223].

5.2.1 Diffraction and Catastrophe Theory

Caustics can be rigorously described and classified by the elementary catastrophes of singularity theory, which describes the abrupt behavior, such as sudden jumps or biforcations, of dynamical systems [224–226]. As an important part of the general singularity theory, *catastrophe theory* was originally developed by René Thom in the 1960s as a powerful mathematical approach to analyze the critical points of multivariable functions, or potential functions $V(x_1, \ldots, x_n; c_1, \ldots, c_m)$, with generally nonisolated (degenerate) critical points. The critical points or the equilibrium points of a multivariable smooth function $V(x_1, \ldots, x_n)$ are the points at which its gradient ∇V vanishes. The critical points at which the determinant of the Hessian matrix of the potential $V_{ij} = \partial^2 V / \partial x_i \partial x_j$ (also called the stability matrix) are nonzero are called isolated, nondegenerate or *Morse critical points* [225–227]. On the other hand, the ones at which the determinant of the Hessian matrix vanishes are called degenerate or nonisolated (also non-Morse critical points). The degenerate critical points of the potential are the ones where not only the first derivatives but also higher-order derivatives vanish. However, besides its dependence on n generalized coordinates or *state variables*, the potential depends additionally on m parameters, called *control variables*. In optics, state variables can be taken as (generalized) coordinates on the rippled surface while control variables specify, for instance, the observation conditions of the caustics. In the language of optical engineering, we can say that state variables are "object-plane coordinates" while state variables are "image-plane coordinates" such as positions or observation angles. The essence of the matter is that since the potential depends on several control parameters $c_1, \ldots, c_m$ so does the Hessian matrix and its eigenvalues $\lambda_1, \ldots, \lambda_m$, which in fact could vanish for certain values of the

---

[7] Remember that an envelope of a family of curves in the plane is a curve that is tangent to each member of the family at some point. All the points of tangency together form the envelope.

[8] Bright fold caustic lines are observed in a coffee cup on a sunny day, earning the not-so-technical designation of "coffee cup caustics."

control parameters, giving rise to nonisolated degenerate critical points. Therefore, the critical behavior of the potential will depend by the choice of the control parameters in a generally very complicated fashion that is the goal of catastrophe theory to precisely understand.

Remarkably, Thom was able to prove that the very complex local structure of general potential functions around degenerate critical points can be analyzed in terms of only a finite number of distinctive and very complex geometrical shapes that are stable with respect to perturbations in the parameters of the potential [227]. These fundamental building blocks account for the general singular behavior of complex functions and are termed "elementary catastrophes." They have direct applications in many physical contexts, including light diffraction from rippled surfaces, which are the essential tool for understanding optical caustics [211, 222, 225, 226]. For instance, the use of catastrophe theory can tame the unphysical divergences that otherwise plague short-wavelength approximations of wave diffraction theory where geometrical optics rays are endowed with amplitude and phase and can interfere where rays cross [222]. Nondiverging integral representations of "interfering rays" called *diffraction catastrophes* can be generally written for the n-dimensional case in the following form:

$$\psi(\tilde{\mathbf{x}}) = \left(\frac{k}{2\pi}\right)^{n/2} \int \cdots \int \psi(\mathbf{x}, \tilde{\mathbf{x}}) e^{jk\phi(\mathbf{x}, \tilde{\mathbf{x}})} d^n \mathbf{x} \qquad (5.36)$$

where n is the number of state variables, $k = 2\pi/\lambda$ the wavenumber, and the potential function $\phi(\mathbf{x}, \tilde{\mathbf{x}})$ is simply the geometrical distance function from a general source point $\mathbf{x} \in \mathbb{R}^n$ on the rippled wavefront (defined on the object hyperplane) to the observation point $\tilde{\mathbf{x}} \in \mathbb{R}^m$ in the image or detection hyper-plane. In the special case of wave propagation between two two-dimensional surfaces, we can specialize expression (5.36) considering $n = m = 2$. The preceding integral represents the superposition of interfering rays and can be evaluated accurately by considering only the contributions from the neighborhoods of the critical points of $\phi(\mathbf{x}, \tilde{\mathbf{x}})$[9]. These are the points where the phase varies slowly and the corresponding terms dominate the integral. For this reason, this approach is called the *stationary phase method,* and its application to the preceding integral in the limit of $\lambda \to 0$ is detailed in [226]. The stationary phase method also allows one to understand that the field amplitude on the caustic diverges with a characteristic power-law scaling $\psi(0) \sim k^\sigma$, where σ is called the *Arnold index.* As expected, the field strength approaches infinity in the geometrical optics limit $\lambda \to 0$.

The connection between light diffraction and catastrophe theory can now be established based on Thom's theorem [225–227], which enables the classification of structurally stable caustics.[10] This profound result shows that for an m-parameter family

---

[9] The contributions to the amplitude from the neighborhood of noncritical points vanish due to the fast-oscillating nature of the integral in this case.

[10] This stability refers for instance to the perturbations of the optical arrangement used to measure them.

of smooth potentials with $m \leq 5$, the potential can be locally described around its singularity points by a quadratic form with isolated "good" critical points (the Morse term) plus a limited number of characteristic terms that carry the "bad" non-Morse singularities. More precisely, when $m \leq 5$, Thom's theorem states that the potential can always be written in canonical form [226]:

$$V = CG(\ell) + \sum_{j=\ell+1}^{n} \lambda_j y_j^2 \qquad (5.37)$$

where ℓ is the number of eigenvalues $\lambda_1, \ldots, \lambda_n$ of the Hessian determinant that vanish at non-Morse critical points, and the y_j are transformed variables obtained via a smooth coordinate transformation from the state variables x_j. The preceding canonical decomposition is valid in an open neighborhood of $\mathbf{x}_0 \in \mathbb{R}^n$ for the fixed value of the control parameter $\mathbf{c}_0 \in \mathbb{R}^m$ that renders $\mathbf{x}_0$ a non-Morse critical point.

The function $CG(\ell)$ is called the *catastrophe germ* since it contains the essence of the singularity. Thom's theorem also specifically prescribes functional forms describing the ways in which the ℓ "bad non-Morse variables" $y_1, \ldots, y_\ell$ occur at degenerate critical points of the potential, after a suitable coordinate change. A more general version of this theorem, also proved by Thom, provides a stronger canonical decomposition of V that does not fix the parameter c_0 and it is valid within open neighborhoods of both ($\mathbf{x}_0$ and $\mathbf{c}_0$):

$$V = CAT(\ell,m) + \sum_{j=\ell+1}^{n} \lambda_j y_j^2 \qquad (5.38)$$

The function $CAT(\ell,m)$ is the *catastrophe function*, which is related to the catastrophe germ as follows [226]:

$$CAT(\ell,m) = CG(\ell) + \wp(\ell,m), \qquad (5.39)$$

where $\wp(\ell,m)$ is called the *perturbation function*, and it contains a number of terms equals to the dimensionality m of the control space, also known as the *codimension*.[11] Perturbations arise by changes in the control parameters such as the detection conditions or experimental configurations in measuring optical caustics regarded as catastrophes. These terms are important because they affect the qualitative feature of the potential functions at their non-Morse critical points while they do not affect them at the Morse critical points (or at noncritical points). In the physics literature, the perturbation terms are often referred to as the *unfolding terms* [211] because they describe how the singularities, i.e., optical caustics, "unfold" as one deviates in control space from the origin. Therefore, we can summarize by saying that the catastrophe function is an ℓ-state variable and m-control variable function that fully describes the local character of the singular potential. The amazing power of Thom's result, which

---

[11] Note that the m parameters are here regarded as free parameters, so they span an m-dimensional space.

also makes its proof extremely complex, is that it is possible to find explicit expressions for the forms of all the elementary catastrophe functions that enter the decomposition of a critical potential. These expressions consist of "universal" polynomial functions that are tabulated, for instance, in references [211, 225–227]. Interestingly, there exist only seven of them in codimension $m \leq 4$. Therefore, there are only seven possible caustics associated to each catastrophe germ. These are called "elementary catastrophes," and in a general sense should be considered as "atomic components" that make up any (stable) singular wave field. In summary, thanks to Thom's theorem it became possible to rigorously investigate structurally stable caustics by replacing the potential that appear in the general diffraction integral in (5.36) with one of the canonical normal forms listed in references [211, 225–227]. Therefore, a scalar field amplitude ψ is be associated to each degenerate (non-Morse) critical point of the potential function according to the integral in (5.36). The resulting diffraction integrals are the "diffraction catastrophes" and provide a characteristic diffraction pattern for each type of caustic. The diffraction integrals corresponding to the fold and the cusp diffraction catastrophes, which are the first two terms in the hierarchy of elementary diffraction catastrophes, are proportional to the *Airy function* Ai(s) and to the *Pearcey function* P(a,b) reported as follows [211]:

$$\mathrm{Ai}(s) = \frac{1}{\pi} \int_0^\infty dt \, \cos(t^3/3 + st) \tag{5.40}$$

$$\mathrm{P}(a,b) = \frac{1}{\sqrt{2\pi}} \int_{-\infty}^{+\infty} e^{j(x^4+ax^2+bx)} dx \tag{5.41}$$

It can be shown by evaluating these integrals with the stationary phase method that they have Arnold index $\sigma = 1/6$ and $\sigma = 1/4$, respectively. More complex closed-form integral expressions have been obtained for higher catastrophes, and we refer the interested reader to the excellent work by Nye for a comprehensive discussion [211]. The Airy and Pearcey integral functions are computed numerically and shown in Figure 5.7b,d.

5.2.2   Engineering Diffraction Caustics

We now show how it is possible to use the rigorous Rayleigh–Sommerfeld diffraction theory, valid beyond the paraxial approximation, to reproduce and study diffraction caustics using planar phase masks instead of dielectric slabs with fluctuating thickness. The proposed approach to "caustics engineering" is suitable for the design of complex focal structures that can be fabricated using planar device technology on the micron scales. In particular, controlled generation of fundamental diffraction catastrophes using traditional Fourier optics methods provides opportunities to engineer miniaturized diffractive devices that exploit the fine-scale structures of focused radiation

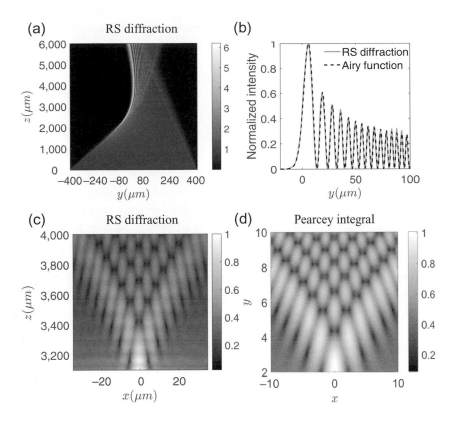

Figure 5.7 (a) Side view of the computed field amplitude of a 1D fold caustic using the Rayleigh–Sommerfeld diffraction theory and a phase mask (see the text) with the coefficient $a = 10/(2f_0)$. (b) normalized 1D intensity profile of the diffraction fold shown in panel (a) at a propagation distance $z = 4,000$ μm and the Airy function. (c) Top view of a 2D cusp caustic computed using the Rayleigh–Sommerfeld diffraction theory with a phase mask (see the text) with the coefficient $a = 6/(2f_0^3)$. (d) Pearcey integral function $P(x, y)$ shown in normalized units.

to implement novel optical concentration and illumination modalities. For example, desired "catastrophe phase masks" could be produced using multilevel lithography on optical chips for applications to novel light sensing and trapping modalities. This vision could be realized by encoding the catastrophe germs into the spatial phase profile $\phi(x, y)$ of a diffracting phase mask with complex transmittance $A = e^{j\phi(x,y)}$ excited at normal incidence by a homogeneous plane wave. Note that the catastrophe germ must be added to a standard parabolic focusing phase because the relevant wavefronts are always specified with respect to a reference paraboloid surface [211]. The diffracted amplitude detected on the observation plane will then display the characteristic caustic for each selected catastrophe germ. This approach, which is valid beyond the usual paraxial approximation, is implemented and applied later to the fold and cusp caustics in Figure 5.7 and validated against the results obtained by computing the corresponding catastrophe integrals. Since the codimension of the fold catastrophe structure is equal to one, we implemented a one-dimensional (1D) phase mask to generate the corresponding diffraction pattern using the following expression:

$$\phi(y) = \frac{2\pi}{\lambda_0}[ay^3 - G(y)], \qquad (5.42)$$

where $\lambda = 633$ nm is the wavelength of the incident light, $G(y)$ is the lens focusing term $G(y) = \sqrt{y^2 + f_0^2} - f_0$ and we have used $f_0 = 4,000$ μm in all our simulations. The coefficient a can be chosen freely since it is a control parameter. Here we choose $a = \frac{5}{3f_0^2}$ in order to make the higher-order phase variation comparable with the quadratic terms. Figure 5.7a shows a cross-sectional view of the propagated field amplitude of a fold caustic, which is remarkably characterized by a maximum field intensity that propagates along a curved trajectory in free space while preserving its transverse structure. This is an example of *accelerating Airy beams* [228], which are weakly diffractive solutions of the wave equation that propagate along parabolic trajectories or, more generally, curved trajectories. Airy beams were first investigated for quantum waves in 1979 by Berry and Balazs, who pointed out how, unlike ordinary quantum wavefunctions, Airy wave packets are nonspreading accelerating solutions to the force-free Schrödinger equation [229]. The discovery of the optical counterparts of these peculiar beams occurred relatively recently [228, 230].

Airy beams are of great interest for applications to particle trapping and micromanipulation because they feature unusual gradient optical forces that can accelerate particles toward their main peaks and propel them along parabolic paths by optical radiation pressure [230, 231]. Figure 5.7c,d compare the plane views of the amplitude distribution of a cusp caustic computed using the Rayleigh–Sommerfeld diffraction integral approach and the Pearcey diffraction integral, respectively. The agreement between the two is remarkable. Since the fold caustics has codimension equal to 2, its diffraction pattern is obtained using the 2D phase profile:

$$\phi(x, y) = \frac{2\pi}{\lambda_0}[ax^4 + by^2 - G(x, y)], \qquad (5.43)$$

where $G(x, y) = \sqrt{x^2 + y^2 + f_0^2} - f_0$ is the 2D focusing term and we choose the control coefficients $\left(a = \frac{6}{f_0^2}, b = 0\right)$. We set the y^2 term equal to zero to avoid conflicting with the quadratic focusing term $G(x, y)$, which yields better focusing quality. Finally, we considered the engineering of diffractive phase masks producing elliptic and hyperbolic umbilic caustics, which are examples of more complex caustics with codimension equal to 3 [211, 222]. We should keep in mind that the corresponding catastrophes unfold in three spatial dimensions, and therefore their full structures and structural stability properties cannot be captured by simply analyzing the propagated field on two-dimensional planes. The phase profile that we have used to generate the elliptic umbilic structure is as follows:

$$\phi(x, y) = \frac{2\pi}{\lambda_0}[ax^3 - bxy^2 - G(x, y)], \qquad (5.44)$$

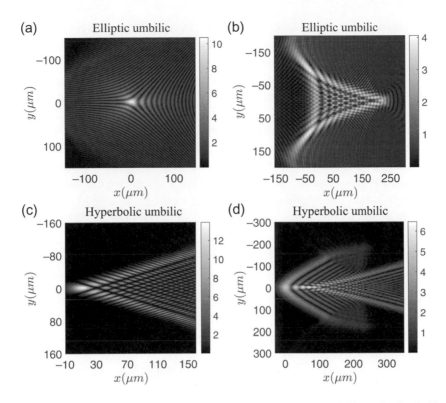

Figure 5.8 Panels (a) and (b) show the front view of the computed field amplitude of elliptic umbilic caustics using the Rayleigh–Sommerfeld diffraction theory and a phase mask (see the text) with the coefficient $a = b = 10/(2f_0^2)(f_0 = 4,000$ μm) sampled at distances $z = 4,500$ μm and $z = 5,700$ μm, respectively. Panels (c) and (d) show the front view of a hyperbolic umbilic caustic obtained using a phase mask (see the text) with coefficients: $a = b = 10/(2f_0^3)(f_0 = 4,000$ μm) at the distances $z = 4,000$ μm and $z = 6,000$ μm, respectively.

where we choose the coefficients $a = b = \frac{5}{f_0^2}$. In the case of the hyperbolic umbilic structure, we used the following phase profile:

$$\phi(x, y) = \frac{2\pi}{\lambda_0}[ax^3 + bxy^2 - G(x, y)] \tag{5.45}$$

with coefficients $a = b = \frac{5}{f_0^2}$. The results of the Rayleigh–Sommerfeld diffraction calculations are shown in Figure 5.8. In particular, in Figure 5.8a,b, we display the computed field amplitudes for an elliptic umbilic caustic probed at distances $z = 4,500$ μm and $z = 5,700$ μm from the phase mask, respectively. Scanning through the focal section a triangular-shaped caustic region begins to appear and then evolves into a distinctive pattern of diffraction maxima and minima within the fold lines of an equilateral triangle. Remarkably, past the focal point it is possible to observe an ordered array of bright and dark spots that are arranged three-dimensionally in space like atoms in a crystal lattice. This fine structure of focused light has been investigated by Berry [222] and Nye [211, 232, 233] and shown to correspond to wave dislocations, which are line singularities in the phase of the waves analogous to dislocations in crystals. In particular, along the dislocation lines the phase of the wave is indeterminate and the amplitude is zero. The theoretical framework for understanding the local phase structure and the dynamics of any dislocation in a scattered scalar wave field is discussed in [232]. The fascinating manifestations of singularities in optical wave fields, including polarization effects, are discussed comprehensively in [234]. Moreover, it is interesting to note that the elliptic umbilic structure naturally encodes an Airy pattern parallel to the fold lines and a Pearcey pattern near the cusps (Figure 5.8b). Such a fine substructure of focused light within the elliptic umbilic can been explained to a good approximation by the interference of four rays, according to the diffraction integral [211, 235]:

$$E(a,b,c) = \frac{1}{2\pi} \int_{-\infty}^{+\infty} \int_{-\infty}^{+\infty} dy e^{j[x^3 - 3xy^2 - c(x^2+y^2) - ax - by]}. \tag{5.46}$$

The integral can be evaluated using the stationary phase method and has an Arnold index $\sigma = 1/3$.

In Figure 5.8c,d, we show front views of an hyperbolic umbilic caustic at two different propagation distances. As the caustic unfolds, its characteristic structure consisting of two intersecting hyperbolic fold surfaces becomes evident in Figure 5.8b. In the interfering-ray picture discussed earlier the hyperbolic umbilic follows from a diffraction (catastrophe) integral similar to the one in equation (5.46) but with a phase equal to: $\varphi(x,y) = j[x^3 + y^3 - cxy - by - ax]$. Since the structural complexity of diffraction catastrophes increases rapidly with the codimension, it becomes progressively more difficult to study them in codimension >3 by considering only their two-dimensional cross-sections. These fascinating patterns of light can be observed by imaging, using a microscope, a drop of water spanning a circular hole in a tape on a vertical glass slide. On the other hand, when the glass slide is mounted horizontally, elliptic umbilical caustics are observed instead [211].

5.2.3   Optical Rogue Waves

The strong phase gradients and fluctuations that drive diffraction caustics and natural focusing phenomena may also result in an abrupt and extreme increase of the field's amplitude in linear systems known as rogue-type behavior. Temporal and spatial *rogue waves* consist of high-amplitude field phenomena occurring with low probability in many oscillating systems due to the superposition of correlated random wave components, such as in the linear concentration of waves occurring in caustics formation, or due to intrinsic nonlinear instabilities in the presence of noise. The term "rogue waves" was originally introduced in oceanography to describe the behavior of "giant waves" of extremely large intensities at low probabilities that have been observed emerging from the ocean and releasing exceptionally destructive power [236, 237]. In 2007, Solli et al. [238] reported the first observation of rogue waves in an optical system using a microstructured optical fiber driven within a noise-sensitive nonlinear regime in which extremely broadband radiation is generated from a narrowband input. The experimental results were modeled using a generalized nonlinear Schrödinger equation demonstrating that rogue waves arise with very low probability starting from initially smooth pulses due to a power transfer process seeded by small noise fluctuations. Although many studies in optics have focused on different nonlinear phenomena as driving mechanisms for the emergence of rogue waves [239], extreme wave amplitudes can also be achieved in the linear regime when a suitable random phase structure is imparted on a coherent optical field [240]. The appearance of extremely intense waves in linear systems reflects the presence of a non-Gaussian statistical distribution with a very long tail for intensity fluctuations, i.e., a heavy-tailed probability distribution in which very large amplitudes appear more often than what would be predicted from a Gaussian distribution. This is the case, for instance, in caustics and in their networks where rogue-type waves have been recently reported using computer-generated smooth random phase masks in combination with spatial light modulators [241, 242].

More generally, rogue waves can be understood in the context of extreme value theory or extreme value analysis, EVA) where extremely rare events are described by non-Gaussian heavy-tailed probability distributions [243]. Heavy-tailed distributions arise in probability theory when limit processes such as the sum of dependent and correlated variates are considered, beyond the validity of the central limit theorem (see additional discussions in Chapter 8). Examples of heavy-tailed distributions are the Pareto distribution, which is a power-law distribution often encountered in econometrics, the log-normal distribution that characterizes a random variable whose logarithm is normally distributed, and the log-Cauchy distribution, which describes a random variable whose logarithm is distributed according to a Cauchy distribution. In the present context, non-Gaussian statistics describes the intensity fluctuations of waves scattered in the far-field zone by correlated or tailored random systems in space and/or time. In particular, the direct connection between far-field light scattering and geometry valid in the single scattering regime, provides opportunities to engineer suitable phase correlations by controlling the aperiodic geometry of deterministic aperiodic media, leading to the formation of interesting rogue waves phenomena [244]. As a representative example, we discuss later the statistics of the far-field intensity distribution obtained when uniformly illuminating aperiodic arrangements of small (with respect to the incident wavelength) scattering particles, referred to as scattering arrays or scattering point patterns. As we make clear in later chapters, the far-field scattered wave intensity is proportional to the structure factor of the arrays, which is essentially its two-dimensional Fourier transform. Therefore, investigating the spatial fluctuations of the structure factor provides information on the statistics of scattered waves that determines the likelihood of rogue wave formation in different structures. We show an example of this approach in Figure 5.9, where we compare the probability density of the intensity fluctuations of waves scattered by an uncorrelated random arrangement of particles on a disk, i.e., a Poisson array, and by a two-dimensional Penrose array that is the primary example of quasicrystal structures introduced in later chapters. The intensity peaks are statistically analyzed to compute histograms, and the intensity

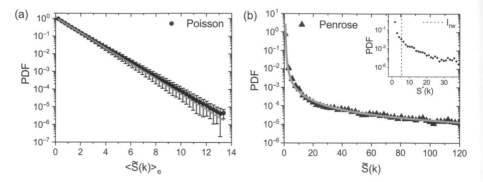

Figure 5.9 (a) Ensemble averaged probability density function of the fluctuations of the normalized structure factor $\tilde{S}(k) = S(k)/\overline{S(k)}$ with respect to its average for a Poisson point pattern. The average is performed over 100 different realizations of the disorder. The continuous gray line is the fit with the negative exponential statistics expected for the uncorrelated speckle behavior of a circular Gaussian random process. (b) Probability density function of the fluctuations of the normalized structure factor of a Penrose quasicrystal. The continuous gray line is fit with respect to the heavy-tailed log-Cauchy distribution with parameters $\mu = -0.41$ and $\sigma = 0.14$. The inset shows the probability density function of the fluctuations of the far-field peak intensity with respect to its mean value $S^*(k) = S(k)/\overline{S(k)}|_{max}$. The dashed black line indicates the threshold value I_{rw} for the formation of rogue waves.

threshold for rogue wave formation is computed after the oceanographic definition as $I_{rw} \geq 2I_{1/3}$, where the $I_{1/3}$ is the mean intensity of the highest third of events [239]. In Figure 5.9a, we observe an inverse-exponential distribution that is characteristic of the interference of uncorrelated random waves as in the case of fully developed speckles (see extended discussions on speckle phenomena in Chapter 8). On the other hand, we found that the intensity fluctuations of the Penrose structure shown in Figure 5.9b are well-described by the log-Cauchy heavy-tailed distribution:

$$p(x; \mu, \sigma) = \frac{1}{\pi x} \frac{\sigma}{[(\log x - \mu)^2 + \sigma^2]} \qquad (5.47)$$

where μ is a real number and $\sigma > 0$. This probability distribution is referred as a "super-heavy tailed" one because its tails decay only logarithmically. As a result, none of the moments of the log-Cauchy distribution are finite. In particular, this distribution does not have a defined mean or standard deviation. The remarkable heavy-tailed nature of the far-field intensity fluctuations in the Penrose structure reflects its aperiodic long-range structural correlations and results in a significantly larger probability for rogue wave formation compared to the Poisson array.

5.2.4   Singular Optics

Singular optics is a rapidly emerging field of modern optics that deals with the analysis of phase singularities in scalar or vector wave fields whose amplitudes vanish [234].

Ray singularities (caustics) are optical catastrophes that support a rich structure of phase dislocations and are of primary interest to singular optics. In addition to phase dislocations, singular optics investigates the behavior of complex beams that produce optical vortices. The study of optical vortices received a tremendous boost when Coullet [245] recognized the mathematical analogy between the description of helically phased beams and superfluid vortices, coining the term "optical vortex." In these situations, the entire phase structure of the light beam rotates and light is said to acquire *orbital angular momentum* (OAM) [246]. The orbital momentum can be many times greater than the spin of circularly polarized beams and gives rise to remarkably new effects that are currently intensively investigated [247]. It was recently realized by Allen et al. [248] that optical beams with an azimuthal phase dependence $\exp(j\ell\phi)$, where ℓ can be any integer value, carry an OAM that is distinct from the spin of photons and that gives rise to an optical vortex along the propagation axis of the beam. These beams have *helical phase fronts* with a number of intertwined helices and their handedness depending on the magnitude and the sign of ℓ. Moreover, we can easily realize that the electromagnetic field transverse to these phase fronts has axial components.

The OAM carried by a wave field can be conveniently analyzed in cylindrical coordinates using modal decomposition into a basis set with azimuthal dependence through Fourier-Hankel decomposition (FHD) according to [249–251]:

$$f(\ell, k_r) = \frac{1}{2\pi} \int_0^\infty \int_0^{2\pi} E_\infty(r, \phi) J_\ell(k_r r) e^{j\ell\phi} r \, dr \, d\phi \qquad (5.48)$$

where $E_\infty(r, \phi)$ is the field distribution in the far-field plane, k_r represents the spatial frequency in the radial direction, and J_ℓ is the ℓth-order Bessel function that carries OAM states with positive and negative ℓ values. This can be directly observed by plotting $|f(\ell, k_r)|^2$, where multiple and well-defined azimuthal components ℓ will appear. Alternatively, integrating over the radial frequency k_r results in the function $F(\ell) = \int |f(\ell, k_r)|^2 k_r dk_r$ which immediately displays the OAM spectrum carried by the analyzed field structure.

Light with OAM can be generated in the laboratory using, for instance, spiral phase plates or engineered diffractive optical elements, enabling novel optical approaches to rotate and manipulate microscopic objects, create new types of imaging and communication systems, and explore the fundamental light–matter interaction properties of OAM beams in both the classical and the quantum regime [252–254].

5.3 What Is Meta-Optics?

Meta-optics or metaphotonics refers to the engineering of optical wave propagation and focusing using metamaterial concepts and both the electric and magnetic interactions of resonant nanostructures. See, for instance, the recent review by Baev et al. [255]. This approach demonstrates unprecedented optical functionalities enabled by devices with subwavelength resonant features and designed responses that originate from material and/or geometrical dispersion controlled by shapes and dimensions. Currently, the meta-optics paradigm is driving a disruptive new vision for optical technology especially in relation to the rapid development of metasurfaces that are suitable for monolithic integration with semiconductor materials and nonlinear devices on optical chips. In particular, applications of metasurfaces to optical beam steering, lensing, optical and thermal imaging, holography, and quantum information processing have been demonstrated beyond what is possible using traditional approaches [256, 257].

5.3.1   Metasurface Engineering

Optical metasurfaces (OMs) are a class of metamaterials with subwavelength thickness that demonstrate exceptional abilities for controlling the flow of light beyond what is offered by conventional, planar interfaces between two naturally occurring media. In particular, OMs are 2D or quasi-2D (flat) artificially structured surfaces consisting of a dense arrangements of resonant optical elements such as nano-antennas with dimensions and relative separations that are much smaller (typically by a factor of 10) than the wavelength. The resonant nature of the light–matter interaction in OMs enables spatial control over the scattering amplitudes and phases of optical waves at the nanoscale level that can be flexibly manipulated through the choice of the resonators' material, size, orientation, morphology, and surrounding environment. Therefore, differently from conventional optical components such as lenses or waveplates that rely on propagation effects to impart a desired phase change, OMs produce *localized changes* in the phase, amplitude, and polarization of a light beam within the length scale of the free-space wavelength. The spatial variation of the geometric parameters across the metasurface results in its spatially varying optical response that provides the ability to impart desired shapes to optical wavefronts and largely control the energy flow [256, 257].

The metasurface concept and its ability to modify the energy flow at an interface between two media were first demonstrated by using V-shaped optical nano-antennas in the midinfrared spectral range [258] and later confirmed in the near-infrared [259]. Due to the spatially varying geometry and orientations of the antennas, these metasurfaces give rise to anomalously refracted and reflected beams satisfying the *generalized Snell's laws* [259]:

$$\sin \theta_t n_t - \sin \theta_i n_i = \nabla \Phi / k_0 \qquad (5.49)$$

$$\sin \theta_r - \sin \theta_i = n_i^{-1} \nabla \Phi / k_0, \qquad (5.50)$$

where the phase gradient $\nabla \Phi$ is the additional momentum contributed by the coupling with the nano-antennas at the interface between two media with refractive indices n_t and n_i. Since the nano-antennas break the symmetry at the interface, light waves must bend in order to conserve the total momentum. In the preceding expressions, θ_t, θ_i, and θ_r are the angles of refraction, incidence, and reflection, respectively. Current optical metasurface designs rely on lossless dielectric materials [260, 261] and more complex phase gradients in both space and time [216] and have demonstrated an impressive range of ultrathin optical devices, including broadband, aberration-free lenses with large numerical aperture [262], engineered structures for the controlled generation of optical vortices, optical beam steerers, as well as adaptive and tunable graphene-based hybrid devices for on-chip ultrafast switching and modulation [257, 263–265].

6   Wave Scattering and Coherence

> After Maxwell people conceived physical reality as represented by continuous fields,
> not mechanically explicable, which are subject to partial differential equations.
>
> *Ideas and Opinions*, Albert Einstein

In this Chapter, we present the framework of the scalar theory of wave scattering, or potential scattering theory, formulated in terms of the perturbative Green function approach. In particular, we establish the fundamental result of potential scattering, i.e., the *Lippman–Schwinger integral equation*, and introduce its self-consistent multiple scattering solution. This approach provides the theoretical underpinning for the multiple scattering formulation of both quantum and classical waves, which later will be applied to different types of aperiodic structures in Chapter 12. The synthetic parameters that characterize radiation and scattering from arbitrary aggregates of particles are introduced and the important optical theorem is discussed for both scalar and vector waves. Finally, we review the basic ideas of optical coherence and address their implications for scattering and diffraction from both deterministic and random media.

6.1   Classical Versus Quantum Scattering

In frequency domain, the Schrödinger's equation for quantum waves and the Helmholtz equation for classical scalar waves have the same mathematical form:

$$\boxed{\nabla^2 \psi + k^2 \psi = 0} \tag{6.1}$$

where ψ is the *complex wavefunction* that denotes either the quantum mechanical or the classical wavefunction such as, for example, any component of the electromagnetic field. The eigenvalue k^2 is simply given by the following:

$$k^2 = \begin{cases} \frac{2m(E-\varphi)}{\hbar^2}, & \text{for quantum waves} \\ \frac{\omega^2}{v^2}, & \text{for classical waves.} \end{cases} \tag{6.2}$$

Here E is the total energy of the quantum particle moving in the potential φ and v is the speed of the classical wave (i.e., acoustic waves, optical waves, etc). Note that $E - \varphi$ is negative for quantum mechanical bound states. Therefore, the propagation of classical waves has a direct correspondence with quantum waves only when $E > \varphi$.

6.1.1 The Classical Scattering Potential

In a general scattering problem, the wavenumber k depends on the position vector $\mathbf{r}$ due to the presence of localized inhomogeneities, such as particles in a discrete medium or fluctuations in the optical parameters of a continuous non-uniform medium (e.g., the turbulent atmosphere) and can be represented as follows:

$$k^2 = k_0^2 + V(\mathbf{r}), \tag{6.3}$$

where k_0^2 is the wavenumber associated to a uniform background and $V(\mathbf{r})$ represents the deviation from the background value due to the scattering process.

For classical waves propagating at a speed $v = c_0/n$ in a dielectric medium with position-dependent refractive index $n(\mathbf{r})$, we obtain that $k_0^2 = \omega^2/c_0^2$ and introduce the *optical scattering potential*:

$$\boxed{V(\mathbf{r}) = k_0^2 \left[\epsilon_r(\mathbf{r}) - 1\right] = \frac{\omega^2}{c_0^2} \left[\epsilon_r(\mathbf{r}) - 1\right]} \tag{6.4}$$

where c_0 is the speed of light in free space and $\epsilon_r(\mathbf{r}) = n(\mathbf{r})^2$ is the relative dielectric function of the inhomogeneous medium. Equation (6.4) indicates that wave scattering originates from the local variations in the dielectric properties of materials and defines the *optical scattering potential* in a way that is analogous to the quantum potential.[1] The optical scattering potential is generally a complex-valued quantity compactly supported over the finite volume of the scattering particles. However, depending whether or not the support of the scattering potential $V(\mathbf{r})$ is a connected or a disconnected region of space, the resulting optical potential describes the wave scattering by a single particle or by a collection of particles (i.e., multiparticle scattering).

The primary goal of *potential scattering theory* consists in finding the field ψ that is generated by the interaction of a known incident wave ψ_0 with a general optical scattering potential $V(\mathbf{r})$, as schematized in Figure 6.1 for the case of a single particle.

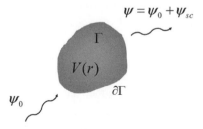

$$\psi = \psi_0 + \psi_{sc}$$

Γ

$V(r)$

$\partial\Gamma$

ψ_0

Figure 6.1 Typical scattering geometry for an object of volume Γ bounded by the surface $\partial\Gamma$. The scattering potential is $V(\mathbf{r})$. The incident (scalar) wave is described by the complex wavefunction ψ_0 and the reradiated (scattered) field is denoted by ψ_{sc}. The total field is given by $\psi = \psi_0 + \psi_{sc}$.

---

[1] Note that the expressions for the quantum and classical scattering potential V as well as the representation of the wavenumber in equation (6.3) must be consistent with the sign convention previously adapted in the definition of the Green function in (2.31).

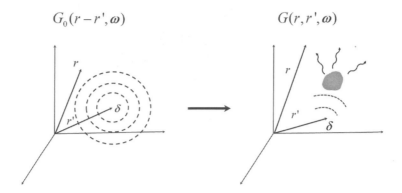

$$G_0(r - r', \omega) \qquad\qquad G(r, r', \omega)$$

Figure 6.2 Schematic representations of (left) the homogeneous medium Green function and (right) total Green function for a nonhomogeneous medium.

6.2   Wave Scattering Formalism

Light scattering by *inhomogeneous linear media* can be rigorously described by the general approach of Green function perturbation theory, which is formally analogous to the many-body perturbation theory of fermions [115, 266]. According to this approach, all the relevant properties of the scattered waves are contained in the retarded Green function associated to the nonhomogeneous medium, also called the full or the *total Green function* $G^+(\mathbf{r}, \mathbf{r}', \omega)$, defined as follows (see Figure 6.2):

$$\boxed{(\nabla^2 + k^2)G^+(\mathbf{r}, \mathbf{r}', \omega) = -\delta(\mathbf{r} - \mathbf{r}')} \tag{6.5}$$

Additionally, in scattering problems the total Green function must satisfy the Sommerfeld radiation condition (SRC) at infinity:

$$\lim_{x \to \infty} r \left[\frac{\partial G^+}{\partial r} - jkG^+ \right] = 0. \tag{6.6}$$

The SRC condition guarantees that no energy is scattered from infinity toward the target object, which is a requirement to solve uniquely the Helmholtz equation.

In Chapter 2, we introduced the free-space retarded Green function $G_0^+(\mathbf{r} - \mathbf{r}', \omega)$ as the field at $\mathbf{r}$ radiated by a point current source (delta function) located at $\mathbf{r}'$ in the infinite homogeneous background medium. The free-space retarded Green function also satisfies the SRC condition at infinity.

Analogously, the *total Green function* $G^+(\mathbf{r}, \mathbf{r}', \omega)$ provides the field at $\mathbf{r}$ radiated by a point source located at $\mathbf{r}'$ *in the nonhomogeneous medium characterized by the scattering potential* $V(\mathbf{r})$. Since the presence of a scattering potential breaks the homogeneity of free-space, the total Green function depends separately on the $\mathbf{r}$ and $\mathbf{r}'$ spatial coordinates.

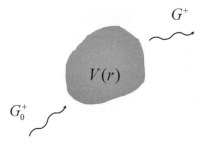

G^+

$V(r)$

G_0^+

Figure 6.3 Schematic representation of the impulse response for a nonhomogeneous medium.

Substituting our definition of the wavenumber (6.3) into equation (6.5), we can write the following:

$$(\nabla^2 + k_0^2)G^+(\mathbf{r},\mathbf{r}',\omega) = -\delta(\mathbf{r} - \mathbf{r}') - V(\mathbf{r})G^+(\mathbf{r},\mathbf{r}',\omega). \tag{6.7}$$

The right-hand side of equation (6.7) contains the contribution of the *primary source* $\delta(\mathbf{r} - \mathbf{r}')$, which corresponds to an incident wave, and of a *secondary source* $V(\mathbf{r})G^+(\mathbf{r},\mathbf{r}',\omega)$ that describes *induced currents in the medium*. These currents act as sources of *secondary radiation* known as the *scattered field*. Using the general Green function technique presented in Chapter 2, we can write the *formal solution* of equation (6.7) as follows:

$$G^+(\mathbf{r},\mathbf{r}',\omega) = G_0^+(\mathbf{r} - \mathbf{r}',\omega) + \int_\Gamma d^3\mathbf{r}_1 G_0^+(\mathbf{r} - \mathbf{r}_1,\omega)V(\mathbf{r}_1)G^+(\mathbf{r}_1,\mathbf{r}',\omega) \tag{6.8}$$

where the integration is over the volume of space Γ occupied by the scattering obstacle rather than over the entire space. This is so because by definition the scattering potential vanishes outside the scattering object. Equation (6.8) is known as the *Lippmann–Schwinger (LS) integral equation* for the total Green function, and can be interpreted as the impulse-response function generated in a scattering experiment whose incident wave is described by the free-space Green function (see Figure 6.3). Clearly, when $\epsilon_r(\mathbf{r}) = 1$ the LS integral equation to $G^+(\mathbf{r},\mathbf{r}',\omega) = G_0^+(\mathbf{r} - \mathbf{r}',\omega)$.

An expression formally analogous to equation (6.8) can be obtained for the *total field* $\psi(\mathbf{r},\omega)$ generated in a scattering experiment by the interaction of the incident wave $\psi_0(\mathbf{r},\omega)$ and the scattering potential $V(\mathbf{r})$. The total field is as follows:

$$\psi(\mathbf{r},\omega) = \psi_0(\mathbf{r},\omega) + \psi_{sc}(\mathbf{r},\omega), \tag{6.9}$$

where $\psi_{sc}(\mathbf{r},\omega)$ denotes the scattered field. *If there are no free sources within the scattering volume* Γ, the total field $\psi(\mathbf{r},\omega)$ must satisfy in Γ the Helmholtz equation:

$$[\nabla^2 + k^2(\mathbf{r})]\psi(\mathbf{r},\omega) = 0, \tag{6.10}$$

and the scattered field must be subjected to the SRC condition at infinity:

$$\lim_{x \to \infty} r \left[\frac{\partial \psi_{sc}}{\partial r} - jk\psi_{sc} \right] = 0. \tag{6.11}$$

Using (6.3), we can rewrite equation (6.10) in the following form:

$$\boxed{\left[\nabla^2 + k_0^2 \right] \psi(\mathbf{r}, \omega) = -V(\mathbf{r})\psi(\mathbf{r}, \omega) \equiv -Q(\mathbf{r}, \omega)} \tag{6.12}$$

which is an *inhomogeneous Helmholtz equation* driven by the unknown *volume source term* $Q(\mathbf{r}, \omega)$. The source term describes, in the case of *penetrable media*, the field-obstacle interaction arising from the penetration of the incident wave into the volume of the scatterer. This approach makes us clearly appreciate the *equivalence between scattering and radiation problems*.[2] According to this picture, we can regard the scattering problem as a *two-step process* where in the first step an incident wave interacts with the scattering potential generating the induced (and unknown) source term $Q(\mathbf{r}, \omega)$, which subsequently radiates the scattered field according to the following:

$$\left[\nabla^2 + k_0^2 \right] \psi_{sc}(\mathbf{r}, \omega) = -Q(\mathbf{r}, \omega). \tag{6.13}$$

This equation follows from the fact that *the incident wave is a homogeneous solution of the Helmholtz equation* and it involves a source term Q that depends on the unknown total field ψ, which can only be obtained by solving the full scattering problem.

Using the standard Green function technique, we can *formally* write the general solution of equation (6.12) as an integral equation *with a convolution kernel*:

$$\boxed{\psi(\mathbf{r}, \omega) = \psi_0(\mathbf{r}, \omega) + \int_\Gamma d^3\mathbf{r}_1 G_0^+(\mathbf{r} - \mathbf{r}_1, \omega)V(\mathbf{r}_1)\psi(\mathbf{r}_1, \omega)} \tag{6.14}$$

where $\psi_0(\mathbf{r}, \omega)$ is a homogeneous solution of the Helmholtz equation corresponding to the incident wave. Equation (6.14) is the *integral equation of potential scattering*. This equation is the LS integral equation for the total field. It must be noted that this equation is a *volume integral equation (VIE)* for the internal field, i.e., valid at points *inside the scattering volume* Γ. Highly specialized and accurate numerical methods are available to solve both scalar and vector integral equations, such as the discrete dipole approximation (DDA) or the method of moments (MoM) [103]. Once the field solution throughout Γ has been found, the solution at points exterior to Γ is obtained by substituting the internal field solution in the integrand of equation (6.14) and then evaluating the volume integral.

The LS integral equations (6.8) and (6.14) govern the scattering from arbitrary *penetrable objects* described by a scattering potential. Solving them for a given incident field and scattering potential defines the *forward scattering problem*. Due to

---

[2] However, in radiation problems the source of the radiated field is known in advance (specified) and it is also assumed to be independent of the radiated field.

the linearity of Maxwell's equations, the forward scattering problem is a linear mapping between the incident wave and the resulting scattered field. On the other hand, the problem of determining the scattering potential from the knowledge of the scattered field, known as the *inverse scattering problem*, is in general a nonlinear one and more difficult to solve. This is so because the scattering field that appears in the LS equation depends in turn on the scattering potential, as shown by equation (6.13). In general, the solution of inverse scattering problems requires the inversion of a set of coupled, nonlinear integral equations for the scattering potential $V(\mathbf{r})$ in terms of the known (i.e., measured) field amplitudes of the total or scattered fields over a given domain or for a set of scattering conditions. Advanced computational techniques have been developed for this task, including optimization and machine learning methods, and we refer interested readers to our cited references on the subject [159, 267–269].

6.2.1 The Multiple Scattering Solution

In the previous section, we have established the two basic equations of potential scattering, which are the LS integral equations for the total Green function and for the total field. Even though these equations can be efficiently solved in very complex three-dimensional geometries by resorting to full-vector numerical methods for integral equations [270], it is still instructive to consider approximate techniques. In particular, we now address in detail the *perturbation approach* originally developed in the context of the quantum scattering theory [90, 115].

According to this method, successive terms in the perturbation expansion of a desired quantity (e.g., the total Green function or the total field) are found by iterations from the preceding ones, until global converge is achieved. Clearly, if the medium scatters weakly (i.e., the refractive index differs only slightly from unity), only few terms will be needed in the perturbative expansion. On the other hand, for strongly scattering media a large number of terms is required, rendering the perturbation method computationally inefficient compared to direct numerical techniques. However, the perturbative approach provides a clear and intuitive physical picture of the general scattering problem and can be formulated on the same footing for both classical and quantum waves. Central to this approach is the use of *diagrammatic representations* that describe different scattering processes and their contributions to desired scattering quantities, similarly to the *Feynman diagrams* widely utilized in the many-body physics of interacting quantum systems and in quantum field theory [113, 115, 271].

We now illustrate how the perturbative method works by focusing, without loss of generality, on the LS field integral equation (6.14). If the scattering is weak, i.e., $|\psi_{sc}(\mathbf{r}, \omega)| \ll |\psi_0(\mathbf{r}, \omega)|$, we can reasonably approximate the total field as follows:

$$\psi(\mathbf{r}, \omega) = \psi_0(\mathbf{r}, \omega) + \psi_{sc}(\mathbf{r}, \omega) \approx \psi_0(\mathbf{r}, \omega) \tag{6.15}$$

and thus obtain the single scattering or the *first-order Born approximation*:

$$\psi_1(\mathbf{r}, \omega) = \psi_0(\mathbf{r}, \omega) + \int_\Gamma d^3\mathbf{r}_1 G_0^+(\mathbf{r} - \mathbf{r}_1, \omega) V(\mathbf{r}_1)\psi_0(\mathbf{r}_1, \omega) \qquad (6.16)$$

This is simply a volume integral that can be computed once the incident wave is known. Moreover, equation (6.12) reduces to the following:

$$\left[\nabla^2 + k_0^2\right]\psi(\mathbf{r}, \omega) = -V(\mathbf{r})\psi_0(\mathbf{r}, \omega). \qquad (6.17)$$

This approximate equation proves the mathematical equivalence between the first-order scattering problem and the radiation with the known *localized source distribution* $\rho(\mathbf{r}, \omega) = V(\mathbf{r}, \omega)\psi_0(\mathbf{r}, \omega)$, where we also made explicit the possible frequency dependence of the scattering potential.

A better approximation of the field solution is obtained if we substitute $\psi_1(\mathbf{r}, \omega)$ for $\psi(\mathbf{r}, \omega)$ in the integrand of the LS equation (6.14). This leads to the *second-order Born approximation*:

$$\psi_2(\mathbf{r}, \omega) = \psi_0(\mathbf{r}, \omega) + \int_\Gamma d^3\mathbf{r}_1 G_0^+(\mathbf{r} - \mathbf{r}_1, \omega) V(\mathbf{r}_1)\psi_1(\mathbf{r}_1, \omega) \qquad (6.18)$$

which can be written explicitly as follows:

$$\psi_2(\mathbf{r}, \omega) = \psi_0(\mathbf{r}, \omega) + \int_\Gamma d^3\mathbf{r}_1 G_0^+(\mathbf{r} - \mathbf{r}_1, \omega) V(\mathbf{r}_1)\psi_0(\mathbf{r}_1, \omega)$$
$$+ \int_\Gamma \int_\Gamma d^3\mathbf{r}_1 d^3\mathbf{r}_2 G_0^+(\mathbf{r} - \mathbf{r}_2, \omega) V(\mathbf{r}_2) G_0^+(\mathbf{r}_1 - \mathbf{r}_2, \omega) V(\mathbf{r}_1)\psi_0(\mathbf{r}_1, \omega). \qquad (6.19)$$

We note that the second term on the right-hand side of this equation is an integral over the scattering volume Γ, while the third term involves an integration performed twice and independently over the same volume. We can continue this iterative procedure by substituting $\psi_2(\mathbf{r}, \omega)$ for $\psi(\mathbf{r}, \omega)$ in equation (6.14) and so on, thus obtaining a sequence of successive (higher-order) approximations:

$$\psi_1(\mathbf{r}, \omega), \psi_2(\mathbf{r}, \omega), \psi_3(\mathbf{r}, \omega), \cdots, \psi_n(\mathbf{r}, \omega), \qquad (6.20)$$

where each term is obtained from the previous one by the *recurrence relation*:

$$\psi_{n+1}(\mathbf{r}, \omega) = \psi_0(\mathbf{r}, \omega) + \int_\Gamma d^3\mathbf{r}_1 G_0^+(\mathbf{r} - \mathbf{r}_1, \omega) V(\mathbf{r}_1)\psi_n(\mathbf{r}_1, \omega) \qquad (6.21)$$

This expression involves integrals performed once, twice, $\cdots$, n times over the scattering volume, resulting in progressively more cumbersome expressions as n increases.

For this reason, such convolution-type integral equations are better expressed in *symbolic form*. As an example, equation (6.19) is rewritten symbolically as follows:

$$\psi_2 = \psi_0 + G_0^+ V \psi_0 + G_0^+ V G_0^+ V \psi_0 \qquad (6.22)$$

where the integrations are implied. The corresponding expression for the general nth term in the field expansion can be written as follows:

$$\psi_n = \psi_0 + \psi_{sc} = \psi_0 + G_0^+ V \psi_0 + G_0^+ V G_0^+ V \psi_0 + \cdots + \underbrace{G_0^+ V \cdots G_0^+ V G_0^+ V}_{n \text{ factors } G_0^+ V} \psi_0.$$

$$(6.23)$$

This solution form is very suggestive because each term in the perturbative expansion has a clear physical significance. For instance, the factor $V(\mathbf{r}_1)\psi_0(\mathbf{r}_1,\omega)d^3\mathbf{r}_1$ that appears in equation (6.19) corresponds to the radiation from the induced currents in the volume element $d^3\mathbf{r}_1$ (centered around point $\mathbf{r}_1$ of the scatterer) created by the interaction of the object with the incident field. The scattering contribution of such an element at a different location $\mathbf{r}$, which can be situated either inside or outside Γ, is obtained by multiplying with the free-space Green function, i.e., by the term $G_0^+(\mathbf{r} - \mathbf{r}_1,\omega)V(\mathbf{r}_1)\psi_0(\mathbf{r}_1,\omega)d^3\mathbf{r}_1$. Therefore, the free-space Green function $G_0^+(\mathbf{r} - \mathbf{r}_1,\omega)$ acts as a *propagator* that transmits the contribution of the *scattered field* from the point $\mathbf{r}_1$ to the observation point $\mathbf{r}$. By integrating over the volume Γ, we thus obtain the total scattering contribution of the particle within the accuracy of the *single-scattering approximation*.

Equation (6.23) shows that in the general case, in addition to the single scattering contribution, there appear higher-order contributions as well. These terms account for the fact that the field scattered once at $\mathbf{r}_1$ can be scattered again from additional points within the object. For instance, the scattering field $G_0^|(\mathbf{r}_2 - \mathbf{r}_1,\omega)V(\mathbf{r}_1)\psi_0(\mathbf{r}_1,\omega)d^3\mathbf{r}_1$, when it reaches another point $\mathbf{r}_2$ inside Γ, can be scattered a second time, giving rise to a contribution at point $\mathbf{r}$, known as *double scattering*, which is $G_0^+(\mathbf{r} - \mathbf{r}_2,\omega)V(\mathbf{r}_2)G_0^+(\mathbf{r}_2 - \mathbf{r}_1,\omega)V(\mathbf{r}_1)\psi_0(\mathbf{r}_1,\omega)d^3\mathbf{r}_1 d^3\mathbf{r}_2$. All the successive terms appearing in equation (6.23) can be similarly interpreted, and the general expression $G_0^+ V \cdots G_0^+ V G_0^+ V$, which consists of n products $G_0^+ V$, represents the multiple scattering processes of order n (see Figure 6.4).

6.2.2   Basic Scattering Parameters

In the analysis of scattering experiments, it is often desirable to describe the rate at which the incident energy is scattered or absorbed by a given obstacle. This can be accomplished by introducing *effective parameters*, known as *cross sections*, that quantify the interaction strength of electromagnetic waves with scattering/absorbing objects.

We will now consider, within the scalar approximation, the *far-field* scattering problem of a dielectric particle, and derive a generally valid expression for the *scattering amplitude*. As we will see, the scattering amplitude contains all the information needed

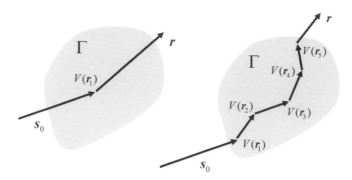

Figure 6.4 Schematic representation of single and multiple scattering events.

to describe the far-field scattering behavior of the particle. Let us start by noticing that the SRC condition (6.11) for the scattered field can equivalently be expressed as follows:

$$\psi_{sc}(\mathbf{r}, \omega) \simeq f(\mathbf{s}, \omega)\frac{e^{jk_0 r}}{r} \qquad r \to \infty \qquad (6.24)$$

where $f(\mathbf{s}, \omega)$ is the *scattering amplitude*[3] (dimensions of a length) in the direction specified by the unit vector $\mathbf{s} \equiv \hat{\mathbf{r}} = \mathbf{r}/r$. A convenient expression for $f(\mathbf{s}, \omega)$ can be immediately found by considering the LS equation (6.14) in the case of an incident plane wave of unit amplitude propagating along the direction $\mathbf{s}_0 = \mathbf{k}_0/k_0$, i.e., $\psi_0(\mathbf{r}) = e^{jk_0\mathbf{s}_0 \cdot \mathbf{r}}$. In this case, we can write equation (6.14) as follows:

$$\psi(\mathbf{s}, \mathbf{s}_0, \omega) = e^{jk_0\mathbf{s}_0 \cdot \mathbf{r}} + \int d^3\mathbf{r}' G_0^+(\mathbf{r} - \mathbf{r}', \omega) V(\mathbf{r}')\psi(\mathbf{r}', \omega), \qquad (6.25)$$

where we have explicitly indicated by $\psi(\mathbf{s}, \mathbf{s}_0, \omega)$ the parametric dependence of the field solution on the direction $\mathbf{s}_0$ of the incident wave. By considering the *far-field region* where $r \gg r'$ (i.e., the $r \to \infty$ limit), we can approximate $|\mathbf{r} - \mathbf{r}'| \sim r - \mathbf{s} \cdot \mathbf{r}'$ (see Figure 6.5) in the exponential of the Green function $G_0^+(\mathbf{r} - \mathbf{r}', \omega)$ and, neglecting $\mathbf{r}'$ in the denominator, we obtain the following:

$$G_0^+(\mathbf{r} - \mathbf{r}', \omega) \sim \frac{e^{jk_0 r}}{4\pi r}e^{-jk_0\mathbf{s} \cdot \mathbf{r}'} \qquad (6.26)$$

Substituting this far-field approximation into the integrand of the LS equation (6.25), we obtain the scattering amplitude for a given incident direction $\mathbf{s}_0$:

$$f(\mathbf{s}, \mathbf{s}_0, \omega) = \frac{1}{4\pi}\int d^3\mathbf{r}' V(\mathbf{r}')\psi(\mathbf{r}', \omega)e^{-jk_0\mathbf{s} \cdot \mathbf{r}'} = \frac{1}{4\pi}\int d^3\mathbf{r}' Q(\mathbf{r}', \omega)e^{-jk_0\mathbf{s} \cdot \mathbf{r}'}$$

$$(6.27)$$

---

[3] Adopting a spherical coordinate system centered on the scatter, the scattering amplitude at a given frequency ω can also be expressed as $f(\theta, \phi)$.

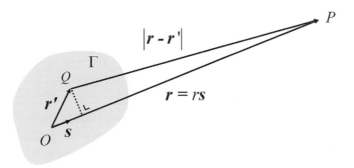

Figure 6.5 Geometry illustrating the far-field approximation for the Green function.

Equation (6.27) shows that, in complete analogy with antenna radiation problems, the scattering amplitude is proportional to the spatial Fourier transform of the induced current source $Q(\mathbf{r}, \omega) = V(\mathbf{r})\psi(\mathbf{r}, \omega)$. The scattering amplitude determines the far-field *radiation pattern* of the sources induced by the incident wave inside the scattering object. Clearly, if the internal field is uniformly distributed inside the object, then the radiation pattern will be directly determined by the geometry of the object (i.e., its shape and dimension), effectively reducing light scattering to Fourier spectral analysis.

Based on the knowledge of the scattering amplitude, other important parameters can be obtained, such as the *differential scattering cross section (dSCS)*:

$$\sigma_d = \frac{d\sigma}{d\Omega} = |f(\mathbf{s}, \mathbf{s}_0, \omega)|^2 \tag{6.28}$$

The dSCS specifies the amount of optical far-field power scattered, per unit incident intensity, within a differential solid angle along the particular scattering direction with angles (θ, ϕ) in a spherical coordinates system centered on the scattering particle. On the other hand, the angularly integrated, or the total *scattering cross section* (in cm^2) is defined as follows:

$$\sigma_s = \int_{4\pi} d\Omega |f(\mathbf{s}, \mathbf{s}_0, \omega)|^2 \tag{6.29}$$

and it represents the ratio of power scattered by the object in the far-field over all the scattering solid angles and the incident power flow per unit area. For electromagnetic waves, the total power P_s scattered in the far-field region can be expressed as follows:

$$P_s = \sigma_s |\mathbf{S}_i|, \tag{6.30}$$

where $\mathbf{S}_i = 1/2\mathrm{Re}(\mathbf{E}_i \times \mathbf{H}_i^*)$ is the time-averaged Poynting vector denoting the incident power flow per unit area.[4] If we divide the scattering cross section of the particle by its

---

[4] The corresponding quantity for scalar wave scattering is the *energy flux vector*
$\mathbf{F} = \alpha\Im\{\psi^*(\mathbf{r}, \omega)\nabla\psi(\mathbf{r}, \omega)\}$, where α is a positive constant and $\Im\{\cdot\}$ denotes the imaginary part of $\{\cdot\}$.

geometrical cross section A (i.e., the projected area perpendicular to the wavevector of the incident wave), we obtain a normalized quantity known as the *scattering efficiency* Q_s:

$$Q_s = \frac{\sigma_s}{A} \tag{6.31}$$

Note that, since the scattering cross section is only an effective area, the scattering efficiency is by no means constrained to assume values smaller than one. On the contrary, due to the possible excitation of the internal resonances of the scattering particles, Q_s may become much larger than one. This is the standard situation in the field of *nanoplasmonics*, where the resonant excitation of surface electron oscillations bound at the metal–dielectric interface of metallic nanoparticles drives values of scattering and absorption cross sections that are much larger than the geometrical cross section of the particles [92, 97–99].

A particle composed of a dispersive material is described by a complex permittivity $\epsilon_p(\mathbf{r}) = \epsilon'_p(\mathbf{r}) + j\epsilon''_p(\mathbf{r})$ and gives rise to power attenuation due to *absorption*. According to Ohm's law, the power absorbed by the particle can be expressed as follows:

$$P_{abs} = \frac{1}{2}\omega \int_\Gamma d^3\epsilon''_p(\mathbf{r})|\psi_{int}(\mathbf{r},\omega)|^2, \tag{6.32}$$

where $\psi_{int}(\mathbf{r},\omega)$ is the internal field that is present within the absorbing particle. Similarly to the scattering cross section, we define the *absorption cross section* as follows:

$$P_{abs} = \sigma_a|\mathbf{S}_i| \tag{6.33}$$

and the *absorption efficiency* Q_a:

$$Q_a = \frac{\sigma_a}{A}. \tag{6.34}$$

For highly transparent particles, the absorption cross section is much smaller than the geometrical cross section, resulting in an absorption efficiency much smaller than one. However, as we noted before in relation to scattering, when internal resonances are excited (e.g., localized surface plasmons in the visible spectral range or phonon-polaritons in the infrared), the values of the absorption efficiency can well exceed unity.

The *extinction cross section* (in cm^2) of the particle accounts for both the effects of scattering and absorption, according to the following:

$$\sigma_{ext} = \sigma_s + \sigma_a, \tag{6.35}$$

and, equivalently, the extinction efficiency Q_{ext} satisfies the following:

$$\boxed{Q_{ext} = Q_s + Q_a} \tag{6.36}$$

Notice that for an object with complex nonspherical shape, the preceding cross sections and efficiencies depend on the direction of incidence.

Finally, in radiative transfer it is useful to define the *albedo* of the particle:

$$\tilde{\omega} = \frac{\sigma_s}{\sigma_{ext}} \tag{6.37}$$

The albedo is a measure of the fraction of scattering compared to total (i.e., extinction) cross section. Energy conservation requires that $0 \leq \tilde{\omega} \leq 1$.

A closely related concept, which is central in the theory of radiative transfer as well as in the statistical description of multiple scattering, is the *phase function* or *phase diagram*, defined as follows:

$$p(\mathbf{s}, \mathbf{s}_0, \omega) = \frac{|f(\mathbf{s}, \mathbf{s}_0, \omega)|^2}{\sigma_{ext}} \tag{6.38}$$

In multiple scattering problems, this function estimates the probability that light incident on a particle from direction $\mathbf{s}_0$ is scattered in the direction $\mathbf{s}$. The phase function satisfies the normalization condition:

$$\int_{4\pi} d\Omega\, p(\mathbf{s}, \mathbf{s}_0, \omega) = \frac{\sigma_s}{\sigma_{ext}} = \tilde{\omega} \tag{6.39}$$

6.2.3   Inverse Scattering in the Born Approximation

We now discuss in more detail the *far-field scattering response of weakly scattering media* within the first-order Born approximation. For media whose refractive index differs only slightly from unity, we can solve analytically the LS integral equations of potential scattering (6.14) by replacing the total field ψ under the integral with the incident field ψ_0. If we consider the incident plane wave $\psi_0(\mathbf{r}) = e^{jk_0\mathbf{s}_0\cdot\mathbf{r}}$, we can write the first-order Born approximation as follows:

$$\psi(\mathbf{s}, \mathbf{s}_0, \omega) = e^{jk_0\mathbf{s}_0\cdot\mathbf{r}} + \int_\Gamma d^3\mathbf{r}'\, V(\mathbf{r}')e^{jk_0\mathbf{s}_0\cdot\mathbf{r}'} G_0^+(\mathbf{r} - \mathbf{r}', \omega). \tag{6.40}$$

Using the far-field approximation for the Green function (6.26), the complex amplitude in the first-order Born approximation for the scattered field ψ_1 becomes

$$\psi(\mathbf{r}) \simeq \psi_1(r\mathbf{s}) = e^{jk_0\mathbf{s}_0\cdot\mathbf{r}} + f_1(\mathbf{s}, \mathbf{s}_0, \omega)\frac{e^{jk_0 r}}{r}, \tag{6.41}$$

where the *first-order scattering amplitude* is defined as follows:

$$f_1(\mathbf{s}, \mathbf{s}_0, \omega) = \int_\Gamma V(\mathbf{r}')e^{-jk_0(\mathbf{s}-\mathbf{s}_0)\cdot\mathbf{r}'} d^3\mathbf{r}' \tag{6.42}$$

In order to clearly understand the consequences of this scattering amplitude, let us introduce the Fourier transform of the scattering potential as follows:

$$\hat{V}(\Delta \mathbf{k}, \omega) = \int_{-\infty}^{\infty} V(\mathbf{r}')e^{-j\Delta \mathbf{k}\cdot\mathbf{r}'}d^3\mathbf{r}', \tag{6.43}$$

where $\Delta \mathbf{k}$ is the Fourier transform variable for reasons that will become apparent in the following. This transform variable corresponds to the *transferred wavevector* (i.e., wavevector difference) according to the following:

$$\boxed{f_1(\mathbf{s}, \mathbf{s}_0, \omega) = \hat{V}(\Delta \mathbf{k}, \omega) = \hat{V}[k_0(\mathbf{s} - \mathbf{s}_0), \omega]} \tag{6.44}$$

Equation (6.44) forms the basis of many inverse scattering techniques [267]. The preceding formula implies that if a scattering object is illuminated using a plane wave directed along $\mathbf{s}_0$ and if the far-field scattered radiation is measured along the direction $\mathbf{s}$, then the complex scattering amplitude $f_1(\mathbf{s}, \mathbf{s}_0, \omega)$ is *uniquely determined by only one Fourier component of the scattering potential*, the one corresponding to the transferred wavevector:

$$\Delta \mathbf{k} = k_0(\mathbf{s} - \mathbf{s}_0) \tag{6.45a}$$

$$|\Delta \mathbf{k}| = 2k_0 \sin(\theta/2), \tag{6.45b}$$

where θ is the angle between the incident and the scattered wavevectors. Therefore, by measuring the scattered complex field (amplitude and phase) along the direction $\mathbf{s}$, it is possible to determine the spectral (Fourier) component of the scattering potential.

On the other hand, when the object is illuminated in direction $\mathbf{s}_0$ and the far-field amplitude measured in all possible directions $\mathbf{s}$, we obtain all the spectral components of $V(\mathbf{r})$ labeled by the wavevectors $\Delta \mathbf{k}$ whose end points lie on a sphere of radius $2\pi/\lambda$ centered at $-k_0\mathbf{s}_0$ (see Figure 6.6). In direct analogy with the theory of X-ray diffraction, this sphere is called the *Ewald's sphere*, and it is illustrated in Figure 6.6.

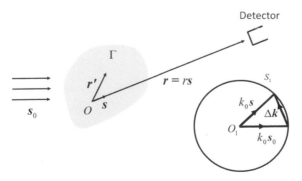

Figure 6.6 Scattering geometry and Ewald's sphere S_1 associated to the incident wavevector $k_0\mathbf{s}_0$. This sphere is the locus of the end points of the vector $\Delta \mathbf{k} = k_0(\mathbf{s} - \mathbf{s}_0)$ for all possible directions $\mathbf{s}$ of the scattered field.

It is interesting to consider now the case where we illuminate the object from another incident direction and measure the scattered far-field *in all possible directions* **s**. With this procedure, we obtain the Fourier components of the scattering potential labeled by wavevectors $\Delta\mathbf{k}$ with end points lying on another Ewald's sphere, indicated by S_2. By scanning all possible incidence directions and measuring the scattered field in all possible directions **s**, we can determine all the Fourier components of the scattering potential labeled by the wavevectors $\Delta\mathbf{k}$ with end points that fill the interior of the sphere S_L of radius $2k_0 = 4\pi/\lambda$, called the *Ewald limiting sphere*, which is shown in Figure 6.7. It is then clear that if we were to measure the scattered far-field for all possible incidence directions and for all possible scattering directions, we would determine all the Fourier components of the scattering potential labeled by the *transferred wavevector* $\Delta\mathbf{k}$ with magnitude:

$$q \equiv |\Delta\mathbf{k}| \leq 2k_0 = \frac{4\pi}{\lambda}, \qquad (6.46)$$

where $\mathbf{q} \equiv \Delta\mathbf{k}$ is also known as the *scattering wavevector*. The preceding inequality sets an upper bound to the achievable spatial resolution in the reconstruction of the scattering potential. In fact, the scattering potential can be approximately reconstructed as follows:

$$\hat{V}_{LP}(\mathbf{r}, \omega) = \frac{1}{(2\pi)^3} \int_{|\mathbf{q}|\leq 2k_0} \hat{V}(\mathbf{q}, \omega) e^{i\mathbf{q}\cdot\mathbf{r}} d^3\mathbf{q}, \qquad (6.47)$$

which is known as the *low-pass filter approximation* of the scattering potential. However, note that the exact scattering potential contains all the Fourier spectral components since the exact inversion of the Fourier transform (6.43) requires us to extend the integration over the entire q-space. We can now clearly appreciate that, due to the assumed far-field approximation of the Green function, expression (6.47)

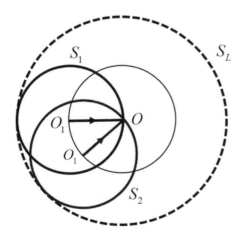

Figure 6.7 Ewald limiting sphere S_L obtained as the envelope of the Ewald's spheres construction showing two spheres $S_1, S_2, \ldots$, associated with all possible incidence directions. The sphere centered at point O is generated by the centers $O_1, O_2, \ldots$, of all Ewald's spheres.

only represents a scattering potential that does not vary appreciably over length scales comparable to the wavelength.[5] The question of extending the spatial bandwidth beyond the *cutoff wavenumber* $2k_0$ in the Fourier synthesis of finitely supported scattering potentials has been addressed intensively using different extrapolation techniques in the context of *super-resolution imaging*. Interested readers can obtain additional information, for instance, in the following papers [189, 272].

6.3   The Optical Theorem

One of the most outstanding results of potential scattering theory is the *optical theorem*, which is encountered in many branches of physics besides optics, such as in mechanics and acoustics, as well as in quantum mechanical, relativistic and nonrelativistic, scattering theory. The theorem is a direct consequence of the conservation of the energy flux in wave scattering and it establishes in very clear terms the importance of *interference effects in scattering phenomena.*

The optical theorem provides a link between the rate at which energy is lost from an incident wave by the combined effects of scattering and absorption processes (i.e., extinction) and the scattering amplitude evaluated in the neighborhood of *only one direction, namely the forward or incidence direction.* For a scattering system illuminated by an incident plane wave, the optical theorem can be stated as follows:

$$\sigma_{ext} = \frac{P_s + P_{abs}}{|\mathbf{S}_i|} = \frac{4\pi}{k}\Im\{f(\mathbf{s}, \mathbf{s}_0, \omega)\} \tag{6.48}$$

where $\Im\{\cdot\}$ as usual denotes the imaginary part of the argument $\{\cdot\}$. This is apparently a surprising result, since extinction is the combined effect of absorption in the particle and scattering by the particle *in all directions.*

The long history of the optical theorem dates back to Lord Rayleigh more than a century ago, and it has been recounted in a classical paper by Roger Newton [273]. In quantum mechanics, a consistent derivation of the optical theorem in the form of equation (6.48) first appeared in a paper by Feenberg [274]. For scalar optical waves, Levine and Schwinger were able to deduce it using a rigorous formulation of diffraction theory [275], while it was van de Hulst [276] who rediscovered the result in 1949 and proposed the intuitive and physically insightful derivation that we summarize as follows.

In van de Hulst's paper, the removal of energy from the incident wave is clearly recognized as a consequence of the *interference* between the incident and the scattered waves "in the neighborhood of the forward direction." In this section, we will consider scalar wavefunctions, which may represent sound waves, electron waves, or any component of the electric or magnetic fields in the scalar optics approximation. However, since light is more generally a vector field described by two orthogonal

---

[5] This result is also consistent with the fact that our analysis has been based on the scalar approximation.

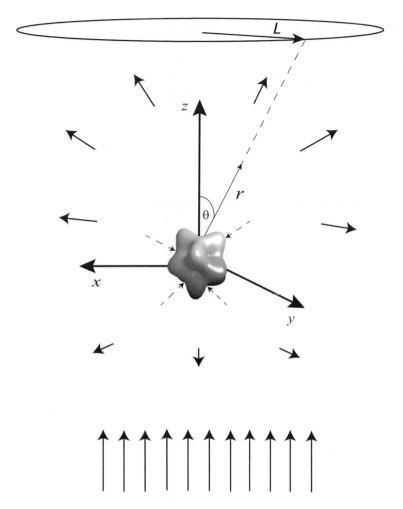

Figure 6.8 Schematics of the scattering geometry of a particle illuminated by an incident plane wave along the z-direction giving rise to scattered waves that are captured by a detector positioned in the far-field zone, several wavelengths away from the object. The energy flow of the incident and scattered waves as well as the energy flow into the particle due to absorption are figuratively indicated by the solid and the dashed arrows.

polarization states, the present approach remains valid only if we assume that the state of polarization is not changed upon scattering and each polarization component can be described independently by the scalar wavefunction.

Let us consider the scattering of scalar waves incident on a single particle, in the presence of a light sensitive detector of circular area located in the far-field zone, as illustrated in Figure 6.8. As we have previously seen, the total scalar field that exists at a distance $r \gg \lambda$ can always be expressed as follows:

$$\psi(r) = \psi_0 + \psi_{sc} = e^{jkz} + f(\theta, \omega)\frac{e^{jkr}}{r}, \tag{6.49}$$

where θ is the angle between the position vector $\mathbf{r}$ and the z-axis. Note that for simplicity we have assumed no azimuthal ϕ dependence. If we restrict our attention to the waves scattered at very small angles (i.e., $\theta \ll 1$) with respect to the incidence direction (i.e., the forward scattering), we can use the well-known *paraxial approximation* $r \simeq z + (x^2 + y^2)/2z$ and express the total far-field intensity as follows:

$$|\psi(r)|^2 = \left| e^{jkz} + f(0, \omega) \frac{e^{jkr}}{r} \right|^2 \simeq 1 + \frac{2}{z} \Re\{ f(0, \omega) e^{jk(x^2 + y^2)/2z} \}, \qquad (6.50)$$

where we neglected the quadratic term in z. Integrating this intensity over the active area A of the detector, we obtain the following:

$$\int |\psi(r)|^2 dx dy = A + \frac{2}{z} \Re\left\{ f(0, \omega) \int e^{jkx^2/2z} dx \int e^{jky^2/2z} dy \right\}. \qquad (6.51)$$

Extending the limits of integration to the interval $(-\infty, +\infty)$, each integral contributes with a term $\sqrt{2\pi z i / k}$ and we obtain the following:

$$\int |\psi(r)|^2 dx dy = A - \frac{4\pi}{k} \Im\{ f(0, \omega) \}. \qquad (6.52)$$

This result shows that the power captured by the active area of the detector is reduced by the presence of the particle. It is clear that this reduction, which is expressed by the second term in equation (6.52), is due to the interference of the incident and the scattered wave in the forward direction, and can be interpreted as if some area of the detector had been covered. Energy conservation dictates that this effect must be accounted for by the extinction cross section σ_{ext} of the particle. Therefore, we can write the following directly:

$$\boxed{ \sigma_{ext} = \frac{4\pi}{k} \Im\{ f(0, \omega) \} } \qquad (6.53)$$

which is the *optical theorem*. Few assumptions are crucial in this derivation of the optical theorem. First, in order for equation (6.52) to be exactly valid, the detector area must be large enough in order to capture all the diffraction peaks[6] and it must be positioned far enough away from the origin in order to guarantee that $\theta \ll 1$. These conditions require that $L^2/z \gg \lambda$ while still $L/z \ll 1$ (i.e., $\theta \ll 1$), where L is the maximum linear dimension of the active area A of the detector. These two conditions can be combined as $\sqrt{\lambda z} \ll L \ll z$, which are compatible since we required $\lambda \ll z$. In addition, this simple derivation requires azimuthal symmetry in the scattering amplitude. Relaxing such a condition does not lead to novel physics, but it involves more complex integrals that can be evaluated based on the Jones lemma, as explained in [43]. Finally, the incident wave must be planar, otherwise the optical theorem as stated in equation (6.53) can fail. This failure was indeed demonstrated in the scattering of Gaussian beams by a spherical particle [277].

---

[6] This is the reason why we extended the range of the integrals to infinity.

6.3.1   The Optical Theorem for Vector Fields

The optical theorem can be extended to include polarization effects in vector scattering and also to multiparticle scattering systems. A particularly clear and insightful physical discussion of this subject is presented in [278, 279]. A rigorous mathematical derivation of the optical theorem for vector waves is based on the Huygens principle and can be found in [132].

The scattering amplitude introduced in Section 6.2.2 must be generalized to include the vector wave nature and polarization effects of the scattered field. Using the far-field expansion of the free-space Green function into the vector form of the LS equation for the total field, it can be rigorously shown [267] that the scattered field at large distances from the object behaves as a transverse spherical wave that propagates outward in the direction of the radial unit vector $\mathbf{s} = \hat{\mathbf{r}}$. In particular, we can express the scattered electric field as follows:

$$\mathbf{E}_{sc}(\mathbf{r}) = \mathbf{F}(\mathbf{s}, \omega)\frac{e^{jk_0 r}}{r} \tag{6.54}$$

where the vector $\mathbf{F}(\mathbf{s}, \omega)$ is independent of r and describes the angular distribution of the scattered radiation (i.e., the radiation pattern) in the far-field zone, and is known as the *vector scattering amplitude*. The vector scattering amplitude satisfies the orthogonality relation $\hat{\mathbf{r}} \cdot \mathbf{F}(\mathbf{s}, \omega) = 0$. In order to describe polarization effects, the incident and the scattered vector fields are separated into two orthogonal polarization components:

$$\mathbf{E}_0(\mathbf{r}) = (\hat{\mathbf{a}}_0 E_{a0} + \hat{\mathbf{b}}_0 E_{b0})e^{jk_0\mathbf{s}_0\cdot\mathbf{r}} \tag{6.55}$$

$$\mathbf{E}_s(\mathbf{r}) = (\hat{\mathbf{a}}_s E_{as} + \hat{\mathbf{b}}_s E_{bs})\frac{e^{jk_0 r}}{r}, \tag{6.56}$$

where $(\mathbf{s}_0, \hat{\mathbf{a}}_0, \hat{\mathbf{b}}_0)$ and $(\mathbf{s}, \hat{\mathbf{a}}_0, \hat{\mathbf{b}}_0)$ form an orthonormal unit vector system that follows the right-hand rule. In any linear scattering experiment, the polarization components of the scattered field must be linearly related to the components of the incident field via the 2×2 *scattering matrix* [132]:

$$\begin{bmatrix} E_{as} \\ E_{bs} \end{bmatrix} = \begin{bmatrix} f_{aa}(\mathbf{s}, \mathbf{s}_0, \omega) & f_{ab}(\mathbf{s}, \mathbf{s}_0, \omega) \\ f_{ba}(\mathbf{s}, \mathbf{s}_0, \omega) & f_{bb}(\mathbf{s}, \mathbf{s}_0, \omega) \end{bmatrix} \begin{bmatrix} E_{a0} \\ E_{b0} \end{bmatrix}. \tag{6.57}$$

In the optical scattering literature [280, 281], the *amplitude scattering matrix* $\mathbf{S}(\mathbf{s}, \mathbf{s}_0, \omega)$ is defined as follows:

$$\begin{bmatrix} E_{as} \\ E_{bs} \end{bmatrix} = \frac{e^{jk_0 r}}{r}\mathbf{S}(\mathbf{s}, \mathbf{s}_0, \omega)\begin{bmatrix} E_{a0} \\ E_{b0} \end{bmatrix}. \tag{6.58}$$

This expression naturally generalizes the scattering amplitude to vector waves. The elements of the amplitude scattering matrix have dimensions of length and can be obtained by contracting the *scattering dyad* with the unit vectors of the chosen orthonormal base [132]. The scattering dyad is the dyadic quantity that relates directly the scattered field with the incident field, and can be expressed in terms of the dyadic transition operator $\overset{\leftrightarrow}{\mathbf{T}}$ using the far-field approximation in equation (12.95). The amplitude scattering matrix generalizes the scattering amplitude $f(\mathbf{s}, \mathbf{s}_0, \omega)$ to the case of vector scattering, and it naturally includes polarization effects. In fact, combinations of its elements can be formed to assemble the 4×4 *Mueller matrix* or *phase matrix* of an arbitrary scattering particle that linearly connects the Stokes parameters of the incident and of the scattered waves [281]. The matrix $\mathbf{S}(\mathbf{s}, \mathbf{s}_0, \omega)$ depends on the size, shape, composition, and orientation of the scattering object with respect to the chosen coordinate system, as well as on the incidence and scattering directions. Typically, when dealing with the scattering of a single particle, two orthonormal unit systems associated to the incident and the scattered fields are defined with respect to the *scattering plane*, which is the plane containing the incident direction $\mathbf{s}_0$ and the scattering direction $\mathbf{s}$. In particular, the incident and scattered vectors can be decomposed into their parallel ($E_{\parallel 0}, E_{\parallel s}$) and perpendicular ($E_{\perp 0}, E_{\perp s}$) components with respect to the scattering plane. The transport of electromagnetic energy can be described by the time-averaged Poynting vector, whose value is directly related to the time-harmonic vector fields as follows:

$$\overline{\mathbf{S}}(\mathbf{r}) = \frac{1}{2} \Re\{\mathbf{E}(\mathbf{r}) \times \mathbf{H}(\mathbf{r})^*\}. \tag{6.59}$$

The component of $\overline{\mathbf{S}}$ directed into the detector and integrated over its sensitive area will provide the detector's time-averaged response to the total electromagnetic wave impinging on it. Since the total fields $\mathbf{E}$ and $\mathbf{H}$ can be decomposed into the sum of their incident ($\mathbf{E}_0, \mathbf{H}_0$) and scattered ($\mathbf{E}_{sc}, \mathbf{H}_{sc}$) waves such as $\mathbf{E} = \mathbf{E}_0 + \mathbf{E}_{sc}$ and $\mathbf{H} = \mathbf{H}_0 + \mathbf{H}_{sc}$, the time-averaged Poynting vector breaks into three distinct components:

$$\boxed{\overline{\mathbf{S}}(\mathbf{r}) = \overline{\mathbf{S}}_0(\mathbf{r}) + \overline{\mathbf{S}}_{sc}(\mathbf{r}) + \overline{\mathbf{S}}_{cross}(\mathbf{r})} \tag{6.60}$$

where we defined the following:

$$\overline{\mathbf{S}}_0(\mathbf{r}) = \frac{1}{2} \Re\{\mathbf{E}_0(\mathbf{r}) \times \mathbf{H}_0^*(\mathbf{r})\} \tag{6.61}$$

$$\overline{\mathbf{S}}_{sc}(\mathbf{r}) = \frac{1}{2} \Re\{\mathbf{E}_{sc}(\mathbf{r}) \times \mathbf{H}_{sc}^*(\mathbf{r})\} \tag{6.62}$$

$$\overline{\mathbf{S}}_{cross}(\mathbf{r}) = \frac{1}{2} \Re\{\mathbf{E}_0(\mathbf{r}) \times \mathbf{H}_{sc}^*(\mathbf{r}) + \mathbf{E}_{sc}(\mathbf{r}) \times \mathbf{H}_0^*(\mathbf{r})\}. \tag{6.63}$$

We observe that the *interference term* $\overline{\mathbf{S}}_{cross}(\mathbf{r})$ describes the fraction of the total power density at $\mathbf{r}$ due to the *interference of the incidence with the scattered waves*.

Let us now take a closer look at the energy balance associated to a homogeneous absorbing particle located (in free space) at the origin of a spherical coordinate system and surrounded by a large imaginary spherical ∂V surface centered at its origin. The surface is located at a large distance (i.e., in the far-field zone) from the particle, so

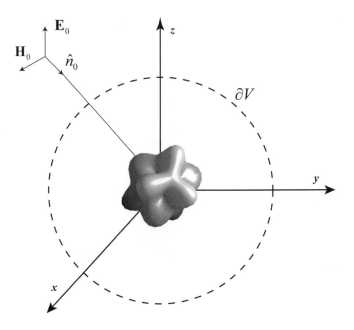

Figure 6.9 Schematics of scattering geometry with an arbitrarily shaped particle centered at the origin of the coordinate system. The particle is assumed to be homogeneous and surrounded by a large spherical surface located in the far-field region.

that the scattered waves can be considered transverse and outward-traveling spherical waves. The typical geometry of light scattering from an arbitrarily shaped particle centered at the origin of the coordinate system is shown in Figure 6.9.

The total power absorbed within the particle is obtained by calculating the *net inward flow* of the radial component of the Poynting vector across ∂V:

$$W_{abs} = -\oint_{\partial V} \overline{\mathbf{S}}(\mathbf{r}) \cdot \hat{\mathbf{n}} dS \qquad (6.64)$$

where $\hat{\mathbf{n}}$ is the unit *outward normal* to the reference sphere ∂V, such that W_{abs} is either positive or zero. Substituting from (6.60), we can equivalently write the following:

$$W_{abs} = W_0 - W_{sc} + W_{cross}, \qquad (6.65)$$

where we defined the following:

$$W_0 = -\oint_{\partial V} \overline{\mathbf{S}}_0(\mathbf{r}) \cdot \hat{\mathbf{n}} dS \qquad (6.66)$$

$$W_{sc} = \oint_{\partial V} \overline{\mathbf{S}}_{sc}(\mathbf{r}) \cdot \hat{\mathbf{n}} dS \qquad (6.67)$$

$$W_{cross} = -\oint_{\partial V} \bar{S}_{cross}(\mathbf{r}) \cdot \hat{n} dS. \tag{6.68}$$

We note that W_0, which gives the net incident power that crosses the surface ∂V, must vanish since the same incident power that enters the volume bounded by ∂V also exits the same volume.[7] Therefore, with $W_0 = 0$ we obtain the following:

$$\boxed{W_{cross} = W_{abs} + W_{sc}} \tag{6.69}$$

which states the *conservation of energy* for the single-particle case. In particular, equation (6.69) shows that W_{cross} yields the *net energy loss*, or extinction, contributed by the power lost due to the absorption within the particle and due to the outward-traveling scattered radiation. As already discussed, we can define the scattering, absorption, and extinction cross sections $\sigma_{sc}, \sigma_{abs}, \sigma_{ext}$ by simply dividing the corresponding (far-field) powers $W_{sc}, W_{abs}, W_{cross}$ by the power density (i.e., the irradiance or intensity) of the incident wave $I_0 = |\bar{S}_0| = (\epsilon_0/2\mu_0)^{1/2}|\mathbf{E}_0(\mathbf{r})|^2$ (time-averaged power flow per unit area). Note that the definition of extinction given by (6.69) relies on surface integrals that explicitly depend on all directions, without any apparent requirement for the energy flow along the forward direction.

The optical theorem for vector waves can be derived directly by evaluating, in the asymptotic limit $kr \to \infty$, the integral follows:

$$W_{cross} = -\frac{1}{2}\Re\oint_{\partial V} \{\mathbf{E}_0(\mathbf{r}) \times \mathbf{H}_{sc}^*(\mathbf{r}) + \mathbf{E}_{sc}(\mathbf{r}) \times \mathbf{H}_0^*(\mathbf{r})\} \cdot \hat{n} dS. \tag{6.70}$$

Assuming plane-wave excitation, the integral above can be asymptotically evaluated using the *method of stationary phase* (i.e., Jones lemma), as detailed in [43, 281], or by using the expansion of plane waves inside the expression (6.70) in terms of the complete base of spherical harmonics, as in [278]. Regardless of the specific derivation method, the resulting expression of the extinction cross section $\sigma_{ext} = W_{cross}/I_0$ in the $kr \to \infty$ limit becomes

$$\boxed{\sigma_{ext} = \frac{4\pi}{k|\mathbf{E}_0(\mathbf{r})|^2}\Im\{\mathbf{E}_0^* \cdot \mathbf{F}(\hat{n}_0, \omega)\}} \tag{6.71}$$

Equation (6.71) is the *optical theorem for vector waves*, which relates the extinction cross section to the imaginary part of the scattering amplitude evaluated in the forward direction $\hat{n}_0$. *Note that this theorem is rigorously valid only for $kr \to \infty$ and approximately valid in the far-field.* Moreover, equation (6.71) holds true only if the incident wave is planar [277]. However, the preceding derivation of the optical theorem clearly illustrates how *the removal of energy in all directions from an incident wave originates from its interference with the scattered wave limited to the neighborhood of the forward direction.*

---

[7] This is true if we assume no absorption in the embedding medium.

Box 6.1 The Ewald–Oseen Extinction Theorem

A very important result of classical optics that further emphasizes the fundamental role of interference in the complex response of materials is the Ewald–Oseen extinction theorem, often referred to as simply the *extinction theorem*. This theorem, first introduced for isotropic media by Paul Peter Ewald and Carl Wilhelm Oseen in 1915 and 1916, rigorously explains light propagation through a transparent medium, where the speed of light is reduced according to the index of refraction of the medium. This is a familiar result when analyzed in terms of macroscopic quantities such as the refractive index, but it is actually very surprising when considered from a microscopic viewpoint. In fact, according to the superposition principle of linear optics, the light field inside any refractive medium is the superposition of the original (incident or primary) wave field and the wave fields radiated by each of the microscopic constituents of the medium. Each wave in the medium is traveling in free space and at the speed of light. It is only when these waves are added that, quite surprisingly, they create a wave that travels at the reduced light speed inside the medium. The Ewald–Oseen extinction theorem explains this intriguing phenomenon by showing how the secondary fields emitted by the microscopic constituent of matter exactly cancel out, or extinguish, the primary incident wave and also produce a component that travels at the slower speed expected for the medium. The net effect is that the wave component that survives in the medium propagates at a slower speed, exactly in agreement with the predictions of the macroscopic theory based on the refractive index concept. The Ewald–Oseen extinction theorem is formulated in terms of integral equations, and it also provides a rigorous foundation for the Lorentz–Lorenz formula as well as for the Fresnel formulae of light refraction/reflection at planar interfaces [43]. Although the mathematical proof of this theorem is too complex to be summarized here, very accessible accounts can be found in the papers by Reali [282] and Ballenegger [283].

The apparent paradoxical nature of this result, which does not explicitly involve absorption, can be resolved by remembering that the dependence of extinction on particle's absorption is hidden in the scattering amplitude $\mathbf{F}(\hat{\mathbf{n}}_0, \omega)$. In fact, the volume integral equation approach developed in Section 6.2.2, and its vector generalization, clearly show that the scattering amplitude depends on the spatial distribution of the particle's internal electric field. In turn, this field is directly affected by the absorptive properties of the scattering particles. In any case, equations (6.68) and (6.69) indicate that the extinction caused by the particle is given by the the interference energy flow component that crosses the far-field reference sphere in all directions. Since the relative phase between waves changes with direction and distance from the scattering particle, the waves interfere over all directions, producing a pattern of alternating radial inward and outward flow of $\overline{S}_{cross}(\mathbf{r})$. In [278, 279], it is clearly shown that this alternating radial energy flow between inward and outward directions has an angular distribution that changes with the distance from the particle in every direction and gives rise to *intensity cancellation when integrated along all directions except in the exact forward direction*. Therefore, the forward-direction dependence implicit in the optical theorem simply reflects the incomplete cancellation of the directions (i.e., angular regions) of opposite radial flow located in the neighborhood of the forward direction.[8] Finally we observe that an analogous interference mechanism is responsible for the extinction of multiparticle systems and particle clusters, where it has been shown that the dominant contribution to the extinction cross section occurs within a much narrower angular region around the forward direction compared to the case of a single particle [279].

6.4   The Description of Partially Coherent Waves

The modern theory of optical coherence consists in the rigorous study of the statistical properties of light. This field of statistical optics, founded on the mathematical theory of stochastic processes and random fields, enables the manipulation of light scattering and interference phenomena in optical environments subject to random fluctuations. Randomness in optical fields can originate due to the unpredictable fluctuations in the light generation processes (e.g., thermal light) or in the medium through which light waves propagate or scatter. For instance, natural light generated by the sun is statistically described as a random field because it originates from a large number of independently emitting atoms with different frequencies and phases. Stochastic light fields are also produced by the scattering from rough surfaces or turbulent media, such as the Earth's atmosphere, which impart random variations to propagating wavefronts. In the next sections, we introduce fundamental concepts in the *theory of optical coherence* with applications to light scattering theory.

6.4.1   Mathematical Description of Optical Coherence

The concept of optical coherence is central to the field of *statistical optics* and is formulated in terms of spatial correlation functions. The basic quantity in the so-called second-order coherence theory[9] is the *mutual coherence function* $\Gamma(\mathbf{r}_1, \mathbf{r}_2, \tau)$, which is defined for a wide-sense stationary random optical field as follows [88, 284, 285]:

$$\Gamma(\mathbf{r}_1, \mathbf{r}_2, \tau) = \langle u^*(\mathbf{r}_1, t_1) u(\mathbf{r}_2, t_2) \rangle, \tag{6.72}$$

---

[8] As a consequence, the angular size of the of the detector in the far-field also plays a central role in determining the cancellation that leads to the correct value of extinction. In fact, whether or not the detector in the far-field receives a reduced amount of power depends sensibly on the detector's angular size.

[9] The term "second-order" refers to the fact that it involves the product of a field variable at two points.

where $\tau = t_2 - t_1$ and the field $u(\mathbf{r},t)$ is assumed to be a scalar analytic signal. The angle bracket denotes time averaging. The approach can be extended to vector fields, as shown in [285] and discussed in Section 6.4.4. It is evident from its definition that the mutual coherence function is a quantitative descriptor of both the spatial and temporal fluctuations of the random field $u(\mathbf{r},t)$ at pairs of points $\mathbf{r}_1$ and $\mathbf{r}_2$ for a fixed time delay τ, i.e., it is a cross-correlation function.

A closely related quantity is the *complex degree of coherence*:

$$\gamma(\mathbf{r}_1, \mathbf{r}_2, \tau) = \frac{\Gamma(\mathbf{r}_1, \mathbf{r}_2, \tau)}{\sqrt{\Gamma(\mathbf{r}_1, \mathbf{r}_1, 0)}\sqrt{\Gamma(\mathbf{r}_2, \mathbf{r}_2, 0)}} = \frac{\Gamma(\mathbf{r}_1, \mathbf{r}_2, \tau)}{\sqrt{I(\mathbf{r}_1)}\sqrt{I(\mathbf{r}_2)}}, \tag{6.73}$$

where $I(\mathbf{r}_1)$ and $I(\mathbf{r}_2)$ are the light intensities at points $\mathbf{r}_1$ and $\mathbf{r}_2$, respectively. It follows from (6.73) and from the general property in Appendix C.24 of the cross-correlation function that the complex degree of coherence is bounded as follows:

$$\boxed{0 \leq |\gamma(\mathbf{r}_1, \mathbf{r}_2, \tau)| \leq 1} \tag{6.74}$$

where the value zero represents a complete absence of correlations (i.e., complete incoherence) and unity represents total correlation (i.e., complete coherence) between the vibrations of the waves at points $\mathbf{r}_1$ and $\mathbf{r}_2$, respectively. Note that the mutual coherence function and the complex degree of coherence are quantitative measures of both temporal and spatial coherence, often discussed as separate concepts. Specifically, the temporal coherence of a light source is described by the *equal-space complex degree of coherence* $\gamma(\mathbf{r}, \mathbf{r}, \tau)$, while the spatial coherence is captured by the *equal-time complex degree of coherence* $\gamma(\mathbf{r}_1, \mathbf{r}_2, \tau)$, where τ is considered as a fixed delay, often assumed to be zero.

The absolute value $|\gamma(\mathbf{r}, \mathbf{r}, \tau)|$ can be directly measured using a Michelson interferometer. The two arms of the interferometer are varied in length to study the field correlation at the same point in space but delayed in time. In particular, $\mathbf{r}$ lies on the central beamsplitter and τ is the time delay introduced between the two interfering beams by displacing, by a distance $c\tau/2$, one of the two mirrors from their original symmetric positions.

On the other hand, the spatial coherence of light at two points $\mathbf{r}_1$ and $\mathbf{r}_2$, which is characterized by $\gamma(\mathbf{r}_1, \mathbf{r}_2, \tau)$, can be measured using the Young's double-slit interference experiment. In fact, the near-axis interference fringes observed on the detection screen manifest the equal-time correlation between the fields at the two slits. A detailed analysis of this experimental configuration [285] leads to conclude that the intensity contrast, or the *visibility* $V(\mathbf{r})$ of the observed interference fringes is directly related to the absolute value of the complex degree of coherence as follows:

$$\boxed{V(\mathbf{r}) \equiv \frac{I_{max}(\mathbf{r}) - I_{min}(\mathbf{r})}{I_{max}(\mathbf{r}) + I_{min}(\mathbf{r})} = |\gamma(\mathbf{r}_1, \mathbf{r}_2, \tau)|} \tag{6.75}$$

Moreover, the phase of the complex degree of coherence can be experimentally determined by measuring the positions of the intensity maxima in the interference fringe pattern [43].

The complex degree of coherence determines the conditions for the observability of interference phenomena between partially coherent waves. In particular, if we consider the geometry of the double-slit Young's interference experiment, we can easily write the field intensity observed on a point P of the diffraction screen located at $\mathbf{r}$ due to the radiation of the two apertures located at $\mathbf{r}_1$ and $\mathbf{r}_2$ as follows:

$$I(\mathbf{r}) = \langle |k_1 u(\mathbf{r}_1, t - R_1/c) + k_2 u(\mathbf{r}_2, t - R_2/c)|^2 \rangle, \tag{6.76}$$

where $k_i = 1/(j \bar{\lambda} R_i)$ and $R_i = |\mathbf{r}_i - \mathbf{r}|$ with $i = 1, 2$. Here $\bar{\lambda}$ denotes the average wavelength of the incident light. Assuming stationary optical fields, it is then possible to compute the ensemble average and obtain the interference law as follows [285]:

$$\boxed{I(\mathbf{r}) = I_1(\mathbf{r}) + I_2(\mathbf{r}) + 2\sqrt{I_1(\mathbf{r})}\sqrt{I_2(\mathbf{r})}\,\Re\{\gamma(\mathbf{r}_1, \mathbf{r}_2, \tau)\}} \tag{6.77}$$

This equation shows that in order to obtain the average intensity at a point $\mathbf{r}$ in the observation plane, one needs to know, in addition to the average intensity of the two interfering beams $I_1(\mathbf{r})$ and $I_2(\mathbf{r})$, also the *statistical correlations* of the fields at the two pinholes, which are described by the real part of the complex degree of coherence at the locations of the pinholes. It also follows that in order to observe interference phenomena the radiation does not need to be purely monochromatic. It is only necessary that the interfering beams possess a nonzero degree of statistical similarity as quantified by the complex degree of coherence.

In situations involving symmetry in the experimental setup, such as in the description of optical images that form near the axis of a centered optical system, the effects of coherence in stationary fields can be described by considering the *mutual intensity function*:

$$J(\mathbf{r}_1, \mathbf{r}_2) = \langle u^*(\mathbf{r}_1, t) u(\mathbf{r}_2, t) \rangle = \Gamma(\mathbf{r}_1, \mathbf{r}_2, 0) \tag{6.78}$$

or its normalized version, the *equal-time complex degree of coherence*:[10]

$$j(\mathbf{r}_1, \mathbf{r}_2) = \gamma(\mathbf{r}_1, \mathbf{r}_2, 0) = \frac{J(\mathbf{r}_1, \mathbf{r}_2)}{\sqrt{I(\mathbf{r}_1)}\sqrt{I(\mathbf{r}_2)}}, \tag{6.79}$$

where $I(\mathbf{r}_i) = \langle u^*(\mathbf{r}_i, t) u(\mathbf{r}_i, t) \rangle$ $(i = 1, 2)$ denotes the average intensities at the points $\mathbf{r}_i$. It follows directly from the mathematical representation of quasi-monochromatic (i.e., narrowband) light that the mutual coherence function and the complex degree of coherence of narrowband light with central frequency ω_0 can be written as follows:

$$\Gamma(\mathbf{r}_1, \mathbf{r}_2, \tau) \simeq J(\mathbf{r}_1, \mathbf{r}_2) e^{-i\omega_0 \tau} \tag{6.80}$$

$$\gamma(\mathbf{r}_1, \mathbf{r}_2, \tau) \simeq j(\mathbf{r}_1, \mathbf{r}_2) e^{-i\omega_0 \tau}, \tag{6.81}$$

provided that the time delay τ is smaller than the coherence time, i.e., for $|\tau| \ll 2\pi/\Delta\omega$. We remark that in general situations, the temporal and spatial coherence of

---

[10] This quantity is also known as the complex coherence factor.

an optical field cannot be considered independently of each other. In fact, as shown by Wolf in 1955 [286], the mutual coherence function satisfies the wave equations:

$$\left(\nabla_1^2 - \frac{1}{c^2}\frac{\partial^2}{\partial\tau^2}\right)\Gamma(\mathbf{r}_1,\mathbf{r}_2,\tau) = 0 \tag{6.82}$$

$$\left(\nabla_2^2 - \frac{1}{c^2}\frac{\partial^2}{\partial\tau^2}\right)\Gamma(\mathbf{r}_1,\mathbf{r}_2,\tau) = 0, \tag{6.83}$$

where ∇_i^2 with $i = 1, 2$ is the Laplacian with respect to the Cartesian coordinates of the position vector $\mathbf{r}_i$ and c is the speed of light. Therefore, the statistical properties of light, i.e., their spatial and temporal behavior, evolve as intrinsically coupled wave quantities upon propagation.

6.4.2 Coherence Time, Coherence Length, and Spectral Width

Based on the coherence functions introduced in the preceding section, we can precisely formulate the temporal coherence properties of a stationary light source. These are measured by the *normalized autocorrelation function*:

$$g(\tau) = \frac{G(\tau)}{G(0)} \equiv \frac{\gamma(\mathbf{r},\mathbf{r},\tau)}{\gamma(\mathbf{r},\mathbf{r},0)} = \frac{\langle u^*(t)u(t+\tau)\rangle}{\langle u^*(t)u(t)\rangle}. \tag{6.84}$$

The nonnormalized temporal autocorrelation function $G(\tau) \equiv \Gamma(\mathbf{r},\mathbf{r},\tau)$ is also known as the *temporal coherence function*. Here we assumed that the fields are detected at a fixed position $\mathbf{r}$, which has been dropped for convenience. This situation is common in many interferometric experiments where two light beams originate from the same source and are superimposed at an arbitrary position $\mathbf{r}$ while differing only by a constant optical path (or delay). This quantity is called the *complex degree of temporal coherence*, and its absolute value cannot exceed unity:

$$0 \le |g(\tau)| \le 1. \tag{6.85}$$

The quantity $|g(\tau)|$ provides a quantitative measure of the degree of temporal correlation between the field $u(t)$ and $u(t + \tau)$. As an example, we can see that for a deterministic and monochromatic field $u = A\exp(-j\omega_0 t)$ (where A is a constant), we have $|g(\tau)| = 1$ for all values of τ. Therefore, this field is completely temporally coherent and the values of $u(t)$ and $u(t + \tau)$ are perfectly correlated for all times. On the other hand, for partially coherent light (i.e., nonmonochromatic light), $|g(\tau)|$ will decrease with τ from its maximum value at $\tau = 0$ because the field fluctuations progressively lose their correlation. A good indicator of temporal coherence is given by the time τ_c at which the $|g(\tau)|$ drops below a prescribed value, typically $1/e$ of the peak value. The time τ_c, which characterizes the width of the function $|g(\tau)|$, is called the *coherence time* and describes the memory time of the fluctuations of the field. When $\tau < \tau_c$, the field fluctuations are strongly correlated, while for $\tau > \tau_c$ they are weakly correlated. Clearly the coherence time of a monochromatic light field is equal to infinity since $|g(\tau)|$ never decreases. When the differences of the time delays experienced by a light field in an optical system are much smaller than the coherence time τ_c we can, to all practical purposes, consider that light field as coherent. This occurs when the distance $\ell_c = c\tau_c$ (where c is the light speed), known as the *coherence length*, is much larger than all optical the path differences encountered by light in the system. For example, in an experiment with two interfering beams, interference appears if the relative path difference does not exceed ℓ_c.

A direct application of the Wiener–Khintchine theorem (C.30) to the temporal coherence function $G(\tau)$ allows us to write the *power spectral density*, or the *power spectrum*, as follows:

$$S(\omega) = \int_{-\infty}^{+\infty} G(\tau)e^{i\omega\tau}d\tau \tag{6.86}$$

Since we are dealing with stationary random functions that extend everywhere in time, the spectrum exists only in the sense discussed in Appendix C.3.

Expression (6.86) allows us to introduce the linewidth, or the *spectral width* of light, as the width of the spectral density in frequency domain. Since $S(\omega)$ and $G(\tau)$ form a Fourier transform pair, their widths are always inversely related. Therefore, a light source with a broad frequency spectrum has a short coherence time, while narrowband light has necessarily a very long coherence time. Since the definition of the width of line shape function is rather arbitrary, the relation between the spectral width and the coherence time depends on the actual spectral profile. When the power spectrum of the field has a Gaussian line shape with frequency bandwidth $\Delta\nu$ (remember that $\omega = 2\pi\nu$) corresponding to its full width at half-maximum (FWHM), we have [78]:

$$\tau_c \simeq \frac{0.664}{\Delta\nu} \tag{6.87}$$

However, if we define the spectral width by

$$\Delta\nu \equiv \frac{\left[\int_0^\infty S(\nu)d\nu\right]^2}{\int_0^\infty S^2(\nu)d\nu}, \tag{6.88}$$

we obtain a simple inverse relation between the spectral width and the coherence time:

$$\Delta\nu = \frac{1}{\tau} \tag{6.89}$$

regardless of the spectral profile of the source. In this case, the coherence time must be defined as follows:

$$\tau_c = \int_{-\infty}^{+\infty} |g(\tau)|^2 d\tau. \tag{6.90}$$

The last expression is known as the *power-equivalent width* of $|g(\tau)|$.

6.4.3 Optical Coherence in the Space-Frequency Domain

The behavior of partially coherent fields can be formulated directly in the frequency domain. This approach also provides the opportunity to clarify few conceptual issues that we left pending from the previous section.

Let us introduce the *cross-spectral density function* $W(\mathbf{r}_1, \mathbf{r}_2, \omega)$, defined as the Fourier transform of the mutual coherence function:

$$W(\mathbf{r}_1, \mathbf{r}_2, \omega) = \frac{1}{2\pi} \int_{-\infty}^{+\infty} \Gamma(\mathbf{r}_1, \mathbf{r}_2, \tau) e^{-j\omega\tau} d\tau. \tag{6.91}$$

The cross-spectral density satisfies the pair of Helmholtz equations:

$$(\nabla_1^2 + k^2) W(\mathbf{r}_1, \mathbf{r}_2, \omega) = 0 \tag{6.92}$$

$$(\nabla_2^2 + k^2) W(\mathbf{r}_1, \mathbf{r}_2, \omega) = 0, \tag{6.93}$$

where the wavenumber $k = \omega/c$.

A direct generalization of the Karhunen–Loéve expansion (discussed in Appendix C.5) allows us to expand the cross-spectral density as [285]:

$$W(\mathbf{r}_1, \mathbf{r}_2, \omega) = \sum_n \lambda_n(\omega) \phi_n^*(\mathbf{r}_1, \omega) \phi_n(\mathbf{r}_2, \omega), \tag{6.94}$$

where the ϕ_n are an orthonormal set of eigenfunctions and λ_n are the eigenvalues of the integral equation:

$$\int_\Omega W(\mathbf{r}_1, \mathbf{r}_2, \omega) \phi_n(\mathbf{r}_1, \omega) d^3 \mathbf{r}_1 = \lambda_n(\omega) \phi_n(\mathbf{r}_2, \omega), \tag{6.95}$$

where Ω is a closed domain in free space. The preceding result is also known as the *coherent mode representation* of the cross-spectral density.

Importantly, the coherent mode representation enables the construction of an ensemble $\{u(\mathbf{r}, \omega)\}$ of sample functions in the frequency domain of the following form:

$$u(\mathbf{r}, \omega) = \sum_n a_n(\omega) \phi_n(\mathbf{r}, \omega), \tag{6.96}$$

where the $a_n(\omega)$ are delta-correlated random coefficients [285]. Computing the correlation function $\langle u^*(\mathbf{r}_1, \omega) u(\mathbf{r}_2, \omega) \rangle$ and using the expansion (6.96) as well as (6.94), we can establish the following:

$$\boxed{W(\mathbf{r}_1, \mathbf{r}_2, \omega) = \langle u^*(\mathbf{r}_1, \omega) u(\mathbf{r}_2, \omega) \rangle} \tag{6.97}$$

where the angle bracket indicates an average taken over an ensemble of space-frequency realizations. Therefore, the cross-spectral density of the field can be expressed as the correlation function of an ensemble $\{u(\mathbf{r}, \omega)\}$ of monochromatic realizations $u(\mathbf{r}, \omega)$ in the space-frequency domain.

An important derived quantity is the *spectral density* $S(\mathbf{r}, \omega)$ of the fluctuating random field, which is defined as follows:

$$S(\mathbf{r}, \omega) \equiv W(\mathbf{r}, \mathbf{r}, \omega) = \langle u^*(\mathbf{r}, \omega) u(\mathbf{r}, \omega) \rangle. \tag{6.98}$$

The spectral density represents the average intensity of the wave field at position $\mathbf{r}$ and at frequency ω. Note that in the preceding expression the $u(\mathbf{r}, \omega)$ should be regarded as sample functions (realizations) of a stationary random process and not as the Fourier frequency components of a random field. Those in fact do not exist since sample functions extend everywhere in time and are not integrable. Therefore, the quantity $u(\mathbf{r}, \omega)$ must be interpreted as the space-dependent envelope of a particular member of the *statistical ensemble of monochromatic realizations*, i.e., $\{u(\mathbf{r}, t) = u(\mathbf{r}, \omega)e^{-j\omega t}\}$, all with the same frequency.

The cross-spectral density is also commonly expressed in terms of the spectral density and the *spectral degree of coherence* $\mu(\mathbf{r}_1, \mathbf{r}_2, \omega)$ as follows:[11]

$$W(\mathbf{r}_1, \mathbf{r}_2, \omega) = \sqrt{S(\mathbf{r}_1, \omega)}\sqrt{S(\mathbf{r}_2, \omega)}\mu(\mathbf{r}_1, \mathbf{r}_2, \omega). \tag{6.99}$$

Schwartz's inequality can be used to show that the absolute value of the spectral degree of coherence is restricted within the following range:

$$0 \leq |\mu(\mathbf{r}_1, \mathbf{r}_2, \omega)| \leq 1, \tag{6.100}$$

and thus provides a measure of the degree of correlation of the field at the two positions $\mathbf{r}_1$ and $\mathbf{r}_1$ at a given frequency ω. The value zero corresponds to the complete lack of coherence, and the value of unity represents full spatial coherence at frequency ω.

We now apply the preceding results to the simple case of a *polychromatic plane wave* propagating along the direction specified by the unit vector $\mathbf{s}_0$. According to our discussion on the coherent mode representation, we can express this wave by the following statistical ensemble:

$$\{u(\mathbf{r}, \omega)\} = \{a(\omega)\}e^{jk\mathbf{s}_0 \cdot \mathbf{r}}, \tag{6.101}$$

where $a(\omega)$ is a random complex envelope. Substituting into (6.97), we obtain the following for the cross-spectral density function:

$$W(\mathbf{r}_1, \mathbf{r}_2, \omega) = S(\omega)e^{jk\mathbf{s}_0 \cdot (\mathbf{r}_2 - \mathbf{r}_1)}, \tag{6.102}$$

where $S(\omega) = \langle a^*(\omega)a(\omega) \rangle$ represents the frequency spectrum of the wave. We deduce from (6.99) that the spectral degree of coherence of a polychromatic plane wave is as follows:

$$\mu(\mathbf{r}_1, \mathbf{r}_2, \omega) = e^{jk\mathbf{s}_0 \cdot (\mathbf{r}_2 - \mathbf{r}_1)}, \tag{6.103}$$

and since $|\mu(\mathbf{r}_1, \mathbf{r}_2, \omega)| = 1$ for all pairs of points, the polychromatic plane wave is completely *spatially coherent* at the frequency ω throughout all space.

---

[11] The quantities $W(\mathbf{r}_1, \mathbf{r}_2, \omega)$, $S(\mathbf{r}, \omega)$ and $\mu(\mathbf{r}_1, \mathbf{r}_2, \omega)$ are always zero for negative frequencies since they are defined for analytic signals.

As we discussed in the previous section, the mutual coherence function contains information on both spatial and temporal coherence effects. However, under special circumstances the two effects can be separated out, and the mutual coherence function can be factored as follows:

$$\Gamma(\mathbf{r}_1, \mathbf{r}_2, \tau) = J(\mathbf{r}_1, \mathbf{r}_2)\gamma(\tau), \tag{6.104}$$

where we introduced the *complex degree of self-coherence* $\gamma(\tau)$, which is independent of the spatial coordinates. It is interesting to ask under which conditions the mutual coherence can be factored in this way. To this purpose, we transform (6.104) into the spectral domain, which yields the following:

$$W(\mathbf{r}_1, \mathbf{r}_2, \omega) = J(\mathbf{r}_1, \mathbf{r}_2)\hat{\gamma}(\omega) = j(\mathbf{r}_1, \mathbf{r}_2)\sqrt{I(\mathbf{r}_1)}\sqrt{I(\mathbf{r}_2)}\hat{\gamma}(\omega), \tag{6.105}$$

where $\hat{\gamma}(\omega)$, which determines the normalized spectral lineshape of $W(\mathbf{r}_1, \mathbf{r}_2, \omega)$, is the Fourier transform of $\gamma(\tau)$. Equation (6.105) implies that the mutual coherence function can be separated into spatial and temporal components provided that the cross-spectral density has a normalized line shape $\hat{\gamma}(\omega)$ that does not vary across the spatial domain of interest.

6.4.4 Unified Theory of Coherence and Polarization

The rich subjects of coherence and polarization of waves have been traditionally considered independent topics in optical physics and engineering. However, it was not until very recently that these two subjects were recognized as manifestations of the same physical phenomenon [287], i.e., the correlation between fluctuations in a light beam.

In this section, we introduce some of the important concepts of the *unified theory of coherence and polarization* developed by Emil Wolf in [287, 288] and beautifully introduced in [285]. According to this theory, the state of coherence and polarization of a random wave are coupled and characterized by the *electric cross-spectral density matrix*:

$$\mathbf{W}(\mathbf{r}_1, \mathbf{r}_2, \omega) = \begin{bmatrix} W_{xx}(\mathbf{r}_1, \mathbf{r}_2, \omega) & W_{xy}(\mathbf{r}_1, \mathbf{r}_2, \omega) \\ W_{yx}(\mathbf{r}_1, \mathbf{r}_2, \omega) & W_{yy}(\mathbf{r}_1, \mathbf{r}_2, \omega) \end{bmatrix} \tag{6.106}$$

where entries defined as follows:

$$W_{ij}(\mathbf{r}_1, \mathbf{r}_2, \omega) = \langle E_i^*(\mathbf{r}_1, \omega)E_j^*(\mathbf{r}_2, \omega)\rangle, \tag{6.107}$$

where $i, j = x, y$ and $E_i(\mathbf{r}, \omega)$ denotes a Cartesian component of the electric field realization at a point $\mathbf{r}$ and frequency ω that belongs to the monochromatic ensemble representing a stochastic beam.[12] The matrix (6.106), which entirely describes both

---

[12] This situation describes a stochastic beam by an ensemble of monochromatic realizations each with two orthogonal components in the far-field region.

the phenomena of coherence and polarization in a light beam, can be obtained from the Fourier transform of the *mutual coherence matrix*:

$$\mathbf{\Gamma}(\mathbf{r}_1, \mathbf{r}_2, \tau) = \begin{bmatrix} \langle E_x^*(\mathbf{r}_1, t) E_x(\mathbf{r}_2, t + \tau) \rangle & \langle E_x^*(\mathbf{r}_1, t) E_y(\mathbf{r}_2, t + \tau) \rangle \\ \langle E_y^*(\mathbf{r}_1, t) E_x(\mathbf{r}_2, t + \tau) \rangle & \langle E_y^*(\mathbf{r}_1, t) E_y(\mathbf{r}_2, t + \tau) \rangle \end{bmatrix}. \tag{6.108}$$

In complete analogy with the case of stochastic scalar fields and equation (6.97), it has been proven in [289] that the electric cross-spectral density matrix of vector fields can be expressed as a correlation matrix in the following form:

$$\mathbf{W}(\mathbf{r}_1, \mathbf{r}_2, \omega) = \begin{bmatrix} \langle E_x^*(\mathbf{r}_1, \omega) E_x(\mathbf{r}_2, \omega) \rangle & \langle E_x^*(\mathbf{r}_1, \omega) E_y(\mathbf{r}_2, \omega) \rangle \\ \langle E_y^*(\mathbf{r}_1, \omega) E_x(\mathbf{r}_2, \omega) \rangle & \langle E_y^*(\mathbf{r}_1, \omega) E_y(\mathbf{r}_2, \omega) \rangle \end{bmatrix}, \tag{6.109}$$

where the field components $E_x(\mathbf{r}, \omega)$ and $E_y(\mathbf{r}, \omega)$ are members of a statistical ensemble in frequency domain [289]. The *spectral degree of coherence* $\nu(\mathbf{r}_1, \mathbf{r}_2, \omega)$ of the field is defined by the following:

$$\mu(\mathbf{r}_1, \mathbf{r}_2, \omega) = \frac{\operatorname{Tr} \mathbf{W}(\mathbf{r}_1, \mathbf{r}_2, \omega)}{[\operatorname{Tr} \mathbf{W}(\mathbf{r}_1, \mathbf{r}_1, \omega) \operatorname{Tr} \mathbf{W}(\mathbf{r}_2, \mathbf{r}_2, \omega)]^{1/2}}, \tag{6.110}$$

where Tr denotes the trace of the matrix and, similarly to the scalar case, we have the following [285]:

$$0 \leq |\mu(\mathbf{r}_1, \mathbf{r}_2, \omega)| \leq 1, \tag{6.111}$$

with the extreme values 0 and 1 corresponding to complete lack of coherence and to complete coherence, respectively. We notice that the absence of off-diagonal components in the definition of the degree of coherence reflects the physical fact that two mutually orthogonal components of the electric field do not produce interference. A rigorous analysis of the Young's interference experiment for a stochastic electromagnetic vector beam [285] leads to establishing a spectral interference law that is the vector analogous of equation (6.77). In particular, it can be shown [285] that the *spectral visibility* of the interference fringes is given by the expression:

$$V(\mathbf{r}, \omega) \equiv \frac{S_{max}(\mathbf{r}, \omega) - S_{min}(\mathbf{r}, \omega)}{S_{max}(\mathbf{r}, \omega) + S_{min}(\mathbf{r}, \omega)} = |\mu(\mathbf{r}_1, \mathbf{r}_2, \omega)|, \tag{6.112}$$

where S_{max} and S_{min} are the maximim/minimum values of the spectral density on the observation screen, and $\mathbf{r}_1$ and $\mathbf{r}_2$ denote the locations of the slits in the diffracting screen.

Based on the knowledge of the electric cross-spectral density matrix, we can derive the spectral density of the beam as follows:

$$S(\mathbf{r}, \omega) = \operatorname{Tr} \mathbf{W}(\mathbf{r}, \mathbf{r}, \omega), \tag{6.113}$$

and the *spectral degree of polarization* at position $\mathbf{r}$ is defined by [285]:

$$P(\mathbf{r}, \omega) = \sqrt{1 - \frac{4 \operatorname{Det} \mathbf{W}(\mathbf{r}, \mathbf{r}, \omega)}{[\operatorname{Tr} \mathbf{W}(\mathbf{r}, \mathbf{r}, \omega)]^2}}, \tag{6.114}$$

where Det denotes the determinant of the matrix $\mathbf{W}$. The preceding definition, which follows naturally from considering the ratio of the spectral density of the polarized part of the beam and its total spectral density [285], clearly demonstrates how polarization phenomena fundamentally originate from the correlations of the electric field components at a given point in space. Note that since the spectral degree of polarization is expressed both in terms of the trace and the determinant of the matrix $\mathbf{W}$, it depends in general on the diagonal as well as the off-diagonal elements of $\mathbf{W}$. As originally realized by Wolf [288], the elements of the electric cross-spectral density matrix change as the beam propagates. This implies that all the quantities $\mu(\mathbf{r}_1, \mathbf{r}_2, \omega)$, $S(\mathbf{r}, \omega)$, and $P(\mathbf{r}, \omega)$ will in general also change upon propagation. Therefore, not only the spectrum of light may change upon propagation, but also its degree of polarization as well as the shape and orientation of the polarization ellipse of the polarized component of the beam. These distinctive modifications in the properties of stochastic electromagnetic waves upon propagation and scattering are generally referred to as *correlation-induced changes* [288] and play a significant role in the modern theory of coherence and its applications to optical engineering [290].

6.5   Coherence Effects in Scattering and Diffraction

Up to this point, we have been concerned with the scattering of coherent radiation from arbitrary objects. We can now employ the general formalism of partial coherence to discuss basic results on the scattering of partially coherent waves (i.e., nonmonochromatic) by a deterministic medium, i.e., a medium with a scattering potential $V(\mathbf{r}, \omega)$ that is a known function of position, and by a random medium with Gaussian disorder. We will also address the diffraction of partially coherent radiation by thin transmitting apertures in opaque screens. We start by presenting a simple derivation of a central result of coherence theory, i.e., the *van Cittert–Zernike theorem*, and discuss its generalization to source with a finite coherence length. Finally, we present the Schell scattering theorem and discuss the diffraction of partially coherent radiation by two-dimensional arrays of small dipole particles both on the far-field and in the Fresnel near-field regions. These results are of great interest in technological applications to scattering, imaging, optical instrumentation design, and radiation engineering. Our analysis is based on the excellent discussions by Wolf [285] and Goodman [284], to which we refer interested readers for additional details.

6.5.1   The van Cittert–Zernike Theorem

We can now build on the formalism introduced in Section 6.4 to derive a central result in the theory of optical coherence known as the *van Cittert–Zernike theorem*. This theorem establishes a Fourier transform relation between the two-point field correlation (i.e., equal-time mutual coherence function or any of its rescaled versions) away from a planar, stationary, spatially incoherent source, and the spatial intensity distribution across the source. It also shows very clearly how the spatial coherence of a wave can be modified in the propagation process from the situation of a completely uncorrelated (i.e., delta-correlated) radiating source. The van Cittert–Zernike theorem plays a central role in astronomical interferometry, where images of celestial bodies can be reconstructed with large angular resolution by Fourier transforming the visibility function of interferometric data collected from multiple telescopes [291]. We first discuss the theorem assuming that the observation points of the field are on a transverse plane located in the far-field region of the source. We will address a more general situation at the end of the section.

Let us consider the source $\sigma(x', y')$ located in a plane transverse to the z-propagation axis at the origin ($z = 0$) and evaluate the equal-time mutual intensity function in a transverse plane located at a certain distance z away from the source and large enough to consider the contributions from the elementary source regions d^2r' and d^2r'' as spherical waves. Under these conditions, we can evaluate the mutual intensity function $\Gamma(\mathbf{r}_1, \mathbf{r}_2, 0)$, alternatively denoted as $\mathbf{J}(\mathbf{r}_1, \mathbf{r}_2)$, at the two points $\mathbf{r}_1$ and $\mathbf{r}_2$ on the observation plane and obtain the following:

$$\Gamma(\mathbf{r}_1, \mathbf{r}_2, z) = \left(\frac{1}{j\bar{\lambda}}\right)^2 \int \int \Gamma(\mathbf{r}', \mathbf{r}'', 0) \frac{e^{-j\bar{k}|\mathbf{r}_1 - \mathbf{r}'|} e^{-j\bar{k}|\mathbf{r}_2 - \mathbf{r}''|}}{|\mathbf{r}_1 - \mathbf{r}'||\mathbf{r}_2 - \mathbf{r}''|} d^2r' d^2r'', \qquad (6.115)$$

where $\bar{\lambda}$ is the mean wavelength of the spectrum of the incident narrowband radiation and $\bar{k}$ is the corresponding wavenumber. If we now consider the case where $\mathbf{r}_1 \gg \mathbf{r}'$ and $\mathbf{r}_2 \gg \mathbf{r}''$, we can use the far-field approximations for the phases:

$$|\mathbf{r}_1 - \mathbf{r}'| \approx |\mathbf{r}_1| - \frac{\mathbf{r}_1 \cdot \mathbf{r}'}{|\mathbf{r}_1|}, \qquad (6.116)$$

and the same for $\mathbf{r}_1$ by substituting $\mathbf{r}' \rightarrow \mathbf{r}''$. Additionally, we can replace the distances in the denominator of equation (6.115) with z. Using the preceding approximations, it becomes possible to explicitly evaluate this integral for a spatially incoherent source described by the mutual intensity function:

$$\Gamma(\mathbf{r}', \mathbf{r}'', 0) = I_0(\mathbf{r}')\delta(\mathbf{r}' - \mathbf{r}''). \qquad (6.117)$$

This form of the mutual coherence corresponds to an extended light source composed of a collection of independent radiators. Using the sampling property of the delta function, we obtain the following:

$$\boxed{\Gamma(\mathbf{r}_1, \mathbf{r}_2, 0) = \frac{\bar{k}^2 e^{-jk(|\mathbf{r}_1| - |\mathbf{r}_2|)}}{4\pi^2 z^2} \int \int I_0(\mathbf{r}') e^{-jk\mathbf{r}' \cdot (\hat{\mathbf{r}}_1 - \hat{\mathbf{r}}_2)} d^2r'} \qquad (6.118)$$

In this expression, z is the distance between the source and the observation planes, while $\hat{\mathbf{r}}_1$ and $\hat{\mathbf{r}}_2$ denote unit vectors along the directions of $\mathbf{r}_1$ and $\mathbf{r}_2$, respectively. This is the far-field form of the van Cittert–Zernike theorem stating that the far-field equal-time mutual correlation function of a spatially incoherent source is directly proportional to the two-dimensional Fourier transform of the intensity distribution across the source. According to this result, the source size determines the coherence of the observed radiation. For instance, this explains why distant stars, whose intensity distribution can be approximated by a delta function, appear to us as perfectly coherent sources.

A more rigorous treatment [285] unveils an even deeper analogy between the equal-time degree of coherence of a light field and the Huygens–Fresnel theory of diffraction [43]. In particular, the equal-time degree of coherence (also known as the complex coherent factor) $j(\mathbf{r}_1, \mathbf{r}_2)$ at two points in the field generated by a spatially incoherent, stationary source can be written as follows [285]:

$$j(\mathbf{r}_1, \mathbf{r}_2) = \frac{1}{\sqrt{I_0(\mathbf{r}_1)I(\mathbf{r}_2)}} \int \int I(\mathbf{r}') \frac{e^{j\bar{k}(R_2 - R_1)}}{R_2 R_1} d^2\mathbf{r}', \tag{6.119}$$

where $R_i = |\mathbf{r}_i - \mathbf{r}'|$ for $i = 1, 2$ are the distances between a typical source point specified by $\mathbf{r}'$ and the field points, and the wavenumber $\bar{k}$ corresponds to the average wavelength of light. Remarkable, the integral in (6.119) has the same mathematical form as the Hyugens–Fresnel diffraction solution to the problem of an aperture σ in an opaque screen illuminated by an incident converging spherical wave positioned at a distance R_2. This surprising similarity is explained by realizing that the mutual coherence function of quasimonochromatic radiation approximately satisfies the Helmholtz wave equation with the average wavenumber $\bar{k}$. The same equation is also obeyed by a purely monochromatic optical field, which points to the identical behavior of these two apparently unrelated quantities. In fact, both the mutual field coherence function and the field dynamics are governed by the Hyugens–Fresnel principle. This also implies that the spatial coherence properties of light must change upon propagation, as quantitatively established by the van Cittert–Zernike theorem.

In the standard derivation of the van Cittert–Zernike theorem, it is assumed that the mutual intensity of the source is delta-correlated, representing a completely incoherent source. However, the theorem can be generalized to account for more general sources with small but nonzero coherence length (i.e., partially coherent radiation). In particular, a generalized van Cittert–Zernike theorem can be established for *quasihomogeneous*, *planar sources* whose mutual intensity function is described as follows:

$$J(\xi_+, \eta_+, \xi_-, \eta_-) = \Gamma(\xi_+, \eta_+, \xi_-, \eta_-, 0) = I(\xi_+, \eta_+)\mu(\xi_-, \eta_-), \tag{6.120}$$

where we defined the source plane variables as follows:

$$\xi_+ = \frac{\xi_1 + \xi_2}{2} \tag{6.121}$$

$$\eta_+ = \frac{\eta_1 + \eta_2}{2} \tag{6.122}$$

$$\xi_- = \xi_2 - \xi_1 \tag{6.123}$$

$$\eta_- = \eta_2 - \eta_1. \tag{6.124}$$

In these expressions, (ξ_i, η_i) for $i = 1, 2$ are the Cartesian coordinates of two arbitrary points in the source plane.[13] Substituting the preceding form of the mutual intensity of the source into the general propagation law directly derived from the Huygens–Fresnel principle [284], it is possible to obtain, under the paraxial approximation, the *generalized form of the van Cittert–Zernike theorem*:

$$J(x_1, y_1; x_2, y_2) = \frac{\kappa e^{-j\psi}}{(\bar{\lambda}z)^2} \int \int I(\xi_+, \eta_+) e^{\left[\frac{2\pi j}{\bar{\lambda}z}(x_-\xi_+ + y_-\eta_+)\right]} d\psi_+ d\eta_+, \quad (6.125)$$

where ψ is a phase factor that can be neglected in the far-field zone and [284]:

$$\kappa(x_+, y_+) = \int \int j(\xi_-, \eta_-) e^{\left[\frac{2\pi j}{\bar{\lambda}z}(x_+\xi_- + y_+\eta_-)\right]} d\psi_- d\eta_-. \quad (6.126)$$

The observation plane relative coordinates $x_\pm$ and $y_\pm$ are defined analogously to their source plane counterparts. In the preceding expressions, $j(\mathbf{r}_1, \mathbf{r}_2)$ denotes the *complex coherence factor* that is related to the complex degree of coherence $\gamma(\mathbf{r}_1, \mathbf{r}_2, \tau)$ as follows:

$$j(\mathbf{r}_1, \mathbf{r}_2) = \gamma(\mathbf{r}_1, \mathbf{r}_2, 0). \quad (6.127)$$

Since confusion with the imaginary unit j can arise in the present context, it is also customary to denote the complex coherent factor with the letter μ. Undoubtedly, the terminology utilized in coherence theory can be quite confusing and is still not completely standardized. An illustrative table summarizing the definitions of the most common quantities used in the theory of optical coherence can be found in [284]. Finally, we observe that since the complex coherence factor of a completely incoherent planar source is a delta function, the generalized van Cittert–Zernike theorem reduces to its standard form in equation (6.118).

6.5.2   Scattering by Deterministic Media

Now that we have established the basic framework of statistical optics, we can address the scattering theory of partially coherence radiation from deterministic random media. Our discussion is based on the approach developed by Wolf and presented in his excellent book [285], to which we refer interested readers for additional details.

We first consider a non-monochromatic, statistically stationary radiation field characterized by a cross-spectral density $W^{(i)}(\mathbf{r}_1, \mathbf{r}_2, \omega)$ that is incident on a static scattering object with scattering potential $V(\mathbf{r}, \omega)$. The coherent mode representation of the cross-spectral density allows us to describe both the incident ψ_0 and the scattered field ψ_{sc} using an ensemble of monochromatic realizations (see equation (6.97)), as follows:

$$W^{(i)}(\mathbf{r}_1, \mathbf{r}_2, \omega) = \langle \psi_0^*(\mathbf{r}_1, \omega) \psi_0(\mathbf{r}_2, \omega) \rangle \quad (6.128)$$

---

[13] The corresponding variables in the observation plane are denoted by (x_i, y_i).

$$W^{(sc)}(\mathbf{r}_1, \mathbf{r}_2, \omega) = \langle \psi_{sc}^*(\mathbf{r}_1, \omega) \psi_{sc}(\mathbf{r}_2, \omega) \rangle. \tag{6.129}$$

Within the accuracy of the Born approximation, it is possible to directly connect the scattered and the incident fields (see equation (6.16)). Therefore, to first order in the scattering theory, we evaluate the cross-spectral density:

$$W^{(sc)}(\mathbf{r}_1, \mathbf{r}_2, \omega)$$

$$= \int_\Gamma \int_\Gamma W^{(i)}(\mathbf{r}_1', \mathbf{r}_2', \omega) V^*(\mathbf{r}_1', \omega) V(\mathbf{r}_2', \omega) G_0^*(|\mathbf{r}_1 - \mathbf{r}_1'|) G_0(|\mathbf{r}_2 - \mathbf{r}_2'|) d^3\mathbf{r}_1' d^3\mathbf{r}_2', \tag{6.130}$$

where $G_0^+(\mathbf{r} - \mathbf{r}', \omega) \equiv G_0(|\mathbf{r} - \mathbf{r}'|)$ is the free-space scalar Green function.[14] Written in terms of the spectral degree of coherence and the previous expression becomes

$$W^{(sc)}(\mathbf{r}_1, \mathbf{r}_2, \omega) = \int_\Gamma \int_\Gamma \sqrt{S^{(i)}(\mathbf{r}_1', \omega)} \sqrt{S^{(i)}(\mathbf{r}_2', \omega)} \mu^{(i)}(\mathbf{r}_1', \mathbf{r}_2', \omega)$$

$$\times V^*(\mathbf{r}_1', \omega) V(\mathbf{r}_2', \omega) G_0^*(|\mathbf{r}_1 - \mathbf{r}_1'|) G_0(|\mathbf{r}_2 - \mathbf{r}_2'|) d^3\mathbf{r}_1' d^3\mathbf{r}_2'. \tag{6.131}$$

When the spectrum of the incident radiation does not depend on the spatial position, i.e., $S^{(i)}(\mathbf{r}_1', \omega) = S^{(i)}(\mathbf{r}_2', \omega) \equiv S^{(i)}(\omega)$, we can obtain the spectral density of the scattered radiation by evaluating this equation at points $\mathbf{r}_1 = \mathbf{r}_2 \equiv \mathbf{r}$ as follows:

$$S^{(sc)}(\mathbf{r}, \omega) = S^{(i)}(\omega) \int_\Gamma \int_\Gamma \mu^{(i)}(\mathbf{r}_1', \mathbf{r}_2', \omega) V^*(\mathbf{r}_1', \omega) V(\mathbf{r}_2', \omega)$$

$$\times G_0^*(|\mathbf{r} - \mathbf{r}_1'|) G_0(|\mathbf{r} - \mathbf{r}_2'|) d^3\mathbf{r}_1' d^3\mathbf{r}_2'. \tag{6.132}$$

This important formula predicts spectral modifications upon scattering of partially coherent radiation, i.e., it shows that the spectrum of the scattered field differs in general from the spectrum of the incident field, even for a static (i.e., time-independent) scattering medium. The spectral modifications arise due to the frequency-dependent terms that appear under the sign of integral in equation (6.132). However, among these factors, the one that usually dominates the spectral changes in the scattering of quasimonochromatic light is associated to the frequency dependence of the spectral degree of coherence $\mu^{(i)}(\mathbf{r}_1', \mathbf{r}_2', \omega)$ of the incident light [285].

Let us now consider as an example the scattering of an incident polychromatic plane wave (i.e., spatially coherent) by a given potential. Making use of the expression (6.103) and of equation (6.131) we obtain the cross-spectral density of the scattered light field as follows:

$$W^{(sc)}(\mathbf{r}_1, \mathbf{r}_2, \omega) = \langle \psi_{sc}^*(\mathbf{r}_1, \omega) \psi_{sc}(\mathbf{r}_2, \omega) \rangle, \tag{6.133}$$

where

$$\psi_{sc}(\mathbf{r}, \omega) = \sqrt{S^{(i)}(\omega)} \int_\Gamma V(\mathbf{r}', \omega) G_0(|\mathbf{r} - \mathbf{r}'|) e^{jk\mathbf{s}_0 \cdot \mathbf{r}'} d^3\mathbf{r}'. \tag{6.134}$$

---

[14] Note that we have assumed the stationarity of the field.

Comparing with the general definition of the spectral degree of coherence in (6.99), we observe that the spectral degree of coherence of the scattered radiation has now a complex modulus equal to unity. This implies that the scattered field at frequency ω remains completely spatially coherent, in contrast with the more general situation described by equation (6.131), where both the spectrum and the state of coherence of the radiation are modified upon scattering.

Using the far-field approximation of the Green function as well as the expression of the first-order scattering amplitude in equation (6.44), we get the cross-spectral density of the scattered far-field:

$$W^{(sc)}(r\mathbf{s}_1, r\mathbf{s}_2, \omega) = \frac{1}{(4\pi r)^2} S^{(i)}(\omega) \hat{V}^*[k(\mathbf{s}_1 - \mathbf{s}_0), \omega] \hat{V}[k(\mathbf{s}_2 - \mathbf{s}_0), \omega]. \quad (6.135)$$

The far-field spectral density of the scattered radiation follows immediately:

$$\boxed{S^{sc}(r\mathbf{s}, \omega) \equiv W^{(sc)}(r\mathbf{s}, r\mathbf{s}, \omega) = \frac{1}{(4\pi r)^2} S^{(i)}(\omega) |\hat{V}^*[k(\mathbf{s} - \mathbf{s}_0), \omega]|^2} \quad (6.136)$$

where $\mathbf{s}_1$ and $\mathbf{s}_2$ are directional unit vectors and $\hat{V}(\mathbf{k}, \omega)$ denotes the three-dimensional Fourier transform of the scattering potential $V(\mathbf{r}, \omega)$. The preceding formulas provide a complete description of first-order scattering of partially coherent light in the space-frequency domain. Equation (6.136) is consistent with the general theory of *stochastic linear filters* introduced in Appendix C.4, as it describes the effect of a linear deterministic filter with transfer function $\hat{V}^*[k(\mathbf{s} - \mathbf{s}_0), \omega]$ on a stochastic incident signal with the power spectrum $S^{(i)}(\omega)$. The equivalent description of the coherence properties of the scattered radiation field in the space-time domain can be obtained based on the Fourier frequency transform of the cross-spectral density.

6.5.3   Scattering by Random Media

The theory of wave scattering by partially coherent radiation in the first-order Born approximation can be naturally extended to random media, which are characterized by a scattering potential $V(\mathbf{r}, \omega)$ that randomly fluctuates as a function of position [285]. The classical example of such media is the Earth's atmosphere, whose refractive index varies randomly in space and time due to the fluctuations in pressure and temperature. At small enough time scales, such temporal fluctuations of the atmospheric parameters can be neglected,[15] and the problem reduces to a static one where only ensemble averages over different realizations of spatial disorder must be considered.

The cross-spectral density and the spectral density of the scattered radiation from a random medium in the static regime can be obtained by ensemble averaging the previous expressions in equations (6.131) and (6.132), respectively. For instance, averaging equation (6.131) over the spatial disorder yields the following:

---

[15] The temporal fluctuations in the Earth's atmosphere can be neglected for time scales shorter than approximately 100 milliseconds.

$$W^{(sc)}(\mathbf{r}_1, \mathbf{r}_2, \omega) = \int_\Gamma \int_\Gamma \sqrt{S^{(i)}(\mathbf{r}_1', \omega)} \sqrt{S^{(i)}(\mathbf{r}_2', \omega)} \mu^{(i)}(\mathbf{r}_1', \mathbf{r}_2', \omega)$$
$$\times C(\mathbf{r}_1', \mathbf{r}_2', \omega) G_0^*(|\mathbf{r}_1 - \mathbf{r}_1'|) G_0(|\mathbf{r}_2 - \mathbf{r}_2'|) d^3\mathbf{r}_1' d^3\mathbf{r}_2', \tag{6.137}$$

where

$$\boxed{C(\mathbf{r}_1', \mathbf{r}_2', \omega) = \langle V^*(\mathbf{r}_1', \omega) V(\mathbf{r}_2', \omega) \rangle} \tag{6.138}$$

is the spatial *correlation function of the scattering potential* and $\langle\langle \cdot \rangle\rangle$ denotes the operation of ensemble averaging over different realizations of the scattering medium. This average involves microscopically different configurations of the random medium (e.g., the atmosphere) as described through its scattering potential.

In practical calculations, the random medium is assumed to obey a Gaussian, homogeneous, and isotropic disorder described by the spatial correlation function:

$$C(\mathbf{r}_1, \mathbf{r}_2, \omega) = \frac{A}{[\sigma\sqrt{(2\pi)}]^3} e^{-|\mathbf{r}_2 - \mathbf{r}_1|^2/(2\sigma^2)}, \tag{6.139}$$

where A and σ are positive constants and σ is considered smaller than the linear dimensions of the scattering medium. In the far zone, the cross-spectral density of the random medium reduces to the following:

$$W^{(sc)}(r\mathbf{s}_1, r\mathbf{s}_2, \omega) = \frac{1}{(4\pi r)^2} \int_\Gamma \int_\Gamma \sqrt{S^{(i)}(\mathbf{r}_1', \omega)} \sqrt{S^{(i)}(\mathbf{r}_2', \omega)} \mu^{(i)}(\mathbf{r}_1', \mathbf{r}_2', \omega)$$
$$\times C(\mathbf{r}_1', \mathbf{r}_2', \omega) e^{-jk(\mathbf{s}_2 \cdot \mathbf{r}_2' - \mathbf{s}_1 \cdot \mathbf{r}_1')} d^3\mathbf{r}_1' d^3\mathbf{r}_2'. \tag{6.140}$$

We now apply this theory to the case where the incident field consists of a spatially coherent polychromatic plane wave. Therefore, the cross-spectral density becomes $W^{(i)}(\mathbf{r}_1, \mathbf{r}_2, \omega) = S^{(i)}(\omega) e^{jk\mathbf{s}_0 \cdot (\mathbf{r}_2 - \mathbf{r}_1)}$, which substituted into the far-field formula (6.140) yields the important result [285]:

$$\boxed{W^{(sc)}(r\mathbf{s}_1, r\mathbf{s}_2, \omega) = \frac{1}{(4\pi r)^2} S^{(i)}(\omega) \hat{C}[-k(\mathbf{s}_1 - \mathbf{s}_0), k(\mathbf{s}_2 - \mathbf{s}_0), \omega]} \tag{6.141}$$

where

$$\hat{C}(\mathbf{k}_1, \mathbf{k}_2, \omega) = \int_\Gamma \int_\Gamma C(\mathbf{r}_1', \mathbf{r}_2', \omega) e^{-j(\mathbf{k}_1 \cdot \mathbf{r}_1' + \mathbf{k}_2 \cdot \mathbf{r}_2')} d^3\mathbf{r}_1' \mathbf{r}_2' \tag{6.142}$$

is the Fourier transform of the spatial correlation function $C(\mathbf{r}_1', \mathbf{r}_2', \omega)$ of the scattering potential.

By setting $\mathbf{s}_1 = \mathbf{s}_2 = \mathbf{s}$, we obtain the spectrum of the scattered field in the far-field zone:

$$S^{(sc)}(r\mathbf{s}, \omega) = \frac{1}{(4\pi r)^2} S^{(i)}(\omega) \hat{C}[-k(\mathbf{s} - \mathbf{s}_0), k(\mathbf{s} - \mathbf{s}_0), \omega]. \tag{6.143}$$

This equation shows very clearly that the spectrum of the scattered field will generally differ from the one of the source and it can be analytically evaluated in closed form for the Gaussian random medium. The resulting spectral density of the scattered polychromatic incident plane wave becomes [285]

$$S^{(sc)}(r\mathbf{s}, \omega) = \frac{A\Omega}{(4\pi r)^2} S^{(i)}(\omega) e^{-2(k\sigma)^2 \sin^2(\theta/2)} \tag{6.144}$$

where Ω is the scattering volume and θ the scattering angle such that $\mathbf{s} \cdot \mathbf{s}_0 = \cos\theta$. On the other hand, when the incident field has an arbitrary degree of spectral coherence, the spectral density in the far-field region for the scattered radiation can be obtained by setting $\mathbf{s}_1 = \mathbf{s}_2 = \mathbf{s}$ in the general formula (6.140). For example, when the incident illumination consists of natural (incoherent) light, the cross-spectral density function of the incident field becomes [285]

$$W^{(i)}(\mathbf{r}_1, \mathbf{r}_2, \omega) = S^{(i)}(\omega) \frac{\sin(k|\mathbf{r}_2 - \mathbf{r}_1|)}{|\mathbf{r}_2 - \mathbf{r}_1|}. \tag{6.145}$$

In this case, after some calculations, it is possible to obtain the following [285]:

$$S^{(sc)}(r\mathbf{s}, \omega) = \frac{A\Omega}{(4\pi r)^2} \frac{1}{2(k\sigma)^2} S^{(i)}(\omega) \left[1 - e^{-2(k\sigma)^2}\right] \tag{6.146}$$

We note that when $k\sigma \ll 1$, corresponding to a strongly uncorrelated medium over the wavelength scale of the incident radiation, the last equation becomes identical to (6.144). In the opposite limit, the angular distributions of the scattered radiation in the far-field zone can be very different for the two cases.

The scattering analysis of partially coherent waves presented in this section can be naturally extended to discrete random media composed by systems of particles with random positions. If each particle has a scattering potential $v(\mathbf{r}, \omega)$, the scattering potential of the discrete random medium becomes

$$V(\mathbf{r}, \omega) = \sum_n v(\mathbf{r} - \mathbf{r}_n, \omega). \tag{6.147}$$

In this case, the correlation function of the discrete scattering medium is as follows:

$$C(\mathbf{r}_1, \mathbf{r}_2, \omega) = \langle V^*(\mathbf{r}_1, \omega) V(\mathbf{r}_2, \omega) \rangle_d = \sum_m \sum_n \langle v^*(\mathbf{r}_1 - \mathbf{r}_m, \omega) v(\mathbf{r}_2 - \mathbf{r}_n, \omega) \rangle, \tag{6.148}$$

and the previously derived formulas can still be utilized formally unchanged.

A particularly simple result can be derived when applying this theory to the scattering of a spatially coherent polychromatic plane wave by a random distribution of identical particles. Evaluating equation (6.143) with the preceding correlation function, it becomes possible to express the spectral density of the scattered light in the far-field region as follows [285]:

$$S^{(sc)}(r\mathbf{s}, \omega) = \frac{1}{(4\pi r)^2} S^{(i)}(\omega) |\hat{v}[k(\mathbf{s} - \mathbf{s}_0), \omega]|^2 \mathbb{S}[k(\mathbf{s} - \mathbf{s}_0)], \tag{6.149}$$

where $\hat{v}(\mathbf{k}, \omega)$ denotes the Fourier transform of the single particle scattering potential $v(\mathbf{r}, \omega)$ and

$$\boxed{\mathbb{S}(\mathbf{k}) = \left\langle \left| \sum_n e^{-j\mathbf{k}\cdot\mathbf{r}_n} \right|^2 \right\rangle} \tag{6.150}$$

is the *disorder-averaged structure factor* of the random system of particles. The fundamental role played by this quantity in scattering theory will be addressed in great detail in the next chapter. Consistently with the general predictions of the theory of partial coherence in the space-frequency domain, equation (6.149) shows that the spectrum of the far-field radiation scattered by a system of particles is also modified with respect to the spectrum $S^{(i)}(\omega)$ of the incident radiation. The spectral modifications induced by systems of scattering particles have been specifically addressed in [290, 292, 293].

6.5.4 The Schell Scattering Theorem

We conclude this chapter by discussing the diffraction of partially coherent, quasi-monochromatic waves by a thin transmitting structure Σ described by the amplitude transmittance:

$$t_\Sigma(\mathbf{r}'_\perp) = \begin{cases} 1 & \text{if } \mathbf{r}'_\perp \in \Sigma \\ 0 & \text{otherwise} \end{cases} \tag{6.151}$$

where $\mathbf{r}'_\perp$ is a two-dimensional position vector that specifies the location of points in the plane of the aperture at $z = 0$.

Following [284], let us denote by $J_0(\mathbf{r}'_\perp; \mathbf{r}''_\perp)$ the mutual intensity function[16] of the incident radiation on the transmitting aperture and by $J_t(\mathbf{r}'_\perp; \mathbf{r}''_\perp)$ the one of the light transmitted immediately after the aperture. The corresponding field amplitudes are connected as follows:

$$\psi_t(\mathbf{r}'_\perp, t) = t_\Sigma(\mathbf{r}'_\perp)\psi_0(\mathbf{r}'_\perp, t - t_0), \tag{6.152}$$

where t_0 is a small delay introduced upon propagation through the aperture. The mutual intensity of the transmitted light can be written as follows:

$$J_t(\mathbf{r}'_\perp; \mathbf{r}''_\perp) = t_\Sigma(\mathbf{r}'_\perp)t_\Sigma^*(\mathbf{r}''_\perp)J_0(\mathbf{r}'_\perp; \mathbf{r}''_\perp). \tag{6.153}$$

Using the expression for the propagation of the mutual intensity that is derived from the Huygens–Fresnel principle under the assumptions that the sizes of both the

---

[16] This quantity is the same as the equal-time mutual coherence function $\Gamma(\mathbf{r}_1, \mathbf{r}_2, 0)$ evaluated on the plane.

diffractive aperture and the observation region are much smaller than the propagation distance, we can obtain the intensity pattern at a transverses position $\mathbf{r}_\perp$ in the (x, y) plane as follows [284]:

$$
I(\mathbf{r}_\perp) \simeq \frac{1}{\tilde{\lambda} z^2} \int\limits_{-\infty}^{+\infty}\!\!\!\int J_t(\mathbf{r}'_\perp; \mathbf{r}''_\perp) e^{-i\frac{2\pi}{\tilde{\lambda}}(r'_2 - r'_1)} d^2\mathbf{r}'_\perp d^2\mathbf{r}''_\perp \tag{6.154}
$$

where $\tilde{\lambda}$ is the average wavelength and r'_2, r'_1 are the distances between the point $\mathbf{r}_\perp = (x, y)$ on the plane at z and the points $\mathbf{r}_{\perp 2}$ and $\mathbf{r}_{\perp 1}$ on the plane at $z = 0$.

When a *Schell model field* is incident on the aperture, we can write the following:

$$
J_0(\mathbf{r}'_\perp; \mathbf{r}''_\perp) = A(\mathbf{r}'_\perp) A(\mathbf{r}''_\perp) j(\mathbf{r}''_\perp - \mathbf{r}'_\perp), \tag{6.155}
$$

where $j(\mathbf{r}''_\perp - \mathbf{r}'_\perp)$ is the equal-time complex degree of coherence (or complex coherence factor) introduced in Section 6.4, $A(\mathbf{r}'_\perp) = \sqrt{I(\mathbf{r}'_\perp)}$, $A(\mathbf{r}''_\perp) = \sqrt{I(\mathbf{r}''_\perp)}$, and $I(\mathbf{r}_\perp)$ is the optical intensity across the aperture wave field. Assuming for simplicity an incident wave on the aperture with constant intensity I_0 and enforcing the paraxial approximation on the distance $r'_2 - r'_1$ we can obtain, setting $\mathbf{r}'_\perp = (\xi_1, \eta_1)$ and $\mathbf{r}''_\perp = (\xi_2, \eta_2)$, the following expression for the diffracted intensity [284]:

$$
I(x, y) = \frac{I_0}{(\tilde{\lambda} z)^2} \int\limits_{-\infty}^{+\infty}\!\!\!\int\!\!\int\!\!\int P(\bar{\xi} - \Delta\xi/2, \bar{\eta} - \Delta\eta/2) P^*(\bar{\xi} + \Delta\xi/2, \bar{\eta} + \Delta\eta/2)
$$

$$
j(\Delta\xi, \Delta\eta) e^{-i\frac{2\pi}{\tilde{\lambda} z}(\bar{\xi}\Delta\xi + \bar{\eta}\Delta\eta)} e^{i\frac{2\pi}{\tilde{\lambda} z}(x\Delta\xi + y\Delta\eta)} d\bar{\xi} d\bar{\eta} d\Delta\xi d\Delta\eta, \tag{6.156}
$$

where a *complex pupil function* $P(\xi, \eta)$ has been introduced that generalizes the amplitude transmittance[17] $t_\Sigma(\xi, \eta)$. The following definitions have also been used: $\bar{\xi} = (\xi_1 + \xi_2)/2$, $\bar{\eta} = (\eta_1 + \eta_2)/2$, $\Delta\xi = \xi_2 - \xi_1$, and $\Delta\eta = \eta_2 - \eta_1$.

Expression (6.156) can be simplified by neglecting the contribution to the integral of the first exponential factor, under the following assumption:

$$
z \gg \frac{\bar{\xi}\Delta\xi + \bar{\eta}\Delta\eta}{\tilde{\lambda}}. \tag{6.157}
$$

This choice is satisfied when a positive lens of focal length z is positioned in contact with the aperture (thus canceling out exactly the quadratic phase term). More generally, in the absence of a lens, condition (6.157) is valid in the Fraunhofer far-field zone, where $z > (2D^2)/\tilde{\lambda}$, with D denoting the maximum width

---

[17] This allows one to account for potential absorption in the aperture or aberration effects if it behaves as a thin lens.

of the aperture. This condition allows us to separate the integral and to rewrite it more compactly as follows [284]:

$$I(x, y) = \frac{I_0}{(\tilde{\lambda}z)^2} \int\int_{-\infty}^{+\infty} \mathbb{P}(\Delta\xi, \Delta\eta) j(\Delta\xi, \Delta\eta) e^{i\frac{2\pi}{\tilde{\lambda}z}(x\Delta\xi + y\Delta\eta)} d\Delta\xi d\Delta\eta \qquad (6.158)$$

where $\mathbb{P}$ denotes the *autocorrelation function* of the complex pupil function P, which is:

$$\mathbb{P}(\Delta\xi, \Delta\eta) \equiv \int\int_{-\infty}^{+\infty} P\left(\tilde{\xi} - \frac{\Delta\xi}{2}, \tilde{\eta} - \frac{\Delta\eta}{2}\right) P^*\left(\tilde{\xi} + \frac{\Delta\xi}{2}, \tilde{\eta} + \frac{\Delta\eta}{2}\right) d\tilde{\xi} d\tilde{\eta}. \qquad (6.159)$$

The expression in (6.158) shows that the intensity distribution $I(x, y)$ is the two-dimensional Fourier transform of the product of $\mathbb{P}$ and j. This important result is referred to as *Schell's theorem* [284, 294] and provides a simple approach for calculating diffraction patterns of arbitrary apertures illuminated by partially coherent radiation.

Schell's theorem generalizes the Fraunhofer diffraction formula of Chapter 4 to the case of partially coherent radiation. This can be immediately appreciated when considering the case of a normally incident coherent plane wave (i.e., $j(\mathbf{r}''_\perp - \mathbf{r}'_\perp) = 1$). In fact, using the Wiener–Khintchine theorem, we can transform the integral (6.158) into the following:

$$I(x, y) = \frac{I_0}{(\tilde{\lambda}z)^2} \left| \int\int_{-\infty}^{+\infty} P(\xi, \eta) e^{i\frac{2\pi}{\tilde{\lambda}z}(x\xi + y\eta)} d\xi d\eta \right|^2. \qquad (6.160)$$

At the opposite extreme, the incoherent radiation has a coherence width much smaller than the aperture size, and we have $\mathbb{P}(0,0) = I_0 \Sigma$ (Σ is the area of the aperture) across the integration range in (6.158), which yields the following:

$$I(x, y) = \frac{I_0 \Sigma}{(\tilde{\lambda}z)^2} \int\int_{-\infty}^{+\infty} j(\Delta\xi, \Delta\eta) e^{i\frac{2\pi}{\tilde{\lambda}z}(x\Delta\xi + y\Delta\eta)} d\Delta\xi d\Delta\eta. \qquad (6.161)$$

In the incoherent limit, the intensity distribution of the diffraction pattern does not depend on the shape of the aperture anymore but rather on the complex coherence factor j alone. In the case of partially coherent incident radiation, both the functions $\mathbb{P}$ and j play a role in determining the diffracted intensity $I(x, y)$. Since the intensity in (6.158) is expressed as the Fourier transform of a product of two functions, using the convolution property of the Fourier transforms we can rewrite Schell's theorem in the alternative form:

$$I(x, y) \propto |\mathfrak{F}\{P(\xi, \eta)\}|^2 * \mathfrak{F}\{j(\Delta\xi, \Delta\eta)\} \qquad (6.162)$$

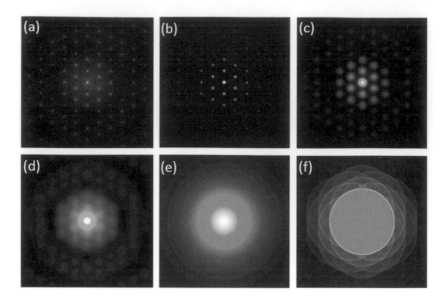

Figure 6.10 Shell's theorem diffraction of partially coherent light by a triangular phase mask with lattice constant $a = 500$ nm and circular holes with radius $R = 100$ nm. The central wavelength is $\lambda = 550$ nm. The panels show the far-field diffraction patterns of radiation with different coherence properties. (a) Perfectly coherent (Fraunhofer) radiation and partially partially coherent radiation with (b) $\sigma_\mu = 10\lambda$; (c) $\sigma_\mu = \lambda$; (d) $\sigma_\mu = 0.1\lambda$; (e) $\sigma_\mu = 0.01\lambda$; (f) Lambertian source.

where $\mathfrak{F}\{\cdot\}$ represents the two-dimensional Fourier transform and $*$ denotes the convolution operator.[18] This form of Shell's theorem clearly demonstrates how the partially coherent nature of radiation results in the smoothing of the diffraction pattern of the aperture compared to the case of coherent illumination. This effect can be clearly observed in Figure 6.10, where we show the calculated diffraction patterns of partially coherent light through a binary phase mask with apertures arranged on a triangular lattice. The computation is based on the Shell theorem formula and different values of the spectral degree of coherence for a source with a Gaussian complex coherence factor $\mu(\Delta\xi, \Delta\eta)$ of width σ_μ are considered. In particular, Figure 6.10a, f shows the two extreme cases of perfectly coherent light (Fraunhofer far-field) and of an incoherent Lambertian source diffracted by the mask, respectively. Figure 6.10b–d displays the diffraction patterns corresponding to progressively decreasing values of the spectral coherence of the source, which significantly broaden the bright diffraction spots observed under coherent illumination.

Schell's theorem has been recently generalized to cases where the detection plane is not in the far-field region of the aperture [295]. In particular, if the detector is

---

[18] Note that due to the even symmetry and real character of the equal-time complex degree of coherence, typically modeled by a Gaussian function, the Fourier transform of j is purely real, which renders the expression for the intensity in (6.162) real as well.

positioned in the Fresnel zone at a distance z from the diffracting aperture, a *generalized Schell theorem* holds true in the following form [295]:

$$\boxed{I(x,y) \propto |\mathbb{F}\{P(\xi,\eta,z)\}|^2 * \mathfrak{F}\{j(\Delta\xi,\Delta\eta)\}} \qquad (6.163)$$

where $\mathbb{F}\{\cdot\}$ denotes the Fresnel transform of the diffracting aperture. Its modulus squared is also known as the *incoherent point spread function*. Note how the standard (far-field) formulation of Schell's theorem is recovered as a special case when when the detector is located in the far-field of the aperture.

7 Structure Factor and Diffraction

Indeed no amount of mere reading will of itself improve one's understanding of anything. What is required in addition to the acquiring of information is some measure of intense reflection upon the several matters thus gathered in.

Wisdom of the West, Bertrand Russell

In this Chapter, we specifically address the link between structural and radiation properties of aggregates of scattering particles within the Born approximation that is valid in the single-scattering regime, i.e., under weak scattering conditions. The central quantity in our discussion is the *structure factor* that bridges the geometrical correlations and radiation properties of scattering media. The significance of the structure factor in different situations is illustrated through a number of examples that address scattering media with periodic and aperiodic geometries. Specifically, our discussion will be focused on one- and two-dimensional arrays of small scattering particles with a size smaller than the wavelength. This choice provides us with the opportunity to introduce the principal categories of aperiodic structures based on their distinctive Fourier spectral properties. Finally, we consider the physics of lattice resonances in two-dimensional arrays of small dielectric and metallic scattering particles.

7.1 Single Scattering and Geometry

We discuss a very interesting and fruitful connection that exists between the theory of elastic light scattering of scalar waves, considered within the first-order Born approximation, and the geometrical structure of scattering objects characterized by their spatial Fourier spectra. For simplicity, we consider the scattering of waves from nondispersive particles that are approximated as radiating scalar point dipoles. Extensions to vector point dipoles will be discussed in Chapter 12. The present assumptions underpins the *kinematic scattering theory*, which was originally developed in the context of the X-ray scattering of materials [296].

The considered scattering geometry is illustrated in Figure 7.1, where a plane wave with wavevector $\mathbf{k}_0$ is incident on the sample. With respect to an arbitrary origin fixed in the sample, a detector D positioned in the far-field receives scattered radiation that propagates with wavevector $\mathbf{k}$. Since we are considering elastic scattering

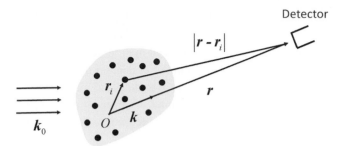

Figure 7.1 Geometry for the elastic wave scattering experiment with a sample composed of discrete point scatterers.

events where total energy is conserved, we have the condition $\Delta \mathbf{k} = \mathbf{k} - \mathbf{k}_0$ with $|\Delta \mathbf{k}| = 2k_0 \sin(\theta/2)$, where θ is the *scattering angle*, i.e., the angle between $\mathbf{k}$ and $\mathbf{k}_0$.

The amplitude of the scattered wave at the detector is proportional to the following:

$$e^{\mathbf{k}_0 \cdot \mathbf{r}_i} \frac{e^{j\mathbf{k} \cdot |\mathbf{r} - \mathbf{r}_i|}}{|\mathbf{r} - \mathbf{r}_i|}. \tag{7.1}$$

If we assume that $\mathbf{r} \gg \mathbf{r}_i$, we can substitute the *far-field approximation* $k|\mathbf{r} - \mathbf{r}_i| \approx kr - \mathbf{k} \cdot \mathbf{r}_i$ in the phase of equation (7.1) and neglect the small term $\mathbf{k} \cdot \mathbf{r}_i$ in the denominator. With these choices, the field amplitude at the detector becomes

$$\frac{e^{jkr}}{r} e^{-j(\mathbf{k} - \mathbf{k}_0) \cdot \mathbf{r}_i} = \frac{e^{jkr}}{r} e^{-j(\Delta \mathbf{k}) \cdot \mathbf{r}_i}. \tag{7.2}$$

The total amplitude at the detector contributed by all the scattering points inside the sample is equal to the following:[1]

$$\boxed{A = A_0 \sum_i e^{-j\Delta \mathbf{k} \cdot \mathbf{r}_i}} \tag{7.3}$$

where A_0 is the constant scattering efficiency of each (identical) particle in the medium. The summation (7.3) is known in antenna engineering as the *array factor*. This is simply a *phasor sum* that accounts for the geometrical phase shifts of the far-field scattered waves that originate at the different dipole positions in a discrete particle medium, or a *dipole array*.

The preceding simple treatment requires exact knowledge of the geometrical configuration of the scattering particles. However, this cannot practically be achieved in situations that involve a very large number of particles or in the presence of randomness where a statistical approach must naturally be employed. Moreover, in a general scenario, the particle positions are not fixed in space and the detection process itself requires a finite acquisition time. In all such cases, the experimentally accessible scattering intensity should be regarded as an appropriate average quantity defined over

---

[1] Note the crucial reliance on the single-scattering approximation here.

a time scale that is longer compared to the thermodynamic fluctuations occurring in the sample, which are typically in the 10^{-12}s-10^{-14}s range. Therefore, a suitable statistical mechanics approach needs to be utilized in order to deal with wave scattering from macroscopic systems with a large number of degrees of freedom.

The goal of the following sections is to introduce the relevant concepts that are involved in the calculation of the scattered intensity from arbitrary dipolar arrays and even with incomplete information on their detailed microscopic arrangements, as in the case of random media.

7.2   The Structure Factor and Its Meaning

We proceed in the calculation of the scattered intensity from samples of arbitrary geometry and establish the connection with important structural parameters of the medium such as the radial distribution function and the correlation function. These quantities describe the microstructure of a scattering medium and are directly proportional to the intensity of scattered waves within the validity of the first-order Born approximation. An extended discussion of correlation functions for random scattering media is provided in Section 8.1.

According to the principles of scalar wave optics [78], the scattered intensity at the detector is obtained by squaring the total complex amplitude, or the array factor, i.e., $|A|^2$. Specifically, the scattered intensity can be expressed as follows:

$$I = \langle |A|^2 \rangle = |A_0|^2 \left\langle \left| \sum_i e^{-j\Delta \mathbf{k} \cdot \mathbf{r}_i} \right|^2 \right\rangle = |A_0|^2 \left\langle \sum_{l,m} e^{-j\Delta \mathbf{k} \cdot (\mathbf{r}_l - \mathbf{r}_m)} \right\rangle, \qquad (7.4)$$

where the angle bracket denotes the appropriate ensemble average operation over the disorder. Clearly, if the sample under consideration is deterministic, meaning that it is composed of known constituents located at fixed positions, the averaging procedure can simply be dropped. However, we believe that it is instructive to consider the more general situation in the following derivations.

Inserting the integral representation of the delta function, we can rewrite the last expression as follows:

$$I = \langle |A|^2 \rangle = |A_0|^2 \left\langle \int d^3 \mathbf{r} \sum_{l,m} e^{j\Delta \mathbf{k} \cdot \mathbf{r}} \delta(\mathbf{r} + \mathbf{r}_m - \mathbf{r}_l) \right\rangle. \qquad (7.5)$$

The exponential factor can be removed from the summation, and since only the term containing the δ function fluctuates across the sample, we obtain the following:

$$I = \langle |A|^2 \rangle = |A_0|^2 \int d^3 \mathbf{r} e^{j\Delta \mathbf{k} \cdot \mathbf{r}} \left\langle \sum_{l,m} \delta(\mathbf{r} + \mathbf{r}_m - \mathbf{r}_l) \right\rangle. \qquad (7.6)$$

We recognize in the expression (7.6) the Fourier transform of the quantity within the angle brackets. Before proceeding further, following Goodstein [297], we extract from the double summation the diagonal terms with $l = m$ as follows:

$$\left\langle \sum_{l,m} \delta(\mathbf{r} + \mathbf{r}_m - \mathbf{r}_l) \right\rangle = \left\langle \sum_{l \neq m} \delta(\mathbf{r} + \mathbf{r}_m - \mathbf{r}_l) \right\rangle + N\delta(\mathbf{r}), \qquad (7.7)$$

where N is the total number of scattering particles. The first term on the right-hand side of this equation allows us to define a key quantity in scattering theory, which is the *radial distribution function*:

$$\boxed{\rho g(\mathbf{r}) = \frac{1}{N} \left\langle \sum_{l \neq m} \delta(\mathbf{r} + \mathbf{r}_m - \mathbf{r}_l) \right\rangle} \qquad (7.8)$$

where $\rho = N/V$ is the density of scatterers. The function $g(\mathbf{r})$ contains information on the local correlations in the positions of the scattering particles (see Section 8.1.2 for an extended discussion).

According to its definition, $g(\mathbf{r})$ records a δ function whenever any two scattering particles in the sample are separated by the vector distance $\mathbf{r}$. If the particles that make up the sample cannot be considered static, as is the case in real solid, liquid or gaseous substances, the recording step is performed for each possible configuration and then thermodynamically averaged. Moreover, if a material is nonisotropic, the $g(\mathbf{r})$ depends on both magnitude and direction of vector $\mathbf{r}$, while for isotropic materials $g(r)$ will only depend on the magnitude r of the separation vector. Therefore, for general crystalline solids the $g(\mathbf{r})$ is strongly peaked around certain separation vectors corresponding to the periodic positions of atoms in a cell, while for ideal random gases $g(r) = 1$, since any two-particle separation is equally likely. Corrections to this intuitive model are discussed in Section 8.1.

Liquids or amorphous structures are characterized by *local correlations* and are somewhat intermediate between the two extreme scenarios of periodic crystals and uniform random gases. In liquids or amorphous media, $g(r)$ features several oscillations of decreasing intensity around the asymptotic value 1. The radial locations of the local oscillation peaks of $g(r)$ unveil different *correlation lengths* associated to the more probable coordination shells in the particles positions, which will be lost completely at large enough separations. More precisely, the positions of the local maxima of the radial distribution function give the average pair distance of the first, second, etc., neighbors in the sample. On the other hand, the widths of the peaks of $g(r)$ quantify the fluctuations in these distances. More precisely, the area under each peak gives the number of particles in successive coordination cells. The distance at which the oscillations around unity become negligible defines the overall *coherence length* of the structure. Figure 7.2 is schematic illustration of the radial distribution functions of periodic and amorphous systems.

We can now continue the derivation of the scattered intensity. Substituting equations (7.8) and (7.7) into (7.6), we obtain the following:

$$I(\Delta\mathbf{k}) = |A_0|^2 N \left[1 + \rho \int e^{j\Delta\mathbf{k}\cdot\mathbf{r}} g(\mathbf{r}) d^3\mathbf{r} \right]. \qquad (7.9)$$

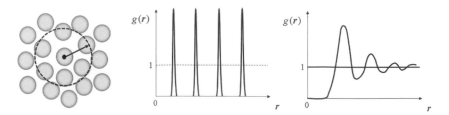

Figure 7.2 Schematic of the radial distribution functions of a periodic solid (sampled along its crystal axis), a perfect random gas, and an amorphous or liquid structure.

It is customary to redefine this expression by subtracting a term proportional to a δ function that vanishes everywhere except when $\Delta \mathbf{k} = 0$[2]. Therefore, we set the following [297]:

$$\rho \delta(\Delta \mathbf{k}) = \rho \int e^{j \Delta \mathbf{k} \cdot \mathbf{r}} d^3 \mathbf{r}. \tag{7.10}$$

This trick allows us to obtain the following important expression:

$$I(\Delta \mathbf{k}) = |A_0|^2 N \left[1 + \rho \int e^{j \Delta \mathbf{k} \cdot \mathbf{r}} h(\mathbf{r}) d^3 \mathbf{r} \right]. \tag{7.11}$$

which defines the following function:

$$\boxed{h(\mathbf{r}) = g(\mathbf{r}) - 1} \tag{7.12}$$

known as the *total correlation function*. We now realize that, according to equation (7.11), the scattered intensity for a uniformly disordered system I_{RG}, such as an ideal random gas with $g(r) = 1$, reduces to the following:

$$I_{RG} = |A_0|^2 N. \tag{7.13}$$

Therefore, we can eliminate the prefactor in equation (7.11) by defining the dimensionless quantity:

$$\boxed{S(\Delta \mathbf{k}) = \frac{I(\Delta \mathbf{k})}{I_{RG}} = 1 + \rho \int e^{j \Delta \mathbf{k} \cdot \mathbf{r}} h(\mathbf{r}) d^3 \mathbf{r}} \tag{7.14}$$

The function $S(\Delta \mathbf{k})$ is called the *static structure factor* or, more simply, the *structure factor*. Its generalization to the dynamic case will be discussed in Section 7.2.4. We can write the last expression more compactly as follows:

$$\boxed{S(\mathbf{q}) = 1 + \rho \hat{h}(\mathbf{q})} \tag{7.15}$$

where we renamed the *scattering wavevector* $\Delta \mathbf{k} \equiv \mathbf{q}$ and denoted by $\hat{h}(\mathbf{q})$ the spatial Fourier transform of the total correlation function $h(\mathbf{r})$. Within the single-scattering

---

[2] This can be done on the basis that $\Delta \mathbf{k} = 0$ corresponds to *forward scattering* that cannot be distinguished experimentally from the unscattered (transmitted) part of the incident beam. Therefore, it makes no physical difference if we rescale the formula at that point.

approximation, the structure factor encodes important information regarding the microstructure of the scattering object. From the previous equation, we realize that by enforcing the condition $\hat{h}(q) = -1/\rho$ it becomes possible to engineer, within the single-scattering approximation, a discrete medium that does not scatter the incoming radiation around the scattering wavevector q. As we fully appreciate in Section 8.4.1, this condition is satisfied in the forward scattering direction (i.e., $q = 0$) by *hyperuniform random media* [298, 299].

Remembering the general relation for the scattered intensity in equation (7.4) and using (7.14), we immediately obtain the general structure factor (ensemble averaged):

$$S(\mathbf{q}) = \frac{1}{N}\left\langle \left|\sum_i e^{-j\mathbf{q}\cdot\mathbf{r}_i}\right|^2\right\rangle = \frac{1}{N}\left\langle \sum_{l,m} e^{-j\mathbf{q}\cdot(\mathbf{r}_l - \mathbf{r}_m)}\right\rangle. \tag{7.16}$$

For arrays of scatterers with N elements at fixed positions, the previous formula reduces to the following:

$$S(\mathbf{q}) = \frac{1}{N}\left|\sum_i e^{-j\mathbf{q}\cdot\mathbf{r}_i}\right|^2 = \frac{1}{N}\sum_{l,m} e^{-j\mathbf{q}\cdot(\mathbf{r}_l - \mathbf{r}_m)} \tag{7.17}$$

This equation provides a clear link between the spatial distribution of scattering points (radiating dipoles) and the static structure factor. For instance, if the array is periodic, then its structure factor is nonzero on a periodic lattice as well, which is the *reciprocal lattice*. See Section 7.2.5 for an extended discussion. The structure factor of an array of point particles can alternatively be obtained by considering the normalized Fourier power spectrum of the *array density*. The array density $\rho(\mathbf{r})$ for an array of point-like scatterers located at positions $\mathbf{r}_i$ is given by the following generalized function:

$$\rho(\mathbf{r}) = \sum_{i=1}^{N} \delta(\mathbf{r} - \mathbf{r}_i) \tag{7.18}$$

Remembering the shifting property of the Fourier transform, we immediately obtain the expression for the Fourier components of the array density that coincide with the previously derived *array factor*:

$$\hat{\rho}(\mathbf{q}) = \sum_{i=1}^{N} e^{-j\mathbf{q}\cdot\mathbf{r}_i}. \tag{7.19}$$

This characteristic "interference function" is a complex trigonometric sum that encodes all the interference properties arising from the distinctive geometrical arrangement of the points in the array. The relation between the array factor and the static structure factor follows immediately:

$$S(\mathbf{q}) = \frac{1}{N}|\hat{\rho}(\mathbf{q})|^2 \tag{7.20}$$

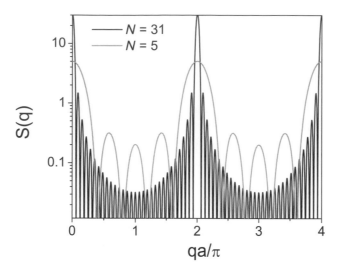

Figure 7.3 Calculated structure factor of a periodic chain of scattering points with period a for different particle numbers N.

This equation establishes a direct connection between the Fourier spectral components of the array density and the structure factor, which is proportional to the measured intensity of the scattered radiation in the far-field. Measurements of the structure factor by scattering experiments can in principle be Fourier antitransformed numerically to yield $h(r)$ or $g(r)$.

In Figure 7.3, we show examples of structure factors corresponding to one-dimensional (1D) periodic chains of point scatterers with an increasing number of elements. We note that the effect of constructive interference effects manifests dramatically when increasing the number N of scattering elements, giving rise to very narrow diffraction peaks separated by an increasing number of side lobes of smaller intensity. More generally, the structure factor provides access to the *q-space* (i.e., reciprocal space with units of inverse length) description of a scattering sample, directly obtained by detecting the far-field single-scattered radiation intensity as a function of the scattering angle. The quantity q^{-1} determines to the *resolution of the scattering process* because the wave scattering cannot probe spatial features that are smaller than q^{-1}. A comparison between the inherent length scale of the scattering process, given by q^{-1}, and the length scales of an array of scatterers determines whether the waves combine in phase (constructively) or randomly at the detector. For a random system of N scatterers, we can distinguish between two limiting situations:

- If the scattering particles are within a q^{-1} distance from each other, their phases can be considered approximately the same, and thus they will add constructively at the detector. The total scattered amplitude will be proportional to N, and the total scattered intensity will be proportional to N^2. This occurs when $\mathbf{q} \cdot (\mathbf{r}_l - \mathbf{r}_m) < 1$ in equation (7.17).

- If the scatterers are pairwise separated by distances $> q^{-1}$, their phases will largely fluctuate and hence they will add randomly at the detector. In this case, the total scattered amplitude will be proportional to $\sqrt{N}$ and the total scattered intensity will be proportional to N. This occurs when $\mathbf{q} \cdot (\mathbf{r}_l - \mathbf{r}_m) > 1$ in equation (7.17), which allows us to regard the summation as an uncorrelated *random walk process* or a *random phasor sum* with average amplitude proportional to the square root of the number of steps, i.e., N (see Chapter 8).

Box 7.1 Light Scattering by Fractal Aggregates

Fractal aggregates, such as colloids or aerosols, have *scale invariant* shapes that, within a limited range, appear the same when viewed over different scales. The scale invariance of fractal aggregates is quantitatively described by the noninteger Hausdorff or fractal dimension D that appears in the scaling:

$$N = \eta \left(\frac{R_g}{a} \right)^D ,$$ (7.21)

where N is the number of component particles in the aggregate (proportional to its mass), R_g is the *gyration radius* that measures the overall size of the aggregate, a is the radius of the individual particles, and η is a proportionality constant of order unity. The fractal dimensions of typical diffusion-limited aggregates (DLA) are $D = 1.7 - 1.9$. Understanding light scattering by fractal aggregates requires us to analyze the scaling of their structure factor $S(q)$ under the assumption of no internal multiple scattering, here known as the Rayleigh-Debye-Gans (RDG) approximation. The essence of the *scaling method* consists in comparing the intrinsic length scale of wave scattering, i.e., q^{-1}, with the scales in the scattering system and then compute their contributions to the intensity based on the linearity principle [300]. For a fractal aggregates, this approach leads to the following [300]:

$$S(q) = 1, \qquad\qquad q < (R_g)^{-1}$$ (7.22)
$$S(q) = C(q R_g)^{-D}, \quad q > (R_g)^{-1},$$ (7.23)

where $C \simeq 1$. These equations do not include the possibility of observing the constituent particles with q, which can happen when $q > a^{-1}$. If the fractal aggregates consists of homogeneous spherical particles, we can express the total structure factor in the following product form:

$$S(q) = S_{FA}(q) S_0(q),$$ (7.24)

where $S_{FA}(q)$ refers to the fractal aggregate and $S_0(q)$ to the individual spherical constituents. Therefore, the hallmark of fractal scattering in the RDG approximation is the power-law scaling $S(q) \propto q^{-D}$ for $q > (R_g)^{-1}$.

The results derived in this section are applicable only to the *single-scattering regime* of arbitrary arrays composed of small scatterers with no internal structure. The validity range of this approximation is difficult to establish under general conditions, and it has been the subject of intense investigation. However, based on theoretical analysis it has been possible to propose the simple approximate condition [301]: for particles that are significantly smaller than the radiation wavelength, the single-scattering regime is achieved when the interparticle distance is larger than one wavelength in the medium. In particular, the free-space volume occupied by a particle in the medium must be larger than λ^3 for small particles. Therefore, for aggregates of small particles the volume fraction should be smaller than $\rho_{max} = 4/3\pi(a/\lambda)^3$. When the volume fraction does not exceed this limit, the power scattered by a unit volume of the system is obtained by independently summing the scattering contributions of each particle in the system. In particular, the amount of scattering in the system will be directly proportional to the concentration of particles, and complicated multiple scattering effects can be neglected. However, the scattering approach based on the structure factor, sometimes referred to as *kinematic scattering theory*, still provides useful physical insights into the properties of many systems, including the scattering and absorption of light by fractal aggregates [300]. Finally, within this approximation, the differential scattering cross section of a collection of N scattering dipoles with polarizability α can be expressed as follows [82]:

$$\frac{d\sigma}{d\Omega} = \left(\frac{\omega}{c}\right)^4 N\alpha^2(1 - \sin^2\theta\cos^2\varphi)[1 + \rho\hat{h}(q)] \qquad (7.25)$$

where $\rho = N/V$ is the number density, θ is the scattering angle, $q = 2k_0\sin(\theta/2)$, and φ is the angle between the scattering plane and the polarization vector. It is evident from the preceding result that the structural correlations of an aggregate modify both the intensity and the angular distribution of the radiation scattered by an isolated particle.

The effect of spatial correlations can be taken into account, within the first-order Born approximation, by correcting the expression of the differential scattering cross section of a single scatterer $d\sigma'/d\Omega$ (which can be separately computed using accurate multipolar theory, e.g., Mie theory for spheres) by the static structure factor $S(q)$ as follows:

$$\frac{d\sigma}{d\Omega} = \frac{d\sigma'}{d\Omega}S(q) \qquad (7.26)$$

where $q = (4\pi/\lambda_{eff})\sin(\theta/2)$ and $\lambda_{eff} = \lambda/n_{eff}$ is the wavelength in a medium with effective refractive index e_{eff}, which can be found using appropriate mixing formulas [123]. The effects of structural correlations in disordered random media have been recently investigated experimentally by Conley et al. [302], resulting in significant control of the transport mean free path[3] of light in planar geometries.

---

[3] This important quantity describes the average distance that light travels in a nonhomogeneous sample before complete randomization of its initial propagation direction. See Chapter 14 for a precise definition.

7.2.1 Form Factor and Kinematic Scattering

The scattering analysis based on the structure factor can be extended to arrays of particles with angle-dependent *scattering amplitude*, or with a radiation diagram described by the angular function $f(\theta, \omega)$. We assume for simplicity azimuthal symmetry in the problem (i.e., no ϕ dependence). In this case, the strength of the field scattered by the particle i at an angle θ with the direction of the incoming wave ψ_0 and at a distance $r = |\mathbf{r}|$ is simply as follows:

$$\psi = \psi_0 f_i(\theta, \omega) \frac{e^{jk_0 r}}{r}, \tag{7.27}$$

where $\psi_{inc} = \psi_0 e^{jk_0 z}$, k_0 is the wavevector of the radially outgoing scattered wave, and the particle is located at the origin of the reference (spherical) coordinate system. In the context of X-ray (atomic) scattering, the function $f_i(\theta, \omega)$ that describes the scattering ability of individual atomic units is called the atomic form factor, or simply the *form factor*. In X-ray and neutron scattering, the form factors are generally real-valued functions that decrease monotonically with increasing θ (or decreasing λ) and have a magnitude that is proportional to the atomic number Z of the investigated elements. The same concept is also utilized in optics, where it is known as the scattering amplitude describing the properties of anisotropic particles. However, as we discussed in relation to the optical theorem, the scattering amplitude of optical systems is generally a complex-valued function. This is invariably the case with dielectric and metallic scattering particles whose linear dimensions are comparable or larger than the wavelength of the excitation field. The explicit form of $f_i(\theta, \omega)$ is determined by both the size and the shape of the scattering objects.

We can write the far-field contribution to the scattering field at $\mathbf{r}$ from a particle i located at position vector $\mathbf{r}_i$ relative to an arbitrarily defined origin:

$$\psi_i = \psi_0 e^{i\mathbf{k}_0 \cdot \mathbf{r}_i} f_i(\theta, \omega) \frac{e^{j\mathbf{k}_{sc} \cdot (\mathbf{r} - \mathbf{r}_i)}}{|\mathbf{r} - \mathbf{r}_i|}. \tag{7.28}$$

The schematic geometry is sketched in Figure 7.4. It is easy to generalize this situation to N scatterers in a diffractive sample since, due to the linear superposition principle, we simply add up the individual contributions and obtain the total scattered wave:

$$\psi_{sc} = \psi_0 e^{i\mathbf{k}_{sc} \cdot \mathbf{r}} \sum_{i=1}^{N} f_i(\theta, \omega) \frac{e^{-j\mathbf{q} \cdot \mathbf{r}_i}}{|\mathbf{r} - \mathbf{r}_i|} \tag{7.29}$$

where $\mathbf{q} = \mathbf{k}_{sc} - \mathbf{k}_0$ and all the terms that do not depend on the index i have been factored outside of the sum. It should also be clear that the validity of equation (7.28) is contingent on the first-order Born approximation since we only considered the coherent superposition of the waves scattered independently by all the particles in the array.[4] Multiple scattering effects where waves are deflected more than once before reaching the far-field detector are completely neglected. An in-depth discussion of a

---

[4] Needless to say, we neglected any possible absorption mechanism as well.

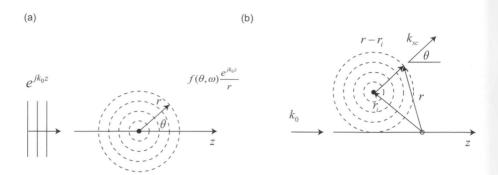

Figure 7.4 Schematic geometry that illustrates the scattering contribution of a plane-wave illuminated particle positioned at the origin of a spherical reference system (a) and positioned at an off-centered location (b). Adapted from reference [303].

more general approach that includes multiple scattering effects in systems of particles will be presented in Chapter 12.

7.2.2   Fundamental Equations of Kinematic Scattering

In many practical applications, the distance of the scattering sample to the measuring detector is much larger than the linear dimensions of the sample, and therefore we can use the following approximation:

$$|\mathbf{r} - \mathbf{r}_i| \approx r. \tag{7.30}$$

We can insert this condition in the denominator of equation (7.29) and consider its squared modulus, which is proportional to the differential scattering cross section [303]:

$$\frac{d\sigma}{d\Omega} = \frac{1}{N}\left|\sum_{i=1}^{N} f_i(\theta, \omega)e^{-j\mathbf{q}\cdot\mathbf{r}_i}\right|^2 \tag{7.31}$$

This equation naturally generalizes the structure factor in equation (7.17), which is recovered when $f_i(\theta, \omega) = 1$, to situations where the individual particles have different radiation patterns.

So far we have been concerned with the scattering from samples composed of discrete particles. However, in many practical situations it is useful to consider scattering objects that are characterized by a continuous distribution of scattering units, such as electron clouds, nuclear and magnetization densities, etc. To deal with these cases, we need a continuum generalization of equation (7.31). This can be achieved by introducing the *scattering density function* $\alpha(\mathbf{r})$ via the relation:

$$f(\theta, \omega) = \alpha(\mathbf{r})d^3\mathbf{r}. \tag{7.32}$$

More generally, the scattering density function depends on both the position and the wavevector, i.e., $\alpha = \alpha(\mathbf{r}, \mathbf{k})$. This quantity describes the ability of an infinitesimally small scattering element of volume $d^3\mathbf{r}$ located at position $\mathbf{r}$ to scatter incoming radiation in a certain direction, and it is proportional to the density of the scattering substance. If we assume that each infinitesimal scattering element scatters identically and isotropically, then the scattering density function can be factored into the product of two terms, i.e., the density of scattering elements $\rho(\mathbf{r})$ and the form factor F (here considered to be independent of $\mathbf{k}$). Substituting the scattering density function into equation (7.31) allows us to immediately obtain its generalization to the continuum case:

$$\frac{d\sigma}{d\Omega} \propto \left| \int_V \alpha(\mathbf{r}) e^{-j\mathbf{q}\cdot\mathbf{r}} d^3\mathbf{r} \right|^2 \qquad (7.33)$$

where V is the total volume for the scattering events to take place. Equations (7.31) and (7.33) are two fundamental results of kinematic scattering theory since they establish a direct connection between the structure of a target material and the resulting scattered intensities. In particular, equation (7.33) shows that the (elastic) differential scattering cross section is linked to the scattering density function via the squared modulus of its three-dimensional Fourier transform. This also implies that the scattered far-field can be expressed as follows:

$$\psi_{sc} = \int_V \alpha(\mathbf{r}) e^{-j\mathbf{q}\cdot\mathbf{r}} d^3\mathbf{r} \qquad (7.34)$$

We note that this formula agrees with the first-order Born approximation of scalar waves discussed in Chapter 6. When a generic kinematic scattering experiment is considered, e.g., X-ray diffraction, it provides information on the three-dimensional Fourier components of the electron density.

7.2.3   Kinematic Diffraction of Periodic Structures

It is often important to apply the kinematic theory of wave diffraction to three-dimensional crystalline structures characterized by a lattice and a basis unit. When the scattering structures can be represented as the geometric repetition of identical patterns, equation (7.34) splits into the product of factors: an integral over the scattering elements contained inside an individual unit cell of the lattice[5] and a periodic summation that defines the reciprocal lattice vectors.

Let us now address in more detail the wave scattering from a three-dimensional crystal structure composed of a periodic lattice and a repeating unit basis element.

---

[5] In the context of X-ray diffraction from periodic systems, the structure factor is often restricted to take into account only the contributions of particles inside the repeating unit basis.

The periodicity of the underlying lattice is described by considering a lattice periodic scattering density function:

$$\alpha(\mathbf{r}) = \alpha(\mathbf{r} + n_1\mathbf{a} + n_2\mathbf{b} + n_3\mathbf{c}), \qquad (7.35)$$

where $\{n_1, n_2, n_3\}$ are integer numbers and $\{\mathbf{a}, \mathbf{b}, \mathbf{c}\}$ are linearly independent vectors, called *primitive translation vectors*, that span the periodic lattice by integer translations (additional details are discussed in Chapter 11). In particular, the parallelepiped with edges $\{\mathbf{a}, \mathbf{b}, \mathbf{c}\}$ is the basic building block of the crystal, or its *unit cell*, and the entire crystal is obtained by periodically translating it in space according to the primitive translation vectors of the underlying lattice. Therefore, the lengths $\{a, b, c\}$ of the primitive translation vectors are called *lattice constants*.

Let us now introduce a three-dimensional array of δ functions with the lattice periodicity:

$$\rho(\mathbf{r}) = \sum_{n_1}\sum_{n_2}\sum_{n_3} \delta[\mathbf{r} - (n_1\mathbf{a} + n_2\mathbf{b} + n_3\mathbf{c})]. \qquad (7.36)$$

This function provides the density of lattice points and generalizes the one-dimensional periodic comb introduced in Section 7.5.2. Using this function and the convolution operation, we can mathematically describe the entire crystal structure as follows:

$$\alpha(\mathbf{r}) = \rho(\mathbf{r}) * \alpha_{cell}(\mathbf{r}), \qquad (7.37)$$

where α_{cell} denotes the scattering density within the unit cell. Applying the convolution theorem in Fourier space and remembering equation (7.34) we obtain the following factorization for the scattered field:

$$\boxed{\psi_{sc} = \hat{\rho}(\mathbf{q}) \int_{V_{cell}} \alpha(\mathbf{r}) e^{-j\mathbf{q}\cdot\mathbf{r}} d^3\mathbf{r}} \qquad (7.38)$$

where the preceding integration is limited to the volume of the unit cell. Using the shifting property of the Fourier transform, we get the following:

$$\boxed{\hat{\rho}(\mathbf{q}) = \sum_{n_1}\sum_{n_2}\sum_{n_3} \exp[-j\mathbf{q}\cdot(n_1\mathbf{a} + n_2\mathbf{b} + n_3\mathbf{c})]} \qquad (7.39)$$

This expression is the *three-dimensional array factor*, or the interference term, associated to the periodic lattice. Equation (7.38) is a statement, limited to periodic structures, of the more general *array theorem* of antenna engineering according to which the far-field radiation diagram of a large array of radiating antennas can be decomposed into the product of its array factor times the radiation pattern associated to a single (isolated from the array) antenna element, which is given by the integral over the unit cell in equation (7.38). We state this important result for a large array of identical scattering particles arranged in an arbitrary geometry as follows:

$$\frac{d\sigma}{d\Omega} \propto |\hat{\rho}(\mathbf{q}) f(\lambda, \theta, \varphi)|^2 \propto S(\mathbf{q})|\bar{\psi}_{sc}|^2, \qquad (7.40)$$

where $f(\lambda, \theta, \varphi)$ is the scattering function, i.e., the radiation diagram, of each individual particle in the array; $S(\mathbf{q})$ is the structure factor of the array; and $\overline{\psi}_{sc}$ denotes the scattered field amplitude due to a single (isolated) element of the array at a reference point. This important theorem separates the array geometry from the single-scattering element, and it is rigorously valid for any structure (periodic or nonperiodic) whose point density can be expressed as an infinite comb of singular components akin to the case of equation (7.36). The relevant case of quasiperiodic densities with nonperiodic singular combs in their scattering spectra will be introduced later in Section 7.5 and rigorously addressed in Chapter 11 using the Poisson summation formula.

7.2.4 The Dynamic Structure Factor

The dynamic structure factor $S(\mathbf{q}, \omega)$ generalizes the previously defined static structure factor to situations in which the scattering particles in the medium are described by time-dependent position vectors. As a result, $S(\mathbf{q}, \omega)$ encodes information about interparticle correlations in both space and time, and it can be experimentally measured by either inelastic neutron scattering or Raman scattering. A convenient stating point for the definition of $S(\mathbf{q}, \omega)$ is to consider the time-dependent correlation function of the density–density Fourier components $F(\mathbf{q}, t)$, known as the *intermediate scattering function*:

$$F(\mathbf{q}, t) = \frac{1}{N} \langle \rho_{\mathbf{q}}(t) \rho_{-\mathbf{q}}(0) \rangle \tag{7.41}$$

where $\rho_{\mathbf{q}}(t)$ denotes the $\mathbf{q}$-spatial Fourier component at time t. The function $F(\mathbf{q}, t)$ is directly related to the cross section measured in inelastic scattering experiments, and it can be expressed as the spatial Fourier transform:

$$F(\mathbf{q}, t) = \int G(\mathbf{r}, t) \exp(-j\mathbf{q} \cdot \mathbf{r}) d\mathbf{r}, \tag{7.42}$$

where $G(\mathbf{r}, t)$ is the *van Hove function* defined as follows:

$$G(\mathbf{r}, t) = \left\langle \frac{1}{N} \int \sum_{i=1}^{N} \sum_{j=1}^{N} \delta[\mathbf{r}' + \mathbf{r} - \mathbf{r}_j(t)] \delta[\mathbf{r}' - \mathbf{r}_i(0)] d\mathbf{r}' \right\rangle. \tag{7.43}$$

The Fourier spectrum of the intermediate scattering function yields the dynamic structure factor [304]:

$$S(\mathbf{q}, \omega) = \frac{1}{2\pi} \int_{-\infty}^{+\infty} F(\mathbf{q}, t) e^{j\omega t} dt. \tag{7.44}$$

Therefore, the dynamic structure factor can be expressed as the spatial and temporal Fourier transform of van Hove's time-dependent pair correlation function. Moreover, it can be shown that the static structure factor $S(\mathbf{q})$ and dynamic structure factor $S(\mathbf{q}, \omega)$ are fundamentally related as follows [304]:

$$\boxed{\int_{-\infty}^{\infty} S(\mathbf{q}, \omega)d\omega = F(\mathbf{q}, 0) = S(\mathbf{q})} \tag{7.45}$$

This result is called the *elastic sum rule*. The differential scattering cross section for inelastic scattering is directly proportional to the dynamic structure factor, and it is linked to the elastic scattering differential cross section as follows:

$$\frac{d\sigma}{d\Omega} = \int \frac{d^2\sigma}{d\Omega d\omega} d\omega. \tag{7.46}$$

This expression is clearly consistent with the elastic sum rule. More specialized discussions and rigorous derivations of the results introduced in this section can be found in the excellent reference [304].

7.2.5 The Origin of Reciprocal Space

A fundamental concept in the wave scattering from periodic structures is the one of *reciprocal lattice*, which follows naturally by considering the nonzero contributions to the oscillatory complex sum in equation (7.39). The array factor summation is equal to zero unless its terms add up coherently, which happens in the following:

$$\boxed{\mathbf{q} \cdot (n_1\mathbf{a} + n_2\mathbf{b} + n_3\mathbf{c}) = 2\pi n} \tag{7.47}$$

where n is an integer number. The preceding condition is met for the vector $\mathbf{q}$ that can be expanded as follows:

$$\mathbf{q} = h\mathbf{A} + k\mathbf{B} + l\mathbf{C}, \tag{7.48}$$

where (h, k, l) are integers and the following relations hold:

$$\boxed{\mathbf{A} = \frac{2\pi\mathbf{b} \times \mathbf{c}}{\mathbf{a} \cdot (\mathbf{b} \times \mathbf{c})} \quad \mathbf{B} = \frac{2\pi\mathbf{c} \times \mathbf{a}}{\mathbf{a} \cdot (\mathbf{b} \times \mathbf{c})} \quad \mathbf{C} = \frac{2\pi\mathbf{a} \times \mathbf{b}}{\mathbf{a} \cdot (\mathbf{b} \times \mathbf{c})}} \tag{7.49}$$

The denominators in these expressions provide the volume V_c of the unit cell. Note that the dimensions of vectors $\{\mathbf{A}, \mathbf{B}, \mathbf{C}\}$ are reciprocal with respect to the ones of the primitive lattice vectors $\{\mathbf{a}, \mathbf{b}, \mathbf{c}\}$ and that, thanks to the well-known properties of scalar and vector products, the two set of vectors satisfy the following conditions:

$$\mathbf{A} \cdot \mathbf{a} = \mathbf{B} \cdot \mathbf{b} = \mathbf{C} \cdot \mathbf{c} = 2\pi \tag{7.50}$$

$$\mathbf{a} \cdot \mathbf{B} = \mathbf{a} \cdot \mathbf{C} = \mathbf{b} \cdot \mathbf{A} = \mathbf{b} \cdot \mathbf{C} = \mathbf{c} \cdot \mathbf{A} = \mathbf{c} \cdot \mathbf{B} = 0. \tag{7.51}$$

Therefore, the set of vectors where the array factor $\hat{\rho}(\mathbf{q})$ does not vanish define a periodic three-dimensional lattice in $\mathbf{q}$-space, known as the *reciprocal lattice*, with $\mathbf{q}$-space density:

$$\hat{\rho}(\mathbf{q}) \propto \sum_h \sum_k \sum_l \delta[\mathbf{q} - (h\mathbf{A} + k\mathbf{B} + l\mathbf{C})], \tag{7.52}$$

where $\{\mathbf{A}, \mathbf{B}, \mathbf{C}\}$ are called *reciprocal lattice vectors*. In general, when $\mathbf{q}$ can be represented as in equation (7.48), the array factor $\hat{\rho}(\mathbf{q})$ sums up to the value V/V_c, which depends on the number of illuminated unit cells and vanishes otherwise. The discrete nature of the support of $\hat{\rho}(\mathbf{q})$ shows that diffraction from periodic structures can only occur for the specific set of $\mathbf{q}$ vectors that belong to the reciprocal lattice. The corresponding scattering peaks are called *Bragg peaks*, after William Henry Bragg who shared the Nobel Prize with his son Lawrence Bragg in 1915 "for their services in the analysis of crystal structure by means of X-rays." These fundamental concepts also will be discussed extensively in the context of Chapter 11.

7.2.6 The Phase Problem and the Autocorrelation Approach

Diffraction experiments are often utilized as the primary tools to recover structural information on different physical systems. As we have seen in the previous sections, the link between the scattering density function and the scattered field is formulated in terms of the Fourier transform, which establishes a one-to-one map between complex functions. However, when the diffracted intensity $I(\mathbf{q}) \propto |\psi_{sc}|^2$ is measured instead of the complex field ψ_{sc}, we lose most of the relevant information contained in Fourier phase. The shifting properties of the Fourier transform should immediately convince the reader that complex functions that differ by arbitrary translations still share the exact same intensity, despite their different phases! The same problem also arises when considering the fact that a real function $f(\mathbf{r})$ and its inversion-symmetric counterpart $f(-\mathbf{r})$ also have the same (intensity) diffraction pattern. Incidentally, this is the reason why the structure factor of an arbitrary point pattern is always centrosymmetric. Therefore, the lack of phase information associated to intensity patterns results in a loss of uniqueness in the recovery of the original function since most of the structural information is actually encoded in the phase. This difficulty makes the recovery of the structure from its measured diffraction pattern generally a very complex problem that requires advanced probabilistic and optimization methods [305]. An alternative approach to retrieve structural information relies on the *autocorrelation function* (ACF), which can be obtained directly from the measured scattered intensity $I(\mathbf{q})$. This follows by expressing the autocorrelation of the function $\rho(\mathbf{r})$ as a convolution, according to the following:

$$\boxed{ACF(\mathbf{r}) \equiv \rho(\mathbf{r}) \otimes \rho(\mathbf{r}) = \rho(\mathbf{r}) * \overline{\rho(-\mathbf{r})}} \tag{7.53}$$

where we used the overline to denote the complex conjugate and the symbol $\otimes$ for the autocorrelation operator. Passing to the Fourier domain and using the convolution theorem, we can write the autocorrelation function as follows:

$$ACF(\mathbf{r}) = \mathbb{F}^{-1}\{|\hat{\rho}(\mathbf{q})|^2\} \propto \mathbb{F}^{-1}\{S(\mathbf{q})\} \tag{7.54}$$

where $\mathbb{F}^{-1}$ denotes the inverse Fourier transform operator, $\hat{\rho}(\mathbf{q})$ is the Fourier transform of the density $\rho(\mathbf{r})$, and $S(\mathbf{q})$ is the structure factor. The preceding equation establishes a one-to-one correspondence between the measured diffraction intensity and the ACF and is the basis of the *Patterson map approach* for the real-space representation of diffraction data. Despite the fact that the loss of phase in the diffraction intensity creates a fundamental ambiguity in the recovery of the correct crystal structure,[6] the Patterson map approach turns out to be a very valuable tool for structure analysis [306].

7.3   Structure Factor and Antenna Theory

The prototypical problem of antenna theory is that of computing the far-zone radiation diagram of arrays of identical radiating elements typically arranged in linear or planar periodic geometries. The kinematic approach to light scattering and radiation discussed in this chapter can be directly utilized to gain fundamental insights into such a problem. Let us illustrate this point by considering first the symmetric arrangement of two radiating isotropic dipoles as shown in Figure 7.5a. If the two antenna elements are excited with the same phase, the fundamental equivalence between radiation and scattering allows us to consider the equivalent problem of two identical point scatterers illuminated by a monochromatic plane wave at normal incidence with respect to the vertical z-axis, as shown in Figure 7.5b. The array density for this simple problem can be written as follows:

$$\rho(z) = \delta(z - d/2) + \delta(z + d/2), \tag{7.55}$$

from which the corresponding array factor is found immediately by Fourier transformation:

$$\hat{\rho}(k_z) = e^{-jk_z d/2} + e^{jk_z d/2} = 2\cos(k_z d/2). \tag{7.56}$$

Noting that $k_z = k\cos\theta$, we deduce the familiar expression for the array factor of a two-element dipole array [175]:

$$AF \equiv \hat{\rho}(k_z) = 2\cos\left[\frac{1}{2}(kd\cos\theta)\right] \tag{7.57}$$

---

[6] Patterson maps display only 24 out of the 230 three-dimensional space groups available to characterize the scattering density function, reflecting the noninvertibility of the scattering problem [303].

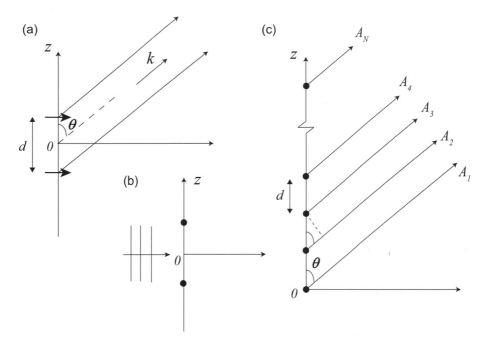

Figure 7.5 (a) Configuration of two radiating antennas forming a two-element linear array and (b) equivalent diffraction model with two point scatterers illuminated by a normally incident plane wave. (c) Configuration of an N-element linear antenna arrays. The vectors $A_1, \ldots, A_N$ denote the phasors corresponding to the scattering wave contributions of each particle in the array.

However, different antennas in an array can be excited with an arbitrary phase. Generally, a constant (uniform) phase lead β due to the relative phase excitation between consecutive antenna elements must be considered in order to shape the radiation pattern. In particular, the ability to electronically control the excitation phase enables the possibility of steering the radiation diagram as achieved by phase scanning arrays [175]. Considering the additional contribution of the excitation phase to the argument of the array factor, we can express the total (excitation plus optical path difference) phase as follows:

$$\psi = kd \cos \theta + \beta. \tag{7.58}$$

We now review some important properties of periodic arrays before discussing their aperiodic counterparts. A basic geometry is the one of the N-element *uniform linear array*, illustrated in Figure 7.5c, characterized by a constant value of β. The corresponding array factor is easily obtained:

$$AF = \sum_{n=1}^{N} e^{j(n-1)\psi} \tag{7.59}$$

and can be summed analytically, yielding the well-known expression [175]:

$$AF = \left[\frac{\sin(N\psi/2)}{\sin(\psi/2)}\right] \tag{7.60}$$

The angles θ_m that yield the directions of maximum radiation are readily computed:

$$\psi/2 = [(kd\cos\theta + \beta)/2]_{\theta=\theta_m} = m\pi, \tag{7.61}$$

where $m = 0, \pm 1, \pm 2, \pm 3, \ldots$. Inserting the wavenumber $k = 2\pi/\lambda$, we rewrite the previous condition as follows:

$$d\cos\theta_m + \beta/k = m\lambda. \tag{7.62}$$

Considering the case of in-phase array excitation ($\beta = 0$) and identifying the directions of maximal radiation by the complementary angles $\bar{\theta}$ measured with respect to the horizontal axis instead of the vertical z-axis, we have the following:

$$\boxed{d\sin\bar{\theta}_m = m\lambda} \tag{7.63}$$

This equation is *Bragg's diffraction law*. This simple derivation shows the formal analogy between the broadside radiation properties of uniform linear arrays with in-phase antenna excitation and the associated scattering problem of periodic linear chains of point-like scattering particles illuminated at normal incidence by a plane wave. The array theorem factorization discussed in Section 7.2.3 can be utilized to obtain the overall radiation diagram of the array when the radiation properties of the individual antennas are not isotropic or, equivalently, when the corresponding scattering particles cannot be considered point-like. However, we note that due to the additional degree of freedom provided by the generally nonzero phase excitation β, uniform linear antenna arrays can also be excited to produce maximum radiation along the axis of the array. This feature corresponds to the so-called *end-fire radiation mode* achieved when $\beta = \pm kd$, which yields a radiation diagram pointing downward or upward with respect to the z-axis, respectively. More generally, by controlling the value of β, the radiation can be shifted in any desired direction. For uniform linear arrays, we have the following [175]:

$$\boxed{\beta = -kd\cos\theta_0} \tag{7.64}$$

where θ_0 is the direction of the desired radiation maximum.

Large planar antenna arrays are used to scan the main directional beam of antennas across any desired point in space, which is ideal for many applications, including radar tracking and searching, remote sensing, and imaging [175]. Neglecting the mutual coupling among the individual antennas, we can express the array factor of a rectangular planar arrays of antennas as the product of two linear arrays [175]:

$$AF(\theta, \phi) = \frac{1}{MN}\left[\frac{\sin(M\psi_x/2)}{\sin(\psi_x/2)}\right]\left[\frac{\sin(N\psi_y/2)}{\sin(\psi_y/2)}\right], \tag{7.65}$$

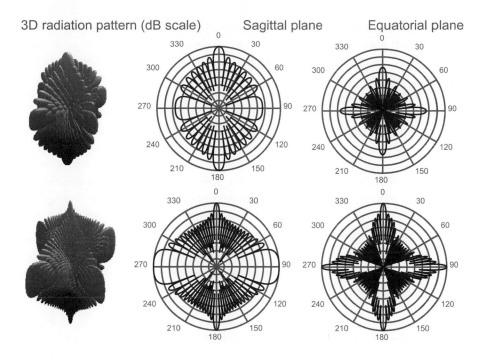

Figure 7.6 Absolute value of the array factor (radiation diagrams) of a 225-element square array of isotropic radiators with uniform amplitude (i.e., $\beta_x = \beta_y = 0$) for two different values of the array's period d. The first row in the figure corresponds to an array with period $d = \lambda/2$. The second row in the figure corresponds to an array with period $d = \lambda$. Three-dimensional radiation patterns as well as polar cuts across the sagittal plane ($\phi = 0°$) and the equatorial plane ($\theta = 90°$) are displayed. The quantity $U = 20\log_{10}(|AF|)$ is plotted in dB scale.

where we defined the total phases as follows:

$$\psi_x = kd_x\hat{\mathbf{x}} + \beta_x = kd_x \sin\theta \cos\phi + \beta_x \qquad (7.66)$$

$$\psi_y = kd_y\hat{\mathbf{y}} + \beta_y = kd_y \sin\theta \sin\phi + \beta_x, \qquad (7.67)$$

and where d_x and d_y are the spatial periods along the x- and y-axis, respectively, and a spherical reference system centered on the array has been adopted. As an example, we show in Figure 7.6 the antenna radiation patterns of a 225-element uniform square array with periodicity $d = \lambda/2$ and $d = \lambda$. The three-dimensional radiation patterns and cross-sectional polar plots for the vertical x-z (sagittal) plane as well as for the horizontal x-y plane (equatorial) are also shown. The principal maximum is referred to as the *major lobe* and the weaker side lobes as the *grating lobes*. A grating lobe is defined as a peak, other than the main one, produced when the spacing of the antenna array is sufficiently large to enable in-phase addition of radiated fields in more than one direction (similarly to the higher-order diffraction modes of a grating). The onset of undesired radiation lobes in the x-z and y-z planes can be avoided when the spacings between the elements in the x- and y-directions satisfy $d_x < \lambda/2$ and $d_y < \lambda/2$. While the radiation patterns of planar arrays can be very complex due to the appearance of many grating lobes if $d_x = d_y = d > \lambda/2$, the radiation in the equatorial plane (i.e., the plane of the array) always reflects the point-group symmetry of the array, i.e., the C_{4v} symmetry for the square lattice. This follows from the general fact that the group-theoretic symmetry properties of classical as well as quantum systems determine the physical properties of the supported eigenmodes [307]. Specifically, the radiation patterns of two-dimensional and three-dimensional antenna arrays have been studied using the rigorous group theory formalism by Kritikos, who utilized it as a new design tool for the synthesis of antennas with orthogonal radiation patterns [308]. Generally, when considering nonperiodic antenna arrays with increased rotational symmetry axes compared to their periodic counterparts, we observe more isotropic equatorial radiation patterns in addition to reduced (sagittal) grating lobes due to the lack of global periodicity. However, this behavior generally comes at the cost of a reduced intensity of the main radiation lobes and a reduced scattering efficiency.

7.4   Radiation Properties of Aperiodic Arrays

The lack of crystallographic rotational symmetries in aperiodic arrays of radiators provides unique opportunities to engineer radiation patterns with unusual directional properties or with enhanced isotropy over a broader range of geometrical parameters compared to traditional periodic arrays. This flexibility can be utilized to design novel active devices with more complex colorimetric responses for optical sensing [309, 310] or to achieve more efficient solar light trapping [311]. The radiation properties of planar antenna arrays based on aperiodic tilings have been recently investigated by Pierro et al. [312]. A significant result of this study is that the interplay between discrete and continuous components in the radiation spectrum depends strongly on the degree of long-range order of the array geometry. In particular, it was found that the more "orderly appearing" tilings, such as the Penrose or the octagonal tiling, display narrow grating lobes manifesting strong collective interference effects while the radiation patterns of more random-looking tilings, such as the Pinwheel or Denzer, are significantly broadened [312]. This behavior reflects the fact that quasiperiodic tilings generated by the projection method (more details in Chapter 11) have discrete diffraction spectra, whereas tilings obtained by more general substitution rules may exhibit either discrete or continuous spectra depending on the eigenvalues of the transformation matrix that implements the substitutions.

Here we focus on radiation properties of aperiodic arrays with spiral symmetry. In Figure 7.7, we show the calculated radiation diagram for a Vogel spiral array. The equatorial plane diagrams feature very remarkable isotropy for both $d = \lambda$ and $d = \lambda/2$, which reflects the distinctive circular isotropy of the Vogel spiral geometry discussed in detail in Section 11.3.5. A similar trend is also observed when analyzing the sagittal planes. When increasing the average separation d between the neighboring particles in the Vogel array, we observe a large number of secondary lobes that are significantly weaker compared to the case of the periodic square array shown in Figure 7.6. This behavior results from the lack of discrete scattering Bragg peaks

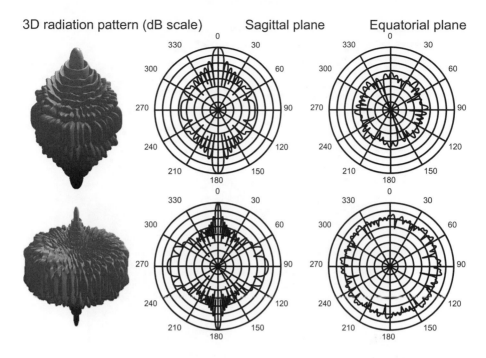

3D radiation pattern (dB scale)   Sagittal plane   Equatorial plane

Figure 7.7 Magnitude of the array factor (radiation diagrams) of a 225-element Vogel spiral golden angle array of isotropic radiators for two different average neighbor distances d in the array. The first row shows the case of $d = \lambda/2$ and the second row the case of $d = \lambda$. The sagittal plane and equatorial plane plots are in dB scale with the outermost circle corresponding 0 dB, and in each inner circle the signal decreases by 10 dB.

in the diffraction spectra of Vogel spiral arrays. The enhanced wide-angle (planar) light scattering properties of engineered Vogel spiral arrays of metallic nanoparticles have been recently investigated for device applications to ultrathin film silicon solar cells [311].

We now discuss two additional examples of radiation diagrams from the aperiodic deterministic arrays shown in Figure 7.8 together with the corresponding structure factors. In particular, in Figure 7.8(a) we illustrate the geometry of an Eisenstein point pattern of radiating elements. This is a diffractive, though nonperiodic, array derived from the number-theoretic distribution of complex prime elements in the Eisenstein's ring of integers (see Chapter 10). Eisenstein integers are complex numbers of the form $a + b\omega$, where a and b are natural integers and $\omega = (-1 + j\sqrt{3})/2$ is one of the cubic roots of one. These numbers form a triangular lattice in the complex plane and a commutative ring of algebraic integers in the imaginary quadratic field $\mathbb{Q}(\sqrt{-3})$. It is known from algebraic number theory [313–315] that the ring of Eisenstein integers is a unique factorization domain (UFD) in which every nonzero and nonunit element can be written as a product of prime elements (or irreducible elements), uniquely up to rearrangement, complex conjugation, and associates (i.e., unit multiples), in strict analogy with the fundamental theorem of arithmetic for the natural integers. Moreover,

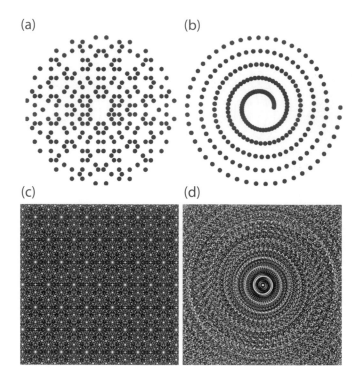

Figure 7.8 Panels (a) and (b) show the pattern of a 259-element Eisenstein array and of a 250-element Archimedean spiral array with structural parameters $(a, b, c) = (1, 0.17, 1)$, respectively. Panels (c) and (d) plot the corresponding structure factors of the Eisenstein array and of the Archimedean spiral array, respectively.

simple characterizations exist to construct Eisenstein primes in the complex plane, which are used to generate the array shown in Figure 7.8(a) [205]. Since multiplication by a unit and complex conjugation both preserve primality, the Eisenstein arrays exhibit characteristic 6×2 or 12-fold symmetry with six different rotational axes. The necessary background on algebraic number fields will be discussed in Chapter 10. Figure 7.8c shows the structure factor of the Eisenstein array that features, in addition to sharp diffraction peaks, a weaker continuous component or a diffuse background than is typically associated to structural disorder in complex media. Moreover, upon closer inspection, the diffraction pattern of Eisenstein primes reveals an hierarchical structure that encodes spatial correlations at multiple length scales, recently characterized using multifractal analysis [10, 317]. Aperiodic structures whose diffraction spectra feature the coexistence of diffractive (singular) and continuous components are referred to as singular continuous. Singular-continuous spectra are often associated to complex systems with chaotic dynamics, fractal and multifractal structures, and are also observed in quasicrystals [39, 316]. Recent investigations of the spectral and scattering properties of complex primes arrays are provided in [10, 205]. The characteristic rotational symmetry of the Eisenstein arrays is displayed by the corresponding radiation diagrams that are shown in Figure 7.9 for both

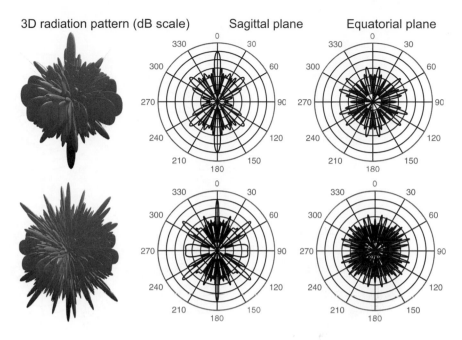

3D radiation pattern (dB scale)   Sagittal plane   Equatorial plane

Figure 7.9 Magnitude of the array factor (radiation diagrams) of a 259-element Eisenstein array of isotropic radiators for two different average neighbor distances d in the array. The first row shows the case of $d = \lambda/2$ and the second row the case of $d = \lambda$. The sagittal plane and equatorial plane plots are in dB scale with the outermost circle corresponding 0 dB, and in each inner circle the signal decreases by 10 dB.

$d = \lambda$ and $d = \lambda/2$. The highly diffractive nature of these arrays is evidenced by the many narrow radiation directions with reduced side lobes compared to the periodic square array.

An interesting example of a structure that combines in a unique way both directional and isotropic properties is provided by the generalized Archimedean spiral array displayed in Figure 7.8b along with its characteristic structure factor shown in Figure 7.9d. Generalized Archimedean spiral arrays are constructed according to the following polar representation:

$$r = a + b\theta^{1/c} \tag{7.68}$$

where (a, b, c) are free parameters and (r, θ) are the polar coordinates. The standard Archimedean spiral is recovered when $c = 1$. Other spiral shapes that are included are the hyperbolic spiral $(c = -1)$ and the Fermat's spiral $(c = 2)$. The structure factor of the generalized Archimedean spiral does not feature sharp diffraction peaks. On the opposite, it encodes a very rich and textured structure in its diffuse scattering component that consists of concentric rings qualitatively similar to the case of periodic circular gratings. The radiation diagrams of the Archimedean arrays are shown in Figure 7.10. Compared to Vogel spirals, these arrays reveal a significantly enhanced directionality in the sagittal plane in combination with azimuthal

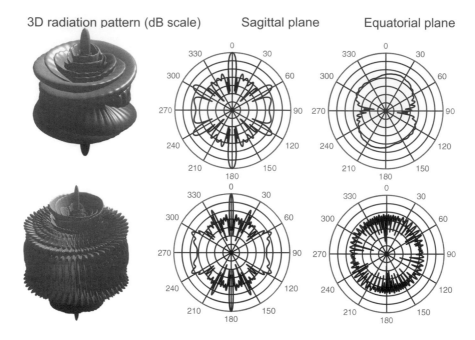

Figure 7.10 Magnitude of the array factor (radiation diagrams) of a 250-element generalized Archimedean spiral array of isotropic radiators for two different average neighbor distances d in the array. The parameters used to generate the array are specified as follows: a = 1, b = 0.17, c = 1, and θ varies in the range $[0, 10\pi]$. The first row shows the case of $d = \lambda/2$ and the second row the case of $d = \lambda$. The sagittal plane and equatorial plane plots are in dB scale with the outermost circle corresponding 0 dB, and in each inner circle the signal decreases by 10 dB.

isotropy (for large values of d/λ), as evidenced by the corresponding equatorial plane diagrams. Archimedean arrays form distinctive Archimedes' screw-shaped radiation diagrams with very unusual azimuthal structures largely controlled by their geometrical parameters, providing opportunities to create novel beams.

7.5   Examples of Structure Factor Calculations

In this section, we introduce fundamental concepts in the spectral classification of deterministic aperiodic structures by presenting additional examples and detailed calculations of their structure factors.

We first consider an array of scatterers with uncorrelated random positions over the length scales of the incident wavelength. In this case, the total correlation function is equal to zero and the corresponding structure factor is constant and equal to one. Therefore, the waves scatter with equal intensity in all directions, giving rise to a perfect isotropic response, which is the hallmark of uncorrelated random media. This idealized regime is often referred to as the *incoherent regime*. In this case, we can neglect on average the detailed phase relationships among the different scattered waves, which reduces its overall optical response to the one of a single particle multiplied by the total number of illuminated particles in the system. As a result, light diffraction by *uncorrelated random arrays* with discrete particles is determined by the properties of the individual particles, i.e., their size, shape, and composition, and does not depend on their particular geometrical arrangements.

In contrast, the *coherent scattering regime* refers to situations in which the spatial positions of the scatterers in the array are correlated over the wavelength range of interest. In this case, the distinctive geometrical relationships among the positions of the particles are captured by a nontrivial correlation function. Many fascinating examples of the coherent single scattering of light in complex media can be found in Nature when diffraction occurs in biological tissues, scales, and other structures characterized by a coherence length of the order of the optical wavelength. Under such conditions, spectacular iridescent phenomena take place similar to the ones displayed by butterflies' wings, the exoskeleton of beetles, or the inside certain shells, just to name a few [318, 319].

The correlation function and the structure factor of perfectly periodic structures that infinitely extend in space consist of infinite trains of delta functions that correspond to ideal Bragg peaks, as previously shown in Figure 7.3 for one-dimensional (1D) periodic arrays (i.e., chains) of point scatterers. More complex and interesting structural correlation effects in the single-scattering regime are displayed by quasiperiodic structures such as the *Fibonacci chain*. In this case, the position of the nth scattering particle along the Fibonacci chain is given by the following:

$$x_n = n + \frac{1}{\tau}[(n+1)/\tau] \qquad (7.69)$$

where $\tau = (1 + \sqrt{5})/2 = 1.618034 \ldots$ is the *golden mean* and $[x]$ indicates the integer part of the real number x. Note that the first term in equation (7.69) describes a periodic chain of particles with unit spacing, while the second term is scaled by a factor of $1/\tau$ every time that n is increased by a factor τ. As a result, the Fibonacci chain can be regarded as the sum of two periodic functions with *incommensurate periods*, i.e., 1 and τ. The structure factor of the Fibonacci chain is compared to that of a single realization of a random chain of scatterers in Figure 7.11. We appreciate that, contrary to the random chain, the Fibonacci quasiperiodic chain supports sharp Bragg peaks that are aperiodically distributed in q-space. Due to the incommensurate nature of the Fibonacci structure, there are no coincidences in q-space between any two diffraction peaks. As a result, the diffraction peaks of a Fibonacci chain densely fill the q-space with a characteristic self-similar intensity distribution pattern and can be indexed using two integers m and m' [11]. This is in stark contrast to the case of periodic 1D structures where only one integer label suffices to label the diffraction peaks. The scenario introduced in the preceding is not limited to Fibonacci structures, but it is characteristic of any 1D incommensurate structure,[7] as we will better appreciate in Chapter 11.

---

[7] This is not the case for higher-dimensional arrays.

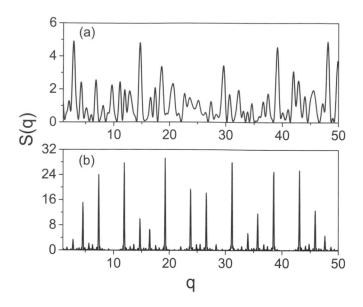

Figure 7.11 Calculated structure factor of (a) a pseudorandom chain of uniformly distributed scattering points and (b) a Fibonacci quasiperiodic chain (scattering points distributed according to the one-dimensional Fibonacci sequence). In both cases, the number of scatterers is $N = 31$.

Much more complex (and aesthetically pleasing!) diffraction patterns arise for two-dimensional arrays even in the case of very simple geometries. This is often the case for point patterns that display circular or spiral symmetry, as illustrated in Figure 7.12. In particular, in Figure 7.12a we show a uniform distribution of N point particles arranged around a circular region at angles $\theta_n = (2\pi/N)n$, where $n = 0, \ldots, N - 1$. It is instructive to analytically compute the corresponding array factor by transforming the Fourier components of the array density to polar coordinates. The spatial frequency components associated to the points' coordinates $x_n = r_n \cos \theta_n$ and $y_n = r_n \sin \theta_n$ can be expressed as follows:

$$v_x = v_\rho \cos \phi \qquad (7.70)$$

$$v_y = v_\rho \sin \phi, \qquad (7.71)$$

where $v_\rho = \sqrt{v_x^2 + v_y^2}$ is the radial spatial frequency conjugate to the polar radius r and ϕ is the polar angle in spatial frequency space that is conjugate to the polar position angle θ. Using the preceding relations, we can transform the Cartesian expression of the array factor into its polar counterpart:

$$\hat{\rho}(\boldsymbol{v}) = \hat{\rho}(v_\rho, \phi) = \sum_{n=0}^{N-1} e^{-2\pi j(v_x x_n + v_y y_n)} = \sum_{n=0}^{N-1} e^{-2\pi j r_n v_\rho \cos(\theta_n - \phi)} \qquad (7.72)$$

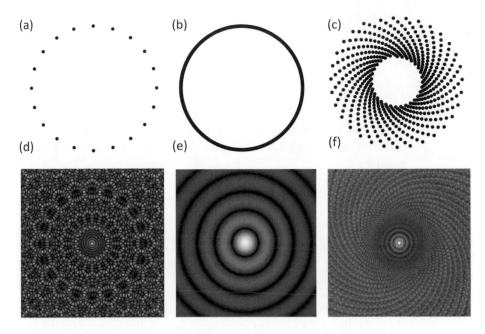

Figure 7.12 Representative two-dimensional arrays of scattering point particles and computed structure factors in panels (a–c). The corresponding structure factors are shown in panels (d–f).

where we made use of the trigonometric identity $\cos\theta\cos\phi + \sin\theta\sin\phi = \cos(\theta - \phi)$. The structure factor obtained by squaring the modulus of the preceding expression is displayed in Figure 7.12d for $N = 20$ particles.

Interestingly, the structure factor of discrete point particle arrays can be utilized as a starting point for the computation of light diffraction from continuous objects. As an example of this approach, we compute the far-field diffraction pattern of the continuous circular region that is obtained in the limit of an infinite number of particles at angular positions $\theta_n = (2\pi/N)n$, each scattering the incoming (normal incident) radiation with identical scattering amplitude F. We will then consider the continuum limit of the previous summation for $N \to \infty$ and $F \to 0$, in such a way that the product NF remains constant (say $NF = 1$). In this limit, the previous summation becomes an integral:

$$\frac{1}{2\pi} \sum_{n=0}^{N-1} NF e^{-2\pi j r_n v_\rho \cos(\theta_n - \phi)} \Delta\theta_n \to \frac{1}{2\pi} \int_0^{2\pi} d\theta' e^{-2\pi j r v_\rho \cos\theta'}, \qquad (7.73)$$

where we introduced the continuum angular variable $\theta' = \theta - \phi$ and used $\Delta\theta_n = (2\pi/N)\Delta n = 2\pi/N$. Remembering the integral representation of the mth-order Bessel function:

$$J_m(x) = \frac{j^{-m}}{2\pi} \int_0^{2\pi} e^{j[x\cos\theta + m\theta]} d\theta, \qquad (7.74)$$

we finally obtain the desired diffraction pattern:

$$S(v_\rho) = |J_0(2\pi r v_\rho)|^2$$

(7.75)

This formula coincides with the expression for the far-field diffraction pattern, shown in Figure 7.12e, of an infinitesimally thin ring pupil of radius r displayed in Figure 7.12b. Given the cylindrical symmetry of the scattering problem, the diffraction pattern of the thin ring can be easily obtained by directly computing its Fourier–Bessel transform:

$$\mathbb{B}[g(r')] \equiv 2\pi \int_0^\infty r' g(r') J_0(2\pi v_\rho r') dr',$$

(7.76)

where

$$g(r') = \frac{\delta(r - r')}{2\pi r'}$$

(7.77)

is the mass density of the circular thin ring of radius r in the two-dimensional plane. Note that the delta function must be scaled by the factor $2\pi r'$ in order to ensure the correct normalization of the two-dimensional polar integral of $g(r)$ over the entire plane.

7.5.1   Single Scattering of Incoherent Radiation

The structure factor of an arbitrary array of small particles can be utilized to build up a simple model for the analysis of the single-scattering properties of spatially incoherent scalar waves diffracted by planar apertures. To this purpose, let us remember that spatially incoherent radiation can be thought of as a statistical ensemble of monochromatic plane waves with a uniform distribution of propagation directions. Considering a two-dimensional geometry for simplicity, each plane-wave component is incident on a planar diffracting structure, here assumed composed of an arbitrary distribution of small apertures, such that the field distribution on the plane of the array is represented as follows:

$$\psi_0 = \psi_{inc} \sum_i \delta(\mathbf{r} - \mathbf{r}_i) = e^{-jk_0 x \sin\theta} \sum_i \delta(\mathbf{r} - \mathbf{r}_i),$$

(7.78)

where $\mathbf{k}_0 = k_0[\sin\theta \hat{\mathbf{x}} + \cos\theta \hat{\mathbf{z}}]$ is the wavevector of the incident wave, $\mathbf{r} = x\hat{\mathbf{x}} + z\hat{\mathbf{z}}$ and the diffracting screen is located on the plane $z = 0$. The unit amplitude incident wave propagates at an angle θ with the longitudinal z-axis. The x-, y-, and z-components of the incoming wavevector are called the transverse $k_\perp$ (in the plane of the screen) and longitudinal $k_\parallel$ components, respectively.

The diffracted far-field intensity distribution is easily obtained by considering the two-dimensional spatial Fourier transform of the field distribution ψ_0 in the plane of the array:

$$\boxed{I = |\mathbb{F}\{\psi_0\}|^2 = S(k_\perp - k_{0\perp}) = S(k_\perp - k_0 \sin\theta)} \tag{7.79}$$

where we used the shifting property of the Fourier transform. We note that the diffraction pattern of a plane wave that is incident on the screen at an angle θ can be simply obtained from the one under normal incidence $S(k_\perp)$ by rigidly translating it by an amount given by the in-plane incident momentum $k_{0x} = k_0 \sin\theta$. If the incoming radiation is spatially incoherent, the plane-wave components that are incident at different angles will be independent from each other (no fixed-phase relations exist), and their intensity contribution can then be simply added together to obtain the total diffracted intensity as follows:

$$I_{tot}(k_\perp) = \int\!\!\!\int_{-\infty}^{+\infty} I_0(k'_\perp) S(k_\perp - k'_\perp) d^2 k'_\perp, \tag{7.80}$$

where $I_0(k'_\perp)$ is a weighting factor describing the angle-dependent amplitudes of the incoming plane waves in the ensemble.

We recognize in the preceding result a convolution integral that we write concisely:

$$\boxed{I_{tot}(k_\perp) = I_0(k_\perp) * S(k_\perp)} \tag{7.81}$$

Note that this equation correctly predicts that the diffracted far-field intensity is given by the structure factor of the screen when all the waves in the ensemble are normally incident to it, i.e., $I_0(k_\perp) = \delta(k_\perp)$. We should also note the similarity of the equation (7.81) with the more general Schell's theorem derived in the previous chapter. This simple yet powerful approach to light diffraction of incoherent radiation has been recently utilized to design aperiodic arrays of scattering nanoparticles with enhanced directional light extraction from light-emitting devices [320, 321].

7.5.2 Periodic and Quasiperiodic Structures

In this section, we return to the application of kinematic scattering theory to the analysis of the spectral properties of 1D quasiperiodic structures. Despite its simplicity, this approach is sufficient to capture many fundamental aspects of aperiodic systems, and it provides the basis for the more rigorous discussion in Section 11.4.

Let us start by considering the Fourier spectral properties of an infinite periodic chain of point scatterers spaced by the lattice constant a. The results that we will obtain in this section can be generalized to finite-size chains of particles of arbitrary shape. The first step is to analytically represent the geometry of the scattering chain by introducing its density function $\rho_P(x)$. This is achieved by the Shah function $\text{Sh}(x)$, also known as the periodic comb or the impulse train function, defined as follows:

$$\boxed{\text{Sh}(x) = \sum_{n=-\infty}^{+\infty} \delta(x - n)} \tag{7.82}$$

The Fourier spectral components of the chain can immediately be obtained using the Fourier transform property of the scaled Shah function:

$$F\left\{\text{Sh}\left(\frac{x}{a}\right)\right\} = \hat{\rho}_P(v) = a\,\text{Sh}(av) = a\sum_{m=-\infty}^{+\infty}\delta(av - m), \qquad (7.83)$$

where F denotes the Fourier transform operator and $v = 1/x$ is the Fourier spatial frequency along the x-axis, which is related to the corresponding wavevector component k by the relation: $k = 2\pi v$. Equation (7.83) shows that the diffraction spectrum of the linear chain is different from zero only on a discrete set of spatial frequencies satisfying the relation $v = m/a$. In terms of diffraction vectors, the Fourier spectrum is nonzero for the discrete set:

$$k_m = m\frac{2\pi}{a}, \qquad (7.84)$$

where m is an integer label. The set of discrete k-vectors defined by equation (7.84) for all integer values of m span the *reciprocal space* of the periodic chain. This simple derivation shows that the Fourier spectrum of a periodic structure consists of a series of delta peaks that can be indexed by a single integer, known as the *Miller index*, according to the following:

$$F_m = \sum_m \delta(k_m - m2\pi/a) \qquad (7.85)$$

Let us now discuss the case of an aperiodic incommensurate structure obtained by superimposing two periodic chains with different spatial periods such that their ratio approximates the irrational number α:

$$\rho_I(x) = \sum_{n,k}[\delta_1(x - na) + \delta_2(x - \alpha ma)], \qquad (7.86)$$

with n and m integers. The chain defined by the density in (7.86) is a deterministic structure, but it is nonperiodic since there cannot be any spatial coincidence between the two periodic components when α is an irrational number. However, each of the periodic chain components δ_1 and δ_2 can be expanded in a Fourier series similar to (7.85), giving rise to the following spectrum:

$$F_{m,m'} = \sum_{m,m'}\delta[k_{m,m'} - 2\pi/a(m + m'/\alpha)] \qquad (7.87)$$

Note that $F_{m,m'}$ is labeled by two independent Miller integers m and m' (even though the structure is one-dimensional), but it still consists of a series of delta peaks similarly to the case of the periodic structure. However, for incommensurate structures, the density of diffraction peaks in reciprocal space is much larger than for periodic structures because the fractional parts of the multiples of the irrational number α cover densely (i.e., uniformly) the unit interval. This is the fundamental mathematical property of uniformly distributed sequences modulo one that we will discuss in detail in Section 8.5 in relation to deterministic sub-randomness. Due to their discreteness, the diffraction spectra of both periodic and incommensurate nonperiodic structures are referred as *pure-point spectra* since they are characterized by the presence of sharp Bragg peaks.

The diffraction spectrum of the quasiperiodic Fibonacci structure can be derived analytically for 1D chains. A detailed quantitative treatment [322] leads to conclude that the nonzero components of the Fibonacci spectrum occur at wavevectors:

$$k_{m,m'} = \frac{2\pi\tau^2}{\tau^2 + 1}(m + m'/\tau), \tag{7.88}$$

where τ is the golden mean. The Fourier transform of the Fibonacci chain can also be calculated analytically in the following form [11]:

$$F(k) = \sum_{m,m'} A_{m,m'}\delta(k - k_{m,m'}) \tag{7.89}$$

where the Fourier amplitude coefficients are explicitly given by the following [11]:

$$A_{m,m'} = \frac{\sin\left[\frac{\pi\tau}{\tau^2+1}(\tau m' - m)\right]}{\left[\frac{\pi\tau}{\tau^2+1}(\tau m' - m)\right]}\exp\left[j\pi\frac{\tau - 2}{\tau + 2}(\tau m' - m)\right]. \tag{7.90}$$

Long-range correlated or deterministic structures without periodicity that nevertheless exhibit an *essentially discrete Fourier spectrum* with peaks that densely fill the reciprocal space are called *quasiperiodic*. Materials with such a property exist in Nature and are called quasiperiodic crystals or *quasicrystals* [38, 323]. Their diffraction spectra are characterized by more Miller indices than the dimensionality of the embedding space, which is a distinctive feature for both incommensurate structures and quasicrystals. For one-dimensional problems, we cannot distinguish between incommensurate structures and quasicrystal structures. However, in higher spatial dimensions, quasicrystals are distinct from incommensurate structures, also known as modulated crystals, since they do not admit an average periodic lattice from which they are obtained by an irrational modulation. These important concepts will be addressed in more depth in Chapter 11. According to equation (7.90) the brightest diffraction spots of a Fibonacci chain appear when $\tau m' - m \simeq 0$, which implies that the ratio m/m' is close to τ. This condition is satisfied when (m,m') are successive Fibonacci numbers such as $(1,1), (2,1), (3,2), (5,3)$, and so on. For labels outside such a Fibonacci sequence, the diffracted intensity strongly decreases. This behavior explains why, despite the fact that the Bragg peaks densely fill the reciprocal space, we mostly observe isolated bright peaks in the diffraction pattern of quasicrystals. Finally, we remark that the Fourier spectrum of Fibonacci quasicrystals is *self-similar*, or fractal. This follows from the fact that the sequence of diffraction vectors in equation (7.88) is invariant when multiplied by any power of τ. Analogous properties hold true also in higher dimensions, as demonstrated by Levine and Steinhardt [322].

7.5.3   More Complex Aperiodic Structures

In the previous section, we have established that the diffraction spectrum of a Fibonacci quasicrystal exhibits a self-similar countable set of delta-like Bragg peaks at incommensurate intervals. Here we introduce more general deterministic structures that display even richer Fourier spectra characterized by the fact that the individual Bragg peaks cluster to form broader bands in reciprocal space. These types of spectra are known as *singular-continuous spectra*. More precisely, we say that singular-continuous structures support Fourier spectra that can be covered by an ensemble of open intervals with arbitrarily small total length, i.e., they are "singularly supported." The chief example of this category of deterministic aperiodic structures is the *binary Thue–Morse (TM) sequence* [51, 324]. The TM sequence of order N has $M = 2^N$ elements composed of two symbols, 0 and 1, defined recursively as follows:

$$s_0 = 0, s_{2n} = s_n, s_{2n+1} = 1 - s_n. \tag{7.91}$$

The preceding equations generate an infinite string of digits that never repeats itself. Interestingly, the TM sequence is self-similar and can also be generated by the simple substitution: $0 \longrightarrow 01, 1 \longrightarrow 10$. Moreover, the Fourier spectrum of the TM sequence s_n can be obtained analytically by first defining the sequence over the two-digit alphabet $(1, -1)$, namely $s_n = (1, -1, -1, 1, -1, 1, 1, -1 \cdots)$, and then considering the associated *generating function*:

$$f(z) = \sum_{n=0}^{+\infty} s_n z^n \tag{7.92}$$

Following [325], we note that the invariance of the TM sequence under the substitution $1 \longrightarrow 1, -1, -1 \longrightarrow -1, 1$ implies the following functional equation:

$$f(z) = (1 - z) f(z^2) \tag{7.93}$$

with the following factorization:

$$f(z) = (1 - z)(1 - z^2)(1 - z^4)(1 - z^8) \cdots \tag{7.94}$$

Since we found an analytical expression for the TM generating function, we easily obtain the unilateral Z-transform of the TM sequence through the substitution $z \longrightarrow 1/z$. This step allows us to compute analytically the Fourier transform of the TM sequence. In fact, we should remember that the Fourier transform of a sequence can be obtained directly by evaluating its Z-transform on the unit circle, i.e., for $z = e^{j\omega}$, where ω is the angular frequency. Therefore, the Fourier spectrum of the TM sequence shown in Figure 7.13 can be written in closed form as follows:

$$F_{TM}(\omega) = \prod_{k=1}^{+\infty} \left[1 - e^{j\omega 2^k} \right] \tag{7.95}$$

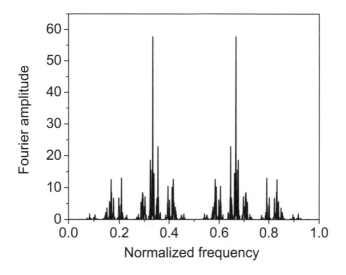

Figure 7.13 Calculated Fourier amplitude spectrum of the Thue–Morse aperiodic chain (shown in normalized units). The chain was truncated to a length of 1,024 elements. The spectrum features prominent scattering peaks that would be absent in the absence of long-range order, such as in a truly random chain with delta-correlated disorder.

Note that this spectrum obeys the scaling law $F_{TM}(\omega) = [1 - \exp(j\omega)]F_{TM}(2\omega)$ and satisfies the symmetries $F_{TM}(-\omega) = F^*_{TM}(\omega)$ and $F_{TM}(\omega + 2\pi) = F_{TM}(\omega)$. No exact labeling scheme exists for the TM diffraction peaks and the reciprocal space is simply identified with the support of its singular-continuous Fourier spectrum, even though its gaps can be predicted based on an extended approach [326]. A detailed mathematical analysis of the TM spectral properties demonstrates the absence of delta peaks in the spectrum [51]. However, the TM spectrum features very intense diffraction peaks of finite widths due to the global scale invariance, which corresponds to periodicity on a logarithmic scale. The peculiar nature of the TM structure makes us clearly realize that long-range ordered aperiodic structures without Bragg peaks still possess numerous symmetries such as scale invariance, conjugation, and mirror symmetry, which give rise to highly structured Fourier spectra and sizable diffraction effects. For example, the infinite TM sequence reproduces itself when the complementary elements are inserted after each element or upon decimation, when every other element is eliminated (implying that the scaling factor of the self-similarity is equal to 2).

It is clear from these preliminary considerations that the connection between the type of long-range order of aperiodic structures and their spectral properties is nontrivial. In fact, deterministic structures exist that display a continuous Fourier spectrum. The Rudin–Shapiro (RS) structure [327, 328] is the primary example of a deterministic system with *absolutely continuous* diffraction spectra and with a discrete (i.e., pure-point) energy spectrum akin to random media in the localization regime. However, the nature of the eigenmodes and their localization properties are not yet fully

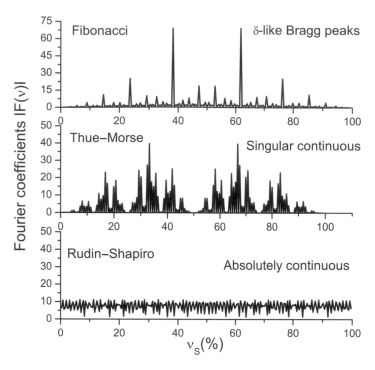

Figure 7.14 (a) Absolute value of the Fourier coefficients of a quasiperiodic (Fibonacci) structure, (b) of an aperiodic (TM) structure with singular continuous spectrum, (c) of an aperiodic structure with absolutely continuous spectrum (RS structure). Adapted from [331]

understood [329]. For instance, differently from fractal structures such as Fibonacci and Thue–Morse ones, the integrated density of states of 1D Rudin–Shapiro structures was found to scale logarithmically and their energy spectra display multifractal behavior for certain values of the scattering strength [330]. RS multilayers are generated from a two-letter alphabet subject to the inflation rule: $AA \rightarrow AAAB$, $BB \rightarrow BBBA$, $AB \rightarrow AABA$, $BA \rightarrow BBAB$, starting from the initial seed AA [71]. The transmission spectra of RS photonic or electronic structures are characterized by a distribution of narrow resonant peaks. The Fourier spectra of the three main examples of one-dimensional aperiodic sequences generated by binary substitutions are displayed in Figure 7.14. Rudin–Shapiro structures are expected to share most of their physical properties with disordered random systems, including the presence of localized optical states (i.e., Anderson-like states). However, the abundance of short-range correlations, whose main effect is to reduce the degree of disorder and localization, favors the existence of resonant extended states in the energy spectra, and significantly complicates the theoretical analysis of Rudin–Shapiro and other deterministic structures with absolutely continuous Fourier spectra. In fact, extended states have been recently discovered in the spectrum of Rudin–Shapiro structures [332, 333].

7.5.4   Uniform Random Structures

We now consider in more detail the kinematic scattering properties of linear chains of scattering particles with uniformly distributed random coordinates. Since this specific model of disorder is characterized by the absence of spatial correlations, i.e., it is a δ-correlated random processes, sharp Bragg peaks cannot appear in the spectrum of the chain, as we will now show.

Let us consider a 1D array of N-point particles with random uncorrelated spacings. In this system, each particle is translated along the x-direction according to a randomly fluctuating distance d_n such that the density of the random chain $\rho_R(x)$ is expressed as follows:

$$\rho_R(x) = \sum_{n=0}^{N-1} \delta(x - d_n). \tag{7.96}$$

Remembering the shifting property of the Fourier transform, we immediately obtain the Fourier transform of the random chain's density:

$$F\{\rho_R(x)\} = \sum_{n=0}^{N-1} e^{-j2\pi\nu d_n} = 1 + e^{-j2\pi\nu d_1} + e^{-j2\pi\nu d_2} + \cdots \tag{7.97}$$

The corresponding Fourier power spectrum is given by the following:

$$I(\nu) = \left(1 + e^{-j2\pi\nu d_1} + e^{-j2\pi\nu d_2} + \cdots\right) \times \left(1 + e^{j2\pi\nu d_1} + e^{j2\pi\nu d_2} + \cdots\right). \tag{7.98}$$

The last expression can alternatively be written as follows:

$$I(\nu) = \left[N + \sum_{j=0}^{N-1}\sum_{k=0}^{N-1} \cos 2\pi\nu(d_j - d_k) \right] \tag{7.99}$$

where in the double sum $j \neq k$. The second term in the preceding equation is a super-position of cosine random functions with uncorrelated random periods, and therefore it sums up to zero for large enough values of N. As a result, the power spectrum $I(\nu)$ of the random chain is equal to a constant determined by the total number of particles in the chain. No sharp diffraction peaks contribute to this spectrum. Indeed, after ensemble averaging, a constant (flat) Fourier spectrum characterizes wave diffraction from arrays of particles with randomly uncorrelated positions.

7.6   Lattice Resonance in Particle Arrays

Wave diffraction and scattering effects are significantly enhanced when the incident wavelength becomes comparable to the correlation length ℓ_c of the considered structures. For example, dielectric media with a spatial periodicity $a \sim \lambda$, known as *photonic crystals*, are known to dramatically modify the dispersion relation and the optical density of photons' states inside controllable frequency regions and can produce a substantial degree of optical confinement. As we will see in Chapter 13, a detailed description of the electromagnetic response of arrays of identical particles with sizes and separations comparable to λ is generally quite complicated due to the interactions of both lattice and site resonances. However, a simple model based on vector dipole interactions is often sufficient to describe the basic physics of arrays in the limit of particles (or holes) that are small compared to both λ and their separations [334, 335]. In this case, a *dynamic scattering approach* enables the understanding of the basic physics demonstrated in many recent experiments with plasmon supporting nanostructures, such as the observation of lattice resonances with narrow line shapes [336, 337] and of extraordinary light transmission effects [338] in periodically structured metallic surfaces. The method, which is systematically reviewed in the excellent paper by García de Abajo [339], considers the linear scattering of an external plane wave on a periodic array of identical and nonmagnetic particles that are small compared to the wavelength and their separations. Each particle is characterized by an induced dipole moment $\mathbf{p}_n = \alpha_E \mathbf{E}(\mathbf{r}_n)$ at position $\mathbf{r}_n$, where α_E is the electric polarizability. The dipole radiates an electric field at a position $\mathbf{r}$ that is obtained by contraction with the dyadic Green function, i.e., $\mathbf{E}(\mathbf{r}) \propto \overleftrightarrow{\mathbf{G}}_0(\mathbf{r} - \mathbf{r}_n)\mathbf{p}_n$. The dipole in the array is excited by the external plane wave $\mathbf{E}_{inc} = \mathbf{E}_0 \exp(j\mathbf{k}_\| \cdot \mathbf{r}_n)$ (where $\mathbf{k}_\|$ is the component of the incoming wave momentum parallel to the array) and by the fields scattered by all the other dipoles in the array at its position $\mathbf{r}_n$, yielding the *self-consistent coupled-dipole equations*:

$$\mathbf{p}_n = \alpha_E \left[\mathbf{E}_{inc}(\mathbf{r}_n) + \sum_{m \neq n} \overleftrightarrow{\mathbf{G}}_0(\mathbf{r}_n - \mathbf{r}_m)\mathbf{p}_m \right]. \qquad (7.100)$$

This coupled dipole model, which is the starting point of the Foldy–Lax formulation of multiple scattering, will be discussed in detail and extended to both electric and magnetic contributions in Chapter 12, where we will be concerned with aperiodic arrays. However, when specialized to periodic arrays, the preceding equations will admit the periodic Bloch-form solution $\mathbf{p}_n = \mathbf{p} \exp(j\mathbf{k}_\| \cdot \mathbf{r}_n)$. This ansatz leads directly to the important relation [334, 339]:

$$\mathbf{p} = \frac{1}{1/\alpha_E - \overleftrightarrow{\mathbf{G}}(\mathbf{k}_\|)} \mathbf{E}_0 \qquad (7.101)$$

that separates the properties of single particles, captured by their polarizability α_E, from the ones of the periodic lattice, described by the *lattice sum*:

$$\overleftrightarrow{\mathbf{G}}(\mathbf{k}_\|) = \sum_{n \neq 0} \mathbf{G}_0(\mathbf{r}_n) \exp(-j\mathbf{k}_\| \cdot \mathbf{r}_n), \qquad (7.102)$$

where $\mathbf{r}_0 = 0$. The lattice sum, which contains the collective response of the array, can be converted into a rapidly converging series using the Ewald's summation method originally developed in the dynamical theory of X-ray diffraction [340]. The lattice sum can be expressed as a summation over the reciprocal lattice vectors of the array, and explicit forms have been obtained by García de Abajo et al. [339] corresponding to normal incidence. Moreover, this formalism has been applied to compute the reflection and the absorption in periodic particle arrays, leading to a particularly simple formula for the zero-order reflection coefficient of perfect conductors under normal incidence [334, 335]:

$$r = \frac{2\pi jk/A}{1/\alpha_E - G_{xx}(0)} \qquad (7.103)$$

This expression refers to the xx dyadic component and A denotes the area of the unit cell. The singularities of the lattice sum, which physically originate from the accumulation of in-phase scattered fields, determine the reflection/transmission behavior of the arrays. In particular, zero reflection can be achieved when $G_{xx}(0) \to \infty$, rendering the array invisible (i.e., with unit transmission) even in the presence of absorbing particles. A detailed analysis performed in [334] for arrays of perfectly conducting particles determines closed-form expressions for the lattice *singularities* under both normal and oblique incidence conditions. In particular, it is shown that the normal-incidence lattice sum for a square array of period a diverges when $\lambda \gtrsim a$, corresponding to the Rayleigh's grazing condition for a diffracted beam.

A similar behavior is also obtained for hole arrays in perfect-conductor screens due to the *Babinet principle* that links the transmitted field of a hole array for a given incident polarization with the reflected field of the complementary array of disks for the orthogonal polarization. Therefore, the reflectance spectra of particle arrays are identical to the transmittance spectra of the complementary perforated screens. This general principle allows one to apply the previous Ewald's summation technique to the case of hole arrays as well and deduce two important features that appear for a square lattice under normal incidence: (i) the lattice sum $G_{xx}(0)$ diverges when $\lambda = a$, leading to vanishing transmission; (ii) the array has a transmission resonance (100% transmission) at a wavelength slightly above a because, no matter how large the term $1/\alpha_E$ becomes for small holes (remember that the polarizability is proportional to the cube of the hole radius), there is always one wavelength at which the diverging lattice sum matches it [334]. These striking phenomena are the manifestation of collective responses driven by multiple scattering paths (i.e., dynamic diffraction effects) among the holes of planar periodic arrays and have a similar origin to *Wood's anomalies* observed in metal gratings [341]. Lattice surface resonances in metallic nanostructures are equivalently interpreted as due to the interference of surface bound states excited at metal–dielectric interfaces with the radiation continuum of transmitted waves, giving rise to the characteristic transmission spectrum with the *Fano line shape* [342, 343]:

$$T = C\frac{(q + \epsilon)^2}{1 + \epsilon^2} \qquad (7.104)$$

where C is a constant, q is a parameter describing the strength of the coupling to the lattice surface resonance, and ϵ is interpreted as the frequency of light.

The coupled-dipole approach introduced here and discussed in more detail in Chapter 12 also provides a good starting point to describe the extraordinary transmission effects observed in two-dimensional quasiperiodic hole arrays [344, 345]. The finite lattice sums for these aperiodic structures exhibit pronounced maxima at the diffraction peaks which, according to the general gap-labeling theorem, reflect the singularities of the Fourier transforms of the corresponding particles/holes distributions [70, 346, 347]. The diffraction spots of quasiperiodic structures span their *Fourier module*, which extends the familiar concept of reciprocal lattice to aperiodic diffractive systems with non-crystallographic symmetries, which are discussed in Chapter 11. The important message here is that collective effects due to the long-range electromagnetic interactions can be very important in quasiperiodic structures with pure-point and singular-continuous diffraction spectra [54, 71], providing novel opportunities to engineer focusing and subwavelength field confinement using deterministic aperiodic structures [348–350].

8 Tailored Disorder

In all chaos there is a cosmos, in all disorder a secret order.

The Archetypes and the Collective Unconscious, Carl Jung

Despite our ability to identify regularities and discover patterns, random structures and processes abound all around us and in fact have stimulated scientific understanding for a long time and greatly expanded its reach. In this Chapter, we introduce the statistical description of random media with a focus on structural correlations and their implications for transport and far-field scattering properties. Within the limits of the kinematic wave scattering theory, we propose a pedagogical excursion that builds on the fundamental connection between random walks and diffusion phenomena. In this context, the analysis of correlated random walks provides the opportunity to introduce anomalous transport within the mathematical framework of fractional equations. We then introduce the fascinating concept of hyperuniform disorder and discuss some of its implications for the scattering of radiation by random structures with engineered correlations. Finally, we introduce the concept of subrandomness and illustrate its applications to wave scattering from aperiodic media that are constructed using low-discrepancy sequences.

8.1 Random Media in the Single-Scattering Regime

In contrast to deterministic structures, the composition, shape, orientation, and position of each constituent particle of a random medium are practically unknown and can only be modeled statistically. As a result, the measurable quantities of random structures are defined through ensemble averaging procedures that rely on appropriate models of disorder. From this perspective, a "random medium" is the physical manifestation of a *statistical random process* that describes the uncertainty of materials properties, such as the dielectric permittivity, or configuration parameters, such as the positions, shapes, and orientations, of component particles. Each set of values assumed by material properties and configuration variables gives rise to a specific *realization of the random medium*. The set of all possible realizations of the random medium is called the *ensemble*, which represents the random process viewed "as a whole."

Random processes can occur in a great variety of forms depending on the nature of the random variables and their correlation properties (see Appendix C). For example, many random processes are completely characterized by their correlation functions; the most notable examples are real and complex Gaussian processes. An important property of Gaussian stationary random processes is that they are completely determined by their first and second statistical moments, which motivates their wide application to science and engineering. However, the assumptions on the model of disorder in random processes always require a careful experimental justification, which is not always an easy task to achieve in practice. In fact, the structural properties of many natural and man-made random media are not always captured by either Gaussian statistics or uncorrelated uniform randomness but are influenced to various degrees by distinctive correlation effects. More broadly, the engineering of correlation effects in random and deterministic aperiodic structures have recently emerged as a promising approach to control wave scattering and transport phenomena in complex media with "tailored disorder" and controllable complexity. In this context, we introduce the basic ideas behind hyperuniform and subrandom point patterns based on low-discrepancy sequences.

8.1.1   Statistical Description of Random Many-Particle Systems

The microstructure of a discrete random medium composed of N identical and nonoverlapping particles at positions $\mathbf{r}_1, \mathbf{r}_2, \ldots, \mathbf{r}_N$ embedded in a background medium of volume V is described in terms of certain *statistical distribution functions*, as we will see in this chapter. When the particles can be considered point-like with no internal degrees of freedom, the discrete random medium can be regarded as a statistical *spatial point process*. Point processes are very important tools in the spatial description or randomness. The mathematical theory of spatial point processes applies to a variety of physical problems encountered in different fields such as ecology, neuroscience, cell biology, materials science, condensed matter physics, and even cosmology [304, 351–354].

The most general statistical approach to characterize a discrete random medium is formulated in terms of the *N-particle probability density function*, which is the normalized joint probability density function $p(\mathbf{r}_1, \mathbf{r}_2, \ldots, \mathbf{r}_N)$ proportional to the probability of finding the N particles in a specified microscopic configuration [304]. More precisely, $p(\mathbf{r}_1, \mathbf{r}_2, \ldots, \mathbf{r}_N)$ is a nonnegative quantity defined in such a way that $p(\mathbf{r}_1, \mathbf{r}_2, \ldots, \mathbf{r}_N) d^3\mathbf{r}_1 d^3\mathbf{r}_2 \ldots d^3\mathbf{r}_N$ is the probability of finding simultaneously the first particle within the volume element $d^3\mathbf{r}_1$ centered at $\mathbf{r}_1$, the second particle within the volume element $d^3\mathbf{r}_2$ centered at $\mathbf{r}_2 \ldots$ and the Nth particle within the volume $d^3\mathbf{r}_N$ centered at $\mathbf{r}_N$. By definition, the probability density function is assumed to be normalized, meaning that its integral over the entire volume V is equal to one. This description of random media originates in statistical mechanics, where it is employed to describe the distribution of point particles (i.e., atoms and molecules) in the corresponding position-momentum phase space [355].

The knowledge of $p(\mathbf{r}_1, \mathbf{r}_2, \ldots, \mathbf{r}_N)$ allows one to rigorously introduce the ensemble average $\langle A \rangle$ of any configuration-dependent quantity $A(\mathbf{r}_1, \mathbf{r}_2, \ldots, \mathbf{r}_N)$ associated to the random medium using the following expression:

$$\langle A \rangle = \int_V A(\mathbf{r}_1, \mathbf{r}_2, \ldots, \mathbf{r}_N) p(\mathbf{r}_1, \mathbf{r}_2, \ldots, \mathbf{r}_N) \prod_{\alpha=1}^{N} d^3\mathbf{r}_\alpha \qquad (8.1)$$

When the total number of particles is very large, the complete statistical information via the N-particle probability density function is usually not practical and generally unavailable. However, one can obtain the probability of a *reduced configuration*, where the position of only $n < N$ particles is considered fixed and the remaining $N-n$ particles are unconstrained. The reduced probability density function for a subsystem with $n < N$ particles is simply obtained from the N-particle joint density by integrating it over the remaining $N-n$ particles' positions. Therefore, the reduced density function is the marginal distribution for the considered subset of particles of the larger N-particle system. For example, the single-particle and the two-particle reduced probability density functions can be explicitly obtained as follows:

$$p(\mathbf{r}_i) = \int_V p(\mathbf{r}_1, \mathbf{r}_2, \ldots, \mathbf{r}_N) \prod_{\substack{\alpha=1 \\ \alpha \neq i}}^{N} d^3\mathbf{r}_\alpha \qquad (8.2)$$

$$p(\mathbf{r}_i, \mathbf{r}_j) = \int_V p(\mathbf{r}_1, \mathbf{r}_2, \ldots, \mathbf{r}_N) \prod_{\substack{\alpha=1 \\ \alpha \neq i,j}}^{N} d^3\mathbf{r}_\alpha. \qquad (8.3)$$

When the particles are identical, it is more relevant to consider the probability that any set of $n < N$ particles are at positions $\mathbf{r}_1, \ldots, \mathbf{r}_n$ in any possible permutation, leading to define the *n-particle density function*:

$$\rho^{(n)}(\mathbf{r}_1, \ldots, \mathbf{r}_N) = \frac{N!}{(N-n)!} \int_V \cdots \int_V p(\mathbf{r}_1, \ldots, \mathbf{r}_N) d\mathbf{r}_{n+1} d\mathbf{r}_{n+2} \cdots d\mathbf{r}_N \qquad (8.4)$$

For *statistically homogeneous media*, the functions $\rho^{(n)}(\mathbf{r}_1, \ldots, \mathbf{r}_N)$ are translationally invariant and depend only on the relative displacement of the particles with respect to a selected site, say with respect to $\mathbf{r}_1$:

$$\rho^{(n)}(\mathbf{r}_1, \ldots, \mathbf{r}_n) = \rho^{(n)}(\mathbf{r}_{12}, \mathbf{r}_{13}, \ldots, \mathbf{r}_{1n}), \qquad (8.5)$$

where $\mathbf{r}_{ij} = \mathbf{r}_j - \mathbf{r}_i$.

The n-particle density function allows us to precisely define the concept of positional correlation. This is achieved by the *n-particle correlation function*:

$$g_n(\mathbf{r}_1, \ldots, \mathbf{r}_n) = \frac{\rho^{(n)}(\mathbf{r}_1, \ldots, \mathbf{r}_n)}{\rho^n} \qquad (8.6)$$

where $\rho = \rho^{(1)}(\mathbf{r}_1) = N/V$ is the constant number density of particles in the thermodynamic limit of $N, V \to \infty$. Note that for a completely uncorrelated random system, we have $\rho^{(n)}(\mathbf{r}_1, \ldots, \mathbf{r}_n) = \rho^n$. It follows that the deviations of g_n from unity provide a measure of statistical correlations of the positional degrees of freedom in the system.

Of particular importance is the *two-particle function*:

$$g_2(\mathbf{r}) = \frac{\rho^{(2)}(\mathbf{r})}{\rho^2} \tag{8.7}$$

that we have introduced for a translation invariant system. This quantity is usually referred to as the *pair-correlation function*, and it plays a central role in scattering theory because it can be measured experimentally (see our previous discussion in Section 7.2). For a random medium that is both statistically homogeneous and isotropic, the pair-correlation function depends only on a radial distance r, and it is known as the *radial distribution function*. If we denote by $s(r)$ the surface area of a sphere of radius r, it follows that $\rho s(r) g_2(r) dr$ is proportional to the conditional probability of finding a particle center in a spherical shell of volume $s(r)dr$, provided that there is another particle at the origin.

Closely related to the pair-correlation function is the *total correlation function*, which for a statistically homogeneous medium is written as follows:

$$h(\mathbf{r}) = g_2(\mathbf{r}) - 1. \tag{8.8}$$

In a homogeneous system that does not possess long-range order, $g_2(r) \to 1$ when $r \to \infty$ and therefore $h(r) \to 0$. This means that $h(r)$ is generally an integrable function and that its Fourier transform $\hat{h}(\mathbf{k})$ is well defined. As previously shown in Section 7.2, the *structure factor* $S(\mathbf{k})$ is related to the total correlation as follows:

$$\boxed{S(\mathbf{k}) = 1 + \rho \hat{h}(\mathbf{k})} \tag{8.9}$$

where $\mathbf{k}$ is the Fourier conjugate variable with respect to the position vector $\mathbf{r}$. Since the structure factor is the Fourier power spectrum of the number density of the point pattern, it is always nonnegative. Clearly for systems that are homogeneous and isotropic, the structure factor depends only on the magnitude of the vector $\mathbf{k}$. In what follows, we will focus on the local density description of discrete random media and illustrate, through detailed analytical calculations, its impact on local correlations and the structure factor.

8.1.2   Particle Densities and Distribution Functions

We define the *one-particle number density function* $n^{(1)}(\mathbf{r})$ for a single realization of a discrete random medium with N particles as follows:

$$\boxed{n^{(1)}(\mathbf{r}) = \sum_{i=1}^{N} \delta(\mathbf{r} - \mathbf{r}_i)} \tag{8.10}$$

where δ denotes the Dirac delta function. Note that, by definition, the integration of $n^{(1)}(\mathbf{r})$ over the entire space of the medium is equal to N. At this point, we can introduce the *average number density* of the random medium as the configurational or ensemble average of its one-particle number density function:

$$\langle n^{(1)}(\mathbf{r}) \rangle = \int_V n^{(1)}(\mathbf{r}) p(\mathbf{r}_1, \mathbf{r}_2, \ldots, \mathbf{r}_N) \prod_{\alpha=1}^{N} d^3 \mathbf{r}_\alpha. \tag{8.11}$$

Using the sampling property of the Dirac delta function and exchanging the summation with integral, we obtain the following:

$$\langle n^{(1)}(\mathbf{r}) \rangle = \sum_{i=1}^{N} p(\mathbf{r}) = N p(\mathbf{r}), \tag{8.12}$$

where $p(\mathbf{r})$ is the single-particle probability density previously defined in (8.2). Realize that the average number density depends on the position vector because the random medium is in general *statistically nonhomogeneous* and therefore the particles are not distributed with equal probability across the volume V. However, for a statistically homogeneous random medium, the particles are distributed with the same probability everywhere in space, and therefore $p(\mathbf{r}) = 1/V$. In this case, the average number density reduces to the familiar constant $\rho - N/V$.

We can extend this approach to capture multiparticle correlations by introducing the *two-particle number density function*:

$$\boxed{n^{(2)}(\mathbf{r}, \mathbf{r}') = \sum_{i=1}^{N} \sum_{\substack{j=1 \\ j \neq i}}^{N} \delta(\mathbf{r} - \mathbf{r}_i) \delta(\mathbf{r}' - \mathbf{r}_j)} \tag{8.13}$$

This function measures the degree of influence or positional correlation between two particles. When the particles of the random medium are densely packed, $n^{(2)}(\mathbf{r}, \mathbf{r}') \neq n^{(1)}(\mathbf{r}) n^{(1)}(\mathbf{r}')$ because the occupation of position $\mathbf{r}$ is strongly influenced by the occupation of position $\mathbf{r}'$.

In analogy with the average number density, we introduce the ensemble average of the two-particle number density function as follows:

$$\langle n^{(2)}(\mathbf{r}, \mathbf{r}') \rangle = \int_V n^{(2)}(\mathbf{r}, \mathbf{r}') p(\mathbf{r}_1, \mathbf{r}_2, \ldots, \mathbf{r}_N) \prod_{\alpha=1}^{N} d^3 \mathbf{r}_\alpha. \tag{8.14}$$

Inserting the definition of the two-particle number density function, we obtain the following:

$$\langle n^{(2)}(\mathbf{r}, \mathbf{r}') \rangle = \sum_{i=1}^{N} \sum_{\substack{j=1 \\ j \neq i}}^{N} p(\mathbf{r}, \mathbf{r}') = N(N-1) p(\mathbf{r}, \mathbf{r}'), \tag{8.15}$$

where $p(\mathbf{r}, \mathbf{r}')$ is the two-particle probability density defined in (8.3).

Following the previous logic, we can now introduce the *pair-distribution function*[1] $g(\mathbf{r}, \mathbf{r}')$ in terms of the two-particle average number density:

$$\langle n^{(2)}(\mathbf{r}, \mathbf{r}')\rangle = \langle n^{(1)}(\mathbf{r})\rangle \langle n^{(1)}(\mathbf{r}')\rangle g(\mathbf{r}, \mathbf{r}') = N(N-1)p(\mathbf{r}, \mathbf{r}'). \tag{8.16}$$

In the case of a *statistically homogeneous random medium*, we obtain the simpler expression:

$$\boxed{\langle n^{(2)}(\mathbf{r}, \mathbf{r}')\rangle = \rho^2 g(\mathbf{r}, \mathbf{r}') = \rho^2 g(\mathbf{r} - \mathbf{r}')} \tag{8.17}$$

Particles that are infinitely apart so that $|\mathbf{r} - \mathbf{r}'| \to \infty$ do not influence each other and their positions are statistically uncorrelated. Therefore, $p(\mathbf{r}, \mathbf{r}') \to p(\mathbf{r})p(\mathbf{r}')$, which implies $g(\mathbf{r}, \mathbf{r}') \to 1$. Again, if the medium is additionally *isotropic*, then the pair distribution function depends only on the the magnitude of the distance $r = |\mathbf{r}_1 - \mathbf{r}_2|$ between two particles and the function $g(r)$ is called the *radial distribution function*. This function describes how the particles' density varies as a function of distance from a reference particle, and it is proportional to the probability of finding a pair of particles separated by a distance $\mathbf{r}$ relative to the probability expected for a uniformly random distribution with the same density.

At this point, it is interesting to ask ourself what kind of mechanism is responsible for establishing particle correlations in a statistically homogeneous random medium in which $p(\mathbf{r}_i) = 1/V$. This is possible because the particles have *finite size,* and if the medium is dense enough, i.e., the particles are closely packed, the nonoverlapping condition directly generates correlations. In fact, for nonoverlapping hard spheres the joint probability density function is zero at distances smaller than the diameters of the spheres. The same applies to the pair distribution function because from equations (8.15) and (8.17), we deduce the following:

$$\boxed{p(\mathbf{r}_i, \mathbf{r}_j) = \frac{N}{N-1} \frac{g(\mathbf{r}_i, \mathbf{r}_j)}{V^2}} \tag{8.18}$$

In the limit of large N, we obtain $p(\mathbf{r}_i, \mathbf{r}_j) \approx g(\mathbf{r}_i, \mathbf{r}_j)/V^2$. Typically, the radial distribution function of a uniform and isotropic random media composed of scattering spherical particles of radius a is approximated by $g(r) = 0$ for $r < 2a$ and $g(r) = 1$ otherwise. This choice is called *hole correction* (HC) approximation and reflects the fact that particles cannot interpenetrate.

8.1.3   The Ornstein–Zernike Integral Equation

How good is the HC approximation for the radial distribution function of a uniform random medium with a large particle density? A simple argument for one-dimensional

---

[1] This function is also known as the *pair-correlation function*.

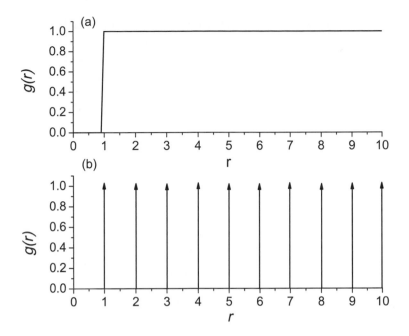

Figure 8.1 Pair distribution functions of a homogeneous and isotropic random medium composed of one-dimensional scattering particles of unit size. (a) HC approximation for small values of the volume fraction $f \ll 1$. (b) Case of complete filling $f = 1$. Adapted from reference [132].

structures shows that it is not a good approximation when the fractional volume f of the particles is large [132].

To convince ourselves, let us imagine an extreme situation where the entire one-dimensional space is homogeneously filled by the scattering particles, i.e., $f = 1$. The particles have all the same linear size a. Since the particles cannot overlap, their centers will be separated by integer multiples of a and the pair distribution function will be zero for $r \neq ma$, where m is a nonzero integer. Otherwise, it consists of a periodic series of Dirac delta functions positioned at the centers of the particles. This situation cannot clearly be described by the HC approximation. On the other hand, when the filling fraction f is small, the HC approximation for $g(r)$ provides a good description. These two limit situations are sketched in Figure 8.1.

We could intuitively expect that in situations where f is not one but still appreciable, the shape of $g(r)$ should interpolate in between the HC approximation and the maximum concentration (i.e., complete filling) situation. The problem, however, is to predict the correct functional form of $g(r)$ that is valid for intermediate values of the filling fractions. This challenge has been thoroughly addressed in statistical mechanics, where the exact form of the distribution functions describing the structural order of different states of materials is of primary importance [304, 356].

The problem was exactly formulated in 1914 by Ornstein and Zernike [357], who introduced the *direct correlation function* $c(\mathbf{r})$. We recall that we previously defined the total correlation function, or total influence function, as follows:

$$h(\mathbf{r}_{ij}) = g(\mathbf{r}_{ij}) - 1, \tag{8.19}$$

where $\mathbf{r}_{ij} = \mathbf{r}_i - \mathbf{r}_j$. This function measures the "influence" of particle i on particle j separated by the distance $\mathbf{r}_{ij}$.

The idea of Ornstein and Zernike was to split the total influence into two contributions, i.e., a *direct and an indirect part*. The direct contribution is provided by the direct correlation function $c(\mathbf{r})$, while the indirect part is contributed by the direct influence of particle i on a third particle, labeled k, which in turn exerts a total influence (i.e., direct and indirect) on particle j. The indirect contribution must be averaged over all the possible positions of the particle k and weighted by the density ρ of particles. This decomposition can be written mathematically in the form of an integral equation:

$$h(\mathbf{r}_{ij}) = c(\mathbf{r}_{ij}) + \rho \int_V d^3\mathbf{r}_k c(\mathbf{r}_{ik}) h(\mathbf{r}_{kj}) \tag{8.20}$$

which is called the *Ornstein–Zernike equation*. The second term in the Ornstein–Zernike equation formally defines the indirect part of the total correlation function $h(\mathbf{r})$. Note that the Ornstein–Zernike equation involves two unknowns, $c(\mathbf{r})$ and $h(\mathbf{r})$, and therefore cannot be solved in the absence of an explicit relation, known as the *closure relation*, which connects the two unknown functions.

Equation (8.20) is in a convolution-type integral relation and therefore it can be transformed into the following algebraic equation by Fourier transformation:

$$\hat{h}(\mathbf{k}) = \hat{c}(\mathbf{k}) + \rho \hat{h}(\mathbf{k}) \hat{c}(\mathbf{k}) \tag{8.21}$$

where we denoted the Fourier transformed quantities by the carot symbol.

The Ornstein–Zernike equation defines the direct correlation function $c(\mathbf{r})$ or $\hat{c}(\mathbf{k})$ in terms of the total correlation function $h(\mathbf{r})$ or $\hat{h}(\mathbf{k})$. The shape of the interaction potential between the scattering particles is taken into account by the choice of the closure relation. In the scattering theory of classical particles it is customary to model their interaction by a hard-potential, which reflects the fact that classical objects do not overlap in space. A commonly used closure relation is the *Percus–Yevick approximation* [358]. In the case of hard spherical particles,[2] the Percus–Yevick approximation leads to an analytical solution of the Ornstein–Zernike equation, which we computed in Figure 8.2 for three different values of the filling fraction. We note that the Percus–Yevick radial distribution functions $g(r) = h(r) + 1$ are equal to zero for distances less than a diameter, naturally implementing the HC. All the curves assume their maximum

---

[2] The hard-sphere potential $u(r)$ is equal to infinity for $r < 2a$ and is equal to zero otherwise.

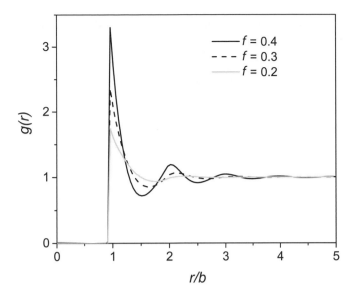

Figure 8.2 Percus–Yevick pair distribution function for hard spheres in units normalized with respect to the diameter of the particles. Analytical results are shown for three different values of the filling fraction f as specified in the legend.

values at $r = 2a$, oscillate as r increases, and asymptotically approach unity. This asymptotic value corresponds to the uniform random gas picture, where the particles, positions are all independent. Since the oscillations of $g(r)$ are more pronounced for the cases with larger values of f, the positions of the particles are less independent in this regime.

Finally, it is important to realize that the maximum volume fraction of solid spheres is limited to values that are significantly less than one. Moreover, the particle volume fraction at random close packing (RCP)[3] depends on the types of objects being packed. In particular, for randomly packed monodisperse hard spheres, the highest possible filling fraction that can be achieved is $f \approx 0.64$. Note that this value is significantly smaller than the maximum theoretical filling fraction of $f \approx 0.74048$ that results from the hexagonal close packing (HCP).

8.1.4   The Percus–Yevick Structure Factor

The Percus–Yevick approximation provides a closed-form analytical solution for the structure factor of a *random packing of hard spheres*. In this section, we discuss more details behind the analytical calculation of the structure factor for an isotropic and

---

[3] This denotes the maximum volume fraction of solid objects obtained when they are packed randomly.

uniform random medium composed of solid spheres. Combining equation (8.21) with the structure factor $S(\mathbf{k}) = 1 + \rho \hat{h}(\mathbf{k})$, we obtain the following:

$$S(k) = 1 + \rho \frac{\hat{c}(k)}{1 - \rho \hat{c}(k)} \tag{8.22}$$

where $\rho = 6f/\pi b^3$ is the number of particles per unit volume and $b = 2a$ is their diameter. Using the Percus–Yevick approximation for hard spheres, the Fourier transform $\hat{c}(k)$ of the direct correlation function $c(r)$ can be analytically computed, and the full solution is obtained as follows [359]:

$$
\begin{aligned}
\hat{c}_{PY}(k) = 4\pi b^3 \Bigg\{ & \frac{(\alpha + \beta + \gamma)}{(2ka)^2} \cos(2ka) - \frac{(\alpha + 2\beta + 4\gamma)}{(2ka)^3} \sin(2ka) \\
& - \frac{2(\beta + 6\delta)}{(2ka)^4} \cos(2ka) + \frac{2\beta}{(2ka)^4} + \frac{24\delta}{(2ka)^5} + \frac{24\delta}{(2ka)^6} [\cos(2ka) - 1] \Bigg\},
\end{aligned}
\tag{8.23}
$$

where all the constants are defined as follows [359, 360]:

$$
\begin{aligned}
\alpha &= \frac{(1 + 2f)^2}{(1 - f)^4} \\
\beta &= -6f \frac{(1 + f/2)^2}{(1 - f)^4} \\
\gamma &= \frac{f(1 + 2f)^2}{2(1 - f)^4}.
\end{aligned}
\tag{8.24}
$$

Note that once $\hat{c}_{PY}(k)$ has been analytically obtained, not only the static factor can be computed through equation (8.22), but also the pair distribution function $g(r)$ can be obtained using the equation (8.21) and the Fourier inversion of the spherically symmetric function $\hat{h}(k)$:

$$h(r) = 4\pi \int_0^\infty dk k^2 \left[\frac{\sin(kr)}{kr} \right] \hat{h}(k), \tag{8.25}$$

along with the relation $g(r) = h(r) + 1$.

Analytical calculations of the static structure factor of random aggregates of spheres with $a = 1$ μm and different values of the filling fraction f are shown in Figure 8.3. We notice that for spheres of small electric size, i.e., $a \ll \lambda$, the light wavelength cannot resolve the fine structure of the random medium, which appears to be homogeneous. In this limit, light diffraction inhibits the optical transmission. On the opposite hand, when $a \gg \lambda$, light can fully probe the random nature of the medium and the structure factor approaches one. For the intermediate situations, the effects of structural correlations in the random medium become important and the resulting structure factor and diffraction intensity feature distinctive oscillations.

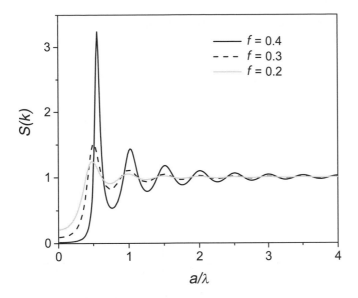

Figure 8.3 Calculated static structure factor within the Percus–Yevick approximation for hard spheres of radius $a = 1$ µm. Analytical results are shown for three different values of the filling fraction f as specified in the legend.

8.2   Speckle Phenomena and Random Walks

Random walks provide very convenient models to understand the single-scattering properties of discrete random media. In this section, we introduce the simplest possible random walk model that allows us to discuss the far-field scattering properties of a discrete random medium comprised of a collection of N fixed scatterers with subwavelength dimensions, such as small spherical particles described as electric dipoles. The medium is assumed to be sufficiently diluted so that we can safely neglect the contributions of multiple-scattering phenomena. Moreover, the small particles are randomly distributed over a uniformly illuminated volume that is much larger than the excitation wavelength. This condition ensures that the path differences of the scattered waves from the random medium to the far-field detector are much greater than the wavelength of light and the corresponding phases are uniformly distributed in the $[0, 2\pi]$ interval.

Under these assumptions, the scattered field observed at the position $\mathbf{r}$ of the detector and at a time instant t_0 can be written as a sum of phasors:

$$\psi(\mathbf{r}, t_0) = |\psi| e^{j\theta} = \sum_{n=1}^{N} a_n e^{j\phi_n} \tag{8.26}$$

where $a_n = a(\mathbf{r}_n; \mathbf{r}, t_0)$ and $\phi_n = \phi(\mathbf{r}_n; \mathbf{r}, t_0)$ are, respectively, the amplitudes and phases of the individual field components that are scattered from different locations of the object to the far-field detector. These quantities generally depend on the positions and orientations of the scatterers within the illuminated target volume. Since the phases ϕ_n and amplitudes a_n randomly fluctuate, the expression in equation (8.26) constitutes a *random phasor sum* and can be graphically represented as a *random walk process* of N steps on the complex plane. The resultant vector from the superposition (summation) of the many randomly phased vector components $\mathbf{a}_n = a_n \exp(j\phi_n)$ of lengths a_n and phases ϕ_n can be very large or small depending on the relative phases of such components, corresponding to constructive or destructive interference.

Box 8.1 Gaussian Random Variables

We recall few basic facts about the Gaussian (normal) random variables. We denote by $X \sim N(\mu, \sigma^2)$ a Gaussian random variable with mean μ and variance σ^2, which is distributed according to the *Gaussian probability density function* (pdf):

$$p(x) = \frac{1}{\sqrt{2\pi}\sigma} \exp\left[-\frac{(x-\mu)^2}{2\sigma^2}\right] \tag{8.27}$$

When $\mu = 0$ and $\sigma = 1$ we call $X \sim N(0,1)$ a standard Gaussian random variable. Any Gaussian random variable can be reduced to standard form by considering $Z = (X - \mu)/\sigma$, which is in fact distributed as $N(0,1)$. This is a straightforward consequence of the following useful property of a Gaussian random variable, valid for any two real numbers α and β:

$$\alpha + \beta N(\mu, \sigma^2) = N(\alpha + \beta\mu, \beta^2\sigma^2) \tag{8.28}$$

In addition, if $N(\mu_1, \sigma_1^2)$ and $N(\mu_2, \sigma_2^2)$ are two mutually independent Gaussian random variables, their linear combination is also a Gaussian (normal) random variable and can be written as:

$$\alpha N(\mu_1, \sigma_1^2) + \beta N(\mu_2, \sigma_2^2) = N(\alpha\mu_1 + \beta\mu_2, \alpha^2\sigma_1^2 + \beta^2\sigma_2^2) \tag{8.29}$$

This important result on the additivity of the means and variances can be generalized to the sum of any number of independent normal variables. In particular, and linear combination of statistically independent normal variables is itself a normal variable. A further generalization of fundamental importance is the *central limit theorem* (CLT), which states that the sum of any N statistically independent and identically distributes (i.i.d.) random variables, but not necessarily normal variables, though finite means and variances will converge to a normal random variable when $N \to \infty$. The CLT establishes Gaussian random variables as objects of primary importance in statistics and in the natural sciences where the sum of a large number of often uncontrollable fluctuations, not necessarily normally distributed, converges to the well-behaved Gaussian distribution (i.e., the celebrated "bell curve").

Two representative walk trajectories obtained by computing the random phasor sum in 8.26 with $N = 10^5$ are displayed in Figure 8.4a in black and gray. In this example, the phase angles ϕ_n are uniformly distributed in the $[-\pi, +\pi)$ interval while

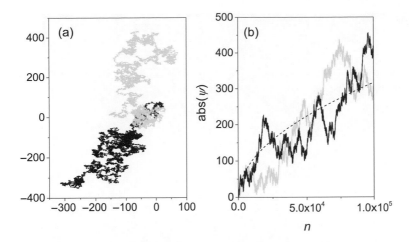

Figure 8.4 (a) Two Gaussian random walk trajectories and (b) corresponding absolute displacement of the walker as a function of the number of terms in the phasor sum up to $N = 10^5$. The dashed line is the expected root square behavior based on the central limit theorem.

the individual amplitudes, or step sizes a_n, are drawn from a Gaussian distribution with zero average and unit variance. The highly fragmented random walk trajectories are self-similar fractal curves with non-integer fractal dimension $d_w < 2$, known as the *walk dimension* [361]. In Figure 8.4b, we show the evolution of the absolute value of the resultant vector $\psi(\mathbf{r}, t_0)$ with respect to the number of terms in the sum for two separate realizations of the random phasor sum. It is expected that as the number of random walks increases, the curves will better and better approximate the square root dashed line drawn in the figure. This behavior reflects the general fact that when the amplitudes (steps) are distributed according to any probability distribution with zero mean and a finite variance, not necessarily a Gaussian distribution, the mean squared displacement of the random walk scales with the number of steps n according to $\mathbb{E}[\psi^2] \propto n$, where $\mathbb{E}$ denotes the expectation value or average value. The addition of random vector components is the hallmark of the *speckle phenomenon* that manifests the coherent nature of the interference of a large number of randomly phased waves. This phenomenon consists in the appearance of a characteristic fine-scale granularity, known as "speckle," which can be easily observed in the far-field transmission/reflection intensity pattern of a laser beam through/from paper, fabric, suspensions of particles, or surfaces with roughness on the scale of the optical wavelength. As an example we show in Figure 8.5 the simulated speckle pattern observed in the far-field due to the scattering of radiation from a randomly roughed square surface that is rough relative to the optical wavelength.

The primary concern of the theory of speckles is to determine the statistical properties of the resultant vector amplitude $|\psi|$ and phase θ describing the total electric field and its intensity $|\psi|^2$ at a single space-time point under various physical situations described by different types of random phasor sums [362].

Figure 8.5 Image of a simulated speckle pattern (i.e., electric field intensity distribution) in the far-field due to scattering by a randomly roughened surface.

8.2.1   Fully Developed Speckle

In this section, we discuss the main statistical properties of phasor sums with an infinite number of independent terms, which model wave scattering from uncorrelated random structures that are extended with respect to the incident wavelength. Following the convention introduced by Goodman [362], in this situation it is convenient to rescale the sum in equation (8.26) by the factor $1/\sqrt{N}$ so that its variance remains finite even when there are an infinite number of phasors. Therefore, we consider the statistical properties of the random phasor sum:

$$\boxed{\mathbf{A} = A e^{j\theta} = \frac{1}{\sqrt{N}} \sum_{n=1}^{N} a_n e^{j\phi_n}} \qquad (8.30)$$

whose real and imaginary parts are given by the following:

$$R = \Re\{\mathbf{A}\} = \frac{1}{\sqrt{N}} \sum_{n=1}^{N} a_n \cos \phi_n \qquad (8.31)$$

$$I = \Im\{\mathbf{A}\} = \frac{1}{\sqrt{N}} \sum_{n=1}^{N} a_n \sin \phi_n. \qquad (8.32)$$

The amplitudes a_n and phases ϕ_n are statistically independent random variables with respect to each other for any n so that knowing the length of a component vectors provides no knowledge on its phase angle and vice versa. It is also assumed that a_n

and ϕ_n are statistically independent of a_m and ϕ_m when $n \neq m$ and that the phases ϕ_n of the contributing phasors are uniformly distributed in the interval $[-\pi, +\pi)$. The approximation of a *uniformly distributed phase* breaks down for an extended system of small scattering particles if we consider directions close to the forward-scattering because small particles do not introduce large path differences.

The immediate consequence of the assumed statistical independence and uniform phase distribution is that the expectation values of both the R and I are identically zero while the variance can be easily evaluated from the general relation:

$$\sigma_X^2 = \mathbb{E}[X^2] - (\mathbb{E}[X])^2 \tag{8.33}$$

where X is a generic random variable. Considering the variance of R we have the following:

$$\sigma_R^2 = \mathbb{E}[R^2] = \frac{1}{N} \sum_n \sum_m \mathbb{E}[a_n a_m] \mathbb{E}[\cos \phi_n \cos \phi_m], \tag{8.34}$$

and analogously for σ_I^2, where we made use of the statistical independence between amplitudes and phases. The preceding expression is equal to zero when $n \neq m$. Therefore, considering only the nonzero $n = m$ terms, we obtain the following:

$$\boxed{\sigma_R^2 = \sigma_I^2 \equiv \sigma^2 = \frac{1}{N} \sum_{n=1}^{N} \frac{\mathbb{E}[a_n^2]}{2} = \frac{\langle a^2 \rangle}{2}} \tag{8.35}$$

where we used the trigonometric identity $\cos^2 \phi_n = 1/2 + (\cos 2\phi_n)/2$ and the fact that the expectation value of the uniformly distributed phase vanishes. We also introduced the angle bracket as an alternative convention to denote the average value of the quantity of interest. Consistently with our statistical assumptions, we can also easily verify that the correlation coefficient between the real and imaginary components of the phasor sum, defined as $\Gamma_{R,I} \equiv \mathbb{E}[RI] = 0$, vanishes.

An important question in the theory of speckle regards the statistical properties of R and I in the $N \to \infty$ limit. Since we are dealing with the sum of many independent random variables, the central limit theorem (CLT) ensures that the joint probability density function for R and I converges to the Gaussian distribution:

$$p_{R,I}(R, I) = \frac{1}{2\pi\sigma^2} \exp\left(-\frac{R^2 + I^2}{2\sigma^2}\right), \tag{8.36}$$

where $\sigma^2 = \sigma_R^2 = \sigma_I^2$. While technically correct only in the $N \to \infty$ limit, the preceding result well approximates the more practical situation of a random walk with a large number ($N \gtrsim 10^3$) of independent steps. Since the contours of equal probability density of the joint distribution are circular in shape, the resultant complex phasor $\mathbf{A}$ is known in statistics as a *circular complex Gaussian variate* or, if the phasors are also time dependent, a circular complex Gaussian process. The joint probability distributions of the amplitude $A \geq 0$ and phase $-\pi \leq \theta < \pi$ of the resultant vector defined in equation (8.30) can be obtained by transforming the variables from Cartesian to polar coordinates according to the following:

$$p_{A,\theta}(A,\theta) = p_{R,I}(R,I)J = p_{R,I}(A\cos\theta, A\sin\theta)|\mathbf{J}|, \qquad (8.37)$$

where $|\mathbf{J}| = A$ is the determinant of the Jacobian matrix[4] of the transformation.

Implementing the change of variables, we easily obtain the joint probability density [362]:

$$p_{A,\theta}(A,\theta) = \frac{A}{2\pi\sigma^2}\exp\left(-\frac{A^2}{2\sigma^2}\right). \qquad (8.38)$$

The *marginal probability distribution* of the resultant amplitude A, i.e., the distribution of the amplitude regardless of the values of the phase, is obtained by integrating the joint density $p_{A,\theta}(A,\theta)$ with respect to $\theta \in [-\pi, \pi)$, yielding the following:

$$\boxed{p_A(A) = \frac{A}{\sigma^2}\exp\left(-\frac{A^2}{2\sigma^2}\right)} \qquad (8.39)$$

This statistical distribution of the length A of the resultant phasor is known as the *Rayleigh probability density function*. The moments of the amplitude can be found analytically as follows [362]:

$$\langle A^q \rangle = \int_0^\infty A^q p_A(A) dA = 2^{q/2}\sigma^q \Gamma\left(1 + \frac{q}{2}\right). \qquad (8.40)$$

The marginal probability density of the phase, obtained by integrating the joint density over $A \in [0,\infty)$, yields $p_\theta(\theta) = 1/(2\pi)$, as expected based on the uniform distribution of the phase in $\theta \in [-\pi, \pi)$. Moreover, we can also verify that $p_{A,\theta}(A,\theta) = p_\theta(\theta)p_A(A)$, consistently with our previous assumption that A and θ are statistically independent random variables. The probability density of the speckle intensity $I = f(A) = |\mathbf{A}|^2 = A^2$ is obtained from the probability transformation law [363]:

$$p_I(I) = p_A[f^{-1}(I)]\left|\frac{dA}{dI}\right| = p_A(\sqrt{I})\left|\frac{dA}{dI}\right|, \qquad (8.41)$$

where $p_A(A)$ is the Rayleigh density. Completing the calculation, we derive the probability density of the intensity in the following form:

$$\boxed{p_I(I) = \frac{1}{2\sigma^2}\exp\left(-\frac{I}{2\sigma^2}\right)} \qquad (8.42)$$

This expression is an inverse exponential distribution (with average intensity $\langle I \rangle = 2\sigma^2$) characteristic of the so-called *fully developed speckle regime*. This is to be contrasted with partially developed speckle that occurs in the presence of phase correlations. This more general regime describes the scattering from correlated particle systems, such as deterministic aperiodic structures and quasicrystals, which generally exhibit nonexponential intensity fluctuations resulting from the significant

---

[4] This is the matrix of all the first-order partial derivatives of the transformation.

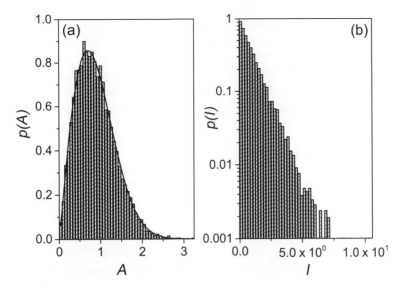

Figure 8.6 (a) Histogram of the resultant amplitudes A of a random phasor sum with $N = 10^5$ terms evaluated 10^4 times. The continuous line is the theoretically predicted Rayleigh distribution. (b) Histograms of the resultant intensities $I = |A|^2$ of a random phasor sum with $N = 10^5$ terms evaluated 10^4 times. The phase angles are uniformly distributed in the $[-\pi, +\pi)$ interval, and the individual amplitudes a_n are taken from a Gaussian distribution with zero average and unit variance.

positional correlations. An example of this behavior was previously shown for the Penrose quasicrystal in relation to the fluctuations of its structure factor discussed in Section 5.2.3. The histograms of the computed amplitudes and intensities of 10^4 random phasor sums each with $N = 10^5$ terms are shown in Figure 8.6. They well approximate the Rayleigh distribution (a continuous line) and the inverse exponential distributions theoretically predicted for the $N \to \infty$ limit based on the application of the central limit theorem.

The moments of the intensity of fully developed speckle have the simple analytical expression $\langle I^q \rangle = (2\sigma^2)^q q!$ from which we deduce that the second moment of the intensity is equal to $\langle I^2 \rangle = 2(2\sigma^2)^2 = 2\langle I \rangle^2$. Note that, without the choice of the $1/\sqrt{N}$ normalization in the phasor sum, the average intensity can be written as $\langle I \rangle = N\mathbb{E}[a_n^2] = Na^2$, when all the amplitudes in the sum are equal. The strength of the intensity fluctuations in a speckle field is usually quantified by the *contrast* C defined as follows:

$$C = \frac{\sigma_I}{\langle I \rangle} \tag{8.43}$$

where $\sigma_I = \sqrt{\langle I^2 \rangle - \langle I \rangle^2}$ is the standard deviation of the intensity. In general, the contrast depends on the nature of phase correlations as well as the number of phasors contributing to the resultant sum. As a consequence, it is sensitive to the structural correlations of the underlying scattering media. In the case of fully developed speckle, $C = 1$ because the variance σ_I^2 of an exponential distribution is equal to the square of the average intensity. In this case, the intensity fluctuations of speckle patterns are of the same order as the average intensity. The degree of sharpness of the speckle pattern is alternatively quantified by the *scintillation index*, which is the normalized variance of intensity fluctuations (i.e., the square of the contrast).

8.2.2   Finite Sums: The Kluyver–Pearson Formula

The previous results were derived for random phasor sums with an infinite number of terms such that the central limit theorem can be directly applied to derive the relevant statistics. However, the statistical understanding of random phasor sums with a finite number N of terms becomes important when we reduce the size of the illuminated region of the samples. Finite-size effects in speckle statistics result in enhanced clutter power that degrades image quality and reduces the performance of radar systems [364].

These more challenging problems can be solved through the calculation of the joint characteristic function of the real and the imaginary parts of $\mathbf{A}$, shown explicitly in equations (8.31) and (8.32), as we discuss in this chapter. The *characteristic function* of a random variable X is the Fourier transform of its joint probability density $p(x)$ or, equivalently, the expectation value of $e^{j\omega x}$:

$$M(\omega) = \mathbb{E}[e^{j\omega x}] = \int_{\mathbb{R}} p(x)e^{j\omega x}dx. \tag{8.44}$$

This concept is very important in probability theory because the characteristic function of the sum of several *independent* random variables is the product of the characteristic functions of the individual components of the sum. Therefore, we can determine the probability density function of the sum of independent random variables through the Fourier inverse transform of the characteristic function in product form. This approach is computationally more efficient than computing the probability density function of the sum of independent random variables from the convolution of the probability density functions of the individual components [363, 365].

We now apply the characteristic function method to compute the probability density of the finite random phasor sum in equation (8.30). First, we express the characteristic function of the phasor sum as follows:

$$M(\omega) = \langle \exp\left[j(\omega_R R + \omega_I I)\right]\rangle = \left\langle \exp\left[j\sum_{n=1}^{N} a_n'(\omega_R \cos\phi_n + \omega_I \sin\phi_n)\right]\right\rangle, \tag{8.45}$$

where $a_n' = a_n/\sqrt{N}$ and we introduce the two-dimensional frequency $\omega = \sqrt{\omega_R^2 + \omega_I^2}$. Here the angle bracket indicates the averaging procedure that needs to be carried out over realizations of the random amplitudes and phases of the N scattering particles.

Using the well-known identity $a \cos \phi + b \sin \phi = c \cos(\phi - \theta)$ with $c = \sqrt{a^2 + b^2}$ and $\theta = \tan^{-1}(b/a)$, we can rewrite the last expression as follows:

$$M(\omega) = \left\langle \prod_{n=1}^{N} \exp\left[ja'_n \omega \cos(\phi_n - \theta_\omega) \right] \right\rangle, \qquad (8.46)$$

where $\theta_\omega = \tan^{-1}(\omega_I/\omega_R)$. When the contributions from the amplitudes and phases of the different scatterers are statistically independent from each another, then the average of the preceding product is equal to the product of the averages. Moreover, if we assume that the scattering amplitudes are statistically independent from the phases and that the phases are uniformly distributed, then averaging each factor of the product over the uniform phase distribution leads to the close-form result:

$$M(\omega) = \langle \exp\left[j(\omega_R R + \omega_I I) \right] \rangle = \prod_{n=1}^{N} \left\langle J_0 \left(\frac{\omega a_n}{\sqrt{N}} \right) \right\rangle, \qquad (8.47)$$

where J_0 is the zero-order Bessel function of the first kind [366]. It is now important to note that since the preceding joint characteristic function depends only on the magnitude of ω, then the corresponding probability density function of the finite random phasor sum must have circular symmetry, i.e., it is uniformly distributed in phase. This is to be expected since we assumed that the individual phases ϕ_n are uniformly distributed in the interval $[-\pi, \pi)$.

The circular symmetry of the joint characteristic function allows one to recover the radial profile of the probability density function $f(A)$ of the resultant amplitude $A = \sqrt{R^2 + I^2}$ of the finite random phasor sum by computing the Fourier–Bessel transform of $M(\omega)$. The probability density function $p_N(A)$ can then be explicitly obtained from the radial profile of the two-dimensional density by multiplying it by the circumference $2\pi A$ of a circle of radius A, resulting in the following:

$$p_N(A) = A \int_0^\infty \omega \left\langle J_0 \left(\frac{\omega a}{\sqrt{N}} \right) \right\rangle^N J_0(\omega A) d\omega \qquad (8.48)$$

where the angle bracket here denotes the averaging over the probability density function $p(a)$ of the random amplitudes a_n. Concrete results can only be obtained when a specific probability density function $p(a)$ is given for the step lengths a_n. However, it can be shown that when all the phasors have the same length, the preceding integral converges to the Rayleigh distribution in the limit $N \to \infty$ and to a delta function centered at $A = 1$ when $N = 1$ [362]. The probability density function $p_N(I)$ for the intensity $I = A^2$ can be obtained using the probability transformation law in (8.41), yielding the following:

$$p_N(I) = \frac{1}{2} \int_0^\infty \omega \left\langle J_0 \left(\frac{\omega a}{\sqrt{N}} \right) \right\rangle^N J_0(\omega \sqrt{I}) d\omega \qquad (8.49)$$

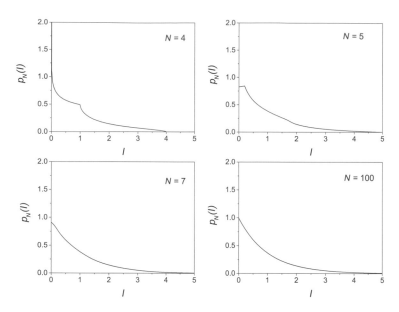

Figure 8.7 Probability density function of the intensity for random phasor sums with N terms (specified in each panel) computed using the Kluyver–Pearson formula. The N phasors have uniformly distributed phases and constant length equal to $1/\sqrt{N}$.

When all the amplitudes of the individual component phasors in the sum are equal to $a/\sqrt{N}$ we can drop the angle bracket in this expression, which then reduces to the *Kluyver–Pearson formula* [362]. Figure 8.7 displays the computed probability density functions of the intensity for random phasor sums with different numbers of terms. We note that when the number of phasors in the sum, i.e., the number of scattering objects, is small ($N \lesssim 10$), the corresponding density functions feature different nonmonotonic and markedly nonexponential behavior. However, when N becomes sufficiently large, the intensity distribution approaches the characteristic negative exponential shape previously derived under the central limit theorem in the $N \to \infty$ limit.

The nonexponential nature of the probability distribution that we have numerically computed for small N affects the values of the speckle contrast C. Interestingly, the speckle contrast can be established analytically by evaluating the moments of the intensity in a finite random phasor sum with randomness arising from the phase distribution. Assuming for simplicity a uniform phase and constant amplitudes ($a_n = a_m = a$), it can be obtained as follows [362]:

$$C = \frac{\sigma_I}{\langle I \rangle} = \sqrt{1 - \frac{1}{N}}. \tag{8.50}$$

This result shows that the fluctuations of the intensity strongly depend on the number of terms in the sum and approach the fully developed regime with $C = 1$ only in the $N \to \infty$ limit.

8.2.3   Correlations in Speckle Statistics

The presence of amplitude and/or phase correlations in disordered systems profoundly modifies their scattering behavior and generally significantly complicates the mathematical analysis. In the context of speckle optics, the effects of *correlated disorder* appear when summing a large number of phasors with nonuniform phase statistics, resulting in partially developed speckle patterns with reduced contrast, i.e., with decreased intensity fluctuations. Generally, the presence of structural correlations in random media reduces their scattering fluctuations and makes it impossible to achieve closed-form solutions for the statistical properties. However, in the presence of focusing effects caused by large-scale inhomogeneities, the contrast (or the scintillation index) can even achieve a maximum value greater than unity. As a side note, we remark that in thick samples with strong enough scattering strength, this focusing effect is normally diminished by multiple scattering effects, and the intensity fluctuations slowly begin to decrease and saturate the scintillation index at a value that approaches unity from above. In a less general context, the effects introduced by phase correlations can be directly appreciated when considering a random phasor sum with a large number $N \to \infty$ of independent phasor components and nonuniform, e.g., Gaussian, phase distribution. Under the usual assumptions that all components are identically distributed and the amplitude and phase of any individual component are independent, the joint probability density function of the complex Gaussian variate $A = R + jI = A \exp(j\theta)$, as obtained by a general form of the central limit theorem, gives rise to nonzero $\langle R \rangle$ and $\langle I \rangle$ as well as *covariance* values:

$$\text{Cov}(R, I) = \langle (R - \langle R \rangle)(I - \langle I \rangle) \rangle, \tag{8.51}$$

where $\text{Cov}(R, R) = \sigma_R^2$ and $\text{Cov}(I, I) = \sigma_I^2$. Specifically, in the presence of phase correlations, it is possible to generalize (8.36) in the following form [362]:

$$p_{R,I}(R, I) = \frac{1}{2\pi \sigma_R \sigma_I \sqrt{1 - \rho^2}} \exp\left[-\frac{\tilde{R}^2 + \tilde{I}^2 - 2\rho \tilde{R}\tilde{I}}{2(1 - \rho^2)} \right] \tag{8.52}$$

where we introduced the following normalized variables:

$$\tilde{R} = \frac{R - \langle R \rangle}{\sigma_R} \tag{8.53}$$

$$\tilde{I} = \frac{I - \langle I \rangle}{\sigma_I} \tag{8.54}$$

and the *correlation coefficient*:

$$\rho = \frac{\text{Cov}(R, I)}{\sigma_R \sigma_I}. \tag{8.55}$$

In the presence of nonzero correlations between R and I, the contours of constant probability density of the corresponding complex Gaussian variate become *elliptical* in the (R, I) plane. Moreover, when the probability density function of the phase is even, the joint distribution is centered on a nonzero point on the real axis, i.e., $\langle R \rangle \neq 0$ and $\langle I \rangle = 0$. In this case, the resultant vector of the random phasor sum deviates significantly from a circular complex random variate. The joint density for the resultant length A and phase θ can in principle be obtained by transforming the general expression (8.52) to polar coordinates and multiplying by the Jacobian determinant, as we did previously in equation (8.37). However, this time the marginal distribution obtained by integrating the resulting expression with respect to θ cannot be found in closed form. Nevertheless, it is possible to readily express the mean amplitude, intensity, and standard deviation in terms of the characteristic function $M_\phi(\omega)$ of the random phase variables, enabling the exact computation of the speckle contrast for a chosen probability density model of the phase. Results obtained for a zero-mean Gaussian phase distribution are explicitly discussed by Goodman [362].

The effect of correlations becomes more complicated in the multiple scattering regime, where light is transmitted or reflected by an optically thick (dense) and structurally disordered medium. In this limit, expressions for the average value and the fluctuations of intensity of classical waves in three-dimensional disordered systems can be derived perturbatively using the microscopic transport theory discussed in Chapter 14, but the calculations are quite difficult. Generally, in the regime of multiple scattering, anomalous long-range correlations develop, and the fluctuations of the total transmission versus the normalized intensity do not follow the inverse exponential law of fully developed speckle but develop instead a stretched exponential tail distribution [367, 368], as shown in Figure 8.8.

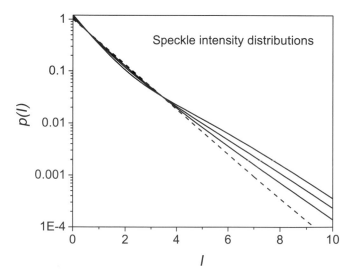

Figure 8.8 Probability distributions of normalized speckle intensity for an incoming plane wave in a random medium and for different values of the dimensionless conductance $g = 3, 5, 10$ (upper to lower curve at $I = 8$). The dashed line corresponds to the Rayleigh distribution ($g = \infty$).

In principle, both spectral (frequency) and spatial correlations in the diffuse regime can be obtained starting from the transport quantity (14.71), with considerable effort [24]. An in-depth analysis performed by Shapiro [369] shows that the light spots in speckle patterns are spatially correlated, no matter how far the waves travel, over the distance ℓ corresponding to the mean free path in the system (see Chapter 14 for an exact definition of this important quantity of transport theory). Moreover, the intensity fluctuations can be separated into a *short-range correlation component* $C^{(1)}$ associated to the rapidly varying speckle pattern, with exponential attenuation proportional to $1/\ell$, and a *long-range one*, known as $C^{(2)}$, due to diffusion relaxation over the scales of the sample thickness. The reason for the slow decay of the transmitted intensity fluctuations traces back to the presence of interference effects that "oppose" the smoothing action of the diffusion process describing the multiple-scattering regime. In addition, a third correlation term $C^{(3)}$ appears that describes the correlations of all incoming modes with all outgoing modes [370]. This term is the optical analogue of the *universal conductance fluctuations* (UFC) that characterize the mesoscopic physics of wires and is the direct consequence of quantum wave interference on a macroscopic scale [371]. Both correlations $C^{(2)}$ and $C^{(3)}$ are lower-order corrections in the expansion parameter $1/g$, where g is the dimensionless conductance that is defined by considering the total (diffuse) transmission under diffuse excitation (i.e., summing all the transmission channels under all the excitation channels). Interference corrections to the classical transport picture based on random walk picture of transport increase with the randomness of the system and can lead to the breakdown of classical (Boltzmann) transport theory. See Chapter 14 for additional discussions. Recently, speckle correlations and light propagation through scattering media have been exploited for wavefront shaping and multispectral imaging applications, resulting in significantly improved focusing and single-shot, lensless imagining techniques with diffraction-limited resolution [372–376].

8.3 Correlated Walks and Anomalous Transport

The fundamental role of correlations in random walks can be better appreciated by considering the simple case of a walk performed on the points of a lattice. At each time step, we assume that the walker jumps from its current position site to one of the other sites of the lattice according to a given probability rule. The simplest walk is performed on a hypercubic d-dimensional lattice of unit spacing when the walker hops at each time step to one of its nearest neighboring sites on the lattice with equal probabilities. After n time steps, the overall vector displacement of the walker is clearly as follows:

$$\mathbf{r}(n) = \sum_{i=1}^{n} \boldsymbol{\ell}_i, \tag{8.56}$$

where ℓ_i is a unit vector pointing to a nearest-neighbor site and representing the ith step of the random walk. In general, we have that $\langle \ell_i \rangle = 0$ when averaged over many realizations of the random walk and therefore $\langle \mathbf{r}(n) \rangle = 0$. However, since the steps are independent (uncorrelated) random variables, we have that $\langle \ell_i \cdot \ell_j \rangle = \delta_{ij}$, where δ_{ij} denotes the Knronecker delta, and the mean squared displacement is as follows:

$$\langle \mathbf{r}^2(n) \rangle = \left\langle \left(\sum_{i=1}^{n} \ell_i \right)^2 \right\rangle = n + 2 \sum_{i>j}^{n} \langle \ell_i \cdot \ell_j \rangle = n. \tag{8.57}$$

In the case when the lattice spacing equal to a, we obtain $\langle \mathbf{r}^2(n) \rangle = (2d)Dt$, where $t = n\Delta t$, the diffusion constant is $D = a^2/(2d)\Delta t$, and d is the dimensionality of the embedding space. The characteristic behavior illustrated in this example is known as regular diffusion, and for this process the speed of the walker vanishes at long times since $v \sim \sqrt{\langle r^2 \rangle}/t \sim 1/\sqrt{t}$. Complex media are characterized by a large diversity in their elementary structural units, which can exhibit interactions at multiple length scales. The onset of nonlocal correlations in space and time (i.e., memory effects) breaks the Markovian character of standard diffusion and results in *anomalous diffusion*. Examples of anomalous transport abound in many different fields of science and engineering, which include the study of liquid crystals, porous and viscoelastic materials, turbulent fluid motion, proteins flow and cells communication, the dynamics of financial markets, the information flow in social media, and across entire ecosystems [377]. Moreover, diffusion through fractal media and, in general, nonuniform structures with loops, bottlenecks, dangling ends, and other spatial inhomogeneities gives rise to anomalous diffusion. First suggested by de Gennes [378], transport properties on percolation clusters can be investigated by performing random walks (the "ant in the labyrinth") to obtain a measure of the diffusion exponent from the time scaling of the *mean squared displacement* (MSD) $\langle r^2(t) \rangle$. The problem of the ant in a labyrinth can easily be extended to fractals, where anomalous diffusion is characterized by $\langle r^2(t) \rangle \sim t^{d/d_w}$, where d is the dimensionality of the embedding space under consideration, and d_w is the fractal random walk dimension corresponding to the fractal trajectories traced by the diffusing walker [361, 379]. In all these situations, the diffusion processes are anomalous and the corresponding MSD scales nonlinearly according to a power law $\sigma^2 \sim D_\gamma t^\gamma$ with $0 \leq \gamma \leq 1$. The standard diffusion process corresponds to the situation when $\gamma = 1$. When $\gamma > 1$, the process is called anomalous *superdiffusion*, and when $\gamma < 1$ it is called anomalous *subdiffusion*. The different regimes of diffusion are summarized in Figure 8.9, where we display the asymptotic behavior of the MSD with respect to time. Subdiffusion is typical for the motion of charge carriers in disordered semiconductors, proteins in cell membranes, cellular transport, and contaminants in underground water. Superdiffusion has been observed in turbulent fluid flows, animal foraging, and even internet traffic [377].

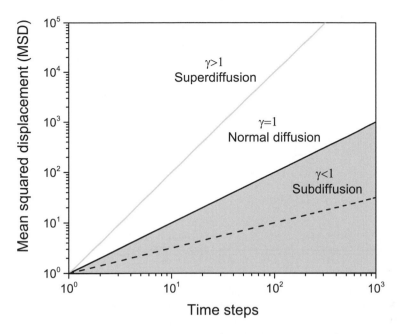

Figure 8.9 Time evolution of the mean squared displacement for different types of diffusion processes. The asymptotic power-law behavior $\sigma^2 \sim D_\gamma t^\gamma$ indicates subdiffusive transport when $\gamma < 1$ and superdiffusive transport when $\gamma > 1$. Normal diffusion occurs when $\gamma = 1$, separating the two regions of anomalous transport.

8.3.1   Beyond Gaussian Statistics: Stable Distributions

Many of the properties of random walks and processes discussed so far rely on Gaussian (or normal) random variables, which play a prominent role in science and engineering due to the central limit theorem. In addition, among all classes of random variables with finite means and variances, only normal variables preserve their class under addition, namely the sum of normal variables is again a normal variable. The problem of identifying additional limiting distributions with such "self-preserving" property upon summation of more general random variables led Paul Lévy to develop the theory of *stable distributions* [380]. The word "stable" is used here because the mathematical form of the limiting distribution is preserved under summation. In general, stable distributions feature slow decaying tails with algebraic decays and generalize the addition property of normal variables to situations where the variances become infinite and the CLT cannot be applied. Stable distributions have been recently rediscovered by Benoit Mandelbrot [381], who analyzed their scaling properties and proposed them, after the pioneering work of Vilfredo Pareto, as more suitable candidates (as opposed to normal distributions) for a more accurate statistical description of the fluctuations of stock and commodity prices [382]. The main ideas that defines the so-called *Lévy alpha-stable distribution* is that a linear combination

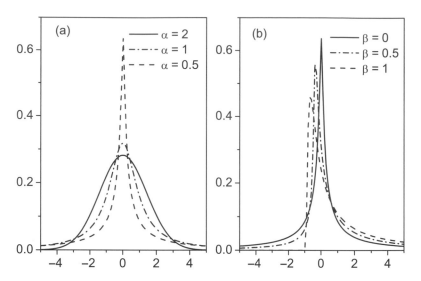

Figure 8.10 Examples of stable probability distribution functions with different values of (a) shape parameter α as listed in the legend. The other parameters in panel (a) assume the values $\beta = 0, \gamma = 1$, and $\delta = 0$. In panel (b), we show stable distributions with different values of the shape parameter β, as listed in the legend. In these examples, the other parameters assume values $\alpha = 0.5, \gamma = 1$, and $\delta = 0$.

of two or more independent random variables with this distribution must have the same distribution, up to location and scale parameters. A random variable is said to be stable if its distribution is stable. Symbolically, if $X_1, X_2, \ldots, X_n$ are independent and identically distributed stable random variables, then for every n we have the following:

$$X_1 + X_2 + \cdots + X_n = c_n X + d_n \tag{8.58}$$

for the real constants d_n and $c_n > 0$. The normal distribution, the Cauchy distribution, and the Lévy distribution all share the preceding property, and are special cases of stable distributions.

Stable distributions form a four-parameter family of continuous probability distributions that are parameterized by the location and scale parameters $\delta \in (-\infty, \infty)$ and $\gamma \in (0, \infty)$, respectively, and by the two shape parameters $\beta \in [-1, 1]$ and $\alpha \in (0, 2]$, which roughly measure the degree of asymmetry and concentration of the distributions. A few notable examples of stable distributions with different shape parameters are shown in Figure 8.10. The probability density function for a general stable distribution, denoted by $p(x; \alpha, \beta, \gamma, \delta)$, cannot be obtained analytically. However, its characteristic function admits an explicit expression. The theory of stable distributions led to the following *generalized central limit theorem* (GCLT) [383]: the sum of a large number N of random variables with symmetric distributions having power-law tails decaying as $|x|^{-\alpha-1}$ with $0 < \alpha < 2$ (having infinite variance) converges to a stable distribution as $N \to \infty$. This theorem plays an important role in the transport theory of inhomogeneous complex media, as we will discuss later in this chapter.

8.3.2   The Continuous Time Random Walk (CTRW)

At the microscopic level, anomalous diffusion processes are described by a more general type of random walk where physical time is considered a continuous real variable. The corresponding random processes are known as *continuous time random walks* (CTRWs) and were originally introduced by E. Montroll and G. H. Weiss [384]. Despite the fact that the original motivation to develop such models was to explain charge transport in amorphous solids [385], CTRWs have found numerous applications in physics and engineering due to their fundamental connection with anomalous transport processes. The main idea behind CTRW is to consider a particle that randomly jumps according to a given probability density for the length of each jumps and according to a probability density of waiting times between subsequent jumps. The picture of a walker jumping instantaneously from one site to another after waiting for a random period of time drawn from a probability density function is suitable to describe the complex dynamics of disordered systems, where particles diffuse through a distribution of trapping sites. In the most general case of a *correlated random walk*, we consider a CTRW where space and time are coupled by the joint probability density function $\psi(x,t)$ of the step's length and duration. This density describes, for a given jump, both the length of the jump and the waiting time. Based on $\psi(x,t)$, we can obtain the respective probability densities for the jump lengths and the waiting times as follows:

$$\lambda(x) = \int_0^\infty \psi(x,t)dt \tag{8.59}$$

$$w(t) = \int_{-\infty}^\infty \psi(x,t)dx. \tag{8.60}$$

By definition, $\lambda(x)dx$ is the probability of observing a jump with length in the interval $(x, x + dx)$ while $w(t)dt$ is the probability of a waiting time for the jump in the time interval $(t, t + dt)$. On the other hand, when the jumping time and length are independent random variables, the temporal and the spatial contributions are not correlated and we have $\psi(x,t) = w(t)\lambda(x)$. In this case, the CTRW is said to be uncoupled.

Our goal is now to obtain an expression for the probability density function $p(x,t)$ of the walker's position at time t in the most general case of a coupled or correlated CTRW process. Let us then define $\eta(x,t)$ to be the probability density function of the positions of walkers that just arrived at site x at time t after completing a jump. This quantity obeys the following *master equation*:

$$\boxed{\eta(x,t) = \int_{-\infty}^\infty dx' \int_0^\infty dt' \eta(x',t')\psi(x - x', t - t') + \delta(x)\delta(t)} \tag{8.61}$$

The first term in the right-hand side of this equation describes the case in which the last completed step was at point x' and was concluded at time $t' < t$ while the actual step of the walker is the one from x' to x that took the time $t - t'$ for completion. The second term in the equation corresponds to the initial condition (i.e., the actual step is the zeroth one). The probability density function $p(x,t)$ of the walker's position at time t can now be expressed as follows:

$$p(x,t) = \int_{-\infty}^{\infty} dx' \int_0^{\infty} dt' \eta(x',t') \Psi(x - x', t - t'), \qquad (8.62)$$

where we introduced the quantity $\Psi(x,t)$ that is the probability density function for the displacement of the random walker during the last, uncompleted step of the walk. The specific form of $\Psi(x,t)$ will depend on the particular process considered. Since the waiting times are all nonnegative quantities, we can express the previous equation in the Fourier–Laplace representation and, when combined with the Fourier–Laplace representation of equation (8.61), finally obtain the following:

$$p(k,s) = \frac{\Psi(k,s)}{1 - \psi(k,s)} \qquad (8.63)$$

where (k,s) are the Fourier–Laplace conjugate variables of (x,t). The last equation is the main result for the time-space coupled CTRWs, also known as the *Montroll–Weiss equation*. We can further simplify this result by considering jumps that are separated by resting periods whose duration is specified by the $w(t)$ probability density function. In this case, the last incomplete step corresponds to the resting period of the particle and we can write the following:

$$\Psi(x,t) = \delta(x) \left[1 - \int_0^t w(t')dt' \right] = \delta(x) \int_t^{\infty} w(t')dt', \qquad (8.64)$$

where the quantity that multiplies the Dirac delta function is called the *survival probability* on a site since it gives the probability that the waiting time on a site exceeds t. Fourier–Laplace transforming the last equation, we get the following:

$$\Psi(k,s) = \Psi(s) = \frac{1 - w(s)}{s}, \qquad (8.65)$$

from which we obtain the Fourier–Laplace transform of the probability density of finding a walker at position r and at time t:

$$p(k,s) = \frac{1 - w(s)}{s} \frac{1}{1 - \psi(k,s)} \qquad (8.66)$$

This important result can be easily extended to three-dimensional walks by replacing $k \to \mathbf{k}$ in this expression and in equation (8.63).

Very different types of anomalous transport processes can be obtained depending of the values of the characteristic waiting time $T = \int_0^{\infty} w(t)t\,dt$ and of the variance of the jump length $\Sigma^2 = \int_{-\infty}^{\infty} \lambda(x)x^2 dx$. In particular, when $w(t)$ and $\lambda(x)$ are independent and Σ^2 is finite, the CTRW model is asymptotically equivalent to the Brownian motion discussed in Appendix C.6.2. More specifically, a continuous time random walk in which the waiting times are exponentially distributed and the jumps are normally distributed coincides with the Brownian motion. However, more complex types of walks can be obtained when considering distributions with divergent T or Σ^2 values. As an example, choosing $w(t) \sim t^{-1-\alpha}$ and $w(s) \sim 1 - s^{\alpha}$ ($0 < \alpha < 1$), the characteristic time T is divergent and the variance Σ^2 of the jump lengths is finite. Under these conditions, it is possible to show [386] that the $MSD \sim t^{\alpha}$ so that the character of the diffusion process is anomalous and subdiffusive. On the other hand, by choosing waiting time distributions that give rise to finite T (e.g., a Poisson waiting time distribution) but jump length distributions that yield a diverging Σ^2 (e.g., an inverse algebraic power law), we obtain the so-called *Lévy flights* that manifest anomalous superdiffusive transport with $MSD \sim t^{\beta}$ with $1 < \beta < 2$. Finally, we note that for $w(t) = \delta(t - \tau)$ the CTRW reduces to the regular random walks over discrete times discussed in the previous section, and any choice of $w(t)$ that falls off fast enough will yield a regular diffusion behavior. On the opposite, anomalous diffusion will be achieved for slow-decaying $w(t)$ functions.

Representative examples of Lévy flight trajectories with $N = 10^5$ steps are shown in Figure 8.11. In these cases, we considered a random walk with a uniform phase distribution and with the values of the jump length drawn from two different stable distributions. In particular, Figure 8.11a displays the flight obtained by considering a jump length (or step-size) that follows the stable Cauchy distribution. It is evident in this case that only a small number of steps dominate the overall transport properties. This is to be contrasted with standard random walks where each step contributes with a similar amount to the macroscopic transport properties. In Figure 8.11b, we plot the absolute displacement of the walker along with the expected trend for a traditional random walk with uniform phase and Gaussian step-size distribution, corresponding to normal diffusion. The significantly larger distances covered by the Lévy flight clearly unveil its anomalous transport character. In Figure 8.11c, we show a Lévy flight trajectory obtained when the step-size is drawn from a stable distribution that is very close to a Gaussian one. In this hybrid situation ($\alpha = 1.8$), we can clearly appreciate how the Lévy flight trajectory resembles the morphology expected for a traditional random walk (compare with Figure 8.4) but also displays a small number of larger jumps that betray its anomalous behavior caused by the underlying stable distribution. The corresponding absolute displacement is shown in Figure 8.11d, where we can clearly appreciate both the similarity with the average trend of normal diffusion but also the presence of enhanced statistical fluctuations, i.e., localized deviations from the quadratic trend of the dashed curve, which are caused by the stable distribution of step-sizes. A similar situation microscopically describes the transport in inhomogeneous media, where the step length distribution may vary significantly depending on the geometry and the spatial distribution of scatterers. In these circumstances, the step length distribution may be characterized by a diverging variance that breaks the usual requirements of the central limit theorem, resulting in anomalous transport. In recent years, inhomogeneous optical materials have been engineered with local fluctuations

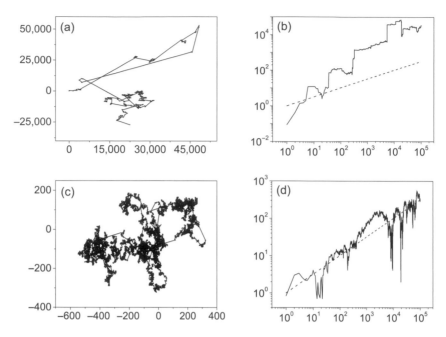

Figure 8.11 Examples of Lévy flight trajectories and the absolute displacements of the corresponding walkers as a function of the numbers of steps ($N = 10^5$). (a) Lévy flight with amplitudes distributed according to the Cauchy distribution ($\alpha = 1, \beta = 0, \gamma = 1, \delta = 0$) and (c) a stable distribution with $\alpha = 1.8, \beta = 0, \gamma = 1, \delta = 0$. Panels (b) and (d) display two absolute displacements of a Lévy walker corresponding to panels (a) and (c), respectively. The dashed lines indicate the expected trend for a Gaussian step-size distribution ($\alpha = 2, \beta = 0, \gamma = 1, \delta = 0$) corresponding to the normal diffusion process.

in the density of scattering centers that result in a heavy-tailed step length distribution and superdiffusive transport of light [387, 388]. In addition, subdiffusive wave transport has been investigated in deterministic aperiodic media and in correlated random media [22, 389, 390].

8.3.3   The Approach of Fractional Transport

In recent years, a powerful mathematical framework was introduced in applied sciences and engineering that makes use of macroscopic transport equations with generalized derivatives and integrals of fractional order [391, 392]. This approach exploits the mathematics of *fractional calculus* to provide an efficient description of anomalous transport phenomena in situations where long-range memory and spatial correlation effects become important. In such cases, the traditional Markovian random systems' approach becomes inadequate. A basic introduction to fractional calculus can be found in Appendix G.

Fractional diffusion equations contain nonlocal integro-differential operators with *power-law kernels* that physically describe space-time correlations in the scattering events through nonhomogeneous or *correlated random environments*. For example, nonlocal effects in multiple light scattering naturally arise because the mean field at a given position inside a complex medium generally depends on the spatial distribution of the surrounding scattering particles as well. Remarkable examples of anomalous transport have been recently discovered in various scientific domains such as turbulent plasmas, viscoelasticity, percolation and transport through fractals and porous media, amorphous solids, and biological systems, and even unveiled in internet traffic [377, 386]. The coarse-grained (i.e., macroscopic) description of a correlated random walk can be formulated as a time-space fractional diffusion equation of the form:

$$\boxed{D_*^\alpha u\,(x,t) = \partial^\beta u\,(x,t)\,/\partial\,|x|^\beta}$$

(8.67)

where $0 \le \alpha \le 1$ and $1 \le \beta \le 2$ are real numbers that are directly connected to the scaling law of the anomalous transport $\langle x(t)^2\rangle \sim t^\gamma$, where $\gamma = \alpha/\beta$. For subdiffusive system, $\alpha/\beta < 1$. The symbol $D_*^\alpha u\,(x,t)$ denotes the Caputo fractional derivative of order α in time while $\partial^\beta u\,(x,t)\,/\partial\,|x|^\beta$ denotes the symmetric Riesz fractional derivative of order β in space [393–395]. The full equivalence between fractional diffusion and continuous-time (correlated) random walks was rigorously established by Mainardi [396, 397]. In optics, superdiffusive optical transport has been intensively investigated, and artificial media that give rise to superdiffusion of photons, called *Lévy glasses*, have been demonstrated. A very comprehensive discussion of anomalous transport and fractional diffusion analysis of Lévy walks is provided in [398, 399]. Interestingly, by engineering the scattering medium it is possible to achieve ultraslow diffusion, or *Sinai subdiffusion*, which is a process characterized by the averaged mean-square displacement that grows logarithmically with time, i.e., $\langle x^2(t)\rangle \approx \ln^4 t$. Ultraslow photon subdiffusion was recently investigated in deterministic aperiodic photonic structures [22] and in random laser systems, where it results in a more efficient lasing behavior [390]. The application of logarithmic subdiffusion in this context is discussed in the last chapter of the book. Logarithmic subdiffusion originates in one spatial dimension from fractional equations of distributed order (DO-FEs). DO-FEs generalize fractional diffusion equations to orders of differentiation that are distributed according to continuous weighting functions [400, 401]. DO-FEs effectively describe transport in multiscale complex media, such as multiscale and multifractal systems. Optical systems that display logarithmic subdiffusion are very interesting because their diffusion constant $D = \lim_{t\to\infty} \langle x\,(t)^2\rangle\,/t \to 0$ may vanish in the absence of strong interference effects. This could lead to a novel pathway for localizing waves over a broader frequency spectrum that by exploiting a fundamentally different photon transport process. The fundamental solutions of different fractional transport models are shown in Section G.2.

8.4   Hyperuniform Point Patterns

In this section, we introduce the basic ideas behind the important field of hyperuniform point patterns, which are tailored disordered systems with a characteristic arrangement of particles that suppresses long-wavelength density fluctuations. These systems, which lie somewhat in between perfect crystals and amorphous solids, were originally introduced by Torquato and Stillinger [298] in 2003 and are currently at the center of intense research activities due to their wide applicability to many physical phenomena [6]. Remarkably, examples of hyperuniform point patterns include classical and quantum equilibrium phases of materials, driven nonequilibrium systems, random jammed particle packings, as well as disordered photonic crystal structures with large and isotropic bandgaps [6, 298, 299, 352, 402–404]. Recently, hyperuniformity has been discovered in the disordered arrangement of the chicken cone photoreceptor system consisting of five different cell types, a property that had never been observed in living organisms before [405]. Moreover, hyperuniformity has also been discovered in the distributions of prime numbers as well as in the locations of the nontrivial zeros of the Riemann zeta function [6, 352, 406, 407]. Additional details on this topic are presented in the next chapter. The hyperuniformity concept has recently been extended to include heterogeneous media with different components and continuous random fields [404, 408].

As a direct application of the previously introduced formalism of density correlation functions, we begin by addressing the density fluctuations of many-particle random systems that are modeled as statistical point patterns. Important information on the nature of point patterns can be obtained by studying the statistical fluctuations of the number of particles that are contained inside a regular domain or observation windows. Indeed, the magnitude of such local density fluctuations has been recently proposed as a possible *order metric* to quantify the degree of structural disorder/order of a given point pattern [298]. Besides its fundamental interest, the behavior of number density fluctuations in complex systems has great relevance to a number of applications in the physical and biological sciences. In particular, detailed knowledge on the exact nature of local fluctuations in point patterns has been utilized to study seemingly disparate phenomena such as the large-scale structures of the Universe (through fluctuations in the density of galaxies) [353], the structure of living cells, and the organization of granular matter [351], as well as the thermodynamics of liquids [304]. In many practical situations, one is interested in characterizing the variance in the local number of points, known as the *number variance*, of a general point pattern. Such a problem extends naturally to higher dimensions with fascinating applications to number theory [409], integrable quantum systems [352], and photonic materials [321, 402, 410]. With the goal of characterizing density fluctuations in many-body systems, Torquato and Stillinger [298] introduced the notion of an hyperuniform point pattern, which attracted significant attention in both fundamental studies and engineering applications.

An *hyperuniform point pattern* is defined as a point pattern with a number variance that, in some local observation domain Ω (e.g., a circular window), grows more

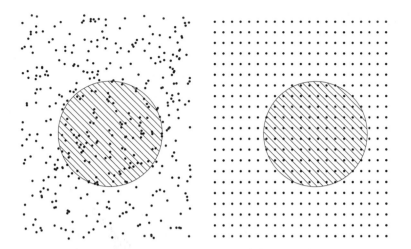

Figure 8.12 Schematics of a circular domain Ω on (left) a disordered point pattern consisting of a single realization of a Poisson point pattern and (right) on a periodic square array. The statistics (number variance) of the points contained in Ω are fundamentally different for these two patterns leading to the notion of hyperuniformity.

slowly than the mean as the size of the domain increases. In Figure 8.12, we show typical periodic and disordered (nonperiodic) two-dimensional point patterns with the corresponding observation windows. In order to better understand the concept of hyperuniformity, let us choose a domain Ω centered at a particular vector position $\mathbf{x}_0$ and investigate the size scaling of its number variance. If the point pattern is random, the number of points N_Ω contained in Ω will fluctuate across different realizations, and ensemble averaging over different configurations is necessary in order to extract the number variance for a given size of Ω. The question that was posed by Torquato and Stillinger [298] is about the dependence of the number variance on the size of Ω, which is assumed fixed at position $\mathbf{x}_0$. Clearly, in dealing with regular periodic or deterministic aperiodic point patterns, there is no need to introduce any ensemble averaging procedure since there is no uncertainty in the positions of the particles. In this case, the fluctuations in N_Ω originate by letting the window Ω uniformly sample the point patterns, i.e., by shifting it across the structure. Exact results on the number of lattice points inside a random window and generalizations in d-dimensional Euclidean space ($d > 3$) were previously established for a cubic (hyper)lattice [411, 412]. Understanding the fluctuations of points in deterministic lattices is a problem of great relevance in number theory with applications to coding and cryptography. A classical number-theoretic problem defined on a lattice is the so-called *Gauss circle problem* of determining the number $N(r)$ of integer lattice points that are contained in a circle of radius r centered at the origin. Gauss showed that $N(r)$ can be expressed as the area of the circle plus a highly fluctuating error term $E(r)$ that he was able to bound as $|E(r)| \leq 2\sqrt{2}\pi r$ [409]. Interestingly, the Gauss circle problem is closely related to the number of ways of writing the integer n as the sum of two squares [413].

Let us now return to random point patterns and consider the number variance $\sigma^2(R)$ for a circular window of fixed radius R by computing the configurational (ensemble averaged) quantity:

$$\boxed{\sigma^2(R) = \langle N^2(R)\rangle - \langle N(R)\rangle^2}$$ (8.68)

For a disordered point pattern, such as the Poisson spatial pattern illustrated in (8.12), the number variance coincides with the average number $\langle N(R)\rangle$, which in turn scales with the area of the window in two spatial dimensions or, in general, with the volume of an hypersphere in d-dimensional Euclidean space. Therefore, for general Poisson point patterns, we immediately obtain the scaling law:

$$\sigma^2(R) \propto R^d$$ (8.69)

The same law is also valid for other random point patterns that are statistically homogeneous. The question naturally arises if there exist some point patterns characterized by a number variance that grows more slowly than the volume of the window Ω. In particular, Torquato and Stillinger [298] asked for what class of point patterns the number variance $\sigma^2(R)$ grows as the surface area of Ω.[5] Considering a spherical domain, they defined a point pattern *hyperuniform* when, for large R values, the number variance obeys the scaling relation [298]:

$$\boxed{\sigma^2(R) \sim R^{d-1}}$$ (8.70)

In the precise sense defined, hyperuniform point patterns, also known as superhomogeneous point patterns [415], show "more uniformity" than standard homogeneous disordered point patterns such as Poisson random point processes. It is important to point out that in addition to the preceding scaling law provided, which identifies a type of hyperuniformity called Class I, there are two other possible types of scaling, as discussed in [6].

In Figure 8.13, we compare the scaling of the variance $\sigma^2(R)$ for three different types of hyperuniform patterns, subrandom patterns (defined in Section 8.5.1) and for a random Poisson point pattern. The variance scaling for the hyperuniform arrays follows a linear trend (in double logarithmic scale) with a slope approximately equal to one, consistent with their hyperuniform nature. We discovered that this is also the case for the more uniform subrandom arrays that we introduce in Section 8.5.1, which are presented in Figure 8.13b. On the other hand, as expected based on the equivalence between its mean and variance, the variance of the Poisson random point pattern grows quadratically. As we will discuss in the next section, hyperuniform point patterns are characterized by *vanishing infinite-wavelength density fluctuations*. This is a property shared by all periodic crystal structures, certain deterministic aperiodic systems such as Vogel spirals and the Pinwheel pattern, and special disordered point

---

[5] It was already established that for homogeneous and statistically isotropic point patterns, the variance cannot grow slower than the surface area of the window Ω [414].

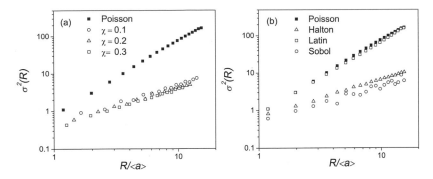

Figure 8.13 Scaling analysis of the variance $\sigma^2(R)$ for different types of aperiodic point patterns. The data are plotted in double logarithmic scale versus the normalized radius $R/<a>$, where $<a>$ is the average interparticle separation. (a) The symbols show the variance scaling with normalized radius for a Poisson random point pattern (black squares) and for three hyperuniform point patterns with $\chi = 0.1$, (open circles), $\chi = 0.2$ (open triangles), and $\chi = 0.3$ (open squares). (b) The symbols show the variance scaling with respect to the normalized radius for a Poisson random point pattern (black squares) and for three main examples of subrandom arrays. .

patterns. The correlation analysis presented at the end of this chapter provides some clues on how to predict the occurrence of hyperuniformity in a given point pattern by considering the nature of third- and fourth-order structural correlations.

A particularly intriguing feature of hyperuniform point patterns is that their direct correlation function $c(\mathbf{r})$, defined by the Ornstein–Zernike integral equation in (8.20), decays slower than r^{-d} and therefore the volume integral of $c(\mathbf{r})$ does not exist. As a result, the direct correlation of hyperuniform point patterns is long-ranged, while their pair correlation function $g(\mathbf{r})$ remains short-ranged. This implies a remarkable type of *hidden local order* that is just opposite to what manifested at their thermodynamic critical points by complex systems, where the pair correlation acquires a long-ranged character but the direct correlation remains short-ranged [304]. A very comprehensive review on the fascinating properties and numerous applications of hyperuniform states of matter can be found in [6] while a description of a general theoretical methodology to very efficiently construct perfectly hyperuniform packings in d-dimensional Euclidean space is provided in [416].

In the next sections, we will gather and discuss the main results, originally presented by Torquato and Stillinger in [298], on the number statistics of general point patterns, and we emphasize their implications for wave scattering in hyperuniform media.

8.4.1 Fluctuations in Random Point Patterns

We will review here the results obtained in [298] for random point patterns. A general variance formula has been derived starting from the expression of the number of points

$N(\mathbf{x}_0, \mathbf{R})$ contained inside an oval window centered at $\mathbf{x}_0$ and characterized by the vector $\mathbf{R}$ of its semi-axes. It is rather easy to verify the following:

$$N(\mathbf{x}_0, \mathbf{R}) = \int_V n^{(1)}(\mathbf{r})w(\mathbf{r} - \mathbf{x}_0; \mathbf{R})d\mathbf{r} = \sum_{i=1}^{N} w(\mathbf{r}_i - \mathbf{x}_0; \mathbf{R}). \qquad (8.71)$$

where $w(\mathbf{r} - \mathbf{x}_0; \mathbf{R})$ is the window *indicator function* of the domain Ω, which is equal to one if $\mathbf{r}$ is inside Ω and zero otherwise. Under the condition of statistical homogeneity of the point pattern, the preceding expression does not depend on the vector position $\mathbf{x}_0$, as is assumed in the rest of this section.

The average number of points contained within Ω is simply obtained by computing the ensemble average over the realizations of the disorder, which according to the general prescription in (8.1) yields the following:

$$\langle N(\mathbf{R}) \rangle = \int_V \sum_{i=1}^{N} w(\mathbf{r}_i - \mathbf{x}_0; \mathbf{R})p(\mathbf{r}_1, \mathbf{r}_1, \dots, \mathbf{r}_N)d\mathbf{r}_1 d\mathbf{r}_2 \dots, d\mathbf{r}_N. \qquad (8.72)$$

Using the definitions of marginal distributions introduced in Section 8.1.1, we can rewrite the average number of points as follows:

$$\langle N(\mathbf{R}) \rangle = \int_V \rho^{(1)}(\mathbf{r}_1)w(\mathbf{r}_1 - \mathbf{x}_0; \mathbf{R})d\mathbf{r}_1 = \rho V_{\Omega(R)}, \qquad (8.73)$$

where we indicated by $V_{\Omega(R)}$ the volume of the general window region with geometrical parameters $\mathbf{R}$ and ρ is the constant number density of the point pattern.[6]

By taking the ensemble average of the square of equation (8.72) and combining it with equation (8.73), we obtain the general formula for the *local number variance*, as originally derived in [298]:

$$\boxed{\langle N^2(\mathbf{R}) \rangle - \langle N(\mathbf{R}) \rangle^2 = \langle N(\mathbf{R}) \rangle \left[1 + \rho \int_V h(\mathbf{r})\alpha(\mathbf{r}; \mathbf{R})d\mathbf{r} \right]} \qquad (8.74)$$

where $h(\mathbf{r})$ is the total correlation function and $\alpha(\mathbf{r}; \mathbf{R}) = V(\mathbf{r}; \mathbf{R})/V_{\Omega(R)}$ is the *scaled intersection volume*, i.e., a purely geometrical quantity that gives the intersection volume between two identical windows whose centers are separated by the vector $\mathbf{r} = \mathbf{r}_1 - \mathbf{r}_2$. Explicit analytical expressions exist for the intersection volume of spherical windows in arbitrary d-dimensional Euclidean space [298].

The formula for the local number variance for an arbitrarily shaped window can be rewritten in the Fourier domain and, using Parseval's theorem and the definition of the structure factor $S(\mathbf{k})$, it is possible to derive the Fourier representation of the number variance [298]:

---

[6] The uniform nature (i.e., translational invariance) of the point pattern is here used to eliminate the dependence on the window position.

$$\langle N^2(\mathbf{R})\rangle - \langle N(\mathbf{R})\rangle^2 = \langle N(\mathbf{R})\rangle \left[\frac{1}{(2\pi)^d} \int_V S(\mathbf{k})\tilde{\alpha}(\mathbf{k};\mathbf{R})d\mathbf{k} \right], \qquad (8.75)$$

where $\tilde{\alpha}(\mathbf{k};\mathbf{R})$ is the Fourier transform of the scaled intersection volume. The preceding formula, which implies that the local number variance is strictly positive, encodes the fundamental properties of number fluctuations in terms of the structure factor and the geometrical parameter $\tilde{\alpha}(\mathbf{k};\mathbf{R})$. Moreover, when considering the formula (8.75) in the limiting case of a window that grows infinitely large, the function $\tilde{\alpha}(\mathbf{k};\mathbf{R})$ tends to a Dirac delta function so that we can write the following:

$$\boxed{\frac{\langle N^2(\mathbf{R})\rangle - \langle N(\mathbf{R})\rangle^2}{\langle N(\mathbf{R})\rangle} = 1 + \rho \int_V h(\mathbf{r})d\mathbf{r} = S(\mathbf{k}=0)} \qquad (8.76)$$

This is a fundamental results that links local density fluctuations with the long-wavelength limit (i.e., static) structure factor of the point pattern.

Using the previous results, an *asymptotic variance formula*, valid for large R, can be obtained in the case of statistically homogeneous point patterns, and simple formulas arise when considering spherical windows [298]. In particular, substituting the first-order expansion of the known analytical expression for the scaled intersection volume function into the local number variance formula (8.74), it is possible to obtain the following expression valid for large R [298]:

$$\langle N^2(\mathbf{R})\rangle - \langle N(\mathbf{R})\rangle^2 = 2^d \phi \left[A \left(\frac{R}{D}\right)^d + B \left(\frac{R}{D}\right)^{d-1} + \ell \left(\frac{R}{D}\right)^{d-1} \right], \qquad (8.77)$$

where ϕ is a dimensionless density defined by the following:

$$\phi = \rho \frac{\pi^{d/2}}{2^d \Gamma(1+d/2)} D^d, \qquad (8.78)$$

and D is a characteristic microscopic length for the system such as the mean first-neighbor distance between points. In the preceding expression, A and B are the asymptotic *volume and surface area coefficients* and ℓ denotes terms of lower order than x.

Of particular interest to our discussion is the expression for the volume coefficient A that can be written as [298]:

$$A = 1 + \rho \int_V h(\mathbf{r})d\mathbf{r}. \qquad (8.79)$$

By comparing this expression with the definition of the structure factor in (8.9), it becomes evident that hyperuniform (i.e., $A = 0$) random point patterns satisfy the following:

$$\boxed{A = \lim_{|\mathbf{k}|\to 0} S(\mathbf{k}) = 1 + \rho \int_V h(\mathbf{r})d\mathbf{r} = 0} \qquad (8.80)$$

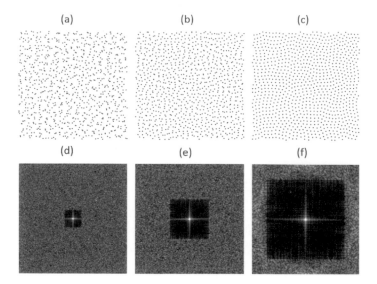

Figure 8.14 (a–c) Hyperuniform point patterns with order parameters $\chi = 0.1$, $\chi = 0.2$, and $\chi = 0.3$, respectively. As χ increases toward one, the system becomes more ordered. (d–f) The corresponding structure factors $S(k)$. Scattering exclusion regions of increasing sizes are clearly visible around $k = 0$.

The result established indicates that hyperuniform point patterns do not possess infinite-wavelength fluctuations and that, in the single-scattering regime, they are characterized by a complete suppression of forward scattered radiation. This last property has been recently exploited to engineer enhanced transparency (in the forward scattering direction) in a random hyperuniform medium carefully designed to suppress scattering within a square domain of the reciprocal space centered at the origin [410] and to design improved directional light emitters [321]. The notion of *stealthy hyperuniform point patterns* has been introduced and discussed in this context to characterize this intriguing consequence of hyperuniformity in homogeneous random point patterns [299, 403]. A general feature of hyperuniform random structures is that under normal incidence excitation, their diffraction patterns feature a compact region around the origin in which there is no scattering. Clearly, the exact forward direction ($\mathbf{k} = 0$) still displays a strong peak due to the transmitted incident beam. This behavior is clearly demonstrated in Figure 8.14, where we have generated hyperuniform point patterns with different degrees of local order and displayed the corresponding structure factors. The "stealth" nature of these arrays is manifested by the presence of a square-shaped exclusion region in the diffraction space that can be controlled by design. Stealth arrays are generated using the process outlined in [410]. Briefly, the algorithm first requires the generation of an initial random point pattern, then a potential function associated with the targeted square exclusion region is minimized using the particle positions as free parameters. The control parameter, χ, with $0 \leq \chi \leq 1$, determines the size of the exclusion region, with the maximum size being limited by the number of free parameters (the number of particles) that are available. The tunable degree of structural correlations that can be achieved in stealthy hyperuniform structures has been recently shown to produce different wave transport regimes ranging from ballistic to photon diffusion, and even Anderson localization has been reported in these systems [417–419].

8.4.2   Fluctuations in Deterministic Point Patterns

In this section, we discuss the notion of hyperuniformity for deterministic point patterns. An exact formula has been derived for the number variance of a single realization of a point pattern consisting of a large number of points N in a system of large volume V [298]. This important result applies to any deterministic point pattern, not necessarily regular or homogeneous (i.e., translationally invariant), including aperiodic structures and quasicrystals [420, 421]. In these situations, the fluctuations in the number density arise because the window $\Omega(R)$ uniformly samples the entire space.[7]

For deterministic systems, the number of points $N(\mathbf{x}_0, \mathbf{R})$ that appear in equation (8.71) needs to be averaged according to the probability of uniform sampling of the volume (i.e., $p = 1/V$). A calculation similar to what led to equation (8.73) yields the average number of points within the window [298]:

$$\overline{N(R)} = \frac{1}{V} \int_V \sum_{i=1}^N w(|\mathbf{r}_i - \mathbf{x}_0|; R) d\mathbf{x}_0 = \rho V_{\Omega(R)} = 2^d \phi \left(\frac{R}{D} \right)^d. \tag{8.81}$$

Squaring the equation (8.71) and averaging provides the following:

$$\overline{N^2(R)} = \frac{1}{V} \left[\int_V \sum_{i=1}^N w(|\mathbf{r}_i - \mathbf{x}_0|; R) d\mathbf{x}_0 + \int_V \sum_{i \neq j}^N w(|\mathbf{r}_i - \mathbf{x}_0|; R) w(|\mathbf{r}_j - \mathbf{x}_0|; R) d\mathbf{x}_0 \right]. \tag{8.82}$$

If we denote $r_{ij} = |\mathbf{r}_i - \mathbf{r}_j|$ and remember the definition of the scaled intersection volume, we can express the equation above as follows:

$$\overline{N^2(R)} = \rho V_{\Omega(R)} + \frac{\rho V_{\Omega(R)}}{N} \sum_{i \neq j}^N \alpha(r_{ij}; R). \tag{8.83}$$

Based on these results, it becomes possible to establish (exercise) the local variance formula for the deterministic case [298]:

$$\boxed{ \sigma^2(R) = 2^d \phi \left(\frac{R}{D} \right)^d \left[1 - 2^d \phi \left(\frac{R}{D} \right)^d + \frac{1}{N} \sum_{i \neq j}^N \alpha(r_{ij}; R) \right] } \tag{8.84}$$

---

[7] For simplicity, $\Omega(R)$ is again assumed to be a spherical window of radius R and the size of the system is much larger than the window radius in order to avoid boundary effects.

This expression becomes identical to the previously derived equation (8.74) in the case of statistically homogeneous infinite systems. However, equation (8.84) contains small-scale fluctuations with respect to R due to the presence within the brackets of the sum of scaled intersection volumes over all the particles pairs. The local variance formula (8.84) is very relevant to engineering applications because it is valid for a single realization enabling, in principle, to identify a particular point pattern that minimizes the variance at a fixed value of R or that achieves a targeted value of the variance. In particular, since for deterministic hyperuniform patterns the variance must grow as R^{d-1}, equation (8.84) implies the following for large R [298]:

$$\sigma^2(R) = \Lambda(R) \left(\frac{R}{D}\right)^{d-1} + O\left(\frac{R}{D}\right)^{d-2}, \tag{8.85}$$

where

$$\Lambda(R) = 2^d \phi \left(\frac{R}{D}\right) \left[1 - 2^d \phi \left(\frac{R}{D}\right)^d + \frac{1}{N} \sum_{i \neq j}^{N} \alpha(r_{ij}; R)\right] \tag{8.86}$$

is the asymptotic *surface area function* which, differently from the case of a random homogeneous point patterns, explicitly depends on R. In order to "smooth out" the small-scale variations of $\Lambda(R)$, it is useful to define a surface area coefficient as the radial average [298]:

$$\overline{\Lambda(R)} = \frac{1}{L} \int_0^L \Lambda(R) dR. \tag{8.87}$$

Since this expression is valid for a single realization, it can be employed to engineer a particular (deterministic) pattern that minimizes the surface area coefficient. It is important to realize that for a deterministic hyperuniform pattern it is not possible to find a simple formula that expresses the volume coefficient in terms of the structure factor, as was done previously in the case of statistically homogeneous point patterns. Therefore, hyperuniform deterministic point patterns do not necessarily suppress the forward scattering of incident waves.

Finally, general order metrics, which are useful for the classification of different hyperuniform point patterns, can be introduced in d-dimensional Euclidean space dividing the surface area coefficient $\overline{\Lambda(R)}$ by $\phi^{(d-1)/d}$ [298]. This parameter is a scalar that quantifies the *degree of randomness* in a point pattern, and it has been numerically computed for a number of one-, two-, and three-dimensional regular as well as aperiodic point patterns, as detailed in [298, 404]. A significant outcome of these studies is that while the smallest values of the order metrics are obtained with Bravais lattices, it does not follow in general that the densest point patterns are the ones that minimize the fluctuations in the number variance. This behavior is simply demonstrated in three spatial dimensions by the fact that the body-centered cubic (BCC) lattice was found to possess a smaller asymptotic (normalized) surface area coefficient than the denser face-centered cubic (FCC) lattice [404].

8.4.3   Classification of Hyperuniform Systems

Hyperuniform structures can be classified based on the scaling of their structure factor $S(k)$, which decays as a power-law decay in the vicinity of the k-space origin [6]:

$$S(k) \sim k^\alpha \quad k \to 0 \tag{8.88}$$

with scaling exponent $\alpha > 0$ roughly describing the degree of short-range order.[8] In the limit $\alpha \to \infty$, hyperuniform systems become stealthy and wave scattering is suppressed within a compact region of k-space. The applicability of the scaling equation (8.88) becomes difficult or not possible for structures whose structure factor is discontinuous and densely supported, as is the case for quasicrystals.

A more general hyperuniformity criterion that is valid in the case of quasicrystals and also for singular measures was recently introduced by Oğuz et al. [420], who proposed to consider the integrated or cumulative *density function* defined as follows:

$$Z(k) = \int_0^k S(\mathbf{q}) a_d q^{d-1} dq \tag{8.89}$$

where $a_d = d\pi^{d/2}/\Gamma(1 + d/2)$ is the surface area of a d-dimensional sphere of unit radius. The preceding integral is a smoother function compared to $S(k)$ and is computed in k-space over a sphere of radius k from the origin. The integrated intensity $Z(k)$ and the number variance $\sigma^2(R)$ are linked by the integral relation [420]:

$$\sigma^2(R) = -\rho v_1(R) \left[\frac{1}{(2\pi)^d} \int_0^\infty Z(k) \frac{\partial \tilde{\alpha}_2(k; R)}{\partial k} dk \right], \tag{8.90}$$

where ρ is the density, $v_1(R) = a_d R^d / d$ and $\tilde{\alpha}_2(k; R)$ is the square of the Fourier transform of the indicator function of a d-dimensional sphere of radius R divided by the volume the sphere, which can be obtained explicitly in terms of the Bessel function $J_\nu(x)$ of the first kind of order ν [420].

For one-dimensional systems, the $Z(k)$ scales according to [420]:

$$Z(k) \sim k^{1+\alpha} \quad k \to 0, \tag{8.91}$$

---

[8] As α increases, so generally does the degree of short-range order in the pattern.

where hyperuniformity again corresponds to $\alpha > 0$. The scaling behavior of the number variance $\sigma^2(R)$ of particles within a spherical observation window of radius R enables the following classification:

$$\sigma^2(R)\bigg|_{R\to\infty} \sim \begin{cases} R^{d-1} & \alpha > 1 & \text{strongly hyperuniform (Class I)} \\ R^{d-1}\ln R & \alpha = 1 & \text{logarithmic hyperuniform (Class II)} \\ R^{d-\alpha} & 0 < \alpha < 1 & \text{weakly hyperuniform (Class III)} \\ R^{d-\alpha} & \alpha < 0 & \text{antihyperuniform} \end{cases}$$

(8.92)

Box 8.2 Sum Rules and Order Metric

The structure factors of a general correlated random system must deviate from unity for some wavevectors. However, from equation (8.9) we can immediately appreciate that the value of the total correlation function at the origin, i.e., $h(\mathbf{r} = 0)$, can be used to obtain a general *sum rule* that must be obeyed by its Fourier transform, proportional to $S(\mathbf{k}) - 1$, as follows [6]:

$$\rho h(\mathbf{r} = 0) = \frac{1}{(2\pi)^d} \int_{\mathbb{R}^d} [S(\mathbf{k}) - 1] d\mathbf{k}$$

(8.93)

It follows that for patterns with a minimum (nonzero) interparticle distance, $h(\mathbf{r} = 0) = -1$. Combining this information with the general definition of the structure factor (8.9) as well as the hyperuniform requirement in equation (8.80), we can obtain the direct-space sum rule obeyed by any hyperuniform system:

$$\rho \int_{\mathbb{R}^d} h(\mathbf{r}) d\mathbf{r} = -1$$

(8.94)

This equation implies that $h(\mathbf{r})$ in hyperuniform structures *must manifest negative correlations* for some values of $\mathbf{r}$.

Integrals over the total correlation function can also be introduced to quantitatively describe the degree of order present in a given point pattern. A convenient scalar positive order metric that measures the degree of translational order in a point pattern with respect to a completely uncorrelated random system was introduced in [298, 299], and it manifests as a sum rule for the square of the total correlation function:

$$\tau = \frac{1}{D^d} \int_{\mathbb{R}^d} h^2(r) d\mathbf{r} = \frac{1}{(2\pi^d)D^d} \int_{\mathbb{R}^d} \tilde{h}^2(k) d\mathbf{k},$$

(8.95)

where D is some characteristic length scale of the system and $\tilde{h}^2$ denotes the Fourier transform of the total correlation function. This quantity diverges in perfect crystals or quasicrystals of infinite size and so it is better suited to characterize the degree of pair correlations in amorphous structures.

Crystal lattices and traditional quasicrystals obtained by the cut and projection method or by substitution rules are found to be strongly hyperuniform systems (Class I), while examples of weakly and logarithmic hyperuniform structures are scarce [421]. When $\alpha < 0$, the systems are identified as antihyperuniform. These peculiar structures are not hyperuniform as they display unbounded intensity fluctuations. The corresponding point patterns are clustered and describe, for instance, the configurations of thermal critical points. However, the scaling behavior of $\sigma^2(R)$ can be directly related to the scaling of $Z(k)$ for small k, and the values of α that identify different structures agree in both definitions [420]. However, the k-space classification of hyperuniform structures through the $Z(k)$ scaling has the advantage to be an intrinsic property of the point pattern itself, while the asymptotic scaling of the variance $\sigma^2(R)$ may depend on the choice of window shape.

8.4.4   Numerical Optimization of Arbitrary Point Patterns

In previous sections, we have considered in detail the implications of structural correlations in the far-field scattering properties of particle arrays described as point patterns. We now discuss a general methodology for the creation of customized point patterns that exhibit a desired structure factor. This can be achieved with optimization approaches that make use of the *multivariate regression method*. Such techniques are widespread in network-based optimization and machine learning [154, 155] and became of central importance to the field of computational imaging [305].

We will discuss the implementation of the optimization method in relation to the problem of finding the coordinates x_i of an N-point one-dimensional pattern with a prescribed structure factor expressed as follows:

$$S(k_m) = \frac{1}{N} \left| \sum_{i=1}^{N} e^{jk_m \cdot x_i} \right|^2, \tag{8.96}$$

where we explicitly indicate the structure factor at some discrete value k_m of the wavevector with $m = 1, \ldots, M$.

Let us consider $N = 250$ for concreteness and target a flat spectrum point pattern with desired structure factor $S_0(k_m) = 1$ for $M = 500$ equally spaced values of k_m ranging in the interval $[0.1, \ldots, 1]$. The multivariate regression method minimizes a properly defined *cost function* C through iterations of the *gradient descent method*. For the problem at hand, we minimize the mean square estimate (MSE):

$$C = \sum_{j=1}^{M} [S(k_j) - S_0(k_j)]^2 \tag{8.97}$$

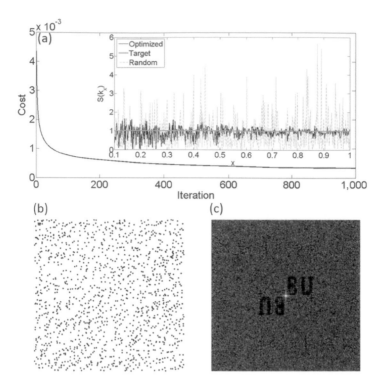

Figure 8.15 (a) The value of cost function C during the minimization process as a function of the iteration number, for constant learning rate $\alpha = 1$. The inset shows the optimized structure factor $S(k)$ (black line) compared with the targeted structure factor (gray line). For comparison, the additional plot of the $S(k)$ corresponding to a single realization of a random array (dashed line). (b) Optimized two-dimensional point pattern and (c) corresponding structure factor.

The gradient descent method minimizes the cost function by updating the coordinates of all the particles at each step according to the following rule:

$$\mathbf{x} \rightarrow \mathbf{x} - \nabla_{\mathbf{x}} C, \tag{8.98}$$

where α is a suitably chosen constant parameter known as the learning rate, $\nabla_{\mathbf{x}}$ is the gradient operator in the N-dimensional space ($N = 250$ in the example of the linear array), and $\mathbf{x}$ is the vector containing the coordinates of all the particles that determine the dimensionality of the optimization problem. This algorithm has been implemented on a laptop (4 GB RAM and Core-i5 CPU), and the results of the optimization obtained in a few-minutes time are shown in Figure 8.15a. As an interesting and more challenging example in higher dimensionality ($N = 2,000$), we consider the design of a two-dimensional point pattern that exhibits an arbitrarily structured diffraction pattern that displays the letters "BU" in the corresponding structure factor. The optimized point pattern and the corresponding structure factor are shown in Figure 8.15b–c, respectively, proving the validity of the approach.

8.5   Equidistributed Sequences and Light Scattering

We have seen in the previous sections that the intensity of waves diffracted by an arbitrary array of point particles is proportional to the associated structure factor. In particular, for a one-dimensional arrangement of scattering points with coordinates x_n we can write the following:

$$S(q) = \frac{1}{N} \left| \sum_{n=1}^{N} e^{2\pi i q x_n} \right|^2 . \tag{8.99}$$

This expression is a complex *exponential sum* whose behavior depends on the specific arrangement of the particles. Exponential sums have a long and fascinating history in number theory [422] and have also been intensively investigated in the mathematical theory of *equidistributed sequences*, which is the foundation of efficient sampling methods such as the quasi–Monte Carlo technique [423]. Moreover, the engineering of novel scattering systems based on equidistributed sequences may result in subdiffusive wave transport and weak localization phenomena, as recently discussed by Sgrignuoli and Dal Negro [389].

In what follows, we will discuss, in the context of kinematic scattering, the main concepts in the theory of uniform sequences, originally developed by Hermann Weyl in 1916 [424]. Generally speaking, this theory is concerned with point sets and sequences having a uniform distribution inside a real interval, such as the distribution of the fractional parts of certain sequences of real numbers $\{x_n\} = x_n - [x_n]$ in the unit interval $I = [0, 1)$. Here $[x_n]$ denotes the integer part of x_n, which is the greatest integer smaller or equal to x_n. Uniform distribution theory investigates well-distributed point sets and sequences. The fundamental notion of this theory is the concept of an equidistributed sequence, or a sequence that is *uniformly distributed modulo one*, abbreviated as u.d. mod(1). A sequence x_n of real numbers is said to be u.d. mod(1) when the proportion of its terms that fall within any half-open subinterval of I is proportional to the length of that interval. More precisely, a sequence x_n of real numbers is said to be u.d. mod(1) if it satisfies the following [425]:

$$\lim_{N \to \infty} \frac{A([a, b); N)}{N} = b - a \tag{8.100}$$

for every pair of numbers a and b with $0 \le a < b \le 1$, where $A([a, b); N)$ denotes the number of terms x_n with $1 \le n \le N$, whose fractional part belongs to the interval $[a, b)$. Informally, a number sequence x_n is u.d. mod(1) in the interval I if every half-open subinterval of I eventually contains its "proper share" of fractional parts.

A central theorem in the theory of equidistributed sequences is the *Weyl's criterion* that provides the necessary and sufficient condition for a general sequence x_n to be u.d. mod(1) in term of the asymptotic behavior of the corresponding exponential sum.

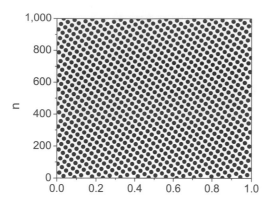

Figure 8.16 Illustration of the filling of the unit interval $I = [0, 1)$ of the fractional part of the first $n = 1,000$ terms of the $n\alpha$ sequence, where $\alpha = (1 + \sqrt{5})/2$.

Specifically, Weyl's theorem [425], which can be generalized in any dimension, states that an arbitrary sequence x_n of real numbers is u.d. mod(1) if and only if [425]

$$\lim_{N \to \infty} \frac{1}{N} \sum_{n=1}^{N} e^{2\pi i q x_n} = 0 \qquad (8.101)$$

for all integers $q \neq 0$. It is important to realize that the trigonometric sum in equation (8.101) in fact coincides with the array factor of kinematic diffraction theory. In particular, its squared modulus is proportional to the far-field diffracted intensity from an array of point scatterers with coordinates x_n. Therefore, Weyl's theorem implies that large-scale arrays of point scatterers arranged according to a u.d. mod(1) sequence will strongly suppress the far-field scattering radiation, except for the forward direction. From this perspective, u.d. mod(1) arrays manifest a complementary behavior with respect to the previously introduced hyperuniform point patterns that suppress wave scattering only in the forward direction.

The direct application of Weyl's theorem allows us to deduce that the sequence $x_n = n\alpha$, where α is any irrational number, is u.d. mod(1) since its fractional parts uniformly fill the unit interval I. Moreover, the sequence of the factional parts $x_n = \{n\alpha\}$ is also dense since the difference between consecutive terms can be made arbitrarily small (see Figure 8.16).[9] A theorem proved by Kronecker established that equidistributed sequences are always dense. However, the converse is not true since dense sequences are not necessarily equidistributed. In fact, consider as an example the sequence of the fractional parts of the logarithms $x_n = \{\log n\}$. This is a dense sequence in I since $\log (n + 1) - \log n \to 0$ when $n \to \infty$. On the other hand, it is easy to show that the associated trigonometric sum, when divided by N, does not converge to zero. Therefore, by Weyl's criterion this dense sequence is not u.d. mod(1). Interestingly, it can be proved based on general trigonometric summation

---

[9] More technically, a sequence $x_n \in I$ is dense if for all $x \in I$ there exists a subsequence x_{n_k} that converges to x.

methods that the sequence $x_n = \{\alpha n \log n\}$ is u.d. mod(1) for all irrational numbers $\alpha \neq 0$, and therefore dense. Additional examples of equidistributed sequences and their relation to spatial randomness, arithmetic chaos, and Poisson behavior will be discussed in Section 10.2.

It is interesting to reflect on the implications of uniform distribution theory on the diffraction properties of point particles. In particular, it is known that the irrational numbers α that are more slowly approximated, in the technical sense later discussed in Section 9.4, by rational numbers (e.g., the golden mean) are the ones for which the sequence $\{n\alpha\}$ is distributed with the smallest discrepancy [426]. This is the fundamental mathematical reason why incommensurate and quasicrystal structures have dense pure-point diffraction spectra, as previously introduced in Section 7.5.2. On the other hand, when α is rational, the sequence $x_n = \{n\alpha\}$ cannot be u.d. mod (1) since there will only be a finite set of values of fractional parts. Therefore, the corresponding point scatterers is characterized by periodically repeating diffraction peaks, i.e., such structures will be periodic.

8.5.1 Low-Discrepancy Sequences and Subrandomness

The degree of uniformity of equidistributed sequences is quantified by the mathematical concept of discrepancy. For a one-dimensional sequence x_n of N real elements, the discrepancy $D_N = D_N(x_1, \cdots, x_n)$ is defined by the following [425]:

$$
D_N = D_n(x_1, \ldots, x_N) = \sup_{0 \leq a < b \leq 1} \left| \frac{A([a,b); N)}{N} - (b-a) \right| \tag{8.102}
$$

For any numerical sequence of N numbers, we have: $1/N \leq D_N \leq 1$. In addition, upper bounds that estimate the discrepancy of any finite sequence of real numbers can be rigorously obtained in terms of the exponential sums that appear in Weyl's criterion, as precisely formulated by LeVeque's inequality and by the Erdös–Turán theorem [425]. These are important results that highlight the relation between discrepancies and exponential sums.

The discrepancy D_N of a sequence x_n with N terms is low if the fraction of points in the sequence falling into an arbitrary subset of the unit interval is close to being proportional to the length of the interval, as would happen in the case of a u.d. mod(1) sequence.[10] An important theorem establishes that a sequence x_n is u.d. mod(1) if and only if $\lim_{N \to \infty} D_N = 0$ [427], thus proving the fundamental equivalence between uniform sequences mod(1) and zero-discrepancy sequences. Finite-length sequences with such asymptotic property are often referred to as *subrandom* or *quasirandom sequences*. These sequences are very useful in numerical applications, where they outperform, at least for low-dimensional problems, the results obtained using uniformly distributed random numbers. In particular, the Koksma–Hlawka theorem links directly the errors in multidimensional integration to the discrepancy of the sequence used to

---

[10] This would also happen on average for a random variable with uniform distribution on the same interval, but not for any specific realization of that variable.

sample the integrand [425]. Low-discrepancy sequences share some basic properties of random variables but also have an advantage over pure random numbers in that they cover a given domain of interest more quickly and more evenly. Sequences of subrandom numbers can also be generated from random numbers by imposing a negative correlation such that clustering of points is avoided. Interestingly, the elements of subrandom sequences can be generated either in a deterministic fashion, as in the case of the Halton, Faure, and Sobol sequences, or by a stochastic algorithm, as for the Latin hypercube sequence [423].

The archetypical low-discrepancy sequence is the one developed by van der Corput, which serves as a basis building block for all the others. Due to its importance, we describe in this section its deterministic construction. Let us start by choosing an integer base b. In order to obtain the nth element in the van der Corput sequence, we first have to express n in base b:[11]

$$n = \sum_{j=0}^{m-1} a_j(n) b^j, \tag{8.103}$$

where m is the smallest integer such that $a_j(n) = 0$ for all $j > m - 1$. We then reverse the number n in order to find the value of the nth element of the sequence, denoted by b_n:

$$b_n = \sum_{j=0}^{m-1} \frac{a_j(n)}{b^{j+1}}. \tag{8.104}$$

In reversing the number representation, we ensure that the values of the sequence always lie in the $[0, 1)$ interval. Moreover, the elements of the van der Corput sequence (in any base) form a dense set in the unit interval. In the limit of a large sequence, the elements will be equidistributed over the unit interval. To put this procedure more explicitly, if we are counting up to integer $n = a_m a_{m-1} \ldots a_0$ then the nth term b_n is given by reversing the digits, thus $b_n = 0.a_0 a_1 \ldots a_m$. The van der Corput sequence has a discrepancy that scales with the number of terms N in the sequence as $\sim \ln(N)/N$.

The *Halton sequence* is a multidimensional extension of the van der Corput sequence. In order to build the Halton sequence, we use the points in the sequence of the van der Corput but change the prime number base for each dimension. Typically base 2 is used for the first dimension, base 3 for the second dimension, base 5 for the third dimension, etc. The *Faure sequence* is obtained similarly to the Halton sequence but with two major differences: (i) the same base is chosen in all dimensions; and (ii) a permutation of the elements is performed for each dimension. The Faure's approach combines the theory of low-discrepancy sequences with the combinatorial theory of vector permutations. However, the speed in the computation of the Faure sequence decreases with increasing dimension, which makes this approach less efficient than

---

[11] In order to achieve a more uniform distribution of the unit interval, the base is often chosen to be a prime number.

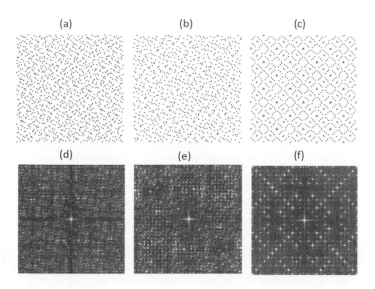

Figure 8.17 Subrandom arrays with $N = 1,000$ particles generated according to two-dimensional (a) Halton (b) Faure, and (c) Sobol sequences. Calculated structure factors for (d) Halton, (e) Faure, and (f) Sobol arrays.

the Halton sequence. The most sophisticated low-discrepancy deterministic sequence is the *Sobol sequence*, which relies heavily on number theory and the properties of primitive polynomials in order to implement permutations of the elements of the van der Corput sequence along each dimension. Additional details on the generation of low-discrepancy sequences can be found in the excellent paper [427]. Finally, it is interesting to realize that low-discrepancy sequences can also be obtained using a stochastic algorithm, known as *Latin hypercube sampling* (LHS) [428]. The LHS method divides each dimensions of an hyperspace (i.e., in two spatial dimensions, this is a simple square region) in N sections and randomly positions one sample in each row and in each column of the grid. The step is repeated in order to distribute random samples in all the sections of the grid with the requirement that there must be only one sample in each row and each column of the hypergrid. This ensures that different random samples are never too closely spaced in each dimension.

In Figure 8.17, we show two-dimensional examples of subrandom arrays ($n = 1,000$ particles) generated according to the Halton, Faure, and Sobol sequence, displayed in Figure 8.17a–c. Figure 8.17d–f shows the corresponding structure factors. A significant degree of spatial correlations is manifested by the bright spots that appear in the Faure and Sobol arrays, while Halton arrays are found to be structurally more complex. We also note the decreasing intensity of the structure factor of the investigated structures near the origin of k-space, which indicates the hyperuniform nature of subrandom structures. Structural correlations in three-dimensional subrandom arrays have been recently investigated by Sgrignuoli and Dal Negro, who found, based on the numerical scaling of the number variance, a connection between deterministic subrandomness and hyperuniformity [389].

8.5.2   Correlations Analysis of Aperiodic Arrays

In Box 8.2, we have derived the direct-space sum rule that must be obeyed by any hyperuniform system and commented that hyperuniform structures manifest negative correlations for some values of the position vector $\mathbf{r}$. Here we propose a general approach to identify regions of negative structural correlations in arbitrary point patterns based on the analysis of their higher-order correlations. This powerful method, which was originally developed to investigate correlations in the energy levels of nuclear spectra [429], will be applied here to the golden-angle Vogel spiral, to deterministic aperiodic patterns, based on the distribution of prime numbers, and to the Pinwheel structure.

Besides providing a precise characterization of level repulsion and long-range order, this approach is sensitive to three- and four-level correlation effects captured by the skewness γ_1 and the excess γ_2 functions [430]. We show in particular that the presence of negative γ_2 regions, which describe particles repulsion or anticlustering, can serve as a good indicator of hyperuniform behavior as well.

Let us first define the moments:

$$\mu_j = \langle (n - \langle n \rangle)^j \rangle \qquad (8.105)$$

where n is the number of elements in a region of size length L and $\langle \cdots \rangle$ represents an average taken over many such intervals throughout the entire system [429, 430]. The moments involve k-level correlations for all $k \leq j$. In particular, the number variance μ_2 is a measure of two-point correlations and its scaling allows one to identify the hyperuniformity of point patterns. Additionally, here we consider the size scaling of the γ_1 and the γ_2 functions, which are defined in terms of the moments as follows [429, 430]:

$$\boxed{\gamma_1 = \mu_3 \mu_2^{-3/2} \qquad \gamma_2 = \mu_4 \mu_2^{-2} - 3} \qquad (8.106)$$

where μ_3 and μ_4 are proportional to third-level and fourth-level correlations, respectively. For a Poisson point pattern, it is easy to show that the functions γ_1 and γ_2 have the following analytical expressions:

$$\gamma_1 = a(R/d_0)^{-d/2} \qquad \gamma_2 = b(R/d_0)^{-d}, \qquad (8.107)$$

where the coefficients a and b are equal to $1/2\sqrt{\rho}$ and $1/3\rho$, respectively. The parameter ρ is the density of scatterers equal to N/L, N/A, and N/V in one-, two-, and three-dimensional systems, respectively. Here, N denotes the number of points and L, A, and V correspond to the linear size, the area, and the volume of the system under consideration. Before considering the scaling analysis of the γ_1 and γ_2 functions, we introduce in this section some background on the two-dimensional aperiodic point patterns that we will later analyze. In particular, we show in Figure 8.18 the Ulam spiral pattern and the coprime array along with the corresponding structure factors. The Ulam spiral array is generated according to the distribution of prime numbers arranged on a square spiral curve, and was originally conceived by the mathematician Stanislaw Ulam in 1963 allegedly while scribbling during the presentation of "a long and very boring paper" at a scientific meeting [431].

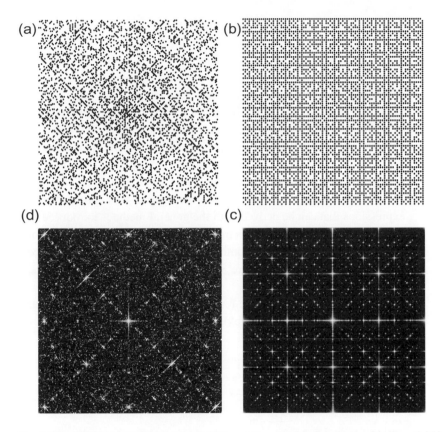

Figure 8.18 (a) Ulam spiral with $N = 7,743$ particles. (b) Coprime array with $N = 6,350$ particles. The corresponding structure factors are shown in (d) and (c), respectively. The cubic root of the structure factors is displayed in order to better emphasize the fine features embedded in the continuous background.

Despite its global aperiodic nature, the Ulam spiral exhibits remarkable regularities such as the appearance of noticeable diagonal, horizontal, and vertical lines containing large numbers of primes generated by quadratic polynomials of the form $f(n) = 4n^2 + bn + c$, where b and c are integer constants, which includes Euler's prime-generating polynomial[12] $x^2 - x + 41$. In particular, "rays" emanating from the central region of the Ulam spiral and making angles of $45°$ with the horizontal and vertical axes correspond to prime numbers of the form $4x^2 + bx + c$ with b even while horizontal and vertical rays correspond to prime numbers of the same form with b odd. A series of conjectures, dating back to Hardy and Littlewood's "Conjecture F," estimate the density of primes along such rays and show that some polynomials are especially rich in primes while others are exceptionally poor [432]. The local order that is characteristic of the Ulam spiral is also manifested by the many peaks appearing in the structure factor shown in Figure 8.18d, reflecting the correlation of the primes

---

[12] It is known that no polynomial $P(n)$ can exist that generates as its roots all the prime numbers.

along diagonal lines. The peaks are also the absence of rigorous spectral results, the singular-continuous nature of the diffraction.

The coprime array shown in Figure 8.18b is generated by plotting the points of the square lattice $\mathbb{Z}^2$ with coprime integer coordinates. This aperiodic structure has dihedral group symmetry D_4 and four-fold rotational symmetry. Rigorous mathematical results exist on the nature of this array and on its spectral properties, discussed in detail in [316]. Interestingly, it can be proven that its asymptotic density (i.e., the large-structure limit) is equal to $6/\pi^2 = 1/\zeta(2)$, where $\zeta(x)$ denotes the Riemann's zeta function. The array contains square-shaped holes of arbitrary size. However, the ones of increasing size are exponentially rare, but repeat with lattice periodicity [316]. The autocorrelation and the diffraction spectrum of the coprime array have analytical expressions that can be derived with the general methods introduced in Section 11.4. In particular, the diffraction spectrum is pure-point and concentrated on the elements of $\mathbb{Q}^2$ with a square-free denominator. These properties can be recognized in the structure factor shown in Figure 8.18c, manifesting the purely diffractive nature of the coprime array.

In Figure 8.19, we show the results of the scaling analysis of γ_1 and γ_2 on (a) a Poisson point pattern; (b) the golden-angle (GA) Vogel spiral; (c) the coprime array;

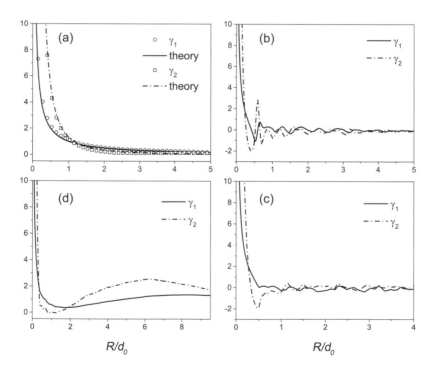

Figure 8.19 High-order correlation analysis on aperiodic point patterns showing the size scaling of the γ_1 and γ_2 coefficients for (a) finite-size Poisson random pattern (symbols) and theoretical expectation (lines); (b) golden-angle Vogel spiral; (c) Coprime array; and (d) Ulam spiral. All arrays have approximately $N = 4,000$ particles. The symbol d_0 denotes the average interparticle separation.

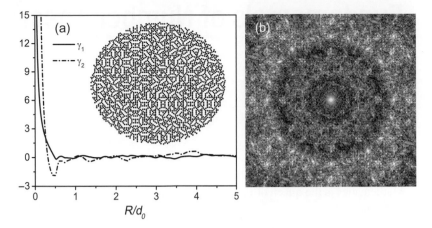

Figure 8.20 High-order correlation analysis of the Pinwheel pattern. (a) Scaling of the γ_1 and γ_2 correlation functions of the Pinwheel point patterns shown in the inset ($N = 2,963$ particles). (b) Structure factor of the analyzed Pinwheel point pattern (cubic root shown).

and (d) the Ulam spiral. The continuous lines shown in Figure 8.19(a) are the analytical results for the uncorrelated Poisson pattern, which are in excellent agreement with the numerical trends. We observe that both the Vogel spiral and the coprime spiral exhibit a region where γ_2 is oscillatory and negative, reflecting strong structural correlations with repulsion behavior. Consistently, we found that the number variance $\sigma^2(R)$ grows slower than quadratically, unveiling their hyperuniform (Class I) nature. Finally, we investigate the Pinwheel structure in Figure 8.20. The pinwheel array is obtained by considering the nodes of the aperiodic Pinwheel tiling, which is constructed from a right-triangular prototile (building block) using a deterministic fractal generation rule (inflation) [39, 316, 433]. The array is aperiodic and nondiffractive. In particular, Radin has shown that there is no discrete component in its diffraction spectrum. However, it is presently unknown if the spectrum is continuous [433, 434]. Upon inflation, its prototile is repeated in infinitely many distinct orientations. As a consequence, the Pinwheel tiling displays continuous ("infinityfold") rotational symmetry. These properties can be clearly recognized in the structure factor shown in Figure 8.20b, where no sharp diffraction peaks are visible and the highly structured diffuse background manifests isotropy. Similarly to the cases of Vogel and coprime array, the scaling of γ_1 and γ_2 in the Pinwheel structure indicates the presence of significant structural correlations with anticlustering behavior shown in Figure 8.20a. We found that the number variance $\sigma^2(R)$ of the Pinwheel array grows slower than quadratically, which indicates its hyperuniform geometry [433].

9   Fundamentals of Number Theory

> Mathematics is the queen of the sciences and number theory is the queen
> of mathematics.
>
> Carl Friedrich Gauss

The present Chapter and the next one are concerned with aperiodic order, unpredictability, and complexity as they appear in number theory, which is the area of mathematics that provides the most compelling examples of the phenomenon of *aperiodic order*, i.e., the emergence of multiscale aperiodic patterns and structural irregularities, or "arithmetic randomness," deeply rooted in the properties of integer numbers. In particular, we introduce fundamental concepts in the theory of numbers and discuss a selection of topics from analytic to algebraic number theory that provide paradigmatic examples of the *interplay between structure and randomness* in number theory, with direct implications to light scattering in aperiodic media. Selected topics include the multiscale aperiodic structure of certain arithmetic functions, the distribution of prime numbers, Riemann's zeta function and its surprising connection with random matrices, generalized prime numbers in complex algebraic fields, the stochastic behavior of some modular operations in prime (Galois) fields, and the structure of elliptic curves over finite fields. Due to the vastness of the subject, our discussion is limited to a self-contained introduction to a selection of topics that bear direct relevance to the study of aperiodic order and structural complexity. Readers motivated to deepen their knowledge of these fascinating subjects are encouraged to engage directly with our extensive list of cited references.

9.1   Arithmetic Functions

Arithmetic functions comprise a vast class of mathematical functions that display extremely complex and aperiodic behavior. Many analytical results exist on the mean value of arithmetic functions, but a general theory capable of predicting their local behavior, size, and approximation in terms of simpler functions is currently missing [436]. Arithmetic functions provide a striking demonstration of deterministic aperiodicity in number theory, where almost-periodic and quasiperiodic behavior is widespread. Our discussion is limited to an informal introduction aided by plots of selected arithmetic functions that illustrate their distinctive complexity. A rigorous discussion of this difficult subject with connections to functional analysis methods can be found in the [436, 437].

An *arithmetic function* is a real or complex-valued function $f(n)$ defined over positive integers. Examples of arithmetic functions are the *divisor functions* $\sigma_k(n)$, whose values at a positive integer n are equal to the sum of the kth powers of the positive divisors of n, including 1 and n:

$$\sigma_k(n) = \sum_{d|n} d^k \qquad (9.1)$$

The notation $\sum_{d|n}$ denotes the sum over all the numbers d that divide n (i.e., the divisors of n). Special cases are the function $\sigma_0 = \sum_{d|n} 1$, which counts the number of positive divisors of n, also denoted by $d(n)$, and the function $\sigma_1 = \sum_{d|n} d$, also denoted by $\sigma(n)$, which is the sum of the positive divisors of n. The reader can simply verify that $\sigma_0(12) = 6$ and that $\sigma_1(12) = 28$. It can also be shown that if a number n has the prime factorization $n = p_1^{\alpha_1} \dots p_r^{\alpha_r}$, then $d(n) = (\alpha_1 + 1)(\alpha_2 + 1) \dots (\alpha_r + 1)$, while

$$\sigma(n) = \frac{p_1^{\alpha_1+1} - 1}{p_1 - 1} \frac{p_2^{\alpha_2+1} - 1}{p_2 - 1} \dots \frac{p_r^{\alpha_r+1} - 1}{p_r - 1}. \qquad (9.2)$$

Divisor functions appear in many important number-theoretic identities and display a very complex aperiodic behavior. Interestingly, they admit explicit series representations in terms of Ramanujan sums, which are highly oscillating trigonometric expansions discovered in 1918 by the Indian mathematician Srinivasa Ramanujan using elementary methods[1] of number theory [438]. The characteristic aperiodic behavior of the divisor functions $d(0)$ and $\sigma(n)$ is displayed in Figure 9.1.

9.1.1 Multiplicative Functions

Divisor functions are primary examples of *multiplicative functions* in number theory. Specifically, an arithmetic function $f(n)$ is said to be *completely multiplicative* if $f(ab) = f(a)f(b)$ for all natural numbers a and b while it is *multiplicative* if $f(ab) = f(a)f(b)$ for all coprime integer numbers a and b. Note that two integers a and b are said to be coprime (or relatively prime) if their greatest common divisor (gcd) is equal to one, i.e., symbolically $(a, b) = 1$. This means that the number 1 is the only positive integer factor that divides both numbers. Another important example of a multiplicative arithmetic function is provided by the *Euler totient function* $\varphi(n)$. This function is defined to be the number of positive integers less than n that are coprime to n. An elegant formula exists to compute $\varphi(n)$ for any positive integer with prime decomposition $n = p_1^{\alpha_1} \dots p_r^{\alpha_r}$. The formula can be obtained by using the multiplicative nature of the totient function, which reduces the problem to the computation of factors of the form $\varphi(p^k)$ for all integers k and primes p. Since p is prime, the only

---

[1] In number theory, "elementary methods" are methods that do not require the use of complex analysis.

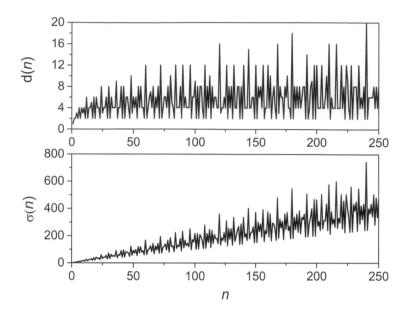

Figure 9.1 Plot of the divisor function $d(n)$ and of the sum of divisors $\sigma(n)$.

numbers in the set $\{1, 2, \ldots p^k - 1, p^k\}$ that are not relatively prime to p are those $k - 1$ that are divisible by p. Therefore, we can write the following:

$$\varphi(p^k) = p^k - p^{k-1} = p^k \left(1 - \frac{1}{p}\right). \tag{9.3}$$

The general product formula for the totient function follows by direct multiplication:

$$\boxed{\varphi(n) = n \prod_{p|n} \left(1 - \frac{1}{p}\right)} \tag{9.4}$$

and this product runs over all the prime factors p of the integer n. The distinctively aperiodic behavior of the Euler $\varphi(n)$ is displayed in Figure 9.2.

The Euler totient function is ubiquitous in number theory and enjoys many interesting properties, including the following one established by Gauss:

$$\sum_{d|n} \varphi(d) = n. \tag{9.5}$$

A simple proof of this fact is provided in [439]. Euler himself proved that if a and n are relatively prime numbers, then

$$\boxed{a^{\varphi(n)} \equiv 1 \bmod n} \tag{9.6}$$

This theorem is known as Euler's (totient) theorem. This result generalizes the so-called *Fermat's little theorem*, which follows from it when n is prime. A brief introduction to modular arithmetic is provided in Box 9.1. Euler's theorem is central to the implementation of modern public-key cryptosystems, particularly RSA

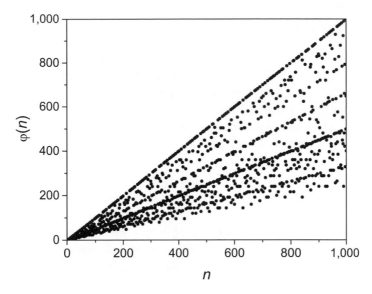

Figure 9.2 Plot of the Euler totient function up to $n = 10^3$.

cryptography, where shared public keys are based on the product of two randomly chosen and very large prime factors that are kept secret.

An additional multiplicative function that appears in number theory in connection with the aperiodic prime number distribution is the *Möbius function* $\mu(n)$, which is defined as follows: it is $\mu(1) = 1$, and when $n > 1$ with factorization into primes $n = p_1^{\alpha_1} \ldots p_r^{\alpha_r}$, we define the following:

$$\mu(n) = \begin{cases} 0 & \text{if any exponent } \alpha_i > 1. \\ (-1)^r & \text{if every exponent } \alpha_i = 1. \end{cases} \qquad (9.7)$$

Alternatively, we see that $\mu(n)$ is equal to 1 if n is a square-free positive integer[2] with an even number of distinct prime factors, it will be equal to -1 if it is a square-free positive integer with an odd number of distinct prime factors, and it will be equal to 0 if n is not square-free (i.e., it has repeated prime factors in its decomposition). A closely related arithmetic function is the Mertens function $M(n) = \sum_{k=1}^{n} \mu(k)$. This function counts the number of square-free integers up to n that have an even number of prime factors, minus the those that have an odd number, and it displays an apparently chaotic behavior similar to the fractional Brownian motion discussed in Appendix C. This function grows very slowly and it was conjectured by Mertens that, when extended to positive real numbers by $M(x) \equiv M(\lfloor x \rfloor)$, its absolute value never exceeds the square root of x (This conjecture was proven wrong in 1985 by Andrew Odlyzko and Herman te Riele). The Möbius and the Mertens functions are plotted in Figure 9.3.

---

[2] A square-free integer is an integer that is divisible by no perfect square other than 1. Therefore, its prime factorization has exactly one factor for each prime that appears in it.

Box 9.1 Modular Arithmetic

Euler's totient theorem gives us the opportunity to introduce the tools of *modular arithmetic*, developed by Gauss in his foundational treatise *Disquisitiones Arithmeticae* published in 1801. In modular arithmetic, it is the remainder modulo a given integer that matters, similarly to the familiar case of arithmetic on the 12-hour clock, where numbers wrap around every 12 hours (the number 12 is called the *modulus*). More precisely, we say that two integer numbers a and b are congruent modulo the positive integer n, and we write $a \equiv b \bmod n$, if n divides their difference $a - b$. Equivalently, there is an integer k such that $a = kn + b$ and a and b leave the same remainder when divided by n. For example, $19 \equiv 8 \bmod 11$ and $a \equiv b \bmod 1$ for every a and b. Using the elementary properties of divisibility, we can establish that the congruence relation is an equivalence relation on the integers, meaning that it is reflexive ($a \equiv a \bmod n$), symmetric ($a \equiv b \bmod n$ implies $b \equiv a \bmod n$), and transitive ($a \equiv b \bmod n$ and $b \equiv c \bmod n$ imply $a \equiv c \bmod n$). Therefore, integer numbers are divided into n distinct equivalence classes and the representative of an integer in a class is equal to its remainder on division by n. The set of all the equivalence classes is denoted by $\mathbb{Z}/n\mathbb{Z} = \{0, 1, 2, \ldots, n - 1\}$. Therefore, a *complete set of residues* is a choice of representatives for each equivalence class in $\mathbb{Z}/n\mathbb{Z}$. This set is called the *ring of integers modulo n*. Congruences with the same module can be added, subtracted, or multiplied member by member as if they were standard equations. The same is true for any finite number of congruences with the same module. However, the common factor k can be canceled from the congruence $ak \equiv bk \bmod n$ only if it is coprime with the modulus, i.e., if $(k, n) = 1$. Otherwise, we have $a \equiv b \bmod n/d$, where $d = (k, n)$. Finally, if $(a, n) = 1$, there exists an *inverse* b of $a \bmod n$, i.e., $ab \equiv 1 \bmod n$. Moreover, such inverse is unique modulo n. We denote by $(\mathbb{Z}/n\mathbb{Z})^*$ the set of all invertible elements of $\mathbb{Z}/n\mathbb{Z}$. Numbers that have inverses are called *units*. Therefore, $(\mathbb{Z}/n\mathbb{Z})^* = \{a \in \mathbb{Z}/n\mathbb{Z}, |, ax \equiv 1 \bmod n\}$. This set is closed under multiplication and forms a group called the *group of units modulo n*. The cardinality of $(\mathbb{Z}/n\mathbb{Z})^*$ is equal to the Euler totient function $\varphi(n)$.

The connection between the Euler totient function and the Möbius function is provided by the formula [439]:

$$\varphi(n) = \sum_{d|n} d\mu\left(\frac{n}{d}\right) \tag{9.8}$$

This type of summation occurs very often in number theory and, together with equation (9.5), provides a specific example of the all-important *Möbius inversion formula*.

More generally, given two arithmetic functions f and g such that $f(n) = \sum_{d|n} g(d)$, the Möbius inversion formula allows us to express $g(n)$ by "inverting through the Möbius function" as follows:

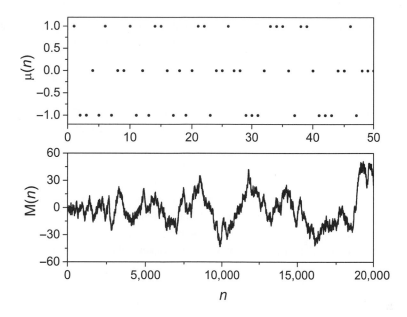

Figure 9.3 Top: plot of the Möbius function $\mu(n)$. Bottom: plot of the Mertens function $M(n)$ up to $n = 2 \times 10^4$.

$$g(n) = \sum_{d|n} f(d)\mu\left(\frac{n}{d}\right) \qquad (9.9)$$

The Möbius inversion formula is a special case of *Dirichlet product*, or the discrete convolution $f * g$ between two arithmetic functions f and g, which is defined as follows:

$$h(n) = f * g \equiv \sum_{d|n} f(d)g\left(\frac{n}{d}\right). \qquad (9.10)$$

Note that equation (9.8) can be written as the Dirichlet convolution $\varphi = \mu * N$ by introducing the arithmetic function $N(n) = n$ for all n. This follows from the alternative form $\varphi(n) = \sum_{d|n} \mu(d)\frac{n}{d}$ that holds true because when d runs over all the divisors of n, so does n/d.

It can be shown that the Dirichlet product of arithmetic functions is commutative and associative, which allows one to develop the algebra of arithmetic functions based on the notion of the Dirichlet product [439].

9.1.2   Multifractal Scaling of Complex Signals

The complex oscillatory behavior that is characteristic of many arithmetic functions bears a qualitative similarity with the individual realizations of stochastic self-similar (fractal) processes, such as the fractional Brownian motion (discussed in Appendix C). This observation can be supported quantitatively by applying the quantitative tools of *multifractal scaling analysis* based on wavelet leaders [440, 441]. This is a general approach that allows one to establish if complex one-dimensional signals of arbitrary nature, such as turbulent wind data, time series, network traffic, and arithmetic functions, display self-similarity or more complex multifractal properties. In this section, we introduce the main ideas behind the method and show its application to the fractional Brownian process, which is a particular type of stochastic signal with fractal scaling properties. Having established the fundamental ideas through a well-established reference example, we will then proceed to analyze the fractal and multifractal properties of different types of arithmetic functions throughout the rest of this chapter. The fundamental goal of the scaling analysis of complex aperiodic signals, such as $1/f$ stochastic noise processes, is the estimation of the characteristic scaling exponents that describe their scale-invariance properties. For a one-dimensional signal $u(t)$ that depends on the parameter t (i.e., time, space, etc.) this can be generally achieved considering the scaling of *multiresolution quantities* $T_u(a,t)$ that describe the signal u around position t at an observation scale a. Typical multiresolution quantities consist of wavelet coefficients, as described later in this chapter. The scale-invariant behavior of the analyzed signal is revealed by the power-law behavior of the ensemble average (in time or space) of the qth power of $T_u(a,t)$, known as the structure functions $S(q,a)$, in a given (large) range of scales [441]:

$$S(q,a) = \frac{1}{n_a} \sum_{k=1}^{n_a} |T_u(a,k)|^q \simeq G_q a^{\zeta(q)} \tag{9.11}$$

Clearly, no ensemble average is needed when analyzing a deterministic signal. In the previous expression, n_a is the number of the multiresolution coefficients at each scale a, which is related to the duration n of the observation by $n_a \approx n/a$. On the other hand, the coefficients G_q depend on the nature of the analyzed process and not on the scales.

From a practical standpoint, multifractal analysis consists in estimating the scaling exponents $\zeta(q)$ from a given set of data, which in general requires sophisticated statistical methods. However, this is not the case for arithmetic functions, since no true randomness is involved. The function $\zeta(q)$ can be formally expanded as follows:

$$\zeta(q) = c_1 q + c_2 q^2/2 + c_3 q^3/6 + \cdots \tag{9.12}$$

When $\zeta(q)$ is a linear function of q, the process $u(t)$ is a monofractal, or a homogeneous fractal characterized by global scale-invariance symmetry, i.e., only one scaling exponent. This is the case for the *fractional Brownian motion* (FBM), which includes the standard Brownian motion as a special case (see Appendix C for more details). On the other hand, the signal $u(t)$ is said to be a multifractal whenever $\zeta(q)$ is a nonlinear function of q. The simplest deviation from the linear model occurs when $\zeta(q) = c_1 q + c_2 q^2/2$, which is referred to as a log-normal multifractal. Therefore, the statistical estimation of the values of the c_n coefficients is of great relevance to quantitatively decide if a given signal is monofractal ($c_2 = 0$) or multifractal ($c_2 \neq 0$), or even to distinguish between possible multifractal models. A statistical test to accomplish this goal has been developed in [441] based on log-cumulants, wavelet leaders' multiresolution, and nonparametric bootstrap statistics.[3] The local fluctuations in the regularity of the sample path of a signal/process versus time/space are described by its singularity or *multifractal spectrum* $D(\alpha)$, where α is called the *Hölder exponent* [442]. The Hölder exponent quantifies the strength of singularity of a signal $u(t)$ around the point t_0 by comparing its local variation around t_0 with a *local power law*:

$$|u(t) - P_{t_0}(t)| \leq C|t - t_0|^{\alpha'}, \tag{9.13}$$

where $P_{t_0}(t)$ is a polynomial of degree less than α' and where $\alpha \geq 0$ and $C \geq 0$ are constants. The Hölder exponent α is defined as the largest of such α' exponents. The variability of the range of the Hölder exponents of a signal/process is captured by its multifractal spectrum $D(\alpha)$ defined as the Hausdorff dimension [443] of the set of points t_i such that $\alpha(t_i) = \alpha$. Empirical multifractal analysis consists of inferring the multifractal spectrum of the available data from a single observation of finite duration. This can be done by estimating the scaling exponents $\zeta(q)$ using directly equation (9.11) and then obtaining $D(\alpha)$ via a Legendre transform [442]. There are currently several approaches that can be utilized to achieve this goal based on the analysis of the continuous wavelet transform [442, 444, 445]. Here we discuss in some detail the one that uses the *wavelet leaders* as the preferred multiresolution quantities [440, 441]. Wavelet leaders are largely used for the analysis of different types of physical data especially in the context of turbulence [440]. A mother wavelet is a zero-average, finite-energy, oscillatory function $\psi_0(t)$ characterized by its number of vanishing moments. It must be appreciated that the wavelet energy remains mostly localized in a narrow support both in the time and in the frequency domains. Importantly, the collection of scaled and translated replicas (templates) $\{\psi_{j,k}(t) = 2^{-j}\psi_0(2^{-j}t - k), j, k \in \mathbb{Z}\}$ form an orthonormal basis of $L^2(\mathbb{R})$. The discrete wavelet transform (DWT) of the analyzed process $u(t)$ is defined by its coefficients $d_u(j,k)$ via the inner product [442]:

$$\boxed{d_u(j,k) = \langle \psi_{j,k} | u \rangle = \int_{\mathbb{R}} u(t) 2^{-j} \psi_0(2^{-j}t - k)dt} \tag{9.14}$$

The DWT provides a complete *time-scale representation* of the signal $u(t)$ where

$$u(t) = \sum_{j,k} d_u(j,k)\psi_{j,k}(t). \tag{9.15}$$

---

[3] In bootstrap statistics, one approximates the unknown probability distribution of a random variable by repeated resampling with replacement from the available data. This is particularly useful when only small datasets are available.

Wavelet leaders $L_u(j,k)$ were introduced by Jaffard [446] by considering dyadic intervals $\lambda_{j,k} = [k2^j, (k+1)2^j)$ and the set union $3\lambda_{j,k} = \lambda_{j,k-1} \cup \lambda_{j,k} \cup \lambda_{j,k+1}$ as follows:

$$L_u(j,k) = \sup_{\lambda' \in 3\lambda_{j,k}} |d_{u,\lambda'}| \tag{9.16}$$

Hence the leader $L_u(j,k)$ is the largest of the discrete wavelet coefficients in the neighborhood $3\lambda_{j,k}$ computed at all finer scales $2^{j'} \leq 2^j$. Following the analysis in [446–448], it can be shown that the wavelet leaders exactly reproduce the Hölder exponent of the signal $u(t)$ and that the structure functions of the wavelet leaders $S_L(q,2^j)$ obey the following local power-law behavior in the limit of fine scales:

$$S_L(q,2^j) = \frac{1}{n_j} \sum_{k=1}^{n_j} |L_u(j,k)|^q = F_q |2^j|^{\zeta(q)} \tag{9.17}$$

This expression is an exact result, and it allows us to extract the multifractal spectrum $D(\alpha)$ via the Legendre transform of the scaling exponents $\zeta(q)$[4]. Moreover, the measurements of the scaling exponents provides access to the coefficients c_n (i.e., the log-cumulants), whose meaning is the following [441]: the coefficient c_1 yields the location of the maximum of $D(\alpha)$, c_2 characterizes its width, and c_3 its asymmetry. Therefore, knowledge of the coefficients c_1, c_2, c_3 provides the most important information on the multifractal spectra of the empirical data under consideration.

We now apply the multifractal scaling analysis to the fractional Brownian motion, which is a continuous-time Gaussian random process with increments that are not independent (see Appendix C for more details). In particular, we consider in Figure 9.4 a sample path of the positively correlated FBM characterized by the *Hurst exponent*[5] $H = 0.4$. The case $H = 1/2$ corresponds to the standard Brownian motion or Wiener process. The fractional Brownian motion is known to be a self-similar fractal process with scaling exponent H. The linear behavior of the scaling exponents $\zeta(q)$ clearly demonstrates that the process is a monofractal, and consistently $c_2 \simeq 0$. The corresponding multifractal spectrum is very narrow, as expected for monofractal signals. By increasing the length of the sample path, the multifractal spectrum is eventually supported by a single point. The spectrum $D(\alpha)$ shows a maximum at the location $\alpha \simeq 0.4$, in close agreement with the estimated value of $c_1 = 0.398$. Deviations of the peak of $D(\alpha)$ from the estimated c_1 coefficient are in fact expected due to the finite

---

[4] More strictly, only an upper bound for the multifractal spectrum can be obtained following this method. However, for most multiplicative processes, whose multifractal spectra are concave, the Legendre transform of the scaling exponents yields exactly the multifractal spectrum [441].

[5] The Hurst exponent is an "index of long-range dependence" or persistence of a time series that quantifies the amount of long-term positive ($0.5 < H < 1$) or negative ($0 < H < 0.5$) correlation among the values of the series. For self-similar time series, H has values between 0 and 1 and it is directly related to fractal dimension $1 < d_f < 2$ by the relation $d_f = 2 - H$.

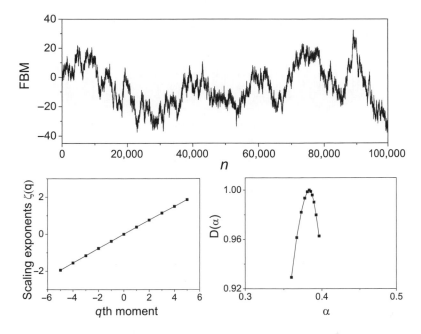

Figure 9.4 Multifractal scaling analysis of the fractional Brownian motion (FBM). One sample path is shown along with the linear scaling of the exponents $\zeta(q)$ and the estimated multifractal spectrum $D(\alpha)$.

size of the analyzed sample path. The multifractal scaling analysis will be applied in later sections to investigate the distinctive structure of primes and related arithmetic functions.

9.1.3 Additive Functions

Another important class of arithmetic functions is the one of additive functions. An additive function is an arithmetic function $f(n)$ for which $f(ab) = f(a) + f(b)$ whenever $(a, b) = 1$. If this property holds for all positive integers a and b, even when they are not coprime, then the function $f(n)$ is called *completely additive*. A simple example is the restriction of the logarithmic function to the natural numbers, since $\log(ab) = \log(a) + \log(b)$. A less elementary example of a completely additive arithmetic function is provided by the so-called *Big Omega function* $\Omega(n)$ defined as the total number of prime factors of n, counted with multiplicity. For example, $\Omega(4) = 2$ and $\Omega(20) = \Omega(2 \cdot 2 \cdot 5) = 3$. An example of an arithmetic function that is additive but not completely additive is the *small omega function* $\omega(n)$ defined as the total number of distinct prime factors of n. Therefore, $\omega(4) = 1$ and $\omega(20) = 2$. The arithmetic functions $\Omega(n)$ and $\omega(n)$ are also known as the *prime omega functions*. These functions obey many important number theoretic relations [409]. Both $\Omega(n)$ and $\omega(n)$ behave very irregularly for large values of n since both functions are equal

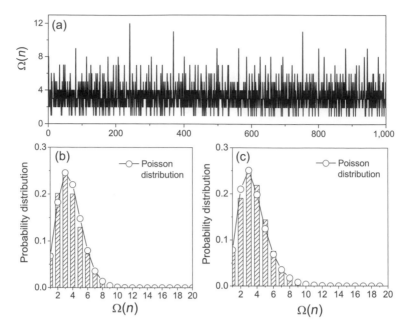

Figure 9.5 (a) Plot of the $\Omega(n)$ function for the integers in the interval $10,000 \leq n \leq 11,000$. (b) The probability distributions (shaded bars) obtained by counting the number of prime factors in the interval $8 \times 10^5 \leq n \leq 9 \times 10^5$ and (c) in the interval $1 \leq n \leq 10^6$. The height of each bar is equal to the probability of selecting an observation within that bin, and the height of all of the bars sums up to 1. The open dot symbols are the values computed according to the exact Poisson distribution.

to 1 when n is prime. This can be seen clearly in Figure 9.5a, where we plot $\Omega(n)$ for the integers in the interval $10,000 \leq n \leq 11,000$.

However, $\Omega(n)$ and $\omega(n)$ have a well-known *average order*, or arithmetic average, that shows a characteristic $\log\log(n)$ asymptotic scaling [409]. This can simply be appreciated for $\omega(n)$ by observing the following:

$$\omega(n) \equiv \sum_{p_i | n} 1 \approx \sum_{p_i \leq n} \frac{1}{p_i}, \tag{9.18}$$

where we converted the sum over the primes that divide n into a summation over all primes up to n weighted with a probability factor $1/p_i$. This factor gives the probability that any randomly selected integer number is divisible by the prime p_i, since every p_ith number is divisible by p_i. We can then convert expression (9.18) into a sum over all integers up to n using the probability $1/\log x$ that a randomly chosen integer x is a prime[6] and obtain the approximated arithmetic average of $\omega(n)$ as follows:

$$\overline{\omega}(n) \approx \sum_{x \leq n} \frac{1}{x \log x}. \tag{9.19}$$

---

[6] See discussion of the prime number theorem (PNT) later in the chapter.

Integrating the last result from $x = 2$ up to $x = n$ we immediately get the average order estimate $\overline{\omega}(n) \approx \log \log n$. Considerations based on probability theory are used in the important part of number theory known as *probabilistic number theory*, whose main idea is that different prime numbers can be considered, in some technical sense, as manifestations of independent random variables [449]. An intuitive argument developed in [8] uses probability theory, under the assumption that divisibility by different primes constitutes an independent property (in the statistical sense), and deduces that the probability distribution of the number of prime factors can be approximated by the shifted *Poisson distribution*:

$$P\{\Omega(n) = k\} \approx \frac{(\overline{\Omega} - 1)^{k-1}}{(k - 1)!} \exp(-\overline{\Omega} + 1) \tag{9.20}$$

where $k = 1, 2, \ldots$ and $\overline{\Omega} \approx \log \log(n)$ denote the average number of prime factors. The same result can also be obtained considering the function $\omega(n)$. Figure 9.5a–b shows the distributions of the number of prime factors in the intervals $8 \times 10^5 \leq n \leq 9 \times 10^5$ and $1 \leq n \leq 10^6$, respectively. The results obtained according to the exact Poisson distributions are also shown by the open dot symbols. It is remarkable that the two distributions appear very similar despite the large difference in the size of the sampling intervals. Moreover, we observe that up to integers of the order of 10^{135} the most probable number of different prime factors is 6 or less [8]. We should appreciate how the Poisson distribution, which is often considered the epitome of random behavior, surprisingly appears behind the elementary divisibility properties of integer numbers. The Poisson behavior in number theory and its relation to spatial randomness will be discussed in more detail in Chapter 10.

9.1.4   The von Mangoldt Function

An important arithmetic function that is neither multiplicative nor additive is the von Mangoldt function Λ (see Table 9.1). This function also plays a very important role in number theory in relation to the distribution of prime numbers. More specifically, its Dirichlet series on the critical line (see the next section) encodes the so-called Riemann spectrum, which consists of a series of spikes at the positions of the imaginary parts of the nontrivial zeros of the Riemann zeta function [450]. The *von Mangoldt function* is formally defined, for every integer $n \geq 1$, as follows:

$$\Lambda(n) = \begin{cases} \log p & \text{if } n = p^k \text{ for some prime } p \text{ and some } k \geq 1. \\ 0 & \text{otherwise.} \end{cases} \tag{9.21}$$

The first few values of the von Mangoldt function are shown in Table 9.1. The von Mangoldt function satisfies the important identity:

$$\log n = \sum_{d \mid n} \Lambda(d). \tag{9.22}$$

Table 9.1 First nine values of the von Mangoldt function $\Lambda(n)$.

n	1	2	3	4	5	6	7	8	9
$\Lambda(n)$	0	$\log 2$	$\log 3$	$\log 2$	$\log 5$	0	$\log 7$	$\log 2$	$\log 3$

Inverting this expression using the Möbius inversion formula, we can establish a simple connection between the von Mongoldt and the Möbius functions:

$$\Lambda(n) = \sum_{d|n} \mu(d) \log \frac{n}{d} = -\sum_{d|n} \mu(d) \log d \qquad (9.23)$$

The summatory of the von Mangoldt function is known as the second *Chebyshev function* $\psi(x)$, defined as follows:

$$\psi(x) = \sum_{n \leq x} \Lambda(n). \qquad (9.24)$$

Von Mangoldt provided a rigorous proof of an explicit formula for $\psi(x)$ involving a sum over the nontrivial zeros of the Riemann zeta function, introduced in Section 9.2. This was an important step toward the first proof of the prime number theorem (PNT). The PNT describes the asymptotic distribution of prime numbers among the positive integers, and it quantitatively states the rate at which primes become less and less frequent as they become larger and larger. The theorem was conjectured by Gauss (at age 16!) and rigorously proved independently by Jacques Hadamard and Charles Jean de la Vallée Poussin in 1896 using ideas of analytic number theory introduced by Bernhard Riemann.

The PNT can be simply stated by saying that for large enough N, the probability that a random integer not exceeding N is prime is asymptotically close to $1/\log N$. Therefore, according to the PNT, a random integer with $2n$ digits (i.e., $N \sim 10^{2n}$) will be approximately half as likely to be a prime compared to a random integer with at most n digits. This also implies that the average gap between consecutive prime numbers among the first N integers is roughly equal to $\log N$. We will come back to this theorem when discussing its fundamental connection with the Riemann zeta function (i.e., the distribution of its nontrivial zeros) later in the chapter. Figure 9.6 plots the first 100 values of the von Mangoldt and the Chebyshev functions.

The aperiodic staircase behavior of the Chebyshev function is connected with the locations of the nontrivial zeros of the Riemann zeta function which, according to the famous *Riemann's hypothesis*, all have real part equal to $1/2$. More specifically, Hans Carl Friedrich von Mangoldt proved in 1895 an important explicit formula that shows how $\psi(x)$ can be obtained as a sum over the nontrivial zeros of the Riemann zeta function [426]:

$$\psi(x) = x - \sum_{\rho} \frac{x^{\rho}}{\rho} - \frac{\zeta'(0)}{\zeta(0)} - \frac{1}{2}\log(1 - x^2) \qquad (9.25)$$

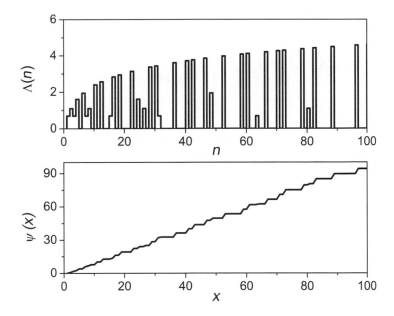

Figure 9.6 Plot of the von Mangoldt $\Lambda(n)$ and the Chebyshev function $\psi(x)$.

In this summation, ρ ranges over all the nontrivial zeros of the Riemann zeta function ζ, which are the zeros with real part in the interval $[0,1]$. We notice at this point that there exists also a first Chebyshev function $\vartheta(x) = \sum_{p \leq x} \log p$ where the sum extends over all prime numbers p that are less than or equal to x. The Chebyshev functions play an important role in the analytic proof of the PNT since they are related to the *prime counting function* $\pi(x)$, defined as the number of primes less than or equal to some real number x, but are much simpler to manipulate. Interestingly, both Chebyshev functions are asymptotic to x, which is a statement equivalent to the PNT [439]. Figure 9.7a shows the function $\psi(x) - x$ for $x < 8 \times 10^4$ along with the two curves $\pm 0.8\sqrt{x}$. It may appear from the plot that this function is always bounded by the two curves in light gray color. However, this is not the case. In fact, a classical result proven by Schmidt [451] in 1903 established that for some positive constant K, there are infinitely many natural numbers x such that $\psi(x) - x > K\sqrt{x}$ and such that $\psi(x) - x < -K\sqrt{x}$. Sharper bounds have been derived in the influential paper by Hardy and Littlewood [452], and more recent estimates can be found in [453]. We investigate the rapidly fluctuating behavior of the $\psi(x) - x$ function using multifractal analysis based on wavelet leaders that reveals a fractal nature similar to the fractional Brownian motion. The observed linear trend of the scaling exponents in Figure 9.7b shows that the analyzed signal is indeed a monofractal, which is confirmed by the very small value of the c_2 logarithmic cumulant. This is also consistent with the narrow support of the singularity spectrum shown in Figure 9.7c. Therefore, our numerical analysis is consistent with a self-similar fractal structure characterized by the scaling exponent $H \simeq 0.4$. These conclusions have been found to be robust with respect to the size of the analyzed signals up to the maximum size explored ($x = 10^8$).

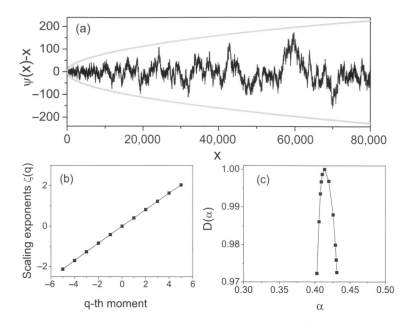

Figure 9.7 (a) Plot of the function $\psi(x) - x$. The light grey curves correspond to the functions $\pm 0.8\sqrt{x}$. Panels (b) and (c) show the results of the wavelet leader multifractal analysis. (b) scaling exponent $\zeta(q)$ and (c) corresponding singularity, or multifractal, spectrum $D(\alpha)$.

9.2   The Riemann Zeta Function

We now introduce one of the most mysterious objects of all mathematics, the Riemann zeta function, which is intimately connected to many fundamental results concerning the aperiodic distribution of the prime numbers and also appears in many different areas of physics [454]. While many properties of this function have been completely understood, there are still several important conjectures, most notably the Riemann hypothesis, that remains unproven to this day [450, 455].

The Riemann zeta function is the principal example of an L-function, which is a complex-valued generating function that encodes in the distribution of its values fundamental information about an underlying arithmetical or algebraic structure, often difficult to retrieve by elementary or algebraic methods. L-functions originate from analytic continuation of so-called Dirichlet series, and play a crucial role in analytic number theory because they admit an Euler product, i.e., they can be written as a product over prime numbers.

More formally, consider any complex-valued series of the following form:

$$F(s) = \sum_{n=1}^{\infty} \frac{f(n)}{n^s} \tag{9.26}$$

where $s \in \mathbb{C}$ is called a *Dirichlet series*. In this context, it is common to write $s = \sigma + jt$, where $(\sigma, t) \in \mathbb{R}$. The function $F(s)$ is the called the Dirichlet generating function for the coefficients $f(n)$. Remarkable examples of Dirichlet series that link many of the arithmetic functions that we have introduced in the previous sections are shown here [409, 439]:

$$\sum_{n=1}^{\infty} \mu(n)n^{-s} = \frac{1}{\zeta(s)} \tag{9.27}$$

$$\sum_{n=1}^{\infty} \varphi(n)n^{-s} = \frac{\zeta(s-1)}{\zeta(s)} \tag{9.28}$$

$$\sum_{n=1}^{\infty} \Lambda(n)n^{-s} = -\frac{\zeta'(s)}{\zeta(s)} \tag{9.29}$$

$$\sum_{n=1}^{\infty} \sigma_k(n)n^{-s} = \zeta(s)\zeta(s-k), \tag{9.30}$$

where $\zeta(s)$ is the *Riemann zeta function*. We note that sums involving the divisor functions can be related to the Riemann zeta function $\zeta(s)$ (as shown in the preceding examples), thus enabling the use of analytic methods for the study of divisor functions as well [409].

The Riemann zeta function $\zeta(s)$ is a special type of Dirichlet series defined as follows:

$$\zeta(s) = \sum_{n=1}^{\infty} \frac{1}{n^s} \tag{9.31}$$

This series converges absolutely in the half-plane where the real part of s is larger than 1. This can be easily verified by expressing $n^{-s} = \exp(-s\log n) = n^{-\sigma}\exp(-jt\log n)$ and realizing that $|n^{-s}| = n^{-\sigma}$, implying that the series expression for $\zeta(s)$ converges absolutely if $\sigma > 1$.

The zeta function has been studied previously by Euler but only as a function of the real variable s. The great contribution of Bernhard Riemann was to regard this function as a function of the complex variable s. This apparently small change of perspective truly opened immense vistas in number theory by showing that arithmetic properties of integers can be deduced from analytic properties of complex functions. In particular, using the tools of complex analysis, Riemann extended the previous definition and proved in his landmark 1859 memoir [456] titled *On the Number of Prime Numbers Less than a Given Quantity*[7] that $\zeta(s)$ can be analytically continued over the entire complex plane except for the simple pole at $s = 1$. Therefore, the extended (analytically continued) $\zeta(s)$ function is defined over the entire complex plane with $s \neq 1$ and agrees with the previous summation formula (9.31) in the complex half-plane $\Re(s) > 1$.

---

[7] Interestingly, this nine-page paper was also the only paper that Riemann published on the subject of number theory.

This analytic continuation is called the *Riemann zeta function*. In the $\sigma > 1$ half-plane, $\zeta(s)$ satisfies the formula:

$$\zeta(s) = \sum_{n=1}^{\infty} \frac{1}{n^s} = \prod_{p} \left(1 - \frac{1}{p^s}\right)^{-1} \tag{9.32}$$

where the infinite product, called the *Euler product*, is taken over all the prime numbers. It can be shown that the preceding expansion formula follows directly from the fundamental theorem or arithmetic [439]. Euler used his product formula with $s = 1$ to prove that the sum of the reciprocals of prime numbers diverges, which implies the infinitude of the set of prime numbers. In the same paper, Riemann also proved the outstanding fact that $\zeta(s)$ obeys the following *functional equation*:

$$\xi(s) \equiv \frac{1}{2}s(s-1)\pi^{-s/2}\Gamma\left(\frac{s}{2}\right)\zeta(s) = \xi(1-s), \tag{9.33}$$

Box 9.2 Dirichlet Series and Euler Products

The Euler product provides the connection between the fundamental theorem of arithmetic and complex analytic functions. More generally, it is true that every multiplicative f satisfies the identity:

$$\sum_{n=1}^{\infty} f(n) = \prod_{p} \left[1 + \sum_{k=1}^{\infty} f(p^k)\right], \tag{9.34}$$

where the product is extended over all primes p. The preceding expansion is called the Euler product of the series, and it is valid if the series on the left is absolutely convergent. Moreover, if $f(n)$ is completely multiplicative, then each factor in the Euler product of the series is a geometric series and the Euler product simplifies to the following:

$$\sum_{n=1}^{\infty} f(n) = \prod_{p} [1 - f(p)]^{-1}. \tag{9.35}$$

These Euler products are used to obtain series that generate many functions of multiplicative number theory. For example, the completely multiplicative function $f(n) = n^{-s}$ yields the Euler product representation of the Riemann zeta function:

$$\zeta(s) = \sum_{n=1}^{\infty} n^{-s} = \prod_{p} [1 - p^{-s}]^{-1}. \tag{9.36}$$

---

where $\Gamma(s)$ denotes the Euler gamma function. The function $\xi(s)$ defined in this equation is called the *completed zeta function*, and it is analytic over the entire complex plane. The functional equation is valid for all complex s values and displays mirror symmetry around the vertical line at $\sigma = 1/2$, called the *critical line*. In view of the Euler product (9.32), it follows that $\zeta(s)$ has no zeros in the $\sigma > 1$ half-plane. Using the functional equation (9.33), it turns out that $\zeta(s)$ vanishes for $\sigma < 0$ exactly at the so-called *trivial zeros* $\zeta(-2n) = 0$ (i.e., over the negative even numbers).[8] All other zeros of $\zeta(s)$ are called *nontrivial* and are all complex. It is known that such zeros must lie in the critical strip $0 \leq \sigma \leq 1$ and, by virtue of the symmetry of the functional equation, must be symmetrically positioned with respect to the real axis and the critical line. The exact location of the nontrivial zeros is an open problem with fundamental implications for the particular aperiodic order of the distribution of the prime numbers.

Figure 9.8 graphically illustrates the complex behavior of the $\zeta(1/2 + jt)$ on the critical line $\sigma = 1/2$. Many important results are now known on the zeros of the zeta function. For example, one of Riemann's famous conjectures, later proved by von Mangoldt, concerned the number $N(T)$ of nontrivial zeros $\rho = \sigma + jt$ (counted with multiplicities) in the interval $0 \leq t \leq T$, which has the following asymptotic behavior:

$$N(T) \sim \frac{T}{2\pi} \log \frac{T}{2\pi} \qquad (9.37)$$

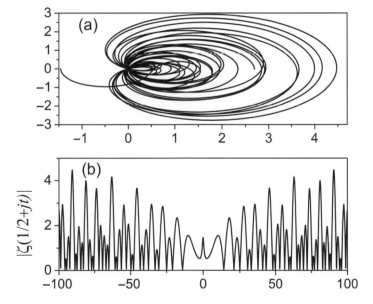

Figure 9.8 (a) Values (real and imaginary part) of the Riemann zeta function $\zeta(1/2 + jt)$ on the critical line for $0 \leq t \leq 100$. (b) Absolute value $|\zeta(1/2 + jt)|$ of the Riemann zeta function as a function of t.

---

[8] This follows directly from the basic properties of the gamma function.

Hence, according to this result, there are infinitely many nontrivial zeros and their frequency of occurrence increases with their imaginary parts, as observed in Figure 9.8. In addition, Hardy showed in 1941 that there are infinitely many zeros of the Riemann zeta function on the critical line. In 1942, Selberg showed that there exists a positive constant $c > 0$ such that the number of nontrivial zeros *on the critical line* satisfies $N_0(T) \geq cN(T)$, thus establishing that a positive proportion of the nontrivial zero of $\zeta(s)$ are indeed on the critical line. Levinson later showed that at least one third of the nontrivial zero lies on the critical line ($c > 1/3$), and the latest record is held by Brian Conrey, who showed $c > 2/5$ [455, 457].

9.2.1   The Hardy Zeta Function

The aperiodic distribution of the nontrivial zeros of the zeta function on the critical line is completely captured by the *Hardy zeta function $Z(t)$*, also known as the Riemann–Siegel θ function, which is a real-valued function defined as follows:

$$Z(t) = \exp j\,\theta(t)\zeta\left(\frac{1}{2} + jt\right)$$

$$\theta(t) = \arg\left[\Gamma\left(\frac{1}{4} + \frac{jt}{2}\right)\right] - \frac{t \ln \pi}{2},$$

(9.38)

where Γ is the Euler gamma function. It follows from the functional equation for $\zeta(s)$ that the Hardy Z function is an infinitely differentiable real-valued function (for real values of t). Moreover, since $|Z(t)| = |\zeta(1/2 + jt)|$, the zeros of $Z(t)$ correspond to the zeros of the Riemann zeta function on the critical line. This function therefore allows one to precisely locate and study the zeros on the critical line by utilizing methods from real analysis. A comparison between the modulus of the zeta function (the black line) on the critical line and the Hardy Z function is shown in Figure 9.9. The function $Z(t)$ shows a single negative local maximum at $t \simeq 2.5$, which is the only one known in the range $t \geq 0$. It is also unknown if there are any positive local maxima. Interestingly, the occurrence of an additional negative local maximum or of a positive local minimum would disprove the Riemann's hypothesis [458]. Edwards showed that the monotonic decrease at large t of the graph of $Z'(t)/Z(t)$, where the prime indicates the derivative, is also equivalent to Riemann's hypothesis [459]. The computational burden of the high-precision computation of the Riemann zeta function on the critical line can be significantly reduced considering the *Riemann–Siegel formula*:

$$Z(t) = 2 \sum_{n \leq \sqrt{t/(2\pi)}} \frac{\cos\left[\theta(t) - t \log n\right]}{n^{1/2}} + O(t^{-1/4})$$

(9.39)

where the error term has a complex asymptotic expansion in terms of trigonometric functions. Since the locations of the zeros of $Z(t)$ coincide with the nontrivial ones

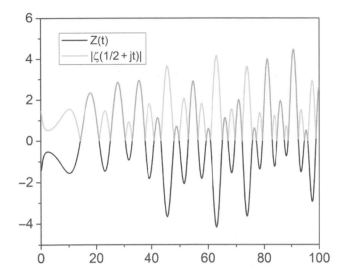

Figure 9.9 The black line: shows the modulus of the Riemann zeta function on the critical line, and the light gray line shows the Hardy Z function.

of the Riemann zeta function, it follows from equation (9.37) that the density of real zeros of the $Z(t)$ up to T is given by the following:

$$\rho_T = \frac{c}{2\pi} \log \frac{T}{2\pi} \tag{9.40}$$

for some constant value $c > 2/5$. Therefore, the number of zeros in an interval of a given size T slowly increases. Clearly, if the Riemann hypothesis is true, all of the zeros within the critical strip are real zeros of $Z(t)$ or, equivalently, the constant c is equal to one.[9]

Despite its apparently erratic behavior, the Hardy zeta function displays a gentle average growth that has been intensively studied in analytic number theory [460]. In particular, its even moments up to the fourth order are known to exhibit logarithmic growth.[10] For example, the second moment, or root mean square value of $Z(t)$, features the following asymptotic scaling:

$$\frac{1}{T} \int_0^T Z^2(t)dt \sim \log T. \tag{9.41}$$

Moments higher than four have also been studied, but little is known. However, it has been conjectured that

$$\frac{1}{T} \int_0^T Z^{2k}(t)dt = o(T^\epsilon) \tag{9.42}$$

---

[9] It is known presently, based on numerical evidence, that the Riemann hypothesis is satisfied by more than a billion zeros and no counterexamples have yet been discovered.

[10] Note that $Z(t)$ is an even function, so there are no odd moments.

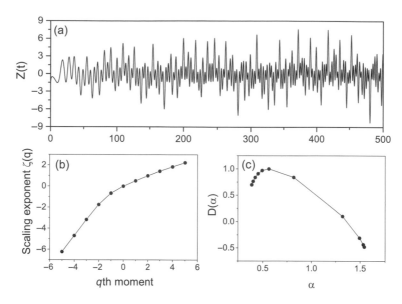

Figure 9.10 (a) The graph of the Hardy Z function up to $t = 500$. In the multifractal analysis shown in (b) and (c), we analyzed a much larger segment up to $t = 10^6$. (b) Scaling exponent behavior and (c) multifractal (singularity) spectrum.

for every positive ϵ. (The little o notation means that in the preceding equation, the left-hand side divided by the right-hand side converges to zero.) This conjecture is called the *Lindelöf hypothesis*, and it is weaker than the Riemann hypothesis. Equivalently, it can be stated in the form $Z(t) = o(t^\epsilon)$, which limits the growth rate of the peaks of $Z(t)$ [459].

9.2.2   Multifractality of the Riemann's Zeros

The complex behavior of the nontrivial zeros of the Riemann zeta function on the critical line is reflected in the rich scaling properties of the Hardy zeta function. In Figure 9.10, we analyze the outstanding complexity of the function $Z(t)$ using the multifractal analysis introduced in Section 9.1.2. Our numerical data unveil strong multifractality in the aperiodic behavior of $Z(t)$. This is evidenced by the nonlinear trend of the scaling exponent in Figure 9.10b as well as by the broad singularity spectrum shown in Figure 9.10c. We remark that, differently from what previously observed in the case of arithmetic functions, the strong multifractal behavior discovered for $Z(t)$ suggests a multiscale structural complexity in the distribution of the nontrivial Riemann's zeros. While we are not aware of any rigorous study on the multifractal nature of $Z(t)$, we believe that our results provide an initial step in this direction that is consistent with recent discoveries in the field of *quantum chaos*. Following a "physics-driven" approach to the Riemann's hypothesis, Wu and Sprung [461] in fact constructed a quantum mechanical model that possesses the Riemann's zeta zeros as its own eigenvalues. The study of spectral statistics demonstrated that these eigenvalues display

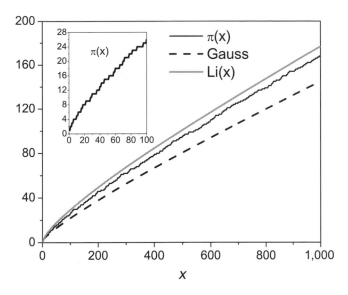

Figure 9.11 The evolution of the prime counting function $\pi(x)$ (black line) and of two important approximations, namely the logarithmic integral (gray line) and Gauss's estimate (dashed line). The inset shows a detail of $\pi(x)$ computed for the first 100 integers.

the same level repulsion behavior observed in quantum chaotic systems. Such novel quantum potentials turned out to be very irregular and multifractal in nature. Subsequently, Schumayer et al. [462] additionally studied quantum potentials whose energy spectra coincide with the prime numbers. Their work also suggested that the potentials calculated for the prime numbers and for the zeros of the Riemann $\zeta(s)$ function are indeed multifractals. Additional aspects on the fundamental connection between $\zeta(s)$ and quantum chaos will be discussed in Section 9.3.1. Finally, the quasiperiodic self-similar fractal nature of $\zeta(s)$ and of more general Dirichlet series was already anticipated in the seminal work of Harald Bohr, who even established a connection with the Riemann hypothesis [463].

9.2.3 The Distribution of Prime Numbers

The search for patterns and regularities in the distribution of prime numbers has fascinated mathematicians for generations. Arguably, the most mysterious and unpredictable behavior in number theory is in fact exhibited by the prime numbers themselves. This is captured by the prime counting function $\pi(x)$, which is defined as the number of prime numbers smaller than or equal to x. This function appears as an aperiodic staircase such that every time the integer x equals a prime, its value is increased by one. However, despite the very irregular series of jumps locally observable on a fine scale (see inset in Figure 9.11), it is evident from Figure 9.11 that the counting function approaches a concave and smoother trend when observed at a larger scale. This unique interplay between apparent randomness at the local level and regularity at a larger scale is characteristic of the most profound problems in number theory that attracted the curious attention of generations of mathematicians and physical scientists. In particular, the young Gauss at age 14, when pondering over a large table of prime numbers, conjectured based on numerical evidence that for large integers x the fraction of primes is approximately $1/\log(x)$. This led him to postulate the celebrated PNT, later demonstrated in 1896 by Jacques Hadamard and La Vallée-Poussin, which states that the relative error tends to zero in the asymptotic approximation (i.e., when $x \to \infty$):

$$\boxed{\pi(x) \sim \frac{x}{\log x}} \tag{9.43}$$

Concerning the rigorous proof of the PNT, it suffices to mention here that, as quantitatively proved by Hadamard and La Vallée-Poussin, the absence of zeros of Riemann's zeta function on the line $\sigma = 1$ implies the validity of the PNT in the form of equation (9.43).

By considering the number of primes in blocks of length 1,000 in order to smooth out their local irregularities, Gauss subsequently refined his original estimate by introducing the *logarithmic integral function*:

$$\mathrm{Li}(x) = \int_2^x \frac{dt}{\log t}. \tag{9.44}$$

The logarithmic integral function indeed produces a good approximation of $\pi(x)$ already over the finite range plotted in Figure 9.11 (the gray line), even though it appears to systematically overestimate the exact values of $\pi(x)$. Understanding the error term $\pi(x) - \mathrm{Li}(x)$ became a very intense research priority in number theory after the PNT was rigorously proved. Littlewood showed in 1914 that despite the apparent dominance of $\mathrm{Li}(x)$ over $\pi(x)$, the two functions will eventually swap in magnitude infinitely often, even though it remains unknown today what is the smallest value of x for this inversion to occur for the first time. Early work by Stanley Skewes in 1933 demonstrated that the first inversion will surely take place before reaching the enormously large upper bound, or "Skewes number":

$$N_s = 10^{10^{10^{34}}}. \tag{9.45}$$

This number stood for years as the *largest useful number* defined in number theory. In 2000, C. Bays and R. Hudson improved Skewes estimate and showed that the first sign change occurs before approximately 10^{316}, which is a number well beyond the reach of present-day computational power. Further refinement to this upper bound for inversion was subsequently provided, but without changing the order of magnitude [464].

One of the most beautiful facts of analytic number theory is that even the fine-scale, detailed structure of the primes encoded in the local fluctuations of $\pi(x)$ can be accounted for by an explicit expansion formula that relies on precise knowledge of the positions of the nontrivial zeros of the Riemann $\zeta(s)$, as we will explicitly show in the next section.

9.2.4 The Riemann–von Mangoldt Explicit Formula

In his famous 1859 article [456], Bernhard Riemann sketched the derivation of a very beautiful formula, later proved in 1895 by von Mangoldt, which enables the exact analytical reconstruction of the prime counting function $\pi(x)$ in terms of the zeros $\{\rho\}$ of the $\zeta(s)$ function. More precisely, Riemann considered the normalized prime counting function $\pi_0(x)$ defined as follows:

$$\pi_0(x) = \frac{\lim_{h \to 0}[\pi(x+h) - \pi(x-h)]}{2}. \tag{9.46}$$

The rigorous derivation of Riemann's explicit formula, as discussed, for instance, by Edwards [459], requires advanced methods of complex analysis and lies beyond the scope of our brief introduction. However, given the great importance of this result in analytic number theory, we highlight in this section some of the key ideas. Riemann started by computing the natural logarithm of the $\zeta(s)$ in the Euler product form:

$$\log \zeta(s) = -\sum_p \log(1 - p^{-s}), \tag{9.47}$$

where we remind that the summation is considered over the prime numbers p. By Taylor expansion of the logarithm in the right-hand side, we obtain the following:

$$\log \zeta(s) = \sum_p \sum_{n=1}^{\infty} \frac{1}{n} p^{-ns} = \sum_p p^{-s} + \frac{1}{2}\sum_p p^{-2s} + \frac{1}{3}\sum_p p^{-3s} + \cdots \tag{9.48}$$

Riemann then implemented the substitutions:

$$p^{-s} = s\int_p^{\infty} x^{-s-1}dx, \quad p^{-2s} = s\int_{p^2}^{\infty} x^{-s-1}dx, \quad \ldots, \tag{9.49}$$

resulting in the following:

$$\log \zeta(s) = \sum_p \sum_{n=1}^{\infty} \frac{s}{n} \int_{p^n}^{\infty} x^{-s-1}dx. \tag{9.50}$$

Now we can interchange the integral and the sum in the last expression and obtain the following:

$$\log \zeta(s) = s\int_1^{\infty} \sum_{\substack{p,\,n \\ p^n < x}} \frac{1}{n} x^{-s-1}dx = s\int_1^{\infty} \Pi(x)x^{-s-1}dx, \tag{9.51}$$

where we defined the following function:

$$\boxed{\Pi(x) = \sum_{p^n < x} \frac{1}{n} = \pi_0(x) + \frac{1}{2}\pi_0(x^{1/2}) + \frac{1}{3}\pi_0(x^{1/3}) + \cdots} \tag{9.52}$$

The first term in the sum simply counts primes below x. Now note that a prime p is less than $x^{1/2}$ when p^2 is less than x. Therefore, the second term of $\Pi(x)$ counts squares of primes smaller than x, with weight $1/2$. Similarly, the third term counts cubes of primes less than x with a weight $1/3$, and so forth. The function $\Pi(x)$ has a staircase graph with a jump by 1 at each prime number and smaller jumps at larger powers of primes. The normalized prime counting function $\pi_0(x)$ can be recovered from $\Pi(x)$ using the Möbius inversion formula:

$$\pi_0(x) = \sum_{n=1}^{\infty} \frac{\mu(n)}{n} \Pi(x^{1/n}) = \Pi(x) - \frac{1}{2}\Pi(x^{1/2}) - \frac{1}{3}f(x^{1/3})$$

$$- \frac{1}{5}\Pi(x^{1/5}) + \frac{1}{6}\Pi(x^{1/6}) - \cdots, \tag{9.53}$$

where $\mu(n)$ is the Möbius function.

The fundamental connection between the zeros of $\zeta(s)$ and the prime counting function $\pi_0(x)$ now starts to appear quite clearly since the term $\log \zeta(s)$ in equation (9.51) can be rewritten in terms of the zeros of $\zeta(s)$ via its *Hadamard product expansion*. Hadamard's theory is concerned with the global characterization of analytic functions by their singularities, and it allows one to define analytic functions by the position of their zeros and singular points. In particular, the function $\log \zeta(s)$ has logarithmic singularities at the roots ρ of $\zeta(s)$, which makes it proportional to the formal summation $\sum_{\rho} \log(1 - s/\rho)$. Hence, by exponentiation, we arrive at the Hadamard product expression for the *completed zeta function*:

$$\xi(s) = \xi(0) \prod_{\rho} \left(1 - \frac{s}{\rho}\right). \tag{9.54}$$

Riemann then used the general Fourier theory to invert equation (9.51) and obtain an integral representation for the function $\Pi(x)$ in the following form:

$$\Pi(x) = \frac{1}{2\pi j} \int_{a-j\infty}^{a+j\infty} \log \zeta(s) \frac{x^s}{s} ds, \tag{9.55}$$

where $a > 1$. By combining the functional equation and the Hadamard product expansion for $\zeta(s)$ Riemann obtained an expression for $\log \zeta(s)$ that he substituted in the preceding integral. Finally, skillfully using complex integration, he was able to analytically derive the main result of his paper, the so-called *Riemann–von Mangoldt explicit formula*:

$$\Pi(x) = \mathrm{Li}(x) - \sum_{\rho} \mathrm{Li}(x^{\rho}) + \int_x^{\infty} \frac{dt}{t(t^2 - 1)\log(t)} - \log(2) \tag{9.56}$$

where ρ are the nontrivial zero of the Riemann $\zeta(s)$. What Riemann masterfully achieved is nothing short of an outstanding feat whose detailed mathematical steps are carefully retraced in Edwards's excellent book [459].

Inserting Riemann's result in (9.53), we can express the normalized prime counting function as follows:

$$\pi_0(x) = R(x) + \sum_\rho R(x^\rho) + \sum_{n=1}^\infty \frac{\mu(n)}{n} \int_{x^{1/n}}^\infty \frac{dt}{t(t^2 - 1)\log t} \qquad (9.57)$$

where we have defined the following:

$$R(x) = \sum_{n=1}^\infty \frac{\mu(n)}{n} \mathrm{Li}(x^{1/n}), \quad R(x^\rho) = -\sum_{n=1}^\infty \frac{\mu(n)}{n} \mathrm{Li}(x^{\rho/n}). \qquad (9.58)$$

Note that the term $\log 2$ is not present in expression (9.57) because it follows from the prime number theorem that $\sum_n \mu(n)/n = 0$. The preceding summations are only conditionally convergent (i.e., the way the terms are summed matters!), requiring complex conjugate pairs of nontrivial zeros to be summed consecutively and in the order of increasing imaginary part. In analogy with conventional Fourier series, the terms $R(x^\rho)$, referred to as *Riemann's harmonics* [450], can be considered basis functions that better approximate the erratic behavior of $\pi_0(x)$ when increasing the number of zeros in the expansion. However, differently from standard Fourier analysis, Riemann's harmonics are highly oscillating and nonperiodic functions [465]. The progressive approximation of $\pi_0(x)$ in terms of an increasing number of Riemann's harmonics is illustrated in Figure 9.12, where we summed up to 60 harmonics. Additional details on the numerical implementation of these delicate series can be found in [465].

It should now be clear that the essence of Riemann's beautiful explicit formula is that if we know exactly where all the nontrivial zeros ρ of $\zeta(s)$ are located, then we also know exactly where all the primes are located. Riemann computed the first three complex zeros $\rho_n = \sigma_n + jt_n$ and noticed that they all have the real part equal to $1/2$. From this observation, Riemann conjectured that all the nontrivial zeros have real part equal to $1/2$. This hypothesis, currently supported by ample numerical evidence up to several hundred billion zeros, is of course the celebrated Riemann's hypothesis (RH) that is considered to be the most important unsolved problem in pure mathematics [455, 466]. For the most pragmatic, it is worth remembering that this is also one of the seven *Millennium Prize Problems* of the Clay Mathematics Institute, whose correct solution will result in a prize of $1 million US awarded by the Institute to the discoverers.

9.2.5 Liouville's λ Function and Riemann's Hypothesis

One of the most interesting arithmetic functions is the Liouville Lambda function, denoted by $\lambda(n)$, which gives the parity of the number of prime factors of n. Specifically, its value is $+1$ if n is the product of an even number of prime numbers, and -1 if it is the product of an odd number of primes. It is easy to show that $\lambda(n)$ can be expressed in terms of the Big Omega function $\Omega(n)$ according to $\lambda(n) = (-1)^{\Omega(n)}$,

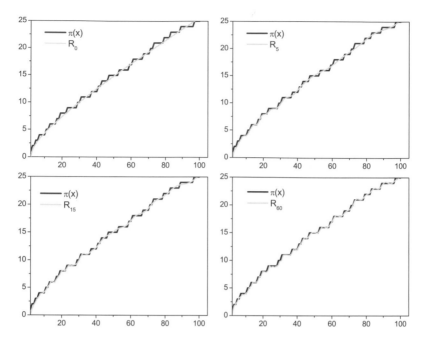

Figure 9.12 The evolution of the prime counting function $\pi(x)$ (the black line) and of two important approximations, namely the logarithmic integral (the gray line) and Gauss's estimate (the dashed line). The inset shows a detail of $\pi(x)$ computed for the first 100 integers.

which shows that $\lambda(n)$ is completely multiplicative since $\Omega(n)$ is completely additive. The Dirichlet series for the Liouville function is related to the Riemann zeta function as follows [467]:

$$\frac{\zeta(2s)}{\zeta(s)} = \sum_{n=1}^{\infty} \frac{\lambda(n)}{n^s}, \tag{9.59}$$

showing how $\lambda(n)$ encodes information on the nontrivial zeros of $\zeta(s)$.

The $\lambda(n)$ function, shown in the inset of Figure 9.13, manifests a complex aperiodic behavior with oscillations at all scales and appears to be an uncorrelated random function with no discernible patterns, in the sense specified as follows:

$$\frac{1}{N} \sum_{n \leq N} \lambda(n) f(n) \to 0. \tag{9.60}$$

Here $N \to \infty$ and $f(n)$ represents any deterministic function or zero-entropy dynamical system. Many fundamental results of number theory can be deduced assuming the uncorrelated randomness of $\lambda(n)$; most importantly the prime number theorem follows from [467]:

$$\lim_{n \to \infty} \frac{\lambda(1) + \lambda(2) + \cdots + \lambda(n)}{n} = 0, \tag{9.61}$$

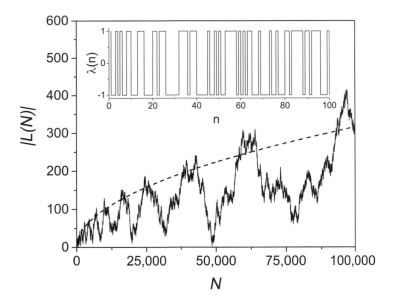

Figure 9.13 Absolute value of the summatory Liouville function $L(N)$ (the solid line) and the quadratic function $\sqrt{n}$ (the dashed line). The oscillations in the absolute value of $L(N)$ reflect the first nontrivial zero of the Riemann zeta function. The inset shows the first 100 values of the Liouville $\lambda(n)$ function.

where $L(N) = \sum_{n \leq N} \lambda(n)$ is the summatory Liouville function. Interestingly, the asymptotic behavior of the summatory Liouville function also provides an *equivalent formulation of the Riemann's hypothesis* (RH):

$$\left| \sum_{n \leq N} \lambda(n) \right| \leq C_\epsilon N^{\frac{1}{2}+\epsilon} \tag{9.62}$$

where C_ϵ and ϵ are positive constants. We recognize that the preceding expression is a statement on the nature of a random walk since the central limit theorem provides the $\sqrt{N}$ upper bound (which would prove the RH). The absolute value of L(n) is shown in Figure 9.13. From this, we can deduce that the RH is equivalent to the statement that an integer number has equal probability of having an odd or an even number of prime factors. This remarkable formulation of the RH shows in very clear terms how "randomness" plays a fundamental role in number theory.

It is interesting to realize that no complex calculus or advanced mathematical concepts are behind this alternative RH formulation besides elementary arithmetic notions (i.e., unique factorization). Establishing the randomness of an arithmetic function provides in this case the solution to the most central unsolved problem of mathematics. However, despite the intuitive appeal of equation (9.62), proving the randomness of a numerical sequence is generally a formidable task (see additional discussions in Section 10.5).

9.3   Riemann's Zeros and Random Matrices

In recent years, a surprising connection between the Riemann zeta function and Random Matrix Theory (RMT) has been discovered. This fruitful line of research in number theory was pioneered by Hugh Montgomery, who in 1972 investigated the spacings between zeros of the zeta function and formulated his now famous *pair-correlation conjecture*. This conjectures states that the two-point correlation function of the zeros of the zeta function $\zeta(s)$ on the critical line is described by the formula:

$$R_2(r) = 1 - \left(\frac{\sin \pi r}{\pi r} \right)^2 \tag{9.63}$$

This expression is exactly equal to the two-point correlation function of the eigenvalues of a random Hermitian matrix from the *Gaussian Unitary Ensemble* (GUE) [430], supporting the conjecture by Polya and Hilbert that the complex zeros of $\zeta(s)$ may correspond to the real eigenvalues of some Hermitian operator. In Figure 9.14, we show the pair-correlation function of the nearest neighbor spacing of the first 10^5 nontrivial zeros of the Riemann zeta function. The continuous line that very well reproduces the behavior of the data corresponds to the analytical pair-correlation formula conjectured by Montgomery. In the inset, we show the histogram distribution of the nearest neighbor spacing of the zeta zeros, which can be modeled considering the probability density function of a Poisson process. In the 1980s, Andrew Odlyzko

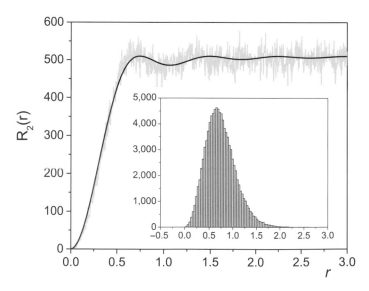

Figure 9.14 The pair-correlation function of the nearest neighbor spacing (normalized to the mean spacing) of the nontrivial zeros of the Riemann zeta function. The inset shows the histogram distribution of the nearest neighbor level spacing. The continuous line through the data is the analytical pair-correlation formula conjectured by Montgomery. The first 10^5 zeros on the critical line are considered in this analysis.

provided empirical support to the Montgomery conjecture by large-scale computer calculations of millions of zeros at values up to 10^{20} [468]. The conjecture has now been extended to higher-order correlations as well as to the zeros of more general zeta functions such as the L-functions that we will briefly introduce in the next section.

The history of Montgomery's discovery is worth being recounted. During a visit to the Institute of Advanced Studies in Princeton, Montgomery showed his numerical results on the level spacing of the zeros of $\zeta(s)$ to the physicist Freeman Dyson, who immediately recognized in Montgomery's conjecture expressed by equation (9.63) the pair-correlation function for the eigenvalues of large random Hermitian matrices with independent and identically distributed (i.i.d) Gaussian entries. These belong to the GUE that is utilized by physicists to statistically compute the distribution of excited energy levels of heavy nucleons or atoms in an external field that breaks time-reversal symmetry.

The random matrix approach in physics was pioneered by Wigner and Dyson, who pictured complex nuclei as black boxes where a large number of constituent particles interact according to extremely complex and largely unknown forces. Similarly to the established approach of statistical mechanics, they tackled this complex problem by considering an ensemble of Hamiltonians, each describing a different nucleus. Wigner postulated that the statistical behavior of the energy levels in the ensemble was entirely captured by the corresponding statistics of the eigenvalues of a random matrix. In particular, in 1957 Wigner proposed that in a sequence of levels with the same spin and parity, the probability density function for the normalized spacing can be described as follows:

$$p_{WS}(s) = \frac{\pi s}{2} \exp\left(-\frac{\pi s^2}{4}\right) \tag{9.64}$$

where s is the spacing between two consecutive levels normalized by the mean spacing. This approximate rule is known as the *Wigner surmise*. Another important result in the theory of random matrices enables the estimation of the fraction of suitably normalized eigenvalues that lie within a given interval $[a, b]$. Remarkably, this can be expressed as the integral of the eigenvalue probability distribution $\mu_{A,N}$ associated to an $N \times N$ random matrix A:

$$\int_a^b \mu_{A,N}(x)dx = \int_a^b \frac{2}{\pi}\sqrt{1 - x^2}dx \tag{9.65}$$

When properly normalized, the curve $\mu_{A,N}(x)$ looks like a semicircle of radius 1. For this reason, the preceding equation is generally known as the *semicircle law*. It is valid in the limit $N \to \infty$ and it is almost independent of the probability density p used to choose the independent and equally distributed entries a_{ij} of the matrix A. Remarkable exceptions are random matrices specified by probability densities with infinite moments, such as the Cauchy distribution, banded matrices, whose eigenvalue density depends on the width of the band, and matrices in Toeplitz form, which exhibit

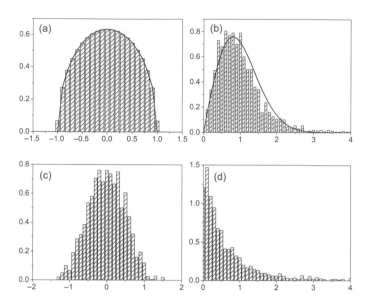

Figure 9.15 Distribution of eigenvalues and level spacing in symmetric random matrices. 50 statistical realizations of $2,000 \times 2,000$ random matrices are considered here, and all eigenvalues are normalized by the factor $2\sqrt{N}$. (a) Eigenvalues distribution of symmetric Gaussian random matrices from the GOE ensemble. The continuous line is the theoretical prediction according to the semicircle law. (b) Level spacing distribution of Gaussian random matrices from the GOE ensemble and theoretical prediction (the continuous line) according to the Wigner surmise. (c) Eigenvalues distribution of symmetric Toeplitz matrices and (d) level spacing distribution.

a Poissonian density, as shown in Figure 9.15. Note that in analyzing the spectral statistics of $N \times N$ random matrices, a frequent choice is to normalize their eigenvalues dividing by the constant $2\sqrt{N}$ since the average eigenvalue has a size approximately equal to $\sqrt{N}$. The semicircle law shown in Figure 9.15 represents a characteristic *universal behavior* akin to the central limit theorem for infinitely large random matrices, as it holds for all mean zero, finite moment distributions. The important principle of universality in random matrix theory states that in the limit of a large matrix dimension, the statistical correlations of the eigenvalue spectra of an ensemble of matrices are independent of the probability distribution that defines the ensemble and only depend on the symmetry properties of the distribution.

Following these ideas, it becomes possible to compute averaged quantities over suitable ensembles of large random matrices, representing nuclear Hamiltonians, in order to statistically describe the observable properties of the system under investigation. Central to this approach is the assumption that the *level density* and the *level spacing* of eigenvalues do not depend, in the limit of large matrices, on the detailed statistical distributions of the individual matrix elements. In fact, the statistical properties of the eigenvalues of random matrices should only depend on general symmetry requirements of the system, such as its time-reversal symmetry. Based on group theoretic arguments, Dyson indeed demonstrated that if an ensemble of random matrices is invariant under a symmetry group, then it must necessarily belong to one of three separate classes, which he named the *orthogonal, unitary*, and *symplectic*. When Gaussian disorder is considered, the corresponding random matrix ensembles are known as the Gaussian Orthogonal Ensemble (GOE), the Gaussian Unitary Ensemble (GUE), and the Gaussian Symplectic Ensemble (GSE). Additional classes of random matrices, such as the MANOVA and the Wishart ensembles, arise in the context of multivariate statistics [430, 469, 470]. However, the GOE and GUE ensembles, which respectively contain real-symmetric and complex Hermitian matrices whose entries are independently chosen from Gaussian distributions, play a crucial role in physics and number theory. It was in fact experimentally observed that the spacings of the energy levels of heavy nuclei are in excellent agreement with those of the eigenvalues of real symmetric matrices from the GOE ensemble, improving on the Wigner surmise model of nuclear physics for systems that obey time-reversal symmetry. On the other hand, substantial numerical evidence exists today to support the *GUE conjecture*, which states that the spacing between consecutive zeros (in the limit of zeros with larger and larger imaginary parts) of *L*-functions is the same as the one between eigenvalues of large complex Hermitian matrices from the GUE ensemble. A very accessible yet rigorous treatment of analytical random matrix methods applied to *L*-functions can be found in [426]. Beyond nuclear physics [471] and number theory [472], random matrix ensembles have found fruitful applications in different areas such as quantum chaos and chaotic dynamics, quantum transport [473–475], and the electromagnetic properties of atomic and optical random media [469, 476–478], which we address specifically in Chapter 12.

9.3.1   Prime Numbers and Quantum Chaos

We previously discussed the surprising connection between random matrices and number theory. We now turn our attention the link between quantum chaos and number theory, leading to the conjecture that the zeta zeros are the eigenvalues of a time-irreversible quantum system that is classically chaotic. Quantum chaos studies the relationship between quantum dynamics and classical chaotic systems. A fundamental result in this field is the connection between the statistics of eigenvalues of certain classes of random matrices and the properties of the corresponding classical dynamics. In particular, it has been established that quantum systems with classically regular (nonchaotic) dynamics typically feature eigenvalues that follow Poisson statistics. Since the most likely spacing for Poisson statistics is $s = 0$, such systems are said to exhibit *level clustering*. In contrast, the eigenvalues of a quantum mechanical system with a chaotic dynamics in the classical limit obey the statistics of different ensembles of random matrices, depending on the symmetries of the system. Such systems, for example the ones with a Wigner surmise level spacing closely approximating the exact GOE distribution, exhibit *level repulsion*.

Timberlake and Tucker [479] numerically investigated the possible connection between prime numbers and quantum physics by applying to the sequence of primes the same statistical methods used to study the spacing of eigenvalues of random

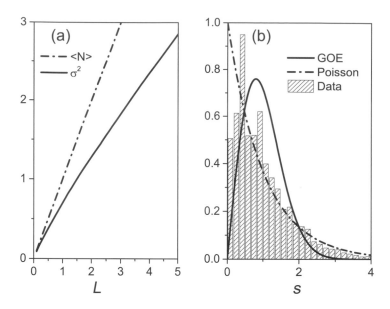

Figure 9.16 (a) Number variance σ^2 (solid line) and average number (dash-dot) as a function of the interval length L for the first 10^4 primes. (b) Nearest neighbor spacing distributions (NNSD) for the sequence of the first 10^4 primes. The curves show the GOE (solid line) and the Poisson (dash-dot) distributions.

matrices. Their results have shown level repulsion among small sequences of primes that turns into level clustering and into Poisson statistics for larger primes. This suggests that primes could be considered energy eigenvalues of a quantum system whose classical dynamics appears to be chaotic at low energies while it becomes increasingly regular at higher energies. The results are in good agreement with the Berry and Robnik statistical model that considers an eigenvalue sequence that is a combination of several independent Poisson sequences and a single independent GOE sequence, thus interpolating between GOE and Poisson statistics [480].

Structural correlations in the distribution of primes are captured by investigating the average number and the number variance of primes that fall within a given interval. Correlations of even higher order are also discussed in [479]. In Figure 9.16a, we consider the scaling of the number variance σ^2 and of the average number of primes as a function of the length L of an interval that slides across the sequence containing the first 10^4 primes. We notice that the variance scales at a smaller rate compared to the mean number of primes $< N >$, showing that the prime numbers form in fact a one-dimensional hyperuniform point set (see Section 8.4 for a general discussion of hyperuniform point patterns). The hyperuniformity of prime numbers was also recently established by Torquato et al. [406] by analyzing the scaling behavior of the structure factor.

In Figure 9.16b we show the histogram of the nearest-neighbor spacing distributions (NNSD) for a sequence of 10^4 primes. Consistent with the findings of Timberlake and Tucker [479], such prime sequences appear to deviate quite significantly from the predictions of both the GOE (solid line) and the Poisson (dash-dot) probability distributions, requiring instead the more flexible Berry and Robnik model that is mixture of Poisson and GOE sequences. In spite of the lack of rigorous theorems, these results strongly suggest a connection between the distribution of primes and the eigenvalues of a quantum system.

9.3.2   L-Functions and the Structure of Primes

L-functions generalize many properties of the Riemann zeta function $\zeta(s)$ in the sense that they admit similar Euler product expansions and encode fundamental arithmetic information in their coefficients. The zeta function $\zeta(s)$ is only the first in a large family of complex analytic L-functions, which can be defined in different mathematical contexts outside analytical number theory, including group representation theory and algebraic topology, elliptic curves on finite fields, as well as dynamical systems theory [458, 481]. For our introductory purposes, it suffices to regard an L-function simply as a Dirichlet series (see Section 9.2) that admits a Euler product and that converges for complex numbers s with sufficiently large real parts. We recall that one of the main facts in the theory of Dirichlet series is that if they converge for $s_0 = \sigma_0 + jt_0$, then they converge in the semiplane with $\Re(s) > \sigma_0$ as well. Moreover, there exists a minimum[11] in this set of converge called the *abscissa of convergence* α such that a Dirichlet series converges if $\sigma > \alpha$ and diverges if $\sigma < \alpha$. However, on the boundary $\sigma = \alpha$ the convergence of the series is not determined and should be analyzed case by case. A simple geometrical interpretation of this property in the context of wave diffraction will be discussed in Section 11.5.

In this section, we introduce the L-functions $L(s, \chi)$ from Dirichlet characters since they contain information on primes in arithmetic progressions. They can be expressed as follows:

$$L(s, \chi) = \sum_{n=1}^{\infty} \frac{\chi(n)}{n^s} \tag{9.66}$$

where $\chi(n)$ is a Dirichlet character. The *Dirichlet characters* are completely multiplicative (see definition in Section 9.1.1) arithmetic functions defined on the positive integers and play a crucial role in number theory and in the theory of finite Abelian groups. Moreover, Dirichlet characters are periodic with period q (a positive integer) and $\chi(n) = 0$ when $(n, q) > 1$, i.e., when not relatively prime. In particular, when two positive integers q and a are relatively prime, Dirichlet proved in 1837 that there are infinitely many primes in the arithmetic progression $\{n \mid n = kq + a\}$ with $k \in \mathbb{N}$, i.e., there are infinitely many primes congruent to a modulo q. For example, the prime numbers $\{5, 13, 17, 29, 37, 41, 53 \ldots\}$ and $\{3, 7, 11, 19, 23, 31, 43 \ldots\}$ are congruent to 1 and to 3 modulo 4, respectively.[12] Equivalently, we say that there are infinite prime

---

[11] This minimum value can be $-\infty$ if the series converges everywhere or $+\infty$ if it converges nowhere.
[12] Notice that this does not mean that the primes form themselves an arithmetic progression. This last aspect will be proven by the Green–Tao theorem discussed later in the text.

numbers of the form $4k + 1$ and $4k + 3$. Despite the fact that special cases like the ones listed previously are readily demonstrated using simple arithmetic arguments, the proof for a general arithmetic progression requires complex analysis and relies on the analytical properties of $L(s, \chi)$. More specifically, it requires that $L(s, \chi) \neq 0$ (and bounded) in the limit $s \to 1$, which is very difficult to rigorously prove [439, 482]. Dirichlet proved that there are infinite primes in arithmetic progression by showing that the series

$$\sum_{p \equiv a \bmod q} \frac{1}{p^s} \tag{9.67}$$

diverges when $s \to 1$. In evaluating the preceding expression, Dirichlet characters play an essential role (see Box 9.3). Since Dirichlet characters are completely multiplicative, $L(s, \chi)$ has a Euler expansion (review Box 9.2):

$$L(s, \chi) = \prod_p \left(1 - \frac{\chi(p)}{p^s}\right)^{-1} \tag{9.68}$$

for $\Re(s) > 1$. Therefore, it is possible to consider the logarithm of the preceding expression. Expanding the logarithm into its Taylor series and using the orthogonality property of the character functions to "filter out" primes of the desired form $a + nq$, we can obtain the relation [483]:

$$\boxed{\frac{1}{\phi(q)} \sum_\chi \overline{\chi(a)} \log L(s, \chi) \sim \sum_{p \equiv a \bmod q} \frac{1}{p^s}} \tag{9.69}$$

where $\overline{\chi}$ denotes complex conjugation and $\phi(q)$ is the Euler totient function. The last equation provides the bridge between the analytical properties of $L(s, \chi)$ and the leading behavior of the summation in (9.67). Since it is easy to establish that $L(s, \chi_0)$ has a simple pole at $s = 1$, where χ_0 is the principal character defined as $\chi_0(a) = 1$ if $(a, q) = 1$ and $\chi_0(a) = 0$ otherwise, the proof of Dirichlet theorem reduces to show that $L(s, \chi)$ is nonzero and finite in the limit $s \to 1$. In this case, the left-hand side of equation (9.69) tends to infinity, implying, in view of its right-hand side, that there are infinitely many positive terms in the sum and therefore infinite primes in any arithmetic progression. Proving that $L(s, \chi) \neq 0$ for $s \to 1$ for $\chi \neq \chi_0$ is a very delicate business that lies well outside our present scope. Interested readers are referred to the very accessible proof, based on the Abel's summation formula, that can be found in the excellent book by G. Everest and T. Ward [483].

Moreover, Dirichlet showed that all prime progressions contain to first order the same fraction of primes, meaning that the sequences of primes are very well distributed in the reduced residue classes modulo q. Therefore, analogously to the PNT, the Dirichlet theorem of primes in arithmetic progressions unveils additional large-scale structure in the distribution of primes despite their random-looking behavior manifested at smaller scales.[13] Specifically, introducing the function $\pi_{q,a}(x)$ that denotes the numbers of primes up to x in the reduced residue class a modulo q, we have, as $x \to \infty$, the following:

$$\pi_{q,a}(x) = \frac{\text{Li}(x)}{\phi(q)} + O(x^{1/2+\epsilon}). \tag{9.70}$$

where the leading term is independent of a. For example, it follows from this theorem that $\pi_{4,1}(x) \sim \pi_{4,3}(x)$ or that half of the primes are of the form $4k + 3$ and half of the form $4k + 1$. The theorem is valid subject to the *generalized Riemann hypothesis* (GRH) for L-functions, which, in analogy with the hypothesis for $\zeta(s)$, states that the nontrivial zeros of L-functions lie on a line with real part equal to $1/2$. However, despite the previous asymptotic result, it was first noticed by Chebyshev that certain residue classes appear to contain a slightly larger number of elements even when considering very large values of x. This phenomenon is called *Chebyshev's bias*. For instance, if we compared the number of primes in the two classes $4q + 1$ and $4q + 3$ up to the first 10^5 primes, we would notice that $\pi_{4,3} = 4,808$, that $\pi_{4,1} = 4,783$, and that a slight yet systematic dominance of the $4q + 3$ class would appear to persist also for other values of x. A more careful numerical observation of these so-called prime races would lead us to discover that occasionally such a lead will be inverted. In fact, from time to time more primes of the form $4q + 1$ than of the form $4q + 3$ will appear, with only a very brief lead quickly relinquished for much longer stretches. Littlewood discovered in 1914 that there exist arbitrarily large values of x for which there are more primes of the form $4q + 1$ up to x than primes of the form $4q + 3$. Moreover, he was able to bound this difference with the incredibly slow growing term [484]:

$$\pi_{4,1}(x) - \pi_{4,3}(x) \geq \frac{\sqrt{x}}{2 \log x} \log \log \log x. \tag{9.71}$$

A fascinating account of different types of prime number races is provided in the very accessible paper by Granville and Martin [485] while Chebyshev's bias found explanation in the rigorous work of Rubinstein and Sarnak [486]. More recent work by Lemke Oliver and Soundararajan [487] considers the patterns of residues modulo q among strings of consecutive primes and introduces additional conjectures based on extensive numerical experimentation. Despite the fact that the fundamental reasons for such unexpected biases are still not fully understood, all these results reinforce that primes are not simply randomly distributed according to a Markovian, i.e., memoryless, process. On the contrary, specific correlations persist even at the larger scales due to the presence of higher-order, slow-growing terms in asymptotic densities, manifesting characteristic structures. In closing, we mention a recent result that directly leverages the fundamental dichotomy between structure and randomness in the distribution of primes and follows from a more advanced *structural theorem*

---

[13] We should notice, however, that primes also possess a local structure. For example, they are all odd numbers except the number 2. Their last digit is 1,3,7, or 9 (with two exceptions), and they are all adjacent to a multiple 6 (with two exceptions).

perspective [488]. This is the Green–Tao theorem, proved in 2004, according to which prime numbers contain arbitrarily long arithmetic progressions. This means that for every natural number k, there exist arithmetic progressions of primes with k terms. Examples are the progressions $\{3,5,7\}$ with $k = 3$ and $\{5,11,17,23,29\}$ with $k = 5$. The largest known progression of primes discovered to date has $k = 26$. However, the Green–Tao theorem guarantees the existence of progressions with arbitrary large k, even though it does not tell us where they are.

Box 9.3 Dirichlet Characters

Dirichlet introduced his characters in order to construct a suitable indicator or *characteristic function* for primes in arithmetic progression (AP), i.e., a function defined over the integers that is nonzero when primes are in AP and zero otherwise. Specifically, Dirichlet first defined the characters as completely multiplicative arithmetic functions $\chi: (\mathbb{Z}/q\mathbb{Z})^* \to \mathbb{C} \setminus \{0\}\}$ between elements of the group of units modulo q (see Box 9.1) to nonzero complex numbers. For all integers m, n, we have that $\chi(mn) = \chi(m)\chi(n)$ and $\chi(n+q) = \chi(n)$, where q ic called the modulus or *conductor* of the character. It is easy to deduce that $\chi(1) = 1$. Furthermore, since Euler's theorem establishes that every element $a \in (\mathbb{Z}/q\mathbb{Z})^*$ is such that $a^{\varphi(q)} \equiv 1 \bmod q$, we have that $\chi(a)^{\varphi(q)} = \chi(a^{\varphi(q)}) = 1$, implying that Dirichlet characters can be obtained as the $\varphi(q)$-th roots of unity $\exp[2k\pi j/\varphi(q)]$, where $0 \leq k < \varphi(q)$. It follows that the number of Dirichlet characters with conductor q is $\varphi(q)$. The set of units modulo q forms an abelian group of order $\varphi(q)$, where $\varphi(q)$ is Euler's totient function. More generally, Dirichlet characters χ can be regarded as homomorphisms between the elements of a finite abelian group G and the nonzero complex numbers. For any group, the map $\chi_0 : \chi_0(g) = 1$ for all $g \in G$ is called the *trivial character*. Dirichlet extended the characters to arithmetic functions $\chi: \mathbb{Z} \to \mathbb{C} \setminus \{0\}$ over all the integers, called Dirichlet characters $\bmod q$, defined by: $\chi(n) = \chi(n \bmod q)$ if n is invertible modulo q (equivalently, if $(n,q) = 1$) and $\chi(n) = 0$ otherwise. Characters $\bmod q$ obey the fundamental *orthogonality relation*:

$$\frac{1}{\varphi(q)} \sum_{\chi} \overline{\chi(a)}\chi(n) = \begin{cases} 1, & n \equiv a \bmod q \\ 0, & \text{otherwise,} \end{cases}$$

where the overbar denotes complex conjugation. The preceding relation is indeed the characterization function for primes in AP since knowing $\chi(p \bmod q)$ for all χ indicates immediately if $p \equiv a \bmod q$ or not.

Since prime numbers have zero density[14] in the integers, the Green–Tao theorem can be regarded as a generalization of *Szemerédi's theorem* [488, 489], which

---

[14] In number theory it is said that a subset Ω of the positive integers has natural density ρ if the proportion of elements of Ω among all natural numbers from 1 to n converges to ρ as n tends to infinity. This can be explicitly evaluated by considering the asymptotic value of $\omega(n)/n$, where $\omega(n)$ is the counting function that gives the number of elements of Ω less than or equal to n.

states that a subset of the natural numbers with positive upper density contains infinitely many arithmetic progressions of length n for all positive integers n. Similar questions related to identifying specific patterns in the distribution of primes have been extended to their complex-valued generalizations such as the Gaussian primes (see Section 10.3). In particular, Terence Tao showed in 2005 that Gaussian primes contain clusters of primes with minimal distances, called constellations, of any given shape [490].

Box 9.4 What Is a Group?

 A group consists of a set G and a rule $\star$ that combines any two elements $a, b \in G$ to obtain an element $a \star b \in G$, and the composition rule is required to satisfy the following properties, or group axioms:

- **Identity Law**: There is a neutral element $e \in G$ such that $e \star a = a \star e = a$ for every $a \in G$.
- **Inverse Law**: For every $a \in G$, there is a unique element $a^{-1} \in G$ that satisfies $a \star a^{-1} = a^{-1} \star a = e$.
- **Associative Law**: For any three elements $a, b, c \in G$, we have the following: $a \star (b \star c) = (a \star b) \star c$.

In addition, if for all $a, b \in G$ we have $a \star b = b \star c$, the group is called commutative or an *abelian group*. When the set G has a finite number of elements, the group $(G, \star)$ is called a *finite group* and its number of elements, called the *order* of G, is denoted by $|G|$. Groups are ubiquitous in mathematics and in physics, where they precisely describe symmetry operations. Simple examples from mathematics include the following:

- $(\mathbb{Z}, \star)$, where $\mathbb{Z}$ is the set of integer numbers and the operation $\star$ is the addition. This is an abelian group whose identity element $e = 0$ and the inverse of every element a is $-a$. Note that the set $\mathbb{Z}$ with the multiplication operation does not form a group, since most elements lack multiplicative inverses inside $\mathbb{Z}$.
- $(\mathbb{F}_p^*, \star)$, where $\mathbb{F}_p^* \equiv (\mathbb{Z}/p\mathbb{Z})^* = \{1, 2, \ldots, p - 1\}$ with p a prime number is the group of the invertible elements, or the units, of the ring of integers modulo p and the operation $\star$ is the multiplication (modulo p). Since p is prime, every element has an inverse in $\mathbb{F}_p^*$ and the group has finite order equal to $p - 1$. A group of infinite order with the usual multiplication operation can be obtained by considering $G = \mathbb{R}^*$, which denotes all the real numbers except the number zero.
- An example of a noncommutative group is the *general linear group* formed by $GL_n(\mathbb{R})$, which is the set of n-by-n matrices with real coefficients and nonzero determinant, with the operation $\star$ given by matrix multiplication. Other groups can be formed by considering other fields in place of $\mathbb{R}$, such as, for instance, the finite field $GL_n(\mathbb{F}_p)$.

> **Box 9.5** Normal Groups and Quotient Groups
>
> A subset H of G that forms a group under the operation $\star$ is called a *subgroup*. Important subgroups in group theory are the *normal subgroups* denoted by N, which are invariant under conjugation. This means that $\forall n \in N, \forall g \in G: gng^{-1} \in N$. Alternatively, a subgroup N of a group G is normal if and only if the coset equality $aN = Na$ holds for all a in G. (Remember that given an element $a \in G$, we define the corresponding *left coset* $aH \equiv \{ah \mid h \in H\}$, and similarly for the right coset). The usual notation for the normality relation is $N \triangleleft G$. Normal subgroups are important because they can be used to construct the *quotient groups* (or factor group) of the given group. This is the group G/N obtained by aggregating similar elements of a larger group using an equivalence relation that preserves the fundamental group structure. More formally, let N be a normal subgroup of a group G. We can then define the set G/N to be the set of all left cosets of N in G, i.e., $G/N = aN \mid a \in G$. The operation on G/N, which is a set of cosets, is naturally defined as follows: for each aN and bN in G/N, the product of aN and bN is $(aN)(bN) \equiv (ab)N$. It can be checked that this operation on G/N is always associative and that N is the identity element in G/N. Moreover, the inverse of an element aN is $a^{-1}N$. Therefore, G/N has the structure of a group, i.e., the quotient group of G by N. As a simple example, consider the subgroup H of $\mathbb{Z}$ containing only the even integers. This is a normal subgroup, because $\mathbb{Z}$ is abelian. Moreover, there are exactly two cosets: $0 + H$, which are the even integers, and $1 + H$, which are the odd integers. Therefore, the quotient group $\mathbb{Z}/2\mathbb{Z}$ is the cyclic group with two elements. This set is isomorphic to the set $\{0, 1\}$ with addition modulo 2. Informally, these two sets are simply identified.

9.3.3   Multifractality and Hyperuniformity of Prime Numbers

Although a probabilistic approach that regards prime numbers as pseudorandom variables can be very useful in number theory, prime numbers are deterministic objects that exhibit considerable structure. This is strikingly demonstrated, for instance, by the Chebyshev's bias or by the Green–Tao theorem. In fact, many more structural patterns have been discovered among the primes. For instance, Goldbach's conjecture, which is one of the oldest and best-known unsolved problems in number theory, states that every even integer greater than 2 is the sum of two primes, and Vinogradov proved in the 1930s that every sufficiently large odd number is the sum of three primes. Moreover, unexpected biases in the distribution of consecutive primes have been recently discovered [487], and Zhang proved that there is some number N (smaller than 70 million) such that there are infinitely many pairs of primes that differ by N [491]. All these important contributions point toward the existence of some "rigidity" in the distribution of primes with distinctive "hidden correlations" that manifest in the most unexpected ways. Prime numbers exhibit in the most surprising ways the interplay

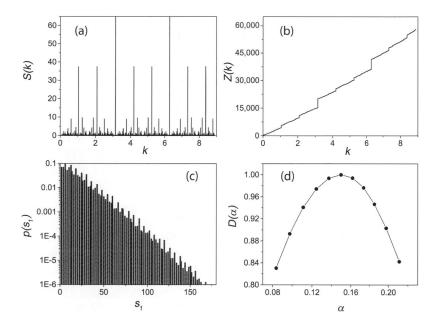

Figure 9.17 (a) Structure factor of the first $N = 9,592$ prime numbers and (b) corresponding counting function $Z(k)$. (c) Histogram of the gap distribution of the first $N = 576,1455$ prime numbers and (d) corresponding multifractal spectrum $D(\alpha)$.

between structure (or rigidity) and randomness that characterizes the most profound problems in number theory as exemplified by the equivalent formulation of Riemann's hypothesis in terms of a random walk (see our discussion in Section 9.2.4). Recently, the characteristic multiscale order of the prime numbers has been investigated using the powerful methods of statistical mechanics, and it was discovered that the prime numbers form an *hyperuniform point set with an effective limit-periodic structure* [406, 407, 492].[15] This implies anomalously suppressed density fluctuations for prime numbers compared to uncorrelated (Poisson) systems at large length scales. Moreover, primes in the interval $M \le p \le M + L$ with M large are described by a scalar order metric that identifies a transition between order, exhibited when L is comparable to M, and uncorrelated disorder, when L is logarithmic in M [407]. The structure factor of prime numbers is shown in Figure 9.17a, and the corresponding density function $Z(k)$, defined in Section 8.4.3, is displayed in Figure 9.17(b). We observe that, unlike the structure factor of quasiperiodic crystals, the diffraction peaks in the spectrum of primes occur at certain rational multiples of π and exhibit a remarkable limit-periodic structure. Indeed, Torquato et al. [406, 407] have derived an analytical formula for

---

[15] Limit-periodic structures are specifically addressed in Chapter 11 and consist of set unions of an infinite number of periodic systems with rational periods. They exhibit a dense spectrum of sharp diffraction peaks with rational ratios between their positions. However, since primes are erratically distributed with the nonuniform asymptotic density $\rho(x) \sim 1/\ln x$, they are only effectively limit-periodic.

the heights of the peaks of the structure factor $S(k)$ that occur at certain rational values of k/π:

$$S(\pi m/n) \sim \frac{N}{\varphi^2(2n)}\mu^2(2n) \qquad (9.72)$$

where $|m| < n$ integers, $\varphi(n)$ is the Euler's totient function and $\mu(n)$ is the Möbius function. This formula is valid for a finite but large number N of primes in the interval $(M, M + L)$. Note that if n is even or has repeated prime factors, then $\mu(2n) = 0$ and the corresponding peak in the structure factor vanishes. On the other hand, if n is odd and square-free, then there will appear a sharp peak in the structure factor of size, $N/\varphi^2(n)$, as it is the case for the peak observed at $k = \pi/3$. In the limit of $N \to \infty$ the peaks in the structure factor will tend to Dirac delta functions located at rational numbers with odd, square-free denominators. In addition to the sharp peaks, the structure factor of primes also contains a diffuse background component, which is indicative of a more disordered, "noisy" structure that decreases in intensity but seems to persist even for large N values [407, 492]. The hyperuniform nature of the primes is demonstrated in Figure 9.17b by the scaling behavior of the $Z(t)$ that clearly converges to zero in the $k \to 0$ limit, as originally discovered by Torquato et al. [406, 407]. Finally, Figure 9.17(c–d) addresses the multiscale order "hidden" in the distribution of *prime gaps* $g_n = p_{n+1} - p_n$, which are the differences between two consecutive prime numbers. The sequence g_n displays a random-looking behavior similar to an uncorrelated lattice gas, and many questions on the nature of this quantity still remain unanswered. Stochastic methods have been used to gain some insights on g_n. For example, the *Cramér stochastic model of primes* considers a sequence of independent random variables $\{X_n\}_{n=2}^{\infty}$ assuming values 0 and 1 with the probability $\mathcal{P}(X_n = 1) = 1/\log n$, from which the random sequence of primes becomes the set [449]:

$$S = \{n \geq 2 : X_n = 1\}. \qquad (9.73)$$

This stochastic approach is consistent with the Riemann's hypothesis and with the *twin primes conjecture* that posits the existence of infinitely many primes p such that $p + 2$ is prime. Moreover, following a probabilistic method, Cramér conjectured in the 1920s the following for large primes:

$$g_n \ll (\log p_n)^2. \qquad (9.74)$$

Note, however, that the sequence $\{g_n\}_{n=2}^{\infty}$ grows without bounds because the numbers $k! + j$ with $2 \leq j \leq k$ are all composite. Studying primes in short intervals, such as from x to $x + \log x$, where x is a large number, Gallagher proved in 1966 that primes have a pseudorandom spatial distribution with gaps following the Poisson statistics, as if they were spatially uncorrelated [493]. This can be appreciated in Figure 9.17c, where we show the probability density function of prime gaps considering approximately 5.8 million primes. However, we must note that differently from a standard Poisson process, the prime gaps exhibit a finer structure of oscillations with recurrent peaks at multiples of 6. We analyze this surprising substructure using the multifractal

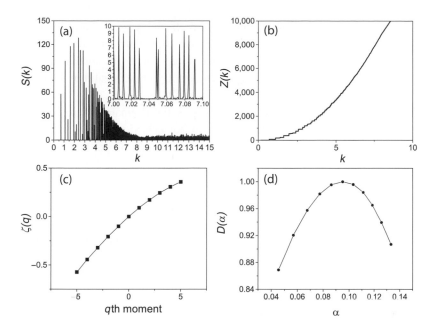

Figure 9.18 (a) Structure factor of the first $N = 10^5$ critical zeros of Riemann's zeta function and (b) corresponding counting function $Z(k)$. (c) Scaling exponent $\zeta(q)$ and (d) corresponding singularity or multifractal spectrum $D(\alpha)$ of Riemann's gaps $d_n = t_{n+1} - t_n$, where t_n are the positive imaginary parts of the critical zeros.

analysis introduced in Section 9.1.2 based on the multiresolution wavelet leaders method. Our results are shown in Figure 9.17d, where a multifractal spectrum $D(\alpha)$ with characteristic downward concavity is clearly visible. This is indicative of the strong heterogeneity exhibited by the set of gaps directly revealing a multiscale complexity. Finally, we observe that $D(\alpha)$ features a Hurst exponent $H < 0.5$, which demonstrates a marked antipersistent behavior in the analyzed sequence of gaps.

We have seen in Section 9.2.4 that prime numbers are directly related to the (nontrivial) zeros of the Riemann zeta function via explicit formulas, of which the Riemann–von Mangoldt explicit formula represents a primary example. Therefore, one might in principle expect to observe multifractality similarly encoded in the spacing of consecutive nontrivial zeros of the Riemann zeta function. However, we must remember that in this case the distribution of gaps (i.e., level spacing) in Riemann's zeros has correlations well described by the GUE statistics, as discussed in Section 9.3. This situation is investigated in Figure 9.18. In Figure 9.18a, we show the structure factor of the first 10^5 zeros of the Riemann's zeta function on the critical line.[16] The spectrum is composed mostly of isolated peaks, as evident from the enlarged view in the inset, with a linearly decreasing intensity toward $k = 0$. The hyperuniform nature of the zeros is demonstrated in Figure 9.18b, where we show

---

[16] This means we are considering here the positive imaginary parts of the zeros arranged in ascending order.

the corresponding counting function $Z(k)$, which tends toward the origin of k-space quadratically. These numerical results are consistent with the analysis obtained by Torquato et al. based on the Fourier transform of the analytical pair correlation function $R_2(r)$ of the zeros (see equation (9.63)) that is valid in the infinite-size limit $N \to \infty$ [494].

Strong multifractality for the spacing of the Riemann's zeros is demonstrated in Figure 9.18c,d where we show the nonlinear behavior of the scaling exponents $\zeta(q)$ and the multifractal spectrum $D(\alpha)$, respectively. However, we should compare these results with the much broader multifractal spectrum of the Hardy $Z(t)$ function, which encodes the positions of nontrivial zeros, as we discussed in Section 9.2.2. We observe that, consistent with the more correlated GUE nature of the spacings of Riemann's zeros (i.e., Riemann's spacings) compared to the Poisson distribution of the prime gaps, the multifractal spectrum $D(\alpha)$ of Riemann's spacings is narrower compared to the one of the prime gaps, pointing toward a less heterogeneous structure of Riemann's spacings. However, since in this case the Hurst exponent is smaller, Riemann's spacings exhibit a stronger antipersistent behavior with shorter-range memory compared to the prime gaps, which is indicative of a significantly less ordered distribution of the zeros compared to the one of the primes. It is worth noting that multifractal analysis was originally applied to the study of the prime number distribution (as opposed to its gaps) by Wolf, who computed the generalized fractal dimensions and the multifractal spectrum of the primes [495]. Moreover, by analyzing the number of primes contained in successive large intervals, Wolf showed that they are distributed according to a power spectrum that follows the $1/f^\beta$ power law with exponent ≈ 1.64 [496]. This behavior indicates that primes appear to be distributed among the natural numbers with all possible gaps between them and with no characteristic length scale, similarly to the critical states of self-organized dynamical systems that exhibit fractal behavior [497]. Indeed, a characteristic fractal structure was discovered in the oscillations of prime gaps by Ares and Castro [498]. All these studies are consistent with the previously mentioned Gallagher's proof that, under the n-tuple Hardy–Littlewood conjecture, the fraction of intervals that contain exactly a given number of primes follows a Poisson distribution. More recently, Sgrignuoli et al. discovered strong multifractality in the complex distribution of primes in quadratic fields and in the irreducible elements of quaternion fields [10]. The fascinating results discussed in this section make us fully appreciate how a rigid structure in the distribution of primes emerges at multiple length scales when considering suitably large intervals. Under these conditions, prime numbers behave as a purely diffractive hyperuniform point pattern that is effectively limit-periodic, in stark contrast to the uncorrelated randomness manifested over short intervals.

9.4   Continued Fractions and Approximation

We now abandon the fascinating field of analytic number theory to introduce important concepts in the approximation theory of numbers based on continued fractions.

The study of continued fractions is an essential part of number theory that not only enables very accurate approximations of irrational numbers by rational ones but also reveals fundamental structural information that is otherwise hidden in the decimal (or binary) expansion of numbers. For instance, irrational numbers, such as $\sqrt{2}$ or e, display a random-looking decimal expansion but an extremely structured and simple expansion in continued fractions. Continued fractions are therefore a powerful tool to learn with many applications that range from approximation theory to number theory and geometry, and are essential in the study of aperiodic order.

Every irrational number can be represented by an infinite series of nested fractions, called a *continued fraction*. This can be achieved following a very simple procedure that we will illustrate in this section considering the irrational number $\pi = 3.1415926535\ldots$ We start by splitting the integer from the fractional part, which we then rewrite "upside down" as follows:

$$\pi = 3 + 0.1415926535\cdots = 3 + \cfrac{1}{\cfrac{1}{0.1415926535\cdots}} = 3 + \cfrac{1}{7.0625133059\cdots}$$

$$(9.75)$$

We now iterate the same reasoning in the last denominator line, which yields the following:

$$\pi = 3 + \cfrac{1}{7 + 0.0625133059\cdots} = 3 + \cfrac{1}{7 + \cfrac{1}{\cfrac{1}{0.0625133059\cdots}}} \qquad (9.76)$$

Continuing the process, we get the following:

$$\pi = 3 + \cfrac{1}{7 + \cfrac{1}{15.9965944066\cdots}} = 3 + \cfrac{1}{7 + \cfrac{1}{15 + 0.9965944066\cdots}} \qquad (9.77)$$

We note along the way that if we replace the number 16 in the last denominator, we obtain the following useful approximation of π:

$$\pi \approx 3 + \cfrac{1}{7 + \cfrac{1}{16}} = \frac{355}{113}, \qquad (9.78)$$

which agrees with the value of π to six decimal places! Continuing the expansion process, we add an additional layer to our fraction:

$$\pi = 3 + \cfrac{1}{7 + \cfrac{1}{15 + \cfrac{1}{1 + 0.0034172310\cdots}}} \qquad (9.79)$$

This expansion process, which consists in repeated flips and separations of the integer and fractional parts, produces better and better approximations as we increase the levels of the resulting continued fraction approximations. Since all the numerators are equal to 1, it is customary to introduce the shorthand notation for the continued fraction expansion of a generic number α in *simple (or regular) form*:

$$\alpha = a_0 + \cfrac{1}{a_1 + \cfrac{1}{a_2 + \cfrac{1}{a_3 + \cfrac{1}{a_4 + \cfrac{1}{\ddots}}}}} = [a_0, a_1, a_2, a_3, a_4, \ldots] \qquad (9.80)$$

where $a_i > 0$ for $i \geq 1$. In a way similar to the infinite decimal expansion, the infinite continued fraction of $\alpha \notin \mathbb{Q}$ exists and converges to α at a great speed. Moreover, this representation is unique. Therefore, every irrational number is the value of a unique infinite continued fraction, whose coefficients can be found using the nonterminating algorithm illustrated (9.80). On the other hand, the continued fraction representation for a rational number is finite and only rational numbers have finite representations. In contrast, the decimal representation of rational numbers may be finite or infinite with a repeating cycle.

9.4.1   Convergents and Recursion Formulas

When the simple continued fraction expansion is known, α can be approximated by a series of rational numbers, called the *convergents to* α, that are simply obtained by truncating the continued fraction at any given level. For example, the nth convergent to α is the rational number:

$$\frac{p_n}{q_n} = [a_0, a_1, \ldots, a_n] \qquad (9.81)$$

that is obtained considering only the terms up to a_n. Retracing the main steps for the continued fraction approximation of any number using symbols rather than numbers, it becomes evident that successive convergents are generated from earlier ones according to the following *fundamental recursion formula*:

$$\boxed{\begin{aligned} p_n &= a_n p_{n-1} + p_{n-2} \\ q_n &= a_n q_{n-1} + q_{n-2} \end{aligned}} \qquad \begin{aligned} (9.82) \\ (9.83) \end{aligned}$$

valid for $n \geq 2$ with $p_0 = a_0$, $q_0 = 1$ and $p_1 = a_1 a_0 + 1$, $q_1 = a_1$. Since the convergents get closer and closer to the number α, it is interesting to investigate the difference between successive convergents, which manifests a simple pattern. In particular, simple numerical exploration reveals that the numerators of the differences of successive convergents are all equal to 1 and that their values alternate between positive

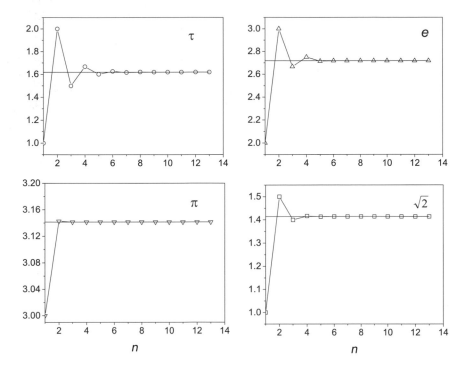

Figure 9.19 Approximations to remarkable irrational numbers by finite continued fractions. In particular, we consider the golden mean $\tau = (1 + \sqrt{5})/2$, e, π, and $\sqrt{2}$. The alternating behavior of successive convergents is clearly shown along with the very fast convergence to π compared to the other numbers.

and negative numbers. In fact, using the recursion formula (9.83) and reasoning by induction it can easily be established that:

$$\frac{p_{n-1}}{q_{n-1}} - \frac{p_n}{q_n} = \frac{(-1)^n}{q_{n-1}q_n} \tag{9.84}$$

The proof of this theorem is very simple. The result is valid for $n = 1$ since $p_0 q_1 - p_1 q_0 = a_0 a_1 - (a_1 a_0 + 1) = -1$. Assuming its validity for $n = N$, a simple calculation shows that it is also valid for $n = N + 1$, completing the induction. In addition, we can deduce that the even convergents p_{2n}/q_{2n} are bounded and increasing while the odd ones p_{2n+1}/q_{2n+1} are bounded and decreasing, as exemplified in Figure 9.19 for representative irrational numbers. In the simple form notation, we can easily obtain the following remarkable continued fraction expansions:

$$\pi = [3, 7, 15, 1, 292, 1, 1, 1, 2, 1, 3, 1, 14, 2, 1, 1, 2, \ldots] \tag{9.85}$$
$$\sqrt{2} = [1, 2, 2, 2, 2, 2, 2, 2, 2, 2, 2, 2, 2, 2, 2, \ldots] \tag{9.86}$$
$$e = [2, 1, 2, 1, 1, 4, 1, 1, 6, 1, 1, 8, 1, 1, 10, 1, 1, 12, 1, \ldots] \tag{9.87}$$
$$\tau = [1, 1, 1, 1, 1, 1, 1, 1, 1, 1, 1, 1, 1, 1, 1, 1, 1, 1, 1, \ldots] \tag{9.88}$$

It is very interesting to compare the preceding continued fractions with the corresponding decimal representations. In doing so, we discover that irrational numbers such as $e = 2.71828182845\ldots$ feature an apparently random decimal expansion with no evident regularity while they admit instead a regularly structured continued fraction representation. Even more strikingly, the continued fraction representations of $\sqrt{2}$ and τ are periodic. In all these cases, the continued fraction expansion brings to light fundamental structural patterns associated to these irrationals. However, the number π does not show any regular pattern in both its decimal and (regular) continued fraction expansion, which points toward its (yet unproven) normality. It is natural to ask what is the reason for this very different behavior. A simple example will suffice to clarify the situation. Consider a continued fraction with the simple repetitive pattern: $\alpha = [a, b, b, b, \ldots]$. By definition, we can rewrite the expression as follows:

$$\alpha = a + \frac{1}{[b, b, b, \ldots]}, \tag{9.89}$$

and we are now left to determine the value of $\beta = [b, b, b, \ldots]$. By applying the same reasoning as before, we can write the following:

$$\beta = b + \frac{1}{[b, b, b, \ldots]}. \tag{9.90}$$

We now recognize that the last denominator is exactly the number β, which yields the quadratic equation:

$$\beta = b + \frac{1}{\beta}. \tag{9.91}$$

Solving for $\beta > 0$ and reinserting into equation (9.90), we can finally compute the value of α as follows:

$$\boxed{\alpha = [a, b, b, b, \ldots] = \frac{2a - b}{2} + \frac{\sqrt{b^2 + 4}}{2}} \tag{9.92}$$

Similar reasoning can be applied also to continued fractions that repeat in a more general fashion, such as $\alpha = [a_1, a_2, \ldots, a_\ell \overline{b_1, b_2, \ldots, b_m}]$, where the overbar identifies the repeating pattern. In this case, the following can easily be shown [499]:

$$\alpha = \frac{r + s\sqrt{D}}{t}, \tag{9.93}$$

where $r, s, t, D > 0$ are integers. In addition, a number $(r + s\sqrt{D})/t$ $(r, s, t, D > 0$ integers) can be represented as $\alpha = [a_1, a_2, \ldots, a_\ell \overline{b_1, b_2, \ldots, b_m}]$.

9.4.2   Periodic Expansions and Quadratic Surds

The real numbers whose continued fraction expansions eventually repeat are precisely the quadratic irrationals, or *quadratic surds*, as shown by Lagrange [426]. These are irrational numbers, such as τ and $\sqrt{2}$, that are the solution to some quadratic equation with integer coefficients. Quadratic irrational numbers, considered as a subset of the complex numbers, are algebraic numbers of degree 2 (see Box 10.1). The square roots of all (positive) integers that are not perfect squares are quadratic irrationals and can be represented by periodic continued fractions. In contrast, the decimal representations of quadratic irrationals are apparently random. Because π or e are transcendental numbers, as opposed to algebraic ones, they cannot have a periodic continued fraction expansion. However, while the nonperiodic continued fraction expansion of e exhibits a very regular pattern, the one of π appears to be random. Interestingly, though, if we relax the regularity requirement that all the numerators of the fractions are equal to 1, it becomes possible to obtain *generalized continued fraction* of transcendental numbers that show simple patterns. For instance, in the case of π we have the following [500]:

$$\frac{4}{\pi} = 1 + \cfrac{1^2}{2 + \cfrac{3^2}{2 + \cfrac{5^2}{2 + \cfrac{7^2}{2 + \cfrac{9^2}{2 + \cdots}}}}}, \quad \frac{\pi}{4} = \cfrac{1}{1 + \cfrac{1^2}{3 + \cfrac{2^2}{5 + \cfrac{3^2}{7 + \cfrac{4^2}{9 + \cdots}}}}} \tag{9.94}$$

These and similar surprising expansions are among the most elegant formulas of mathematics. Most irrational numbers do not have any periodic or regular behavior in their continued fraction expansion. Nevertheless, Khinchin proved that for almost all real numbers α, the coefficients a_i of their continued fraction representations have a geometric mean that is a constant, known as Khinchin's constant, $K = \lim_{n \to \infty}(a_1 a_2, \ldots, a_n)^{1/n} \approx 2.6854520010\ldots$, independent of the value of α. Other regularities associated to the denominators of the convergents of continued fractions were discovered by Paul Lévy and are also exemplified by Lochs's theorem stating that each additional term in the continued fraction representation of a typical real number increases the approximation accuracy by approximately one decimal place. This is consistent with the fact that among numbers with bounded denominators, the continued fractions in fact produce the best approximation [426].

9.4.3   Diophantine Approximation

Diophantine approximation theory deals with the approximation of real numbers in terms of rational numbers. The central question is therefore, given any real number θ, how closely can we approximate it using fractions? More precisely, if we consider any given positive ϵ, can we find a rational number m/n within ϵ from θ that satisfies $|\theta - m/n| < \epsilon$? The answer is certainly yes because the rational numbers are dense over the real ones. It follows that for any real number θ and positive ϵ there are indeed an infinity of rational numbers m/n that satisfy the previous inequality. A better way to approach the problem is to consider all rational numbers with a fixed denominator

n, assumed to be a positive integer. An important result of Diophantine theory asserts that there are infinitely many rational numbers m/n (with $n > 0$) such that

$$\left|\theta - \frac{m}{n}\right| < \frac{1}{n^2}, \tag{9.95}$$

and this result cannot be improved by replacing, for instance, $1/n^2$ by $1/n^3$. More precisely, the preceding inequality becomes false even when $1/n^2$ is replaced by $1/n^{2+\epsilon}$ for any positive ϵ. However, even if the exponent cannot be improved, the result can be strengthened by a suitable constant factor. A famous result proved by Hurwitz states that $1/n^2$ can be replaced by $1/(\sqrt{5}n^2)$ and that no larger constant than $\sqrt{5}$ can be used. Therefore, the inequality

$$\boxed{\left|\theta - \frac{m}{n}\right| < \frac{1}{\sqrt{5}n^2}} \tag{9.96}$$

is the best possible approximation since it becomes false when $\sqrt{5}$ is replaced by any larger constant. The preceding statement, referred to as *Hurwitz's theorem*, is valid for any irrational number θ. Specifically, this theorem states that to each irrational number θ there corresponds infinitely many rational numbers m/n that satisfy the preceding inequality. However, the inequality can be improved if only a finite number or rational numbers is considered or if some restrictions are imposed. For instance, when certain numbers are skipped, better approximations can be achieved, as we explain next. Let us first introduce the concept of *equivalent real numbers*: two real numbers θ and θ' are equivalent if there exist integers p, q, r, s as follows:

$$\theta' = \frac{p\theta + q}{r\theta + s}, \tag{9.97}$$

with $ps - qr = \pm 1$. The preceding expression defines an equivalent relation and the set of all numbers that are equivalent to a fixed θ is a countable set. If the irrational number $(\sqrt{5} - 1)/2$ and all its equivalents are eliminated, then the remaining irrational numbers can be approximated by an infinity of rational numbers that satisfy equation (9.96) with $\sqrt{8}$ replacing $\sqrt{5}$. Next, if the number $\sqrt{2}$ and all equivalents are eliminated, then each remaining irrational number can be approximated by the infinity of rational numbers satisfying equation (9.96) with $\sqrt{221}/5$ replacing $\sqrt{5}$. This process of elimination can be continued, leading to the so-called *Markov spectrum* of irrational numbers that contains the constants $\{\sqrt{5}, \sqrt{8}, \sqrt{221}/5, \sqrt{1517}/13, \ldots\}$ converging to the number 3. The Markov spectrum, and the closely related *Lagrange spectrum* obtained by the formula $L_n = \sqrt{9 - 4/m_n^2}$, where m_n is the nth Markov number,[17] are identical discrete sets in the interval $[\sqrt{5}, 3)$ and they are also equal, but become continuous sets after the so-called Freiman's constant $F = 4.527829\ldots$ As sets of numbers, the Markov and Lagrange spectra have very rich geometrical

---

[17] The nth Markov number is the nth smallest integer m such that the equation $m^2 + x^2 + y^2 = 3mxy$ has a solution in positive integers x and y.

properties [501]. In particular, it was established recently that the transition region between the discrete initial part and the continuous final part of these spectra has a fractal structure [502].

Finally, we point out an important result proved by Joseph Liouville, who obtained in 1844 the first lower bound for the approximation of algebraic numbers (see Box 10.1). In particular, Liouville's theorem states that if θ is a real *algebraic number* of degree $d \geq 2$, then there is a nonzero constant $C > 0$, depending on θ, such that the inequality

$$\left| \theta - \frac{m}{n} \right| \geq \frac{C}{n^d} \tag{9.98}$$

holds for all rational numbers m/n with $n > 0$, and in this case we say that θ is of type d. For rational values θ, we have $d = 1$, and θ is irrational if and only if $d \geq 2$. Thus, Liouville's theorem established that algebraic numbers cannot be well approximated by the rational numbers. An important example is the algebraic number $(1 + \sqrt{5})/2$, known as the golden mean, which plays a central role in the theory of aperiodic order, specifically in relation to quasicrystals. As a consequence, a real number θ can be well approximated by rationals, in the sense that $|\theta - m/n| < C/n^d$, only if it is a *transcendental number* such as π or e. Liouville's result also provides a criterion for the transcendence of a number and enabled to construct the first examples of transcendental numbers. Indeed, almost all real numbers are transcendental, since Georg Cantor proved in 1874 that the algebraic numbers form a countable set. Thue, Siegel, and Dyson have subsequently improved on Liouville's original exponent d that appears in the approximation of algebraic numbers by rationals. In particular, Roth proved in 1955 that there are only finitely many fractions m/n satisfying $|\theta - m/n| < 1/n^{2+\epsilon}$ for every $\epsilon > 0$, previously conjectured by Siegel in 1921, and received the Fields Medal for this work.

10 Aperiodic Order in Number Theory

Order is the pleasure of reason; but disorder is the delight of imagination.

Le Soulier de Satin, Paul Claudel

This Chapter focuses on the phenomena of quasiperiodicity and arithmetic randomness by analyzing the characteristic Poisson behavior in a number of situations ranging from equidistributed sequences (modulo one) to elliptic curves over finite fields. This naturally leads to the investigation of spatial Poisson processes and their fundamental properties, which are central tools for the structural analysis of random point patterns. We introduce the necessary field-theoretic background to address generalized prime numbers in algebraic rings, which consist in aperiodic point patterns with fascinating geometrical and wave scattering properties that will be presented in Chapters 12 and 13.

In this Chapter, we also address the aperiodic Poisson behavior in the context of finite Galois fields and introduce some of the most important applications to cryptographic engineering, the algorithmic generation of pesudorandomness, as well as the construction of deterministic aperiodic structures with constant (i.e., flat) Fourier spectra. Finally, we characterize the aperiodic order of elliptic curves over finite fields that are primary examples of deterministic structures with ideally Poisson behavior up to third-order structural correlations. The chapter concludes with an outlook on the foundational concepts of randomness, structure, and information from the perspective of algorithmic information theory.

10.1   Almost Periodic and Quasiperiodic Functions

Almost periodic and quasiperiodic functions are essential tools in the description of aperiodic order as it appears in number theory and geometry. The theory of almost periodic functions was originally formulated by Harald Bohr while working on the convergence properties of Dirichlet series [463, 503–505]. In this context, Bohr was able to prove that every Dirichlet series is almost periodic in its half-plane of absolute convergence [481].

The key insight to understand the concept of almost periodicity comes when rewriting the terms that appear in the series expression of the Riemann zeta function as follows:

$$n^{-s} = e^{\log n^{-s}} = n^{-\sigma} e^{-jt(\log n)} = n^{-\sigma}[\cos(t \log n) - j \sin(t \log n)], \qquad (10.1)$$

where $s = \sigma + jt$. Therefore, the complex behavior produced on the imaginary axis (i.e., with σ fixed) when summing a finite number of such terms stems from the nature of trigonometric polynomials with $\log n$ (real) frequencies that are not all commensurable. More generally, all complex trigonometric polynomials in the form

$$\boxed{P(x) = a_1 e^{j\lambda_1 x} + a_2 e^{j\lambda_2 x} + \cdots + a_n e^{j\lambda_n x}} \qquad (10.2)$$

with real frequencies λ_n and complex coefficients a_n define almost periodic functions in Bohr's sense [506]. More precisely, a function $f(x)$ is almost periodic if it is the uniform limit of a sequence of trigonometric polynomials $\{P_n(x)\}$ of the form (10.2), i.e., if for each $\epsilon > 0$, we can find $N(\epsilon) > 0$ such that $|f(x) - P_n(x)| < \epsilon$ for $n > N(\epsilon)$ and $x \in \mathbb{R}$. In other words, a function is *almost periodic* if for every $\epsilon > 0$ there exists a linear combination of sine and cosine waves that is within a distance less than ϵ from $f(x)$[1]. Bohr was able to prove that this definition is equivalent to the existence of a relatively dense[2] set of *almost-periods*, meaning that there are "enough" translations that make the shifted function differ arbitrarily little from the function itself. This property is displayed in Figure 10.1, in which we consider three different translates of the function $f(x) = \cos(2\pi x) + \cos(2\pi \alpha x)$, where $\alpha = (1 + \sqrt{5})/2$ is the golden ratio. This continuous function is nonperiodic. In fact, a continuous periodic function attains its maximum value, which here is 2, once in every period, but there is no x for which $f(x) = 2$. On the other hand, there are many choices of translations a that make $f(x + a) = f(x)$ approximately true for all x. One is clearly displayed in Figure 10.1c.

The concept of almost periodic functions that we have introduced has a well-known counterpart in the time domain, where a function of time is called *quasiperiodic* when it contains two or more mutually incommensurate frequencies. For example, we consider the following function:

$$f(t) = \cos\left(\frac{2\pi t}{a}\right) + \cos\left(\frac{2\pi t}{b}\right) \qquad (10.3)$$

with Fourier components $(\pm 1/a \pm 1/b)$. If a/b is a rational number, the preceding function will be periodic and its period will be the least common multiple of a and b. On the other hand, if a/b is an irrational number, the function cannot be periodic.

---

[1] The original 1925 Bohr's definition [504] more precisely requires convergence in the uniform $\|f(x)\| = \sup_x |f(x)|$.

[2] Relatively dense means that for each ϵ, there are radii r_1 and r_2 such that every disk of radius r_2 contains at least one translation a satisfying the condition $|f(x + a) - f(x)| < \epsilon$ and within a disk of radius r_1 from a there are no other translations that satisfy the preceding condition.

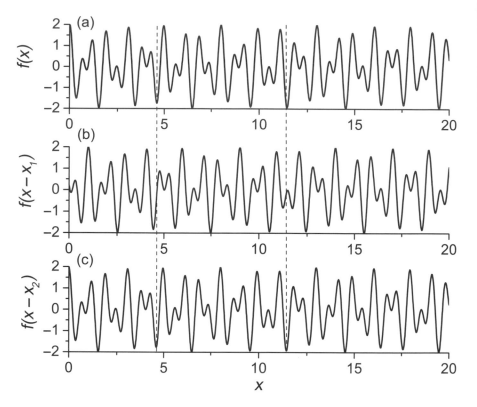

Figure 10.1 (a) The function $f(x) = \cos(2\pi x) + \cos(2\pi\alpha x)$, where $\alpha = (1 + \sqrt{5})/2$ is the golden ratio. (b) The function $f(x)$ is translated by $x_1 = 33$ units and (c) the function $f(x)$ translated by $x_2 = 34$ units. Notice the great similarity between the functions in (a) and (c), which is due to the rational approximation $\alpha \approx 55/34$.

However, its behavior will not be completely chaotic either. On the contrary, it will manifest a specific type of aperiodic order that is called quasiperiodic. This notion of ordered, though aperiodic, motion is well known in the theory of dynamical systems and in classical mechanics. For instance, the orbital periods of the Earth around the Sun and the Moon around the Earth are incommensurate, which renders their dynamics quasiperiodic. Quasiperiodic oscillations are also manifested in the forced oscillations on a simple pendulum that obey the following familiar dynamic equation:

$$\ddot{x}(t) + \omega_0^2 x(t) = f(t), \tag{10.4}$$

in which $f(t)$ is the force (per unit mass) applied to a moving particle. Choosing, for instance, $f(t) = A \sin \gamma t$, we can write the general solution of (10.4) as follows:

$$\boxed{f(t) = C \cos(\omega t + \alpha) + A(\omega^2 - \gamma^2)^{-1} \sin \gamma t} \tag{10.5}$$

When γ is a number incommensurable with ω_0 (i.e., the ratio γ/ω_0 is irrational), the solution $x(t)$ is a nonperiodic oscillatory solution that belongs to the class of quasiperiodic motions. These types of forced oscillations are special cases of almost periodic oscillations that develop when $f(t)$ in (10.4) takes the following form:

$$f(t) = \sum_{i=1}^{\infty} A_i \sin \gamma_i t, \tag{10.6}$$

with γ_i arbitrary real numbers and $\sum_i |A_i| < \infty$. In this case, the general part of the solution to be added to $C \cos(\omega t + \alpha)$ will be the following:

$$\sum_{i=1}^{\infty} A_i (\omega^2 - \gamma_i^2)^{-1} \sin \gamma_i t. \tag{10.7}$$

For values of $\gamma_i \in \mathbb{R}$ such that this series and its second derivative converge absolutely, the function represented by (10.7) is an almost periodic solution of equation (10.4). This dynamical picture also suggests that almost periodic functions, similarly to periodic functions, can be expanded in absolutely convergent Fourier series. This important aspect will be discussed in the next section.

10.1.1 Spectral Representation

The example of forced harmonic oscillations that we have discussed in the previous section motivates the introduction of the spectral representation of an almost periodic function $f(t)$ $(t \in \mathbb{R})$ in the following form:

$$f(t) = \sum_{i=1}^{\infty} a_i e^{j\lambda_i t} \tag{10.8}$$

with $a_i \in \mathbb{C}$ and $\lambda_i \in \mathbb{R}$ such that $\sum_{i=1}^{\infty} |a_i|^2 < \infty^3$. Note that the preceding expression differs from the familiar Fourier series of periodic functions because it involves the real numbers λ_i in its exponent instead of integer numbers. Moreover, the λ_i's are arbitrary and are not the multiples of a positive number, i.e., there are no fundamental frequencies and overtones. Almost periodic functions represented as in (10.8) were investigated by Besicovich [505] and consist of infinitely many harmonic terms:

$$\Re\{f(t)\} = \sum_{i=1}^{\infty} (\alpha_i \cos \lambda_i t - \beta_i \sin \lambda_i t) \tag{10.9}$$

$$\Im\{f(t)\} = \sum_{i=1}^{\infty} (\alpha_i \sin \lambda_i t + \beta_i \cos \lambda_i t) \tag{10.10}$$

---

[3] This condition implies the absolute and the uniform convergence of the series in (10.8) for all $t \in \mathbb{R}$.

The relationships between an almost periodic function $f(t)$ and the coefficients a_i in its trigonometric series representation can be obtained in terms of its *mean value*:

$$M\{f\} = \lim_{\ell \to \infty} \frac{1}{2\ell} \int_{-\ell}^{+\ell} f(t)dt. \tag{10.11}$$

The mean value exists for any almost periodic function and enables one to obtain the Fourier coefficients [436]:

$$a_i = M\{f(t)e^{-j\lambda_i t}\} = \lim_{\ell \to \infty} \frac{1}{2\ell} \int_{-\ell}^{+\ell} f(t)e^{-j\lambda_i t}dt \tag{10.12}$$

Naturally, the numbers $\lambda_i \in \mathbb{R}$, with $i = 1, 2, \ldots$, are called the *Fourier exponents* of f and $a_i \in \mathbb{C}$ are the Fourier coefficients of f. In the particular case when the number of Fourier exponents is finite, the function f reduces to a trigonometric polynomial.

Basic theorems and properties related to Fourier series of periodic functions, such as the shift, modulation, and differentiation properties of Fourier series, can be generalized to almost periodic functions as well. For instance, the Parseval formula in the case of almost periodic functions is given by [505]:

$$\sum_{i=0}^{\infty} |a_i|^2 = M\{|f|^2\}. \tag{10.13}$$

Finally, we caution the readers that in the current mathematical literature there are many additional types (and definitions) of almost periodic functions associated to various specialized normed spaces besides the ones originally introduced by Bohr and Besicovich [507]. The rationale for such numerous variations resides in their specific merits with respect to application fields that are beyond the scope of our introductory exposition. Interested readers are encouraged to consult our cited references, particularly the classic book by Schwarz and Spilker [436], where profound connections between almost periodic functions and arithmetic functions are discussed in great mathematical detail. The application of space-dependent quasiperiodic functions to aperiodic geometry and wave diffraction from quasicrystals will be discussed in Chapter 11.

10.2   Poisson Behavior in Number Theory

The epitome of the disordered behavior of discrete random systems is often identified with the Poisson distribution that governs the probability of a given number of independent events occurring with a known constant mean rate in a fixed time or space interval. For example, this statistical behavior is the one followed by the number of phone calls received by a call center per hour or by the number of nuclear disintegration events per second from a radioactive material. Therefore, it may at first appear surprising that the Poisson behavior frequently appears in number theory, where it is often discovered behind a large range of purely arithmetic phenomena. In the present chapter, we have already encountered the Poisson behavior in the prime omega function Ω that registers the total number of prime factors (with multiplicity) of a given number n, in the level spacings of the eigenvalues of certain random matrices, and among prime numbers, whose gaps, i.e., the differences between adjacent prime numbers, are also described by the Poisson distribution. The appearance of the Poisson distribution in the difference of random variables (i.e., in the level spacing) occurs because the levels themselves are Poisson distributed. In fact, a fundamental theorem of statistics states that the probability distribution of the sum (difference) of two or more independent random variables is the convolution (correlation) of their individual distributions, which, for the case of Poisson variables, is again a Poisson distribution. Note that this does not happen, for instance, when considering the difference of random variables that are uniformly distributed. In this case, we deduce that their difference will have a probability distribution with triangular shape, as it happens, for example, to the spacing of the decimal digits of π.

In the following section, we discuss the Poisson phenomenon in the context of uniformly distributed sequences (review Section 8.5), which is probably the simplest arithmetic setting in which it originates. Our presentation is strongly influenced by the analysis provided in the book by Miller and Takloo-Bighash, to which we refer interested readers for additional details [426].

10.2.1 Equidistributed Sequences and Poisson Behavior

We now investigate the sequence of the fractional parts $x_n = \{n^k \alpha\}$, where $k \geq 1$ is a fixed integer and $\alpha \in \mathbb{R}$ is a fixed real number. It is easy to realize that when $\alpha = p/q \in \mathbb{Q}$ the sequence x_n for $n = 1, \ldots, \infty$ is periodic with a period q. A more interesting situation arises when α is irrational. In this case, no two $\{n^k \alpha\}$ are equal and the x_n for $n = 1, \ldots, \infty$ forms a sequence of numbers that is dense[4] and equidistributed in the interval $[0, 1]$, or uniformly distributed modulo one. This follows immediately from the Weyl's criterion discussed in Section 8.5 since the associated exponential sums, which can be explicitly calculated using geometric series, are bounded [424]. A much more difficult question is related to the nature of the spacings between members of the sequence $x_n = \{n^k \alpha\}$. This problem originated in the context of quantum chaos in relation to the distribution of the energy levels of integrable systems [508]. We approach this problem by first considering as a reference case independent random variables uniformly distributed in $[0, 1]$, which also form equidistributed sequences. Arranging their values in increasing order, the distribution of their mth neighbor spacings displays the Poisson behavior described by [426]:

$$p(t) = \frac{t^{m-1}}{(m-1)!} e^{-t} \qquad (10.14)$$

---

[4] Remember that a sequence in $[0, 1]$ is dense if for all $x \in [0, 1]$ there is a converging subsequence. Equidistributed sequences are dense.

Given that the sequence $x_n = \{n^k \alpha\}$ and one of the uniform random numbers are both equidistributed in $[0, 1]$, we could ask if the spacings between their members also share the same Poissonian behavior characterized by the distribution in (10.14). It turns out that this is indeed the case but only for certain values of the parameters (k, α), depending on the Diophantine approximation properties of α [509, 510]. For instance, when $\alpha \in \mathbb{Q}$, the sequence of points becomes periodic and clearly their spacing does not feature Poissonian behavior. Therefore, we consider only irrational values of α and ask what distribution governs the spacings (wrapped in the unit interval) between adjacent y_n, which are the $x_n = \{n^k \alpha\}$ arranged in increasing order. Despite its apparent simplicity, this is still an open problem in number theory, which led to formulate the following *fundamental conjecture* [426]: for any $k \geq 2$ and for almost all (in the sense of measure theory), α the sequence $\{n^k \alpha\}$ features Poisson behavior in the limit $n \to \infty$. In particular, we see that for $\alpha \notin \mathbb{Q}$ the fractional parts $x_n = \{n\alpha\}$ are uniformly distributed modulo one but not Poissonian. Moreover, it can be shown that in this case there are only three possible values of nearest-neighbor distances, for each number N of terms in the sequence (they depend on α and N). Partial results toward this conjecture are proved by Rudnick et al. [509] and are beyond the limited scope of our discussion. This characteristic behavior is clearly displayed in Figure 10.2, where we compare the distributions of first- and second-neighbor spacings for $N = 10^4$ points of the sequences $x_n = \{n^k \alpha\}$ for $k = 2$ and $k = 1$. When $k = 2$, the normalized distributions are very well reproduced by the Poisson model in equation (10.14).

10.2.2   Arithmetic Randomness and Chaos

The type of erratic, random-like behavior exemplified in Section 10.2.1 appears universally in number theory and geometry, where it represents a distinctive route to aperiodic order, which we refer to as *arithmetic randomness*. Let us keep in mind, however, that no real randomness is involved in these situations since equidistributed sequences, or any other number-theoretic constructs displaying complex aperiodic behavior, are ultimately deterministic in nature since they always produce the same outputs given precisely defined initial conditions and model parameters. However, the numerical values of the outputs vary in a very complex, totally unpredictable way even for very small variations of parameters, producing a situation similar to the sensitive (exponential) dependence on the initial conditions that characterizes the unpredictable behavior of *deterministic chaotic dynamical systems*. However, while for continuous time systems chaotic behavior requires at least a three-dimensional phase space (e.g., the non-linear pendulum with periodic forcing), discrete-time systems can already exhibit chaos in one dimension, as it is the case even for the iterations of the fractional part of "mod 1" sequences [511, 512].

This peculiar behavior can be directly appreciated by considering the differences $\Delta_n(x)$ between the values of two equidistributed sequences with irrational parameters that differ by a very small amount. In particular, we consider the sequences $x_n = \{n^k \alpha\}$ and $x'_n = \{n^k \beta\}$, where α is equal to the golden mean and $\beta = \alpha + \epsilon$. For both

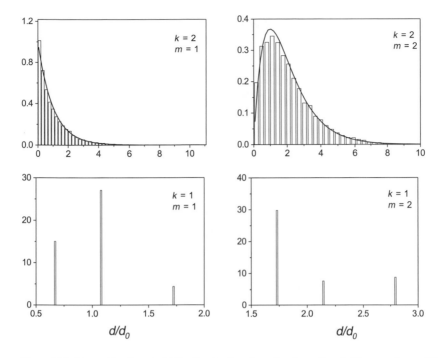

Figure 10.2 Normalized probability density functions for the mth neighbor spacings of the equidistributed sequence $x_n = \{n^k \alpha\}$ for two different values of k and m, specified in the figure. In these examples, α is equal to the golden mean. The continuous lines are the predictions from the equation (10.14). The quantity d_0 denotes the mean spacing between adjacent points.

sequences, we have fixed $k = 2$ in order to obtain the Poissonian behavior. Generating the first 10^5 elements of the sequences and sorting their values in ascending order, we plot the differences $\Delta_n(x)$ in Figure 10.3. We notice that the values $\Delta_n(x)$ fluctuate unpredictably even for very small values of $\epsilon = 10^{-8}$, with a characteristic behavior that resembles closely different realizations of Brownian motion (see Appendix C). Therefore, despite the fact that the two different trajectories of x_n and x'_n cannot diverge exponentially because their difference is bounded within the unit interval, this difference still fluctuates in an unpredictable way when the "initial parameters" α and β of the corresponding sequences are slightly perturbed. Since no randomness is present behind the peculiar unpredictability that we have observed in many contexts of number theory, we speak of "*arithmetic chaos*" instead of "arithmetic randomness," which is informally utilized by number theorists [510].

Finally, more complex sequences can be studied. However, no qualitatively novel phenomena have been discovered with respect to what is observed in the prototypical sequence $x_n = \{n^k \alpha\}$. For instance, the sequences of the fractional parts of $f(n)\alpha$ have been intensively investigated for certain types of $f(n)$. In particular, definite results exist for *lacunary* integer sequences, defined by the property that $\liminf \frac{f(n+1)}{f(n)} > 1$, and characterized by the presence of large gaps between adjacent values. A simple

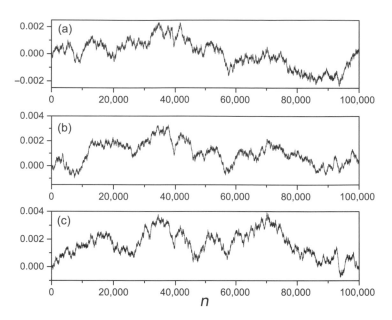

Figure 10.3 Differences $\Delta_n(x)$ between the values, sorted in ascending order, of $x_n = \{n^k \alpha\}$ and $x'_n = \{n^k \beta\}$, where α is the golden mean and $\beta = \alpha + \epsilon$. The values of ϵ are as follows: (a) $\epsilon = 10^{-8}$; (b) $\epsilon = 10^{-6}$; and (c) $\epsilon = 10^{-3}$. $k = 2$ in all cases.

example is to consider an integer $g \geq 2$ and set $f(n) = g^n$. Rudnick and Zaharescu [513] recently proved that for almost all values of α (in the sense of measure theory), the sequences $\{f(n)\alpha\}$ of fractional parts, with $f(n)$ lacunary integer sequences, are uniform modulo one and exhibit Poissonian behavior [513, 514].

10.2.3   Deterministic Arrays with Poisson Behavior

We now show how the arithmetic theory of equidistributed sequences can be naturally utilized to generate deterministic structures that nevertheless exhibit ideal Poisson behavior like the one of uncorrelated random systems. Let us point out again that the term "deterministic" characterizes systems in which randomness plays no role in the determination of future states, as is the case for the aperiodic order that appears naturally in number theory. The general phenomenon of "arithmetic randomness" discussed in the previous section is illustrated in Figure 10.4, where we compare the geometrical structures and the corresponding first-neighbor spacings statistics for a homogeneous spatial *Poisson point process* and for the deterministic planar point pattern generated by the equidistributed sequences $x_n = \{n^m \alpha\}$ and $y_n = \{n^k \beta\}$. In this case, we choose α equal to the golden mean and $\beta = \pi$, while $m = k = 2$ in order to achieve Poisson behavior along the x- and y-coordinates. The composition Poisson behavior along two orthogonal spatial directions produces an aperiodic distribution of points that uniformly fill the square interval $[0, 1] \times [0, 1]$, as shown in Figure 10.4.

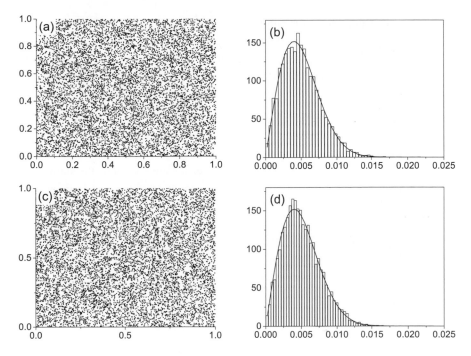

Figure 10.4 (a) Homogeneous Poisson point process (single realization). (b) Normalized probability density for the NNSD of the pattern shown in (a). The continuous line is the prediction based on equation (10.15). (c) Aperiodic point pattern with $N = 10^4$ points whose coordinates are generated based on the equidistributed sequences discussed in the text. (d) Normalized probability density for the NNSD of the pattern shown in (c). The continuous line is the prediction based on equation (10.15). Both patterns have $N = 10^4$ points.

Poisson point processes such as the one illustrated in Figure 10.4a are the hallmark of point process spatial statistics and found numerous applications in different branches of science and technology, including geostatistics, biology, neuroscience, ecology, astronomy, forestry, etc., where they serve as prototypical models for the random distributions of points in space [351]. Homogeneous Poisson point processes are also fundamental building blocks to construct more complex spatial processes and serve as ideal models to define *complete spatial randomness* (CSR), i.e., a uniform distribution of randomly scattered points in space without any correlations. Before proceeding further with our discussion of arithmetic randomness, let us first recall some basic facts about Poisson point processes, starting from the following definition: a homogeneous Poisson process N is a random process with the two properties:

- **Poisson distribution of counts**: the number of points within any bounded set Ω obeys the Poisson distribution with mean $\lambda \mu(\Omega)$, where $\mu(\Omega)$ denotes the measure (area) of the spatial region Ω and λ is a characteristic constant parameter called the *intensity* of the point process and describes the mean number of points within a unit volume.

- **Independence**: the number of points of $N(\Omega_n)$ in n disjoint sets Ω_n is n independent Poisson random variables with means $\lambda \mu(\Omega_n)$, for any arbitrary n. This property conveys the attribute of complete or pure spatial randomness and guarantees that there are no interactions (correlations) between the points of the pattern.

Since these properties are invariant under rotation and translation, homogeneous Poisson processes are stationary and isotropic random processes. In addition, Poisson processes are also invariant under linear transformations, with a suitable scaling of their intensity [351]. Many analytical results are known on the statistical properties of homogeneous Poisson point patterns, which are discussed in detail in the excellent books by Janine Illian et al. [351] and by Kallenberg [365]. The most important result for our purpose is the distribution function of the nearest-neighbor distance between points, or the nearest-neighbor spacing distribution (NNSD). In particular, the probability density function of the distances to the kth nearest neighbors for a Poisson process with intensity λ are given by the following:

$$d_k(r) = \frac{2(\lambda \pi r^2)^k}{r(k-1)!} e^{-\lambda \pi r^2} \qquad (10.15)$$

A detailed derivation of this important result can be found in [351]. We note that when $k = 1$, the resulting first-neighbor spacing distribution coincides with the *Rayleigh distribution* that we previously encountered in the theory of random walks. In particular, this distribution describes the statistics of the resultant sum of a large number of complex vectors (i.e., phasors) whose real and imaginary components are independently and identically distributed Gaussian variables with equal variance and zero mean [362].

We use the analytical result in Figure 10.4 to model the first-neighbor spacing distributions of the Poisson point process (a single statistical realization is shown in Figure 10.4b) and of the deterministic point pattern generated based on the equidistributed sequences introduced previously (Figure 10.4d). The agreement between the first-neighbor spacing data and the analytical Poisson model is remarkable, demonstrating how complete spatial randomness can be generated simply using uniform sequences modulo one. It is also clear that while Poisson point patterns require ensemble averaging over different realizations of disorder, this is not needed for the aperiodic patterns generated by uniform sequences modulo one due to their deterministic character.

10.2.4   Visualizing "Arithmetic Randomness"

The analysis of the aperiodic behavior of a real-valued sequence x_n can be greatly aided by the visualization of the corresponding phasor sum:

$$z_n = \sum_{n=1}^{N} e^{2\pi j x_n} \qquad (10.16)$$

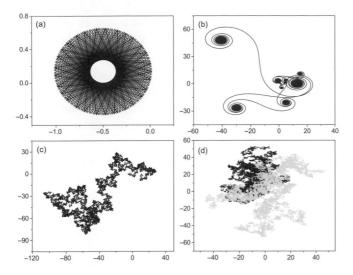

Figure 10.5 Walks ($N = 10^4$) on the following sequences: (a) $x_n = \{\sqrt{2}n\}$; (b) $x_n = \{\sqrt{2}n \log n\}$; (c) $x_n = \{\sqrt{2}n^2\}$; and (d) $x_n = \{\sqrt{2}p_n\}$, where p_n are prime numbers (the black line) and $x_n = \{\sqrt{2}u_n\}$, where u_n are zero-mean Gaussian random variables with unit variance (the gray line). As usual, curly brackets indicate the fractional parts.

The idea is to generate "walks" in the complex plane traced by the phasors in the preceding summation that encodes fundamental structural information on the associated real-valued sequence [457]. This can be achieved by constructing the polygonal line $L(u_n)$ of unit-length segments whose successive vertices are the points z_N in the complex plane defined by $z_N = \sum_{n \le N} \exp(2\pi j x_n)$. This direct visualization method is often used in number theory when calculations are too difficult to perform explicitly and can reveal hidden structural information. Moreover, this method is directly linked to Weyl's criterion, previously discussed in Section 8.5, according to which a sequence $\{x_n\}$ is uniformly distributed modulo one if and only if the associated phasor sum in (10.16) tends to zero when divided by N in the limit $N \to \infty$. As a consequence, the two-dimensional walks associated to equidistributed sequences cannot deviate significantly from the origin of the complex plane.

In Figure 10.5, we apply this geometric approach to sequences that exhibit profoundly different arithmetic phenomena. We first consider in Figure 10.5a walks on the equidistributed sequence $x_n = \{\sqrt{2}n\}$. We know already that the points x_n densely fill the unit interval because this sequence is uniformly distributed modulo one. Moreover, since $k = 1$, it does not manifest the random-like Poissonian behavior. This nonrandom character is immediately reflected by the very structured "star-like appearance" of the associated walk $L(u_n)$. The general shape of this walk depends only weakly on the choice of the irrational number α that appears in the sequence $x_n = \{\alpha n\}$. In fact, in the limit of $N \to \infty$, this walk consists of infinitely many branches confined within a dense annular region centered at $(-1 + j/\tan \pi\alpha)/2$ and bounded between the limiting circles with radii $1/|2 \tan \pi\alpha|$ and $1/|2 \sin \pi\alpha|$ [457]. In Figure 10.5b we show the walk associated to the sequence $x_n = \{\sqrt{2}n \log n\}$, also known to be uniformly distributed modulo one [457]. Despite the fact that its values appear to be more randomly scattered compared to the previous case, the walk shown in Figure 10.5b still exhibits a highly structured, regular shape consisting of several annular regions joined by almost straight segments. As discussed originally in [457], this behavior originates from the slow asymptotic growth of the logarithmic function that makes the walk behave locally as $x_n = \{\alpha_N n\}$, where $\alpha_N \approx \{\alpha \log N\}$ around $n = N$. The jumps connecting the different annular regions are due to the values of N for which α_N is approximately an integer.

A qualitatively novel behavior is observed when studying walks on purely Poissonian sequences or on the distribution of primes. An example is illustrated in Figure 10.5c for the sequence $x_n = \{\sqrt{2}n^2\}$. The corresponding walk has no spiral arms and resembles a single realization of a random walk process, revealing the presence of arithmetic randomness. Furthermore, in Figure 10.5d we show the walk over the sequence $x_n = \{\sqrt{2}p_n\}$, where p_n are prime numbers. This walk (the black line) is also qualitatively similar to the single realization of a random walk process, shown by the gray line for a direct comparison. Note in particular the remarkable difference between the curves in Figure 10.5b,d. According to the prime number theorem (PNT), we have $p_n \sim n \log n$, which would lead us to think (erroneously) that the two curves should behave very similarly. The reason why this is not the case is that the large-scale, global (i.e., asymptotic) regularity of primes does not sufficiently constrain their irregular behavior at smaller scales. Primes, similar to gas molecules confined within a geometrical enclosure, appear to behave as randomly as possible given their arithmetical constraints.

10.3   Prime Numbers and Aperiodic Arrays

Let us now turn our attention again to the distinctive aperiodic complexity of prime number distributions. In particular, we discuss higher-dimensional generalizations of the familiar concept of prime number because they enable deterministic constructions of point patterns that support strongly localized optical resonances, spectral gaps, as well as distinctive critical modes with multifractal intensity fluctuations [10, 205]. In the next section, we introduce fundamental concepts related to prime numbers in quadratic fields and the corresponding point patterns. Additional information can be found in the classic algebraic number theory books by Lang [313] and Cohen [314, 515] or in the paper by Dekker [315].

10.3.1   Quadratic Fields and Integer Rings

The first attempts at extending the properties of integers and rationals to more abstract number-theoretic systems can be found in Euler's work on the solutions of Diophantine equations and in Gauss's work on quadratic reciprocity, quadratic forms, and cyclotomic fields, which mark the early history of algebraic number theory [516]. Stimulated to by the search for higher reciprocity laws and the solution of Fermat's last theorem, *algebraic field theory* grew into a vast area of number theory with fundamental connections to many other branches of mathematics [482, 517]. Given our introductory scope, we will limit our discuss to few elementary aspects with impact on the lattice theory of aperiodic systems. Readers interested in a systematic discussion of algebraic number theory and its relation to contemporary research in mathematics are referred to the excellent textbooks [313, 314, 482, 515, 517]. Quadratic fields and their rings of integers can be naturally represented in the two-dimensional plane, giving rise to very complex aperiodic patterns whose (Fourier) spectral properties still defy analytical comprehension [315]. Specifically, prime arrays are based on the aperiodic distribution of the prime elements in the rings of imaginary quadratic integers, as we will explain in this section.

Quadratic fields play a fundamental role in *algebraic number theory* (ANT) where number-theoretic questions are generalized to more abstract algebraic structures such as rings and ideals. Let us introduce these important concepts. A quadratic field is a degree 2 extension of the field of rational numbers $\mathbb{Q}$. This means that if we consider $d \neq 1$ (called the *radicand*) to be a specified square-free integer, then the sets of numbers

$$\mathbb{Q}(\sqrt{d}) = \{a + b\sqrt{d} \mid a, b \in \mathbb{Q}\} \tag{10.17}$$

form a *quadratic field* with basis $\{1, \sqrt{d}\}$. Note that this set of points is also a linear vector space over $\mathbb{Q}$, spanned by the linear combinations of the basis $\{1, \sqrt{d}\}$ with rational coefficients. The dimension of such a vector space over the field $\mathbb{Q}$ is called the *degree* of the field extension, which is equal to 2 in the case of quadratic fields. Quadratic fields with $d > 0$ are called *real quadratic fields* while the ones with $d < 0$ *imaginary quadratic fields*. For example, if $d = -1$, the corresponding field is known as the field of *Gaussian rationals*. In our discussions, we will deal exclusively with imaginary quadratic fields because their integer elements can be naturally interpreted as the coordinates of discrete points forming remarkable two-dimensional lattices.

Since $\mathbb{Q}(\sqrt{d})$ forms a field, we can add and subtract its elements and obtain additional elements in $\mathbb{Q}(\sqrt{d})$. Moreover, we can multiply and divide (by nonzero numbers) the elements of $\mathbb{Q}(\sqrt{d})$ without ever stepping outside it. Moreover, due to its field nature the sum and product operations in $\mathbb{Q}(\sqrt{d})$ obey the usual distributive, associative, and commutative arithmetic laws that are satisfied over the familiar field of rational numbers. In order to deal with discrete point sets, we address the *ring structures* associated to quadratic fields because they naturally generalize the properties of integer numbers to higher dimensions.

Box 10.1 What Are Algebraic Numbers?

We introduce the concept of an *algebraic number* as any complex number that is the root of a nonzero polynomial in one variable with integer coefficients:

$$a_n X^n + a_{n-1} X^{n-1} + \cdots a_1 X + a_0. \tag{10.18}$$

Therefore, all integers and rational numbers are algebraic numbers, as well as are all roots of integers. Algebraic numbers are closed under standard operations, i.e., the sum, difference, product, and quotient (with a nonzero denominator) of two algebraic numbers are again algebraic. Therefore, algebraic numbers form a *field*. However, there are real and complex numbers, such as π and e, that are not algebraic. Instead, they are *transcendental numbers*. The set of algebraic numbers is countable and has measure zero (in the sense of the Lebesgue measure) as a subset of the complex numbers. Therefore, almost all complex numbers are transcendental.

We can now define the important subset of the *algebraic integers*, as those complex numbers are the roots of a polynomial in the following form:

$$X^n + a_{n-1} X^{n-1} + \cdots a_1 X + a_0 \tag{10.19}$$

for some $n \geq 1$ and a_i are integer coefficients. Note that here the leading coefficient in the polynomial must be equal to 1. The sum, difference, and product of algebraic integers are again algebraic integers, implying that the algebraic integers form a ring. Therefore, we can say that algebraic numbers are to algebraic integers what rational numbers are to standard integers. For example, any integer number n is also an algebraic integer because it is a root of the linear, integer-coefficient polynomial $X - n$. The complex number $2 + 5j$ is also an algebraic integer because it solves the equation $X^2 - 4X + 29 = 0$. A noteworthy example of a real algebraic integer that plays a fundamental role in Diophantine approximation and in diffraction theory is the Pisot–Vijayaraghavan number, also called the *Pisot number* or PV number (see discussion in Section 11.5). In general, we can determine which elements of $\mathbb{Q}(\sqrt{d})$ are algebraic integers and which form they have by studying the mod4 divisibility properties of d. Finally, it is important to realize that algebraic numbers and integers can be defined in any number field. For instance, if K is a number field, its ring of integers is the subring of algebraic integers in K, which is often denoted as $\mathcal{O}_K$.

One of the goals of ANT is to investigate algebraic structures with interesting properties starting from quadratic fields $\mathbb{Q}(\sqrt{d})$. It is precisely this goal that led mathematicians in the nineteenth century to develop the concept of *algebraic integers*, which share many properties with the usual set of integers (see Box 10.1). However, it is important to realize that the rings associated to a quadratic field $\mathbb{Q}(\sqrt{d})$ do not generally admit a well-defined notion of primality and lack the possibility of unique factorization, i.e. the fundamental theorem of arithmetic may fail in the rings of generalized integers! A classical example is provided by the integer ring associated to the quadratic field $\mathbb{Q}(\sqrt{-5})$, which comprises all the complex numbers of the form $a + jb\sqrt{5}$ where a and b are integers. In this set there is no unique factorization because, for instance, the number 6 can be written in more than one way as $6 = 3 \cdot 2 = (1 + \sqrt{-5}) \cdot (1 - \sqrt{-5})$ and these are two different factorizations in the ring.

Box 10.2 What Are Algebraic Rings and Fields?

Consider a set of integer numbers $\mathbb{Z}$ with the usual operations of summation and multiplication. In $\mathbb{Z}$, we can add, subtract, and multiply elements (but not divide!) and always obtain elements of $\mathbb{Z}$. Moreover, the sum and product operations are connected by the distributive law. These familiar properties define any ring structure. More generally, a (commutative) *algebraic ring* (with identity) is a set R with two binary operations, denoted by $+$ (addition) and $\cdot$ (product), such that

- $\{R,+\}$ forms an *Abelian group*.

In addition, the product operation $\cdot$ satisfies the following properties:

- There is a multiplicative identity $1 \in R$ such that $1 \cdot a = a \cdot 1 = a$ for every element $a \in R$.
- The product operation is *associative*, i.e., for all $a, b, c \in R$, we have the following: $a \cdot (b \cdot c) = (a \cdot b) \cdot c$.
- The product operation is *commutative*, i.e., for all $a, b \in R$ we have the following: $a \cdot b = b \cdot a$.

Finally, in a ring the $+$ and $\cdot$ operation are linked by the *distributive law*, i.e., for all $a, b, c \in R$ we have: $a \cdot (b+c) = a \cdot b + a \cdot c$. We should note that every element in a ring has an additive inverse but not every nonzero element is required to have a multiplicative inverse. For instance, in the ring of integers $\mathbb{Z}$, only the elements ± 1 have a multiplicative inverse. Another important example is the set $\mathbb{Z}/n\mathbb{Z} = \{0, 1, 2 \ldots, n - 1\}$, which, with the usual addition of a multiplication modulo n, forms a ring is called the ring of integers modulo n. Moreover, the collection of all polynomials with coefficients in $\mathbb{Z}$ forms a ring with the polynomial addition and multiplication denoted by $\mathbb{Z}[x]$. More generally, if R is any ring (finite or infinite), we can form the ring of polynomials whose coefficients are in R. We now introduce the concept of an *algebraic field*. A commutative ring in which every nonzero element has a multiplicative inverse is called a field. For example, $\mathbb{Q}$ with the usual multiplication and addition operations forms a field. An important finite field in number theory is $\mathbb{F}_p \equiv (\mathbb{Z}/p\mathbb{Z}) = \{0, 1, 2, \ldots, p - 1\}$ with p a prime number (with addition and multiplication as usual). Important fields for cryptography applications are the ones obtained from rings of polynomial with coefficients in a field, particularity in a finite field. In these cases, all the familiar properties of $\mathbb{Z}$ carry over to the polynomial ring $\mathbb{F}[x]$. Polynomial fields and their role in the generation of aperiodic order will be addressed in Section 10.4.

The realization that the fundamental theorem of arithmetic could fail in certain algebraic rings created a significant crisis in nineteenth century mathematics. Historically, this problem became manifest when Lamé announced in 1847 a proof of Fermat's last theorem, (wrongly) assuming the validity of the fundamental theorem of arithmetic in the cyclotomic ring $\mathbb{Z}[e^{2\pi j/n}]$ for every $n \geq 1$. We recall that this famous theorem, which greatly contributed to the development of modern ANT, states that the equation $x^n + y^n = z^n$ has no integer solutions in x, y, z for $n \geq 3$. It was the German mathematician Ernst Kummer that settled this issue by proving that indeed unique factorization fails in some cases (the first being $n = 23$, implying that Lamé's solution was wrong), but it can be recovered when working with more abstract elements that he called ideal numbers. This was the first step in establishing the modern theory of *ideals* (see Box 10.3), which plays a prominent role in ANT. In particular, the pioneering contributions by Kummer and Dedekind, largely stimulated by the many unfruitful attempts to prove Fermat's last theorem,[5] culminated in the formulation of a generalized version of the fundamental theorem of arithmetic that is valid for ideals, giving birth to the modern ANT [517].

For the purpose of generating aperiodic point patterns of scattering particles that reflect the distributions of prime elements in the integer rings of quadratic fields, we are naturally led to consider the rings of algebraic integers $\mathcal{O}_{\mathbb{Q}(\sqrt{d})} \subset \mathbb{Q}(\sqrt{d})$. An important question is how to properly construct a ring of algebraic integers starting from an algebraic field. An important result of number theory [482] provides a convenient characterization for the integer rings of quadratic fields as follows: if $d \equiv 1 \bmod 4$, then $\mathcal{O}_{\mathbb{Q}(\sqrt{d})} = \mathbb{Z}\left[\frac{1+\sqrt{d}}{2}\right]$, otherwise $\mathcal{O}_{\mathbb{Q}(\sqrt{d})} = \mathbb{Z}[\sqrt{d}]$. As important examples, we consider $\mathcal{O}_{\mathbb{Q}(\sqrt{-1})} = \mathbb{Z}[i] \subset \mathbb{Q}(\sqrt{-1})$, explicitly defined by the following set:

$$\mathbb{Z}[i] = \{a + jb \mid a, b \in \mathbb{Z}\} \tag{10.20}$$

where $j = \sqrt{-1}$ denotes the imaginary unit. This set of complex numbers, with the familiar complex summation and multiplication operations, forms the ring of *Gaussian integers*. These were originally introduced by Gauss in relation to the problem of *biquadratic reciprocity*, which establishes a relation between the solutions of the two biquadratic congruences $x^4 \equiv q \bmod p$ and $x^4 \equiv p \bmod q$. Generalized reciprocity laws can also be obtained for cubic congruences, leading to the cubic reciprocity law that is more naturally proven in the ring $\mathcal{O}_{\mathbb{Q}(\sqrt{-3})} = \mathbb{Z}\left[\frac{1+\sqrt{-3}}{2}\right] \subset \mathbb{Q}(\sqrt{-3})$, often denoted as $\mathbb{Z}[\omega]$, defined by the following:

$$\mathbb{Z}[\omega] = \{a + e^{2\pi j/3}b \mid a, b \in \mathbb{Z}\} \tag{10.21}$$

This important ring is known as the ring of the *Eisenstein integers*. Gauss also proved that unique factorization occurs in these rings.

---

[5] In 1847, Kummer ended up proving the validity of Fermat's last theorem for $n = p$ a prime number exponent. The full solution of this theorem had to wait until 1995 when Andrew Wile's proof, originally announced in a lecture in Cambridge in 1993, was finally published.

Box 10.3 What Are Ideals?

A central result of ANT is that *every ideal is a unique product of prime ideals*. But what are ideals in the first place? Here we introduce the important ring-theoretic concepts that underpin the factorization theory of algebraic integers by focusing on the key concept of an *ideal*, as precisely formulated by Richard Dedekind following the initial work of Kummer.

An ideal I is a special subset of a ring that generalizes the structural properties of certain subsets of the integers, such as the even numbers or the multiples of a given integer number. Note that the addition and subtraction of even numbers results in even numbers, thus preserving the property of *evenness*. Moreover, multiplying an even number by any other integer number (not necessarily even) always results in an even number. So the set of even numbers is closed with respect to addition/subtraction and also *absorbs* odd numbers via multiplication. These two simple observations exemplify the defining properties of a general ideal. More formally now, an ideal is a nonempty subset I of a ring R (here assumed commutative) such that if $r, s \in I$, then $r - s \in I$ and, if $r \in R$ and $s \in I$, then $rs \in I$. As we have seen, the even integers form an ideal in the ring of all integers. Similarly, the set of all integers divisible by a fixed integer n is an ideal, which is denoted by $n\mathbb{Z}$. There are many types of ideals. *Principal ideals* consist of the multiples of a single nonnegative number. Of paramount importance in ANT are *prime ideals* because they share many important properties of the prime numbers in the ring of integers (i.e., the familiar primes). An ideal I of a commutative ring R is prime whenever the two following properties are satisfied: (i) if $a, b \in R$ such that their product ab is an element of I, then a is in I or b is in I and (ii) I is not the whole R. You should note that the preceding definition generalizes the well-known property of prime numbers: if p is a prime number that divides the product ab of two integers, then p divides a or p divides b. Finally, ideals are used to construct the *quotient ring* R/I of the ring R by the ideal I. The elements of R/I are the *cosets* $I + r$ of the additive group of I in R. For example, if $n\mathbb{Z}$ denotes the set of integer multiples of $n \in \mathbb{Z}$, then $\mathbb{Z}/n\mathbb{Z}$ is isomorphic to the ring of integers modulo n, denoted by $\mathbb{Z}_n$. This situation is completely analogous to the way in which, in group theory, a normal subgroup can be used to construct a quotient group (see Box 9.5).

These two rings are examples of imaginary quadratic rings ($d < 0$) and are useful structures to investigate in relation to light scattering from two-dimensional aperiodic arrays. In fact, they naturally give rise to interesting geometrical patterns in the two-dimensional complex plane whose metric properties are described by the multiplicative *norm function* $N(\alpha) = \alpha\alpha'$, where α' is the complex conjugate of α in $\mathbb{Q}(\sqrt{d})$. In a quadratic field the *conjugate* of the element $\alpha = a + b\sqrt{d}$ is defined as $\alpha' = a - b\sqrt{d}$ and the norm becomes $N(\alpha) = |a^2 - db^2|$, where the absolute value ensures the positivity. In the rings of integers, the norms are given as follows:

$$N(\alpha) = \begin{cases} \left| a^2 + ab - \dfrac{d-1}{4}b^2 \right| & \text{when } d \equiv 1 \bmod 4. \\[2mm] \left| a^2 - db^2 \right| & \text{when } d \equiv 2, 3 \bmod 4. \end{cases} \qquad (10.22)$$

Note in particular that the norm for a Gaussian integer $\alpha = a + jb$ is given by $N(\alpha) = a^2 + b^2$, while for an Eisenstein integer $\alpha = a + \omega b$, with $\omega = (1 + \sqrt{-3})/2$, it reduces to $N(\alpha) = a^2 + ab + b^2$.

Since the norm is a multiplicative function, the norm of a product of different elements equals the product of their norms. Note that the elements of the rings associated to real quadratic fields $(d > 0)$ form dense sets on the real line, i.e., they are not discrete lattices. However, we can always represent the element $a + b\sqrt{d}$ of any quadratic field by the point with coordinates $(a, b\sqrt{|d|})$ in the plane. In particular, the set of points corresponding to the quadratic integers form a discrete lattice with basis vectors $[(1,0), (0, \sqrt{|d|})]$ if $d \equiv 2, 3 \bmod 4$ and $[(1,0), (1/2, 1/2\sqrt{|d|})]$ if $d \equiv 1 \bmod 4$. Here we focus on imaginary quadratic rings and ask which one admits a unique factorization into irreducible or prime elements, similarly to the Gaussian and Eisenstein cases. Equivalently, we ask which imaginary quadratic ring is also a *unique factorization domain (UFD)*. Unique factorization domains are commutative rings in which every nonzero element different from a unit[6] can be written as a product of irreducible elements (see the definition in Section 10.3.2) uniquely up to ordering and multiplication by units, in a way that is formally analogous to the fundamental theorem of arithmetic for standard integer numbers. Remarkably, it is possible to show that there are only nine (complex) quadratic integer rings that are also UFDs. These are the ones corresponding to the following radicand values $d = -1, -2, -3, -7, -11, -19, -43, -67, -163$ [482]. This important result is known as the *Baker–Stark–Heegner theorem* and the preceding nine values of d are known as the *Heegner numbers*. The first four Heegner numbers identify more specialized algebraic structures called *Euclidean domains*, which are UFDs additionally characterized by the existence of a division algorithm with respect to their norm. On the other hand, when considering real $(d > 0)$ quadratic fields, it is presently unknown if there exists a finite or an infinite number of corresponding rings of integers with the unique factorization property.

10.3.2   Primes and Irreducibles

When dealing with algebraic number fields, the usual concept of primality requires a more precise definition. In particular, we need to distinguish between irreducible elements and prime elements.

---

[6] A unit of an algebraic number field or of its ring of integers is an integer of the field whose inverse is also an integer of the field. This is the set of numbers with unit norm. Two integers α and β of an algebraic number field are *associates* of each other when $\beta = u\alpha$, where u is a unit of the field.

Box 10.4 Quadratic Residues, Legendre Symbol, and the Reciprocity Law

Consider a nonzero rational integer q (i.e., a nonzero element of $\mathbb{Z}$) and a rational prime p that does not divide q. We say that q is a quadratic residue modulo p if there exists a natural number n such that: $n^2 \equiv q \bmod p$. Otherwise, the number q is called a quadratic nonresidue modulo p. For odd p, there are exactly $(p-1)/2$ quadratic residues and the same number of quadratic nonresidues. The product of two quadratic residues is a quadratic residue, the product of a quadratic residue and a quadratic nonresidue is a nonresidue, and of two quadratic nonresidues is a quadratic residue. The distribution of quadratic residues modulo a prime has correlation properties that closely resemble those of white-noise random sequences [8]. However, it also respects the "fundamental symmetry" due the reciprocity law discussed later. For an odd prime p and any rational integer q, we introduce the *Legendre symbol* $\left(\frac{q}{p}\right)$ with value $+1$ if q is a quadratic residue and with value -1 if q is a quadratic nonresidue modulo p. The Legendre symbol allows one to express very concisely the *quadratic reciprocity* theorem:

$$\boxed{\left(\frac{p}{q}\right)\left(\frac{q}{p}\right) = (-1)^{[(p-1)/2][(q-1)/2]}} \tag{10.23}$$

where p and q are distinct, odd natural primes. Notice that if at least one of them is of the form $4n + 1$ (i.e., it is congruent 1 mod 4), then

$$\left(\frac{p}{q}\right) = \left(\frac{q}{p}\right), \tag{10.24}$$

and p is a quadratic residue modulo q if and only if q is a quadratic residue modulo p. When p and q are of the form $4n + 3$, we deduce that p is a quadratic residue modulo q if and only if q is a quadratic nonresidue modulo p. These profound statements, both summarized in the equation (10.23), provide conditions for the solutions of quadratic congruences and were originally proved by Gauss, who referred to this elegant result in his *Disquisitiones Arithmeticae* as to his "fundamental theorem." Precisely the search for "higher reciprocity laws" in other rings stimulated the development of class field theory and culminated in the discovery of the most general reciprocity law by Emil Artin in 1923, for which all other laws follow as special cases.

An integer of a quadratic field is a composite if it is the product of two non-units integers of the field. On the other hand, an integer of an algebraic number field is called *irreducible* if it is not composite and not a unit. The irreducible elements are therefore only divisible[7] by themselves and their associates in addition to the units of the field. Importantly, a nonzero element π of the ring of integers of a given quadratic field is called a *prime* if it is not a unit and, whenever it divides the product of any two integers α and β, then it is a divisor of α or a divisor of β. Irreducible elements and primes coincide in the usual ring of integers $\mathbb{Z}$ but denote different objects in rings

---

[7] Divisibility is implicitly understood within the ring of integers of the considered field.

that are not UFDs. In particular, every prime is irreducible, but the converse is not generally true. Only when working in a UFD it becomes true that every irreducible is also a prime.

Generally, the integers of an algebraic number fields do not form a UFD and the fundamental theorem of arithmetic remains valid only at the level of ideals. In these cases, instead of numbers factorized in terms of irreducible or prime elements, we should instead consider ideals factorized in terms of prime ideals (see Box 10.3). The norm of an integer α of a quadratic field entirely determines if it is an irreducible element (or a prime element in case of UFDs). In fact, an important result of algebraic number theory states that if $N(\alpha)$ is a rational prime, i.e., a standard integer prime in $\mathbb{Z}$, then α is a prime in $\mathbb{Q}(\sqrt{d})$ (or, more generally, an irreducible element if not in a UFD) [313]. On the other hand, if the integer p is a rational prime (standard prime), it may or may not be a prime in the a UFD $\mathbb{Q}(\sqrt{d})$. If it is, it is called an *inert prime*. To determine which rational prime is inert depends on the divisibility properties of the *discriminant* δ of the quadratic field, defined by the following:

$$
\delta = \begin{cases} d, & \text{if } d \equiv 1 \bmod 4. \\ 4d, & \text{otherwise.} \end{cases}
\tag{10.25}
$$

In particular, an odd rational prime p will be inert if it does not divide the discriminant δ and the Legendre symbol $\left(\frac{\delta}{p}\right)$ equals -1 (see Box 10.4 for a definition of the Legendre symbol). The rational prime 2 is dealt with separately [482]. Precise characterization criteria for the prime elements in UFDs require the important concept of *quadratic residue*, which we introduce in Box 10.4. For instance, in the ring of Gaussian integers ($d = -1, \delta = -4$) the primes are (i) the numbers α whose norm is a natural prime congruent 1 mod 4; (ii) the natural primes p congruent 3 mod 4 (p is inert in the Gaussian integers) and associates; and (iii) the numbers α whose norm is 2 and their associates. Similar criteria hold for other quadratic fields [315]. The units of quadratic fields form a multiplicative group. The Gaussian integers have the four units $\mathbb{Z}[i]^{\times} = \{\pm 1, \pm i\}$, while the Eisenstein integers have the six units $\mathbb{Z}[\omega]^{\times} = \{\pm 1, \pm \omega, \pm \omega^2\}$. The rings of integers associated to complex quadratic fields with $d = -2$ and $d < -3$ have two units, i.e., the set $\mathcal{O}^{\times}_{\mathbb{Q}(\sqrt{d})} = \{\pm 1\}$ [482]. On the other hand, real quadratic fields ($d > 0$) have an infinite number of units! Examples of the distinctively aperiodic point patterns corresponding to the complex plane distributions of prime elements in complex quadratic fields are illustrated in Figure 10.6, where we plot finite patches of the Eisenstein and Gaussian primes. The structural and light scattering properties of electric point dipoles arranged according to these aperiodic patterns have been recently investigated by Wang et al. [205] using Green's matrix spectral theory, which we thoroughly discuss in Chapter 12. In particular, this study demonstrated that existence of critical modes in such structures, which are spatially extended fractal modes with long lifetimes, and discovered spectral gaps in the distribution of scattering resonances reflecting the distinctive degree of structural order in such aperiodic systems. More recently, Sgrignuoli et al. [10] experimentally investigated the intensity fluctuations in the optical resonances of primes in quadratic

(a)   (b)

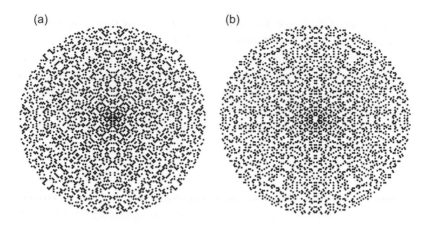

Figure 10.6 Distribution on the complex plane of the (a) $N = 3,129$ Gaussian primes and
(b) $N = 3,276$ Eisenstein primes. Notice that since multiplication by a unit and complex
conjugation both preserve primality, the arrays exhibit characteristic $4 \times 2 = 8$-fold symmetry
and $6 \times 3 = 12$-fold symmetry. However, these arrays display a very regular structure that
coexists with their lack of periodicity.

and quaternion rings using TiO_2 nanopillars deposited atop transparent substrates
and demonstrated multifractal scaling of classical waves in the optical regime. In
addition, the connection between these structures and the formation of rogue wave
phenomena in the linear optical regime has been addressed in [350]. Besides its fundamental interest, the fruitful connubium between algebraic number theory and photonic
structures provides a novel mechanism to tailor wave transport and photon states at
multiple length scales with applications to active photonic devices and broadband
nonlinear components [317]. Moreover, such an approach can readily be transferred
to semiconductor electronics and artificial atomic lattices, enabling the exploration
of novel quantum phases and many-body phenomena that emerge directly from the
fundamental structures of algebraic number theory.

10.3.3   Moat Problems and Primes Constellations

Understanding the structural properties of aperiodic arrays constructed based on prime
number distributions is not only relevant to number-theoretic questions but can also
provide information on the nature of wave transport through these complex systems.
Of particular interest are results on asymptotic density, the distribution of local patterns, and their spatial correlations. In the following, we address some examples beginning with the "moat problem" for the Gaussian and Eisenstein primes. The moat
problem asks whether it is possible to walk to infinity by taking steps of bounded
length on primes. When considering rational primes (prime numbers in $\mathbb{Z}$), this is
not possible since $\{(n + 1), ! + i\}$ with $i = 2 \ldots, n + 1$ are n consecutive composite
integers. In other words $\lim \sup\{p_{n+1} - p_n\} = \infty$, i.e., there are arbitrarily large gaps
in the primes. The moat problem can be extended to Gaussian primes or to integers

in other quadratic fields, where it is concerned with the characterization of prime-free regions that encircle the origin of the complex plane similarly to the moats that surround castles. The *Gaussian moat problem* was posed originally by Basil Gordon in 1962 at the International Congress of Mathematicians held in Stockholm, when he asked whether or not there exist k-moats (i.e., moats of width k) in the Gaussian integers for arbitrarily large k [518]. The Gaussian moat problem remains unsolved today. Gethner et al. [518] showed in 1998 that there is no walk of step length $\leq\sqrt{26}$ starting from the origin. More recently, Tsuchimura provided computational results that showed the existence of moats up to size $\leq \sqrt{36}$ [519].

Ilan Vardi developed a probabalistic model for Gaussian primes that generalizes Cramér's model for rational primes and then applied percolation theory to deduce the existence of a critical step size for percolation to infinity [520]. This gives a heuristic reason to believe that a walk to infinity on the Gaussian primes is not possible since the probability for a lattice point to be a prime, which is proportional to the asymptotic *density of Gaussian primes* $\rho_G \approx 2/(\pi \log x)$ in a disk or radius x, becomes smaller than any fixed value. On the other hand, Eisenstein primes have a higher asymptotic density $\rho_E \approx 3\sqrt{3}/(2\pi \log x)$ than Gaussian primes, possibly resulting in smaller moats. The moat problem for the Eisenstein primes has been recently investigated by West and Sittinger, who computationally demonstrated the existence of moats with width $\leq\sqrt{16}$ [521].

The distinctive aperiodic geometry of the primes in complex quadratic fields has direct implications on the properties of wave transport. In particular, the existence of arbitrarily large moats in the Gaussian and Eisenstein primes would result in light trapping (cavity effects) in physical systems realized by replacing each prime with a scattering center [10, 205]. The localization behavior of optical waves in such systems is discussed in Chapter 12.

It is also very interesting to explore the appearance of local patterns of primes, called *constellations*. In the familiar situation of rational primes, we say that a given pattern of primes is *admissible* if in principle it can occur infinitely often.[8] A famous conjecture due to Hardy and Littlewood, known as the *prime k-tuples conjecture*, states that any admissible pattern of primes appears infinitely often. For example, the pattern $(n, n + 2)$ is admissible while the triplet $(n, n + 2, n + 4)$ is not because one of its members is always divisible by 3. Asserting that there are infinitely many integer values n such that the pattern $(n, n + 2)$ contains two primes is equivalent to the still-unproven twin prime conjecture (TPC). A similar set of questions on the admissibility of local (two-dimensional) clusters of primes can be extended to the Gaussian primes. In this context, connected clusters of n primes are named *n-animals* [522]. Maximal admissible animals are those that become inadmissible by adding just an extra element anywhere, and are called *lions*. Using computer search algorithms

---

[8] This does not imply that it will actually occur infinitely often. It only means that it might occur.

and heuristic arguments, Renze et al. [523] were able to identify all the possible lions of the Gaussian primes, which, up to symmetry, are exactly 52. All other admissible animals are congruent to a subset of these 52 fundamental clusters.

Once prime constellations have been identified as admissible, it becomes important to understand what is their probability of occurrence. For rational primes, a classical derivation hinges on the important theorem of Mertens:

$$\prod_{p \leq x} \left(1 - \frac{1}{p}\right) \sim \frac{e^{-\gamma}}{\log x} \tag{10.26}$$

where p denote prime numbers and $\gamma = 0.5772\ldots$ is Euler–Mascheroni constant that is defined as the limit of the difference between the harmonic series and the natural logarithm.[9] Using Mertens's theorem within a simple random model for primes, it is possible to obtain the asymptotic probability for the occurrence of an n-tuple of primes below x as follows [8]:

$$W_n(x) \sim \frac{C_n}{\log^n x}, \tag{10.27}$$

where C_n denotes the Hardy–Littlewood constant for the n-tuple. In the special case of $n = 2$, we obtain the probability of twin primes with $C_2 \approx 0.66016\ldots$ and the counting function, i.e., the number of primes $p \leq x$ such that $p + 2$ is also a prime:

$$\pi_2(x) \sim 2C_2 \frac{x}{(\log x)^2} \sim 2C_2 \int_2^x \frac{dt}{(\log t)^2}. \tag{10.28}$$

Mertens's theorem (10.26) has been extended to Gaussian integers enabling the heuristic computation of the probability that an admissible Gaussian constellation of size n occurs in the Gaussian primes. The detailed calculations, which are quite straightforward, can be found in the paper by Renze et al. [523] that generalizes the classical prime k-tuples conjecture to $\mathbb{Z}[i]$. The main result is that the asymptotic number of occurrences of a connected n-element set of Gaussian primes within a disk of radius R from the origin in the first octant is as follows:

$$\frac{C_{Gn} 2^{n-3} R^2}{\pi^{n-1} \log^n R}, \tag{10.29}$$

where C_{Gn} is the Gaussian Hardy–Littlewood constant for the n-element constellation. The validity of the preceding formula for the number of occurrences of Gaussian primes' constellations has been tested numerically up to $R \leq 2 \times 10^5$, showing excellent accuracy for the two smallest lions, i.e., the "diamond cluster" ($n = 4$) and the 12-prime lion called "the castle."

---

[9] Despite its appearance in many important equations of mathematics and physics, the nature of the Euler–Mascheroni constant eluded centuries of investigations. Presently, we do not even know if it is a rational or an irrational number.

10.4   Finite Fields

In this section, we discuss the peculiar type of aperiodic order and spatial randomness that originate in certain algebraic *field structures* with a finite number of elements, also called *Galois fields*. As previously introduced in Box 10.2, a field is a set of elements with addition, subtraction, multiplication, and division (except by 0) operations that satisfy the usual commutative, associative, and distributive laws. Familiar examples of fields with an infinite number of elements are the real, complex, and the rational numbers. On the other hand, any residue system modulo a prime number p, which we indicate by $\mathbb{F}_p \equiv (\mathbb{Z}/p\mathbb{Z})$, forms a field with p elements with respect to the usual operations of addition and multiplication modulo p. This follows from the basic fact that when p is a prime, every nonzero element has a multiplicative inverse in $\mathbb{F}_p$. Finite fields of prime order p are also called *prime fields* and are denoted by $GF(p)$. Their number of elements p is called the *characteristic* of the field.[10] It is easy to prove that every finite field has prime characteristic.[11]

Uniform randomness strikingly appears as a fundamental attribute of finite fields that is often exploited for the generation of pseudorandom numbers. This can already be understood in $\mathbb{F}_p$, when looking at the distribution of its modular inverses, i.e., the values $\bar{x} = x^{-1} \bmod p$ as x advances linearly through $1, 2, \ldots, p-1$. The erratic distribution of the modular values is shown as an example for $\mathbb{F}_{701}$ in Figure 10.7a. Figure 10.7b shows the histogram plot of the probability density function for the computed values, indicating a distribution similar to the one of a uniform random variable. The lack of correlations between the successive inverses $\bmod p$ can be generally tested considering the sum:

$$S(p) = \sum_{x=1}^{p-1} e^{2\pi j(x+\bar{x})/p}. \tag{10.30}$$

Such a summation quantifies the amount of correlation that exists between x and $\bar{x}$ and is expected to be of size $\sqrt{p}$ for a uniform random variable due to random cancellations, consistent with the central limit theorem. Indeed, it has been shown by Weil in 1948 that $|S(p)| \leq 2\sqrt{p}$, establishing the arithmetic randomness of the modular inverses [524]. This property is exploited by *inversive congruential generators* that produce strings of pseudorandom numbers with good uniformity properties [525]. Of particular importance for the generation of pseudorandom sequences are also finite fields of order equal to a prime power p^m, where p is a prime number and m is a positive integer. These fields are also called *extension fields*. Interestingly, the elements of extension fields are not usual numbers but are often represented using polynomials.

---

[10] More precisely a field has characteristic p if $p \cdot a = 0$ for every element a in the field, with p the smallest positive integer for which this happens.

[11] Assume that $p \cdot 1 = (r \cdot 1)(s \cdot 1)$ is a composite (here $a = 1$). From the definition of characteristic, it follows that $r \cdot 1 \neq 0$, since $n > r$. Therefore, it has a multiplicative inverse, yielding $s \cdot 1 = 0$. But this is a contradiction since $n > s$.

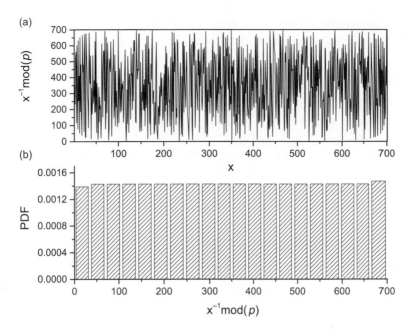

(a)

(b)

Figure 10.7 (a) Distribution of the multiplicative inverses mod p for $p = 701$ and (b) histogram of the normalized probability density function (pdf) of the values.

We will later discuss how these finite fields are constructed. Finite fields with p^m elements are generally denoted as $GF(p^m)$ after the French mathematician Évariste Galois, and have found numerous applications in physics, communication theory, error-correcting codes, cryptography, and even artistic design [8]. In particular, Galois extension fields play a crucial role in the widespread Advanced Encryption Standard (AES) symmetric block-cipher algorithm that secures internet transactions and communications by manipulating bytes encoded as elements of $GF(2^8)$ [526, 527]. It is a fundamental result that all realizations of $GF(p^m)$ are isomorphic, so that we can say that there is effectively one finite field with p^n elements. Therefore, the only possible fields with a finite number of elements are prime fields $GF(p)$ and extension fields $GF(p^m)$ with $m > 1$. We will later discuss in Section 10.4.2 how these fields are constructed. Let us first introduce the important concepts of order and primitive roots. We remember that, as previously discussed in Section 9.1.1, if a is an integer not divisible by the prime p, then the Euler's totient theorem reduces to the Fermat's little theorem $a^{p-1} \equiv 1 \mod p$. We can now fully appreciate that this elementary result, first proved by Gottfried Leibniz and published in 1683 [528], describes a special property of the units in the field $\mathbb{F}_p$. However, there could still be smaller powers of a that are congruent to 1. This observation allows us to introduce the important notion of the *order* of a modulo p as the smallest exponent k such that $a^k \equiv 1 \mod p$, denoted by $k = \text{ord}_a \mod p$.

Another fundamental property of the finite field $\mathbb{F}_p$ is that its group of units $\mathbb{F}_p^*$ is generated by its *primitive roots* of $\mathbb{F}_p$ or *generators*, which are elements $g \in \mathbb{F}_p^*$ of maximum order, namely $p - 1$. Thus, it is always possible to write all the elements of $\mathbb{F}_p^*$ as powers of its primitive roots as follows:

$$\mathbb{F}_p^* = \{1, g, g^2, \ldots, g^{p-2}\} \tag{10.31}$$

In other words, if g is a primitive root (modulo p) then the period of the integer sequence $g^n \bmod p$ (with $n = 1, 2, \ldots$ a positive integer) is equal to $p - 1$, which is the largest possible period since for $n > p - 1$ the sequence will repeat. Therefore, the elements of the sequence $g^n \bmod p$ are a *permutation* of all the nonzero remainders modulo p. Moreover, they are special kinds of permutations, called *perfect permutations*, with the property that each delay difference δ between their elements occurs at most once. Perfect permutations play an important role in the engineering of sound diffusers and in cryptography and are examples of pseudorandom sequences with constant (Fourier) power spectrum.[12]

The arithmetic concept of order introduced in this section also plays a fundamental role in the theory of decimal fractions since it can be shown that the fraction $1/p$ (with p a prime) has a period of length precisely $k = \mathrm{ord}_p \bmod 10$ [8]. Therefore, the period of $1/p$ will equal $p - 1$ whenever 10 is a primitive root modulo p. For example, since 10 turns out to be a primitive root of $1, 301$, the fraction $1/1, 301$ will be periodic with a decimal fraction of period length $1, 300$. Moreover, the digits in each period will occur asymptotically equally often (as is the case for the digits of π!) forming an aperiodic sequence with constant power spectrum. The same arguments carry over to other number bases, providing a convenient method to generate pseudorandom binary sequences when base 2 is chosen [8].

As originally remarked by Gauss in his *Disquisitiones*, the distribution of primitive roots among the field elements is very mysterious and there is currently no way to predict where they will appear. In fact, the distribution and spacing of primitive roots modulo p are believed to exhibit uncorrelated Poisson behavior [529, 530]. This behavior can be directly appreciated in Figure 10.8, where we plot in Figure 10.8a the difference between the consecutive primitive roots of $\mathbb{F}_p$ with $p = 9871$ and in Figure 10.8b the corresponding distribution of nearest-neighbor spacings, which is approximately described by the exponential function $p(s) = \exp(-s)$. However, when g is a primitive root modulo a large prime p, the discrete exponential map $n \to g^n \bmod p$ behaves as a random variable with uniform distribution. This type of "arithmetic randomness," or pseudorandomness, is routinely utilized in the construction of pseudorandom number generators. Interestingly, the number of primitive roots is known exactly and is equal to $\varphi(p - 1)$, where φ is the Euler's totient function. For example, in $\mathbb{F}_{11}$ there are $\varphi(10) = 4$ primitive roots, which, as it is easy to see, are the elements $2, 6, 7, 8$. We could also flip this question around and, fixing a number a, ask for which prime p the number a will be a primitive root. Not surprisingly, this list of primes p will not display any recognizable pattern! Emil Artin conjectured in 1920 that if $a \neq 1$ is not a perfect square, there are infinitely many

---

[12] This situation must be contrasted with random "white noise" where the power spectrum is flat, or constant, only in the statistical sense.

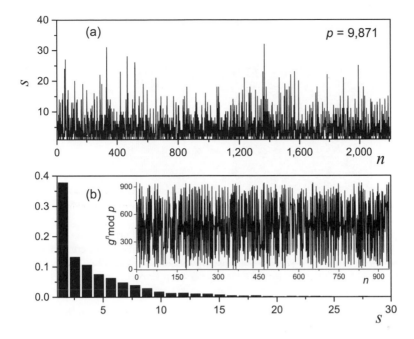

Figure 10.8 (a) Differences between consecutive primitive roots $s = g_{i+1} - g_i$ modulo $p = 9,871$. (b) Normalized probability density function (histogram distribution) of the differences between consecutive primitive roots modulo $p = 9,871$. The inset displays the first $p - 1$ values of the discrete map $n \rightarrow g^n \bmod p$ with $g = 627$ and $p = 941$.

primes p such that a is a primitive root modulo p. Artin's conjecture is still unsolved, but its validity has been reduced to the one of the generalized Riemann hypothesis [499]. The connection between primitive roots and modular exponentiation has been exploited for the creation of aperiodic numeric sequences and arrays called *Costas arrays*. Specifically, a binary Costas array A is created by selecting a primitive root g modulo a prime p and then setting $A_{i,j} = 1$ if $j \equiv g^i \bmod p$ and $A_{i,j} = 0$ otherwise. As an example, we show in Figure 10.9a the finite portion of the Costas array generated considering $6133^i \bmod 9871$ for $i = 1, 2, 3, \ldots, 9,800$. In Figure 10.8b, we display the histogram of the (normalized) nearest-neighbor spacing between particles and the analytical prediction for the probability density distribution of an equal-density Poisson point pattern model, obtained from equation (10.15). The agreement between the numerical data and the analytical prediction is remarkable, demonstrating the Poisson nature of the Costas arrays that reflects the underlying pseudorandomness of the modular exponentiation.

10.4.1 The Discrete Logarithm

The complex aperiodic behavior of modular exponentiation in finite fields constitutes an example of a *one-way function*. This is a function that is easy to compute in one direction but extremely difficult to compute in the opposite direction. With respect to

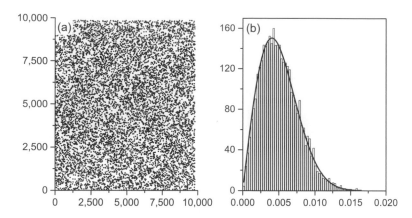

Figure 10.9 (a) Costas array generated with $p = 9,871$ and primitive root $g = 6,133$ for $i = 1, 2, \ldots, 9800$. (b) NNSD of the Costas array shown in (a). The continuous line is the prediction based on the Poisson point pattern model.

our previous example, it is easy enough to compute $g^x \bmod p$, but, due to the highly irregular behavior of the modular exponentiation with the exponent, it is generally difficult to solve the *discrete logarithm problem (DLP)* that consists in finding the exponent x such that $g^x \equiv h \bmod p$, where g is a primitive root and h is a nonzero element of $\mathbb{F}_p$. In fact, while for small p it is possible to obtain the solution by an exhaustive search through all the possible exponents, at large p this task becomes unfeasible. The "intrinsic asymmetry" of the DLP is at the origin of its many applications to *public-key cryptography*, most notably the Diffie–Hellman key exchange algorithm that enables two persons to share a secret key while communicating publicly over an insecure channel [526, 531]. The number x between 0 and $p - 2$ is called the discrete logarithm of h to the base g, denoted by $\log_g h$, or equivalently the *index* of h modulo p, denoted by $\text{ind}_g h$. It is important to realize that according to Fermat's little theorem we have $g^{p-1} \equiv 1 \bmod p$, which implies that $x = \log_g h$ cannot be unique, but it is defined instead only up to the addition/subtraction of multiples of $p - 1$, i.e., it is defined modulo $p - 1$. The discrete logarithm satisfies many properties of the usual logarithm, despite the fact that the modular exponentiation varies very irregularly with the exponent contrary to its continuous counterpart. For example, it is not difficult to prove that $\log_g ab = \log_g a + \log_g b$ for all $a, b \in \mathbb{F}^*$.

Finally, we remark that a generalized DLP can be defined in any cyclic group generated by the element g where the composition group law is used instead of the multiplication modulo p. A particularly important case is the DLP defined for elliptic curves over finite fields, which we will discuss in Section 10.4.3.

10.4.2 Galois Extension Fields and Pseudorandomness

The field elements of $GF(p^m)$ are vector m-tuples, matrices, or polynomials with components or coefficients from $GF(p)$, depending on the actual field representation.

Addition and subtraction of m-tuples are defined columnwise modulo p, while the multiplication is defined by the product of polynomials of degree up to $m - 1$ associated to the m-tuples, reduced modulo a suitable polynomial called *irreducible polynomial* $\pi(x)$ of degree m with coefficients in $GF(p)$, i.e., over $GF(p)$. Therefore, when considering multiplications in Galois extension fields, irreducible polynomials play a role analogous to the one of prime numbers p with respect to modular reduction in $GF(p) = \mathbb{Z}/p\mathbb{Z}$. This fundamental property follows from the fact that given a general finite field $\mathbb{F}$ with p elements and an irreducible polynomial $\pi(x) \in \mathbb{F}[x]$ of degree $d \geq 1$, where $\mathbb{F}[x]$ denotes the ring of polynomials with coefficients in the field $\mathbb{F}$, the quotient ring $\mathbb{F}[x]/\pi(x)$ forms a field with p^d elements [531]. The elements of $GF(2^m)$ are often identified by the polynomials $A(x) = a_{m-1}x^{m-1} + \cdots, a_1 x + a_0$, where the binary coefficients $\{0, 1\}$ belong to the prime field $GF(2)$. As a result, in polynomial addition/subtraction the coefficients must be added/subtracted modulo 2. However, polynomial multiplication/division in $GF(p^m)$ is less intuitive. In fact, in order for the multiplication to possess an inverse, it must be defined modulo a given irreducible polynomial $GF(p^m)$ over $GF(p)$ (of degree m). This situation is entirely analogous to the multiplication in the residue system $\mathbb{Z}/n\mathbb{Z}$, where the integers $1, 2, \ldots, n - 1$ have a multiplicative inverse modulo n only when n is prime, i.e., irreducible.

Let us now exemplify these abstract concepts by considering the field $GF(2^3)$, whose elements are polynomials of the form $A(x) = a_2 x^2 + a_1 x + a_0$ with components taken from $GF(2)$. We can encode these polynomials through the binary values of their coefficients, i.e., by specifying the binary components of 3-bit vectors as the triplets (a_2, a_1, a_0). Using this encoding scheme, we can explicitly list all the elements of $GF(2^3)$ as follows:

$$GF(2^3) = \{0, 1, x, x + 1, x^2, x^2 + 1, x^2 + x, x^2 + x + 1\}, \qquad (10.32)$$

where each element in $GF(2^3)$ corresponds to an ordered triplet from the list:

$$\{000, 001, 010, 011, 100, 101, 110, 111\} \qquad (10.33)$$

with the least significant digit on the left. To illustrate how addition and subtraction actually work, let us consider, for instance, the sum of the two elements $A(x) = x^2 + x + 1$ and $B(x) = x^2 + 1$. We then proceed as in the standard polynomial sum by adding the corresponding coefficients, but remembering that modulo 2 operations must be considered here. We will then obtain the following:

$$C(x) = A(x) + B(x) = x^2(1 + 1) + x + (1 + 1) = x \qquad (10.34)$$

because $1 + 1 = 0$ modulo 2. We also notice $A(x) - B(x) = x$ as well, and that adding or subtracting in $GF(2^m)$ always yields the same results!

A naive application of these rules to polynomial multiplication would produce the following:

$$D(x) = A(x) \cdot B(x) = x^4 + x^3 + x^2 + x^2 + x + 1 = x^4 + x^3 + x + 1. \qquad (10.35)$$

However, the polynomial $D(x)$ cannot be the correct result of the multiplication since it does not even belong to the field $GF(2^3)$! In order to obtain the correct answer, we need to additionally reduce the result modulo a properly defined polynomial. This problem is solved by reducing $D(x)$ with respect to an irreducible polynomial $\pi(x)$, which is a polynomial that does not factor into polynomials of lower degree. Irreducible polynomials are analogous in polynomial rings of prime numbers in the ring of integers $\mathbb{Z}$. In fact, just as for the integers, every polynomial *with coefficients in a field* has an essentially unique factorization as a product of irreducible polynomials [531]. However, given a field $GF(2^m)$, there are several irreducible polynomials, and the result of the computation will in general depend on its particular choice, which needs to be specified from the outset. For instance, the encryption standard AES, which operates in $GF(2^8)$, utilizes in its implementation the irreducible polynomial $\pi(x) = x^8 + x^4 + x^3 + x + 1$.

Let us continue our example by considering the irreducible polynomial $\pi(x) = x^3 + x + 1$. Modular reduction with respect to $\pi(x)$ means that we need to calculate $P(x) \equiv D(x) \bmod P(x)$. Using high school polynomial division and remembering that coefficients must be added/subtracted in $GF(2)$, the reader can easily obtain: $x^2 + x \equiv D(x) \bmod P(x)$. Note that the polynomial $x^2 + x$ belongs this time to $GF(2^3)$ and is in fact the correct result of the polynomial multiplication. We should recognize that the quotient ring $GF(2)/\pi(x)$ is in fact equal to the Galois extension field $GF(2^3)$.

Let us now recall that the nonzero elements of any finite field $\mathbb{F}$ are generated by a primitive root, i.e., the elements of the multiplicative group $\mathbb{F}^*$ form a cyclic group closed under multiplication. In particular, denoting the nonzero elements of $GF(p^n)$ by $GF(p^n)^*$, which is a multiplicative group with $p^n - 1$ elements, it is possible to show that there is a *generating polynomial* or *primitive element* $g(x)$ such that every element in $GF(p^n)^*$ can be expressed as a power of $g(x)$. There are $\varphi(p^n - 1)$ generating polynomials, where φ is the Euler's totient function. This property, which is the analog for polynomials of the primitive roots for primes, allows us to generate all the elements of Galois extension fields starting from a primitive element $g(x)$, after choosing an irreducible polynomial $\pi(x)$. For example, if we choose to represent the elements of $GF(2^4)$ using 4-tuples with components from $GF(2)$, we can obtain all the 16 elements by successive powers of the primitive element $g(x) = 0100$ modulo the irreducible polynomial $\pi(x) = x^4 + x + 1$. More generally, we can arrange all the generated elements of $GF(2^n)$ into a $(2^n \times n)$ *element matrix* whose columns contain n periodic binary sequences. However, the distributions of ones and zeros along the columns (i.e., within each period) display the "random-like" or pseudorandom Poisson behavior manifested by the map $k \to g^k \bmod p$ already discussed in Section 10.4. Galois sequences with binary values have been used in many engineering applications due to their distinctive spectral properties [8]. In particular, Galois sequences derived from $GF(2^n)$ are uncorrelated (i.e., delta-correlated) pseudorandom binary sequences with a constant Fourier power spectrum. However, in contrast to other pseudorandom binary sequences, they can be efficiently generated using a simple linear recursion. This provides a very convenient way to generate aperiodic systems starting from

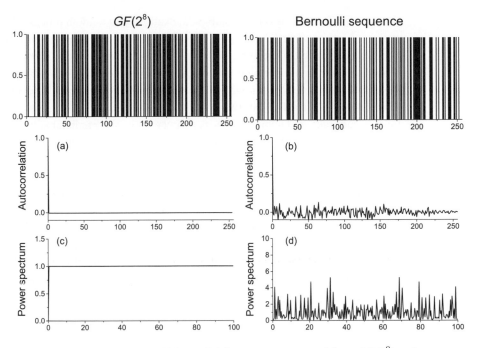

Figure 10.10 Top row: bar-coded binary Galois sequence generated from $GF(2^8)$ and single-realization of a binary Bernoulli random process with probability $p = 0.5$. The elements of the Galois sequence have been modified by replacing 0's by 1's and 1's by -1's. (a,b) The autocorrelation functions. (c,d) The Fourier power spectra of the two sequences displayed in (a) and (b), respectively.

binary valued sequences. For instance, taking the first column of the element matrix associated to $GF(2^4)^*$, we get the binary *Galois sequence*:

$$\{s_i\} = 1000100110101111. \tag{10.36}$$

This sequence can be generated by the linear recursion:

$$s_{i+4} = s_{i+1} + s_i \tag{10.37}$$

subject to the initial conditions $a_1 = 1, a_2 = a_3 = a_4 = 0$.

These sequences can be efficiently generated using a finite-state machine, or a *linear feedback shift register (LFSR)*, which produces $p^n - 1$ distinct nonzero states that repeat periodically using straightforward hardware implementations [526, 527].

In Figure 10.10, we compare the spectral and correlation properties of a random binary sequence with the Galois sequence defined by the first column of the element matrix of $GF(2^8)$. The elements of $GF(2^8)$ are generated with respect to its minimal primitive polynomial, which is the primitive element with the smallest possible number of nonzero terms. For the random sequence, we choose a *Bernoulli p-process*, which models the familiar coin tossing[13] in a series of independent trials

---

[13] Remember that for a fair or balanced coin tossing experiment, $p = 1/2$.

(i.e., experiments). This is the prototypical example of a random process that produces strings of *independent binary digits* (i.e., zeros and ones) with a given probability p, here set to $p = 1/2$. The structural similarity between the two binary sequences is remarkable at a first inspection, as demonstrated by their respective bar plots. Nevertheless, we recognize that these two sequences have profoundly different spectral and structural properties. The flat (two-valued) correlations and power spectra characteristic of pseudorandom Galois sequences reflect an inherent predictable behavior in their bits distribution even before any periodic repetitions. This "predictability pitfall" of Galois sequences, which is also exhibited by other pseudorandom sequences generated by linear and polynomial feedback shift registers, makes them unsafe for realistic cryptographic applications. In fact, the generation of private random keys based on Galois sequences, while computationally very efficient, would compromise security by enabling eavesdroppers to use their knowledge of some bits in the key to predict, with a high probability, all the future values [526].

Pseudorandom Galois sequences can naturally be extended to two- or higher-dimensions if their period length $2^n - 1$ can be factored into two or more coprime factors, using the so-called *Sino representation* [8]. For instance, since the period length in $GF(2^4)$ is $15 = 3 \times 5$, the corresponding aperiodic sequence of elements $\{k\}$ can be converted into a two-dimensional 3×5 array whose rows contain the (least-positive) residues modulo 5 (denoted $\langle k \rangle_5$) and whose columns the (least-positive) residues modulo 3 (denoted $\langle k \rangle_3$). The resulting array has the desired spectral and correlation properties in two spatial dimensions. Alternatively, by using the preceding recursion relation repeatedly, the first column of a 2D Galois array can be generated starting from the $2^n - 1$ elements of the 1D Galois sequence. Subsequent columns are generated by permutations of the elements determined by the binary values in the first column. The binary value of the nth column will be the same as the ones in the first one if the nth value in the first column is 1; otherwise, the binary value of the nth column will be inverted ($1 \rightarrow 0, 0 \rightarrow 1$) with respect to the first column, where $2 \leq n \leq 2^n - 1$ (remember that the zero element is disregarded). By following this deterministic linear recursion and permutation scheme, 2D aperiodic Galois arrays of scattering nanoparticles have been demonstrated that are diagonally symmetric and feature flat Fourier spectra. The integration of these aperiodic Galois structures with microfluidics technology for optical biosensing based on the modifications of their complex colorimetric responses to small refractive index variations was recently demonstrated by Lee et al. [532].

10.4.3   Elliptic Curves over Finite Fields

In this section, we introduce a class of deterministic aperiodic structures based on elliptic curves and the associated discrete logarithm problem. Besides its profound ramifications to diverse mathematical areas, the study of elliptic curves may offer novel opportunities for the engineering of light scattering and localization phenomena in aperiodic environments beyond the limitations of traditional random media. The scattering and transport properties of electromagnetic waves through systems of electric and magnetic dipoles arranged according to elliptic curves over finite fields have been recently explored by Dal Negro et al. [9]. An important outcome of this study, which addresses structure–property relationships for a large number of systems, is that arrays based on elliptic curves manifest an extremely rich spectrum of scattering and localization properties that can outperform uniform random structures in terms of optical confinement and the strength of light–matter interaction. An elliptic curve $E(\mathbb{K})$ over a number field $\mathbb{K}$ is a nonsingular curve (i.e., with a unique tangent at every point) with points in $\mathbb{K}$ that are the solutions of a cubic equation. Therefore, elliptic curves can be thought of as the set of solutions in the field $\mathbb{K}$ of equations in the form [531]:

$$y^2 = x^3 + Ax + B, \tag{10.38}$$

where the coefficients A and B belong to $\mathbb{K}$ and satisfy the nonsingular condition $\Delta_E = 4A^3 + 27B^2 \neq 0$ for the discriminant Δ_E that excludes cusps or self-intersections (i.e., knots) [531, 533]. Elliptic curves specified as in the preceding equation are said to be given in the Weierstrass normal form. When $\mathbb{K}$ coincides with the set of real numbers $\mathbb{R}$, we can graph $E(\mathbb{R})$ and view its solutions (x, y) as actual points of a plane curve. An example is shown in Figure 10.11a for a representative elliptic curve (EC) over the real numbers defined by the parameters $A = 27$ and $B = 4$. Clearly, different choices for the field $\mathbb{K}$ will result in different sets of solutions for the same cubic equation, since elliptic curves can also be regarded as particular examples of algebraic varieties. Algebraic varieties over the field of rational numbers $\mathbb{Q}$ have been investigated already by postclassical Greek mathematicians, most notably by Diophantus, who lived around 270 CE in Alexandria, Egypt. In his honor, we refer to a polynomial equation in one or more variables whose solutions are sought among the integers or rational numbers as a "Diophantine equation." The history of Diophantine equations and elliptic curves runs central to the development of the most advanced ideas of number theory that led to the proof of the celebrated Fermat's last theorem by the British mathematician Andrew Wiles in 1995 [517]. The study of elliptic curves constitutes a major area of current research in number theory with important applications to cryptography and integer factorization. Interestingly, when endowed with an extra point $\mathcal{O}$ at infinity, the points of elliptic curves acquire the structure of an Abelian group with the point $\mathcal{O}$ serving as the neutral group element. In particular, the group of rational points (solutions in $\mathbb{Q}$) of the elliptic curve $E(\mathbb{Q})$ is finitely generated (Mordell's theorem) and can be decomposed into the direct sum of $\mathbb{Z}$ with finite cyclic groups [531, 533]. More specifically, one can also show the group of rational points has the form: $E(\mathbb{Q}) \cong T \bigoplus \mathbb{Z}^r$, where T is a finite group consisting of torsion points (i.e., a point $P \in E$ satisfying $mP = \mathcal{O}$ is called a point of order m in the group E; all points of finite order form an Abelian subgroup called the torsion group of E) and r is a nonnegative number, called the algebraic rank of the elliptic curve E, which somehow characterizes its size [533, 534].

An example of the composition group law for the previously introduced elliptic curve over the real numbers is illustrated in Figure 10.11a, where two points with real-valued coordinates P and Q are summed to obtain the point R'. The simplest

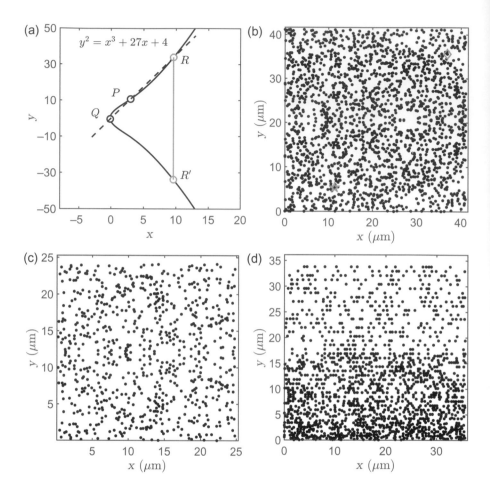

Figure 10.11 (a) Elliptic curve over the real numbers generated by (10.38) with coefficients $A = 27$ and $B = 4$. The sum operation of points P and Q on the elliptic curve over the rationals is also shown. (b) Elliptic curve from (a) defined over the finite field $\mathbb{F}_{2111}$. The coordinates are rescaled in order to display an interparticle separation equal to 450 nm. The gray markers identify two representative points W (rhombus symbol) and M (star symbol) on the curve, respectively. (c,d) The point patterns created by solving the discrete log problem $W = kM$ on the curve.

way to introduce the group composition law, or *elliptic addition*, is to implement the following geometrical construction [531, 533]: we first draw the line that intersects P and Q. This line will generally intersect the cubic at a third point, called R. We then define the addition $P + Q$ as the point $-R$, i.e., the point opposite R. It is possible to prove that this definition for addition works except in a few special cases related to the point at infinity and intersection multiplicity [531]. The type of elliptic curves that we will investigate in this book are defined over the finite field $\mathbb{F}_p \equiv \mathbb{Z}/p\mathbb{Z}$ where p is an odd prime number. This is the set of integers modulo p, which is an algebraic field when p is prime. An elliptic curve over $\mathbb{F}_p$ is still defined by equation (10.38), where the equal sign is replaced by the congruence operation:

$$y^2 \equiv x^3 + Ax + B \mod p, \tag{10.39}$$

where the coefficients $A, B \in \mathbb{F}_p$ and the discriminant Δ_E in this case must be incongruent to 0 when reduced modulo the prime p. Since $\mathbb{F}_p$ is a finite group with p elements, the elliptic curve defined (10.39) has only a finite number of points that we expect to be approximately $p + 1$ in number (remember the necessity to add the extra point at infinity). It turns out that the actual number of points N_p of the curve $E(\mathbb{F}_p)$ fluctuates from $p + 1$ within a bound $2\sqrt{p}$, which is a result proved in 1933 by Helmut Hesse. More precisely, if we define the quantity $a_p = p + 1 - N_p$, Hesse's theorem states that $|a_p| \leq 2\sqrt{p}$ [531, 533]. One of the most challenging yet unsolved problems in mathematics, which is also a millennium prize problem of the Clay Mathematics Institute [535], is the *Birch and Swinnerton-Dyer (BSD) conjecture* that identifies the algebraic and the analytic rank of an elliptic curve [533]. The analytic rank of a curve E is equal to the order of vanishing of the associated Dirichlet L-function $L(E, s)$ at $s = 1$. The aforementioned L-function $L(E, s)$ is a complex-valued function that is constructed based on the numbers a_p. This function, which is analogous to the Riemann zeta function ζ and the Dirichlet L-series, can be analytically continued over the whole complex plane and it encodes information on the number of solutions of E modulo a prime onto the properties of the associated complex function $L(E, s)$. Moreover, $L(E, s)$ satisfies a Riemann-type functional equation connecting its values $L(E, s)$ and $L(E, 2 - s)$ for any s. According to the Sato–Tate conjecture, the random-looking fluctuations observed in the "error term" a_p when the prime p is varied are captured by a "sine-squared" probability distribution. This conjecture was proved in 2008 by Richard Taylor limited to particular types of elliptic curves [536]. In Figure 10.11b, we show the elliptic curve over $\mathbb{F}_p$ with $p = 2111$ that has the same parameters as the curve $E(\mathbb{R})$ previously shown in Figure 10.11a. We note that the curve has been rescaled by a constant parameter so that the average separation between points equals 450 nm, which enables resonant scattering responses across the visible spectrum. Apart from this irrelevant scaling, the points on this curve appear to be randomly distributed in stark contrast with its counterpart defined over the field of real numbers. Moreover, working with EC over finite fields allows one to define the associated discrete logarithm problem that plays as essential role in elliptic curves cryptography due to its nonpolynomial complexity [531, 534]. Let EC be an elliptic curve over $\mathbb{F}_p$ (see Figure 10.11b) and M (star symbol) and W (rhombus symbol) two points on the curve. The discrete logarithm problem is the problem of finding an integer k such that $W = kM$. By fixing a starting point W and applying this group operation repeatedly to all the points M on the curve E in Figure 10.11b, we can obtain the point patterns shown in Figure 10.11c,d, which are the physical representations of the abstract discrete logarithm problem on the original curve E. Specifically, Figure 10.11c,d displays curves characterized by the coordinates $(x_M; k)$

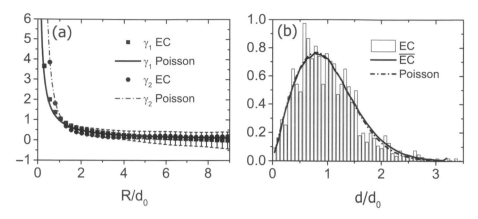

Figure 10.12 Structural properties of elliptic curves compared to random Poisson point patterns. (a) Scaling analysis of the skewness and excess structural correlation parameters introduced in section 8.5.2 (b) Histogram distribution of the first-neighbor spacing for a single elliptic curve (white rectangles), averaged over 900 elliptic curves (continuous line) and for a random Poisson point pattern (dashed-dot line).

and $(y_M; k)$ rescaled in order to have an average interparticles separtion equal to 450 nm, respectively. These types of aperiodic deterministic structures are referred to as elliptic curve discrete logarithm ($EC\ DL$). Clearly, the distribution of points in $EC\ DL$ strongly depends on the choice of the initial point W on the starting EC. In our work, we have uniformly sampled nine different starting points on E. We have found that the resulting $EC\ DL$ curves can be divided into two main categories: $EC\ DL$ point patterns that are symmetric with respect to the x-axis (Figure 10.11c) and others that do not show this structural symmetry and are generally less homogeneous (Figure 10.11d). Moreover, the number of elements $EC\ DL$ cannot be controlled exactly because it depends on the value of the integer k. The complexity of the discrete logarithm problem for elliptic curves over finite fields is at the heart elliptic curve cryptography (ECC), which is the most advanced approach to public-key cryptography that protects highly secure communications (up to top-secret classification) using keys that are significantly smaller compared to alternative methods such as RSA-based cryptosystems [531, 534]. The structural properties of elliptic curves with respect to random Poisson point patterns are analyzed in Figure 10.12, where in Figure 10.12a we compare the higher-order correlation parameters introduced in (8.5.2). The data points refer to an elliptic curve of the form (10.39) with $p = 2,111$ and randomly selected A and B parameters. The continuous and dash-dot lines are the results for a random Poisson point pattern. In Figure 10.12b, we show the histogram distribution of the first-neighbor spacing for a single elliptic curve and for the average (continuous line) over 900 randomly selected A and B parameters. The averaged results perfectly match a random Poisson point pattern.

10.4.4 Generating Weak and Strong Pseudorandomness

The problem of generating pseudorandom numbers that closely approximate sequences of truly random ones using *deterministic methods* has a long history that dates back to the very beginning of modern computing (around the 1940s). Pseudorandom numbers are very important in many branches of applied mathematics and engineering such as in stochastic simulations and for Monte Carlo numerical methods. However, only the statistical properties of long sequences of pseudorandom numbers are relevant to such applications while the distinctive correlations between successive pseudorandom numbers do not play a significant role. On the other hand, the unpredictable nature of truly random numbers becomes essential for cryptographic applications, where they are often used to generate encryption keys. A simple example is provided by the stream cipher, which is a symmetric key cryptographic algorithm that encrypts each plaintext digits one at a time by performing an exclusive-or (XOR) operation with the corresponding digit of a keystream, i.e., a stream of random or pseudorandom numbers. In this context, pseudorandom number generators (PRNGs) are generally considered cryptographically insecure with only a few remarkable exceptions, as we will show later in this section. Besides efficiency, the main requirement of a high-quality PRNG is to produce sequences that are uniformly distributed. However, statistical randomness does not necessarily imply "true randomness," which is instead related to the objective unpredictability manifested by the absence of any discernible pattern. In general, sophisticated algorithms and a very careful choice of parameters are needed in order to develop PRNGs whose outputs are sufficiently close to random sequences to successfully pass a number of accepted *randomness tests*. In the 1950s, John von Neumann jokingly cautioned about confusing pseudorandomness and genuine randomness by remarking that "Anyone who considers arithmetical methods of producing random digits is, of course, in a state of sin."

The simplest PRNGs is the *linear congruential generator* defined by the following recursion:

$$x_{n+1} = (ax_n + b) \bmod M \tag{10.40}$$

where a, b, M are integer constants and the recursion is initiated by a seed value x_0. Each integer iterate x_n can then be converted to a fraction within the unit interval by dividing x_n by the modulus M. Recurrences such as the preceding one are clearly periodic with the period at most M, which can be achieved by ensuring the validity of certain constraints on the parameters [464]. The linear congruential generator, similarly to other simple PRNGs, can be easily broken by observing a number of outputs much smaller than the period, despite its apparent randomness, and the same weakness applies even if the linear congruence is replaced with higher-order polynomials [526]. However, linear congruential generators are more secure when implemented over elliptic curves [464], i.e., with the elliptic addition introduced in the previous section. A more sophisticated general-purpose PRNG is the *Mersenne Twister*, so named because its period $2^{19937} - 1$ is a Mersenne prime, which is a prime of the form $M_p = 2^p - 1$, for some prime number p[14]. The Mersenne twister is based on a matrix linear recurrence over the binary finite field $\mathbb{F}_2$ and generalizes simple linear feedback shift registers.

Another class of iterative methods relies on the *discrete exponential generator* defined by the following:

$$\boxed{x_{n+1} = g^{x_n} \bmod M} \tag{10.41}$$

for given values of g, M, and the seed x_0. It is clear that the security of these PRNGs depends on the complexity of the discrete logarithm problem, which makes them suitable for certain cryptographic applications if a chosen group of bits is kept from x_n [464]. Recently, PRNGs capable of producing unpredictable bitstreams in a cryptographically secure way have been demonstrated by leveraging the number-theoretic properties of quadratic residues (see Box 10.4). We denote by *strong pseudorandomness* the one generated by cryptographically secure PRNGs in contrast to the weak pseudorandomness generated by standard (noncryptographically secure) methods. One of the most popular algorithms to create strong pseudorandomness is based on the *Blum Blum Shub (BBS)* pseudorandom bit generator [537, 538]. The BBS generator outputs a bitstream based on the following recursion relation:

$$\boxed{x_{n+1} \equiv x_n^2 \bmod M} \tag{10.42}$$

where $M = pq$ is the product of two large prime numbers that need to satisfy certain special properties and $x_0 < M$ is an initializing bit, called *seed*, which is coprime to M. Taking the least significant digit of each of the output values, which can easily be done by checking whether they are odd or even, produces a binary bitstream. The preceding recursion eventually repeats itself after a large period given by $\lambda(\lambda(M))$, where $\lambda(M)$ here denotes the Carmichael's λ-function [538]. We note that the squaring operation implements a one-way permutation of the quadratic residues. The security of this approach relies on the intractability of the *quadratic residuosity problem*, which relates to the difficulty of finding square roots modulo a number of a specified form. It can be proven that the BBS generator is *cryptographically secure* in the sense that, if someone ignores the seed and the factors of M, it is not possible (computationally infeasible) to guess with better than 50% chances the next output from the observations of all the previous ones [537, 538].

---

[14] At the time of writing, only 51 Mersenne primes are known and many fundamental questions about them remain unresolved. For instance, it is not known whether there are a finite or an infinite number of them. In general, they are very large numbers and difficult to find because they are close to a power of 2. In fact, all recently discovered Mersenne primes after 1997 have been found by the Great Internet Mersenne Prime Search (GIMPS), which is a distributed computing collaborative project that uses freely available software on the Internet.

A natural question that we address at this point concerns the nature of two-dimensional point patterns obtained by plotting successive iterated values of PRNGs, i.e., by plotting the points with coordinates (n, x_n). We refer to the resulting deterministic aperiodic point patterns as *PRNG-arrays*. Such arrays generally feature Fourier spectral properties and statistical distributions that are very similar (practically indistinguishable) from the ones obtained in uniform random arrays, which is why they are so relevant in the context of Monte Carlo sampling. In fact, despite the deterministic nature, PRNG-arrays have nearest-neighbor distribution functions that closely approximate random Poisson point patterns. This situation is exemplified in Figures 10.13a,b, where we consider PRNG-arrays constructed from a linear congruential generator and a BBS generator, respectively. However, higher-order correlation analysis can reveal significant differences between truly random and unsophisticated PRNG-arrays such as the linear congruential generators.

The deterministic nature of PRNGs is sometimes recognized by plotting successively generated values as pairs or triplets of points in the plane or in space. In this case, the correlations between subsequent output values of PRNGs can give rise to ordered spatial structures and discernible geometric patterns. An interesting result established by George Marsaglia [539] states that n-tuples with coordinates from the consecutive output values of linear congruential generators are sequentially correlated and produce a *lattice structure*, i.e., they lie on equally spaced hyperplanes in n-dimensional space, as shown in Figure 10.13d for triplets of outputs. In contrast, the uncorrelated values from a truly random sequence always give rise to a featureless cloud of scattered points. On the other hand, the deterministic nature of more advanced PRNGs, such as the Mersenne Twister or cryptographically secure PRNGs, cannot simply be revealed by such correlation plots, as we show in Figure 10.13c for the case of the BBS generator. High-quality PRNGs are engineered to pass a large number of randomness tests so that distinguishing strong pseudorandomness from genuine randomness is a very difficult problem that we discuss in Section 10.5.

10.4.5 Pseudorandomness and Coded Apertures

A direct application of spatial pseudorandomness to optical technology is provided by *coded aperture imaging* [540–542], which relies on wave diffraction from engineered pseudorandom binary arrays with prescribed correlations. The radiation from an illuminated object is "blocked" according to a known pattern casting a "coded shadow" from which it is possible to reconstruct, using computational algorithms, the original object. The main idea can be understood considering the following image formation model [541]:

$$I(x, y) = O(x, y) \star A(x, y) + N, \tag{10.43}$$

where $I(x, y)$ is the recorded intensity after the coded aperture with transmission function $A(x, y)$, $O(x, y)$ represents the intensity distribution of the object, N is the

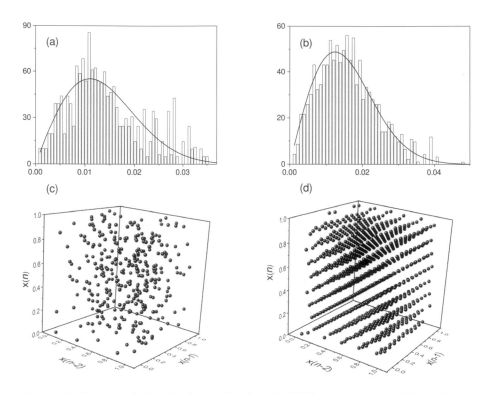

Figure 10.13 Nearest-neighbor distribution functions for RNG-arrays constructed from a linear congruential generator (a) and a BBS generator (b). The continuous lines are predictions based on the Poisson statistics. Correlation plots for triplets of outputs corresponding to the BBS generator (c) and the linear congruential generator (d).

noise, and $\star$ denotes the correlation operator. The reconstructed object is defined as follows [541]:

$$\hat{O}(x, y) = I(x, y) \star G(x, y) = RO \star (A \star G) + N \star G, \qquad (10.44)$$

where R is the reflection operator, and we introduced the postprocessing array $G(x, y)$, which is derived from A with the crucial property $A \star G \approx \delta$. Under these conditions, the original object can be perfectly reconstructed (except for the noise) as follows:

$$\hat{O} = O + N \star G. \qquad (10.45)$$

Therefore, the imaging problem has been reduced to the construction of binary arrays A and G, which are approximately delta-correlated. This is conveniently achieved by binary apertures derived from the pseudorandom distribution of quadratic residues modulo a prime, resulting in so-called uniformly redundant arrays (URA) and their modified version (MURA), whose construction algorithms are detailed in [541].

10.5 Randomness, Complexity, and Information

The quest to develop robust and quantitative criteria to decide whether a set of available data contains some recognizable structure or pattern has a long history in pure and applied mathematics and has very practical ramifications in computer science, statistics, cryptographic research, as well as many engineering disciplines (evolutionary biology, brain science, finance, etc.). Moreover, the general understanding of randomness is per se an important theoretical question that has often been addressed by philosophers, mathematicians, and practitioners in the natural sciences since the very beginning of probability theory. However, it was only in the second half of the last century that algorithmic and computational concepts were fruitfully introduced in order to precisely define and quantify the randomness and complexity contained in binary strings irrespective of the statistical or deterministic processes from which they originate. The distinction between the statistical randomness of a stochastic process that generates a binary string and the randomness of the string itself motivated Andrei Kolmogorov's formal definition of complexity of a mathematical object as the length of the shortest computer program that produces the object. According to this approach, which is a central tenet of *algorithmic information theory (AIT)*,[15] the *algorithmic complexity* of an object is always a measure of the computational resources that are needed to completely describe it [543]. This simple idea is used to introduce the concept of *Kolmogorov or algorithmic randomness*, according to which a sequence of bits $A = a_1 a_2, \ldots, a_k$ is considered random if and only if the length of the shortest computer program (running on a universal Turing machine) outputting A is at least k. Equivalently, a random string has maximal algorithmic complexity, which is comparable to its length. As a result, random strings cannot be *algorithmically compressed* using any algorithm or description shorter in length than the string itself. Therefore, at the core of the computational approach pioneered by Martin-Löf and Kolmogorov is the realization that *randomness is incompressibility*. We remark that in this context, "randomness" has a different meaning with respect to the common usage in statistics. In fact, statistical randomness refers to the process that produces the string, such as the flipping of a coin, while algorithmic randomness is a measurable property of the string itself. This concept also contrasts with the traditional idea of randomness that we encounter in probability theory because no specific element of a sample space can be declared random. As an example, consider the binary sequence 100101110101 generated from a Bernoulli 1/2-process. This sequence is more complex (and possibly more random) than the sequence 010101010101 because it cannot be described more efficiently than simply writing out its digits while the last sequence can be efficiently compressed by the concise description "alternate 0 and 1 for six times."

---

[15] AIT fruitfully combines information theory and computer science to study the complexity and randomness of strings. Besides Andrei Kolmogorov, pioneering figures in this field include Per Martin-Löf, Ray Solomonoff, and Gregory Chaitin.

Box 10.5 Ω, Computable, and Random Numbers

Algorithmic information theory has produced many stimulating results, such as Chaitin's incompleteness theorem, that challenge common mathematical and philosophical intuitions. Most notable is the construction of Chaitin's constant Ω, which is a real number that encodes the probability that a universal Turing machine will halt when fed as its input a random computer program, generated, for example, by flipping a fair coin [544, 545]. Although Ω is precisely and easily defined, one can only compute finitely many digits in any consistent axiomatic theory, making it fundamentally unknowable. However, despite the fact that the digits of Ω cannot be determined, there are still many properties of Ω that can be known. For example, the binary represntation of Ω is an algorithmically random sequence and thus a normal number. In addition, Ω is a noncomputable number in the sense that there is no algorithm that can be used to compute its digits. This is in stark contrast with computable numbers, such as algebraic numbers or many transcendental numbers, including e, π, which in fact can be computed to within any desired accuracy by a terminating algorithm of finite size.

According to the AIT viewpoint, the *information content* of a string is measured by the length of its most-compressed representation. This representation is essentially a computer program, written using some definite but otherwise irrelevant programming language, which can output the original string. This definition may appear very puzzling at first sight. In fact, from this point of view, a 1,000-page edition of Dante's *Divine Comedy* contains less information than 1,000 pages filled with completely random letters, despite their uninspiring appearance. However, this is so because if we wanted to reconstruct the entire sequence of random letters we would need to reproduce, more or less, each single letter in the sequence. On the other hand, when presented the string "*Nl mzzo de camin d nostr vit*," someone with reasonable knowledge of Dante's masterpiece (and of the Italian language) could reconstruct it immediately. A fundamental result of AIT, proven by Gregory Chaitin, establishes that the complexity of a specific string, and therefore its algorithmic randomness, cannot be formally proven if the complexity of the string is above a certain threshold [544, 546]. The essence of Chaitin's result is that, despite most strings are expected to be random, the randomness of a given (sufficiently long) string is ultimately *undecidible*. This theorem, which is known as *Chaitin's incompleteness theorem*, sets a fundamental limit to our knowledge of complexity and shares many ideas and techniques with Gödel's famous first incompleteness theorem that established in 1931 the intrinsic limitations of any formal axiomatic system capable of expressing basic arithmetic. In demonstrating his theorem, Gödel introduced an ingenious numbering scheme according to which certain natural numbers are assigned to terms, formulas, and proofs of a formal system that is powerful enough to express true arithmetic facts, i.e., to prove theorems about natural numbers. Based on the arithmetization of the formal system, Gödel rigorously constructed a number-theoretic assertion that says of itself that it is unprovable. This situation is reminiscent of the famous *liar paradox* first introduced by the Greek philosopher Eubulides of Miletus in the fourth century BCE, which is conveyed by the simple statement: "This statement is false." Thanks to his numbering method, Gödel turned the liar's paradox into a mathematical theorem by constructing a statement that says: "This statement is unprovable." He then noticed that if the last statement is unprovable, then it is true (since it says that it is unprovable), implying that the original formal system in which it was expressed is incomplete, since it contains a true statement that cannot be proven in that system. On the contrary, if the statement is provable, then it is false, and the original formal system is inconsistent, since it allowed us to logically deduce a false statement in it. Therefore, in any consistent formal system there will always be statements that are true but cannot be proven within the system, no matter how ingeniously its axioms are chosen (as long as the formal system is powerful enough to express arithmetic statements). Nontechnical expositions of Gödel's incompleteness theorem can be found in the now classic book by E. Nagel and J. Newman [547] or in the clear introduction to mathematical logic by Raymond Smullyan [548].

Returning now to algorithmic theory, we remark that the Kolmogorov complexity is *uncomputable*, meaning that there is no program that takes a string as its input and produces the value of the corresponding Kolmogorov complexity as the output. Therefore, there remains in practice the hard problem of establishing if a specific (and sufficiently complex) sequence is indeed random or not and how to quantify its degree of randomness. This task will lead us to the information-theoretic concepts of entropy and approximate entropy (ApEn) on one side, and to statistical randomness testing on the other, as we will introduce in the next sections.

10.5.1 Randomness and Information Theory

In 1948, Claude Shannon published a landmark paper entitled "A Mathematical Theory of Communication" [549] which, based on probability theory and statistics, formulated the fundamental principles of *information theory* that led to countless applications for communication engineering, signal processing, coding theory, cryptography, and bioinformatics, only to name a few. Central to Shannon's vision was the idea of measuring the *information content* of possible messages or set of symbols, often encoded using binary digits, which are exchanged with given probabilities over a noise channel. In this context, he considered information as resulting from the resolution of an uncertainty, which naturally led to quantifying it through the entropy of its source. In Shannon's theory, the concept of entropy, which normally describes the degree of randomness in a random variable, is reinterpreted as the average content of information that we acquire when presented with one of the variable's possible outcomes. Note that according to this picture it is the "element of surprise" resulting from uncovering an unexpected outcome among many possible ones that constitutes the information content of the source. As a result, there is more entropy, and therefore more information is gained, when revealing the outcome from a rolling die with six equiprobable faces than from a coin flip, which only provides one bit of information.

More precisely, consider n independent and mutually exclusive choices of symbols that are generated by a source, each with probability p_i, with $i = 1, \ldots, n$. The average information content of this source, measured in bits per symbol, was defined the Shannon as follows:

$$H = -\sum_{i=1}^{n} p_i \log_2 p_i \qquad (10.46)$$

This expression, known as the *Shannon entropy of the source*, measures the amount of uncertainty associated with the values of a discrete random variable when only its distribution is known.[16] Note that when $n = 1$, there is no choice and the information content (or entropy) is equal to zero. The entropy is maximized when all the independent symbols are equiprobable ($p_i = p = 1/n$) and its maximum value is $\log_2 n$, which characterizes maximum randomness. Also note that when a source transmits a sequence of N independent and identically distributed symbols, then its entropy is NH bits (per N-symbols message). On the other hand, if the source symbols are identically distributed but not independent, their correlation decreases the entropy to a value less than NH. We now discuss the useful tool of the *approximate entropy (ApEn)*, which is a more sophisticated metric developed to characterize the chaotic behavior of complex systems [550]. We introduce this useful concept by considering again a source that generates n independent symbols with probabilities p_i, with $i = 1, \ldots, n$. A powerful new perspective is gained when computing the entropy associated to each block of consecutive k-symbols that appear in a given string. For instance, we could consider blocks of two consecutive symbols, or diagrams, and define the entropy per diagram still using equation (10.46), where now p_i refers to the probability of obtaining one among the n^2 possible diagrams and the summation runs over the range $i = 1, \ldots, n^2$. Triplets of consecutive symbols, or trigrams, contribute with n^3 possibilities. The same idea applies to k-grams, or blocks of consecutive k-symbols, for every k. When dealing with a finite-size string of length m, we can estimate the probability $p_i = n_i/N$ from the number n_i of occurrences of a particular i-block of length k in the sequence, with $i = 1, \ldots, n^k$, among the $N = m + 1 - k$ blocks of length k in the string.[17]

We can now define the *block-entropy* $H(k)$ corresponding to a block of size k in the usual way:

$$H(k) = -\sum_{i=1}^{n^k} \frac{n_i}{N} \log_2 \frac{n_i}{N}. \qquad (10.47)$$

---

[16] Note that since we often do not know the source, only the a posteriori probabilities p_i can be empirically estimated from the frequencies of symbols appearing in a (large) particular sequence before us, according to the Law of Large Numbers.

[17] It is not difficult to convince yourself that there are indeed $N = m + 1 - k$ blocks of length k in the string since these are exactly the blocks beginning at the first, second, ..., $m + 1 - k$th position along the string. For example, if $m = 6$ and $k = 3$, the blocks of length k in the string NUMBER are NUM,UMB,MBE,BER, which in fact are $N = m - k + 1 = 4$.

Note that while $H(1)$ is exactly equal to the previously defined Shannon entropy of the source, $H(k)$ with $k \neq 1$ encodes instead the correlations due to possible sequential dependencies of blocks of symbols in a string. Moreover, in this case we ignore the generating source and its probabilities can only be estimated a posteriori based on the frequencies of symbols in the generated string. Based on block entropy, we finally introduce the concept of *approximate entropy* ApEn(k) as follows:

$$\boxed{\mathrm{ApEn}\,(k) = H(k) - H(k-1)} \tag{10.48}$$

where by definition ApEn $(1) = H(1)$. The approximate entropy estimates the amount of new information contributed by the last symbol in a block given that we know already its predecessors in the block. This concept is also related to the conditional entropy of a k-block with respect to a block of length $k - 1$.

In the case of strings that are highly correlated or predictable, the knowledge of a given block already determines the succeeding symbol, and therefore adding new symbols into the block does not provide additional information. In all these cases, the ApEn will be very small, indicating that the string has certain structure that makes it highly redundant. In contrast, a random string of length n has maximum entropy and no discernible structure. In this case, we expect $H(1) = \log_2 n$ and $H(2) = \log_2 n^2 = 2 \log_2 n$ (since there are n^2 equally probable diagrams) and therefore ApEn $(2) = \log_2 n = H(1)$. More generally, the approximate entropy is maximized whenever any of the k-block appears in the string with equal frequency, defining a *normal sequence* or, for strings of bits, a *normal number*. Therefore, if a sequence of binary digits has maximal ApEn, then it is random and the real number represented by that random binary sequence is a normal number.[18] We recall that a real number is said to be normal to the base b if its infinite sequence of digits is distributed uniformly in the sense that each of the b digits appears with frequency (i.e., has density) $1/b$, and all possible blocks of digits of length k appear with frequency $1/b^k$ for all $k > 1$. A number is called absolutely normal if it is normal to all bases $b \geq 2$. Intuitively, the notion of absolute normality means that no digit, or finite block of digits, occurs more frequently than any other, and that this is true regardless of the base used to represent the number. Incidentally, almost all real numbers turn out to be absolutely normal and there is a link with u.d. mod(1) sequences established by the following important result of discrepancy theory [425]: a number α is normal to the base b if and only if the sequence $\{b^n \alpha\}$ is u.d. mod(1).

In Figure 10.14, we demonstrate the relevance of the approximate entropy concept in the quantification of the structural complexity of several prototypical deterministic aperiodic binary sequences compared against a periodic string, a random Bernoulli process with $p = 1/2$, and the output binary string of a cryptographically secure BBS generator. Note that all these systems have Shannon entropy $H(1) = 1$ but greatly differ in terms of the consecutive ApEn values, demonstrating varying degrees of

---

[18] However, the converse is not true, as there are normal numbers that are not random. A simple example us Champernowne's normal number generated by writing 0 and 1 followed by all pairs 00, 01, 10, 11, followed by all triplets 000, 001, etc.

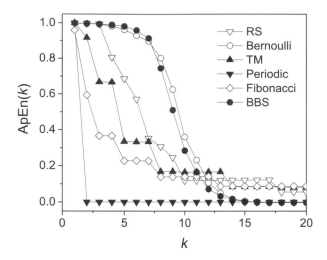

Figure 10.14 Approximate entropy ApEn(k) of different types of deterministic aperiodic binary sequences and of a 512-digit Bernoulli 1/2-process. In particular, we analyzed 512-digit Rudin–Shapiro (RS), Thue–Morse (TM), and periodic sequences as well as a 610-digit Fibonacci sequence. Results for a 420-digit binary sequence obtained by a Blum Blum Shub (BBS) generator are also shown.

redundancy and block correlations. In particular, the ApEn of the periodic sequence drops to zero dramatically already at the diagram correlation level ($k = 2$) because the second digit of any diagram is completely determined by the first digit. On the other hand, the ApEn(k) curves of the Fibonacci, Thue–Morse, and Rudin–Shapiro binary strings drop more gently, indicating a progressively weaker degree of structural correlation, consistent with their spectral properties. Interestingly, the nonrandom nature of the RS sequence is clearly unveiled by the sudden drop of its ApEn that begins at $k = 4$, where ApEn$(4) = 0.8$. When conducted beyond the $k = 8$ block level, the approximate entropy analysis indicates the presence of latent regularities and block correlations that appear even for the more structurally complex BBS and Bernoulli strings, clearly exposing their pseudorandom nature.

The ApEn quantifies the conditional uncertainty of new data/observations given the past history of the system under study. Therefore, it is very sensitive to dependencies that occur over the shortest scales. However, many complex signals, particularly in the biological sciences, are characterized by structural organizations at multiple scales. Recently, the concept of *multiscale entropy (MSE)* was introduced in order to address these more complex situations [551, 552]. In the MSE analysis of a one-dimensional discrete time series $\{x_1, x_2, \ldots, x_N\}$, we consider coarse-grained versions $\{y^{(\alpha)}\}$ scaled by the factor α that is constructed by dividing the original signal into nonoverlapping consecutive windows of length α and then averaging the data points inside each window. In general, each element in the coarse-grained series is computed by averaging over groups of elements in the original signal according to the following:

$$y_j^{(\alpha)} = \frac{1}{\alpha} \sum_{i=(j-1)\alpha+1}^{j\alpha} x_i \qquad (10.49)$$

with $1 \leq j \leq N/\alpha$. The scale $\alpha = 1$ corresponds to the original time series while the length of each coarse-grained series will be rescaled by the factor α. Finally. the entropy calculated for each coarse-grained series is plotted as a function of the scale factor α. If, in comparing two time series, we find that for the majority of scales the entropy is larger for one series than for the other, then the former is more complex than the latter. On the other hand, a monotonic decrease of the multiscale entropy indicates that the analyzed signal contains information localized only at the smallest scales. This is indeed the case when comparing white noise time series with $1/f$-noise time series, which contain complex structures across multiple scales and new information is revealed at all scales [552].

In the next section, we introduce the approach of statistical inference for testing the likelihood that a particular finite-size binary string results from chance.

10.5.2 Testing Randomness

At the beginning of the twentieth century, the applied mathematician and philosopher Richard von Mises attempted to develop a precise "test for randomness" from which to define operationally a random sequence as one that passes all the available tests. Many sophisticated tests for randomness have been developed based on statistical inference, which include the *Kolmogorov–Smirnov test* to assess the discrepancy between an empirical distribution function and an hypothesized (or reference) distribution function or for comparing two samples from the same distribution. The discrepancy between an hypothesized and the empirical (measured from data) distribution, known as the Kolmogorov–Smirnov test statistic, is evaluated against the *null hypothesis* H_0 that the empirical sample is drawn from the hypothesized distribution [553].

Before dealing with a simple application of this statistical method, we consider again the Bernoulli 1/2-process that generates independent binary digits with probability 1/2. In such random binary processes, the proportion of ones obtained in n independent trials (i.e., the relative frequency of ones) should approach the probability p as n increases. Precisely, if we denote by S_n the number of ones obtained in n trials, the random quantity $m_n = S_n/n$, called the *sample average*, will converge in probability to p as n increases. This familiar observation was first precisely formulated by Jakob Bernoulli and became known as the *Law of Large Numbers*, which states that the probability of the event "m_n is within a (small) preassigned absolute distance from p" (i.e., $|S_n/n - p|$ in repeated trials) tends to one as n grows without bounds. In particular, the Law of Large Numbers implies that in tossing a fair coin (where $p = 1/2$), the randomly fluctuating sample average $m_n = S_n/n$ will settle down arbitrarily close to $1/2$ in a sufficiently long sequence of flips. This is a very important result because it allows us to estimate the exact or a priori probability p of the process from the observed (a posteriori) behavior of the sample average for large values of n. However, it should be clearly appreciated that the probability p is defined by considering (large) *ensembles* of n independent trials. These are sets of independent experiments containing all possible binary strings of a given length n. The confusion between the level of the individual realization of a random process (i.e., the given string of length n) and its ensemble lies at the heart of many paradoxes and misconceptions in probability theory. A prominent one is the wrong belief that observing unusually long strings of zeros increases the probability of observing a flipping into ones in the next trial. In reality, due to the independence of coin tossing, the probability of observing another zero remains equal to $p = 1/2$ at every toss (with often unpleasant consequences for naive gamblers!). The truth of the matter is that we need to consider the ensemble of all the binary strings of a given length. In this ensemble, the probability of observing unusually long blocks of zeros is indeed small. Meaningful ensemble averaged quantities (estimators such as m_n or the variance) are obtained only when considering many identical trials on strings with the same length n. The Bernoulli $1/2$-process is the primary example of a binary random process that produces a succession of zeros and ones independently and with each digits having the same chance of occurring. However, although the Law of Large Numbers guarantees that when n is big enough the sample average as a random variable has a small probability of deviating from $1/2$ less than some small preassigned value, it says nothing about the possibility that S_n/n will fail to be close to $1/2$ in subsequent trials. In other words, there will still be fluctuations of the quantity S_n/n from its expected value p, although they will occur less and less frequently as n increases. The likelihood of such fluctuations in Bernoulli p-processes is exactly quantified by the *de Moivre's theorem*. This foundational result of inferential statistics allows one to estimate the probability that the sample average S_n/n lies within a specified distance, conventionally measured in terms of $\sigma/\sqrt{n}$, from the exact probability p of the process. Here σ denotes the standard deviation of the process. Specifically, the theorem states that when n is large, the percentage of sample averages m_n that fall within the so-called *confidence interval* $(p - t\sigma/\sqrt{n}, p + t\sigma/\sqrt{n})$ for a given value of t is approximately equal to the fraction of the total area under the normal (Gaussian) curve between $-t$ and t. The error made in this approximation decreases as n gets larger, and the smaller will be the interval around p in which to expect 95% of the sample averages to fall.[19] The area under the normal curve can be readily computed numerically for any values of t. One choice of t that is particularly used in applications is $t = 1.96$, which yields a probability of 0.95. For this particular choice,

---

[19] A Bernoulli process obeys the binomial distribution with $\sigma = \sqrt{p(1-p)}$. Therefore, de Moivre's theorem states that the normal distribution may be used as an approximation to the binomial distribution under certain conditions.

according to de Moivre's theorem we expect that the sample average S_n/n lies within the confidence interval with 95% probability, or 19 out of 20 cases. Finally, for a Bernoulli 1/2-process the 95% confidence interval can be approximated as follows:

$$\boxed{\left(\frac{1}{2} \pm \frac{1}{\sqrt{n}}\right)_{0.95}} \tag{10.50}$$

where we used the approximation $1.96/2 \simeq 1$.

We illustrate how a simple application of de Moivre's theorem allows us to reject the null hypothesis that a given binary string originates from a Bernoulli 1/2-process random process within a specified level of confidence. Despite its simplicity, the following example fully highlights the typical reasoning employed in hypothesis testing or statistical inference. We select the typical value of 0.95 for the confidence interval so that we have a 95% probability that the sample average $m_n = S_n/n$ deviates from $1/2$ in absolute magnitude by less than $1/\sqrt{n}$. We deduce that for a binary string with $n = 100$ digits the confidence interval is approximately $(0.5 \pm 0.1)_{0.95}$, implying that in a series of trials with strings of $n = 100$ digits, the sample average should lie within this interval with 95% probability, or in 19 out of 20 cases. On the other hand, if we observed a particular binary string containing 70 ones and 30 zeros, the corresponding sample average ($S_n/n = 0.7$) is outside the bounds of the confidence interval and the null hypothesis that the string originates from a 1/2-Bernoulli random process must be rejected at the specified 0.95 confidence level. If we observed instead a string with a sample average $S_n/n = 0.45$, which is inside the 0.95 confidence interval, we would not reject the null hypothesis but, importantly, we will not be able to accept it either. In fact, even when the sample average falls inside the confidence interval, there remains always the possibility that such a string was algorithmically constructed by a different random process (e.g., a biased one) with respect to the one specified by the H_0. In addition, it is also possible to reject a sequence generated by a Bernoulli-1/2 random process because there remains a 5% chance of error from our chosen confidence interval level. A better accuracy could be achieved by employing a more demanding confidence level, say a 99% level for more reliable inferences, although a 95% level is typically accurate enough for most practical applications. Therefore, in statistical inference the relevant decision is always between rejecting the null hypothesis H_0 or avoiding its rejection. The option of accepting it is simply unwarranted in this context.

While pedagogically relevant, the preceding approach is very simplistic because it considers only the relative frequency of zeros and ones in the string as an indicator of its randomness. More powerful methods test the randomness of a binary sequence by considering the estimated distribution of elements appearing in a sequence of runs, which are subsegments of the original sequence consisting of consecutive equal elements (e.g., the sequence $AABAAAABBA$ contains five runs, three of which consisting of the element "A" and two of the element "B"). In the so-called *runs tests*, under the null hypothesis, the number of runs in a sequence is a random variable whose conditional distribution given the observed A-type and B-type runs is approximately Gaussian. If the number of observed runs is significantly higher or lower than what was expected, the hypothesis of statistical independence of the elements can be rejected. More detailed information on statistical testing and inference can be found in the excellent book [553].

10.5.3   Randomness and Structure

We have seen in the previous section that in order for an unlimited binary sequence to be considered random (i.e., originated from a random process), it must feature maximal ApEn at all k-block levels (for any $k > 1$), implying its normality. However, this mathematical property cannot certainly be established for sequences of finite length. In fact, while we can always exclude with a given confidence level that a finite-size string originates from a random process, for instance using the ApEn analysis or statistical testing, we cannot conclusively prove its randomness. This situation is due to the presence of statistical fluctuations in random processes, which may well output snippets of random sequences that show highly organized patterns with arbitrarily small algorithmic randomness (e.g., a snippet consisting of a periodic string of symbols). More generally, by analyzing a specific finite pattern, it is not possible to establish if it originated from a random process. This is established by a theorem published by F. P. Ramsey in 1930 when working on foundational problems of formal logic [554], and now a part of the broader *Ramsey theory*. According to this result of combinatorics, a sufficiently large structure (represented as a graph) must necessarily contain any given substructure. As a consequence, complete randomness is impossible for finite structures. Moreover, thanks to Chaitin's incompleteness theorem, we know that within any (powerful enough) axiomatic system we cannot formally prove whether a binary string has complexity larger than or equal to a sufficiently large number. Since random strings have algorithmic complexity comparable to their lengths, it follows that no formal proof of the randomness is even possible. Regardless of the axioms and rules of inference chosen in the system, there will always be some long string whose randomness cannot be established within that system, and it may in fact be a truly random one! Establishing randomness is ultimately an *undecidable problem* and an answer to the natural question, "Is it random?" can never be provided with complete certainty.

11 Aperiodic Order and Geometry

Perfect symmetry is aesthetically boring, and complete randomness is not very appealing either.

Number Theory in Science and Communication, Manfred R. Schroeder

The geometrical shapes of gems and minerals have intrigued naturalists since the beginning of recorded history. The Greek philosopher Aristotle (384–322 BC) wrote about minerals and their properties and Pliny the Elder devoted five volumes of his monumental work *Historia Naturalis* (77 AD) to the "classification of earths, metals, stones, and gems." However, the scientific study of minerals and rocks flourished in post-Renaissance Europe and was founded on the principles of crystallography in combination with the morphological study of rocks after the invention of the microscope in the seventeenth century.

It was probably the French priest and mineralogist René-Just Haüy (1743–1822) who first conceived a "building block model" for the description of crystals with different external shapes. According to his ideas, crystals were composed of *integral molecules* that can be arranged in different ways in order to construct a large number of macroscopic structures. By assuming the existence of a single "elementary shape," Haüy was able to build macroscopic wooden models that explained a large variety of crystal forms (see Figure 11.1). Haüy's block model remains the correct starting point for the definition of *space lattices* and their diffraction properties that we discuss in this Chapter. Building on the concepts of lattices and $\mathbb{Z}$-modules, we analyze the fundamental classes of deterministic aperiodic structures and present the ideas of mathematical diffraction theory.

11.1 The Mathematics of Lattices

The lattice concept can naturally be extended to the n-dimensional Euclidean space E^n, which plays a central role in the modern theory of periodic and aperiodic structures (particularity quasicrystals).

Lattices began to be rigorously studied as more abstract mathematical notions in the eighteenth century by Gauss before they became primary objects in the number-theoretic field known as the "geometry of numbers," pioneered by Minkowski [409, 516]. More recently, lattices have attracted significant attention in computer

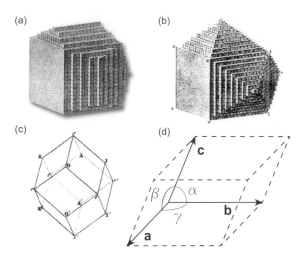

Figure 11.1 (a,b) Regular shapes of crystals obtained by stacking layers of cubic building blocks. Following this approach, Haüy built wooden models that reproduced many crystal forms, including the rhombic dodecahedron shown in (c). Adapted from Haüy's *Traité de Crystallographie* (Paris, Delance 1822). (d) Illustration of the general unit cell formed by the three basis vectors **a**, **b**, and **c**. The angles between the basis vectors are also indicated. The volume of the unit cell is $(\mathbf{a} \times \mathbf{b}) \cdot \mathbf{c}$.

science, coding theory, and cryptography, where they offer a large number algorithms and computationally hard problems for the design of better cryptographic functions and error-correcting codes [555, 556].

Consider E^n and a set of n linearly independent basis vectors $\{\mathbf{b}_1, \mathbf{b}_1, \ldots \mathbf{b}_n\}$. We can form the set of all the integral linear combinations:

$$\boxed{\mathbf{x} = m_1\mathbf{b}_1 + \cdots m_k\mathbf{b}_n} \tag{11.1}$$

where $\{m_i\} \in \mathbb{Z}$ are integer coefficients. Spanning over all the integer values $\{m_i\}$ and representing the resulting vectors as points in E^n gives rise to the *n-dimensional lattice* $\mathcal{L}$. According to the more general approach based on group theory, a lattice can be regarded as the *discrete orbit* of the group formed by the addition/subtraction of the set of linearly independent base vectors with integer coefficients. More generally, any set of vectors in the Euclidean space E^n (i.e., $\mathbb{R}^d$ with the Euclidean metric) generates a countably infinite Abelian group under addition and subtraction that is called a $\mathbb{Z}$-*module*.

The **norm** of a vector measures its squared length, defined as follows:

$$N(\mathbf{b}_i) = |\mathbf{b}_i|^2 = \mathbf{b}_i \cdot \mathbf{b}_i, \tag{11.2}$$

where the symbol $\cdot$ denotes the standard scalar product in E^n. We say that a lattice is an *integral lattice* if every vector in the lattice has integer norm.

To every n-dimensional lattice we can associate a matrix $\mathcal{B}$, called the *generator matrix* or *basis matrix*, whose columns are the components of the basis vectors $\{\mathbf{b}_i\}$. The relation between generator matrices and lattices is not unique. In fact, if $\mathcal{B}$ is the generator matrix of a lattice $\mathcal{L}$, also the matrix $\mathcal{C} = \mathcal{R}\mathcal{B}$ (with right-multiplication by

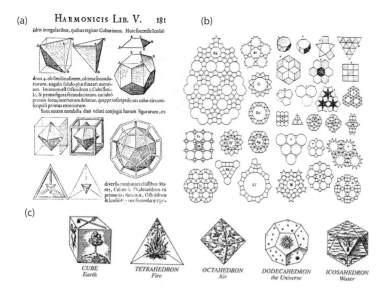

Figure 11.2 (a) A page from Johannes Kepler's *Harmonices Mundi*, book V, Linz, 1619. (b) Kepler's drawings of polygon packing showing the 11 Archimedean tilings. (c) Kepler's drawing of the five Platonic solids (regular convex polyhedra) and Plato's association of each shape to the four elements and to the entire universe. Adapted from Kepler's *Harmonices Mundi*

$\mathcal{B}$ only) will generate the same lattice whenever the matrix $\mathcal{R}$ belongs to the general linear group $GL(n, \mathbb{Z})$ of unimodular $n \times n$ matrices with integer coefficients. This property reflects the obvious fact that every lattice can be spanned by a countable infinity of different bases. However, the determinant of the generator matrix is independent of the choice of the basis vectors and is therefore a property of the lattice $\mathcal{L}$. Moreover, we can construct the symmetric matrix:

$$\mathcal{G} = \mathcal{B}^T \mathcal{B} \tag{11.3}$$

Obviously, when the Gram matrix $\mathcal{G}$ is the $n \times n$ identity matrix, the basis vectors form an orthonormal basis set and the associated lattice coincides with the *unit hypercubic lattice* $\mathcal{I}_n$, also known as the standard or the Cartesian lattice. When considering n-dimensional spaces, it remains true that every lattice $\mathcal{L}$ has an associated dual lattice $\mathcal{L}^*$ that consists of the set of vectors $\mathbf{y} \in E^n$ whose scalar product (in E^n) with every vector $\mathbf{x}$ of $\mathcal{L}$ is an integer. Note that the inessential 2π factor has been disregarded. It can be shown that if $\mathcal{B}$ is a generator matrix for $\mathcal{L}$, then $[\mathcal{B}^{-1}]^T$ is a generator matrix for the dual lattice $\mathcal{L}^*$ and its Gram matrix $\mathcal{G}^* = \mathcal{G}^{-1}$.

11.1.1 Lattices and Tilings

Tilings and patterns have fascinated humankind since pre-historic times. Early examples of tessellations that use clay tiles to realize decorative patterns were already used by the Sumerians around 4000 BC, and tessellations with colored tiles were discovered in the Sumerian city of Uruk, dating back to approximately 3400–3100 BC [557]. Decorative tessellations were also used frequently in Ancient Rome [558] and in Islamic art [559], where impressive examples are displayed in the geometric motives of the Alhambra palace in Granada, Spain. However, the scientific study of planar tilings as set of tiles that cover the plane without gaps or overlaps dates back to the work by Johannes Kepler, published in his 1619 book *Harmonice Mundi* [560] (see Figure 11.2). This work was largely neglected for almost three centuries until the Russian crystallographer Yevgraf Fedorov (alternative spelling Fyodorov) proved that the periodic tiling of the plane can be grouped into only 17 different classes according to their symmetries [561]. In the twentieth century, the Dutch graphic artist M. C. Escher explored many tessellations, both in ordinary Euclidean geometry and in the hyperbolic one, producing intriguing artistic effects. Excellent introductions to the mathematics of tilings can be found in the books *Tilings and Patterns* [562] and *Quasicrystals and Geometry* [39], to which we refer interested readers.

There is an obvious connection between lattices and tilings: to every lattice, we can associate the tiling of the n-dimensional space created by the unit cells of the lattice, i.e., parallelepipedal blocks whose edges are parallel to the basis vectors of the lattice. In the mathematical theory of lattices, unit cells are referred to as the *fundamental regions* associated to the generator matrix. Any unit cells can be considered, but often the primitive unit cells are selected. In the case of $n \times n$ matrices, the volume of the (primitive) cell is simply given by the determinant of the generator matrix of the lattice, which we indicate as $|\mathcal{L}| = \det \mathcal{B}$. Alternatively, this volume can also be obtained as the square root of the *Gram determinant*, which is the determinant of the Gram matrix. Moreover, the vectors $\{\mathbf{v}_1, \mathbf{v}_2, \ldots, \mathbf{v}_n\}$ are linearly independent if and only if the Gram determinant is nonzero, or equivalently, if and only if the Gram matrix is nonsingular. The Gram determinant is a fundamental quantity associated to any lattice, since it does not depend on the choice of the basis, besides possessing a clear geometrical meaning as the n-dimensional volume of the primitive unit cell of the n-dimensional lattice. Since every lattice admits an infinity of different bases, it follows that it can be partitioned into unit cells in an infinity of different ways, one corresponding to each choice of the basis. Moreover, as we already discussed, unit cells do not always display the symmetry of the lattice. Nevertheless, a primitive unit cell that displays the symmetry of two- and three-dimensional lattices can be easily constructed, resulting in the **Wigner–Seitz cell**. The Wigner–Seitz cell is a very important object because it does not depend on the choice of basis vectors and it has the same volume as any other primitive unit cell. In the more general n-dimensional case, the Wigner–Seitz cell, referred to as the **Voronoï polytope** of the lattice $V(\mathbf{x})$, is similarly defined for each lattice point $\mathbf{x}$ as the smallest convex region around $\mathbf{x}$ bounded by the $(n-1)$-dimensional hyperplanes that perpendicularly bisect the straight line segments connecting $\mathbf{x}$ to its neighboring points on the lattice. Since $V(\mathbf{x})$ is defined separately for each point, it can be constructed also for finite or irregular point patterns in E^n, where each point will correspond in general a different $V(\mathbf{x})$. We refer to such point patterns as Delone sets, which will be discussed in more detail in Section 11.4. Considering the general case of a Delone set $\Lambda \in E^n$

(a)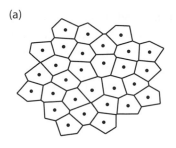
(b)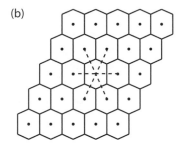

Figure 11.3 Examples of Voronoï cells for (a) a nonperiodic point pattern (i.e., a golden angle Vogel spiral pattern) and (b) an oblique lattice.

(i.e., not necessarily a lattice), the Voronoï cell of point $\mathbf{x} \in \Lambda$ is the set of points in E^n defined by the following:

$$V(\mathbf{x}) = \{\mathbf{v} \in E^n \,|\, |\mathbf{x} - \mathbf{v}| \leq |\mathbf{y} - \mathbf{v}|, \text{ for all } \mathbf{y} \in \Lambda\} \qquad (11.4)$$

This definition identifies the set of points of E^n that lie at least as close to $\mathbf{x}$ as to any other point of Λ. By carrying out this construction for every point of Λ, we create a partition of E^n called the *Voronoï tessellation* induced by Λ. When Λ is a point lattice, then its Voronoï cells are congruent and centrosymmetric polytopes (see Figure 11.3).

Polytopes are geometrical figures bounded by portions of lines, planes, or more generally hyperplanes. In two spatial dimensions, they are polygons, and in three dimensions they are polyhedra. A clear and engaging introduction to the general theory of polytopes can be found in the classic book by Coxeter [563]. The $(n - 1)$-dimensional faces of a Voronoï cell are referred to as *facets*, and the vectors joining $\mathbf{x}$ to its neighboring lattice points whose Voronoï cells share a facet with $V(\mathbf{x})$ are called *facet vectors*. An important result due to Minkowski [564] states that the Voronoï cell of any point lattice in E^n has at least $2n$ and at most $2(2^n - 1)$ facet vectors. This implies that the Voronoï cells of two-dimensional point lattices can only consist of rectangles and hexagons. There are five types of Voronoï cells for three-dimensional point lattices, which were discovered by Fedorov in his studies of the properties of polyhedra that can fill the space by translations [561, 565], now called Fedorov's parallelohedra. There are 52 Voronoï cells in E^4 but their number in higher-dimensional spaces is an extremely challenging problem that currently has no general solution.

11.2 Crystalline Symmetries

Periodic structures display a specific kind of long-range order characterized by translational invariance symmetry along certain spatial directions. More generally, a symmetry of an object is any transformation that leaves it unchanged. More precisely, we say that an object remains "unchanged" after a symmetry transformation if all the distances between its points do not change and thus the object can be brought into coincidence with itself (no stretching or twisting allowed). These transformations are distance-preserving maps also called *isometries* (rigid motions). It can be shown that any isomeric transformation can be reduced to either a translation, a rotation, a reflection, or a combination of these operations. Note that the identity operation, which does nothing to the object, is also considered as one of its symmetry operations. Since any operation can be undone by its inverse, the symmetry operations of an object form a group that includes translations that carry one lattice point into another and all the symmetry operations that fix a lattice point.

It is also useful to distinguish between transformations that are physically realizable, such as translations, rotations or combinations known as proper transformations, and improper transformations, such as reflections or inversions, which change the handedness of an object and cannot be physically realized without deforming it.[1] Rotations are represented by the group of orthogonal matrices, which are matrices whose inverse is equal to their transpose. The determinant of a proper rotation matrix is always equal to $+1$. On the contrary, all the transformations that change the handedness of an object are represented by orthogonal matrices with determinant equal to -1.

Crystal structures in three-dimensional space are mathematically described by 32 crystallographic point–group symmetries, which are combinations of pure rotations, reflection, and their combinations (i.e., mirrors, and rotoinversion operations) compatible with the overall translational symmetry of the 14 Bravais lattices. The addition of translation operations produces *crystallographic space groups*, which have been completely enumerated in 230 different types by Fedorov and Schönflies in 1891 [561].

11.2.1   Point and Space Groups

In order to understand the role of symmetry in crystal structures, it is important to distinguish between two types of isometries, the ones that keep at least one point fixed and the ones that do not. A *point group* is a group of geometric symmetries (i.e., isometries) that leave a point fixed. Point groups can exist in a Euclidean space of any dimension. The discrete point groups in two dimensions, also called *rosette groups*, are used to describe the symmetries of ornaments. There are infinitely many discrete point groups in each number of dimensions. General point groups can be divided into crystallographic point groups and noncrystallographic point groups. The crystallographic point groups are the ones compatible with the translational symmetry of Bravais lattices. In three spatial dimensions, there are only 32 possible crystallographic point groups, while this number reduces to 10 in two spatial dimensions. On the other hand, *space groups* contain all the symmetry operations of the crystal, including both point-symmetry operations, such as rotations, as well as discrete translations.

---

[1] We are all familiar with the fact that in order to change the handedness of an object we need to deform it. In fact, to turn a right-handed glove into a left-handed one, we must turn it inside out, which is not an isometry transformation.

Therefore, the point group of a given space group is the subgroup of symmetry operations that leave one point fixed (i.e., proper and improper rotations). The pioneering works of Federov [561] and Schönflies in 1891 [566], and of Barlow in 1894 [567] led to the complete enumeration of the 230 space group symmetries of three-dimensional crystal structures. On the other hand, in two spatial dimensions, the combination of the 10 point groups with the five different Bravais lattices yields a total of 17 two-dimensional space groups, also known as *plane groups*. The elements of space groups are denoted by the *Seitz symbol* $\{R|\mathbf{t}\}$ that consists of a pair made of an orthogonal transformation R and a translation $\mathbf{t}$ acting on a position vector $\mathbf{r}$ such that $\{R|\mathbf{t}\}\mathbf{r} = R\mathbf{r} + \mathbf{t}$. In three dimensions, Seitz symbols can be represented by the 4×4 matrices composed by a 3×3 rotation submatrix D in the upper-left corner, a 3×1 translation vector $\mathbf{t}$ in the top-rightmost column, and always the 1×4 row vector $(0, 0, 0, 1)$ at the bottom. As a concrete example, let us consider the Seitz symbol for a pure rotation, denoted as $\{R|\mathbf{0}\}$, which in matrix form is expressed as follows:

$$\{R|\mathbf{0}\} = \begin{pmatrix} D_{1,1} & D_{1,2} & D_{1,3} & 0 \\ D_{2,1} & D_{2,2} & D_{2,3} & 0 \\ D_{3,1} & D_{3,2} & D_{3,3} & 0 \\ 0 & 0 & 0 & 1 \end{pmatrix}. \tag{11.5}$$

The set of all orthogonal transformations R in the space group elements forms the point group, which is isomorphic to the quotient of the space group and its translation group $P = S/T$, where S is the space group, T is the translation group, and P is the point group. It is also important to realize that the translation part of space group elements may or may not belong to the translation group. If it belongs to the translation group, it is called a *primitive translation*. Otherwise, it is called a nonprimitive translation. Three-dimensional group elements with nonprimitive translations are linear transformations with either a screw axis (i.e., composition of a rotation by an angle about an axis, called the screw axis, with a translation along this axis) or a glide plane (i.e., a symmetry operation consisting of a reflection in a plane, followed by a translation parallel with that plane). This allows one to further subdivide the 230 crystallographic space groups into the 73 *symmorphic space groups*, which do not contain any screw rotations or glide planes, and the 157 nonsymmorphic groups, which contain those special symmetry operations.

11.2.2 The Crystallographic Restriction

The use of matrices to represent the elements of symmetry groups enables a very elegant demonstration of the crystallographic restriction theorem that sets a fundamental limit to the possible rotation symmetries of periodic lattices. In order to prove this theorem, we must first realize that a point-group transformation R is a symmetry of the lattice, i.e., $\mathbf{r}' = R\mathbf{r}$ for lattice points $\mathbf{r}$ and $\mathbf{r}'$. It follows that the matrix elements of R with respect to the basis of the lattice vectors are integers i.e., R is an integer matrix on a lattice basis.[2] As a consequence, the trace of R is an integer, or $\mathrm{Tr}\, R \in \mathbb{Z}$. Choosing an orthogonal basis, the matrices representing the point-group transformations R are orthogonal matrices and can be written as follows:

$$M(R) = \pm \begin{pmatrix} \cos\phi & -\sin\phi & 0 \\ \sin\phi & \cos\phi & 0 \\ 0 & 0 & 1 \end{pmatrix}. \tag{11.6}$$

Remembering that that the trace of a matrix is invariant under a basis transformation, we deduce that $\mathrm{Tr}\, M = 1 + 2\cos\phi \in \mathbb{Z}$. However, this condition can only be satisfied when the rotation angle $\phi = \pi, 2\pi/3, \pi/2, \pi/3$ or 0. Therefore, the only possible rotations are two-, three-, four-, or sixfold rotations, or the identity operation. The condition

$$\boxed{1 + 2\cos\phi \in \mathbb{Z}} \tag{11.7}$$

is called the *crystallographic condition*, and it restricts the number of possible rotations in a point group of a lattice. The proof can be generalized to higher dimensions. Analogously, the group-theoretic approach to symmetry transformations can be extended to the n-dimensional space E^n in which isometries generalize group operations. Isometries are transformations ϕ of E^n onto itself that preserve the Euclidean distances between points $|\phi(\mathbf{x} - \mathbf{y})| = |\mathbf{x} - \mathbf{y}|$ for all $\mathbf{x} \in E^n$. Linear isometries in E^n can be representant by $n \times n$ unimodular matrices, i.e., matrices with determinant equal to ± 1 corresponding to proper and improper rotations. The isometries that fix the origin are linear isometries [39]. The ones that have no fixed points are generally translations. By iterating linear isometries and their inverses, we can generate a cyclic group of transformation with elements: $I, \phi^{\pm 1}, \phi^{\pm 2}, \ldots$, where I denotes the identity transformation. If there exists a positive integer k such that $\phi^{\pm k} = I$ (and no smaller integer has this property), we say that the isometry ϕ has order k. As an example, reflections are isometries of order $k = 2$ while a rotation by $2\pi/n$ has order $k = n$. In contrast, a translation is an isometry of infinite order. The characterization of isometries in E^n is discussed in the book by Senechal [39], which we recommend to interested readers.

It is possible to establish a link between the possible symmetry operations and the dimensionality of lattices in E^n. Specifically, we can predict the minimal dimensionality of the Euclidean space for which we will have a linear isometry of order k. In particular, if we denote by $n(k)$ the least value of the dimensionality of the Euclidean space for which an element of order k appears in the group of linear isometries, then we have the following [39]:

$$\boxed{n(k) = \Phi(k)} \tag{11.8}$$

---

[2] In the lattice basis, the rotation operation maps every lattice point into an integer number of lattice vectors, so the entries of the rotation matrix in the lattice basis must necessarily be integers.

where the function Φ is defined as follows:

$$\Phi(n) = \begin{cases} \phi(n), & \text{if } n = p^\alpha \text{ where } p \text{ is prime and } \alpha \in \mathbb{N}. \\ \Phi(n_1) + \Phi(n_2), & \text{if } n = n_1 n_2 \text{ and } n_1 \text{ and } n_2 \text{ are coprime.} \end{cases} \tag{11.9}$$

In this formulation, $\phi(n)$ is the Euler totient function. An alternative formulation of this result due to Hiller [568] states that k can occur as the order of an element of the point group of an n-dimensional space group if and only if $n \geq \Phi(k)$. Therefore, since $\Phi(4) = 2$, a fourfold rotation is crystallographic (it leaves the lattice invariant) in two dimensions. On the other hand, since $\Phi(5) = 4$, rotations of order 5 first appear in E^4. This shows how rotations' axis that are noncrystallographic in dimension $n = 3$ may become crystallographic in higher dimensions. A proof of this result is provided in [568].

11.3 From Lattices to $\mathbb{Z}$-Modules

The diffraction peaks of periodic crystals can be labeled in reciprocal space using three integer indices, i.e., the Miller indices. Therefore, the positions of the diffraction peaks can be specified as follows:

$$\mathbf{G} = m_1\mathbf{g}_1 + m_2\mathbf{g}_2 + m_3\mathbf{g}_3, \tag{11.10}$$

where (m_1, m_2, m_3) are integers. However, in certain complex materials, such as the γ-phase of Na_2CO_3, there appear satellite spots and a general diffraction vector is instead expressed as follows:

$$\mathbf{G} = m_1\mathbf{g}_1 + m_2\mathbf{g}_2 + m_3\mathbf{g}_3 + m_4(\alpha\mathbf{g}_1 + \gamma\mathbf{g}_3), \tag{11.11}$$

where the coefficients α and γ are irrational numbers. In such structures, the diffraction peaks corresponding to $m_4 = 0$ are called main reflections and the ones with nonzero m_4 are called satellites. Since α and γ are irrational numbers, there is no three-dimensional basis for which $\mathbf{g}_1$, $\mathbf{g}_2$, $\mathbf{g}_3$, and the vector $\mathbf{q} = \alpha\mathbf{g}_1 + \gamma\mathbf{g}_3$ have simultaneously integer coordinates. As a consequence, this structure cannot display lattice periodicity. However, we may regard its Bragg peaks as projections on the three-dimensional space of a four-dimensional reciprocal lattice. This is in fact the central idea of *quasicrystallography*, which utilizes higher-dimensional space groups to describe more complex structures that lack lattice periodicity. The example provided in (11.11) is typical of an ordered structure with sharp diffraction Bragg peaks without spatial periodicity. More generally, a structure is called *quasiperiodic* if its Bragg diffraction peaks satisfy the following equation:

$$\boxed{\mathbf{G} = \sum_{i=1}^{n} m_i \mathbf{b}_i^*} \tag{11.12}$$

where $m_i \in \mathbb{N}$ are integers and n exceeds the dimensionality of the three-dimensional physical space. A basis for its lattice is provided as usual by the integer span of the vectors $\mathbf{b}_i$, where

$$\mathbf{b}_i^* \cdot \mathbf{b}_j = 2\pi\delta_{ij}, \tag{11.13}$$

and we used an asterisk to denote the basis vectors of the dual space $\{\mathbf{b}_i^*\}$. It is important to realize that when $n > d$ (where d is the dimension of the space spanned by the primitive lattice vectors), the set of vectors defined in (11.12) can still be linearly independent because we are considering only linear combinations of vectors with integer (or rational) coefficients.[3] The number of rationally independent (i.e., linearly independent with rational coefficients) basis vectors is called the *rank* of the reciprocal space, and the set of all the vectors in (11.12) is called the *Fourier module*. Since only integral or rational linear independence matters here, in general the rank of a $\mathbb{Z}$-module can exceed the dimensionality of the physical space. As a result, Fourier modules naturally generalize the familiar concept of the reciprocal space to the case of quasiperiodic structures.

We now address $\mathbb{Z}$-modules from the more general group-theoretic perspective that enables a better understanding of the diffraction space of aperiodic structures. Consider the n-dimensional Euclidean space E^n with an orthonormal basis $\{\mathbf{e}_1, \mathbf{e}_1, \ldots \mathbf{e}_n\}$. Any set of vectors $\{\mathbf{b}_1, \mathbf{b}_1, \ldots \mathbf{b}_k\}$ with $k \neq n$ generates a countably infinite group under addition and subtraction called a $\mathbb{Z}$-*module*. The elements of the $\mathbb{Z}$-module are vectors of the form $\mathbf{x} = m_1\mathbf{b}_1 + \cdots m_k\mathbf{b}_k$ with $\{m_i\} \in \mathbb{Z}$ a set of integer numbers. Spanning over all the integer values $\{m_i\}$ and representing the resulting vectors as points in E^n produces the so-called *orbit* Ω of the $\mathbb{Z}$-module.

The number of *linearly independent*[4] vectors $\{\mathbf{b}_j\}$ is the *rank* of the $\mathbb{Z}$-module. These linearly independent vectors act as the *generators* of the orbits. In general, the orbits of $\mathbb{Z}$-modules in E^n consist of densely filled regions of points, or translates of them. Representatives examples of orbits in E^2 for different $\mathbb{Z}$-modules are shown in Figure 11.4. The nature of the orbits varies dramatically from periodic sets of points, as in the case of the discrete orbits in Figure 11.4a,b, to dense orbits with points arranged along parallel lines (subspace of dimension $d = 1$) in Figure 11.4c to a two-dimensional dense subspace of E^2 as in Figure 11.4d. When the rank of the $\mathbb{Z}$-module equals the dimension of the subspace spanned by $\{\mathbf{b}_1, \mathbf{b}_1, \ldots \mathbf{b}_k\}$ (i.e., the span), then the orbit Ω is called a *lattice*, which is the case in Figure 11.4a,b. In our example with $n = 2$ and $k = 3$, a lattice is obtained whenever the generating vector $\mathbf{b}_3$ can be expressed as a rational linear combination (i.e., a linear combination with rational number coefficients) of the vectors $\mathbf{b}_1$ and $\mathbf{b}_2$. Using the parameters indicated in the caption of Figure 11.4, we see that this is indeed the case for Figure 11.4a,

---

[3] This will not be the case if we considered the usual definition of linear independence where a linear vector space is considered over the field of real numbers.

[4] More properly, we should say **integrally independent** since the linear combination coefficients are all integer numbers.

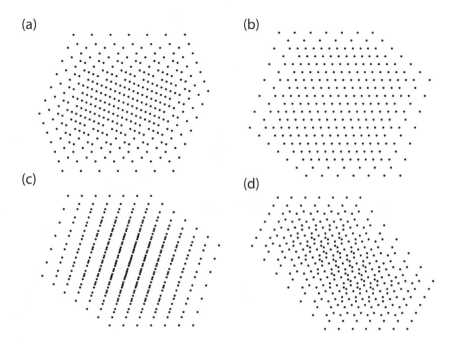

Figure 11.4 Examples of finite portions of the orbits of $\mathbb{Z}$-modules with $n = 2$ and $k = 3$. In all four examples, we considered $\mathbf{b}_1 = [1, 0]$ and $\mathbf{b}_2 = [\cos(\alpha), \sin(\alpha)]$ with $\alpha = 7\pi/5$. In (a) $7\mathbf{b}_3 = 3\mathbf{b}_1 + 5\mathbf{b}_2$; (b) $4\mathbf{b}_3 = 3\mathbf{b}_1 + 2\mathbf{b}_2$; (c) $\mathbf{b}_3 = [\cos(2\alpha), \sin(2\alpha)]$; (d) $\mathbf{b}_3 = \sqrt{2}\mathbf{b}_1 + \sqrt{3}\mathbf{b}_2$. In examples (a) and (b), the rank of the $\mathbb{Z}$-modules is equal to two while for examples (c) and (d) the rank is equal to three.

$\mathbf{b}_3 = (3/7)\mathbf{b}_1 + (5/7)\mathbf{b}_2$, and for Figure 11.4b, $\mathbf{b}_3 = (3/4)\mathbf{b}_1 + (1/2)\mathbf{b}_2$. All the lattices studied in crystallography are special cases of $\mathbb{Z}$-modules with discrete orbits. However, when the vector $\mathbf{b}_3$ is the sum of a rational multiple of $\mathbf{b}_1$ and of an irrational multiple of $\mathbf{b}_2$, as is the case for Figure 11.4c, the orbits of the $\mathbb{Z}$-module become dense along a line, i.e., they stratify in different families of densely filled parallel lines. Finally, in Figure 11.4d we show the case when the coefficients of $\mathbf{b}_3$ are both independent irrational numbers with respect to the $\mathbf{b}_1, \mathbf{b}_2$ basis. In this situation, the orbit becomes dense in two dimensions (not all of it can be visualized). The three situations exemplified in Figure 11.4 are the only ones that can possibly occur, meaning that the orbits of any $\mathbb{Z}$-module in E^n always consist of densely filled subspaces of E^n with dimensions $d \in \{0, 1, 2, \ldots, n\}$ [39]. In the specific case of Figure 11.4, these dense subspaces are zero, one, and two dimensional. Despite their more general nature, $\mathbb{Z}$-modules can be described completely in terms of orthogonal projections of point lattices from higher-dimensional spaces. In particular, it can be proven that every orbit Ω of a $\mathbb{Z}$-module in E^n is the orthogonal projection (onto E^n) of a point lattice in E^k with $k \geq n$ [39]. In general, we can construct many point lattices that project to Ω, so that the orbit Ω can always be lifted to some point lattice. A simple constructive proof of this fundamental fact is provided, for instance, in [39]. However, it is not generally possible to lift Ω to a specified point lattice and, in particular, to the hypercubic or standard point lattice spanned by pairwise orthogonal vectors of unit length. The precise condition under which this operation is possible is specified by *Hadwiger's theorem*. An accessible proof of this theorem can be found, for instance, in chapter 13 of Coxeter's book on regular polytopes [563].

The idea of geometrically projecting from a higher-dimensional lattice onto the physical space enables the generation of more general point sets, called Delone sets, that are essential tools for the description of quasicrystals, as later discussed in Section 11.4. For example, a Delone set with fivefold rotational symmetry can be constructed by projecting a subset of the five-dimensional hypercubic lattice $\mathcal{I}_5$ onto a plane. In this context, Hadwiger's theorem establishes when a set of vectors can be lifted to an orthonormal cross of a higher-dimensional hypercubic lattice, from which a quasicrystal can be formed by a suitable projection scheme. An important approach to the so-called *canonical projection method* was developed by de Bruijn in 1981 [569] and is discussed in the next sections and in the [39, 316]. Alternative methods for the generation quasicrystals based on algebraic number theory and root lattices (as opposed to hypercubic lattices) have also been developed, providing a convenient coordinate representation scheme [316, 570].

11.3.1   Incommensurate Modulated Structures

Incommensurate modulated structures are quasiperiodic systems characterized by a basic periodic structure (or unit cell) with three-dimensional space group symmetry to which it is superimposed a periodic modulation of some physical property, such as the atomic position, the spin, the magnetic moment, etc. The main idea is that modulated structures can be regarded as a lattice periodic *basic structure* and a modulation with a periodicity that is in general incommensurate with the one of the basic structure. In particular, for *incommensurate displacively modulated structures*, the atomic positions are periodically modulated along a given direction at an incommensurate frequency with respect to the periodic equilibrium positions in the underlying periodic basic structure. Therefore, in a modulated structure, the positions of the scattering particles (e.g., atoms) can be expressed as follows:

$$\boxed{\mathbf{r}_{ti} = \mathbf{t} + \mathbf{r}_i + f_i(\mathbf{t} \cdot \mathbf{q})}$$ (11.14)

for some wave vector $\mathbf{q}$. Here $\mathbf{r}_i$ indicates the position of the ith particle in the unit cell defined by the lattice vector $\mathbf{t}$ and $f_i(x)$ are periodic *modulation functions*. If the components of the modulation wavevector $\mathbf{q} = \sigma_1 \mathbf{b}_1^* + \sigma_2 \mathbf{b}_2^* + \sigma_3 \mathbf{b}_3^*$ with respect to the reciprocal lattice are rational, the resulting modulated structure remains periodic. On the other hand, the modulated structure becomes quasiperiodic if one of these components is irrational. Typically sinusoidal displacive modulation functions are considered, such as the following:

$$f_i(x) = A_i \sin(2\pi x + \varphi_i)$$ (11.15)

(a)

(b)

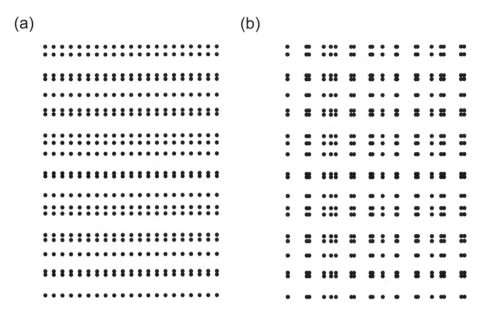

Figure 11.5 Examples of displacively modulated structures. (a) Modulation of a square array along the vertical axis according to the function $f_y(n) = C_2 \sin(2\pi n\alpha)$, where $\alpha = (1 + \sqrt{5})/2$ is the golden mean and $C_2 = 0.4$. (b) Doubly modulated square array with a modulation function along the vertical axis as in (a) and along the horizontal axis according to $f_x(n) = C_1 \sin(2\pi n\nu)$, where $\nu = 1 + \sqrt{2}$ is the silver mean and $C_1 = 0.7$.

More generally, the displacement field can admit more than one modulation wave vector:

$$\mathbf{r}_{ti} = \mathbf{t} + \mathbf{r}_i + f_i(\mathbf{t} \cdot \mathbf{q}_1, \mathbf{t} \cdot \mathbf{q}_2, \ldots, \mathbf{t} \cdot \mathbf{q}_n), \tag{11.16}$$

where the modulation functions are periodic in each of the n arguments. Examples of modulated square lattices in two spatial dimensions are provided in Figure 11.5, where two orthogonal modulation functions $f_x(n)$ and $f_y(n)$ are utilized with irrational frequencies equal to the silver mean $\nu = 1 + \sqrt{2} \approx 2.414213562$ and the golden mean $\alpha = (1 + \sqrt{5})/2 \approx 1.618033989$, respectively. These two irrational numbers are elements of the so-called *metallic means* expressed in a continued fraction as follows:

$$n + \cfrac{1}{n + \cfrac{1}{n + \cfrac{1}{n + \cfrac{1}{n + \cdots}}}} = [n; n, n, n, n, \ldots] = \frac{n + \sqrt{n^2 + 4}}{2} \tag{11.17}$$

The golden ratio is the metallic mean between 1 and 2, while the silver ratio is the metallic mean between 2 and 3. The term *bronze ratio* and additional terms using other names of metals (copper, nickel, platinum, etc.) are used to denote subsequent metallic means. We notice that each metallic mean is a root of the simple quadratic equation $x^2 - nx = 1$, where n is any positive natural number.

Incommensurate modulated structures do not possess three-dimensional lattice periodicity and cannot be described by one of the 230 space groups of classical crystallography. Moreover, the lack of periodicity of incommensurate modulated structures is manifested by the *dense nature of their diffraction patterns*, characterized by the fact that no minimum distance exists between different diffraction spots. Consistently, the diffraction peaks of incommensurate structures are found at vector positions that form, in general, a Fourier module of rank n. As we already observed in Section 11.3, if $n > 3$ the vectors in the Fourier module form lines, planes, or volumes in which every point is arbitrarily close to a vector in the module, i.e., they consist of dense sets of vectors.[5] It also naturally follows that since there is no minimum distance in their diffraction space, quasiperiodic structures are not limited by the crystallographic restriction theorem and can admit noncrystallographic point-group symmetries such as the icosahedral symmetry. In general, the diffraction patterns of incommensurately modulated structures consist of a set of bright or main scattering peaks, which correspond to the Bragg peaks of the reciprocal lattice of the basic structure, and a generally weaker set of *satellite Bragg peaks* that interpolate between the main peaks and have the periodicity of the modulation. This property can be deduced quite easily by considering the density of particles at positions $\mathbf{r}_{ti}$ in the following basic structure:

$$\rho(\mathbf{r}) = \sum_{ti} \delta(\mathbf{r} - \mathbf{r}_{ti}) \tag{11.18}$$

where $\mathbf{r}_{ti} = \mathbf{t} + \mathbf{r}_i + f_i(\mathbf{t} \cdot \mathbf{q})$. The Fourier transform of this function is called the *interference function*, or the array factor. The amplitude of this function provides fundamental insights into the nature of wave interference phenomena associated to a given geometrical structure, and it is related to the structure factor $S(\mathbf{H})$.

The Fourier coefficients of the density function can be expressed as follows:

$$S(\mathbf{G}) = \sum_{ti} \exp(j\mathbf{G} \cdot \mathbf{r}_{ti}) = \sum_i \exp(j\mathbf{G} \cdot \mathbf{r}_i) \sum_t \exp[j\mathbf{G} \cdot (\mathbf{t} + f_i(\mathbf{t} \cdot \mathbf{q}))]. \tag{11.19}$$

Using the Poisson summation formula (see Section 11.4.3) and remembering the periodic character of the modulation function, we can rewrite the previous expression as follows:

$$S(\mathbf{G}) = \sum_m \delta(\mathbf{G} - \mathbf{K} - m\mathbf{q}) \sum_i \exp(j\mathbf{G} \cdot \mathbf{r}_i) f_i(\mathbf{K} + m\mathbf{q}) \tag{11.20}$$

This function is contributed by a set of delta peaks at positions $\mathbf{G} = \mathbf{K} + m\mathbf{q}$. If $m = 0$, the corresponding diffraction peaks are the main Bragg reflections. When $m \neq 0$, the components at $\mathbf{K} + m\mathbf{q}$ are the satellite Bragg reflections.

---

[5] Strictly speaking, there is no minimum distance between the peaks. However, only a finite number of diffraction points are visible (i.e., above the noise level) since the amplitudes of the diffraction peaks are not all constant.

When the modulation function is harmonic, the Fourier coefficients can be written as follows:

$$S(\mathbf{G}) \propto \sum_{\mathbf{t}} \exp[j\mathbf{G} \cdot (\mathbf{t} + \mathbf{A}\cos(\mathbf{t} \cdot \mathbf{q}))]. \tag{11.21}$$

Using the following expansion [366]:

$$\exp(jz\cos\theta) = \sum_{m=-\infty}^{m=+\infty} j^m J_m(z)\exp(jm\theta) \tag{11.22}$$

we explicitly obtain the following:

$$\boxed{S(\mathbf{G}) = \sum_{\mathbf{K},m} \delta(\mathbf{G} - \mathbf{K} - m\mathbf{q})j^{-m}J_m(\mathbf{A} \cdot \mathbf{K})} \tag{11.23}$$

where the summation runs over the reciprocal lattice vectors of the basic structure. The result derived in (11.23) is the structure factor of an aperiodic function of rank 4, provided $\mathbf{q}$ is not a rational combination of the reciprocal lattice vectors of the three-dimensional basic structure. Under these conditions, the diffraction patterns will display peaks that are arranged to form a Fourier module of rank 4. Therefore, four Miller indices will be needed to uniquely label all the diffraction peaks. Two examples of displacively modulated square lattices with irrational modulation along one and two spatial dimensions as well as the corresponding structure factors are shown in Figure 11.6.

11.3.2 Incommensurate Composite Structures

Composite structures consist of two or more interpenetrating crystal structures, or subsystems, with lattice parameters that are mutually incommensurate to each other. Therefore, the full system does not possess any lattice periodicity. However, each subsystem can still be characterized by diffraction vectors that belong to its reciprocal lattice. As a result, the full system has diffraction vectors that can be expressed as linear combinations of the reciprocal vectors of the subsystems. Since the lattice parameters of the subsystems are incommensurate, their reciprocal lattices do not have a common basis and we need more than three Miller indices to label (in three spatial dimensions) the diffraction spots of the full system. More generally, the two subsystems will consist of distinct incommensurate modulated structures, each possessing a basic structure with reciprocal lattices $\Lambda_{n\nu}^*$, where $\nu = 1, 2$. In electronic structures, the atoms of system 1 will be modulated by a function with the periodicity of subsystem 2, and vice versa. The diffraction of atoms of subsystem 1 is characterized by diffraction vectors $h\mathbf{a}_{11}^* + k\mathbf{a}_{12}^* + l\mathbf{a}_{13}^* + m\mathbf{q}_1$ and those of subsystem 2 by the diffraction vectors $h\mathbf{a}_{21}^* + k\mathbf{a}_{22}^* + l\mathbf{a}_{23}^* + m\mathbf{q}_2$, where $\mathbf{q}_1$ belongs to the reciprocal lattice Λ_2^* of subsystem 2 and $\mathbf{q}_2$ to the reciprocal lattice Λ_1^* of subsystem 1. As a result, all the diffraction peaks will be at positions that are linear combinations of the basis vectors $\mathbf{a}_{\nu i}^*$ with $\nu = 1, 2$ and $i = 1, 2, 3$. Generally, the six vectors $\mathbf{a}_{\nu i}^*$ are not

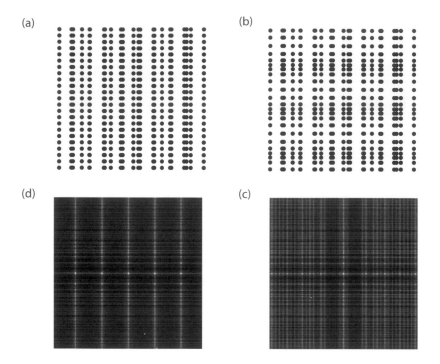

Figure 11.6 Examples of displacively modulated structures with irrational modulation along the horizontal direction (a) according to the function $f_x(n) = C_1 \sin(2\pi n\alpha)$, where α is the golden mean and $C_1 = 0.7$, and (b) along both directions with $f_x(n) = C_1 \sin(2\pi n\alpha)$ and $f_y(n) = C_2 \sin(2\pi n)$ with $C_2 = 0.4$. (d) and (c) show the corresponding structure factors.

all independent and it is possible to choose a basis of rank $n \leq 6$ with vectors $\mathbf{b}_j^*$ such that each diffraction wavevector belongs to the Fourier module spanned by these n basis vectors. Therefore, it is customary to write the following [40]:

$$\mathbf{a}_{\nu i}^* = \sum_{j=1}^{n} Z_{ij}^\nu \mathbf{b}_j^* \qquad (11.24)$$

for certain matrices Z_{ij}^ν. A simple two-dimensional model of incommensurate composites is constructed by considering the following two-particle chains [40]:

$$x_{n\ell} = x_0 + na + f(na), \qquad y_{n\ell} = \ell c \qquad (11.25)$$
$$X_{m\ell} = X_0 + mb + g(mb), \qquad y_{m\ell} = (\ell + 1/2)c, \qquad (11.26)$$

where $f(x) = f(x + b)$ and $g(y) = g(y + a)$ are two arbitrary modulation functions. The Fourier module of this system is generated by the set of vectors $\mathbf{k} = (i/a, j/b, k/c)$, where (i, j, k) are integers. Clearly when a/b is irrational, the resulting structure is aperiodic and the rank of the Fourier module is $n = 3$.

In Figure 11.7, we show two representative composite structures and the corresponding structure factors. These patterns are created by interpenetrating two distinct

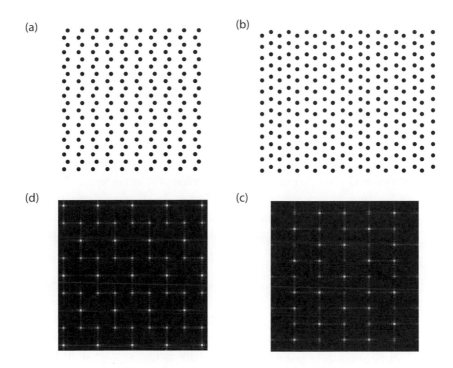

Figure 11.7 Examples of composite particle arrays. (a) Two interpenetrating square arrays with coordinates $(x_{01} + na_1, y_1 = mb_1)$, and $(x_{02} + na_2, y_2 = y_{02} + mb_2)$, where $x_{01} = 0, x_{02} = 1/4, y_{01} = 0, y_{02} = 1/2$ and $a_1 = a_2 = b_1 = b_2 = 1$. (b) Incommensurate composite structure with coordinates generated as in (a) but with the following parameters: $x_{01} = 0, x_{02} = a_1/2, y_{01} = 0, y_{02} = b_1/3$ and $a_1 = 1.6, b_1 = 1, b_2 = 1$, and $a_2 = (1 + \sqrt{5})/2$. (d) and (c) show the corresponding structure factors.

square lattices. The presence is weaker satellite diffraction peaks on top of the fundamental Bragg peaks of a square lattice is evident in Figure 11.7c,d. Incommensurate composites are quite frequently encountered, for example on atomic monolayers on a substrate such as Ar on graphite. For the adsorbed monolayer, the relative distances of the absorbed atoms depend on external factors such as temperature and pressure and are incommensurate with the periodicity of the substrate. Another example is provided by the $Hg_{3-\epsilon}AsF_6$ composite whose host lattice is formed by AsF_6 octahedra. However, in the direction parallel to the x- and y-axis, there are chains of Hg atoms with an atomic distance that is incommensurate with the lattice constant of the host lattice. Wave diffraction through incommensurate composites has been thoroughly discussed by van Smaalen [571].

11.3.3 Wang Tiles: Where Logic Meets Geometry

Arguably the first example of aperiodic tilings are the *Wang tiles* developed by the logician Hao Wang in 1961. Wang tiles can be visualized as square tiles with a color associated to each side. When a finite set of such tiles is selected, it is natural to ask if a tiling of the plane can be obtained by arranging copies of the tiles side by side with matching colors, without rotating or reflecting them. In 1961, Hao Wang conjectured that if this is possible, then there exist also a periodic tiling of the plane (similar to a wallpaper pattern) and an algorithm that decides whether a given finite set of Wang tiles can indeed tile the plane. The problem of determining a mathematical procedure for determining whether a given set of tiles can tile the plane became known as the *domino problem*.

This problem was solved in the negative in 1966 by Wang's student Robert Berger, who proved that no algorithm for the domino problem can exist [572]. Berger's fundamental idea was to reformulate the domino problem in terms of the halting problem of Turing machines, which is the problem of deciding whether a Turing machine will eventually halt. In particular, he sowed that a set of Wang tiles can tile the plane if and only if the corresponding Turing machine does not halt. The well-established undecidability of the halting problem for the Turing machines then implies the undecidability of Wang's tiling problem.

Hao Wang's original conjecture combined with Berger's undecidability result suggested that there must exist a finite set of Wang tiles that tile the plane but *only aperiodically*, known as aperiodic sets, anticipating Penrose's aperiodic tilings discovered by the mathematician Roger Penrose in the 1970s. However, although the first aperiodic set originally discovered by Berger contained 20,426 elements,[6] his discovery was of fundamental importance since Hao Wang conjectured that aperiodic set of tiles could not even exist [562]. In later years, aperiodic sets with increasingly smaller tiles were found. At the time of writing, it is believed that the smallest known set contains only 11 tiles and 4 colors [573].

11.3.4   Aperiodic Tilings and Quasicrystals

The mathematical study of symmetry, planar tilings, and discrete point sets (i.e., Delone sets) paved the way to the discovery of aperiodic order in geometry, leading to the first application in physics of quasiperiodic functions. *Tilings or tessellations* are collections of plane figures called *tiles* that fill the plane without leaving any empty space. Early attempts to tile planar regions of finite size using a combinations of pentagonal and decagonal tiles were already pursued by Johannes Kepler, arguably the founder of the mathematical theory of tilings, in his book *Harmonices Mundi*, published in 1619 [560]. However, it was not until 1974 that the mathematician Roger Penrose discovered the existence of two simple polygonal shapes, called *prototiles*, capable of tiling the infinite Euclidean plane without spatial periodicity [574]. Different types of Penrose tilings can be realized depending on the specific inflation rules utilized [39, 316]. For instance, a simple set of prototiles consists of two *golden*

---

[6] Berger ultimately managed to reduce the aperiodic set to 104 and 92 elements [562].

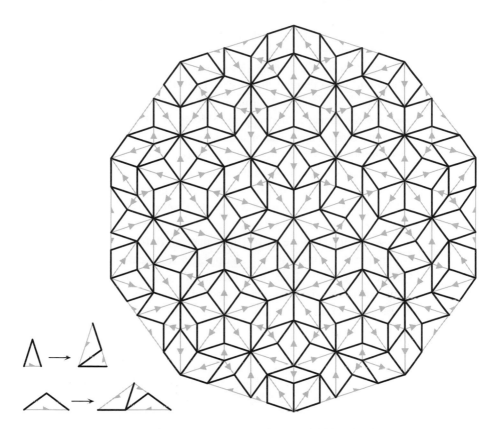

Figure 11.8 A fivefold symmetric patch of the rhombic Penrose tiling (right) obtained by Robinson's stone inflation rule (left). The edge matching rule decorations are indicated by the arrows. Adapted from [316]. Courtesy of Professor Uwe Grimm

triangles, which are triangles whose distinct edge lengths are in the ratio $\tau : 1$ (where τ is the golden mean) and the angles are integer multiples of $\pi/5$. In Figure 11.8, we show a finite patch of the fivefold symmetric rhombic Penrose tiling generated according to Robinson's stone inflation rule, also illustrated in the figure. Note that the prototiles include edge markers which, upon putting them together, must form complete single or double arrows. It is not difficult from this this rule to derive the kites and darts prototiles originally introduced by Penrose [316]. Similar to the case of one-dimensional substitution sequences, we can associate an *inflation matrix* to each inflation rule, which registers the number of prototiles involved by each step of the transformation. For instance, the inflation matrix associated to all the construction rules of the rhombic Penrose tiling that involve the golden triangles is as follows [316]:

$$M_{GT} = \begin{bmatrix} 2 & 1 \\ 1 & 1 \end{bmatrix} \tag{11.27}$$

Since all the entries in this matrix are positive integers, the matrix is a *primitive matrix* describing substitution rule. According to the *Perron–Frobenius theorem*, primitive matrices have a leading eigenvalue, called the *PF eigenvalue*, strictly greater in absolute value than all the others. The PF eigenvalue of the preceding matrix is equal to τ^2, which also determines the ratio of the number of occurrences of the two prototile shapes as they appear in the limit tiling, i.e., after an infinite number of inflations. Since τ is an irrational number, a central theorem in the theory of symbolic substitutions directly implies the aperiodicity of the tiling [316]. Remember that the largest eigenvalue of the substitutional matrix of the one-dimensional Fibonacci substitution $A \rightarrow AB$ and $B \rightarrow A$ is equal to τ, and therefore from this perspective the Penrose tiling can be regarded as the two-dimensional generalization of the Fibonacci sequence.

It was realized in the early 1980s that the diffraction patterns of Penrose tilings is pure point i.e., it consists of sharp diffraction peaks with icosahedral point-group symmetry, which includes the "forbidden" pentagonal symmetry. These structures can no longer be viewed as superimposed modulations, and in order to describe their diffraction patterns completely, we have to introduce more basis vectors, and labeling indices, than the dimensionality of physical space. Figure 11.9 shows the diffraction

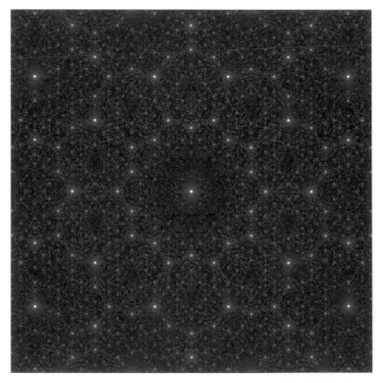

Figure 11.9 Structure factor (diffracted light intensity) of the point pattern dual to the rhombic Penrose tiling. The point pattern was generated using de Bruijin's pentagrid method with all the γ_j parameters equal to $1/2$.

of the point pattern corresponding to the vertices of the triangles in the rhombic Penrose tiling, constructed by the de Bruijn grid method [569]. This approach is purely algebraic and enables the direct computation of the vertex points of the rhombuses in the Penrose tilings. The main geometrical concept behind de Bruijn's method for the determination of the vertex set of the Penrose rhombic tiling is the *pentagrid*, which is the union of five grids in the complex plane specified as follows:

$$\{z \in \mathbb{C} | \Re(z\xi^{-j}) + \gamma_j \in \mathbb{Z}\} \tag{11.28}$$

where γ_j with $j = 0, 1, \ldots, 4$ are real numbers and $\xi = \exp(2\pi i/5)$ is the primitive fifth root of unity. This equation defines an infinite family of equidistant parallel lines (i.e., a grid) in the complex plane. The pentagrid is called regular if no more than two lines intersect at any given point, and it is called singular otherwise. We then construct the projection function:

$$f(z) = \sum_{j=0}^{4} K_j(z)\xi^j \tag{11.29}$$

where $K_j(z)$ is the ceiling of the expression $\Re(z\xi^{-j}) + \gamma_j$. The function defined projects from the five-dimensional lattice $\mathbb{Z}^5$ to the complex plane and, provided the pentagrid s regular, the set of points $\{f(z)|z \in \mathbb{C}\}$ is exactly the vertex set of the Penrose tiling generated by two rhombuses. This simple method can also be extended to other tilings and point patterns with a prescribed degree of noncrystallographic rotational symmetry.

In 1984, Dan Shechtman et al. [37] experimentally demonstrated for the first time the existence of physical structures with noncrystallographic rotational symmetries. In the study of the electron diffraction from certain metallic alloys (Al_6Mn), they discovered sharp diffraction peaks arranged according to the icosahedral point group symmetry consisting of 2-, 3-, 5-, and 10-fold rotation axes for a total of 60 symmetry elements. The sharpness and the intensity of the measured diffraction peaks, which are related to the coherence of the spatial interference effect, turned out to be comparable with the ones observed in ordinary periodic crystals. Moreover, the presence of some disorder did not eliminate the sharp Bragg peaks as long as long-range order was preserved. Stimulated by these findings, Dov Levine and Paul Steinhardt promptly formulated the notion of quasicrystals in a seminal paper titled "Quasicrystals: A New Class of Ordered Structures" [38]. The characteristic long-range correlations of quasicrystals are also reflected in an interesting mathematical property discovered by John Horton Conway in relation to the Penrose tilings. In essence, *Conways' theorem* states that, despite the lack of global periodicity, if we selected a local pattern of an arbitrary size within a quasicrystal, then an identical pattern will surely be found within a distance of twice the selected size. It was subsequently discovered that three-dimensional icosahedral structures can be obtained by projecting periodic crystals from an abstract six-dimensional *superspace*, starting the field of *quasicrystallography* [11, 39, 40].

For example, the particles' density $\rho(\mathbf{r})$ for a structure with icosahedral symmetry can be written as follows:

$$\rho(\mathbf{r}) = \sum_{\mathbf{G}_n} \rho_{\mathbf{G}_n} \exp(j\mathbf{G}_n \cdot \mathbf{r}) = \sum_{\mathbf{G}_n} |\rho_{\mathbf{G}_n}| \exp(j\mathbf{G}_n \cdot \mathbf{r} + \phi_{\mathbf{G}_n}), \qquad (11.30)$$

where $\mathbf{G}_n$ with $n = 0, 1, \ldots, 5$ are the primitive reciprocal vectors of the six-dimensional lattice, which correspond to the positions of the vertices of an icosahedron:

$$\mathbf{G}_0 = \frac{G}{\sqrt{1+\tau^2}}(1,0,\tau) \qquad \mathbf{G}_1 = \frac{G}{\sqrt{1+\tau^2}}(0,\tau,1) \qquad \mathbf{G}_2 = \frac{G}{\sqrt{1+\tau^2}}(-\tau,1,0),$$

$$\mathbf{G}_3 = \frac{G}{\sqrt{1+\tau^2}}(-\tau,-1,0) \qquad \mathbf{G}_4 = \frac{G}{\sqrt{1+\tau^2}}(0,-\tau,1) \qquad \mathbf{G}_5 = \frac{G}{\sqrt{1+\tau^2}}(-1,0,\tau).$$

More generally, the *projection method* regards a k-dimensional quasiperiodic structure as the projection, within a suitable projection window or stripe, of the points of the hypercubic lattice in the $n > k$-dimensional space E^n. According to this approach, the n-dimensional space E^n is decomposed into two orthogonal subspaces:

$$\boxed{E^n = E^{\|} + E^{\perp}} \qquad (11.31)$$

where the subspace $E^{\|}$ is the k-dimensional physical space or *parallel space* and $E^{\perp}$ is called the *perpendicular space*. Once a projection window W is chosen, a lattice point in E^n located inside W is projected onto $E^{\|}$ to obtain a vertex of the quasiperiodic structure in physical space. This will surely be the case when the slope of the hyperplane $E^{\|}$ is an irrational number. On the other hand, when this slope is rational, the projected lattice becomes periodic. We illustrate the projection method in the special case of the one-dimensional Fibonacci quasicrystal, shown in Figure 11.10, where the $E^{\|}$-axis is oriented at an angle $\theta = \arctan(1/\tau)$.

This cut and project procedure can be generalized to higher dimensions. In particular, a pentagonal quasicrystals can be obtained by projecting from the five-dimensional hypercubic lattice to the two-dimensional space and the three-dimensional icosahedral quasicrystal by projecting from the six-dimensional hypercubic lattice to three-dimensional space. The space group symmetries of the resulting quasicrystals are then the ones of the higher-dimensional periodic lattice involved in the projection. The superspace idea was not entirely new. In fact, already in 1981 Nicolaas de Bruijn showed that Penrose tilings can be obtained as the projection of a periodic object in five-dimensional space [569]. Even before, in 1972, de Wolff introduced the higher-dimensional crystallography approach to quantitatively describe the structure of incommensurate modulated phases. According to his work, a modulated crystal with no lattice periodicity can be described in usual crystallographic terms if one employees a four-dimensional space group [571, 575, 576]. The mathematical theory of higher-dimensional crystallographic groups has also been largely explored since the beginning of 1900, when Bieberbach studied the structure of higher-dimensional space groups and answered in the affirmative *Hilbert's 18th problem* on the finite number of groups in every dimension [577]. Seminal work on the algebraic properties

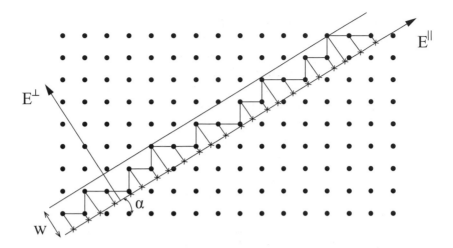

Figure 11.10 Illustration of the strip projection method to generate a one-dimensional Fibonacci quasicrystal by the projection of the two-dimensional square lattice on a line with irrational slope. Only the points of the lattice that are within the projection window W (the stripe) are projected in the parallel space and give rise to the Fibonacci chain.

of four-dimensional groups and on the computational aspects of n-dimensional space groups were conducted by Asher and Janner [578, 579] and by Fast and Janssen [580] in more recent years. We refer the readers with an interest in the rich scientific history of this subject to the excellent books by Janssen [40] and Senechal [39]. The International Union of Crystallography (IUCr) established in 1992 a special commission with the goal of redefining the concept of crystal structures. According to the IUCr report, the term "crystal" should be used in relation to "any solid having an essentially discrete diffraction diagram," *irrespective of spatial periodicity* [581]. The essential attribute of crystalline order (both periodic and quasiperiodic) is to display an essentially discrete spectrum of Bragg peaks. As observed by Maciá Barber [327], such a definition shifted the main attribute of crystalline structures from the real space to the reciprocal space.

11.3.5 Aperiodic Order from Phyllotaxis

Spiral arrays are interesting examples of long-range ordered systems where both translational and orientational symmetries are missing. Spiral curves can be generated by simple mathematical rules, which can adopt many forms in polar coordinates, such as $r = a\theta$ (Archimedean), $r = a\sqrt{\theta}$ (Fermat's or parabolic), $r = a\exp(\theta)$ (logarithmic), etc. A spiral array is generated from a spiral curve simply by restricting r and θ according to an integer quantization condition $\theta_n = \phi n$ and $r_n = f(n)$, where $\phi = 2\pi/\gamma$ is the divergence angle that measures the angular separation between consecutive radius vectors and $f(n)$ is a prescribed function (linear, quadratic, etc.). When γ is an irrational number (different from π), the divergence angle yields an irrational multiple of 2π, and the resulting spiral array entirely lacks rotational symmetry. The spiral geometry is manifested by a large number of physical systems (Figure 11.5) and it plays an important role in the morphological studies of plants and animals [582].

A very interesting example of aperiodic spiral order is provided by the so-called Vogel's spirals, which have been largely investigated by mathematicians, botanists, and theoretical biologists [583] in relation to the fascinating geometrical problems posed by phyllotaxis [582, 584–587]. Phyllotaxis (from the Greek *phullon*, leaf, and *taxis*, arrangement) is concerned with understanding the spatial patterns of leaves, bracts, and florets on plant stems (e.g., the spiral arrangement of florets in the *capituli* of sunflowers and daisies). Charles Bonnet (1720–1793) is usually credited with initiating observational phyllotaxis [588]. However, the field of phyllotaxis goes back to Leonardo Da Vinci (1452–1519), who already described in one of his notebooks the spiral arrangements of leaves, and Kepler (1571–1639), who observed the frequent occurrence of the number 5 in plants [589].

The history of mathematical phyllotaxis began in the first half of the nineteenth century when the brothers L. and A. Bravais presented the first systematic treatment of the phenomenon and recognized the relevance of the theory of continued fractions in this area (i.e., the so-called cylindrical representation of phyllotaxis) [587, 589]. Since then, a long list of notable scientists have contributed to the mathematical development of phyllotaxis, including the mathematician Alan Turing, who proposed a chemical model based on reaction-diffusion equation to account for the main phenomena of morphogenesis [590]. The Fundamental Theorem of Phyllotaxis was first discovered by Adler in 1974 [591]. This important result, formulated in terms of the Bravais and Bravais cylindrical lattice representation, orders the seemingly confusing multiplicity of spiral arms observed on plants. The science of phyllotaxis, born as a branch of botany, has subsequently evolved into a multidisciplinary research field that combines mathematical crystallography, L-systems[7] theory, and computer graphics with anatomical, cellular, physiological, and paleontological observations to explain the outstanding patterns encountered in plant morphogenesis and, more generally, in mathematical biology (patterns found in the structures of polymers, viruses, jellyfishes, proteins, etc.).

Vogel spiral point patterns are defined in polar coordinates (r, θ) by the following equations:

$$r_n = a_0 \sqrt{n} \tag{11.32a}$$

$$\theta_n = n\alpha, \tag{11.32b}$$

---

[7] L-systems or Lindenmayer systems were introduced and developed in 1968 by the Hungarian theoretical biologist and botanist Aristid Lindenmayer (1925–1989). L-systems are used to model the growth processes of plant development, but also the morphology of a variety of organisms. In addition, L-systems can be used to generate self-similar fractals and certain classes of aperiodic tilings such as the Penrose lattice.

where $n = 0, 1, 2, \ldots$ is an integer, a_0 is a positive constant called scaling factor, and α is an irrational number known as the divergence angle. This angle specifies the constant aperture between successive point particles in the array. Since α is irrational, Vogel spiral point patterns lack both translational and rotational symmetry. Accordingly, the diffraction spectrum of Vogel does not exhibit well-defined Bragg peaks, as for standard photonic crystals and quasicrystals, but rather it features a broad and diffuse circular ring whose spectral position is determined by the particles' geometry. Any irrational number ξ can be used to produce the divergence angle ($\alpha°$, in degrees) according to the relationship $\alpha° = 360° - \{\xi\} \cdot 360°$, where $\{\xi\}$ denotes the fractional part of ξ. When ξ is equal to the golden mean ($\xi = (1 + \sqrt{5})/2$), the resulting divergence angle $\alpha \sim 137.508°$ is called the "golden angle" while the resulting Vogel spiral structure is called the golden-angle spiral, or GA spiral. This structure can be decomposed into clockwise (CW) and counterclockwise (CCW) families of out-spiraling particles lines, known as *parastichies*, which stretch out from the center of the structure. Remarkably, the number of spiral arms in each family of parastichies is given by consecutive Fibonacci numbers [592]. A pair of spiral families (i.e., parastichy pair) formed by m spirals in one direction and n spirals in the opposite direction is denoted by (m, n). As an interesting example, a large sunflower's head (i.e., the *capitulum*) features two sets of parastichies (i.e., spirals) running in opposite directions with Fibonacci numbers $(34, 55)$. *Adler's Fundamental Theorem of Phyllotaxis* relates the numbers of visible parastichy pairs (m, n) with the divergence angle of the spiral [587, 591]. Different types of aperiodic Vogel spirals can be generated by choosing other irrational values of the divergence angle, giving rise to vastly different spiral geometries, all characterized by isotropic Fourier spectra, as we can appreciate in Figure 11.11. These examples of Vogel spirals, which have been recently introduced [251, 593], were named μ-, π- and τ-spiral and can be generated by considering $\xi = (5 + \sqrt{29})/2$, $\xi = \pi$ and $\xi = (2 + \sqrt{8})/2$, respectively.

Using analytical Fourier–Hankel decomposition (FHD), Dal Negro et al. [593] recently derived closed-form expressions for the diffraction spectrum of Vogel spiral arrays with an arbitrary divergence angle α. In particular, the Fourier–Hankel transform of the Vogel spiral density:

$$\rho(r, \theta) = \sum_{n=1}^{N} [\delta(r - \sqrt{n}a_0)\delta(\theta - n\alpha)]/r \tag{11.33}$$

can be compactly written as follows [593]:

$$f(m, k_r) = \frac{1}{2\pi} \sum_{n=1}^{N} J_m(k_r a_0 \sqrt{n})e^{inm\alpha}, \tag{11.34}$$

where $k_r = 2\pi q_r$ is the radial wavevector associated to the radial spatial frequency q_r and J_m is the mth-order cylindrical Bessel function of the first kind. The important thing to notice in the previous equation is that the exponential factor contributes with

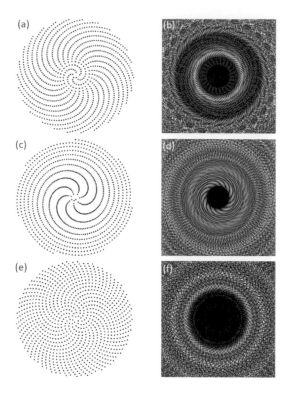

Figure 11.11 $N = 1,000$ μ-spiral array (a), and its diffraction pattern (b). $N = 1,000$ π-spiral array (c), and its diffraction pattern (d). $N = 1,000$ τ-spiral array (e), and its diffraction pattern (f). The diffraction patterns are created with $S\lambda^2 = 0.314$, in an angular range of $80°$.

azimuthal peaks only when the product $m\alpha$ is an integer. Since α is an irrational number, this condition will never be met exactly but it can be approximated using the convergents of the continued fraction expansion:

$$\xi \equiv [\xi_0; \xi_1, \xi_2, \xi_3, \ldots] = \xi_0 + \cfrac{1}{\xi_1 + \cfrac{1}{\xi_2 + \cfrac{1}{\xi_3 + \cfrac{1}{\xi_4 + \cdots}}}}. \tag{11.35}$$

It is clear that for spirals generated using an arbitrary irrational number ξ azimuthal peaks of order m (i.e., Bessel order m) will appear in the FHD due to the denominators q_n of the rational approximations (i.e., the convergents) of $\xi \sim q_n/p_n$. In fact, for all integer Bessel orders $m = p_n$, the exponential sum in equation (11.34) will produce in-phase contributions that yield strong azimuthal peaks. The number-theoretic prediction of distinctive azimuthal peaks that appear in the geometry of Vogel spirals' point patterns is of great importance because the geometry of the pattern encodes its symmetries on the resonant states of the system. An identical azimuthal structure carries over the solution for the diffraction spectrum of Vogel spirals, which can be written as follows [593]:

$$f(m, k_r) = \sum_{n=1}^{N} A_m(k_r)e^{inm\alpha}. \tag{11.36}$$

This result enables one to "encode" well-defined values of discrete components of the orbital angular momentum (OAM) of light. These components are uniquely determined by the rational approximations of the irrational divergence angles (or of ξ) that appear in simple exponential sums. The superposition of many OAM modes characteristic of Vogel spirals' scattered fields was experimentally established in [251], and it creates exciting opportunities for secure optical communication [594].

11.3.6 Multifractal Analysis of Vogel Spirals

The intricate geometry of complex structures and multiscale phenomena such as turbulence cannot be entirely captured by homogeneous fractals characterized by a single fractal dimension, or monofractals. In these more general cases, a spectrum of local scaling exponents associated to different spatial regions needs to be considered in order to achieve a more complete characterization. For this purpose, the concept of multifractals, or inhomogeneous fractals, has been introduced [595, 596] together with a multifractal formalism, which, adapted from statistical physics, quantitatively describes the local fractal scaling of multifractals [595, 597]. Here we apply this approach to the structural characterization of Vogel spiral point patterns that pose significant challenges to the conventional methods of spatial statistics due to their strongly inhomogeneous character leading to distinctively non-Gaussian fluctuations of the neighboring particles [598, 599]. In these cases, it becomes important to consider the distribution of local scaling exponents associated to different spatial regions of the pattern, which is achieved using the multifractal formalism detailed in Appendix E.

Multifractal analysis characterizes the structural complexity of arbitrary point patterns in terms of the corresponding multifractal spectra. From a physical point of view, the multifractal spectrum describes the multiscale distribution of singularities that appear at any scale ℓ in a multifractal signal, providing a quantitative characterization of its degree structural of inhomogeneity.

As for any fractal system, we can describe the scaling behavior of the system's density ρ around any point x by assigning the local power law:

$$\boxed{\rho(x + h) - \rho(x) \sim h^{\alpha(x)}} \tag{11.37}$$

The exponent $\alpha(x)$ is called the singularity exponent, since it describes the local degree of singularity of the density ρ around the point x. The set of all the points that share the same singularity exponent forms a fractal set of dimension $D(\alpha)$. The behavior of the curve $D(\alpha)$ versus α, which is called the singularity spectrum or the multifractal spectrum, describes the statistical distribution of the multifractal quantity ρ. Generally, multifractals are specified by assigning a local measure μ, such as a mass density, a velocity, an electrical signal, or some other scalar physical parameter, to a fractal object. Then the singularity strength $\alpha(x)$ of the multifractal measure μ obeys the local scaling law:

$$\mu[B_x(\ell)] \approx \ell^{\alpha(x)}, \tag{11.38}$$

where $B_x(\ell)$ denotes a ball (i.e., an interval) centered at x and of linear size ℓ. The smaller the exponent $\alpha(x)$, the more singular it will be the measure around x. The multifractal spectrum $D(\alpha)$ therefore characterizes the statistical distribution of the singularity exponent $\alpha(x)$ of the multifractal measure. As a result, if we cover the support of the measure μ with balls of size ℓ, the number of balls $N_\alpha(\ell)$ that, for a given α, scales like ℓ^α, can be written as follows:

$$\boxed{N_\alpha(\ell) \approx \ell^{-D(\alpha)}} \tag{11.39}$$

Note that in the limit of vanishingly small ℓ, $D(\alpha)$ is the fractal dimension of the set of all points x with scaling index α. In the case of singular measures with a recursive multiplicative structure (e.g., the devil's staircase), the multifractal spectrum can be calculated analytically [600, 601].

Multifractal analysis has been applied by Trevino et al. [598] to investigate the distinctive geometrical structure of general Vogel spiral arrays. Here we summarize the results of this analysis by focusing on Vogel spirals point patterns generated with divergence angles equispaced between the α_1 spiral (137.3°) and the GA spiral (137.5077641°) and also between the GA and the β_4 spiral (137.6°), which are displayed in Figure 11.12. These structures are obtained from the GA spiral by one-parameter structural perturbations and possess fascinating geometrical features that are responsible for significant mode localization properties, and optical gaps with complex fractal behavior [250, 251, 349, 593, 598, 602]. We notice that as the divergence angle is varied either above (supra-GA or β-series) or below the golden angle (sub-GA or α-series), the center region of the spiral where both sets of parastichies (CW and CCW) exist shrinks to a point. The outer regions are left with parastichies that rotate only CW for divergence angles greater than the golden angle and CCW for those below. For the spirals with larger deviation from the golden angle (α_1 and β_4), gaps appear in the center head of the spirals and the resulting point patterns mostly consist of either CW or CCW spiraling arms. Stronger structural perturbations (i.e., a further increase in the diverge angle) eventually lead to less interesting spiral structures containing only radially diverging parastichies.

The calculated multifractal spectra and generalized fractal dimensions are shown in Figure 11.13a,c for the spiral arrays in the α and β series, respectively. All spirals exhibit a clear multifractal behavior with singularity spectra of characteristic downward concavity, demonstrating the multifractal nature of the geometrical structure of Vogel's spirals. Moreover, we notice in Figure 11.13a that the GA spiral features the largest fractal dimensionality (D_f, $= 1.873$), which is consistent with its more regular structure. We also notice that $\Delta\alpha$ for the GA-spiral is the largest, consistently with the diffuse nature (absolutely continuous) of its Fourier spectrum. On the other hand, the less homogeneous α_1 spiral structure displays the lowest fractal dimensionality

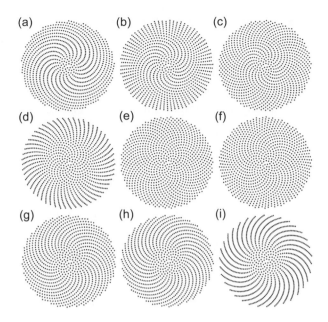

Figure 11.12 Different types of Vogel spiral arrays ($N = 1,000$ particles) generated using different divergence angles (a) 137.3°, α_1; (b) 137.3692546°, α_2; (c) 137.4038819°, α_3; (d) 137.4731367°, α_4; (e) 137.5077641°, GA; (f) 137.5231367°, β_1; (g) 137.553882°, β_2; (h) 137.5692547°, β_3; (i) 137.6°, β_4.

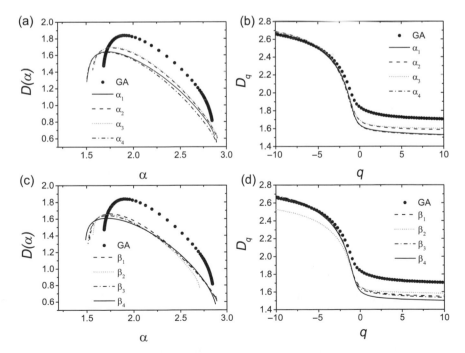

Figure 11.13 Multifractal singularity spectra $D(\alpha)$ for different Vogel spiral arrays ($N = 1,000$ particles) with divergence angles ranging between (a) α_1 and GA and (c) between the GA and β_4. The corresponding generalized fractal dimensions are displayed in (b) and (d), respectively.

$(D_f, = 1.706)$, which is consistent with increased structural disorder. All other spirals in the α series were found to vary in between these two extremes. On the other hand, the results shown in Figure 11.13c demonstrate smaller differences in the singularity spectra of the spirals in the β series, due to the much smaller variation of the perturbing divergence angle α (137.5–137.6). Figure 11.13b,d displays the generalized fractal dimensions of the two set of spirals corresponding to Figure 11.13a,b, respectively.

11.4   Concepts of Mathematical Diffraction Theory

While the diffraction theory of periodic crystals has a long history, the rigorous mathematical study of diffraction from aperiodic crystals is a very recent subject. The first systematic treatment of diffraction by aperiodic structures is due to Hof [603]. In this section, we introduce the basic setup of mathematical diffraction theory closely following references [316, 604, 605]. The key idea of the mathematical theory of diffraction is the so-called *diffraction measure* associated to certain infinite point sets, from which (upon proper normalization) the outcomes of any diffraction experiments can be computed. As we will discuss in this section, such a mathematical investigation reduces to the following two steps: (a) calculate the autocorrelation (or Patterson function) of the point set under consideration; and (b) evaluate its Fourier transform. However, these steps can present formidable mathematical difficulties in general. In what follows, we will illustrate this procedure through several noteworthy worked-out examples. Our discussion follows the approach presented in the excellent book by Michael Baake and Uwe Grimm [316], where interested readers can find more material.

11.4.1   Delone Sets

The analysis of kinematic wave diffraction by aperiodic point sets can be formulated as a rigorous mathematical theory starting from the the basic notions of Delone sets and Dirac combs, from which all the relevant spectral properties can be systematically obtained [316, 604]. We now consider in more detail the fundamental concept of a Delone set Λ that is the mathematical model for the description of very general sets of points in space that satisfy two simple requirements: (a) there is a minimum distance (called r_0) between any two points; and (b) there are no arbitrarily large regions with no points in them, i.e., there is a maximum radius R for "empty holes." More precisely, there exists a radius R such that any sphere (or circle in the plane) of radius R contains at least one point of the system (see Figure 11.14). This property of Delone sets is known equivalently as the homogeneity condition or the relative density condition. Amazingly, the two simple postulates (a) and (b), originally introduced by Delone in the 1930s to formalize the notions of "discreteness" and "homogeneity" of point sets, are sufficient to logically deduce many additional properties and establish a rigorous mathematical foundation for the analysis of crystals as well as aperiodic structures. For instance, both crystals and quasicrystals can be regarded as special

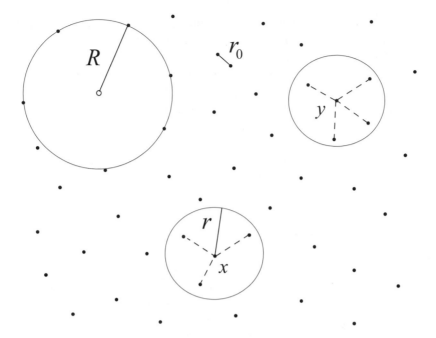

Figure 11.14 A Delone point set Λ with two representative r-stars at points x and y. Delone sets are discrete and relatively dense point sets also denoted by (r_0, R).

Delone sets with enough *local uniformity* to give rise to sharply defined spots in their diffraction patterns. Different from the usual approach to crystallography, in the point set theory developed by Delone, no global symmetry is explicitly enforced and crystalline structures originate from local regularity requirements. In particular, the local behavior of a general (nonperiodic) Delone set can be characterized by the r-stars (or "hedgehog") associated to its points (see Figure 11.14). The r-star at a point $x \in \Lambda$ is the finite set of points of Λ located inside a closed sphere of radius r with center at x, i.e., the points of the Delone set that are also inside a closed ball $B_x(r)$. A Delone set is said to be a *regular system of points* if its r-stars are congruent (i.e., can be brought to overlap) for every value of $r > 0$. Regular system of points are point sets that can be described in terms of crystallographic groups. The r-stars of such systems can be considered as copies of basic patterns that have been interlocked together to form the crystal structure. More intuitively, we can regard regular system of points as Delone sets with local point configurations specified by a set of "protostars" that occur everywhere in the set when appropriately "copied" (i.e., connected by some congruence operation), implying the existence of global symmetries. A profound result proved by Delone shows how regular systems of points arise from a "local regularity" criterion in terms of the r-stars of a Delone set Λ as follows: there exists a critical radius $r_c > 0$ such that if the r_c-stars of the points of Λ are congruent, then the r-stars are congruent for every $r > 0$, and therefore Λ is regular system of points [39]. This theorem clearly establishes a *local order threshold* for a Delone set to be a regular system of points, and it determines how certain local clusters configurations result in a global regularity behavior (i.e., periodicity).

Due to their lack of periodicity, aperiodic crystals cannot be modeled by regular set of points. However, aperiodic crystals that diffract (i.e., with sharp diffraction peaks) are modeled by *repetitive sets* (the converse is not always true since there are repetitive patterns that do not diffract!). A Delone set is said to be repetitive if for every real $r > 0$, its r-stars satisfy the homogeneity condition stating that for all $r > 0$, no two r-stars can be arbitrarily separated. In such aperiodic diffracting structures, any given pattern must necessarily be found somewhere else in the pattern. A simple example of a repetitive aperiodic Delone set is the the one associated to the vertices of Penrose quasicrystals. For such structures, the *Conway's theorem* ensures that if we select a local pattern of an arbitrary size within the structure, an identical pattern will be found within a distance of twice the selected size.

11.4.2   Dirac Comb and Autocorrelation Measures

We now introduce the general characterization of a Delone point set through the summation of Dirac delta functions. The approach can be extended to a higher-dimensional Delone set $\Lambda \subset \mathbb{R}^d$ by introducing the associated *Dirac comb* [316, 604]:

$$\delta_\Lambda = \sum_{x \in \Lambda} \delta_x \tag{11.40}$$

where δ_x is the normalized *Dirac point measure* at point x. Such a measure acts formally on a test function g according to the usual definition:

$$\int_{\mathbb{R}^d} g(y)\delta(y - x)dy = g(x), \tag{11.41}$$

which, expressed symbolically, is equivalent to $\delta_x(g) = g(x)$. Therefore, in this context, measures are introduced as linear functionals on continuous test functions. More details on this approach can be found in [605]. It should be clear that from a physical standpoint the Dirac comb is a rigorous representation of the distribution of matter in a scattering medium. Generally, we consider complex scattering weights $w(x)$ at position vectors $x \in \mathbb{R}^d$ according to the following:

$$\omega = w\delta_\Lambda = \sum_{x \in \Lambda} w(x)\delta_x. \tag{11.42}$$

As a specific example, let us consider a perfect (infinite) crystal defined by its periodic lattice point set $\Gamma \in \mathbb{R}^d$ and the "decoration" (i.e., the crystal's basis) of its *fundamental domain*, or unit cell. Using the repetition property of the convolution with Dirac's combs, the associated *crystallographic measure* becomes the following:

$$\omega = \mu * \delta_\Gamma \tag{11.43}$$

where μ is a finite measure (associated to the basis) that describes the distribution of scatterers inside the fundamental domain of Γ and δ_Γ implements the lattice periodicity. The convolution of the basis measure μ with the Dirac comb δ_Γ produces the crystal's structure by generalizing the known "step-and-repeat" property of the one-dimensional Shah function to the d-dimensional space.

Depending on the nature of μ, the resulting crystallographic measure ω will be pure point, continuous (e.g., when μ is the constant measure in the fundamental domain), or a mixture of the two. For a general Delone set, the diffraction measure $\hat{\gamma}$ associated to ω can be obtained by the Fourier transform of the *autocorrelation measure* γ of ω. The normalized autocorrelation measure is precisely defined as follows [605]:

$$\gamma = \gamma_\omega = \omega \odot \tilde{\omega} \equiv \lim_{R \to \infty} \frac{\omega|_R * \tilde{\omega}|_R}{\text{vol}(B_R)} \tag{11.44}$$

where $\text{vol}(B_R)$ denotes the volume of the open ball with radius R centered around the origin of $\mathbb{R}^d$; $\omega|_R$ is the restriction of the measure ω to the ball B_R; and the symbol $\tilde{\omega}$ extends to measures the "flipping operation" on the function $\tilde{g}(x) = \overline{g(-x)}$ by defining $\tilde{\omega}(g) = \overline{\omega(\tilde{g})}$, where the overline denotes complex conjugation. The $*$ symbol designates the direct convolution of measures, which generalizes the standard convolution of integrable functions [316, 605], while the symbol $\odot$ designates the "volume-averaged" convolution. Notice that in the last equation the volume normalization is needed because ω is generally an unbounded measure. Such a "volume-averaged" convolution of measures is known as the *Eberlein convolution* [316]. The following properties of the Eberlein convolution, which can be easily verified by a direct calculation, turn out to be very useful in the practical computation of the diffraction spectra of aperiodic point sets [316]:

$$\delta_\mathbb{Z} \odot \delta_\mathbb{Z} = \delta_\mathbb{Z} \tag{11.45}$$

$$\delta_\Gamma \odot \delta_\Gamma = (\Omega^{-1})\delta_\Gamma \tag{11.46}$$

$$\delta_{\alpha\mathbb{Z}} \odot \delta_{\alpha\mathbb{Z}} = \frac{1}{\alpha}\delta_{\alpha\mathbb{Z}} \tag{11.47}$$

$$\delta_\Gamma \odot \lambda = (\Omega^{-1})\lambda \tag{11.48}$$

$$\lambda \odot \lambda = \lambda \tag{11.49}$$

where λ denotes the standard Lebesgue measure and $\delta_\mathbb{Z}$ is the distribution form of the Shah or step-and-repeat function previously introduced in Chapter 7. In addition, we have the following:

$$\delta_\mathbb{Z} \odot \delta_{\alpha\mathbb{Z}} = \begin{cases} \frac{1}{p}\delta_{\mathbb{Z}/q}, & \text{if } \alpha = p/q \text{ with } p,q \text{ coprime.} \\ \frac{\lambda}{\alpha}, & \text{if } \alpha \text{ is irrational.} \end{cases} \tag{11.50}$$

These important relations will be applied in the next sections to derive the diffraction spectra of a number of aperiodic structures of great interest in the theory of aperiodic systems. Finally, we notice that when the autocorrelation measure γ of ω exists, then the *diffraction measure* $\hat{\gamma}$ also exists and is always a positive measure. The quantity $\hat{\gamma}$ corresponds to the *kinematic diffraction intensity* that provides the radiation intensity scattered in a given volume of the d-dimensional space.

11.4.3   Crystallographic Structures

Let us now apply the measure-theoretic formalism introduced in the previous section to the computation of the spectrum of arbitrary periodic crystal structures. We first proceed by calculating the autocorrelation of the crystallographic measure ω in equation (11.43). This quantity can be easily obtained by remembering the associativity property of the convolution, which leads to the following expression:

$$\gamma = (\Omega^{-1})(\mu * \tilde{\mu}) * \delta_\Gamma, \tag{11.51}$$

where Ω is the volume of the fundamental domain (i.e., the primitive unit cell), and we used the properties $\delta_\Gamma * \delta_\Gamma = (\Omega^{-1})\delta_\Gamma$ and $\delta_\Gamma = \tilde{\delta}_\Gamma$. Notice that the crystallographic autocorrelation (11.51) is clearly a measure with the lattice periodicity. In order to obtain the corresponding diffraction measure, we need to compute the Fourier transform of the crystallographic autocorrelation measure. This task can be simplified using the *Poisson summation formula* (PSF), which is a central result in the analysis of lattice periodic measures [316]. Given a lattice $\Gamma \subset \mathbb{R}^d$, the PSF allows us to express the Fourier transform associated to the Dirac comb of the lattice in the following form:

$$\boxed{\hat{\delta}_\Gamma = (\Omega^{-1})\delta_{\Gamma^*}} \tag{11.52}$$

where Γ^* is the reciprocal lattice or *dual lattice* of Γ. The dual lattice is spanned by the dual lattice vectors defined by the following set:

$$\boxed{\Gamma^* = \{v \in \mathbb{R}^d | (v, w) \in \mathbb{Z}, \text{for all } w \in \Gamma\}} \tag{11.53}$$

where (v, w) denotes the scalar product of vectors $v, w \in \mathbb{R}^d$. A factor 2π is usually included in the definition of the dual or reciprocal lattice, but it is omitted here. This factor is eliminated in the mathematical literature where the vectors v, w are reinterpreted as spatial frequency vectors (i.e., k-vectors divided by 2π).

Using the PSF and the convolution theorem for measures [316], we can immediately calculate the Fourier transform of the crystallographic autocorrelation measure γ and obtain the following:

$$\boxed{\hat{\gamma} = (\Omega^{-2})|\hat{\mu}|^2 \delta_{\Gamma^*}} \tag{11.54}$$

The last expression is a general result valid for arbitrary crystallographic structures associated to the measures δ_Γ and μ. As expected, the diffraction measure derived in (11.54) is pure point and concentrated on the dual lattice Γ^*.

11.4.4   Incommensurately Modulated Structures

Let us now consider simple examples of diffraction from incommensurate structures, which can be regarded as the "precursors" of quasicrystals. In fact, their study led to the development of the powerful "superspace approach" to crystallography that will be briefly introduced at the end of this section. The mathematical analysis of incommensurate systems was pioneered by de Wolff, Janner, and Janssen [40, 576, 606]. An in-depth discussion of the crystallography of incommensurate structures can be found in the monograph by van Smaalen [571]. As a simple pedagogical example, we will focus specifically on *modulated structures* that provide a simple way to avoid the problem of arbitrarily close points, giving rise to incommensurate Delone sets. In one spatial dimension, such systems are in fact entirely equivalent to quasicrystals, for which they provide a simple model. Here we will introduce a one-dimensional structure consisting of the integer lattice $\mathbb{Z}$ with a positional modulation of its points described by a real-valued displacement function h. The resulting displaced point set is given by the following:

$$\Lambda_h = \{n + h(n)|n \in \mathbb{Z}\}, \tag{11.55}$$

with corresponding Dirac comb denoted as δ_{Λ_h}.

A particularly useful deterministic displacement model that leads to analytical results consists of the following function $h(n) = \epsilon\{\alpha n\}$, where $\{x\}$ denotes the fractional part of x and α and ϵ are real numbers (typically α is an irrational number; this is because when $\alpha = p/q$ with p and q coprime numbers, we obtain a periodic lattice with period q and the case is uninteresting). Such a model is called the *deterministic displacement model* [316]. With this choice, the modulated point set Λ_h respects a minimum distance between points, since for ϵ sufficiently small we have $|h(n)| < \epsilon^8$.

We now summarize basic results on the spectra of incommensurate structures that enable more practical calculations. The autocorrelation γ_h of the Dirac comb on Λ_h can be explicitly calculated and written in the following form [316]:

$$\gamma_h = \sum_{m \in \mathbb{Z}}[(1 - \{\alpha m\})\delta_{m+\epsilon\{\alpha m\}} + \{\alpha m\}\delta_{m-\epsilon(1-\{\alpha m\})}]. \tag{11.56}$$

The corresponding diffraction measure is as follows:

$$\hat{\gamma}_h = \sum_{k \in \mathbb{Z}[\alpha]} |A(k)|^2 \delta_k \tag{11.57}$$

where the set

$$\mathbb{Z}[\alpha] \equiv \langle 1, \alpha \rangle_{\mathbb{Z}} = \{m + n\alpha|m, n \in \mathbb{Z}\} \tag{11.58}$$

defines the Fourier module. We should recognize that the Fourier module of an incommensurate structure has a direct number-theoretic interpretation since it is related to

---

[8] We recall that, as we already discussed in Chapter 7, the fractional part $\{\alpha n\}$ is dense (and equidistributed) in the unit interval $I = [0, 1)$.

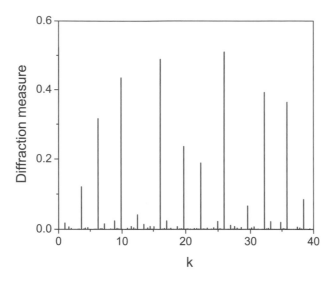

Figure 11.15 Aperiodic diffraction measure (scattered intensity) of an aperiodically modulated point set with α equal to the golden mean and $\epsilon = 0.3$. The diffraction pattern is purely discrete (it does not contain any continuous component) and symmetric with respect to k.

the ring of integers of a real quadratic field. For example, when the irrational number α is equal to the golden mean, the underlying quadratic field is $\mathbb{Q}(\sqrt{5})$ and its integers are given by $\mathbb{Z}\left[\frac{1+\sqrt{5}}{2}\right]$ because $5 \equiv 1 \bmod 4$ (see our general discussion of quadratic fields in Section 10.3.1). Therefore, we should expect that the underlying Galois field structure of the quadratic number field plays an important role in determining the aperiodic diffraction properties of the associated aperiodic point patterns, as recognized originally by Bombieri and Taylor [607]. It turns out that this is exactly the case when the proper field automorphism,[9] referred to as the *star-map* in the present context, is taken into consideration [316]. Typically, this can be identified with the algebraic conjugation in the appropriate quadratic field. The computed diffraction measure for the deterministic displacement model is shown in Figure 11.15. This is a pure-point spectral measure supported on a dense set of points and modulated by a slowly varying envelop $A(k)$. The aperiodic nature of the diffraction is a direct consequence of the basic fact that integer multiples of irrational numbers densely fill the unit interval over the reals, as prescribed by the Fourier module in equation (11.58). However, despite the dense character of the support, not all the diffraction peaks in the spectrum can be distinctly observed since they are modulated by complex weights according to the following complex amplitude:

$$A(k) = e^{-\pi j k^\star} \operatorname{Sinc}(k^\star) \tag{11.59}$$

---

[9] A field automorphism is a one-to-one correspondence $\phi : F \to F$ in a field F that preserves all the algebraic properties of the field. Sometimes it is called an isomorphism. For example, complex conjugation is an automorphism in the field of complex numbers $\mathbb{C}$.

where $\operatorname{sinc} x = \sin(\pi x)/(\pi x)$ when $x \neq 0$ and it is equal to 1 when x vanishes. The star-map $k \mapsto k^\star$ acts on the elements of $\mathbb{Z}[\alpha]$ in the following way: $(m + n\alpha) \mapsto (m\epsilon + n(1 + \epsilon\alpha))$ for any $m, n \in \mathbb{Z}$ [604]. Specifically, the diffraction measure shown in Figure 11.15 corresponds to an incommensurate modulated structure with $\alpha = (1 + \sqrt{5})/2$ calculated according to the general diffraction formula in (11.57) with the amplitude function in equation (11.59). The rigorous proof for the diffraction formula (11.59) and of the absence of continuous components in the diffraction spectrum are not simple and beyond the scope of our introductory discussion [316]. Finally, this result can be regarded as a special case of the diffraction from *model sets*, which are aperiodic point sets (or point patterns) obtained by the *cut and project method*. We discuss next this more general situation by providing examples of diffraction from model sets.

11.4.5 Diffraction from Model Sets

The essential idea behind the model set approach (MSA) is that certain aperiodic patterns can be constructed by simply projecting lattice-periodic structures from a higher-dimensional lattice. These are called *model sets* or *cut and project sets*. Such systems are known as cut and project sets or model sets and they can be produced in many equivalent ways, including via the purely algebraic de Bruijn's grid method [569]. The MSA for the creation of aperiodic structures generalizes the simple observation that quasiperiodic functions can be regarded as projections from periodic higher-dimensional ones. In one real variable, quasiperiodic functions are aperiodic trigonometric polynomials. Such functions are special cases of almost periodic functions, which are functions that "miss the opportunity" of being periodic by an arbitrary small number ϵ (see Section 10.1). As a simple example, we can write a quasiperiodic function in the simple form $f(x) = \cos(2\pi x) + \cos(2\pi x\alpha)$, where α is an irrational number. This function is nonperiodic due to the irrationality of α. However, it can be obtained by projection from the two-dimensional periodic function as follows:

$$f(x) = \cos(2\pi x) + \cos(2\pi y)|_{y=\alpha x}. \tag{11.60}$$

This expression involves the projection from the $\mathbb{Z}^2$ lattice on a line with irrational slope. An analogous mechanism is behind the theory of model sets. Specifically, in the cut and projection scheme (CPS), aperiodic structures naturally emerge by considering a cut across a higher-dimensional periodic structure when the cutting direction is incommensurate with the lattice. In this scheme $\mathbb{R}^d$ is referred to as the *physical* (or parallel) space while $\mathbb{R}^m$ is referred to as the *internal* (or perpendicular) space. Moreover, $\mathcal{L} \in \mathbb{R}^{d+m}$ is the higher-dimensional periodic lattice in $d + m$ dimension from which the aperiodic d-dimensional structure $L = \pi(\mathcal{L}) \in \mathbb{R}^d$ is obtained by the natural projection operator π onto the physical space. Alongside $\mathcal{L}$, we also have to consider the dense set $L^\star = \pi_{int}(\mathcal{L}) \in \mathbb{R}^m$ and the $\star$-map $x \to x^\star$ between the elements of L and the ones of $L^\star$ [316, 608].

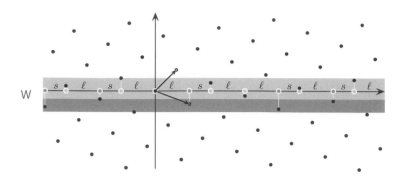

Figure 11.16 Cut and project setting for the silver mean chain. The two-dimensional periodic lattice $\mathcal{L}$ is the so-called Minkowski embedding of the one-dimensional aperiodic sequence L. The points of the lattice contained within the strip W are projected to the horizontal axis (i.e., the parallel space), where they give rise to the silver mean chain, as indicated by the white circled points. From [316]. Courtesy of Professor Uwe Grimm

A model set for a given CPS can now be precisely defined as the following set:

$$\Lambda = \{x \in L \,|\, x^\star \in W\} \tag{11.61}$$

where $W \in \mathbb{R}^m$, called the *window*, is a nonempty subset of $\mathbb{R}^m$. The points of the model set Λ lie in the projected lattice L in physical space and the window in internal space determines which elements of L are selected. This "windowing procedure" is essential in order to avoid overlaps between points. Technical conditions are always imposed on the window in order to ensure that the model set Λ is indeed a Delone set. Often, *regular model sets* are considered, characterized by the fact that the boundary of the window W has zero measure [316]. A central result of the theory of model sets states that regular model sets are pure-point diffractive [603, 609]. The condition $x^\star \in W$ guarantees that all the lattice points involved in the physical space projection lie in the strip W shown in Figure 11.16, where we illustrate the case of the *silver mean chain*. As a point pattern, the silver mean is a Delone set in $\mathbb{R}$. The relevant Z-module for the silver mean chain is as follows:

$$L = \mathbb{Z}[\sqrt{2}] = \{m + n\sqrt{2} \,|\, m, n \in \mathbb{Z}\}, \tag{11.62}$$

which is a dense subset of $\mathbb{R}$ that corresponds to the ring of integers in the quadratic field $\mathbb{Q}(\sqrt{2})$. In this case, the aperiodic order of the silver mean sequence reflects the distribution of integers in the field $\mathbb{Q}(\sqrt{2})$. The diagonal embedding of L in $\mathbb{R}^2$ is given by $\mathcal{L} = \{(x, x^\star) \,|\, x \in \mathbb{Z}[\sqrt{2}]\}$. This is the rectangular lattice spanned by the basis vectors $e_1 = (\sqrt{2}, -\sqrt{2})$ and $e_2 = (1, 1)$, which is shown in Figure 11.16. It can also be proven that the window W consists of the real interval $[-\sqrt{2}/2, \sqrt{2}/2]$ [316]. The Fourier module of the silver mean chain is as follows [316]:

$$L^\odot = \pi(\mathcal{L}^*) = \left\{ \frac{1}{2}(m + n/\sqrt{2}) \,|\, m, n \in \mathbb{Z} \right\}, \tag{11.63}$$

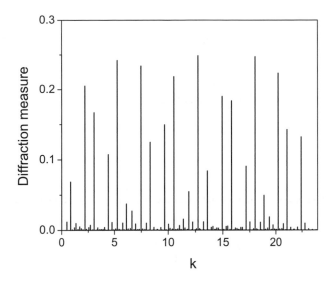

Figure 11.17 Diffraction spectrum of the silver mean chain. A singular peak is represented as a vertical line with a height equal to the intensity function provided in equation (11.64).

and the field automorphism (i.e., the star-map) is defined by the conjugation $\sqrt{2} \mapsto -\sqrt{2}$. The diffraction spectral measure of the silver mean chain can finally be obtained in the general form (11.57) with amplitude function given by the following [316]:

$$A(k) = \frac{1}{2}\,\mathrm{Sinc}\,(\sqrt{2}k^{\star}) \qquad (11.64)$$

The silver mean chain has a peculiar type of quasiperiodic order that is manifested by a substitution rule acting on a two-letter alphabet $\{A, B\}$ according to the following rule: $A \mapsto ABA, B \mapsto A$. We notice that the relative frequencies of the letters of the limit sequence (infinite in size) are equal to $\sqrt{2}/2$ and $(2 - \sqrt{2})/2$. Since both these numbers are irrational numbers, the sequence cannot be periodic. A quasiperiodic chain based on the silver mean substitution can be easily generated by representing the letter A by an interval of length $1 + \sqrt{2}$ and B by one of length 1. This is an example of a *Pisot-type sequence*. Therefore, the corresponding dynamical system is pure point, which in turn implies the pure-point quasiperiodic nature of the diffraction spectrum of the silver mean chain, shown in Figure 11.17. The silver mean chain can be approximated by *periodic approximants* by employing the continued fraction expansion of $\sqrt{2}$, which are the fractions $1, 3/2, 7/5, 17/12, \ldots$ A simple geometrical approach enables the approximation, which is summarized as follows [316]: let us call $r_i = p_i/q_i$ (with $r_0 = 1$) and rotate the window W in Figure 11.16 around the origin of the lattice such that it is made parallel to the lattice direction $(t_i, t_i^{\star}) \in \mathcal{L}$ with $t_i = p_i + q_i\sqrt{2}$. The projection of all the lattice points in the rotated window orthogonally to the horizontal axis results in a periodic point set Δ_i of period t_i that will converge to the silver mean point set in the limit $i \to \infty$.

In the next section, we will discuss the quasiperiodic Fibonacci chain and its diffraction properties as an important application of the general cut and projection scheme.

11.4.6   Diffraction from Quasiperiodic Chains

Before addressing the specific aspects of one-dimensional Fibonacci chains, let us consider again the general diffraction measure $\hat{\gamma}$ associated to the Dirac comb δ_Λ for the model set Λ. As we have discussed before, the corresponding diffraction spectrum can always be written in the following form [604]:

$$\hat{\gamma} = \sum_{k \in L^\circ} |A(k)|^2 \delta_k, \tag{11.65}$$

where $L^\circ = \pi(\mathcal{L}^*)$ is a suitable Fourier module generated from the projection of the higher-dimensional dual lattice $\mathcal{L}^*$ (see Figure 11.18). General mathematical results exist for the computation of the corresponding diffraction amplitude $A(k)$ in terms of the Fourier transform of the characteristic function of the window function W [316]. As an important example with applications to photonic and electronics technology, we address the one-dimensional Fibonacci chain that is a model set obtained from the

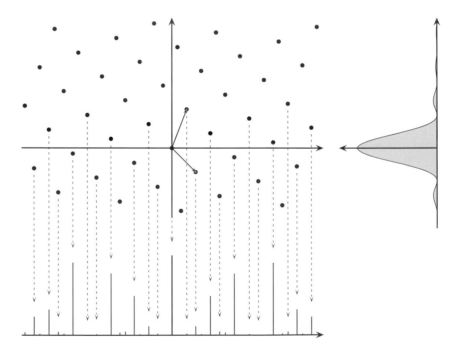

Figure 11.18 Schematic illustration of the embedding method for the computation of the diffraction spectrum of one-dimensional quasiperiodic structures. The arrows indicate some of the points of the dual lattice $\mathcal{L}^* = \{y \in \mathbb{R}^2 | xy \in \mathbb{Z} \text{ for all } x \in \mathcal{L}\}$ that give rise to a point measure (a Dirac delta) at k. The intensity function is shown on the right. From [316]. Courtesy of Professor Uwe Grimm

CPS when $d = m = 1$ and $W = [0, -1)$. The $\mathbb{Z}$-module of the Fibonacci chain is the rank-2 module $L = \mathbb{Z}[\tau] = \{m + n\tau | m, n \in \mathbb{Z}\}$, where $\tau = (1 + \sqrt{5})/2$ is the golden mean. This $\mathbb{Z}$-module obeys the inflation symmetry $\tau\mathbb{Z}[\tau] = \mathbb{Z}[\tau]$ and can also be regarded as the projection of a two-dimensional lattice in several different ways [316].

The most natural way is provided by the Galois structure of the underlying quadratic number field $\mathbb{Q}(\sqrt{5})$, whose quadratic integers coincide with the $\mathbb{Z}$-module $\mathbb{Z}[\sim]$. The corresponding diagonal embedding is $\mathcal{L} = \{(x, x^{\star}) | x \in \mathbb{Z}[\tau]\}$ with a star-map (nontrivial field automorphism) given by the algebraic conjugation $\sqrt{5} \mapsto -\sqrt{5}$. The planar lattice $\mathcal{L}$ is spanned by the basis vectors $e_1 = (1, 1)$ and $e_2 = (\tau, 1 - \tau)$. In this CPS construction, the Fibonacci structure (model set) is constructed by projecting the lattice points within the window $w = (-1, \tau - 1)$. We note that these two basis vectors can be scaled with respect to each other in order to obtain a maximally symmetric embedding to a square, as often presented in the literature. The corresponding Gram matrix, which becomes proportional to the identity matrix $\mathbf{1}$, assumes the simpler form $\mathbf{G} = (\tau + 2)\mathbf{1}$ and describes a scaled square lattice.

It is simple to compute the dual basis matrix for the Fibonacci chain and verify directly that the Fourier module $L^{\odot} = \pi(\mathcal{L}^{*}) = L/\sqrt{5} \subset \mathbb{Q}(\sqrt{5})$. In this case, the Fourier amplitude can be explicitly calculated as follows [316]:

$$|A(k)|^2 = \left[\frac{\tau}{\sqrt{5}} \ \text{Sinc} \ (\tau k^{\star}) \right]^2 \tag{11.66}$$

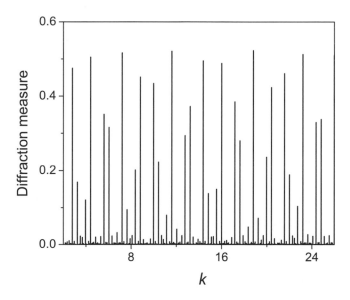

Figure 11.19 Diffraction spectrum of a Fibonacci point set. The spectrum is discrete (pure-point measure). A singular peak of the Fibonacci pure-point measure is represented as a vertical line with a height equal to the intensity function provided in equation (11.66).

The diffraction measure of the Fibonacci chain follows from the general expression (11.65) with $k \in L/\sqrt{5}$, and it is shown in Figure 11.19.

The Fibonacci diffraction spectrum is an aperiodic Dirac comb with intensity function $|A(k)|^2$ that vanishes when $\tau k^\star$ is an integer number different from zero or, equivalently, when $k = m\tau$ with $m \in \mathbb{Z}\backslash\{0\}$. These points of zero diffraction are called *extinctions*. Similarly, the extinctions of the silver chain are the points of its Fourier module such that $k^\star = m\sqrt{2}/2$, where $m \in \mathbb{Z}\backslash\{0\}$.

11.4.7   Diffraction from Limit-Periodic Structures

Limit-periodic point patterns are deterministic structures that consist of a union of an infinite set of distinct periodic structures with different rational periods. Similarly to quasiperiodic systems, they are characterized by a dense set of diffraction Bragg peaks. However, they differ from quasiperiodic structures in the fact that the ratio between the positions of any two scattering peaks is expressed by a rational number. More generally, a sequence (a_n) with $n \in \mathbb{Z}$ of real numbers is said to be limit-periodic if it is the uniform limit of a family of periodic sequences. Limit-periodic point sets display pure-point diffraction spectra supported on a countably infinite, but not finitely generated, Fourier module given by [610]:

$$L^{\odot} = \bigcup_{j \geq 1} \mathbb{Z}/2^j = \{m/2^r \,|\, (r = 0, m \in \mathbb{Z}) \text{ or } (r \geq 1, m \text{ odd})\} = \mathbb{Z}[1/2], \quad (11.67)$$

where (r, m) label the elements uniquely. The most well-known example is the period doubling sequence in one spatial dimension, which is based on the period doubling substitution rule: $A \mapsto AB, B \mapsto AA$. By iterating this rule an infinite number of times, we obtain an aperiodic limit sequence with relative letter frequency equal to $2/3$ for the letter A and equal to $1/3$ for the letter B. This sequence originated in the theory of dynamical systems [611]. We should realize that terminating the previous infinite unions at a finite integer N leads to two periodic point sets. This set can still be described as a model set with the internal space consisting of 2-adic integers. Consequently, the diffraction measure of the associated Dirac comb is pure point. The diffraction formula of a period-doubling chain can be explicitly calculated and its amplitudes are as follows [610]:

$$\boxed{A(k) = \frac{2}{3}\frac{(-1)^r}{2^r}e^{2\pi jk}} \quad (11.68)$$

where $k \in L^{\odot}$. Figure 11.20 shows the amplitude values $|A(k)|$ for $k \in L^{\odot}$ restricted to the interval $[0, 1]$.

11.4.8   Singular Continuous Measures

We will now turn our attention to diffraction from more complex aperiodic systems beyond model sets. The subject is very technical and our discussion will be limited to only providing the essential concepts. Interested readers are encouraged to survey the

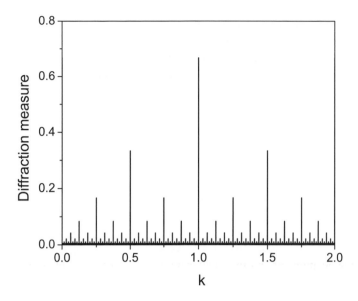

Figure 11.20 Diffraction measure of the period doubling chain with $k \in [0,2]$. This diffraction pattern repeats periodically along the real line with integer period (i.e., with $\mathbb{Z}$-periodicity).

excellent book by Michael Baake and Uwe Grimm and its extensive list of references for a comprehensive mathematical analysis [316].

Undoubtedly one of the foundational results of mathematical diffraction is the Lebesgue decomposition theorem [612]. In our context, this theorem states that, relative to the Lebesgue measure λ, the diffraction measure can be uniquely decomposed as follows [316, 612]:

$$\boxed{\hat{\gamma} = \hat{\gamma}_P + \hat{\gamma}_{SC} + \hat{\gamma}_{AC}}$$ (11.69)

where its pure-point $\hat{\gamma}_P$ part corresponds to singular Bragg peaks (countably many at most), the absolutely continuous part $\hat{\gamma}_{AC}$ corresponds to diffuse background scattering (with a locally integrable density relative to the Lebesgue measure λ), and the singular continuous part $\hat{\gamma}_{SC}$ denotes the remaining contribution that may bridge the entire range between a Dirac distribution and a continuous (and integrable) function. Therefore, there is no simple characterization of the singular continuous part of a diffraction measure. However, in order to distinguish the respective contributions of the three components of a general diffraction measure, an approximate criterion has been proposed [613, 614] based on the scaling behavior of the diffraction intensity with the system size N. Specifically, it was conjectured that $\hat{\gamma}_{SC} \sim N^\alpha$ with $0 < \alpha < 1$ (while $\hat{\gamma}_P \sim N$ and $\hat{\gamma}_{AC}$ remains constant for one-dimensional sequences). This criterion is indeed satisfied for the Thue–Morse sequence, whose singular continuous diffraction spectrum is well established [324], but fails to be valid in general [615]. Singular continuous spectra are observed in structures that are intermediate between quasiperiodic and random. In such systems, the constructive interference responsible for the formation of Bragg peaks is impaired by randomness, but there is still enough long-range order to prevent these peaks from becoming completely diffuse. However, different from periodic and quasiperiodic systems, the singular peaks in structures with singular continuous spectra are not isolated. Besides the well-known Thue–Morse sequence, another one-dimensional example of an aperiodic structure with a singular continuous spectrum is the so-called *circle sequence* [614, 615]. A higher-dimensional Thue–Morse sequence can be generated by following the simple two-dimensional substitution rule [7, 316, 616] acting on a two-letter alphabet:

$$\rho_{TM}: \quad a \mapsto \begin{matrix} b & a \\ a & b \end{matrix}; \quad b \mapsto \begin{matrix} a & b \\ b & a \end{matrix}. \tag{11.70}$$

The resulting two-dimensional Thue–Morse structure is the Cartesian product of Thue–Morse chains, and its diffraction spectrum can be factored in terms of the product of two one-dimensional Thue–Morse spectra (and therefore is singular continuous). A rigorous mathematical analysis of the Thue–Morse spectrum can be found in [316].

A useful quantity that enables one to understand the nature of the different spectral components that contribute to arbitrary diffraction measures is provided by the *integrated intensity function* (IIF), which is simply the distribution function (DF) of the diffraction measure [316, 327]. For one-dimensional structures, the IIF is defined as follows:

$$H(q) = \lim_{N \to \infty} \int_0^q \frac{I_N(q')}{N} dq' \tag{11.71}$$

where $I_N(q)$ denotes the diffracted intensity at a given point (q) in the appropriate conjugate space, which is proportional to the spectral measure $\hat{\gamma}$, and N represents the system size. It should be realized that the exact nature of the diffraction spectrum is only determined by the asymptotic limit of the IIF for a system of infinite size, and therefore only heuristic information can be extracted for any finite-size structure.

In both periodic and quasiperiodic structures, there will be regions where the diffracted intensity vanishes due to the pure-point nature of their diffraction measure (i.e., the discrete nature of the spectra). Over such regions, $H(q)$ remains constant, giving rise to characteristic plateaus regions, and it presents jump discontinuities every time an isolated Bragg peak is integrated. On the opposite, for structures with absolutely continuous spectral measures, the function $H(q)$ will be continuous and differentiable. However, in structures with singular-continuous spectra, Bragg peaks cannot be isolated and cluster into a hierarchy of self-similar patterns that give rise to a continuous, strictly increasing function with no proper plateaus regions. The DF of the Thue–Morse one-dimensional structure is obtained numerically considering the cumulative sum of the *Radon–Nikodym density* [604]:

$$h_N(x) = \prod_{\ell=1}^{N} [1 - \cos(2^\ell \pi x)] \tag{11.72}$$

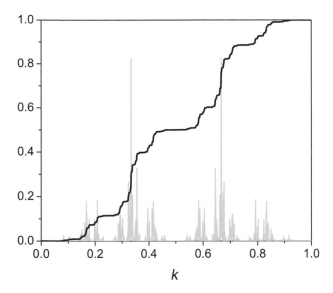

Figure 11.21 The singular-continuous diffraction measure of the Thue–Morse chain (light gray) and its strictly increasing distribution function (black line) on a normalized wavenumber scale.

A simple derivation of this formula that exploits the symmetry properties of the one-dimensional Thue–Morse sequence was provided earlier in Section 7.5.3. The resulting limit distribution function and the singular-continuous diffraction spectrum of the Thue–Morse chain computed from the Radon–Nikodym density are shown in Figure 11.21. Mathematically, the distribution function is in the form of a *Riesz product*, which generally describes a *multiscale signal* with oscillations at every length scale. A rigorous mathematical treatment of the spectral and topological properties of the Thue–Morse sequence and its generalizations can be found in [617]. It is also generally believed that singular-continuous diffraction spectra display multifractal geometry [10].

The distribution function, which is defined for any aperiodic system, can be generalized to higher-dimensional aperiodic structures as well [205]. Finally, we remark that a completely analogous distribution function can be introduced in the study the energy spectra by replacing the normalized intensity distribution appearing in (11.71) with the appropriate spectral measure associated to the density of electronic, vibrational, or optical states. In the study of energy spectra the resulting distribution, known as the integrated density of states (IDOS),[10] plays a central role in the difficult problem of "gap-labeling" of aperiodic structures. According to the important *gap-labeling theorem*, spectral gaps are labeled by the singularities of the Fourier intensity of the scattering potential, independently of the potential strength [326]. In particular, for periodic and quasiperiodic structures it can be rigorously proven that each

---

[10] The IDOS function $H(E)$ at energy E is defined as the function that gives the fraction of eigenvalues less than E. This function is always well defined and particularly convenient to describe aperiodic structures where the standard density of states picture cannot be applied.

δ-function peak appearing in the Fourier transform of the structure, or more precisely each pure-point component in the associated measure, is responsible for a spectral gap. A rigorous mathematical analysis of the gap-labeling theorem has been provided by Bellissard et al. [346].

11.4.9   Engineering Diffuse Scattering

The diffuse part of the diffraction spectrum is typically associated with the presence of disorder in the analyzed structure and it is believed to play only a minor role in the theory of diffraction. However, this perception is inaccurate, as shown by the revealing (and surprising) analysis of Höffe and Baake [616] that provided examples of random and deterministic structures with exactly the same diffuse spectrum. This is best illustrated for one-dimensional structures by comparing the stochastic Bernoulli structure and the deterministic Rudin–Shapiro (RS) structure.

The Bernoulli structure is the point pattern characterized by the stochastic Dirac comb:

$$\boxed{\omega_B = \sum_{n \in \mathbb{N}} \eta_B(n) \delta_n} \tag{11.73}$$

where $\{\eta_B(n)\}$ is a family of random variables that assume the values h_1 and h_2 with probabilities p_1 and p_2. It has been proven that the diffraction measure of the Bernoulli structure consists of the sum of a pure point and an absolutely continuous part [616, 618]:

$$\hat{\gamma}_B = |\langle \mathbf{h} \rangle|^2 \sum_{n \in \mathbb{N}} \delta_n + \sigma^2(\mathbf{h}), \tag{11.74}$$

with mean $\langle \mathbf{h} \rangle = p_1 h_1 + p_2 h_2$ and variance $\sigma^2 = \langle |\mathbf{h}|^2 \rangle - |\langle \mathbf{h} \rangle|^2$.

On the other hand, in the RS sequence, the scattering particles are distributed deterministically and form the Dirac comb:

$$\omega_{RS} = \sum_{n \in \mathbb{N}} \eta_{RS}(n) \delta_n, \tag{11.75}$$

where the scattering strengths are obtained as follows:

$$\eta_{RS}(0) = 1, \quad \eta_{RS}(-1) = 0$$

$$\eta_{RS}(4n) = \eta_{RS}(4n + 1) = \eta_{RS}(n)$$

$$\eta_{RS}(4n + \ell) = \frac{1}{2}\left(1 - (-1)^{n+\ell+\eta_{RS}(n)}\right),$$

and $\ell = 2, 3$. The preceding recurrence produces the aperiodic RS sequence of digits $1, 1, 10, 1, 1, \ldots$ that can alternatively be generated using the substitution rule already introduced in Chapter 7. The diffraction spectrum of the RS structure has been computed rigorously [324]:

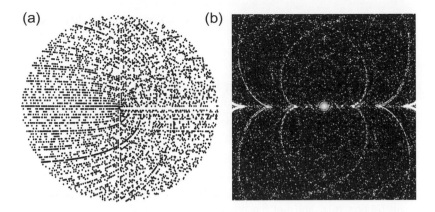

Figure 11.22 (a) Sacks spiral of prime numbers ($N = 4,456$ primes) and (b) a corresponding structure factor (far-field diffraction).

$$\hat{\gamma}_{RS} = \frac{1}{4} + \frac{1}{4}\sum_{n\in\mathbb{N}} \delta_n \qquad (11.76)$$

It is important to remark that, despite its deterministic nature, this RS spectrum becomes identical to the one of the stochastic Bernoulli structure with the following parameters: $h_1 = 1$, $h_2 = 0$, and $p_1 = p_2 = 1/2$. Two measures with the same autocorrelation, or with the same diffraction measure, are called *homometric*. Therefore, this example shows that the Bernoulli sequence and the RS sequence are homometric with probability one, despite the Bernoulli sequence has *configuration entropy* $e = \log 2$ (the maximum value for a binary systems) and the RS sequence has zero entropy. It is argued in [616] that one can find additional examples of structures with the same diffraction spectrum and a configuration entropy in between these extreme values. A particularly striking example of a two-dimensional structure that displays a nontrivial diffuse spectrum is the prime number spiral introduced by Robert Sacks as a slight modification of the Ulam spiral. In the so-called *Sacks spiral*, the natural numbers are plotted along an Archimedean spiral (rather than on a square spiral as in the case of the Ulam spiral), and numbers are spaced so that one perfect square occurs in each full rotation. Then only the prime numbers are marked on the spiral by back dots, as shown in Figure 11.22a. The corresponding diffraction spectrum, shown in Figure 11.22b, features a highly structured diffuse component.

The preceding analysis should dispel the frequent misconception according to which diffuse spectral components do not carry relevant structural information since they essentially reflect structural disorder, and are therefore of limited value to engineering applications. On the opposite, [616] presents concrete examples of two-dimensional structures that encode relevant structural information, such as the degree of rotational symmetry, only in the diffuse components of their spectra. Furthermore, examples of structures that share the same pure-point spectral components and can only be distinguished by the diffuse components in their spectrum are also discussed in [616].

11.5   Which Distributions of Matter Diffract?

The title of this section is borrowed from the seminal paper by Bombieri and Taylor [619] where the authors established a fundamental connection between one-dimensional (1D) quasiperiodic structures and algebraic number theory that allows one to determine which one-dimensional tilings obtained by inflation rules are quasiperiodic. The authors demonstrated that it is the algebraic nature of the substitution rule, and not the relative sizes of the tiles, that is important. We recall that it is possible to associate to each substitution rule a primitive *substitution matrix* whose elements are the number of times a given symbol (i.e., $i = A, B$) appears after the substitution rule, irrespective of the order in which it occurs.[11] The dimension of this matrix is determined by the size of the letter alphabet of the substitution. For example, the substitution matrix of the Fibonacci sequence generated by the two-letter substitution $A \rightarrow AB, B \rightarrow A$ is the 2×2 matrix with column vectors $(1, 1)^T$ and $(1, 0)^T$. Interestingly, relevant information on the nature of the diffraction spectra can be directly obtained from the properties of the substitution matrix. Specifically, Bombieri and Taylor discovered that the characteristic polynomial of the substitution/inflation matrix must have only one root greater than unity in absolute value for the corresponding structure to be quasiperiodic. More precisely, according to the *Bombieri–Taylor theorem* [619], a structure is quasiperiodic if and only if the spectrum of the associated substitution matrix contains a *Pisot–Vijayaraghavan number* (PV number). Moreover, in this case the quasiperiodic distribution of scatterers can be obtained by a suitable the cut and project construction in the same way as continuous quasiperiodic functions are obtained as line cuts of periodic functions in higher-dimensional spaces. Remember that a PV number is a positive algebraic number, i.e., a number that is obtained from the solution of an algebraic equation that is greater than one in modulus and such that all its conjugate elements (the other solutions of the algebraic equation) have an absolute value less than one. For example, the golden mean ϕ, satisfying the algebraic equation $x^2 - x - 1 = 0$, is a PV number.

In general a structure is *diffractive* when its spectrum has countably infinite discrete components. Since quasiperiodic structures are diffractive, the PV property also implies diffraction, i.e., it is a sufficient condition for diffraction phenomena to occur in aperiodic structures. However, the PV property is not a necessary condition for diffraction, since there are diffractive aperiodic structures whose spectra are mixed, i.e., they contain diffractive peaks embedded in a diffuse background that encodes important structural information.[12] These systems are also diffractive but are not

---

[11] Note that different substitution rules can have the same substitution matrix.
[12] See discussion in Section 11.4.

strictly quasiperiodic. Structures with mixed spectra and correlated random systems pose many open questions to mathematical diffraction theory. An informative and rigorous analysis can be found in [316, 620, 621].

The broader question of characterizing all diffractive structures still remains open even in one spatial dimension. This difficulty originates from a hard problem of number theory, to which we now turn our attention. By looking back at equation (7.17), we recognize that the diffraction condition is ultimately a statement on the asymptotic behavior of an infinite sum of exponential functions. As originally recognized in [39], this is a mathematically difficult problem related to the asymptotic behavior of the generalized Dirichlet series on the line of convergence. This series has the following form [622]:

$$f(s) = \sum_{n=1}^{\infty} a_n \exp(-\lambda_n s) \tag{11.77}$$

where λ_n is a sequence of increasing real numbers and $s = \sigma + it$ is a complex variable. Note that when $\lambda_n = n$, this series turns into the power series for $\exp(-s)$, and when $\lambda_n = \log(n)$ it becomes the ordinary Dirichlet series. Dirichlet series play a very important role in analytic number theory in relation to the distribution of prime number since both the Riemann zeta function and the Dirichlet L-functions are special cases of Dirichlet series introduced earlier.

As we discussed in Chapter 9, the generalized Dirichlet's series converges in the complex half-plane for $\sigma > \sigma_0$, where [622]

$$\sigma_0 = \limsup \frac{\ln |\sum_1^n a_n|}{\lambda_n} \tag{11.78}$$

and it diverges when $\sigma < \sigma_0$. The line $s = \sigma_0$ is called the *line of convergence*. Direct comparison with the expression for the diffraction spectrum in (7.17) reveals that $a_n = 1$ and $\lambda_n \sim n$ for structures with scattering particles located at positions x_n that are relatively dense in $\mathbb{R}$ [39]. Therefore, in this case the diffraction condition for 1D structures, or equivalently the asymptotic convergence of the spectrum $\hat{\rho}(q)$, reduces to an analytic statement on the behavior of the generalized Dirichlet series on the line of convergence $\sigma_0 = 0$. Unfortunately, as Hardy notes, "On the line of convergence the question of the convergence of the series remains open, and requires considerations of a much more delicate character" [622]. However, significant progress has since been made by the mathematics community to elucidate this difficult problem. In particular, a fundamental connection between the complex analytical behavior of generalized Dirichlet series and *quasiperiodicity*[13] has been recently established [623–625]. In this context, a complex-valued function $f(s)$ is quasiperiodic on a vertical line if there exists an infinite series of real numbers $\{\tau_n\}$, called quasiperiods, such that for an arbitrarily small $\epsilon > 0$ and for every s in the domain of uniform convergence of the

---

[13] The theorems discussed in the text apply to the more general class of almost-periodic functions. However, almost-periodic functions are necessarily quasiperiodic as well.

series we have $|f(s + i\tau_n) - f(s)| < \epsilon$. Moreover, the quasiperiodicity of Dirichlet functions has been linked to the denseness property of the image of vertical lines by Dirichlet functions, and a necessary and sufficient condition for the quasiperiodicty of $f(s)$ on the line with $s = \sigma_0 + it$ has been related to the geometrical properties of the fundamental domains of the function [625]. These important results generalize seminal work by Harald Bohr, who discovered the quasiperiodic behavior of the standard Dirichlet series, the special case of generalized Dirichlet series when $\lambda_n = \log(n)$, in its half-plane of absolute convergence [463]. Interestingly, Bohr also discovered a connection of the Riemann hypothesis (RH) and quasiperiodicity by proving that the RH for the Dirichlet L-function $L(s, \chi)$ (for χ a nonprincipal character) is equivalent to the quasiperiodicity of $L(s, \chi)$ in the half-plane $\sigma > 1/2$. These results bring us closer to the rigorous classification of 1D quasicrystals entirely based on the number-theoretic properties of generalized L-functions.

12 Multiple Scattering

A clear analysis leads to perspicuous representations.

Philosophical Investigations, Ludwig Wittgenstein

This Chapter further develops the general framework of scattering theory and illustrates its applications to complex multiparticle systems and vector scattering problems. We start by introducing a diagrammatic notation that facilitates the physical interpretation of the scattering equations and their perturbative solutions, in analogy with Feynman's diagrams used in the many-body theory of interacting quantum systems. The single-particle T-matrix and scattering matrix operators will be rigorously defined and their physical meaning discussed for arbitrary multiparticle systems. In particular, we will show that the multiple scattering solution of a general multiparticle system can always be obtained in terms of the single-particle T matrices through an appropriate expansion series. The Foldy–Lax equations of multiple scattering for both scalar and vector waves will be derived. These provide the foundation for coupled-dipoles and Green's matrix spectral methods that are used in the study of complex scattering media. In particular, based on these methods we will discuss the multiple scattering properties of particle arrays with random and deterministic aperiodic geometries.

12.1 Green Function Perturbation Theory

In order to obtain a self-consistent solution of the Lippmann–Schwinger (LS) equation (6.8) for the total Green function, it is convenient to rewrite it using the following symbolic form [132]:

$$\mathbf{G}^+ = \mathbf{G}_0^+ + \mathbf{G}_0^+ \mathbf{V} \mathbf{G}^+, \tag{12.1}$$

where the bold letters are used to denote operators associated to the quantities appearing in (6.8). The operators can be constructed from the corresponding functions regarded as discretized on a computational grid (either in space or wavenumber representation), which provides a convenient matrix representation [24]. Equivalently, the abstract Dirac's bra and ket notation can be used to represent operators in the infinite dimensional Hilbert space associated to the (continuum) coordinate or the momentum representations. For instance, in the coordinate representation, we have the following:

$$G_0^+(\mathbf{r} - \mathbf{r}', \omega) = \langle \mathbf{r}|G_0^+|\mathbf{r}'\rangle \tag{12.2}$$

$$G^+(\mathbf{r}, \mathbf{r}', \omega) = \langle \mathbf{r}|G^+|\mathbf{r}'\rangle, \tag{12.3}$$

where the position operator $\tilde{\mathbf{r}}$ is defined as follows:

$$\tilde{\mathbf{r}}|\mathbf{r}\rangle = \mathbf{r}|\mathbf{r}\rangle \tag{12.4}$$

In addition, we have the usual orthonormal relation between ket vectors:

$$\langle \mathbf{r}|\mathbf{r}'\rangle = \delta(\mathbf{r} - \mathbf{r}') \tag{12.5}$$

and the closure (i.e., completeness) relation:

$$\mathbf{I} = \int d\mathbf{r}|\mathbf{r}\rangle\langle\mathbf{r}|, \tag{12.6}$$

where $\mathbf{I}$ is the identity operator. Since the scattering potential $\sigma(\mathbf{r})$ depends only on one position variable, the corresponding scattering potential operator $\mathbf{V}$ is diagonal [132].

Moreover, the operator symbol $\mathbf{G}_0^+\mathbf{V}$ is equivalent to the integral operator:

$$\mathbf{G}_0^+\mathbf{V} = \int d^3\mathbf{r}_1 G_0^+(\mathbf{r} - \mathbf{r}_1, \omega)V(\mathbf{r}_1). \tag{12.7}$$

Now using the symbolic operator notations' convolution-type integral equations, such as the LS one, can be more simply written as algebraic equations for the corresponding operator quantities.

12.1.1   Self-Consistent Solution of the Scattering Problem

The LS equation (12.1) for the total Green function can be solved self-consistently by successive iterations, yielding an expansion in terms of free-space Green functions. In the compact operator notation, we obtain the following:

$$\begin{aligned}\mathbf{G}^+ &= \mathbf{G}_0^+ + \mathbf{G}_0^+\mathbf{V}\mathbf{G}^+ \\ &= \mathbf{G}_0^+ + \mathbf{G}_0^+\mathbf{V}\left[\mathbf{G}_0^+ + \mathbf{G}_0^+\mathbf{V}(\mathbf{G}_0^+ + \mathbf{G}_0^+\mathbf{V}\cdots]\right. \\ &= \mathbf{G}_0^+ + \mathbf{G}_0^+\mathbf{V}\,\mathbf{G}_0^+ + \mathbf{G}_0^+\mathbf{V}\mathbf{G}_0^+\mathbf{V}\mathbf{G}_0^+ + \cdots\end{aligned} \tag{12.8}$$

The infinite expansion appearing in equation (12.8) is called *Neumann expansion series*. This expression provides the total Green function for a scattering potential in terms of a contribution that comes directly from the source without experiencing any scattering (i.e., the first term), plus a contribution that corresponds to fields scattered only once by the material (i.e., the second term), plus the contributions of fields recurrently scattered, as illustrated in Figure 12.1. It is important to realize that even in the case of a single particle, the recurrent scattering terms describe the fact that the electromagnetic polarization of the scatterer is driven by the total local field, which is the sum of the incident field and the field radiated by the scatterer itself

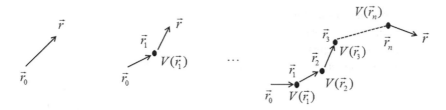

Figure 12.1 Schematic representation of zero-order, first-order, and nth-order scattering process inside an inhomogeneous medium.

(i.e., the scattered field). The incident field polarizes the scatterer that in turn emits a field that modifies the polarization, which subsequently modifies the scattered field, and so on, giving rise to the *self-consistent process of recurrent scattering* even for a single particle (see Figure 12.2).

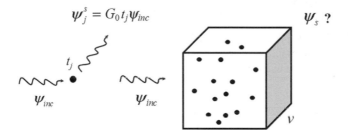

Figure 12.2 Schematic representations of single-particle scattering and of multiple scattering by many particles. In such a medium, the individual scattered fields are interdependent and the electric field exciting each particle is given by the superposition of the external field and the sum of the individual fields scattered by all other particles.

This self-consistent process is contained in the so-called *transition operator* **T**, defined by the following:

$$\mathbf{G}^+ = \mathbf{G}_0^+ + \mathbf{G}_0^+ \mathbf{T} \mathbf{G}_0^+ \tag{12.9}$$

where we denote by the following definition:

$$\mathbf{T} = \mathbf{V} + \mathbf{V}\mathbf{G}_0^+\mathbf{V} + \mathbf{V}\mathbf{G}_0^+\mathbf{V}\mathbf{G}_0^+\mathbf{V} + \cdots$$
$$= \mathbf{V}\left(\mathbf{I} - \mathbf{G}_0^+\mathbf{V}\right)^{-1} = \left(\mathbf{I} - \mathbf{V}\mathbf{G}_0^+\right)^{-1}\mathbf{V}. \tag{12.10}$$

In equation (12.10), we performed a summation over an *operator geometric series*, resulting in a closed-form expression for the transition operator. However, operator expressions are just shorthand compact notations for infinite series of convolution-type integral equations. As a result, in most practical situations the inversion of the operator **T** is not a simple task. Starting from the seminal work of Waterman [626], very powerful numerical approaches generically referred to as the *null-field method* [280, 627], have been developed. These provide suitable representations of the **T** operator, usually expressed in a spherical coordinate basis and known as the *T-matrix*, for a scattering object of arbitrary shape, size, and composition.

We now show that the knowledge of the T-matrix is sufficient to completely solve the inhomogeneous wave equation for the total field (6.10) entirely, or equivalently, to solve the LS equation (6.14) for the field. In order to see this, let us rewrite equation (6.14) in operator notation:

$$|\psi\rangle = |\psi_0\rangle + \mathbf{G}_0^+ \mathbf{V} |\psi\rangle, \tag{12.11}$$

where $|\psi\rangle$ is the Hilbert-space vector corresponding to the field solution ψ. We self-consistently iterate equation (12.11) and obtain the following:[1]

$$\begin{aligned}
|\psi\rangle &= |\psi_0\rangle + \mathbf{G}_0^+ \mathbf{V} |\psi\rangle \\
&= |\psi_0\rangle + \mathbf{G}_0^+ \mathbf{V} |\psi_0\rangle + \mathbf{G}_0^+ \mathbf{V} \mathbf{G}_0^+ \mathbf{V} |\psi_0\rangle + \cdots \\
&= |\psi_0\rangle + \mathbf{G}_0^+ \mathbf{T} |\psi_0\rangle.
\end{aligned} \tag{12.12}$$

This equation makes it very clear that the desired field solution is linear in the incident field $|\psi_0\rangle$ but highly nonlinear in the scattering potential operator $\mathbf{V}$. The infinite expansion appearing in (12.12) is the *Born series* for the field solution.[2] The recurrent solution of the single-scattering problem has been completely solved in terms of the scatterer's T-matrix. It follows from equation (12.12) that $\mathbf{T} |\psi_0\rangle = \mathbf{V} |\psi\rangle$ and therefore from (12.12) we have the following:

$$\mathbf{T} |\psi_0\rangle = \mathbf{V} |\psi_0\rangle + \mathbf{V} \mathbf{G}_0^+ \mathbf{T} |\psi_0\rangle. \tag{12.13}$$

In the case of incident plane waves, the transition operator $\mathbf{T}$ also satisfies a Lippmann–Schwinger equation:

$$\mathbf{T} = \mathbf{V} + \mathbf{V} \mathbf{G}_0^+ \mathbf{T}, \tag{12.14}$$

from which the Born expansion series of the $\mathbf{T}$ operator in equation (12.10) follows directly. Introducing the *scattering operator* $\mathbf{S}$, or the S-matrix:[3]

$$|\psi\rangle = \mathbf{S} |\psi_0\rangle, \tag{12.15}$$

and remembering the definition of the $\mathbf{T}$ operator, we immediately connect the two operators:

$$\boxed{\mathbf{S} = \mathbf{1} + \mathbf{G}_0^+ \mathbf{T}} \tag{12.16}$$

where $\mathbf{1}$ denotes the unit operator. The S-matrix relates the initial state and the final state of a physical system undergoing a scattering process. However, differently from the $\mathbf{T}$ operator,[4] it is evident that the $\mathbf{S}$ operator additionally contains the unscattered (i.e., incident) waves. Combining equation (12.16) with the LS equation for the T operator (12.14), we obtain the following expansion for the S-matrix:

$$\mathbf{S} = \mathbf{1} + \mathbf{G}_0^+ \mathbf{V} + \mathbf{G}_0^+ \mathbf{V} \mathbf{G}_0^+ \mathbf{V} + \cdots \tag{12.17}$$

---

[1] This method of solving integral equations iteratively is also known as the Liouville–Neumann perturbation expansion method.

[2] Born employed this method for the first time in quantum scattering theory [628].

[3] It can be demonstrated that the scattering operator $\mathbf{S}$ is a unitary operator in the absence of absorption.

[4] The T operator directly provides the scattered field solution in terms of the incident field (see equation (12.12)).

Alternatively, we see from this expansion that the S-matrix satisfies the following Dyson equation:

$$\mathbf{S} = \mathbf{1} + \mathbf{G}_0^+ \mathbf{VS}, \tag{12.18}$$

from which we derive the following, by comparing with (12.16):

$$\mathbf{T} = \mathbf{VS}. \tag{12.19}$$

In Section (12.2), we extend this formalism and consider the self-consistent solution for many scatterers, which is properly a multiparticle multiple scattering problem.

12.1.2   Physical Meaning of the S-Matrix

The S-matrix approach is particulary convenient when considering the spectral behavior of structures that support *scattering resonances* (photonic gratings, photonic/plasmonic arrays, open resonator cavities, etc.). Let us rewrite equation (12.15) in the following form:

$$|\psi_{out}\rangle = \mathbf{S}\,|\psi_{in}\rangle, \tag{12.20}$$

where we denoted by $|\psi_{in}\rangle$ and $|\psi_{out}\rangle$ the complex amplitudes of incident and outgoing waves, respectively. The connection of the S-matrix with the resonant behavior of the scattering system becomes evident if we remember that *eigenmodes* can be generally defined as states that exist without requiring any external excitation. As a result, their dispersion relation is directly found through the S-matrix by requiring the following:

$$\mathbf{S}^{-1}(\omega)\,|\psi_{out}\rangle = |\psi_{in}\rangle = 0, \tag{12.21}$$

where we explicitly indicated the frequency dependence of the S-matrix. The nontrivial solutions of (12.21) are obtained numerically by solving for the following:

$$\det \mathbf{S}^{-1}(\omega) = \frac{1}{\det \mathbf{S}(\omega)} = 0. \tag{12.22}$$

This equation shows that the eigenmodes of a general scattering structure are described in terms of the *complex poles of the determinant of the S-matrix*. In fact, it can be rigorously proven that the scattering resonances coincide with the poles of the analytic continuation of $\mathbf{S}(\omega)$ in the complex plane [90, 629, 630]. In particular, the real part of a pole corresponds to the frequency of an incident wave that can excite an eigenmode of the structure, while the inverse of the imaginary part of the pole provides the lifetime of the resonance. This is analogous to what can be obtained using the Green's matrix spectral method presented in Section 12.4. However, the calculation of the poles of the scattering matrix is computationally more challenging. This is so because the calculation of matrix determinants becomes extremely inefficient and unstable in the case of large matrices. In order to overcome this difficulty, several iterative methods have been recently developed that enabled the accurate study of the scattering resonances of metal gratings and plasmonic nanostructures [631–634]. Of particular interest is a Cauchy integral method that allows us to find all the scattering poles located within a given domain of interest [631, 632]. This method, first developed to find an efficient solution of the eigenvalue equations of optical waveguides [635], leverages the so-called *resonance representation of the S-matrix*, which in the general case of N poles within a region Ω can be written as follows [631]:

$$S(\omega) = A(\omega) + \sum_{i=1}^{N} \frac{B_i}{\omega - \omega_p^{(i)}} \qquad (12.23)$$

where B_i are the *residue matrices* of $S(\omega)$ computed at the (isolated) poles $\omega_p^{(i)}$. For instance, in the case of a domain γ that contains only a single pole at ω_p, the residue matrix B of $S(\omega)$ is simply as follows:

$$B \equiv \operatorname*{Res}_{\omega=\omega_p} S(\omega) = \frac{1}{2\pi i} \oint_\gamma S(\omega)d\omega, \qquad (12.24)$$

where this equation must be understood as an elementwise operation. The representation of equation (12.23) shows clearly that the scattering response of a general system can be split in two separate contributions. On one side, the matrix-valued function $A(\omega)$ that has no poles in Ω and describes nonresonant scattering; on the other side, the sum over the complex poles of the S-matrix that takes into account all the contributions of resonant scattering.

Box 12.1 Bound States in the Continuum

Traditionally, the optical resonances of open-scattering systems have a frequencies inside the continuous part of the spectrum and therefore leak out or radiate to infinity. In contrast, the bound states of closed systems populate the discrete part of the spectrum and appear as perfectly confined modes and are characterized by infinite lifetimes. This conventional wisdom is challenged by the existence of exceptional bound states that are present in the radiation continuum of open-scattering systems, known as *bound states in the continuum* (BICs). Such intriguing states, as known as *embedded eigenvalues*, coexist with extended states but are perfectly localized resonances with infinite lifetime and zero width. BICs were originally envisioned in 1929 by von Neumann and Wigner for quantum waves in a specifically constructed 3D potential [636]. However, BICs are very general wave phenomena and have since been discovered in electromagnetic waves [637], acoustic waves, water waves, and a wide range of structures that include optical fibers, photonic crystals, and even in graphene [638]. BICs physically occur due to several different mechanisms, including symmetry and topological protection in periodic structures [637] and parameter tuning via coupled resonances [639–641]. Since photonic structures with BICs behave as very high-Q resonators, they became very appealing for device applications to lasing, optical sensing, and frequency filtering [639, 642–644].

12.1.3 The Diagrammatic Notation

We have seen that the calculation of field amplitudes through the total Green function of a scattering system requires the use of rather complicated integrals to be evaluated over a large number of variables. However, these integrals have a well-defined structure and can be represented graphically in terms of Feynman diagrams. A Feynman diagram gives the contribution of a particular class of scattering paths and corresponds to a graphical representation of a *perturbative contribution* to the total field amplitude or Green's function. This calculation scheme is named after its inventor, Richard Feynman, who first introduced it in 1948 to more simply visualize the complex interactions of subatomic particles. Feynman diagrams became a primary tool in quantum field theory and condensed-matter theory and, as we will discover soon, they also play an important role in the perturbative theory of multiple scattering of classical waves in random media. However, we should always remember that Feynman diagrams are just convenient notations that represent cumbersome LS-type integral equations appearing in scattering theory (e.g., the Neumann's series) and they establish a one-to-one correspondence between every term in the equation and its graphical representation.

As a simple example, we introduce the following "vocabulary":

$$\mathbf{G}^+ = \quad \sim\!\sim \tag{12.25}$$

$$\mathbf{G}_0^+ = \quad \text{——} \tag{12.26}$$

$$\mathbf{V} = \bullet \tag{12.27}$$

Based on the preceding correspondence, we can immediately translate the LS operator equation for the total Green function (12.1) into its *diagrammatic representation*:

$$\sim\!\sim = \text{——} + \text{——}\!\bullet\!\sim\!\sim \tag{12.28}$$

This symbolic expression is clearly a shorthand notation for the LS integral equation:

$$G^+(\mathbf{r},\mathbf{r}',\omega) = G_0^+(\mathbf{r}-\mathbf{r}',\omega) + \int_\Gamma d^3\mathbf{r}_1 G_0^+(\mathbf{r}-\mathbf{r}_1,\omega)V(\mathbf{r}_1)G^+(\mathbf{r}_1,\mathbf{r}',\omega) \tag{12.29}$$

and we remark that in the diagrammatic representation we always imply a spatial integration for each *vertex* (i.e., closed circles) of the diagram.

Iterating (12.28) yields directly the diagrammatic representation of the Neumann series:

$$\sim\!\sim = \text{——} + \text{——}\!\bullet\!\text{——} + \text{——}\!\bullet\!\text{——}\!\bullet\!\text{——} + \text{——}\!\bullet\!\text{——}\!\bullet\!\text{——}\!\bullet\!\text{——} + \cdots \tag{12.30}$$

where each vertex in the diagram corresponds to a characteristic scattering order and the number of vertices determines the dimensionality of the corresponding integral. By comparing the Neumann series (12.8) and its diagrammatic representation (12.30), we see clearly that each diagram stands for a convolution integral whose kernel is the algebraic product of the values of its parts. According to the general definition given in (12.10), the **T** matrix operator, which contains all the contributions of multiple scattering events, can also be diagrammatically represented as follows:

$$\mathbf{T} = \bullet + \bullet\!\!-\!\!\bullet + \bullet\!\!-\!\!\bullet\!\!-\!\!\bullet + \bullet\!\!-\!\!\bullet\!\!-\!\!\bullet\!\!-\!\!\bullet + \cdots$$

$$= \bullet + \bullet\!\!-\!\! \; (\bullet + \bullet\!\!-\!\!\bullet + \bullet\!\!-\!\!\bullet\!\!-\!\!\bullet + \cdots) \tag{12.31}$$

We recognize that the term in parentheses defines again the **T** matrix operator, so that we establish by inspection that the **T** matrix operator satisfies *Dyson's equation*:

$$\boxed{\mathbf{T} = \mathbf{V} + \mathbf{V}\mathbf{G}_0^+\mathbf{T}} \tag{12.32}$$

which coincides with the result derived in the previous section.

The full scope of the Feynman diagrams notation will be better appreciated in the next chapter when we will deal with multiple scattering of waves in random media, requiring *ensemble averaging* of diagrammatic expansions such as (12.30) over different realizations of disorder.

12.2   Multiple Scattering of Scalar Waves

Up to this point, we have described the multiple scattering of light interacting with just one particle, and we have seen that the solution of this problem, which is provided in full by the T-matrix operator (or equivalently by the total Green function) can always be expressed as an infinite series that sums over increasing scattering orders and includes recurrent scattering. We will now discuss the more general situation of a system containing a large number N of scattering particles, and show that its general multiple scattering solution can always be expressed in terms of the T matrices of its components. In particular, following the derivation in [24], we can prove a beautiful result of multiple scattering theory that allows one to express the global T-matrix of a system of arbitrary scattering particles as an expansion series with terms containing the T-matrices of each individual particle. We start by remembering expression (12.10) derived previously for the T-matrix of an arbitrary object described by a scattering potential **V**:

$$\mathbf{T} = \left(\mathbf{I} - \mathbf{V}\mathbf{G}_0^+\right)^{-1}\mathbf{V}. \tag{12.33}$$

If we now left-multiply equation (12.33) by $\left(\mathbf{I} - \mathbf{V}\mathbf{G}_0^+\right)$, we obtain the following:

$$\mathbf{T} = \mathbf{V}\left(\mathbf{I} + \mathbf{G}_0^+\mathbf{T}\right), \tag{12.34}$$

which will turn out to be useful later on in the derivation. Let us now consider a scattering system composed of multiple particles described by the overall scattering potential:

$$\mathbf{V} = \sum_i \mathbf{V}_i, \tag{12.35}$$

where $\mathbf{V}_i$ denotes the scattering potential of an individual particle and the sum extends over the total number of scatterers in the system. Using (12.34), we can express the global T-matrix of the collection of scatterers as the following sum:

$$\mathbf{T} = \sum_i \mathbf{U}_i \tag{12.36}$$

where

$$\mathbf{U}_i = \mathbf{V}_i \left(\mathbf{I} + \mathbf{G}_0^+ \mathbf{T} \right) = \mathbf{V}_i \left(\mathbf{I} + \mathbf{G}_0^+ \sum_j \mathbf{U}_j \right). \tag{12.37}$$

By subtracting $\mathbf{V}_i \mathbf{G}_0^+ \mathbf{U}_i$ from both sides we get the following:

$$\mathbf{U}_i - \mathbf{V}_i \mathbf{G}_0^+ \mathbf{U}_i = \left(\mathbf{I} - \mathbf{V}_i \mathbf{G}_0^+ \right) \mathbf{U}_i = \mathbf{V}_i \left(\mathbf{I} + \mathbf{G}_0^+ \sum_{j \neq i} \mathbf{U}_j \right) \tag{12.38}$$

Solving for $\mathbf{U}_i$, we obtain the following:

$$\mathbf{U}_i = \left(\mathbf{I} - \mathbf{V}_i \mathbf{G}_0^+ \right)^{-1} \mathbf{V}_i \left(\mathbf{I} + \mathbf{G}_0^+ \sum_{j \neq i} \mathbf{U}_j \right) = \mathbf{t}_i \left(\mathbf{I} + \mathbf{G}_0^+ \sum_{j \neq i} \mathbf{U}_j \right), \tag{12.39}$$

where we have indicated with $\mathbf{t}_i$ the T-matrix of the ith single particle according to (12.33). Now we can solve equation (12.39) self-consistently by successive iterations, resulting in the infinite expansion:

$$\mathbf{U}_i = \mathbf{t}_i + \mathbf{t}_i \mathbf{G}_0^+ \sum_{j \neq i} \mathbf{t}_j \left(\mathbf{I} + \mathbf{G}_0^+ \sum_{k \neq j} \mathbf{U}_k \right)$$
$$= \mathbf{t}_i + \sum_{j \neq i} \mathbf{t}_i \mathbf{G}_0^+ \mathbf{t}_j + \sum_{j \neq i} \sum_{k \neq j} \mathbf{t}_i \mathbf{G}_0^+ \mathbf{t}_j \mathbf{G}_0^+ \mathbf{t}_k + \cdots \tag{12.40}$$

By remembering the definition of the multiparticle system's T-matrix in (12.36), we finally get the following:

$$\boxed{\mathbf{T} = \sum_i \mathbf{t}_i + \sum_i \sum_{j \neq i} \mathbf{t}_i \mathbf{G}_0^+ \mathbf{t}_j + \sum_i \sum_{j \neq i} \sum_{k \neq j} \mathbf{t}_i \mathbf{G}_0^+ \mathbf{t}_j \mathbf{G}_0^+ \mathbf{t}_k + \cdots} \tag{12.41}$$

Equation (12.41) is much more complex than simply the sum of the individual particles T-matrices $\mathbf{t}_i$, and provides an expression that includes all the contributions of *recurrent and multiple scattering events* in the many-particle systems, each expressed as a function of the individual particles' t-matrices. Recurrent scattering events are the ones where a field scattered by a specific particle gets scattered by at least another scatterer in the systems, and then returns again to the specific scatterer. This situation only occurs in systems with very strong scattering, and was demonstrated experimentally by Wiersma et al. using TiO_2, ZnO, and $BaSO_4$ strongly scattering samples [645]. The first term in the expansion corresponds to the *independent scattering approximation (ISA)*, where each particle only scatters the field once. Note that each consecutive index value in the summation must differ from the one immediately before it, but there is no restriction on the values of two indices that are not immediately next to each other (e.g., $i = k$ is allowed in the third term of (12.41)).

Now that we have been able to express the T-matrix of a system of particles in terms of each particle's T-matrix, we can write the multiple scattering Born series as in equation (12.12), where $\mathbf{T}$ denotes this time the system's T-matrix (12.41). As a result, the total field solution that includes all recurrent and multiple scattering contributions becomes

$$
\begin{aligned}
|\psi\rangle &= |\psi_0\rangle + \mathbf{G}_0^+ \mathbf{V} |\psi_0\rangle + \mathbf{G}_0^+ \mathbf{V} \mathbf{G}_0^+ \mathbf{V} |\psi_0\rangle + \cdots \\
&= |\psi_0\rangle + \mathbf{G}_0^+ \mathbf{T} |\psi_0\rangle \\
&= |\psi_0\rangle + \mathbf{G}_0^+ \sum_i \mathbf{t}_i |\psi_0\rangle + \sum_i \sum_{j \neq i} \mathbf{G}_0^+ \mathbf{t}_i \mathbf{G}_0^+ \mathbf{t}_j |\psi_0\rangle + \cdots
\end{aligned}
\tag{12.42}
$$

Even if formally identical to the result in (12.12), these equations contain many more terms due to the fields being scattered multiple times and recurrently by all the particles in the system. In particular, the second term provides the first-order Born approximation of *single scattering*, which is a good approximation only for weak scattering at small particle densities (in diluted systems). The third term represents *double scattering* (two-particle) events, and so on, where each successive term increases the order of integration of multiple scattering events.

12.2.1   Foldy–Lax Multiple Scattering Equations

In pioneering papers dealing with multiple scattering in a system of discrete random scatterers, Foldy [646] and Lax [647] developed a general scheme to solve self-consistently the multiple scattering problem in the case of idealized point dipoles. Using the T-matrix approach, this method can be easily generalized to a system of arbitrary scattering particles with known T-matrices $\mathbf{t}_i$. The basic idea behind the general Foldy–Lax self-consistent scheme is to reduce the complex multiple scattering of a system of particles to an equivalent single-scattering problem driven by unknown *local exciting fields* $|\psi_i^{exc}\rangle$ that are defined at each particle's position as follows:

$$
|\psi\rangle = |\psi_0\rangle + \sum_i \mathbf{G}_0^+ \mathbf{t}_i |\psi_i^{exc}\rangle .
\tag{12.43}
$$

The local exciting field $|\psi_i^{exc}\rangle$ is the field that is driving the particle i as a result of the multiple scattering from all the other particles. In a system of N scattering particles with individual T-matrices $\mathbf{t}_i$, we can express the local exciting field that drives particle j as follows:

$$
\boxed{|\psi_j^{exc}\rangle = |\psi_0\rangle + \sum_{i \neq j} \mathbf{G}_0^+ \mathbf{t}_i |\psi_i^{exc}\rangle \qquad j = 1, \ldots, N}
\tag{12.44}
$$

Note that equation (12.44) defines a self-consistent system of **N** coupled equations in the N unknown local exciting fields, which can in principle be solved numerically. Once the local exciting fields have been determined for all the particles, then by substituting them into equation (12.43) we can obtain the full solution of the multiple scattering problem. The coupled field equations (12.43) and (12.44) are the rigorous *Foldy–Lax multiple scattering equations* for a system of N particles. This system can be solved self-consistently once the single particles' T-matrices are known. It is also instructive to realize that the Foldy–Lax system is completely equivalent to the expansion of the N-particles' T-matrix in terms of the single-particle T-matrix $\mathbf{t}_i$, as can be shown by the following iterative recursions:

$$|\psi\rangle = |\psi_0\rangle + \sum_i \mathbf{G}_0^+ \mathbf{t}_i \, |\psi_i^{exc}\rangle$$

$$= |\psi_0\rangle + \sum_i \mathbf{G}_0^+ \mathbf{t}_i \left[|\psi_0\rangle + \mathbf{G}_0^+ \sum_{j\neq i} \mathbf{t}_j \, |\psi_j^{exc}\rangle \right] \tag{12.45}$$

$$= |\psi_0\rangle + \mathbf{G}_0^+ \sum_i \mathbf{t}_i \, |\psi_0\rangle + \sum_i \sum_{j\neq i} \mathbf{G}_0^+ \mathbf{t}_i \mathbf{G}_0^+ \mathbf{t}_j \, |\psi_j^{exc}\rangle .$$

Now we can write the exciting field $|\psi_j^{exc}\rangle$ as follows:

$$|\psi_j^{exc}\rangle = |\psi_0\rangle + \mathbf{G}_0^+ \sum_{k\neq j} \mathbf{t}_k \, |\psi_k^{exc}\rangle \tag{12.46}$$

and substitute in the last term of equation (12.45). By continuing this process iteratively, we generate a series of terms that are identical to field solution in terms of the T-matrix expansion previously derived in equation (12.42). A full derivation of the Foldy–Lax equations is provided in Appendix D.

12.2.2 Foldy–Lax Equations for Point Scatterers

In the case of a system of point scatterers, the T-matrix of each particle has the simple analytical expression:

$$\mathbf{t}_j(\mathbf{r}', \mathbf{r}'') = 4\pi f \delta(\mathbf{r}' - \mathbf{r}_j)\delta(\mathbf{r}'' - \mathbf{r}_j), \tag{12.47}$$

where $f = f' + jf''$ is the isotropic complex scattering amplitude. We restrict our attention to isotropic scatterers that are small compared to the wavelength so that the incoming wave does not probe any internal structure of the scatterer. As a result, to lowest order one has only a spherical symmetric outgoing wave, the so-called *s-wave scattering* characterized by angular momentum $l = 0$. In general, for particles that are not very small compared to the incoming wavelength, one also needs to consider *p and d-wave scattering* with $l = 1, 2$, and so on. Notice that under our assumptions, the scattering amplitude depends only on the wavenumber $k = |\mathbf{k}|$, and it can also be expressed as follows [90]:

$$f(k) = \frac{e^{2j\delta_s(k)} - 1}{2jk}, \tag{12.48}$$

where we introduced the s-wave scattering *phase shift* $\delta_s(k)$. This quantity describe the resonant scattering properties of the single scatterers. In the case of a point particle with one internal *Breit–Wigner resonance*[5] with energy E_0 and width Γ_0, the phase shift can be written explicitly as follows:

$$\cot \delta = -\frac{E - E_0}{\Gamma_0}. \tag{12.49}$$

Moreover, the total scattering cross section for s-waves, which is generally related to the scattering phase shift by the expression [648]:

$$\sigma_s = \frac{4\pi}{k^2} \sin^2 \delta_s(k), \tag{12.50}$$

takes the *Lorentzian shape*:

$$\boxed{\sigma_s = \frac{1}{k^2} \frac{4\pi \Gamma_0^2}{(E - E_0)^2 + \Gamma_0^2}} \tag{12.51}$$

Finally, considering the isotropic elastic scattering (i.e., no absorption) of scalar waves, the scattering amplitude satisfies the optical theorem in the following form:

$$|f(k)|^2 = \frac{\Im\{f(k)\}}{k}. \tag{12.52}$$

Using the T-matrix expression in (12.47), the Foldy–Lax multiple scattering equations of a system of point scatterers are obtained immediately from equation (12.46) in integral form:

$$\psi_i^{exc}(\mathbf{r}) = \psi_0(\mathbf{r}) + \sum_{j \neq i}^{N} \int d^3\mathbf{r}' \int d^3\mathbf{r}'' G_0^+(\mathbf{r}, \mathbf{r}') t_j(\mathbf{r}', \mathbf{r}'') \psi_j^{exc}(\mathbf{r}''). \tag{12.53}$$

Using the analytical expression of the point-scatterer T-matrix, we can compute the preceding integral and obtain the Foldy–Lax equations immediately:

$$\psi_i^{exc}(\mathbf{r}) = \psi_0(\mathbf{r}) + \sum_{m \neq i}^{N} f \frac{e^{jk|\mathbf{r} - \mathbf{r}_m|}}{|\mathbf{r} - \mathbf{r}_m|} \psi_m^{exc}(\mathbf{r}_m) \tag{12.54a}$$

$$\psi(\mathbf{r}) = \psi_0(\mathbf{r}) + \psi_s(\mathbf{r}), \tag{12.54b}$$

where $i - 1, \cdots, N$ and the scattered field $\psi_s(\mathbf{r})$ can be calculated everywhere once the local exciting fields ψ_i^{exc} have been determined:

$$\psi_s(\mathbf{r}) = \sum_{i=1}^{N} f \frac{e^{jk|\mathbf{r} - \mathbf{r}_i|}}{|\mathbf{r} - \mathbf{r}_i|} \psi_i^{exc}(\mathbf{r}_i). \tag{12.55}$$

Note that the preceding Foldy–Lax multiple scattering equations can be solved iteratively in a Born series to any desired level of accuracy and can be reduced to a convenient matrix form as shown in [132].

---

[5] The Breit–Wigner resonance formula is the quantum-mechanical analog of the electromagnetic Lorentz resonance formula discussed in Chapter 3. The two lineshapes are in fact identical.

Substituting (12.48) into the Foldy–Lax equation (12.54a), we can rewrite the local exciting field at the scatterer's position $\mathbf{r} = \mathbf{r}_a$ as follows:

$$\psi^{exc}(\mathbf{r}_a) = \psi_0(\mathbf{r}_a) + \sum_{b \neq a}^{N} \psi^{exc}(\mathbf{r}_b) \frac{e^{2j\delta} - 1}{2} G_{ab}, \tag{12.56}$$

where we introduced the scalar *Green's matrix* G_{ab}, whose elements are equal to [649]:

$$G_{ab} = \begin{cases} \frac{e^{jk|\mathbf{r}_a - \mathbf{r}_b|}}{jk|\mathbf{r}_a - \mathbf{r}_b|} & \text{for } a \neq b \\ 0 & \text{for } a = b. \end{cases} \tag{12.57}$$

We caution that alternative conventions for Green's matrix can be found in the literature [470, 476, 477, 650]. Solving the system (12.56) for $a = 1 \dots, N$ and using equations (12.55) and (12.54b), we can rigorously obtain the scalar wavefunction everywhere in space.

12.2.3 Resonances and Green's Matrix

We will now show, following the papers [649, 651, 652], that the approach introduced in the previous section provides a simple yet rigorous method to study the *scattering resonances* of an arbitrary system of N point scatterers. Within this model, qualitative information about the scattering resonances of the system of scatterers can be extracted just from the spectrum of Green's matrix without looking for resonance poles in the complex energy plane. In particular, the real and imaginary parts of the eigenvalues of Green's matrix can be considered first-order approximations to the relative widths and positions of the system's resonances.

We recall that resonance poles are values of k for which it is possible to solve equations (12.56) in the absence of the incoming wave, i.e., for $\psi_0 = 0$. In this case, the system on equations (12.56) reduces to the following eigenvalue problem for Green's matrix:

$$\boxed{\sum_{b=1}^{N} G_{ab} \psi^{exc}(\mathbf{r}_b) = \xi \psi^{exc}(\mathbf{r}_a)} \tag{12.58}$$

for $a = 1, \dots, N$, where the eigenvalue ξ is given by the following:

$$\xi = -1 - j \cot \delta_s. \tag{12.59}$$

Note here that Green's matrix is complex and symmetric, but not Hermitian, and therefore its eigenvalues are in general complex.

Using an explicit model of the scattering phase shift δ_s and solving (12.59) in the complex energy plane, where $E \rightarrow E - j\Gamma$ and $E_0 \rightarrow E_0 - j\Gamma_0$, it is possible to connect the spectral properties of Green's matrix, expressed through its complex eigenvalues, with the energy positions and widths of the resonances. Substituting the Breit–Wigner phase shift (12.49) for s-wave scatterers into equation (12.59), we obtain two coupled nonlinear equations for the the energy poles. Solving them iteratively, it is possible to obtain up to first order in Γ_0 the following solutions [649]:

$$\Re\{\xi\} \simeq \frac{\Gamma - \Gamma_0}{\Gamma_0} \tag{12.60a}$$

$$\Im\{\xi\} \simeq \frac{E - E_0}{\Gamma_0}. \tag{12.60b}$$

These results show that the eigenvalues of Green's matrix provide important physical information on the resonances of the system of scattering particles. In particular, the real and imaginary parts of the complex eigenvalues of G_{ab} are equal to the relative widths, or decay rates, and to the positions of the scattering resonances, respectively. Moreover, the eigenvectors of Green's matrix with small decay rates closely reproduce the localized optical mode patterns of an arbitrary system of small scalar scatterers [653]. Finally, we remark that the scalar model presented here has been extensively investigated in the literature in relation to the light localization properties of random distributions of point-like scatterers. In this context, universal properties of the spectra of the resulting random Green's matrix have been discovered and applied to many relevant problems of waves in random media, atomic physics, and random lasers [469, 470, 476–478, 650, 654]. The extension of this method to vector wave scattering will be discussed in the next sections.

12.2.4   Physical Meaning of the T-Matrix

For a collection of scalar point scatterers with polarizability α embedded in a surrounding medium with dielectric constant ϵ_0, the scattering potential can be expressed as follows:

$$V(\mathbf{r}) = \sum_i V_i \delta(\mathbf{r} - \mathbf{r}_i) = \frac{\omega^2}{c_0^2} \alpha \sum_i \delta(\mathbf{r} - \mathbf{r}_i), \tag{12.61}$$

where $\mathbf{r}_i$ are the positions of the scatterers.[6] The Green function of the system of scatterers can be written as a general Neumann perturbation series using the system's scattering potential in (12.61). Such a series can be simplified by introducing the *single particle T-matrix* $\mathbf{t}(\mathbf{r}_1, \mathbf{r}_2; \omega)$, which in turn can be expressed, according to the general result (12.31), as an infinite perturbation series that includes all the repeated scattering events from the scatterer.

In order to understand the physical meaning of the single-particle T-matrix, let us remember that the electric polarization $P(\mathbf{r})$ induced by the total (i.e., incident plus scattered) electric field $E(\mathbf{r})$ can be written in terms of the single-particle scattering potential $V_i = \omega^2(\epsilon_r - 1)/c_0^2$ as follows:

---

[6] Notice that in the literature it is often assumed that $\epsilon_0 = 1$ for convenience, which has the effect of transforming the polarizability α into the polarizability volume $\alpha' = \alpha/\epsilon_0$. Here we follow this convention by identifying $\alpha \rightarrow \alpha'$.

$$P(\mathbf{r}) = \epsilon_0(\epsilon_r - 1)E(\mathbf{r}) = \epsilon_0 \frac{c_0^2}{\omega^2} V_i E. \tag{12.62}$$

By applying the general operator identity $\mathbf{T} = \mathbf{VS}$ to the incident electric field E_{inc}, we deduce the following immediately:

$$P(\mathbf{r}) = \epsilon_0 \frac{c_0^2}{\omega^2} t(\omega) E_{inc}, \tag{12.63}$$

where $\mathbf{t}(\mathbf{r}_1, \mathbf{r}_2; \omega) = t(\omega)\delta(\mathbf{r}_2 - \mathbf{r}_i)\delta(\mathbf{r}_1 - \mathbf{r}_i)$. Equation (12.63) implies that the T-matrix of a single particle is related to the polarizability of the particle according to the following:

$$\boxed{\alpha(\omega) = \frac{c_0^2}{\omega^2} t(\omega)} \tag{12.64}$$

where we used the relationship $p = \epsilon_0 \alpha E_{inc}$ valid for a single particle.

This important result unveils the physical meaning of the Neumann perturbation series that rigorously defines the T-matrix. In particular, the incoming field induces a polarization, which corresponds to the first term in the series. This polarization modifies the field around the scatterer, which in turn influences the polarization (second term in the series), etc. The T-matrix for a point scatterer located at $\mathbf{r}_i$ can be approximated to first order (i.e., Born approximation) as $\mathbf{t}(\mathbf{r}_1, \mathbf{r}_2; \omega) = (\omega/c_0)^2 \alpha_0 \delta(\mathbf{r}_2 - \mathbf{r}_i)\delta(\mathbf{r}_1 - \mathbf{r}_i)$, where α_0 is the static polarizability of a spherical particle that neglects all the radiation effects (see discussion in Section 3.4.2).

The effect of resonance that occurs when the particle's size becomes comparable to the wavelength can be consistently incorporated within the point scatterer's model by adding a resonance frequency ω_0 to the retarded polarizability:

$$t(\omega) = \frac{\alpha_0(\omega/c_0)^2}{1 - \omega^2/\omega_0^2 - \frac{2}{3}j\alpha_0(\omega/c_0)^3}. \tag{12.65}$$

This expression guarantees the convergence of the Neumann series beyond the first-order Born approximation. The scattering cross section and extinction cross section are obtained from the optical theorem and can be expressed in terms of the T-matrix as follows:

$$\sigma_s = \frac{|t(\omega)|^2}{4\pi} \tag{12.66}$$

$$\sigma_{ext} = \frac{c_0 \Im\{t(\omega)\}}{\omega}. \tag{12.67}$$

For non-absorbing particles, the extinction cross section equals the scattering cross section and the two preceding expressions can be equated directly providing a relation between the imaginary part of the T-matrix and its absolute value.

12.2.5   Multiple Scattering and Refractive Index

The multiple scattering framework developed in the previous sections provides a physically transparent way to understand the origin of the refractive index in heterogeneous materials composed of an ensemble of radiating dipoles. As a simple application example, we will use the scalar Foldy–Lax equations to derive the refractive index of a *scalar dielectric* consisting of point scatterers with three-dimensional positions $\mathbf{r}_1, \mathbf{r}_2, \ldots, \mathbf{r}_n$. The dipoles are excited by an incident plane wave that will induce multiple scattering of scalar waves in the medium such that the scattered wave produced by the ith particle will be proportional to its complex-valued scattering amplitude f and to the total excitation wave amplitude $\psi(\mathbf{r}_i)$. This amplitude is due to the incident wave plus the scattering contributions of all the other scatterers. An important and implicit assumption behind this model is that the scatterers are separated enough that the spherical wave emitted by one is practically a plane wave when it excites a neighboring scatterer. Such an assumption is necessary because we use a constant scattering amplitude f, which is a quantity usually defined with respect to a plane wave excitation. We can now express the field amplitude $\psi(\mathbf{r})$ at some observation point $\mathbf{r}$ in the usual way:

$$\psi(\mathbf{r}) = e^{jkz} + f \sum_i \psi^{exc}(\mathbf{r}_i) \frac{e^{jk|\mathbf{r}-\mathbf{r}_i|}}{|\mathbf{r} - \mathbf{r}_i|}. \tag{12.68}$$

If we assume that the wavelength is large compared to the interparticle spacings, the sum in these multiple scattering equations can be replaced by an integral over a continuous distribution of scatterers with the volume density $N(\mathbf{r}')$. The amplitude at position $\mathbf{r}$ becomes

$$\psi(\mathbf{r}) = e^{jkz} + f \int \frac{N(\mathbf{r}')\psi^{exc}(\mathbf{r}')e^{jk|\mathbf{r}-\mathbf{r}'|}}{|\mathbf{r} - \mathbf{r}'|} d^3\mathbf{r}'. \tag{12.69}$$

Operating on both sides of this equation with the Helmholtz operator $\nabla^2 + k^2$ (which acts on the unprimed coordinates only) and recalling the definition of the scalar Green function, we get a delta function inside the integral and obtain the following:

$$\left[\nabla^2 + k^2 \left(1 + \frac{Nf}{k^2} \right) \right] \psi = 0. \tag{12.70}$$

We immediately recognize the equation for the propagation of a wave with wavenumber $\tilde{k}$ given by the following:

$$\tilde{k}^2 = k^2 \left(1 + \frac{Nf}{k^2} \right). \tag{12.71}$$

Note that under the weak scattering condition $Nf \ll k^2$, we can approximate this expression as follows:

$$\tilde{k} = k \left(1 + \frac{Nf}{2k^2} \right). \tag{12.72}$$

In terms of the refractive index of the scattering medium, we obtain the following:

$$\tilde{n} = 1 + \frac{Nf}{2k^2} \tag{12.73}$$

The qualitative model introduced here provides a good physical description of the origin of the refractive index in the phenomenological case of scalar dielectrics. However, by extending multiple wave scattering to vector electromagnetic waves, this approach can be used to describe more realistic dielectric media within the *effective medium theory*, further discussed in Chapter 14.

12.3   Extension to Vector Wave Scattering

The perturbative approach for the scalar complex amplitudes discussed previously can be easily generalized to vector waves using the dyadic Green function formalism introduced in Chapter 2. We recall here that the dyadic notation allows us to linearly relate two arbitrarily oriented vectors and to express complex vector equations in simple algebraic form. Therefore, using the dyadic notation, we can write linear vector equations as if they were simpler scalar ones and preserve their mathematical structure.

We will now discuss the multiple scattering of vector waves making use of the source–field dyadic notation introduced in Section 2.8.3. Consider first the case of a single scatterer centered at the origin. We define the *total dyadic Green function* of an inhomogeneous medium with position-dependent wavenumber as follows [132]:

$$\nabla \times \nabla \times \overleftrightarrow{\mathbf{G}}(\mathbf{r}, \mathbf{r}') - k^2(\mathbf{r}) \overleftrightarrow{\mathbf{G}}(\mathbf{r}, \mathbf{r}') = \overleftrightarrow{\mathbf{I}} \delta(\mathbf{r} - \mathbf{r}'). \tag{12.74}$$

Similarly, the equation satisfied by the dyadic Green's function of a homogeneous medium is as follows:

$$\nabla \times \nabla \times \overleftrightarrow{\mathbf{G}_0}(\mathbf{r}, \mathbf{r}') - k_0^2(\mathbf{r}) \overleftrightarrow{\mathbf{G}_0}(\mathbf{r}, \mathbf{r}') = \overleftrightarrow{\mathbf{I}} \delta(\mathbf{r} - \mathbf{r}'), \tag{12.75}$$

where k_0 is the constant wavenumber of the homogeneous medium, here assumed to be the free space. We can formally rewrite equation (12.74) as follows:

$$\nabla \times \nabla \times \overleftrightarrow{\mathbf{G}}(\mathbf{r}, \mathbf{r}') - k_0^2(\mathbf{r}) \overleftrightarrow{\mathbf{G}}(\mathbf{r}, \mathbf{r}') = \overleftrightarrow{\mathbf{I}} \delta(\mathbf{r} - \mathbf{r}') + [k^2(\mathbf{r}) - k_0^2] \overleftrightarrow{\mathbf{G}}(\mathbf{r}, \mathbf{r}'), \tag{12.76}$$

which is the vector generalization of the scalar equation (6.7). Similarly, the second term on the right-hand side can be interpreted as an equivalent vector source function defined over the volume Γ of the scatterer. Using again Greens's superposition theorem, we can formally solve for the dyadic total Green function as follows:

$$\overleftrightarrow{\mathbf{G}}(\mathbf{r}, \mathbf{r}') = \overleftrightarrow{\mathbf{G}_0}(\mathbf{r}, \mathbf{r}') + \int_\Gamma d^3\mathbf{r}_1 \overleftrightarrow{\mathbf{G}_0}(\mathbf{r}, \mathbf{r}_1)[k^2(\mathbf{r}_1) - k_0^2] \overleftrightarrow{\mathbf{G}}(\mathbf{r}_1, \mathbf{r}') \tag{12.77}$$

which is the vector generalization of the scalar Lippmann–Schwinger (LS) integral equation (6.8) in the case of the total dyadic Green function. This equation is also known as the *vector integral equation.*

Let us now define the scattering potential as follows:

$$V(\mathbf{r}) = \begin{cases} 0, \mathbf{r} \notin \Gamma \\ k^2(\mathbf{r}) - k_0^2, \mathbf{r} \in \Gamma \end{cases} \qquad (12.78)$$

and recall that in linear problems the scattered vector field is related to its current source by the following dyadic relation:

$$\mathbf{E}_{sc}(\mathbf{r}) = j\omega\mu_0 \int_\Gamma \overleftrightarrow{\mathbf{G}_0}(\mathbf{r}, \mathbf{r}') \cdot \mathbf{J}(\mathbf{r}') d^3\mathbf{r}'. \qquad (12.79)$$

By expressing the current density in terms of the induced polarization vector using the general relation $\mathbf{J} = \partial \mathbf{P}/\partial t$, we can rewrite equation (12.79) as follows:

$$\mathbf{E}_{sc}(\mathbf{r}) = \int_\Gamma \overleftrightarrow{\mathbf{G}_0}(\mathbf{r}, \mathbf{r}') \cdot V(\mathbf{r}')\mathbf{E}(\mathbf{r}') d^3\mathbf{r}'. \qquad (12.80)$$

A direct comparison of the last two equations establishes a relation between the scattering potential and the induced volume current density:

$$\mathbf{J} = \frac{V\mathbf{E}}{j\omega\mu_0} = \frac{k_0^2 \left[\epsilon_r(\mathbf{r}) - 1\right]\mathbf{E}}{j\omega\mu_0}. \qquad (12.81)$$

In order to simplify these expressions, we will employ Dirac's abstract notation and the dyadic operator formalism. In particular, using the coordinate representation, we can define, in analogy with the scalar case, the following:

$$\overleftrightarrow{\mathbf{G}_0}(\mathbf{r}, \mathbf{r}') = \langle \mathbf{r} | \overleftrightarrow{\mathbf{G}_0} | \mathbf{r}' \rangle \qquad (12.82a)$$

$$\overleftrightarrow{\mathbf{G}}(\mathbf{r}, \mathbf{r}') = \langle \mathbf{r} | \overleftrightarrow{\mathbf{G}} | \mathbf{r}' \rangle, \qquad (12.82b)$$

where the quantities in between the bra and the ket vectors denote dyadic operator quantities. The diagonal operator corresponding to the scattering potential V is known as the *impurity potential operator* and it is given in coordinate representation as follows:

$$\langle \mathbf{r} | \mathbf{V} | \mathbf{r}' \rangle = V(\mathbf{r})\delta(\mathbf{r} - \mathbf{r}') \overleftrightarrow{\mathbf{I}}. \qquad (12.83)$$

Expressing equation (12.80) in dyadic operator notation and adding the incident field, we can immediately obtain the dyadic version of the LS integral equation for the total field:

$$|\mathbf{E}\rangle = |\mathbf{E}_0\rangle + \overleftrightarrow{\mathbf{G}_0} \mathbf{V} |\mathbf{E}\rangle, \qquad (12.84)$$

which generalizes equation (12.11) to the vector scattering case. Following this approach, we can express the LS equation for the total dyadic Green function in a form that is formally identical to the case of scalar waves, and obtain the following:

$$\overleftrightarrow{G} = \overleftrightarrow{G_0} + \overleftrightarrow{G_0} V \overleftrightarrow{G}, \tag{12.85}$$

where the product of operators means dyadic contraction. By iterating this process, we recover the Neumann series expansion for the total dyadic Green function. All the formal results previously obtained for scalar waves naturally carry over to the case of vector waves when using the operator dyadic notation. In particular, the dyadic transition operator $\overleftrightarrow{T}$ satisfies the *dyadic Dyson's equation* [132]:

$$\overleftrightarrow{T} = V + V \overleftrightarrow{G_0} \overleftrightarrow{T}. \tag{12.86}$$

Transforming back to its integral form, Dyson's equation[7] reads as follows:

$$\overleftrightarrow{T}(\mathbf{r},\mathbf{r}') = V(\mathbf{r})\delta(\mathbf{r} - \mathbf{r}') \overleftrightarrow{I} + V(\mathbf{r}) \int_\Gamma d\mathbf{r}'' \overleftrightarrow{G_0}(\mathbf{r},\mathbf{r}'') \cdot \overleftrightarrow{T}(\mathbf{r}'',\mathbf{r}'), \tag{12.87}$$

where as usual Γ denotes the volume occupied by the scatterer.

By iterating the operator dyadic equation (12.86) we obtain a perturbative expansion that can be diagrammatically expressed as follows:

$$\overleftrightarrow{T} = \bullet + \bullet\!-\!\!\bullet + \bullet\!-\!\!\bullet\!-\!\!\bullet + \bullet\!-\!\!\bullet\!-\!\!\bullet\!-\!\!\bullet + \cdots \tag{12.88}$$

Comparing the diagrammatic Neumann series for the total Green function with the preceding dyadic $\overleftrightarrow{T}$ operator, we obtain immediately the following important relation:

$$\overleftrightarrow{G} = \overleftrightarrow{G_0} + \overleftrightarrow{G_0} \overleftrightarrow{T} \overleftrightarrow{G_0}, \tag{12.89}$$

which generalizes (12.9) to the case of vector waves.

In the operator notation, the vector current density becomes a source ket $|\mathbf{J}\rangle$, which we can apply to the right-hand side of equation (12.89) to obtain the following:

$$\overleftrightarrow{G} |\mathbf{J}\rangle = \overleftrightarrow{G_0} |\mathbf{J}\rangle + \overleftrightarrow{G_0} \overleftrightarrow{T} \overleftrightarrow{G_0} |\mathbf{J}\rangle, \tag{12.90}$$

By recognizing

$$\overleftrightarrow{G} |\mathbf{J}\rangle = |\mathbf{E}\rangle \tag{12.91}$$

as the ket associated to the total electric field vector and

$$\overleftrightarrow{G_0} |\mathbf{J}\rangle = |\mathbf{E_0}\rangle \tag{12.92}$$

as the one associated to the incident field vector, we can write the following:

$$|\mathbf{E}\rangle = |\mathbf{E_0}\rangle + \overleftrightarrow{G_0} \overleftrightarrow{T} |\mathbf{E_0}\rangle. \tag{12.93}$$

This equation provides the total vector field in terms of the dyadic operator $\overleftrightarrow{T}$ of the scatterer. The scattered field of a single particle is clearly given by the following:

$$|\mathbf{E}_{sc}\rangle = \overleftrightarrow{G_0} \overleftrightarrow{T} |\mathbf{E_0}\rangle. \tag{12.94}$$

---

[7] In the scattering literature, this equation is also sometimes referred to as the Lippmann–Schwinger equation [90], creating some confusion.

This equation can be written in integral form as follows:

$$\mathbf{E}_{sc}(\mathbf{r}) = \int_{\Gamma} d\mathbf{r}' \overleftrightarrow{\mathbf{G}_0}(\mathbf{r},\mathbf{r}') \cdot \int_{\Gamma} d\mathbf{r}'' \overleftrightarrow{\mathbf{T}}(\mathbf{r}',\mathbf{r}'') \cdot \mathbf{E}_0(\mathbf{r}''), \tag{12.95}$$

where $\mathbf{E}_0$ is the incident field.

Having established the connection between scalar and vector wave equations through the use of the dyadic operator notation, it becomes straightforward to generalize the results to the case of vector scattering from multiple particles. Thanks to the formal equivalence between the scalar and dyadic notations, this can be accomplished by simply rewriting the Foldy–Lax multiple scattering equations obtained for scalar waves (12.44) in the more general dyadic operator notation:

$$|\mathbf{E}\rangle = |\mathbf{E}_0\rangle + \sum_i \overleftrightarrow{\mathbf{G}_0} \overleftrightarrow{\mathbf{T}_i} \, |\mathbf{E}_i^{exc}\rangle \tag{12.96a}$$

$$|\mathbf{E}_j^{exc}\rangle = |\mathbf{E}_0\rangle + \overleftrightarrow{\mathbf{G}_0} \sum_{i \neq j} \overleftrightarrow{\mathbf{T}_i} \, |\mathbf{E}_i^{exc}\rangle, \tag{12.96b}$$

where

$$|\mathbf{E}_j^{exc}\rangle = \overleftrightarrow{\mathbf{G}_j} \, |\mathbf{J}\rangle \tag{12.97}$$

is the dyadic expression for the local exciting field of particle j. Notice that the operators $\overleftrightarrow{\mathbf{T}_i}$ are the exact transition operators for a single particle in the system, and include near-field and far-field contributions. The dyadic Foldy–Lax equations for the electric field can also be derived directly from Maxwell's equations and provide the foundation of vector multiple-scattering and radiative transfer theory [132, 655].

12.3.1   Vector Point Scattering Model

Using the Foldy–Lax equations established in the previous section, it is possible to set up a useful model based on point particles for *vector light scattering* that naturally generalizes the previous scalar approach. In what follows, we will establish the correct multiple scattering equations following an intuitive approach that neglects many of the subtleties behind the more rigorous and self-consistent derivation of the vector point scattering model. The interested reader can find extensive discussions of the many physical aspects of this model in the excellent reviews on the subject [367, 656].

Let us recall that the time derivative of a polarization produces a current density. Therefore, the complex current density associated to a generic time-harmonic electric dipole located at position $\mathbf{r} = \mathbf{r}_i$ can be expressed as follows:

$$\mathbf{J}(\mathbf{r}) = -j\omega\mathbf{p}\delta(\mathbf{r} - \mathbf{r}_i), \tag{12.98}$$

where $\mathbf{p}$ is the complex dipole moment. The electromagnetic field radiated by the oscillating dipole can be found by substituting the preceding expression into the general volume integral equations (2.189) and (2.190), which produce the following:

$$\mathbf{E}(\mathbf{r}) = \omega^2 \mu_0 \overleftrightarrow{\mathbf{G}_0}(\mathbf{r}, \mathbf{r}_0)\mathbf{p} \tag{12.99}$$

$$\mathbf{H}(\mathbf{r}) = -j\omega \left[\nabla \times \overleftrightarrow{\mathbf{G}_0}(\mathbf{r}, \mathbf{r}_0) \right] \mathbf{p}. \tag{12.100}$$

If we regard the dipole as embedded into a system of scattering dipoles, a comparison between the general form of the Foldy–Lax equations (12.96a) and (12.99) makes us recognize the following:

$$\overleftrightarrow{\mathbf{T}}_i \, \mathbf{E}_i^{exc} \equiv \omega^2 \mu_0 \epsilon_0 \alpha \mathbf{E}_i, \tag{12.101}$$

where we introduced the dipole polarizability α relating the dipole moment with the local field acting on it, i.e., $\mathbf{p} = \epsilon_0 \alpha \mathbf{E}_i$. Note that, according to the preceding expression, the transition matrix operator for a point vector dipole corresponds with its dynamic Clausius–Mossotti polarizability provided in (3.69). We remark that a number of other expressions for the dynamical polarizability have been proposed [123, 126, 131], but the difference between their imaginary parts is negligible for objects of small size parameters.

More formally, for a vector point scatterer located at $\mathbf{r} = \mathbf{r}_s$ the T-matrix operator can be written as follows:

$$\overleftrightarrow{\mathbf{T}}_s(\omega, \mathbf{r}, \mathbf{r}') = t(\omega)\delta(\mathbf{r} - \mathbf{r}_s)\delta(\mathbf{r}' - \mathbf{r}_s) \overleftrightarrow{\mathbf{I}}, \tag{12.102}$$

and the T matrix element $t(\omega)$ is related to the dynamic polarizability of the point scatterer by as follows:

$$\alpha(\omega) = \frac{c_0^2}{\omega^2} t(\omega). \tag{12.103}$$

Substituting (12.101) into the Foldy–Lax equations for the local exciting fields (12.96a) allows us to immediately obtain the correct form of the *coupled-dipole equations* of multiple scattering as:

$$\boxed{\mathbf{E}_j = \mathbf{E}_{0j} + t(\omega) \sum_{i \neq j} \overleftrightarrow{\mathbf{G}_0}(\mathbf{r}_{ij})\mathbf{E}_i} \tag{12.104}$$

where $\mathbf{r}_{ij} \equiv \mathbf{r}_i - \mathbf{r}_j$ and $\mathbf{E}_j \equiv \mathbf{E}(\mathbf{r}_j)$ and the same convention for the incident field. The preceding equation shows how the single-particle T-matrix acts as the coupling strength of the local excitation fields. For $r \neq 0$, we use the dyadic Green function in the form (2.197) that we rewrite here for convenience:

$$\overleftrightarrow{\mathbf{G}_0}(\mathbf{r}, \mathbf{r}_i) = \left[\left(\frac{3}{k_0^2 R^2} - \frac{3j}{k_0 R} - 1 \right) \hat{\mathbf{R}}\hat{\mathbf{R}} + \left(1 + \frac{j}{k_0 R} - \frac{1}{k_0^2 R^2} \right) \overleftrightarrow{\mathbf{I}} \right] G_0^+(R), \tag{12.105}$$

where $R = |\mathbf{r} - \mathbf{r}_i|$ and $\hat{\mathbf{R}}$ is the unit vector in the direction of $\mathbf{R} = \mathbf{r} - \mathbf{r}_i$.

The scattered field at position $\mathbf{r}$ is then given as follows:

$$\mathbf{E}_s(\mathbf{r}) = t(\omega) \sum_{i=1}^{N} \overleftrightarrow{\mathbf{G}_0}(\mathbf{r} - \mathbf{r}_i)\mathbf{E}_i. \tag{12.106}$$

We can approximate the expression of the dyadic Green function in order to consider only the *far-field* contributions to the scattering problem. This amounts to neglecting the R-dependent terms that appear in the parentheses of (12.105), which leads to the following expression for the scattered amplitude:

$$\mathbf{E}_s(\mathbf{r}) = \frac{t(\omega)}{4\pi r} \sum_{i=1}^{N} \left(\overleftrightarrow{\mathbf{I}} - \hat{\mathbf{R}}\hat{\mathbf{R}} \right) \mathbf{E}_i e^{-jk_0\hat{\mathbf{R}}\cdot\mathbf{r}_i}. \tag{12.107}$$

This formula demonstrates that the scattered far-field behaves as the coherent sum of outgoing transverse spherical waves centered at the origin and propagating in the direction of the unit vector $\hat{\mathbf{R}}$ [280].

12.3.2   The Discrete Coupled Dipoles Model

We now show how the general Foldy–Lax formulation of multiple scattering can be utilized to obtain the *discrete dipole approximation* (DDA), which is a numerical method often utilized to study the scattering and absorption properties of clusters of small (i.e., dipolar) metallic nanoparticles [657, 658]. For our purposes, let us consider N particles approximated as dipoles with polarizability α_j and complex dipole moment $\mathbf{p}_j = \alpha_j E_{ext,j}$, where $E_{ext,j}$ is the complex local electric field at position $\mathbf{r}_j$ due to all the sources that are external to the j-dipole. In other words, this is the local exciting field at $\mathbf{r}_j$ due to the incident radiation plus the scattered radiation from all the other $N-1$ dipoles. Replacing $E_{ext,j}$ in $\mathbf{p}_j = \alpha_j E_{ext,j}$ with the right-hand side of the coupled-dipole equations (12.104), we obtain N simultaneous complex vector equations for the self-consistent set of dipole moments:

$$\mathbf{p}_j = \alpha_j \left(\mathbf{E}_{0j} + \frac{k_0^2}{\epsilon_0} \sum_{i \neq j} \overleftrightarrow{\mathbf{G}}_0{}_{ji} \mathbf{p}_i \right) \equiv \alpha_j \left(\mathbf{E}_{0j} - \sum_{i \neq j} \overleftrightarrow{\mathbf{A}}_{ji} \mathbf{p}_i \right), \tag{12.108}$$

where we defined the *DDA interaction dyadic* $\overleftrightarrow{\mathbf{A}}_{ji}$, which is related to the dyadic Green function as follows:

$$\overleftrightarrow{\mathbf{A}}_{ji} \equiv -k_0^2 \overleftrightarrow{\mathbf{G}}_0(\mathbf{r}_{ij}). \tag{12.109}$$

In the context of the coupled dipole method, it is customary to represent the dyadic $\overleftrightarrow{\mathbf{A}}_{ji}$ in the following form:

$$\overleftrightarrow{\mathbf{A}}_{mn} = \left[k_0^2 \left(\hat{\mathbf{r}}_{mn}\hat{\mathbf{r}}_{mn} - \overleftrightarrow{\mathbf{I}} \right) + \frac{jk_0 r_{mn} - 1}{r_{mn}^2} \left(3\hat{\mathbf{r}}_{mn}\hat{\mathbf{r}}_{mn} - \overleftrightarrow{\mathbf{I}} \right) \right] G_0^+ \tag{12.110}$$

for $m \neq n$, and where $\mathbf{r}_{mn} = \mathbf{r}_m - \mathbf{r}_n$ and $\hat{\mathbf{r}}_{mn} = (\mathbf{r}_m - \mathbf{r}_n)/r_{mn}$. Defining $\overleftrightarrow{\mathbf{A}}_{ii} \equiv \alpha_i^{-1}$ allows us to write (12.108) as a set of $3N$ inhomogeneous linear equations:

$$\sum_{i=1}^{N} \overleftrightarrow{\mathbf{A}}_{ji} \mathbf{p}_i = \mathbf{E}_{0j} \tag{12.111}$$

for $j = 1 \ldots N$. Moreover, equations (12.111) can be reduced to a symmetric matrix inversion problem $\mathbf{AP} = \mathbf{E}_0$ if we define the $3N$-dimensional vectors $\mathbf{P} \equiv (\mathbf{p}_1, \mathbf{p}_2, \ldots, \mathbf{p}_N)$ and $\mathbf{E}_0 \equiv (\mathbf{E}_{01}, \mathbf{E}_{02}, \ldots, \mathbf{E}_{0N})$ along with a symmetric $3N \times 3N$ block-diagonal interaction matrix $\mathbf{A}$ constructed from the 3×3 symmetric interaction matrices $\mathbf{A}_{ij}$ with the additional terms α^{-1} along the diagonals. Fast computational techniques for the efficient solution of the coupled-dipole equations are based on the complex-conjugate gradient algorithm and the fast Fourier transform method, as discussed in [657, 659].

Once equations (12.111) have been solved for the unknown polarization vectors, the excitation, scattering, and absorption cross sections can be calculated. In particular, the extinction cross section can be directly obtained from the forward scattering amplitude using the optical theorem for vector waves in (6.71), which, using equation (12.108) to identify the scattering amplitude, results in the following expression:

$$\sigma_{ext} = \frac{4\pi k}{|\mathbf{E}_0|^2} \sum_{i=1}^{N} \Im\{\mathbf{E}_{0j}^* \cdot \mathbf{p}_j\} \tag{12.112}$$

The absorption cross section is obtained by considering the energy dissipation of the dipoles in the system. The rate at which energy is removed from a beam is due to both absorption and scattering by a single dipole. This can easily be obtained from the definition of the extinction cross section:

$$P_{abs} + P_s = \left(\frac{dW}{dt}\right)_{ext} = \sigma_{ext}|\bar{\mathbf{S}}|, \tag{12.113}$$

where P_{abs} and P_s are the absorbed and scattered powers, while $\bar{\mathbf{S}}$ denotes the time-averaged Poynting vector that corresponds to the incident intensity.

In Gaussian units,[8] $|\bar{\mathbf{S}}| = (c/8\pi)|\mathbf{E}_0|^2$, which allows us to write the following:

$$\left(\frac{dW}{dt}\right)_{ext} = \frac{\omega}{2}\Im\{\mathbf{E}^* \cdot \mathbf{p}\} = \frac{\omega}{2}\Im\{\mathbf{p} \cdot (\alpha^{-1})^* \mathbf{p}^*\} \tag{12.114}$$

The absorption rate of the dipole is obtained by the difference of its extinction and radiation rates. Note that the important result in (12.114) can equivalently be derived using the expression for the energy dissipation rate, which includes both energy sources and sinks, given by the following:

$$\left(\frac{dW}{dt}\right)_{ext} = -\frac{1}{2}\int_V \Re\{\mathbf{E}^* \cdot \mathbf{J}\}dV = -\frac{1}{2}\int_V \Re\{\mathbf{J}^* \cdot \mathbf{E}\}dV \tag{12.115}$$

---

[8] Here we find it convenient to switch to Gaussian units in order to facilitate the comparison of the DDA equations with the many results published in the literature.

when evaluated for the current density $\mathbf{J}(\mathbf{r}) = -j\omega\mathbf{p}\delta(\mathbf{r} - \mathbf{r}_0)$ associated to a time-harmonic point dipole. Remembering the expression for the average radiated power of an harmonically oscillating dipole:

$$\left(\frac{dW}{dt}\right)_s = \frac{\omega^4}{3c^3}|\mathbf{p}|^2. \tag{12.116}$$

We can express the absorption rate of the dipole by the following difference:

$$\left(\frac{dW}{dt}\right)_{abs} = \frac{\omega}{2}\left\{\Im\{\mathbf{p}\cdot(\alpha^{-1})^*\mathbf{p}^*\} - \frac{2}{3}k_0^3\mathbf{p}\cdot\mathbf{p}^*\right\}. \tag{12.117}$$

Note that in SI units this expression will appear slightly modified as follows:

$$\left(\frac{dW}{dt}\right)_{abs} = \frac{\omega}{2}\left\{\Im\{\mathbf{p}\cdot(\alpha^{-1})^*\mathbf{p}^*\} - \frac{2}{3}\frac{k_0^3}{4\pi\epsilon_0\epsilon_r}\mathbf{p}\cdot\mathbf{p}^*\right\}. \tag{12.118}$$

Returning now to the expression in Gaussian units, the absorption cross section of a cluster of N dipoles is readily obtained by summing over all the contributions of the dipoles:

$$\sigma_{abs} = \frac{4\pi k}{|\mathbf{E}_0|^2}\sum_{j=1}^{N}\left\{\Im\{\mathbf{p}_j\cdot(\alpha_j^{-1})^*\mathbf{p}_j^*\} - \frac{2}{3}k_0^3\mathbf{p}_j\cdot\mathbf{p}_j^*\right\}. \tag{12.119}$$

In principle, the scattering cross section σ_s can always be obtained by the difference of the extinction and the absorption cross section. However, this requires high accuracy in the computation of both cross sections. To avoid this problem, it is possible to directly calculate the scattering cross section by considering the power radiated by the oscillating dipoles, which is as follows:

$$\sigma_s = \frac{k^4}{|\mathbf{E}_0|^2}\int d\Omega \left|\sum_{i=1}^{N}[\mathbf{p}_i - \hat{\mathbf{r}}(\hat{\mathbf{r}}\cdot\mathbf{p}_i)]e^{-jk\hat{\mathbf{r}}\cdot\mathbf{r}_i}\right|^2, \tag{12.120}$$

where $\hat{\mathbf{r}}$ is the unit vector in the direction of scattering and $d\Omega$ is the element of solid angle.

In Figure 12.3, we show results of the efficiency calculations for a square array and for different types of aperiodic arrays of Ag nanoparticles with a diameter of 45 nm, including a Poisson random point pattern. A narrow lattice resonance is clearly visible for the square lattice shown in Figure 12.3a, which is driven by radiative field coupling (multiple scattering) enhanced by the plasmonic response of the particles. This phenomenon was introduced already at the end of Chapter 7. Plasmon-enhanced lattice resonances are also observed for the aperiodic Hurwitz and Lifschitz arrays shown in Figure 12.3g,h, which are the most "regular" of the considered aperiodic geometries. However, all the aperiodic structures display broader and weaker spectral features compared to the plasmonic square lattice due to their larger spatial frequency content that results in a less coherent wave scattering at multiple length scales. Interestingly, an intermediate situation was found in the Penrose array, where several lattice resonances are emerging from a broader incoherent background. Comparing these spectra with

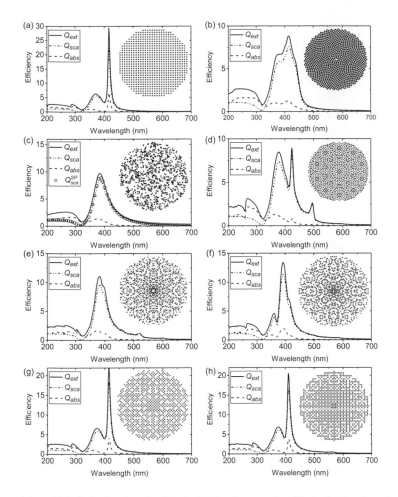

Figure 12.3 Calculated extinction (solid lines), scattering (dash-dot lines), and absorption (dashed lines) efficiency spectra of 45 nm silver nanoparticles arranged in a square lattice (a), golden-angle Vogel spiral (b), Poisson (c), Penrose (d), Eisenstein prime (e), Gaussian prime (f), Hurwitz array (g), and Lifschitz array (h), respectively. The center-to-center average interparticle separation for all the structures is 400 nm. The material dispersion is taken from [121]

the single particle and the uncorrelated random pattern results shown in Figure 12.2c, we directly appreciate the strength of the coherent (i.e., interference-based) lattice resonances in all the investigated aperiodic geometries.

12.3.3   Extension to Coupled Electric and Magnetic Dipoles

In this section, we extend the previous method to arrays of nanoparticles that support both electric and magnetic modes. The ability to deal with both electric and magnetic dipolar excitations is particularly important in the analysis of dielectric nanoparticles with large refractive index, such as Si, since the scattering radiation is strongly contributed by the induced magnetic dipole contribution for particles whose size is comparable to the wavelength [660–663]. Moreover, the analysis of anisotropic magnetodielectric point dipoles is central to the theory and design of novel plasmonic and metamaterial structures [664–666].

The simple model that we introduce here considers coupled electric and magnetic dipoles arranged in arbitrary geometries. We refer to this method as the electric and magnetic coupled dipole approximation (EMCDA). Let us now address the EMCDA method in some detail, remembering that all the expressions appearing in this section are given in the centigram-gram-second (cgs) system of units. The electric and magnetic fields at the ith particle ($\mathbf{E}_i$ and $\mathbf{H}_i$ respectively) resulting from the electric ($\mathbf{p}$) and magnetic ($\mathbf{m}$) dipole moments at the jth particle are as follows [667, 668]:

$$\mathbf{E}_i = a_{ij}\alpha_E\mathbf{E}_j + b_{ij}\alpha_E(\mathbf{E}_j \cdot \mathbf{n}_{ji})\mathbf{n}_{ji} - d_{ij}\alpha_H(\mathbf{n}_{ji} \times \mathbf{H}_j) \quad (12.121)$$

$$\mathbf{H}_i = a_{ij}\alpha_H\mathbf{H}_j + b_{ij}\alpha_H(\mathbf{H}_j \cdot \mathbf{n}_{ji})\mathbf{n}_{ji} + d_{ij}\alpha_E(\mathbf{n}_{ji} \times \mathbf{E}_j), \quad (12.122)$$

where

$$a_{ij} = \frac{e^{ikr_{ij}}}{r_{ij}}\left(k^2 - \frac{1}{r_{ij}^2} + \frac{ik}{r_{ij}}\right) \quad (12.123)$$

$$b_{ij} = \frac{e^{ikr_{ij}}}{r_{ij}}\left(-k^2 + \frac{3}{r_{ij}^2} - \frac{3ik}{r_{ij}}\right) \quad (12.124)$$

$$d_{ij} = \frac{e^{ikr_{ij}}}{r_{ij}}\left(k^2 + \frac{ik}{r_{ij}}\right). \quad (12.125)$$

The electric and magnetic polarizability volumes are as follows:

$$\alpha_E = \frac{3i}{2k^3}a_1 \quad (12.126)$$

and

$$\alpha_H = \frac{3i}{2k^3}b_1, \quad (12.127)$$

and k is the wavenumber of the background medium. The polarizability volumes have units of volume (e.g., m^3).

In equations (12.126) and (12.127), a_1 and b_1 are the Mie–Lorentz coefficients for electric and magnetic dipoles, respectively. The νth-order Mie–Lorentz coefficients can be expressed as follows:

$$a_\nu = \frac{n\psi_\nu(nkr)\psi_\nu'(kr) - \psi_\nu(kr)\psi_\nu'(nkr)}{n\psi_\nu(nkr)\xi_\nu'(kr) - \xi_\nu(kr)\psi_\nu'(nkr)} \quad (12.128)$$

and

$$b_\nu = \frac{\psi_\nu(nkr)\psi_\nu'(kr) - n\psi_\nu(kr)\psi_\nu'(nkr)}{\psi_\nu(nkr)\xi_\nu'(kr) - n\xi_\nu(kr)\psi_\nu'(nkr)}, \quad (12.129)$$

where r is the radius of the spherical scatterer, and n is the relative refractive index of the nanosphere with respect to the background medium. $\psi_\nu(x)$ and $\xi_\nu(x)$ are Riccati–Bessel functions constructed from spherical bessel functions via $\psi_\nu(x) = xj_\nu(x)$ and $\xi_\nu(x) = xh_\nu^{(1)}(x)$. In addition, $j_\nu(x)$ is the spherical bessel function of the first type, and $h_\nu^{(1)}(x)$ is the spherical Hankel function of the first type (more details in Chapter 13).

In order to evaluate the total electric and magnetic fields at the ith particle position resulting from the electric and magnetic dipole moments of the jth one, it is convenient to express the various vector products in equations (12.121)–(12.122) as matrix products. In detail, by introducing the 3×3 matrixes C_{ij} and f_{ij}, defined as follows:

$$
C_{ij} = \begin{bmatrix} a_{ij} + b_{ij}(n_{ij}^x)^2 & b_{ij}n_{ij}^x n_{ij}^y & b_{ij}n_{ij}^x n_{ij}^z \\ b_{ij}n_{ij}^y n_{ij}^x & a_{ij} + b_{ij}(n_{ij}^y)^2 & b_{ij}n_{ij}^y n_{ij}^z \\ b_{ij}n_{ij}^z n_{ij}^x & b_{ij}n_{ij}^z n_{ij}^y & a_{ij} + b_{ij}(n_{ij}^z)^2 \end{bmatrix}
\tag{12.130}
$$

$$
f_{ij} = \begin{bmatrix} 0 & -d_{ij}n_{ij}^z & d_{ij}n_{ij}^y \\ d_{ij}n_{ij}^z & 0 & -d_{ij}n_{ij}^x \\ -d_{ij}n_{ij}^y & d_{ij}n_{ij}^x & 0 \end{bmatrix},
\tag{12.131}
$$

where $n_{ij}^\beta = \beta_i - \beta_j$ ($\beta = x, y,$ and z) are the components of the direction vector from the jth to the ith particle, we can rewrite equation (12.121) and equation (12.122) in the following form:

$$
\mathbf{E}_i = \alpha_E \, C_{ij}\mathbf{E}_j - \alpha_H \, f_{ij}\mathbf{H}_j
\tag{12.132}
$$

$$
\mathbf{H}_i = \alpha_E \, f_{ij}\mathbf{E}_j + \alpha_H \, C_{ij}\mathbf{H}_j
\tag{12.133}
$$

or in the compact notation

$$
\begin{bmatrix} \mathbf{E}_i \\ \mathbf{H}_i \end{bmatrix} = \begin{bmatrix} C_{ij} & -f_{ij} \\ f_{ij} & C_{ij} \end{bmatrix} \begin{bmatrix} \tilde{\alpha}_E & 0 \\ 0 & \tilde{\alpha}_H \end{bmatrix} \begin{bmatrix} \mathbf{E}_j \\ \mathbf{H}_j \end{bmatrix},
\tag{12.134}
$$

where $\tilde{\alpha}_E$ and $\tilde{\alpha}_H$ are 3×3 diagonal matrixes containing the polarizability α_E and α_H defined by the equations (12.126)–(12.127) in the case of isotropic materials. Equation (12.134) defines the dyadic Green's matrix $\overleftrightarrow{G}_{ij}$ that links the electromagnetic field of the ith-particle with the electromagnetic field of the jth particle. Specifically, $\overleftrightarrow{G}_{ij}$ is defined as follows:

$$
\overleftrightarrow{G}_{ij} = \begin{bmatrix} C_{ij} & -f_{ij} \\ f_{ij} & C_{ij} \end{bmatrix} = \begin{bmatrix} \overleftrightarrow{G}_{ij}^{ee} & \overleftrightarrow{G}_{ij}^{eh} \\ \overleftrightarrow{G}_{ij}^{he} & \overleftrightarrow{G}_{ij}^{hh} \end{bmatrix}.
\tag{12.135}
$$

The symbol $\overleftrightarrow{\{\cdots\}}$ is used to underline the fact that we are taking into account all the field components. Therefore, $\overleftrightarrow{G}_{ij}$ is a 6×6 matrix. Moreover, one of the advantages of using a cgs system unit is that the symmetry rules between electric and magnetic quantities are preserved, i.e., $\overleftrightarrow{G}^{ee} = \overleftrightarrow{G}^{hh}$ and $\overleftrightarrow{G}^{eh} = -\overleftrightarrow{G}^{he}$ [667].

The generalization to N scatterers is straightforward. Equation (12.132)–(12.133) can be assembled into a Foldy–Lax scheme such that the total local fields at the position of the ith scatterer $\mathbf{E}_i^{tot}$ and $\mathbf{H}_i^{tot}$ are the sum of the scattered term of all the other particles plus the incident field $(\mathbf{E}_{i,0}; \mathbf{H}_{i,0})$ on the ith particle

$$
\mathbf{E}_i^{tot} = \mathbf{E}_{i,0} + \sum_{\substack{j=1 \\ j\neq i}}^{N} \alpha_{E,i} C_{ij} \mathbf{E}_j - \sum_{\substack{j=1 \\ j\neq i}}^{N} \alpha_{H_i} f_{ij} \mathbf{H}_j
$$

$$
= \mathbf{E}_{i,0} + \sum_{\substack{j=1 \\ j\neq i}}^{N} \left[\alpha_{E,i} \overleftrightarrow{G}_{ij}^{ee} \mathbf{E}_j + \alpha_{H_i} \overleftrightarrow{G}_{ij}^{eh} \mathbf{H}_j \right]
$$

(12.136)

$$
\mathbf{H}_i^{tot} = \mathbf{H}_{i,0} + \sum_{\substack{j=1 \\ j\neq i}}^{N} \alpha_{H,i} \tilde{C}_{ij} \mathbf{H}_j + \sum_{\substack{j=1 \\ j\neq i}}^{N} \alpha_{E,i} \tilde{f}_{ij} \mathbf{E}_j
$$

$$
= \mathbf{H}_{i,0} + \sum_{\substack{j=1 \\ j\neq i}}^{N} \left[\alpha_{H,i} \overleftrightarrow{G}_{ij}^{hh} \mathbf{H}_j + \alpha_{E_i} \overleftrightarrow{G}_{ij}^{he} \mathbf{E}_j \right].
$$

(12.137)

These last two equations can be rewritten in a shorthand notation as follows:

$$
\Xi(\mathbf{r}) = \Xi_{inc}(\mathbf{r}) + \mathbf{M}\, \Xi(\mathbf{r}),
$$

(12.138)

where Ξ is the vector containing the electric $\mathbf{E}_i$ and magnetic field $\mathbf{H}_i$, while $\mathbf{M}$ is a liner integral operator describing the interactions between the scatterers. To solve equation (12.138), successive approximations must be used. The first step is characterized by the Rayleigh–Gans–Debye (RGD) approximation [658, 668]. Within this approximation, $\Xi_{inc}(\mathbf{r})$ is equal to $\Xi(\mathbf{r})$. In this way, we can compute the first-order estimation for every particle. After that, the iterative scheme is obtained by inserting the jth interaction of the fields $\Xi^j(\mathbf{r})$ into the right side of equation (12.138) and evaluating the next interaction in the left side. The solution of equation (12.138) is as follows:

$$
\Xi(\mathbf{r}) = \sum_{l=0}^{\infty} \mathbf{M}^l\, \Xi_{inc}(\mathbf{r}),
$$

(12.139)

which is a direct implementation of the Neumann series:

$$
(\mathbf{I} - \mathbf{M})^{-1} = \sum_{l=0}^{\infty} \mathbf{M}^l,
$$

where $\mathbf{I}$ is the unitary operator. It is very instructive to write the compact matrix form of the equations (12.136)–(12.137) because it defines the full Green's matrix $\overleftrightarrow{G}$. Explicitly, the full Green's matrix has the following form:

$$
\overset{\leftrightarrow}{G} =
\begin{bmatrix}
\hat{0} & \overset{\leftrightarrow}{G}_{12} & \overset{\leftrightarrow}{G}_{13} & \cdots & \overset{\leftrightarrow}{G}_{1N} \\
\vdots & \ddots & \vdots & \vdots & \vdots \\
\overset{\leftrightarrow}{G}_{j1} & \cdots & \hat{0} & \cdots & \overset{\leftrightarrow}{G}_{jN} \\
\vdots & \vdots & \vdots & \ddots & \vdots \\
\overset{\leftrightarrow}{G}_{N1} & \overset{\leftrightarrow}{G}_{N2} & \overset{\leftrightarrow}{G}_{N3} & \cdots & \hat{0}
\end{bmatrix},
\tag{12.140}
$$

where $\hat{0}$ represents the 6×6 zeros matrix, while the 6×6 subblock is expressed by matrix (12.135). $\overset{\leftrightarrow}{G}$ is a $6N \times 6N$ elements, where N expresses the total number of scatterers. In this matrix formulation, the $\mathbf{E}$ and $\mathbf{H}$ fields are assembled in components as a single column:

$$
\begin{pmatrix}
E_{1x} \\
E_{1y} \\
E_{1z} \\
H_{1x} \\
H_{1y} \\
H_{1z} \\
E_{2x} \\
\vdots
\end{pmatrix}.
$$

Within this formalism, the extinction efficiency can be directly obtained from the forward-scattering amplitude using the optical theorem for vector waves for both electric and magnetic polarizations, which results in the following [667]:

$$
Q_{ext} = \frac{4\pi k_0}{\pi |E_{inc}|^2 N R^2} \sum_{i=1}^{N} \Im[\boldsymbol{p}(\boldsymbol{r}_i) \cdot \boldsymbol{E}^*_{inc}(\boldsymbol{r}_i) + \boldsymbol{m}(\boldsymbol{r}_i) \cdot \boldsymbol{H}^*_{inc}(\boldsymbol{r}_i)],
\tag{12.141}
$$

where $\boldsymbol{p}(\boldsymbol{r}_i) = \alpha_e(\boldsymbol{r}_i)\boldsymbol{E}(\boldsymbol{r}_i)$, $\boldsymbol{m}(\boldsymbol{r}_i) = \alpha_m(\boldsymbol{r}_i)\boldsymbol{H}(\boldsymbol{r}_i)$, R is the particle radius, N the particle number, and the asterisk denotes the complex conjugate. Similarly, the absorption efficiency can be obtained by considering the energy dissipation of both dipoles in the system, producing the following:

$$
Q_{abs} = \frac{4\pi k_0}{\pi |E_{inc}|^2 N R^2} \sum_{i=1}^{N} |\boldsymbol{E}(\boldsymbol{r}_i)|^2 \left(\Im[\alpha_e(\boldsymbol{r}_i)] - \frac{2}{3} k_0^3 |\alpha_e(\boldsymbol{r}_i)|^2 \right)
$$
$$
+ \frac{4\pi k_0}{|E_{inc}|^2} \sum_{i=1}^{N} |\boldsymbol{H}(\boldsymbol{r}_i)|^2 \left(\Im[\alpha_m(\boldsymbol{r}_i)] - \frac{2}{3} k_0^3 |\alpha_m(\boldsymbol{r}_i)|^2 \right).
\tag{12.142}
$$

The scattering efficiency can always be evaluated by the difference of the extinction and the absorption efficiency, i.e., $Q_{sca} = Q_{ext} - Q_{abs}$. However, this operation requires high numerical accuracy in the computation of both Q_{ext} and Q_{abs} [667]. To avoid this problem, it is possible also in this case to directly calculate the scattering efficiency Q_{scat} by evaluating the power radiated in the far-field by the oscillating electric and magnetic dipoles, which is as follows [82]:

$$
Q_{sca} = \frac{k_0^4}{\pi |E_{inc}|^2 N R^2} \int \left(\frac{d\sigma}{d\Omega} \right) d\Omega
$$

$$
= \frac{k_0^4}{\pi |E_{inc}|^2 N R^2} \int \left| \sum_{i=1}^{N} e^{ik_0 \mathbf{n} \cdot \mathbf{r}_i} \{ \mathbf{p}(\mathbf{r}_i) - [\mathbf{n} \cdot \mathbf{p}(\mathbf{r}_i)]\mathbf{n} - \mathbf{n} \times \mathbf{m}(\mathbf{r}_i) \right|^2 d\Omega,
$$

(12.143)

where $\mathbf{n}$ is an unit vector in the direction of scattering. Therefore, equation (12.143) defines the differential scattering efficiency in the backward and forward direction when $\mathbf{n}$ is equal to the term $(0, 0, -1)$ and $(0, 0, 1)$, respectively, when the excitation is along the z-axis. Explicitly, the forward and backward scattering efficiencies are as follows:

$$
Q_{bs} = \frac{4k_0^4}{|E_{inc}|^2 N R^2} \frac{d\sigma}{d\Omega} \bigg|_{\theta=\pi}
$$

$$
= \frac{4k_0^4}{|E_{inc}|^2 N R^2} \left| \sum_{i=1}^{N} e^{ik_0 \mathbf{n} \cdot \mathbf{r}_i} \{ \mathbf{p}(\mathbf{r}_i) - [\mathbf{n} \cdot \mathbf{p}(\mathbf{r}_i)]\mathbf{n}_1 - \mathbf{n} \times \mathbf{m}(\mathbf{r}_i) \} \right|^2
$$

(12.144)

$$
Q_{fd} = \frac{4k_0^4}{|E_{inc}|^2 N R^2} \frac{d\sigma}{d\Omega} \bigg|_{\theta=0}
$$

$$
= \frac{4k_0^4}{|E_{inc}|^2 N R^2} \left| \sum_{i=1}^{N} e^{ik_0 \mathbf{n} \cdot \mathbf{r}_i} \{ \mathbf{p}(\mathbf{r}_i) - [\mathbf{n} \cdot \mathbf{p}(\mathbf{r}_i)]\mathbf{n} - \mathbf{n} \times \mathbf{m}(\mathbf{r}_i) \} \right|^2,
$$

(12.145)

where θ is the azimuthal angle.

We now apply this theory to the study of aperiodic planar arrays of small (60 nm in radius) dielectric nanoparticles. In particular, we can directly compare in Figure 12.4 the efficiency spectra for the extinction, scattering, and absorption of periodic, random, and the previously introduced deterministic aperiodic arrays of spherical Si nanoparticles. Also in this case, we observe several resonant scattering peaks due to lattice interference effects that appear in the visible spectral range at similar spectral locations compared to their metallic counterparts. Radiative coupling effects are particularly strong in the Penrose and Gaussian prime arrays shown in Figure 12.4d,e. The response of the single-particle and the Poisson random arrays are shown in Figure 12.4c for comparison. We also found that the GA-spiral, Hurwitz, and Lifschits arrays feature very narrow lattice resonances that are distributed over a broader spectrum compared to the periodic square array. This behavior follows from their more regular geometric structures compared to the other aperiodic systems. However, a novel feature appears in all the dielectric arrays due to the magnetic

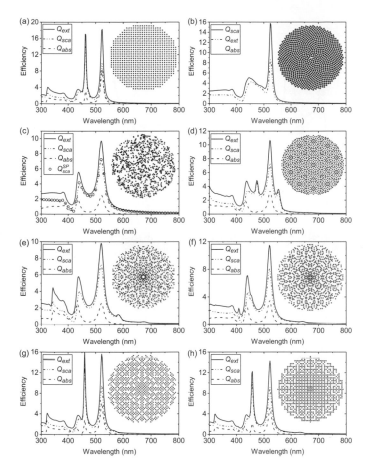

Figure 12.4 Calculated extinction (solid lines), scattering (dash-dot lines), and absorption (dashed lines) efficiency spectra of 60 nm Silicon nanoparticles arranged in a square lattice (a), golden-angle Vogel spiral (b), Poisson (c), Penrose (d), Eisenstein prime (e), Gaussian prime (f), Hurwitz array (g), and Lifschitz array (h), respectively. The center-to-center interparticle separation is 420 nm. The material dispersion is taken from [669]

dipole resonance at long wavelength that is characteristic of high refractive index nanoparticles. On the other hand, the efficiency peaks located approximately between 400–500 nm are due to the contribution of electric dipole scattering and can be very strong for the square and Hurwitz arrays. Their intensity, however, remains always smaller than the one observed for the lattice resonances of the metallic arrays previously discussed in relation to Figure 12.3.

The structural resonances demonstrated in Figure 12.4 for the aperiodic structures can be more clearly understood when comparing them with the behavior of an isolated (single) Si nanosphere. This analysis is shown in Figure 12.5a, where we report the spectra of the scattering efficiency computed considering the contribution of both electric and magnetic scattering modes (continuous line), along with the ones where only the electric dipole (dashed line) or the mangetic dipole are included (dashed-dot

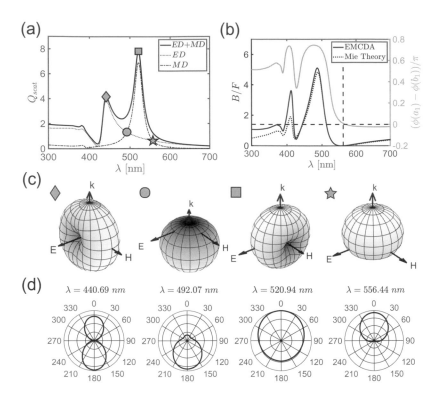

Figure 12.5 (a) Scattering efficiency spectrum of a 60 nm radius silicon nanosphere with material dispersion according to [669]. The dotted and dash-dot lines indicate the electric (ED) and magnetic (MD) dipole contributions, respectively. The different gray markers denote the ED peak (diamond marker), the MD peak (square marker) and the two points where the electric and magnetic contributions interfere, respectively. (b) Ratio of the backward and forward scattering efficiency as a function of the wavelength calculated by truncating the multipolar expansion up to the convergence order provided by $\ell = kr + 4.05^{1/3}$ (dotted line) as compared to the numerical EMCDA formalism (i.e., considering only the dipolar contributions). The gray right axis displays the phase difference of the electric (a_1) and magnetic (b_1) dipoles normalized with respect to π. The black dashed lines highlight the minimum in the backscattered field that occurs when the electric and magnetic dipole are oscillating in phase ($|\phi(a_1) - \phi(b_1)| = 0$). (c) The 3D radiation diagrams and (d) their cuts along the sagittal plane (i.e., the E-k plane) at 440 nm, 492 nm, 520 nm, and 556 nm.

line). It is clear that a strong single-particle magnetic scattering occurs for wavelengths around 550 nm, while strong electric dipole scattering dominates the spectrum at shorter wavelengths with a peak around 450 nm. The symbols on the spectra identify interesting regions where electric, magnetic, or a coherent superposition of the two contribute to the overall scattering efficiency with very different angular distribution properties. The scattering efficiency of a dielectric sphere can be analytically computed based on the Mie theory, which is discussed in detail in Chapter 13.

What is also important in the context of the present discussion is that the interference between the electric and magnetic modes in a dielectric nanoparticle can significantly modify the particle's scattering properties, leading to remarkable reshaping of the directional radiation patterns [662, 670–672]. Directional effects in light scattering from single particles become possible because the fundamental mode of a high-refractive index dielectric sphere or cube is always of the magnetic dipole type [661] and it occurs at a different frequency from the electric one. In fact, according to the Mie scattering theory for spherical particles, the magnetic dipole resonance takes place when the effective wavelength of light inside the particle equals its diameter. However, by engineering the aspect ratio of the scattering particles, it becomes possible to tailor the respective spectral positions of the electric and magnetic modes and create highly directional scattering responses, e.g., to completely suppress the backscattered radiation [672]. This can be easily understood when limiting our analysis to consider only the electric and the magnetic scattering contributions where we can write explicit expressions for the the backscattering and the forward scattering efficiency in the following form:

$$Q_{bs} = \frac{1}{(k_0 r)^2} \left| \sum_{n=1}^{2} (2n+1)(-1)^n (a_n - b_n) \right|^2 \tag{12.146}$$

$$Q_{fd} = \frac{1}{(k_0 r)^2} \left| \sum_{n=1}^{2} (2n+1)(a_n + b_n) \right|^2 . \tag{12.147}$$

The expansion coefficients a_n and b_n are the complex Mie–Lorentz coefficients that describe the electric dipole and the magnetic dipole scattering contributions. We realize by inspecting the preceding expressions that both Q_{bs} and Q_{fd} are written as coherent sums involving the superposition of electric and magnetic contributions. Therefore, the forward and the backward scattering intensities are sensitive to the coherent overlap, i.e., the interference, of the electric and magnetic scattering modes. Note that this is not the case for the total scattering efficiency since no interference effects between different modes can be observed when light is collected over all the scattering angles (see Chapter 13 for more discussions). To demonstrate the importance of mode interference in the radiation properties of the particle, we show in Figure 12.5b the wavelength dependence of the ratio between the backward (B) and the forward (F) scattered intensity along with the phase of the electric scattering coefficient a_1 and of the magnetic scattering coefficient b_1. We can clearly appreciate from the plot that in the region (around 580 nm) where the backscattering is suppressed, the phase difference between the modes is equal to zero while their amplitudes are approximately the same, as shown by the star symbol in Figure 12.5a. Therefore, according to equation (12.146) the backscattering is completely canceled. An hybridized situation can be observed corresponding to the circle in Figure 12.5a, where the electric and magnetic coefficients have the same amplitude but with almost opposite signs, resulting in an enhanced backscattering peak instead. On the other hand, the intermediate regions marked by the square and rhombus symbols correspond to the scattering diagrams of isolated electric and the magnetic modes. The computed radiation diagrams corresponding to the four regions of interest are shown as three-dimensional plots in Figure 12.5c and as two-dimensional cuts along the sagittal plane containing the E field and the wavevector k in Figure 12.5d. The typical electric and magnetic dipole radiation diagrams are consistently observed in correspondence to the rhombus and the square symbols, while the complete absence of backscattering can be clearly appreciated in the diagram labeled by the star symbol consistently with the cancellation of the in-phase electric and magnetic contributions. Finally, we remark that strong magnetic response has also been demonstrated in small clusters of metallic nanoparticles, giving rise to extended coupled modes with substantially lower losses compared to traditional electric modes, providing novel opportunities for optical transport and manipulation inspired by the behavior of complex molecules [673].

12.3.4   Spontaneous Decay Rate of a Dipole Emitter

In the previous sections, we showed how the general framework of the Foldy–Lax multiple-scattering vector equations, when specialized to clusters of resonant electric and magnetic vector dipoles, provides important information on their light-scattering and absorption properties. Here we discuss how the same approach permits us to accurately investigate the fluorescence decay rate of a dipole emitter embedded in a strongly scattering three-dimensional medium with arbitrary geometry. In the weak-coupling regime, quantum electrodynamics shows that the spontaneous decay rate of a dipole emitter aligned along the $\hat{\mathbf{u}}$-direction can be expressed as follows [674]:

$$\Gamma_{\hat{\mathbf{u}}} = \frac{2}{\hbar} \mu_0 \omega_0^2 |\mathbf{p}|^2 \Im\{\hat{\mathbf{u}} \cdot \overleftrightarrow{\mathbf{G}}(\mathbf{r}_0, \mathbf{r}_0) \cdot \hat{\mathbf{u}}\} \tag{12.148}$$

where $\mathbf{p} = p\hat{\mathbf{u}}$ is the dipole moment and $\hat{\mathbf{u}}$ is a direction unit vector. Remember that the field scattered at position $\mathbf{r}$ by an infinitesimal dipole $\mathbf{p}(t)\delta(\mathbf{r} - \mathbf{r}_0)$ located at $\mathbf{r}_0$ is related dyadic Green function as follows:

$$\mathbf{E}(\mathbf{r}) = \mu_0 \omega_0^2 \overleftrightarrow{\mathbf{G}}(\mathbf{r}, \mathbf{r}_0) \cdot \mathbf{p}. \tag{12.149}$$

However, it is crucial to realize that due to the inhomogeneous nature of the medium, $\overleftrightarrow{\mathbf{G}} \neq \overleftrightarrow{\mathbf{G}}_0$ and therefore the field at $\mathbf{r}$ is the self-consistent multiple-scattering field that consists of the sum of the primary field radiated by the dipole and the secondary field scattered by the medium. Therefore, equation (12.148) describes the modification of the decay rate of a dipole in term of the action of the *secondary field* that returns back to the dipole's position after undergoing a series of multiple scattering events in the inhomogeneous medium. Interestingly, this physical picture can also be clearly understood classically[9] by computing, via the electromagnetic Poynting's theorem,[10] the rate of energy dissipation in inhomogeneous environments [92].

---

[9] In this case, however, the classical dipole moment $\mathbf{p}$ should not be confused with the quantum mechanical transition dipole matrix element $\langle g| \hat{\mathbf{p}} |e\rangle$ between the ground $\langle g|$ and the excited $|e\rangle$ state. With proper reinterpretation of the dipole moment, the classical and quantum formulas for the decay rates appear identical.

[10] Assuming time-harmonic fields in linear and nondispersive media.

In free space where $\overleftrightarrow{\mathbf{G}} = \overleftrightarrow{\mathbf{G}_0}$, the decay rate of a dipole simplifies to the following [92]:

$$\Gamma_0 = \frac{\omega_0^3 |\mathbf{p}|^2}{3\pi\epsilon_0 \hbar c^3}. \tag{12.150}$$

We deduce from the preceding equations that the enhancement of the decay rate of a dipole embedded in the inhomogeneous medium can be expressed as follows:

$$\boxed{\Gamma_{\hat{\mathbf{u}}} = \frac{\pi\omega_0}{3\hbar\epsilon_0} |\mathbf{p}|^2 \rho_p(\mathbf{r}_0)} \tag{12.151}$$

where

$$\rho_p(\mathbf{r}_0) = \frac{6\omega_0}{\pi c^2} \Im\{\hat{\mathbf{u}} \cdot \overleftrightarrow{\mathbf{G}}(\mathbf{r}_0, \mathbf{r}_0) \cdot \hat{\mathbf{u}}\} \tag{12.152}$$

is the *partial local density of states*. The formula (12.151) enables the calculation of the spontaneous decay rate of a classical or of a quantum two-level system embedded in a nonhomogeneous scattering environment in terms of the imaginary part of the dyadic Green function evaluated at the dipole's position. Classically, this expression measures the *back action* at the dipole of the secondary field scattered by the medium.

In many practical situations, it is important to consider the decay rate Γ averaged over the orientation $\hat{\mathbf{u}}$ of the transition dipole moment, which can be expressed as follows [675]:

$$\Gamma = \frac{2\mu_0\omega_0^2}{3\hbar} |\mathbf{p}|^2 \Im\{\mathrm{Tr}[\overleftrightarrow{\mathbf{G}}(\mathbf{r}_0, \mathbf{r}_0)]\}, \tag{12.153}$$

where $\mathrm{Tr}[]$ denotes the trace of the tensor in brackets. The polarization-averaged decay rate therefore becomes

$$\Gamma = \frac{\pi\omega_0}{3\hbar\epsilon_0} |\mathbf{p}|^2 \rho(\mathbf{r}_0), \tag{12.154}$$

where we define

$$\rho(\mathbf{r}_0) = \frac{2\omega_0}{\pi c^2} \Im\{\mathrm{Tr}[\overleftrightarrow{\mathbf{G}}(\mathbf{r}_0, \mathbf{r}_0)]\} \tag{12.155}$$

as the *total local density of states*, or the LDOS in the medium [92], which coincides with the previously derived formula (2.56). Therefore, after orientation averaging, the spontaneous decay rate depends on the transition dipole moment and is proportional to the total LDOS in the medium. Using the preceding results above within the discrete dipole model formulated in the previous section, we can compute the decay rate and the LDOS for an arbitrary scatting medium composed of N resonant point scatterers. The approach that we will now discuss was originally developed in [675] and applied to the study of the spontaneous decay rate of a dipole in a strongly scattering random medium. The main idea is to split the dyadic Green function of the inhomogeneous medium as follows:

$$\overleftrightarrow{\mathbf{G}}(\mathbf{r}, \mathbf{r}_0) = \overleftrightarrow{\mathbf{G}_0}(\mathbf{r}, \mathbf{r}_0) + \overleftrightarrow{\mathbf{R}}(\mathbf{r}, \mathbf{r}_0), \tag{12.156}$$

where $\overleftrightarrow{\mathbf{R}}$ corresponds to the scattered field. This can be accomplished by the self-consistent $3N$ Foldy–Lax coupled-dipole equations that provide the local exciting fields on each particle:

$$\mathbf{E}_j = \mu_0 \omega_0^2 \overleftrightarrow{\mathbf{G}_0}(\mathbf{r}_{0j})\mathbf{p} + t(\omega) \sum_{i \neq j} \overleftrightarrow{\mathbf{G}_0}(\mathbf{r}_{ij})\mathbf{E}_i. \tag{12.157}$$

The second term on the right-hand side of equation (12.157) represents the modification of the free-space dyadic Green function due to the presence of the scatterers. Once the exciting electric field at each scatterer is known, it is possible to compute the scattered field $\mathbf{E}_s$ at the source dipole's position $\mathbf{r}_0$:

$$\mathbf{E}_s(\mathbf{r}_0) = t(\omega) \sum_1^N \overleftrightarrow{\mathbf{G}}_0(\mathbf{r}_{j0})\mathbf{E}(\mathbf{r}_j) \tag{12.158}$$

and deduce the Green dyadic $\overleftrightarrow{\mathbf{R}}(\mathbf{r}_0, \mathbf{r}_0)$.

With the knowledge of the scattered field, the normalized spontaneous decay rate in the weak-coupling regime can be readily obtained as follows [92]:

$$\boxed{\frac{\Gamma}{\Gamma_0} = 1 + \frac{6\pi\epsilon_0}{|\mathbf{p}|^2 k^3} \Im\{\mathbf{p}^* \cdot \mathbf{E}_s(\mathbf{r}_0)\}} \tag{12.159}$$

In this numerical approach, near-field and far-field dipole–dipole interactions and multiple scattering are rigorously taken into account.

12.4   The Dyadic Green's Matrix Method

In Section 12.2.3, we discussed the Green's matrix method for arbitrary arrays of scalar point dipoles. In this section, we extend the treatment to vector waves and introduce the *dyadic Green's matrix method*. This method has been recently applied to study the wave transport and the structure–property relations in various types of scattering arrays with deterministic aperiodic geometries considering nanoparticles with electric or magnetic response [9, 10, 205, 349, 599]. Here we will detail the main features of this rigorous technique and show some applications to random and deterministic aperiodic media.

The dyadic Green's matrix method is a powerful tool that has been primarily applied to the study of wave propagation in random media. The method relies on the analysis of the spectra of the dyadic Green's matrix, which is an important class of the so-called *Euclidean random matrices* that appear in Random Matrix Theory (RMT) [430]. In general terms, the elements of a Euclidean random matrix are determined by a function of the positions of pairs of randomly distributed points in the Euclidean space. In particular, Hermitian Euclidean random matrices have important applications in disordered superconductors [676], supercooled liquids [677], diffusion of particles, and scalar phonon localization [678]. In recent years, the interest in non-Hermitian random matrices such as Green's matrix has significantly increased due to their applications in the theoretical description of *open and dissipative systems*. In particular, the study of the spectra of Green's matrices unveiled important information about structural resonances in disordered systems. Indeed, the numerical analysis of Green's matrices has been successfully employed to address different aspects of light propagation in open random media, such as Anderson localization of light [476, 649, 650, 679, 680], and matter waves [681], random lasing [682, 683], light transport in nonlinear media [684], and superradiance in atomic random systems [685–688]. An analytical theory has also been developed for the eigenvalue density of random Green's matrices, providing fundamental insights into light–matter interactions in disordered media [470, 477, 683]. However, the applications of the Green's matrix method have been mostly confined to random media so far. In our work, we apply this approach to random as well as deterministic arrays and shed some light on their unique spectral properties [9, 349, 389, 599]. Before discussing some results, we first review the Green's matrix method in the general context of the electromagnetic multiple scattering formulation for discrete media (i.e., point patterns).

The Green's matrix method describes the vector light propagation in a system of identical, point-like dipoles arbitrarily positioned inside a homogeneous medium (typically in vacuum). This approach is perfectly suitable to investigate the electromagnetic scattering properties of both random and deterministic point patterns. The 3×3 T-matrix $\mathbf{t}$ describes light scattering by one particle and exhibits a scattering resonance at frequency ω_0 with linewidth Γ_0 [689]. The total $3N \times 3N$ T-matrix of an assembly of N scatterers at the positions $\mathbf{r}_1, \mathbf{r}_2, \ldots, \mathbf{r}_N$ is given as follows:

$$
\mathbf{T}_{\mathbf{k},\mathbf{k}'} = \begin{pmatrix} e^{i\mathbf{k}\cdot r_1} \\ \vdots \\ e^{i\mathbf{k}\cdot r_N} \end{pmatrix}^{*} \mathbf{t}\cdot(\mathbf{I} - \mathbf{G}\cdot\mathbf{t})^{-1}\cdot \begin{pmatrix} e^{i\mathbf{k}'\cdot r_1} \\ \vdots \\ e^{i\mathbf{k}'\cdot r_N} \end{pmatrix}, \tag{12.160}
$$

where $\mathbf{I}$ is the unit matrix, $\mathbf{k}$ and $\mathbf{k}'$ are, respectively, the incident and the scattered wavevectors, and in the far-field $|\mathbf{k}'| = |\mathbf{k}| = k = \omega/c_0$. The elements of the $3N \times 3N$ $\mathbf{G}$-matrix represent the general electromagnetic dyadic Green's function that is calculated from the relative positions of the N dipoles. In particular, the diagonal elements of the $\mathbf{G}$-matrix are zero and for $n \neq m$ we can write the following [690]:

$$
\mathbf{G}_{nm}(\mathbf{r}_{nm}) = \frac{3}{2}\frac{e^{jkr_{nm}}}{jkr_{nm}}\left\{[\mathbf{I} - \hat{\mathbf{r}}_{nm}\hat{\mathbf{r}}_{nm}] - \left[\frac{1}{jkr_{nm}} + \frac{1}{(kr_{nm})^2}\right][\mathbf{I} - 3\hat{\mathbf{r}}_{nm}\hat{\mathbf{r}}_{nm}]\right\}. \tag{12.161}
$$

It is important to emphasize that multiple scattering contributions (i.e., all scattering orders) are treated exactly within the Green's matrix method in equation (12.160) and the only approximation involved is the description of the scatterers as vector point dipoles. Hence this method is particularly suited to treat light propagation in cold atomic clouds, where it has been extensively applied [685–688], but it also provides fundamental insights into the transport properties and photonic bandgap structure of periodic and disordered scattering media with a large number of components [691].

Equation (12.160) contains all the information on the electric field scattered by the ensemble of dipoles, and it requires the diagonalization of the $3N \times 3N$ scattering matrix:

$$\mathbf{M}(\omega) \equiv \mathbf{t}(\omega) \cdot [\mathbf{I} - \mathbf{G}(\omega) \cdot \mathbf{t}(\omega)]^{-1}. \tag{12.162}$$

The T-matrix describing one single scatterer $\mathbf{t}(\omega)$ can be expressed as a Neumann–Born series [656]:

$$\mathbf{t}(\omega) = \mathbf{V}(\omega) + \mathbf{V}(\omega) \cdot \mathbf{G}_0(\omega) \cdot \mathbf{V}(\omega) + \cdots = \left(\frac{1}{\mathbf{V}(\omega)} - \mathbf{G}_0(\omega) \right)^{-1}, \tag{12.163}$$

where $\mathbf{V}(\omega)$ is the optical scattering potential for a point scatterer and $\mathbf{G}_0(\omega)$ is the return Green function, i.e., the dyadic Green function $\mathbf{G}_0(\omega, \mathbf{r})$ evaluated at $\mathbf{r} = 0$. In contrast to the analogous scattering formulation for quantum mechanical waves, the optical scattering potential is frequency dependent and for a point dipole at position $\mathbf{r}_i$ takes the following form:

$$\mathbf{V}(\omega, \mathbf{r}) = \left(\frac{\omega}{c_0} \right)^2 \alpha_0 \delta(\mathbf{r} - \mathbf{r}_i), \tag{12.164}$$

where α_0 is the polarizability of the scatterer [656].

Equation (12.161) shows that the matrix elements of the Green function $\mathbf{G}_0(\omega)$ exhibit singularities at $r = 0$. However, this feature can be easily handled by regularization [656]:

$$\tilde{\mathbf{G}}_0(k, \mathbf{r} = 0) = \left(\frac{\Lambda'}{6\pi} + i \frac{k}{6\pi} \right) \mathbf{I}, \tag{12.165}$$

where the inverse length scale parameter Λ' provides the resonance position ω_0 and resonance width Γ_0 of a single dipole by the relations $1/\Lambda' = (\omega_0/c_0)^2 (\alpha_0/6\pi)$ and $1/\Lambda' = (\Gamma_0 c_0/\omega_0^2)$, respectively. Using equations (12.165) and (12.163), we can write the following:

$$\mathbf{t}^{-1}(\omega) = \mathbf{V}^{-1}(\omega) - \mathbf{G}_0(\omega) = \frac{\Lambda'}{6\pi} \left(1 - \frac{\omega_0^2}{\omega^2} \right) \mathbf{I} + i \frac{\omega}{6\pi c_0}, \tag{12.166}$$

where equation (12.164) has been used. Assuming that $\omega_0 \Gamma_0 \ll 1$ (for an atom, typically $\omega_0 \Gamma \sim 10^{-6}$), we can consider, close to the resonance, $\omega \approx \omega_0$ and introduce the detuning parameter $\Delta \equiv \omega - \omega_0/\Gamma_0$. We can then rewrite the total scattering matrix $\mathbf{M}(\omega)$ in equation (12.162) as follows:

$$\mathbf{M}(\omega) = \frac{4\pi c_0}{\omega_0} \left[\mathbf{I}\Delta + \frac{2}{3}\mathbf{I}i - \frac{4\pi c_0}{\omega_0} \mathbf{G}(\omega_0) \right]^{-1}. \tag{12.167}$$

Notice that the only frequency dependence of the total scattering matrix (12.167) near resonance is contained in Δ. Very importantly, we can now realize from equation (12.167) that in order to obtain the global T-matrix in equation (12.160), we need to diagonalize Green's matrix $G(\omega_0)$, which is independent of the frequency detuning Δ. Note that equation (12.167) *depends on the geometric structure of the scattering system*, which makes the Green's matrix method ideally suited to investigate structure–property relationships in aperiodic structures [9, 599]. Green's matrix equation (12.161) fully takes into account the vector character of light. However, its scalar approximation

$$
G_{nm}(r_{nm}) = \begin{cases} \frac{\exp(ikr_{nm})}{ikr_{nm}} & \text{for } n \neq m, \\ 0 & \text{for } n = m, \end{cases}
$$

(12.168)

has been frequently used to simplify the problem [470, 649]. Although the scalar version of the Green's matrix method has been successfully applied to describe several aspects of light propagation in disordered media, it seems to fail to treat the regime of Anderson localization, when light diffusion comes to a halt due interference effects in the strong multiple-scattering regime. Indeed, there is numerical evidence of absence of Anderson localization for vector waves in three dimensions for an ensemble of point scatterers at the same optical densities where Anderson localization is expected to occur for scalar waves [476]. This result is related to the near-field coupling between scatterers that becomes important at high optical densities, and which is more relevant for vector waves. Indeed, note that in the vectorial case, Green function equation (12.161) has a $1/r_{nm}^3$ singularity for $r_{nm} \to 0$ related to the intrinsic transverse nature of electromagnetic waves. In contrast, in the scalar case the singularity of the Green function 12.168 at $r_{nm} \to 0$ is weaker, a fact that may play a crucial role at high optical densities, where Anderson localization is expected to occur [476]. Interestingly, this near-field dipole–dipole coupling was shown to be suppressed by the application of an external magnetic field, which can induce a transition from extended to localized states for light in an ensemble of point scatterers [688]. More recently, the important role of engineered structural correlations in arrays of nanoparticles and the implications on wave transport and localization phenomena were systematically addressed in a number of aperiodic geometries using the dyadic Green's matrix approach [9, 205, 349, 389, 599, 653].

12.4.1 Spectral Properties of Random Arrays

Similarly to the scalar case, it is possible to extract important physical information on scattering arrays from the distribution of eigenvalues Λ of the dyadic Green's matrix in the complex plane. Indeed, the real and imaginary parts of Λ are related to the relative widths $(\Gamma - \Gamma_0)/\Gamma_0$ and frequency positions $(\omega - \omega_0)/\Gamma_0$ of the scattering resonances of the arrays, respectively [649]:

$$
\text{Re}\Lambda(\omega_0) \simeq (\Gamma - \Gamma_0)/\Gamma_0,
$$

$$
\text{Im}\Lambda(\omega_0) \simeq (\omega - \omega_0)/\Gamma_0.
$$

(12.169)

From equations (12.169), we can see that a clustering of eigenvalues around the value $\text{Re}\Lambda = -1$ evidences the formation of a band of long-lived modes (i.e., with very small Γ). We caution the readers that there exists in the literature an alternative convention for Green's matrix that is more suitable for the direct visualization of the long-lived modes of a scattering system. In this case, it is convenient to modify the previous definition of Green's matrix as follows:

$$\boxed{\tilde{\mathbf{G}}_{nm} = i(\delta_{nm} + \mathbf{G}_{nm})} \tag{12.170}$$

where $\mathbf{G}_{nm}$ is defined by equation (12.161) [349, 640]. In this formalism the real and the imaginary parts of Λ are associated to the detuned scattering frequency $(\omega_0 - \omega)$ and to the scattering resonance decay Γ, both normalized with respect to the resonant width Γ_0 of an isolated dipole, respectively [476]. In the rest of our discussion, we will employ the alternative definition of Green's matrix in the form (12.170) and study its spectral properties. In this way, the long-lived modes can be easily identified by $\Im\{\Lambda\} \ll 1$.

In Figure 12.6, we show the eigenvalue Λ distribution of Green's matrix for a Poisson point patterns and representative types of deterministic aperiodic planar arrays with approximately $N = 1,000$ scatterers. In particular, we compare a random distribution of point dipoles located inside a circular region (Figure 12.6a) with several deterministic aperiodic arrangements consisting of a Penrose quasicrystal (b), a Vogel spiral (c), Eisenstein primes (d), Gaussian primes (e), and Hurwitz primes (f). In our analysis, we consider different values of the *optical density* $\rho\lambda^2$, where ρ is the areal density of particles in the arrays and λ the optical wavelength. Therefore, the optical density provides a measure of the "scattering strength" in the system. The results shown in Figure 12.6 correspond to the situation of weak scattering, or small optical density. All the eigenvalues are shaded with respect to the logarithm of the mode spatial extend (MSE) of the corresponding eigenvectors, so that their spatial localization character becomes immediately apparent at small MSE values. The MSE of a given state is defined as follows:

$$\text{MSE}(\omega_k) = \frac{\left(\sum_{n=1}^{3N} |\mathbf{e}_n(\omega_k)|\right)^2}{\sum_{n=1}^{3N} |\mathbf{e}_n(\omega_k)|^4}, \tag{12.171}$$

where $\mathbf{e}_n(\omega_k)$ is the (unit) eigenvector of Green's matrix for the mode of frequency ω_k. The eigenvectors are normalized to obey the condition $\sum_n |\mathbf{e}_n(\omega_k)|^2 = 1$. We note that the inverse of the MSE parameter is known as the **inverse participation ratio** (IPR). The IPR is an alternative measure of the degree of spatial localization of the eigenvectors of the system since it provides the number of sites covered by the eigenvector of a given energy. In particular, an eigenvector that extends over all the N scatterers is characterized by $IPR \simeq 1/N$, whereas an eigenvector localized on a single point has $IPR = 1$. We can observe that the eigenvalue distributions converge to circular disk regions. However, we note the presence of distinctive spiral arms that are visible even at low values of the optical density. For random systems, these spectral regions are typically populated by long-lived resonances, localized over

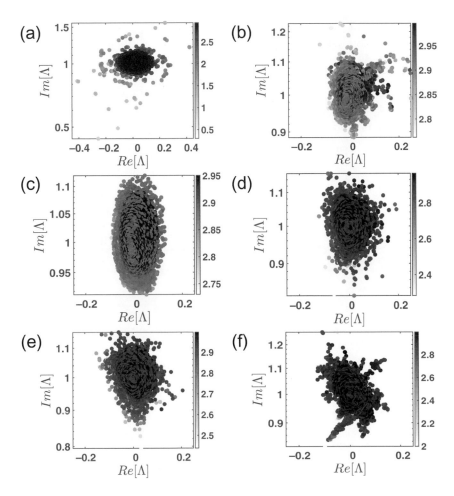

Figure 12.6 Eigenvalues of Green's matrix (12.161) are shown by points on the complex plane for almost 1,000 electric point dipoles arranged in a Poisson (a), Penrose (b), golden-angle Vogel spiral (c), Eisenstein prime (d), Gaussian prime (e), and Hurwitz prime (f) geometry, respectively. These data are produced by an optical density $\rho\lambda^2 = 0.01$ and are shaded according to the log_{10} of the mode spatial extent (MSE).

a small number of scatterers, down to only two particles. It is known that already for systems of only two scatterers randomly placed within the wavelength of the scattered wave field, extremely narrow resonances can appear [649]. Furthermore, in the case of three or few scatterers collective resonances may appear in the spectrum in close analogy to the quantum mechanical Efimov's effect [692]. Simulations based on the scalar Green function approximation have shown that even for a large system of many particles, such electromagnetic resonances can occur [649]. These subradiant quasimodes are called *proximity resonances*. However, they are not related to Anderson wave localization because they do not require multiple scattering in order to occur, as they are driven by the coupling of the reactive (nonpropagating) near-fields [470, 649]. We notice that at small optical density no proximity resonances can be

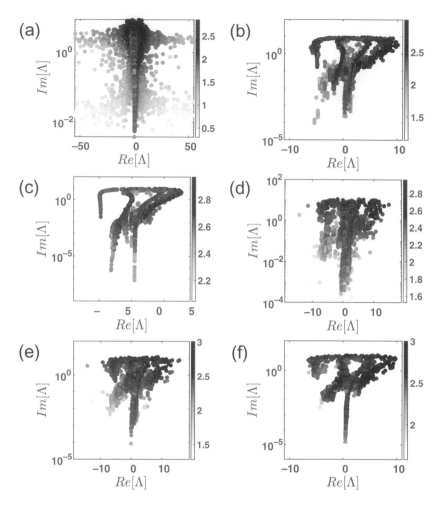

Figure 12.7 Eigenvalues of Green's matrix (12.161) are shown by points on the complex plane for almost 1,000 electric point dipoles arranged in a Poisson (a), Penrose (b), golden-angle Vogel spiral (c), Eisenstein prime (d), Gaussian prime (e), and Hurwitz prime (f) geometry, respectively. These data are produced by an optical density $\rho \lambda^2 = 15$ and are *color coded* according to the log_{10} of the MSE.

observed in the deterministic aperiodic systems investigated in Figure 12.6. This is a consequence of the non-Poissonian nature of the spatial distributions of the scattering dipoles. This scenario is drastically changed for larger values of the optical density, i.e., in the strong scattering regime shown in Figure 12.7, where large spectral gaps appear for all the nonrandom systems.

In particular, despite the lack of periodicity the Penrose array supports spectral gap regions at large enough values of the optical density, i.e., in the presence of strong resonant scattering. However, these gap regions disappear when the value of the optical density is reduced, and the eigenvalue distribution approaches the one of an uncorrelated random system (i.e., circular distribution). This behavior is indicative of a characteristic threshold for bandgap formation in quasiperiodic structures, which has indeed investigated experimentally in the case Penrose quasicrystals [693]. The fine-structured distribution of eigenvalues observed at larger densities reflects the correlated nature of the Penrose array, and it manifests a nested substructure of smaller spectral gaps consistent with the singular continuous energy spectrum of Penrose quasicrystals. These intriguing spectral features point toward a richer physics that is currently well beyond the predictions of conventional random matrix theory [430, 470]. A similar scenario is manifested also by the other investigated aperiodic structures, whose spectral properties have been systematically investigated by Dal Negro and collaborators [599] and by Wang and collaborators [205].

The spectral gap behavior of the considered deterministic aperiodic arrays is consistent with what previously reported on Vogel spiral structures made of dielectric pillars using the finite element method (FEM). In those studies, large bandgap regions were in fact observed in the optical density of states of different Vogel spiral structures [250, 598]. In addition, the presence of strongly localized dipolar eigenmodes in Vogel spiral arrays is confirmed by the analysis of the eigenvectors of Green's matrix. The spatially localized modes associated to the spiral branches extend across few particles along the parastichy lines of Vogel spirals. Similarly to the case of random systems, when the scattering is weakened by lowering the optical density, all the eigenvalue distributions converge to a circular disk region populated by resonant states with large MSE values. However, differently from the random case, the central disk of the Vogel spiral's distribution is surrounded by a very well-defined band of modes with lower MSE values.

12.4.2   Level Statistics in Complex Media

The knowledge of the positions and widths of scattering resonances allows one to investigate level statistics of a disordered medium via Green's matrix method. Level statistics provides important information about electromagnetic propagation in both closed and open systems. Indeed, from RMT it is possible to identify the presence of localized and extended states by investigating level statistics in closed systems.

In closed systems, a well-known result of RMT in the extended regime of wave transport, where almost all modes overlap in space, is the existence of level repulsion [430]. This means that the probability density function for the spacing between adjacent energy levels E_i and E_{i+1}, $|\Delta E| = |E_{i+1} - E_i|$, tends to zero for $|\Delta E| \to 0$ [694]. This fact results in the following expression for the probability density function of normalized energy spacings $s \equiv |\Delta E|/\langle|\Delta E|\rangle$ (for systems with time-reversal symmetry) [694]:

$$P(s) \propto s \exp\left(\frac{-\pi s^2}{4}\right). \qquad (12.172)$$

In contrast, for closed systems, level repulsion does not occur in the localized regime. In this case, two spatially distant, exponentially localized states hardly influence each other, resulting in a Poissonian level-spacing distribution [694]:

$$P(s) \propto \exp(-s) \tag{12.173}$$

so that two states with infinitely close energies become possible.

On the other hand, in dissipative or *open systems*, the eigenenergies are complex with a finite decay rate, as specified by equation (12.169). The concept of level repulsion for extended states can be generalized to this case [694]. Indeed, for each complex energy E_i the nearest quasimode[11] E_{i+1} is the one that minimizes the Euclidean distance between two quasimodes in the complex plane $|\Delta E| = |E_{i+1} - E_i|$. For non-Hermitian matrices belonging to Ginibre's ensemble of random matrices, $P(s)$ reads as follows [694]:

$$P(s) = \frac{3^4 \pi^2}{2^7} s^3 \exp\left(-\frac{3^2 \pi}{2^4} s^2\right), \tag{12.174}$$

so that $P(s = 0) = 0$, which means that two adjacent quasimodes repeal each other in the complex plane. This generalized level repulsion rule for open disordered systems has been recently verified for vectorial waves in the extended regime [688]. As in the case of closed systems, the appearance of localized modes due to increasing disorder leads to the suppression of the (generalized) level repulsion [688]. These results on the spectral statistics of the complex eigenvalues of random systems fully support this picture and even suggest its generalization to the more general deterministic aperiodic structures. In the case of random systems, we can also write $k\ell = 2\pi/(\rho\lambda^2)$, where ℓ is the *scattering mean free path* of the system, which measures the average distance between two successive scattering events. The value of first-level spacing, S_1, is computed at the nearest-neighbor distance between eigenvalues in the complex plane for each eigenvalue, and is normalized by the average S_1 for each case. In particular, the data show the computed probability density function (PDF) of the level spacing $|\Delta E| = |E_{i+1} - E_i|$ in the complex plane. The probability density is normalized such that the total probability equals to one.

We have ensured by performing additional simulations (not shown here) that the overall size of this random system is large enough so that an ensemble averaging procedure that considers different realizations will not alter in any relevant way our main conclusions for the random structures. As expected for random media, the transition toward the level repulsion regime can be distinctly observed by increasing the optical density of the array. However, it is important to keep in mind that a comprehensive theory of level statistics in random open systems of vector point scatterers is currently not available for arbitrary values of the optical density. In random systems, a link has been established between spectral statistics and wave function multifractality [695]. On the other hand, a direct link between spectral multifractality and spectral fluctuations in

---

[11] In our terminology, we use the term "quasimode" to denote an eigevector of Green's matrix with a complex eigenvalue. We will also interchangeably use the term "scattering resonance".

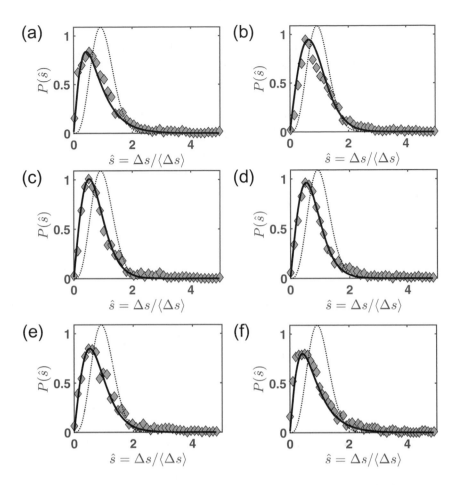

Figure 12.8 Level-spacing statistics of Green's matrix eigenvalues obtained in the low scattering regime ($\rho\lambda^2 = 0.01$) for Penrose (a), golden-angle Vogel spiral (b), Eisenstein prime (c), Gaussian prime (d), Hurwitz prime (e), and Lifschitz prime (f) geometry, respectively. While the traditional Poisson configuration is well reproduced by Ginibre's statistics (the dashed black line), all the deterministic aperiodic structures are compatible with critical distributions.

quasiperiodic structures has been recently established for one-dimensional systems based on the Harper model, where the appearance of a semi-Poisson bandwidth distribution $P(s) \propto s\exp(-s)$ was demonstrated [696] that is characterized by level repulsion. However, to our knowledge, little is currently known on the connections between spectral statistics and level spacing in more general quasiperiodic and deterministic aperiodic two-dimensional structures. We address this issue directly using Green's matrix method by investigating the statistics of first-neighbor eigenvalues in the complex plane for aperiodic systems such as those introduced earlier.

Figures 12.8 and 12.9 summarize our findings for the level-spacing statistics of the random and the aperiodic arrays at two different values of the optical density. For the deterministic aperiodic systems, a clear transition from the absence of level repulsion

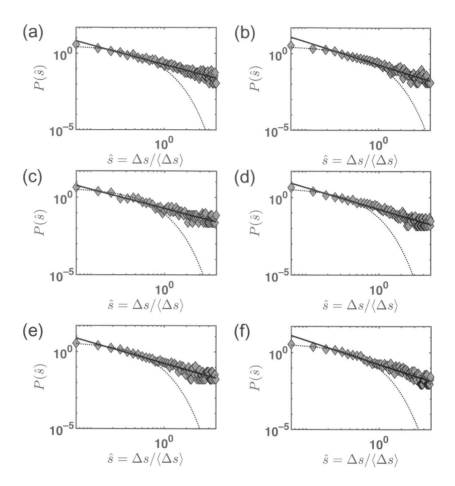

Figure 12.9 Level-spacing statistics of Green's matrix eigenvalues obtained in the large scattering regime ($\rho\lambda^2 = 15$) for Penrose (a), golden-angle Vogel spiral (b), Eisenstein prime (c), Gaussian prime (d), Hurwitz prime (e), and Lifschitz prime (f) geometry, respectively. While the traditional Poisson configuration is well reproduced by a Poisson statistics (the dashed black line), all the deterministic aperiodic structures are compatible with an inverse power-law distributions.

at large optical densities, indicative of localization in agreement with the predictions on the one-dimensional Harper model [696], and level repulsion at small optical densities can be observed. Interestingly, this behavior is similar to what is observed in random systems [688], but has not been previously reported for more general deterministic aperiodic structures with a two-dimensional geometrical support. It is also interesting to note that the level-spacing distributions of the investigated deterministic aperiodic structures feature slowly decaying *power-law tails* that indicate a markedly non-Gaussian spectral statistics. This behavior is consistent with the fractal nature of their density of states [20, 598] and cannot be observed in uniform random media, where the level spacings are uncorrelated and the corresponding probability decays very quickly with their separation.

12.4.3   Resonances of Deterministic Aperiodic Arrays

A very relevant problem in the study of aperiodically ordered structures is the understanding of the nature of their distinctive scattering resonances. This task can be systematically addressed using the Green's matrix method to efficiently compute the polarization eigenvectors and corresponding electric optical modes by contracting the dyadic Green's matrix according to equations (12.99) and (12.100). Therefore, within the validity of the point-particle vector model, the spatial structure of the distinctive optical resonances of aperiodic systems can be investigated. In this section, we study the spatial distribution of the eigenvectors of Green's matrix, which corresponds to the resonant modes of polarization (or current modes) of the arrays. Clearly, once the resonant modes of polarization have been found, the electric field eigenmodes can be obtained everywhere in space by multiplication with the dyadic Green's matrix.

An interesting open problem in the study of deterministic aperiodic structures is the rigorous characterization of their critical modes and scattering resonances. Critical modes are believed to exhibit general characteristics [327]. For instance, in contrast with disordered media, where exponentially localized Anderson modes may occur in two dimensions, critically localized states decay weaker than exponentially, most likely by a power law, and have a rich self-similar structure described by fractal or multifractal scaling. The rich physical behavior of critical states, including the presence of extended fractal wavefunctions at the band-edge energy of the Fibonacci spectrum, and their relation to optical transport has been analytically studied by Kohmoto [59] and Macía Barber [65] using the tight binding method and a transfer matrix renormalization technique, respectively. In quasiperiodic systems, it is believed that the characteristic spatial oscillations of critical modes originate through a series of tunneling events involving the overlap of different substructures that repeat over different length scales [327, 697]. However, for general aperiodic structures, there is no clear definition of critical modes, leading to somewhat confusing situations. For instance, the notion of a slowly decaying envelope function is clearly inadequate to describe the strongly fluctuating scattering resonances that occur in more general aperiodic systems. In addition, precise criteria need to be developed in order to identify critical modes in the spectra of aperiodic arrays. This ability would enable the engineering of critical states that leverages unique transport and localization properties.

We have addressed this problem by investigating all the Green's matrix eigenvectors and the corresponding MSE values for a number of aperiodic systems [9, 10, 205, 349, 389, 599]. This rigorous numerical analysis permits us to distinguish among three main types of polarization eigenvectors. The first type corresponds to long-lived scattering resonances with $Re\Lambda \simeq -1$ and large IPR values, which feature strong spatial localization. For instance, in random systems, proximity resonances involving only two particles, which correspond to $IPR = 0.5$, can be found. Similarly, long-lived resonances can be found in deterministic aperiodic structures.

Representative long-lived scattering resonances with the largest possible IPR values are shown in Figure 12.10 for all the investigated systems. In the case of the Penrose arrays, such resonances extend across few particles and manifest the distinctive

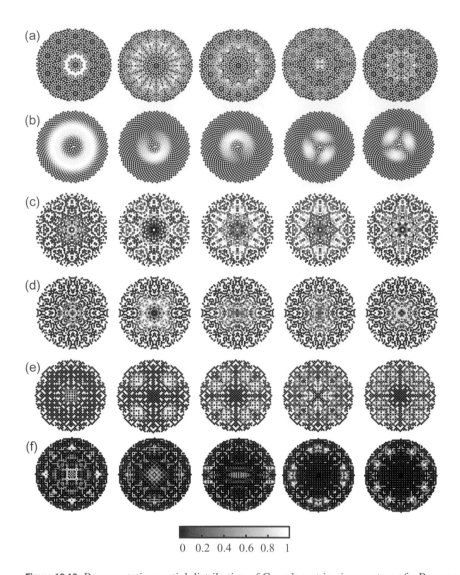

Figure 12.10 Representative spatial distribution of Green's matrix eigenvectors of a Penrose (a), golden-angle Vogel spiral (b), Eisenstein prime (c), Gaussian prime (d), Hurwitz prime (e), and Lifschitz prime (f) array of electric dipoles. These scattering resonances are critical modes, i.e., extended long-lived resonances characterized by an hierarchical structure of self-similar oscillations unfolding at every length scale.

decagonal rotational symmetry of the array. Remarkably, similar resonances can be clearly identified already via the graph Laplacian method discussed in [599], and are therefore *geometrical resonances* of the aperiodic arrays. See Appendix F for a discussion of spectral graph theory applied to aperiodic arrays. In contrast, the long-lived resonances of the Hurwitz array are spatially more extended and characterized by smaller IPR values. The long-lived resonances of the golden-angle Vogel's spiral are also shown in Figure 12.10b. These polarization eigenmodes are strongly localized in the central region of the array and spread across only few neighboring particles along the parastichies, similarly to the behavior of few-body Efimov's resonances in random systems. Vogel spirals exhibit a class of long-lived resonances that are extended azimuthally but localized in the radial direction. Interestingly, these modes also have been found to correspond to the geometrical resonances predicted by the graph Laplacian approach [599]. A remarkable output of this study is that in the investigated deterministic aperiodic arrays, long-lived delocalized modes coexist with both localized and critical modes [599]. The rich spectral diversity of these aperiodic systems makes them very attractive for active device applications that rely on enhanced light–matter coupling over large surfaces, such as nonlinear converters, optical sensors, and novel high-power light sources.

12.4.4   Green's Matrix with Electric and Magnetic Dipoles

In the previous sections, we showed how to extend the CDA approximation to coupled electric and magnetic dipoles. Here, we will discuss how to include in Green's spectral method also the magnetic dipole contribution. Specifically, we will isolate the electric Green's matrix G_{ij}, defined by equation (12.161) replacing the label nm with the pair ij, from the matrix (12.135) by rewriting the matrix (12.130) in the following compact form:

$$C_{ij} = a_{ij}\mathbf{I} + b_{ij}\hat{r}_{ij}\hat{r}_{ij}, \tag{12.175}$$

where a_{ij} and b_{ij} are defined by the equations (12.123) and (12.124) respectively, $\mathbf{I}$ is the 3×3 unity matrix, and the term $\hat{r}_{ij}\hat{r}_{ij}$ has the explicit form:

$$\frac{1}{r_{ij}^2}\begin{bmatrix} (x_i - x_j)^2 & (x_i - x_j)(y_i - y_j) & (x_i - x_j)(z_i - z_j) \\ (x_i - x_j)(y_i - y_j) & (y_i - y_j)^2 & (y_i - y_j)(z_i - z_j) \\ (x_i - x_j)(z_i - z_j) & (y_i - y_j)(z_i - z_j) & (z_i - z_j)^2 \end{bmatrix}.$$

Equation (12.175) is then equivalent to

$$C_{ij} = (1 - \delta_{ij})\frac{e^{ik_0 r_{ij}}}{r_{ij}}k_0^2\left[(\mathbf{I} - \hat{r}_{ij}\hat{r}_{ij}) - \left(\frac{1}{(k_0 r_{ij})^2} + \frac{1}{ik_0 r_{ij}}\right)(\mathbf{I} - 3\hat{r}_{ij}\hat{r}_{ij})\right]$$

$$= \frac{2}{3}ik_0^3 G_{ij}. \tag{12.176}$$

In the same way, matrix (12.131) can be rewritten in the following compact form:

$$f_{ij} = \frac{e^{ik_0 r_{ij}}}{r_{ij}}k_0^2\left(1 - \frac{1}{ik_0 r_{ij}}\right)\hat{r}_{ij}, \tag{12.177}$$

where

$$
\hat{r}_{ij} = \begin{bmatrix} 0 & -\hat{r}_{ij}^z & \hat{r}_{ij}^y \\ \hat{r}_{ij}^z & 0 & -\hat{r}_{ij}^x \\ -\hat{r}_{ij}^y & \hat{r}_{ij}^x & 0 \end{bmatrix}
$$

denotes the Cartesian components of the unit directional vector $\hat{r}_{ij} = r_{ij}/|r_{ij}|$.

Summarizing, the 6×6 subblock of the full Green's matrix defined by matrix (12.135) assumes the following explicit form:

$$
\overset{\leftrightarrow}{G}_{ij} = \begin{bmatrix} \overset{\leftrightarrow}{G}_{ij}^{ee} & \overset{\leftrightarrow}{G}_{ij}^{eh} \\ \overset{\leftrightarrow}{G}_{ij}^{he} & \overset{\leftrightarrow}{G}_{ij}^{hh} \end{bmatrix}
$$
$$
= \begin{bmatrix} \frac{2}{3} i k_0^3 G_{ij} & -\frac{e^{ik_0 r_{ij}}}{r_{ij}} k_0^2 \left(1 - \frac{1}{ik_0 r_{ij}}\right) \hat{r}_{ij} \\ \frac{e^{ik_0 r_{ij}}}{r_{ij}} k_0^2 \left(1 - \frac{1}{ik_0 r_{ij}}\right) \hat{r}_{ij} & \frac{2}{3} i k_0^3 G_{ij} \end{bmatrix}.
$$

(12.178)

In order to be consistent with the notation of equation (12.161), we must multiply the matrix $\overset{\leftrightarrow}{G}_{ij}$ by the factor $3/2ik_0^3$ such that

$$
\frac{3}{2} \frac{1}{ik_0^3} \overset{\leftrightarrow}{G}_{ij} = \begin{bmatrix} G_{ij}^{ee} & G_{ij}^{eh} \\ G_{ij}^{he} & G_{ij}^{hh} \end{bmatrix},
$$

(12.179)

where $G_{ij}^{ee} = G_{ij}^{hh} = G_{ij}$ are defined by equation (12.161), while the off-diagonal terms $G_{ij}^{eh} = -G_{ij}^{he}$ become

$$
G_{ij}^{he} = \frac{3}{2} \frac{e^{ik_0 r_{ij}}}{ik_0 r_{ij}} \left(1 - \frac{1}{ik_0 r_{ij}}\right) \hat{r}_{ij}.
$$

(12.180)

Equation (12.179) allows one to analyze the spectral properties of an arbitrary array of scatterers beyond the usual electric dipole approximation and extend the Green's matrix spectral method to take into account both first-order Mie–Lorentz coefficients. In this more general approach, each particle is described by two dipoles (electric and magnetic dipole). It is instructive to discuss here the effect of the magnetic dipole term on the light localization properties of point scatterers arranged in a planar GA Vogel spiral geometry, which was recently shown to support a delocalization–localization transition for vector waves [349]. We recall that in this situation the resonant electromagnetic field is not only spatially confined in the plane due to the aperiodic geometry of the geometrical support of the array, but it also leaks out into the far-field with a characteristic time scale proportional to the quality factor of the scattering resonance.

Our results are summarized in Figure 12.11, where we considered the optical density $\rho\lambda^2 = 10$. We compare in Figure 12.11a,b the complex eigenvalue distributions obtained by numerically diagonalizing the non-Hermitian matrices (12.161) and (12.179), respectively. Figure 12.11a shows two well-separated spectral regions, near $Re[\Lambda]$ equal to -2 and 0 respectively, that are populated by a significant fraction

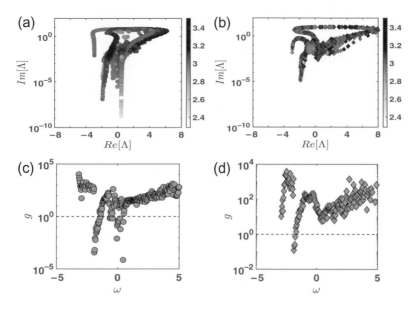

Figure 12.11 (a) Complex eigenvalue distributions of Green's matrix for 1,000 electric dipoles arranged in a golden-angle Vogel spiral geometry. (b) Complex eigenvalue distributions of the electromagnetic Green's matrix for 1,000 electric and magnetic dipoles arranged in a golden-angle Vogel spiral geometry. (c,d) Thouless number g as a function of the frequency ω produced by the complex eigenvalues distributions reported in panel (a,b), respectively.

of the scattering resonances with very small decay rates ($\Gamma_n / \Gamma_0 \ll 1$). On the other hand, we observe that the inclusion of the magnetic contribution, which is coupled with its electric counterpart, not only reduces the fraction of long-lived scattering resonances around $Re[\Lambda] = -2$, but also completely suppresses the localization properties when $Re[\Lambda] \approx 0$. This reflects the fact that the scattering resonances generated by matrix (12.179) leak out from the plane of the array with much smaller characteristic lifetimes compared to the ones produced by matrix (12.161). This "localization loss" effect is better quantified by studying the behavior of the *Thouless number* as a function of ω reported in Figure 12.11c,d and evaluated as follows [476]:

$$g(\omega) = \frac{\overline{\delta\omega}}{\Delta\omega} = \frac{\overline{(1/\Im[\Lambda_n])^{-1}}}{\Re[\Lambda_n] - \Re[\Lambda_{n-1}]} \qquad (12.181)$$

where the symbol $\overline{\{\cdots\}}$ indicates the average of g over different frequency sub-intervals [9, 349, 389]. The Thouless number is a parameter that characterizes the degree of spectral overlap between different optical scattering resonances. In order to demonstrate significant wave localization effects, the Thouless number must decrease below the value 1 when increasing the scattering strength.

The presence of the magnetic dipole introduces an additional loss channel resulting in the reduction of the Thouless number g and the strength of light–matter interaction in the GA Vogel spiral. However, Figure 12.11d shows that there still exists a spectral range in which $g < 1$, unveiling the importance of structural correlations in deterministic aperiodic photonic media to achieve strong light–matter coupling even beyond the traditional electric dipole approximation.

In concluding this chapter, we point out that important experiments on cold atomic gases nicely demonstrate similar scattering and localization phenomena to the ones discussed in this section [698, 699]. Indeed, cold atomic systems can be accurately modeled using a scalar point scatterer model similar to the one introduced in this chapter and provide a formidable model for the study of novel collective effects and phenomena enabled by the multiple-scattering regime.

13 Analytical Scattering Theory

Problems are solved not through vague generalities, or picturesque descriptions of the relation between man and the world, but through technical work.

The Rise of Scientific Philosophy, Hans Reichenbach

In the previous Chapter, starting from Maxwell's wave equations for scalar and vector fields, we established a very general framework to rigorously formulate the solution of multiple scattering problems by arbitrary objects and aggregates in terms of the system's transition operator, or the T-matrix. In particular, we showed how the general multiple-scattering solution for a collection of particles can be expressed in terms of the single-particle T-matrix operators of each individual particle. However, we have not yet shown how the T-matrix can be explicitly obtained. In the following sections, we address rigorous analytical approaches for the computation of the T-matrix of multisphere clusters with arbitrary geometries and nonspherical particles. In particular, we will discuss semianalytical techniques, such as the generalized Mie theory GMT) and the null-field method (NFM), which enable the solution of light scattering by aggregates of particles of arbitrary size and composition. The chapter opens with a pedagogical overview of the chief analytical tool in light-scattering theory, i.e., the Mie scattering of an homogeneous sphere, with emphasis on polariton resonances in small metal and dielectric nanoparticles. Finally, we present applications to metamaterials, resonant field design in clusters of metal nanoparticles, as well as photonic bandgap structures with deterministic aperiodic geometries.

13.1 The Mie–Lorenz Scattering Theory

The problem of vector wave scattering and absorption by spherical particles of arbitrary dimensions and compositions was formally solved in 1908 by Gustav Mie. In his classic paper [700], he applied Maxwell's electromagnetic theory in an attempt to explain the intriguing coloration effects displayed by colloidal suspensions of small gold particles in water, later interpreted in terms of surface plasmon resonances. Since then, the electromagnetic scattering by an homogeneous and isotropic sphere is commonly referred to as the *Mie theory*. However, it is worth noting that earlier studies on elasticity by the German mathematician Alfred Clebsch [701] and on electrodynamics by the Danish physicist Ludvig Lorenz [702] independently contributed to addressing this problem, as well as the 1909 seminal work on radiation pressure by spherical particles by the Dutch-American physicist (and Nobel laureate in Chemistry) Peter Debye [703]. For all these reasons, in the current literature on light scattering the Mie theory is sometimes referred to as the Mie–Lorenz theory or, more correctly, as the Lorenz–Mie–Debye theory. Readers interested in the long and fascinating history of light scattering and absorption by spherical particles are encouraged to read the recent account by Hergert and Wriedt [704]. In this section, following the treatment and conventions provided in [281], we will introduce the main concepts of the Mie scattering theory and discuss some illustrative examples of scattering/absorption of light by spherical nanoparticles and their aggregates. Finally, we will introduce the fundamental ideas behind the GMT and the superposition T-matrix method [627, 705, 706].

13.1.1   Separation of Variable Solution

The key idea behind the Mie theory is the formulation of a vector wave problem in terms of a scalar wave problem, whose solution can be obtained much more easily using the well-known separation of variables technique in spherical coordinates. To appreciate this fact, let us start by remembering that if $(\mathbf{E},\mathbf{H})$ is a physically acceptable solution of Maxwell's equations in a source-free, linear, isotropic, and homogeneous medium, then the fields $\mathbf{E}$ and $\mathbf{H}$ must satisfy the following vector wave equations:

$$\nabla^2 \mathbf{E} + k^2 \mathbf{E} = 0 \tag{13.1}$$

$$\nabla^2 \mathbf{H} + k^2 \mathbf{H} = 0, \tag{13.2}$$

where $k = \omega\sqrt{\epsilon\mu}$. In addition, the fields must be solenoidal (i.e., divergence-free) and coupled to each other by the following dynamic Maxwell's equations:

$$\nabla \times \mathbf{E} = j\omega\mu\mathbf{H} \tag{13.3}$$

$$\nabla \times \mathbf{H} = -j\omega\epsilon\mathbf{E}. \tag{13.4}$$

We then introduce a scalar function ψ that generates the vector fields $(\mathbf{M},\mathbf{N})$ that automatically satisfy all the required properties of the acceptable solutions of the vector wave equations. This is achieved by defining the auxiliary vector fields $\mathbf{M}$ and $\mathbf{N}$ in terms of the yet-unknown scalar function ψ as follows:

$$\boxed{\mathbf{M} = \nabla \times (\mathbf{r}\psi)} \tag{13.5}$$

$$\boxed{\mathbf{N} = \frac{\nabla \times \mathbf{M}}{k}} \tag{13.6}$$

where $\mathbf{r}$ is the position radius vector in a spherical reference system. We note that by definition the vector fields $\mathbf{M}$ and $\mathbf{N}$ are divergence-free and that the curl of $\mathbf{N}$ is proportional to $\mathbf{M}$ while the curl of $\mathbf{M}$ is proportional to $\mathbf{N}$. Moreover, $\mathbf{N}$ satisfies the vector wave equation and using elementary vector identities, we can prove that $\mathbf{M}$ satisfies the following equation:

$$\nabla^2 \mathbf{M} + k^2 \mathbf{M} = \nabla \times \left[\mathbf{r} (\nabla^2 \psi + k^2 \psi) \right].$$ (13.7)

In addition, $\mathbf{M}$ can alternatively be written as $\mathbf{M} = -\mathbf{r} \times \nabla \psi$, which implies that $\mathbf{M}$ is perpendicular to the vector $\mathbf{r}$. Therefore, as we can appreciate from (13.7), $\mathbf{M}$ satisfies the vector wave equation when ψ satisfies the scalar wave equation, and the solution of the original vector equation is reduced to the one of the associated scalar problem. In the context of the Mie theory, the scalar function ψ is called the *generating function* for the vector fields $\mathbf{M}$ and $\mathbf{N}$[1] are the *vector spherical harmonics*. We also note that vector spherical harmonics can be generated by choosing an arbitrary vector $\mathbf{v}$ in place of the position vector $\mathbf{r}$. However, the choice of the so-called *pilot vector* is dictated by the specific symmetry of the problem at hand. Since Mie scattering is formulated in spherical symmetry (i.e., scattering by a sphere), an obvious choice is to choose $\mathbf{r}$ as the pilot vector, which implies that the generated $\mathbf{M}$ satisfies the vector wave equation in spherical polar coordinates.

Considering an arbitrary sphere of radius $r = a$ centered at the origin of a spherical reference system, we can express the scalar Helmholtz wave equation in spherical polar coordinates as follows:

$$\frac{1}{r^2} \frac{\partial}{\partial r} \left(r^2 \frac{\partial \psi}{\partial r} \right) + \frac{1}{r^2 \sin \theta} \frac{\partial}{\partial \theta} \left(\sin \theta \frac{\partial \psi}{\partial \theta} \right) + \frac{1}{r^2 \sin^2 \theta} \left(\frac{\partial^2 \psi}{\partial \phi^2} \right) + k^2 \psi = 0$$

(13.8)

The particular solutions of equation (13.8) are called *spherical wave functions* and can be obtained using the method of separation of variables, as explained later. Let us first assume that the solution can be expressed in the following product form:

$$\psi(\mathbf{r}) = \psi(r, \theta, \phi) = R(r) \Theta(\theta) \Phi(\phi).$$ (13.9)

Substituting this product form into the equation (13.8), dividing the entire equation by $R(r)\Theta(\theta)\Phi(\phi)$, and multiplying it by $r^2 \sin^2 \theta$, we immediately obtain the following:

$$\frac{\sin^2 \theta}{R} \frac{d}{dr} \left(r^2 \frac{dR}{dr} \right) + \frac{\sin \theta}{\Theta} \frac{d}{d\theta} \left(\sin \theta \frac{d\Theta}{d\theta} \right) + \frac{1}{\Phi} \frac{d^2 \Phi}{d\phi^2} + k^2 r^2 \sin^2 \theta = 0.$$ (13.10)

Now observe that the third term of the last equation contains the derivative with respect to the azimuthal angle ϕ while all other terms are independent of the angle ϕ. Therefore, in order for the equation to hold true, the third term itself cannot depend on ϕ and must be a constant. If we indicate this constant by $-m^2$, we can write the azimuthal equation:

$$\frac{d^2 \Phi}{d\phi^2} + m^2 \Phi = 0,$$ (13.11)

---

[1] Note that alternative definitions of the vector spherical harmonics can be found in the literature, as, for instance, in [132].

whose solutions consist of linear combinations of even and odd trigonometric functions $\Phi_e = \cos m\phi$ and $\Phi_o = \sin m\phi$, respectively. Moreover, the requirement that ψ is a single-valued function of the azimuthal angle ϕ necessarily forces m to assume integers or zero values, and positive values of m are sufficient to generate all the linearly independent solutions of (13.11).

If we now insert this separation into equation (13.10) and divide it by $\sin^2 \theta$, we find the following:

$$\frac{1}{R}\frac{d}{dr}\left(r^2\frac{dR}{dr}\right) + \frac{1}{\Theta \sin\theta}\frac{d}{d\theta}\left(\sin\theta\frac{d\Theta}{d\theta}\right) + k^2 r^2 - \frac{m^2}{\sin^2\theta} = 0. \tag{13.12}$$

By applying the same logic as before, since the first and the third terms of equation (13.12) depend only on r while the other two terms depend on θ, we can further separate using the arbitrary constant $n(n + 1)$ into the following two equations:

$$\frac{1}{R}\frac{d}{dr}\left(r^2\frac{dR}{dr}\right) + k^2 r^2 = n(n + 1) \tag{13.13}$$

$$\frac{1}{\Theta \sin\theta}\frac{d}{d\theta}\left(\sin\theta\frac{d\Theta}{d\theta}\right) - \frac{m^2}{\sin^2\theta} = -n(n + 1), \tag{13.14}$$

which we can rewrite as follows:

$$\frac{d}{dr}\left(r^2\frac{dR}{dr}\right) + \left[k^2 r^2 - n(n + 1)\right]R = 0 \tag{13.15}$$

$$\frac{1}{\sin\theta}\frac{d}{d\theta}\left(\sin\theta\frac{d\Theta}{d\theta}\right) + \left[n(n + 1) - \frac{m^2}{\sin^2\theta}\right]\Theta = 0 \tag{13.16}$$

The radial equation (13.15) is known as the *spherical Bessel equation*, whose two linearly independent solutions are denoted by $j_n(kr)$ and $y_n(kr)$, or the nth-order *spherical Bessel functions of the first and second kind*, respectively. The spherical Bessel functions are related to the standard (cylindrical) *Bessel functions* of first and second kind J_m and Y_m with half-integer order $m = n + 1/2$ according to the following identities:

$$j_n(kr) = \sqrt{\frac{\pi}{2kr}}J_{n+\frac{1}{2}}(kr) \tag{13.17}$$

$$y_n(kr) = \sqrt{\frac{\pi}{2kr}}Y_{n+\frac{1}{2}}(kr) \tag{13.18}$$

Despite their quite complicated oscillatory appearance, these functions and their derivatives satisfy simple recurrence relations and possess a number of very useful properties, as discussed in [366]. However, for our purpose it is sufficient to remember

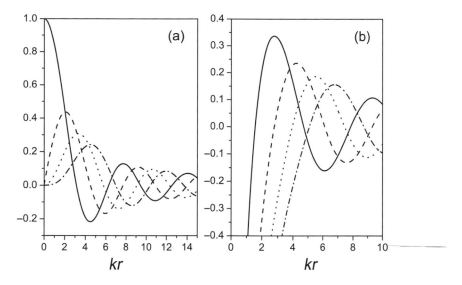

Figure 13.1 (a) Bessel functions of the first kind: $j_0(kr)$ (solid line), $j_1(kr)$ (dashed line), $j_2(kr)$ (dotted line), and $j_3(kr)$ (dash-dotted line). (b) Bessel functions of the second kind: $y_0(kr)$ (solid line), $y_1(kr)$ (dashed line), $y_2(kr)$ (dotted line), $y_3(kr)$ (dash-dotted line).

that $y_n(kr) \to -\infty$ when $kr \to 0$, while $j_n(kr)$ is regular (i.e., it remains finite) when $kr \to 0$. Other fundamental solutions of (13.15) are obtained by considering linearly independent combinations of the spherical Bessel functions, the so-called spherical Bessel functions of the third kind or *spherical Hankel functions*:

$$h_n^{(1)}(kr) = j_n(kr) + j y_n(kr) \tag{13.19}$$

$$h_n^{(2)}(kr) = j_n(kr) - j y_n(kr), \tag{13.20}$$

where j denotes, as usual, the imaginary unit.

The general solutions of equation (13.15) can therefore be obtained in terms of linear combinations of the spherical Bessel functions:

$$R(r) = a_n j_n(kr) + b_n y_n(kr), \tag{13.21}$$

where a_n and b_n are constant coefficients that depend on the specific nature of the problem or on the specific choice of the four spherical Bessel functions j_n, y_n, $h_n^{(1)}$, and $h_n^{(2)}$, generally denoted by $z_n(kr)$. Illustrative plots of Bessel functions are shown in Figure 13.1.

Let us now discuss the solutions of equation (13.16), known as *Legendre's equation*. Its two linearly independent solutions, denoted by $P_n^m(\cos\theta)$ and $Q_n^m(\cos\theta)$, are respectively the *associated Legendre functions of the first and second kind* of order m and degree n. For an integer value of the degree n, the associated Legendre function

$P_n^m(\cos\theta)$ is also called the *associated Legendre polynomial*, which can be expressed as follows:

$$P_n^m(x) = (-1)^m (1-x^2)^{\frac{m}{2}} \frac{d^m}{dx^m} P_n(x), \tag{13.22}$$

where $x = \cos\theta$ and $P_n(x)$ is called *Legendre polynomial*:

$$P_n(x) = \frac{1}{2^n n!} \frac{d^n}{dx^n}(x^2-1)^n \tag{13.23}$$

which is a polynomial of order n. Clearly $P_n^m(\cos\theta) = 0$ when $m > n$.

The general solution of equation (13.16) is provided by a linear combination of these two solutions:

$$\Theta(\theta) = c_{mn} P_n^m(\cos\theta) + d_{mn} Q_n^m(\cos\theta), \tag{13.24}$$

where c_{mn} and d_{mn} are constant coefficients that depend on the nature of the problem. The Legendre functions have also many special properties [366], and it is important to remember that for $\theta = 0, \pi$ only $P_n^m(\cos\theta)$ is regular while $Q_n^m(\cos\theta) \to \infty$.

Based on the knowledge of the preceding solutions, we can finally write down the general solution of the spherical Helmholtz equation (13.8) as follows:

$$\psi(r,\theta,\phi) = \sum_{mn} \big[a_n j_n(kr) + b_n y_n(kr)\big] \times \big[c_{mn} P_n^m(\cos\theta) + d_{mn} Q_n^m(\cos\theta)\big]$$
$$\times \big[e_m \cos m\phi + f_m \sin m\phi\big], \tag{13.25}$$

where the sum is extended over all possible values of the indices m and n and the Hankel functions could have also been used in place of j_n and y_n just depending on the nature of the specific problem.

A key point is that the functions $\cos m\phi$, $\sin m\phi$, $P_n^m(\cos\theta)$, and $z_n(kr)$ are an *orthogonal and complete* set of basis functions. This implies that any other function that is a solution of the homogeneous spherical wave equation in spherical polar coordinates can be expanded as an infinite series over the set of functions:

$$\psi_{emn} = \cos(m\phi) P_n^m(\cos\theta) z_n(kr) \tag{13.26}$$

$$\psi_{omn} = \sin(m\phi) P_n^m(\cos\theta) z_n(kr), \tag{13.27}$$

where the subscripts e and o denote *even* and *odd* functions.[2]

The vector spherical harmonics generated by ψ_{emn} and ψ_{omn} can then finally be obtained, according to equations (13.5) and (13.6), as follows:

$$\mathbf{M}_{emn} = \nabla \times (\mathbf{r}\psi_{emn}) \qquad\qquad \mathbf{M}_{omn} = \nabla \times (\mathbf{r}\psi_{omn}) \tag{13.28}$$

$$\mathbf{N}_{emn} = \frac{\nabla \times \mathbf{M}_{emn}}{k} \qquad\qquad \mathbf{N}_{omn} = \frac{\nabla \times \mathbf{M}_{omn}}{k}. \tag{13.29}$$

These vector functions are the basic building blocks that can be used to expand any other vector function, as we discuss in the following subsection.

---

[2] Since the solution region includes both the positive z-axis as well as the negative z-axis, i.e., the points $\theta = 0, \pi$ are included, we must omit $Q_n^m(\cos\theta)$ because it is singular at these points.

13.1.2 Field Expansions

Any solution of the vector wave equation can be expanded in an infinite series of the vector functions introduced earlier, which are the *normal modes of the spherical particle*. As discussed in this subsection, this very remarkable property allows us to readily solve the problem of vector field scattering by an arbitrary sphere using essentially a generalization of Fourier analysis on the surface of a sphere.

In particular, let us consider an incident plane wave $\mathbf{E}_i = E_0 e^{jkr \cos \theta} \hat{\mathbf{e}}_x$ propagating along the z-direction and polarized along the arbitrary x-direction. This wave can be expanded in vector spherical harmonics as follows:

$$\mathbf{E}_i = \sum_{m=0}^{\infty} \sum_{n=m}^{\infty} (B_{emn}\mathbf{M}_{emn} + B_{omn}\mathbf{M}_{omn} + A_{emn}\mathbf{N}_{emn} + A_{omn}\mathbf{N}_{omn}), \quad (13.30)$$

where, because of the orthogonality of the vector spherical harmonics, the expansion coefficients can be explicitly obtained as follows:

$$B_{emn} = \frac{\int_0^{2\pi} \mathbf{E}_i \cdot \mathbf{M}_{emn} \sin\theta d\theta d\phi}{\int_0^{2\pi} |\mathbf{M}_{emn}|^2 \sin\theta d\theta d\phi}, \quad (13.31)$$

and similar expressions hold for B_{omn}, A_{emn}, and A_{omn}. Moreover, it can be deduced from the orthogonality of the sine and cosine functions and from the component form of the vector spherical harmonics that $B_{emn} = A_{omn} = 0$ for all m and n and that the remaining expansion coefficients vanish unless $m = 1$. Thus the expansion for the incident field can be written as follows:

$$\mathbf{E}_i = \sum_{n=1}^{\infty} \left(B_{o1n}\mathbf{M}_{o1n}^{(1)} + A_{e1n}\mathbf{N}_{e1n}^{(1)} \right) \quad (13.32)$$

where we appended the superscript (1) to the vector spherical harmonics to indicate that the radial dependence of the generating functions ψ_{o1n} and ψ_{e1n} is specified by the Bessel function of the first kind (i.e., $j_n(kr)$), since the incident field must be finite at the origin. It turns out based on the integral expression (13.31) that the expansion coefficients for the incident plane wave can be analytically computed [281] and are given by the following:

$$B_{o1n} = j^n E_0 \frac{2n+1}{n(n+1)} \quad (13.33)$$

$$A_{e1n} = -j E_0 j^n \frac{2n+1}{n(n+1)}. \quad (13.34)$$

The expansion of the incident plane wave in spherical harmonics is therefore:

$$\mathbf{E}_i = E_0 \sum_{n=1}^{\infty} j^n \frac{2n+1}{n(n+1)} \left(\mathbf{M}_{o1n}^{(1)} - j\mathbf{N}_{e1n}^{(1)} \right) \quad (13.35)$$

The corresponding incident magnetic field can be easily obtained from Maxwell's equations taking the curl of the electric field (i.e., $\nabla \times \mathbf{E} = j\omega\mu\mathbf{H}$) and remembering that $\nabla \times \mathbf{M} = k\mathbf{N}$ and $\nabla \times \mathbf{N} = k\mathbf{M}$, resulting in the following:

$$\mathbf{H}_i = \frac{-k}{\omega\mu} E_0 \sum_{n=1}^{\infty} j^n \frac{2n+1}{n(n+1)} \left(\mathbf{M}_{e1n}^{(1)} + j\mathbf{N}_{o1n}^{(1)} \right). \tag{13.36}$$

We can similarly expand the field internal to the scattering sphere (i.e., the internal field) $(\mathbf{E}_{int}, \mathbf{H}_{int})$ in spherical vector harmonics as follows:

$$\mathbf{E}_{int} = \sum_{n=1}^{\infty} E_n \left(c_n \mathbf{M}_{o1n}^{(1)} - j d_n \mathbf{N}_{e1n}^{(1)} \right) \tag{13.37}$$

$$\mathbf{H}_{int} = \frac{-k_{int}}{\omega\mu_{int}} \sum_{n=1}^{\infty} E_n \left(d_n \mathbf{M}_{e1n}^{(1)} + j c_n \mathbf{N}_{o1n}^{(1)} \right) \tag{13.38}$$

where $E_n = j^n E_0 (2n+1)/n(n+1)$. In the region outside the sphere, both the spherical Bessel functions $j_n(kr)$ and $y_n(kr)$ are regular and can be used to expand the scattered field $(\mathbf{E}_s, \mathbf{H}_s)$. However, it is more convenient to use here their linear combinations, or the Hankel functions $h_n^{(1)}(kr)$ and $h_n^{(2)}(kr)$, since their asymptotic behavior corresponds to outgoing and incoming spherical waves, respectively.[3] The expansion of the scattered field can therefore be written as follows:

$$\mathbf{E}_s = \sum_{n=1}^{\infty} E_n \left(j a_n \mathbf{N}_{e1n}^{(3)} - b_n \mathbf{M}_{o1n}^{(3)} \right) \tag{13.39}$$

$$\mathbf{H}_s = \frac{k}{\omega\mu} \sum_{n=1}^{\infty} E_n \left(j b_n \mathbf{N}_{o1n}^{(3)} + a_n \mathbf{M}_{e1n}^{(3)} \right) \tag{13.40}$$

where we used the superscript (3) on the vector spherical harmonics to indicate that the radial dependence of the generating functions is specified by $h_n^{(1)}(kr)$. For each n, the part of the scattered field specified by the $\mathbf{N}^{(3)}$ corresponds to *electric-type modes*, or TM modes, while the part that is specified by the functions $\mathbf{M}^{(3)}$ correspond to *magnetic-type modes*, or TE modes, characterized by vanishing radial components (i.e., they are purely transverse modes in spherical coordinates). In the preceding *multipolar expansion* of the scattered field, each multipole contributes to scattering and absorption by electric and magnetic modes, corresponding to the excitation of *surface-polariton modes* and *eddy currents*. The term $n = 1$ describes dipole fields, $n = 2$ quadrupolar, $n = 3$ octupolar fields, and so on. Note, however, that when the magnetic permeability of the sphere is equal to one (i.e., no magnetic material response), magnetic multipoles can still arise due to electronic excitations in the form of induced current loops that depend on the size and dielectric function of the sphere, as we have seen already in the previous chapter when we considered arrays of

---

[3] Note that if the alternative frequency convention for monochromatic time-harmonic waves is used (i.e., $e^{j\omega t}$), then the role of the two Hankel functions is reversed and $h_n^{(1)}(kr)$ would correspond to an incoming spherical wave.

small Si nanoparticles. A purely magnetic dipole scattering without the involvement of an electric dipole response has been recently demonstrated in a nonmagnetic subwavelength nanoparticle by spectrally overlapping the magnetic dipole resonance with a nonradiative anapole [660]. Electric and magnetic mode interference is also exploited by optical metasurfaces and Huygens surfaces to engineer directional radiation diagrams [707, 708].

The component form of the vector spherical harmonics corresponding to magnetic- and to electric-type modes can be written concisely as follows [281]:

$$\mathbf{M}_{o1n} = \cos\phi\,\pi_n(\mu)z_n(\rho)\hat{\mathbf{e}}_\theta - \sin\phi\,\tau_n(\mu)z_n(\rho)\hat{\mathbf{e}}_\phi \tag{13.41}$$

$$\mathbf{M}_{e1n} = -\sin\phi\,\pi_n(\mu)z_n(\rho)\hat{\mathbf{e}}_\theta - \cos\phi\,\tau_n(\mu)z_n(\rho)\hat{\mathbf{e}}_\phi \tag{13.42}$$

$$\mathbf{N}_{o1n} = n(n+1)\sin\phi\sin\theta\,\pi_n(\mu)\frac{z_n(\rho)}{\rho}\hat{\mathbf{e}}_r + \sin\phi\,\tau_n(\mu)\frac{[\rho z_n(\rho)]'}{\rho}\hat{\mathbf{e}}_\theta \tag{13.43}$$

$$+ \cos\phi\,\pi_n(\mu)\frac{[\rho z_n(\rho)]'}{\rho}\hat{\mathbf{e}}_\phi$$

$$\mathbf{N}_{e1n} = n(n+1)\cos\phi\sin\theta\,\pi_n(\mu)\frac{z_n(\rho)}{\rho}\hat{\mathbf{e}}_r + \cos\phi\,\tau_n(\mu)\frac{[\rho z_n(\rho)]'}{\rho}\hat{\mathbf{e}}_\theta \tag{13.44}$$

$$- \sin\phi\,\pi_n(\mu)\frac{[\rho z_n(\rho)]'}{\rho}\hat{\mathbf{e}}_\phi,$$

where $\mu = \cos\theta$, $\rho = kr$, and the prime indicates differentiation with respect to the argument in parentheses. We remark again that the superscripts (1) or (3) are appended to the $\mathbf{M}$ and $\mathbf{N}$ to denote the type of spherical Bessel functions z_n used. In particular, the superscript (1) is associated with $j_n(k_{int}r)$ and (3) with $h_n^{(1)}(kr)$. We note that the function $\mathbf{M}$ has no radial component and that, for sufficiently large values of kr, the radial component of $\mathbf{N}$ for the scattered field is negligible compared to its transverse component. In the expressions (13.41–13.44), there appear also the polar functions $\pi_n(\mu)$ and $\tau_n(\mu)$ that describe the *angle-dependence of the scattering process* defined as follows:

$$\pi_n = \frac{P_n^1}{\sin\theta} \tag{13.45}$$

$$\tau_n = \frac{dP_n^1}{d\theta} \tag{13.46}$$

These functions can be computed recursively from the following relations:

$$\pi_n = \frac{2n-1}{n-1}\mu\pi_{n-1} - \frac{n}{n-1}\pi_{n-2} \tag{13.47}$$

$$\tau_n = n\mu\pi_n - (n+1)\pi_{n-1}, \tag{13.48}$$

starting from the initial values $\pi_0 = 0$ and $\pi_1 = 1$. Polar plots of the functions $\pi_n = 0$ and τ_n plotted on the 0–360° range for $n = 3 - 5$ are shown in Figure 13.2. The angle-dependent functions $\pi_n(\theta)$ and $\tau_n(\theta)$ are alternatively even and odd functions of μ and assume both positive and negative values. More interestingly, when n increases, the number of their angular lobes also increases and the forward-directed

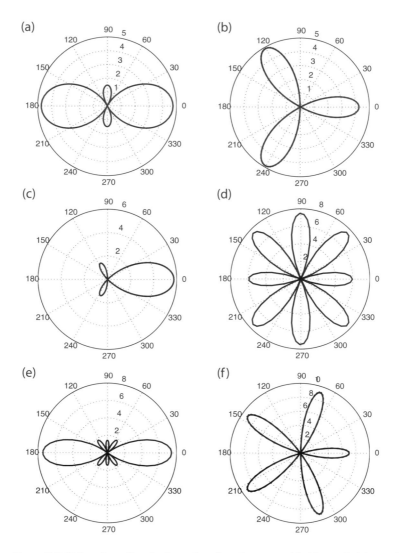

Figure 13.2 Polar plots of angle-dependent functions π_n with (a) $n = 3$, (c) $n = 4$, and (c) $n = 5$ and τ_n with (b) $n = 3$, (d) $n = 4$, and (f) $n = 5$.

lobe becomes narrower. In addition, note that all the functions have forward-directed lobes (i.e., they assume positive values in the forward direction), but the backward lobes disappear for alternating values of n (i.e., these functions assume negative values for backward directions). Due to the behavior of these functions, we understand that spheres of increasing size will give rise to scattering prevalently in the forward direction with narrower and narrower peaks, due to the cancellation of the positive and negative values of $\pi_n(\theta)$ and $\tau_n(\theta)$ that occurs in the backscattering direction. This peculiar forward-scattering behavior is the hallmark of Mie scattering.

Let us now introduce the *Riccati–Bessel functions*:

$$\psi_n(x) = x j_n(x) \tag{13.49}$$

$$\xi_n(x) = x h_n^{(1)}(x). \tag{13.50}$$

Based on the expressions (13.41)–(13.44) and the Riccati–Bessel functions, we can explicitly write down all the components of the electromagnetic field, which we report in Appendix H.

13.1.3 The Mie–Lorenz Coefficients

The original vector wave problem has now been reduced to find the expressions for the coefficients (a_n, b_n, c_n, d_n). Note that for a given multipolar order n, there are four unknown coefficients to be determined. Therefore, they can be obtained by solving four independent equations (for each multiple), which are provided by enforcing the electromagnetic boundary conditions on the surface of the sphere (i.e., $r = a$):

$$E_{i,\theta} + E_{s,\theta} = E_{int,\theta} \qquad E_{i,\phi} + E_{s,\phi} = E_{int,\phi} \tag{13.51}$$

$$H_{i,\theta} + H_{s,\theta} = H_{int,\theta} \qquad H_{i,\phi} + H_{s,\phi} = H_{int,\phi}. \tag{13.52}$$

Using the orthogonality properties of the vector harmonics, it is possible to solve for the scattering and the internal field coefficients in closed form. If we introduce the *size parameter x* and the *relative refractive index m* as follows:

$$x = ka = \frac{2\pi N a}{\lambda} \tag{13.53}$$

$$m = \frac{k_{int}}{k} = \frac{N_{int}}{N}, \tag{13.54}$$

where N_{int} denotes the refractive index of the sphere and N the refractive index of the host medium, the *Mie–Lorenz coefficients* can be written as follows:

$$a_n = \frac{m\psi_n(mx)\psi_n'(x) - \psi_n(x)\psi_n'(mx)}{m\psi_n(mx)\xi_n'(x) - \xi_n(x)\psi_n'(mx)} \tag{13.55}$$

$$b_n = \frac{\psi_n(mx)\psi_n'(x) - m\psi_n(x)\psi_n'(mx)}{\psi_n(mx)\xi_n'(x) - m\xi_n(x)\psi_n'(mx)} \tag{13.56}$$

$$c_n = \frac{-jm}{\psi_n(mx)\xi_n'(x) - m\xi_n(x)\psi_n'(mx)} \tag{13.57}$$

$$d_n = \frac{-jm}{m\psi_n(mx)\xi_n'(x) - \xi_n(x)\psi_n'(mx)} \tag{13.58}$$

In these expressions, the prime indicates differentiation with respect to the argument in parentheses and we have employed the Wronskian relation for the Riccati- Bessel functions $\psi_n(x)\xi_n'(x) - \psi_n'(x)\xi_n(x) = -j$ and considered a non-magnetic sphere. Note that the denominators of the coefficients (a_n, d_n) and (b_n, c_n) are identical. It is also important to realize that if there is a value of the size parameter (i.e., a wavelength or a sphere radius) such that, for a given mode order n, the denominators of the Mie–Lorenz coefficients are very small (or even vanish), then the corresponding normal modes n will dominate the optical response of the sphere and produce resonant scattering effects. It is generally useful to think about the Mie–Lorenz coefficients as complex response functions of independent oscillators whose zero and pole structure determines the interesting resonant features of the radiating system [709].

The frequencies at which the denominators of the Mie–Lorenz coefficients exactly vanish are called the *natural frequencies of the sphere*. These frequencies in general are complex, and called *virtual frequencies*. However, if their imaginary parts are small compared to the real ones, they approximately indicate the frequencies at which an incident wave excites the different electromagnetic modes of the sphere.

All the relevant synthetic field parameters, such as the absorption and scattering cross sections as well as the components of the scattering matrix, can readily be expressed in terms of the known Mie–Lorenz coefficients. In particular, the cross sections are obtained by integrating the analytical far-fields over the imaginary spherical surface that surrounds the scattering particle. Simple closed-form results can be achieved exploiting the orthogonality of the involved functions. In particular, the scattering and extinction cross sections are given by the following [281]:

$$\sigma_{sc} = \frac{2\pi}{k^2} \sum_{n=1}^{\infty} (2n+1) \left(|a_n|^2 + |b_n|^2 \right) \tag{13.59}$$

$$\sigma_{ext} = \frac{2\pi}{k^2} \sum_{n=1}^{\infty} (2n+1) \Re\{a_n + b_n\} \tag{13.60}$$

(where $\Re\{\cdot\}$ denotes the real part of the argument in parentheses) and satisfy the usual requirement $\sigma_{ext} = \sigma_{abs} + \sigma_{sc}$, where σ_{abs} denotes the absorption cross section. The scattering and extinction efficiencies Q_{sc} and Q_{ext} are defined by dividing the corresponding cross sections by the area πa^2 of the sphere.

It is instructive to look at the calculated extinction efficiency shown in Figure 13.3, which corresponds to a dielectric particle in air. Since there is no absorption, the extinction efficiency features a remarkably rich structure consisting of regularly spaced broad oscillations, known as *interference structure*, and an irregular structure of sharper peaks called the *ripple structure*. Note that the interference structure oscillates around the value 2, which is the correct asymptotic limit predicted by scalar diffraction theory [281]. This value is twice as large as what is predicted by geometrical optics in the absence of the unavoidable sharp-edge diffraction effects, motivating what historically became known as the *extinction paradox*. This puzzling result can be understood, and the apparent paradox completely eliminated, using a combination of geometrical optics and rigorous diffraction theory [281]. The interference structure of the extinction efficiency can be explained by the interference between the incident light and the forward-scattered light [281], which is fundamentally connected with extinction, as we already discussed in Section 6.3.1. On the other hand, the ripple structure, which can be observed in weakly absorbing and monodisperse spheres, originates from the resonance behavior of the Mie–Lorenz scattering coefficients associated to the excitation of electromagnetic normal

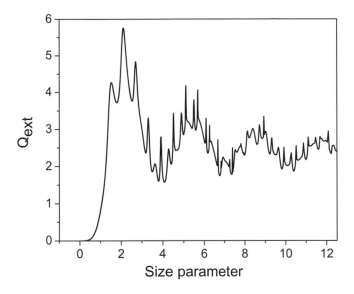

Figure 13.3 Calculated extinction efficiency for a dielectric sphere of refractive index $n = 2$ embedded in air as a function of the size parameter $x = 2\pi a/\lambda$.

modes. Since the coefficients (a_n, d_n) and (b_n, c_n) share the same denominator, both light scattering and light absorption inside the sphere are sharply peaked at its virtual resonance frequencies, giving rise to the ripple structure. The presence of an interference and a ripple structure and the asymptotic limit of the extinction efficiency (equal to 2) are general features of single scattering by weakly absorbing particles that can also be observed in objects with irregular shapes and non-spherical particles (e.g., spheroids).

An important parameter in scattering theory is the *backscattering efficiency*, which is defined as the differential scattering cross section into a unit solid angle around the backscattering direction (i.e., $\theta = 180°$). The backscattering efficiency Q_b can be obtained based on the knowledge of the far-field transverse components, which determine the scattering matrix, and is given in closed form in terms of the Mie–Lorenz scattering coefficients [281]:

$$Q_b = \frac{1}{x^2} \left| \sum_{n=1}^{\infty} (2n + 1)(-1)^n (a_n - b_n) \right|^2 \tag{13.61}$$

Calculation results showing the wavelength spectra of the backscattering efficiencies of gold and silver nanospheres of increasing radii are shown in Figure 13.4. Interestingly, for both materials an optimal range of particle size exists where the backscattering efficiencies are maximized and the spectral width of the backscattering attains a minimum value. Moreover, we observe that the spectral shape of the backscattering is distinctively different from the one of the scattering and extinction spectra.

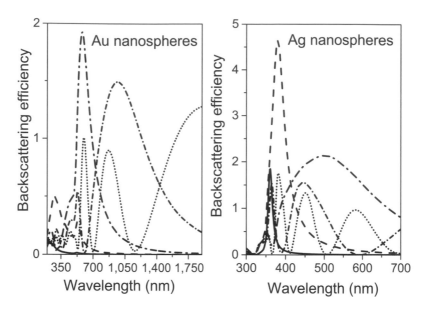

Figure 13.4 Calculated backscattering efficiency of gold and silver nanoparticles for different values of their radius a. In both cases, the solid lines correspond to $a = 20$ nm, dash lines to $a = 40$ nm, dash-dot lines to $a = 80$ nm, short-dash-dot to $a = 160$ nm, and short-dot to $a = 320$ nm. The materials dispersion data (i.e., complex refractive indices) utilized are taken from [112]

13.1.4   Nanoplasmonics and Surface Modes

The interest in the Mie theory and its generalizations has been rekindled in the last few decades thanks to plasmonics and metamaterials technologies, where Mie theory calculations often provide accurate and efficient predictive tools for the design of resonant scattering phenomena [97–99, 139, 140, 710, 711]. Therefore, it is important to address in more detail some of the most current applications of the Mie theory in relation to the resonant optical properties of small (i.e., subwavelength) nanoparticles. In particular, subwavelength metallic particles can exhibit fascinating optical properties that originate from the excitation of *surface modes* that can dramatically increase the efficiency of light absorption and scattering processes. A similar behavior can also occur in certain dielectric materials when excited in a frequency region of negative permittivity. It should be clear from our previous discussion in Section 3.1 that negative permittivity regions always exist in optical materials with dispersion data phenomenologically described by the Drude–Sommerfeld model, such as many noble metals, as well as by the Lorentz model for insulators, such as silicon carbide (SiC). To better understand the resonant optical phenomena in such materials, we look at the conditions for the vanishing of the denominators of the Mie–Lorenz coefficients (13.55) and (13.56). For instance, the condition that the denominator of a_n vanishes is given by the following:

$$m\psi_n(mx)\xi'_n(x) - \xi_n(x)\psi'_n(mx) = 0. \tag{13.62}$$

Using the power series expansions of the spherical Bessel functions of order n as provided in [281] and assuming a nonmagnetic sphere, it is possible to obtain a very simple expression for the vanishing a_n coefficient in the limit of particles with small size parameter:

$$m^2 = -\frac{n+1}{n}$$ (13.63)

where $n = 1, 2, \ldots$ is the mode number. On the other hand, we see that in the limit $x \to 0$ there is no possibility for the b_n coefficients to vanish. For the real frequencies that approximately satisfy (13.63), the Mie–Lorenz scattering coefficient a_n will be very large and there will appear a maximum (i.e., a resonance) in the corresponding optical cross sections. Moreover, the radial dependence of the normal modes associated to these resonant frequencies is characterized by radial (internal) electric fields that scale as r^{n-1}. Therefore, higher-order field solutions will be strongly localized near the surface of the sphere. This is the reason why such modes are termed surface modes of the sphere. For a sufficiently small sphere, so that only the $n = 1$ dipole mode can be considered, we rewrite the expression (13.63) in terms of the complex dielectric function $\epsilon = \epsilon' + j\epsilon''$:

$$\epsilon'(\omega) = -2\epsilon_m$$ (13.64)

where ϵ_m is the dielectric constant of the embedding medium and we have assumed that the sphere is weakly absorbing (i.e., $\epsilon'' \cong 0$). The frequency at which equation (13.64) is exactly satisfied is called the *Fröhlich frequency* ω_F, and the associated normal mode is the Fröhlich mode of the sphere. It is interesting to realize that at the Fröhlich frequency the real refractive index of the sphere vanishes and its extinction coefficient (i.e., the imaginary part of the complex refractive index) assumes the value $\kappa = \sqrt{2}$, as the reader can easily verify. Therefore, in the context of the localized dipole ($n = 1$) oscillations of metallic nanoparticles, the vanishing refractive index is a direct consequences of the Fröhlich oscillation condition.

Assuming negligible damping losses in a metallic nanosphere with Drude dispersion and using equation (13.64), we obtain the Fröhlich frequency:

$$\omega_F = \frac{\omega_p}{\sqrt{1 + 2\epsilon_m}}$$ (13.65)

At approximately the Fröhlich frequency, both the absorption and scattering cross sections of a small spherical particle are significantly enhanced by the internal field resonance associated to the oscillation of conduction electrons. Therefore, in metallic nanoparticles, Fröhlich modes always reflect the excitation of resonant collective oscillations of surface electrons bound at the metal–dielectric interface, known as **localized surface plasmon resonances** (LSPRs). The approximate relation (13.64) shows that a resonant response can be achieved only in small spheres composed of a material with negative dielectric function, since $\epsilon_m > 0$ for an embedding dielectric

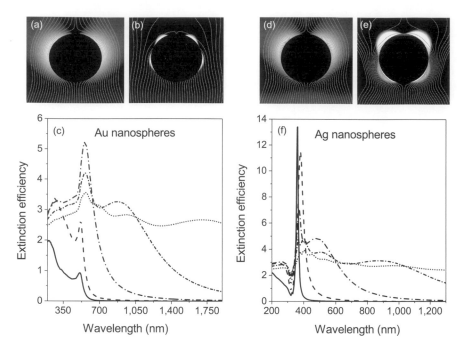

Figure 13.5 (a,b) and (d,e) show the scattered electric field distributions of the dipole and quadrupole resonances of Au and Ag nanospheres with radius a = 80 nm, respectively. (c,f) show the extinction efficiencies for increasing values of a. Solid lines a = 20 nm, dash lines a = 40 nm, Dash-dot lines a = 80 nm, short-dash-dot lines a = 160 nm, short-dot a = 320 nm. Materials dispersion data from [112].

medium. As we have discussed in Section 3.1, noble metals provide unique opportunities to engineer such resonances over large frequency spectra, since their permittivity is negative for all the frequencies that are smaller than the plasma frequency, which typically occurs across the visible and near-infrared spectral range. This is the reason why metal nanoparticles support resonant spectra that can be largely tuned over broadband frequency spectra by varying their sizes, shapes, and dielectric environment. The science and engineering of subwavelength localized resonant fields and enhanced light–matter interactions using metallic nanostructures are the primary focus of the field of *nanoplasmonics* [97–99, 712].

In Figure 13.5, we show the calculated extinction efficiency for gold (Au) and silver (Ag) nanospheres of increasing size. We note that in both cases a resonant peak appears that shifts toward longer wavelengths as the radius of the sphere is increased. This is the dipolar resonance due to the localized surface plasmon (LSP) excitation in small particles. Due to the reduced optical losses of Ag compared to Au, the peak is significantly narrower in the case of Ag nanospheres. When the radius of the spheres is approximately larger than 40 nm, the overall extinction spectra sizably broaden due to the contributions of higher-order electric modes that appear at shorter wavelengths. At the same time, the dipolar resonant peaks significantly red-shift. However, in the

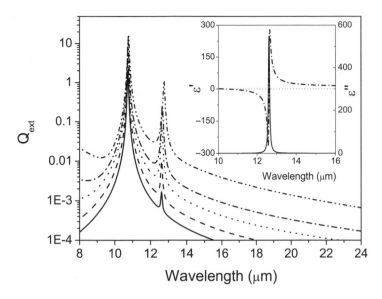

Figure 13.6 Calculated extinction efficiencies of α-SiC nanospheres with progressively increasing values of radius a. The solid line corresponds to $a = 50$ nm, the dash line to $a = 100$ nm, the dotted line to $a = 200$ nm, the dash-dot to $a = 300$ nm, dash-dot-dot to $a = 500$ nm. The material dispersion data (i.e., complex refractive indices) utilized for the calculation are taken from [121]. The inset shows the real and imaginary parts of the relative dielectric function of the bulk α-SiC.

case of dielectric nanospheres (e.g., Si) of diameter comparable with the radiation wavelength in the medium, magnetic mode resonances start to contribute as well [662]. These effects can be further enhanced using multicoated spheres where an additional degree of freedom, the fraction of the total particle volume occupied by the core, is available to control their resonant response [281, 660, 709, 713]. The marked size- and environment-sensitivity of the optical cross sections of subwavelength metal particles is at the origin of their numerous technological applications in nanoplasmonics [97–99]. Let us look more closely at how the relation (13.64) can give rise to spectacular consequences for dielectric media whose optical constants are described by the Lorentz dispersion model for bound electrons. As we know already from our discussion in Section 3.1, a generic Lorentz medium always features a range of negative permittivity for frequencies $\omega_T < \omega < \omega_L$, where ω_T denotes the transverse oscillation frequency and ω_L the longitudinal oscillation frequency. Within this (usually narrow) infrared range known as the *Reststrahlen band*, phonon-polariton modes can be excited where electromagnetic waves strongly couple to the charged lattice vibrations in an ionic crystal. Analogously to the case of surface electron oscillations in metals, small spherical particles composed of dielectric materials such as SiC or MgO can support *localized surface phonon polaritons* (LSPPs) whenever equation (13.64) is satisfied, giving rise to strongly resonant optical responses. This effect is illustrated in Figure 13.6, where we show the calculated extinction efficiency of α-SiC nanospheres with progressively increasing values of radius a.

The one-oscillator dielectric response of this material is provided by equation (3.9), where ω_p corresponds lattice vibrations (ionic oscillators) rather than electronic ones. The inset in the figure shows the real and imaginary parts of the dielectric function that feature a very high-quality resonance around 12.6 μm. Using the Lorentz dispersion (with negligible damping losses) in the condition (13.64), we deduce the Fröhlich frequency for the localized polaritons case:

$$\omega_F^2 = \omega_T^2 \frac{\epsilon_v + 2\epsilon_m}{\epsilon_b + 2\epsilon_m} \tag{13.66}$$

where $\epsilon_v = \epsilon_b + \omega_p^2/\omega_T^2$ is the finite value assumed by the dielectric function (3.9) at frequencies smaller than the transverse optical one. This frequency accurately predicts the position of the lowest-order surface mode of the insulating nanosphere. Note that at resonance ($\omega \approx \omega_F$), the extinction efficiency can be several orders of magnitude larger than the one at $\omega = \omega_T$ and several times bigger than the geometrical cross section. We can also appreciate that the surface mode peak shifts to longer wavelengths as the size of the sphere is increased, in close analogy with the surface plasmon case but with significantly reduced tunability. Moreover, the peak broadens as higher-order surface modes begin to contribute to the overall extinction spectrum. Finally, differently from the surface plasmon counterpart, in the polariton case a series of modes appear slightly below $\omega = \omega_T$. These *low-frequency modes* have the same nature of the *ripple modes* previously shown in Figure 13.3 for a much larger dielectric sphere, i.e., they are the virtual resonances of the spherical particle. In the case of a small sphere of α-SiC, such modes can already be appreciated even at small size parameters because in a Lorentz medium the refractive index is very large immediately below $\omega = \omega_T$.

13.1.5   Scattering from Magnetic Spheres

The light-scattering properties of small magnetic spheres have recently attracted great interest due to the possibility of obtaining zero backscattering from nanostructures when the relative dielectric permittivity and magnetic permeability satisfy the following condition [714]:

$$\epsilon_r = \mu_r \tag{13.67}$$

where $\epsilon_r = \epsilon_1/\epsilon_0$ and $\mu_r = \mu_1/\mu_0$. More generally, the engineering of the magnetic response of a sphere offers the intriguing possibility to achieve highly asymmetric forward-scattering to backscattering ratios and to modify the degree of polarization of the scattered field. The vanishing of the backscattering can be immediately understood by inspecting the general expressions for the Mie–Lorenz scattering coefficients for a magnetic sphere [281]:

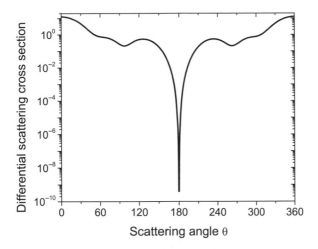

Figure 13.7 Differential scattering cross section of a magnetic sphere with $\epsilon_r = \mu_r = (1.33)^2$. The radius of the sphere is $r = 260$ nm and the incident wavelength is $\lambda = 550$ nm.

$$a_n = \frac{\mu m^2 j_n(mx)[xj_n(x)]' - \mu_1 j_n(x)[mxj_n(mx)]'}{\mu m^2 j_n(mx)[xh_n^{(1)}(x)]' - \mu_1 h_n^{(1)}(x)[mxj_n(mx)]'} \tag{13.68}$$

$$b_n = \frac{\mu_1 j_n(mx)[xj_n(x)]' - \mu j_n(x)[mxj_n(mx)]'}{\mu_1 j_n(mx)[xh_n^{(1)}(x)]' - \mu h_n^{(1)}(x)[mxj_n(mx)]'} \tag{13.69}$$

Observing that $m^2 = \epsilon_r$, we recognize that the condition $\epsilon_r = \mu_r$, known as the *Kerker condition*, implies $a_n = b_n$. Remembering the backscattering efficiency in equation (13.61), we deduce that it vanishes at the Kerker condition, as shown in Figure 13.7. This is a general result valid for spheres of arbitrary size under plane-wave excitation. The same conclusion applies to any axially symmetric particle that is illuminated along its axis of symmetry [714].

In addition to the modification of the angular profile of scattered radiation discussed previously, the Kerker condition also significantly influences the polarization properties of the scattered radiation. This fact can be understood by remembering the expressions for the scattered intensity associated to the transverse and longitudinal components of the scattered field, which are as follows:

$$I_\perp = I_0 \left(\frac{\lambda^2}{4\pi^2 r^2} \right) |S_1(\cos\theta)|^2 \sin^2\phi \tag{13.70}$$

$$I_\| = I_0 \left(\frac{\lambda^2}{4\pi^2 r^2} \right) |S_2(\cos\theta)|^2 \cos^2\phi, \tag{13.71}$$

where $I_\perp$ and $I_\|$ are the scattered intensities perpendicular and parallel to the scattering plane, associated to the ϕ and θ components of the scattered electric field, respectively. Moreover, I_0 is the incident intensity, and ϕ is the angle between the electric field vector of the incident wave and the scattering plane. The nonzero elements of the *amplitude scattering matrix* are as follows [281]:

$$S_1 = \sum_{n=1}^{\infty} \frac{2n+1}{n(n+1)} [a_n \pi_n(\cos\theta) + b_n \tau_n(\cos\theta)] \tag{13.72}$$

$$S_2 = \sum_{n=1}^{\infty} \frac{2n+1}{n(n+1)} [a_n \tau_n(\cos\theta) + b_n \pi_n(\cos\theta)] \tag{13.73}$$

We can immediately realize that $S_1 = S_2$ at the Kerker condition, which implies that the state of polarization of the scattered and incident radiation will be the same irrespective of the scattering angle (i.e., the angle between the incident wavevector and the one of the scattered wave). This situation is very different from the well-known case of the scattering from a small nonmagnetic sphere in which, for a scattering angle $\theta = 90°$, the radiation is completely polarized perpendicularly to the scattering plane.

Additional intriguing effects can be observed in a magnetic sphere, such as the vanishing of the forward-scattering field, which is determined by the amplitude scattering matrix evaluated in the forward ($\theta = 0$) direction [281]:

$$S_1(0°) = S_2(0°) = S(0°) = \frac{1}{2} \sum_{n=1}^{\infty} (2n+1)(a_n + b_n) \tag{13.74}$$

Therefore, the vanishing of the forward-scattering intensity is achieved whenever $a_1 = -b_1$. If we consider a small magnetic sphere, we can use the small argument expansion of the dipolar Mie–Lorenz coefficients a_1 and b_1 and show that zero-forward scattering will be achieved when the optical constants of the sphere are related according to the following [714]:

$$\epsilon_r = \frac{4 - \mu_r}{2\mu_r + 1}. \tag{13.75}$$

In summary, while the scattering intensity from a small dielectric sphere is always symmetric around the scattering angle $\theta = 90°$, for the more general case of a magnetic sphere, a largely tunable asymmetric scattering can be obtained with preferential backscattering or forward scattering achieved under specific conditions on the optical constants [714]. Finally, we point out that in small dielectric nanoparticles with large enough refractive index (e.g., in silicon), it is possible to obtain significant magnetic dipole response, which is important for applications to all-dielectric resonant meta-optics and nanophotonics devices. In this context, the engineering of Mie resonances allows one to achieve the Kirker condition for nonmagnetic, high-index materials[4] and offers novel strategies for the enhancement of optical effects near magnetic and

---

[4] We point out that the traditional Kerker conditions cannot be achieved for magnetic particles at optical frequencies as $\mu_r \approx 1$ for high frequencies.

electric multipolar resonances, giving rise to a variety of interference phenomena in the emerging field of "Mie-tronics" [663, 715, 716].

13.1.6   Mie Theory and Metamaterials Design

The Mie scattering theory can be employed to engineer the effective optical constants of metamaterials composed of lattices of spheres [133, 134, 717]. Following this approach, coated nonmagnetic spheres with a negative index of refraction at infrared frequencies have been demonstrated [717].

In general, the problem of the homogenization of a scattering medium into an effective bulk one, or a metamaterial, is a very complex challenge that can only be rigorously addressed within the general framework of multiple scattering theory, discussed later in Chapter 14. However, the task of obtaining effective bulk optical parameters simplifies considerably when considering metamaterials made of small dipolar spheres at modest density and even closed-form expressions for the effective dielectric function and magnetic permeability of metamaterials can be found. This approach relies on a straightforward generalization of the Clausius–Mossotti equation derived in Chapter 3 to spheres of finite size [717, 718]. In a small sphere, the electric and magnetic dipole resonances are determined by the poles of the a_1 and b_1 Mie–Lorenz scattering coefficients. To these coefficients, it is possible to associate an effective electric polarizability α_e and magnetic polarizability α_m according to the following [717, 718]:

$$\alpha_E = \frac{6\pi j a_1}{k_0^3} \tag{13.76}$$

$$\alpha_M = \frac{6\pi j b_1}{k_0^3}. \tag{13.77}$$

Note that these expressions depend on the radius of the spheres through implicit dependence on the size parameter in the scattering coefficients. Substituting the new polarizability expressions into the standard Clausius–Mossotti equation (see Chapter 3) allows us to obtain simple expressions for the effective (relative) permittivity $\bar{\epsilon}_r$ and permeability $\bar{\mu}_r$ in the following form [717]:

$$\boxed{\begin{aligned} \bar{\epsilon}_r &= \frac{k_0^3 + 4\pi j N a_1}{k_0^3 - 2\pi j N a_1} \\[2mm] \bar{\mu}_r &= \frac{k_0^3 + 4\pi j N b_1}{k_0^3 - 2\pi j N b_1} \end{aligned}} \tag{13.78}$$

$$\tag{13.79}$$

where N is the volume density of the spheres of radius r_0, and the volume filling fraction of the metamaterial is $f = 4\pi N r_0^3/3$.

The preceding expressions for the effective permittivity and permeability completely determine the response of the bulk composite or metamaterial. They can

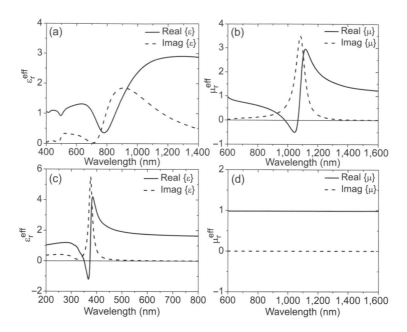

Figure 13.8 (a,b) Metamaterial made of Si nanospheres with radius $r = 140$ nm. The unit cell length is $a = 300$ nm, and the filing fraction is $f = 0.425$. (c,d) Metamaterial made of Ag nanospheres with radius $r = 25$ nm and unit cell length $a = 75$ nm. The filing fraction is $f = 0.155$.

assume almost arbitrary values and depend on the frequency of the wave as well as the radius, density, and composition of the nonmagnetic spheres. However, only small values of the filling fractions should be considered in this approach since large values require corrections due to higher-order multipole terms leading to complicated structure-dependent lattice sums. Negative permittivity metamaterials using dielectric spheres were recently demonstrated following this approach [719] as well as implementations using polaritonic crystals at infrared and optical frequencies [717]. An example of metamaterials homogenization using dielectric and metallic Mie resonators is shown in Figure 13.8, where we considered dense arrays of Si and Ag nanospheres.

This simple method of homogenization can also be applied to more complex structures such as a collection of nonmagnetic coated spheres. In contrast to the previously discussed lattice of simple spheres, an assembly of coated spheres is particularly interesting for the realization of negative index metamaterials because both a negative permeability and a negative permittivity can be obtained at the same frequency [717]. In particular, the core of a spherical particle can be tuned to provide a magnetic dipole resonance characterized by $\bar{\mu}_r \ll 0$ while the coating can be engineered to provide an electric dipole resonance with $\bar{\epsilon}_r$. The appropriate Mie–Lorenz scattering coefficients for this coated sphere geometry are well known [281, 709], and the effective media optical constants for the homogenized coated spheres medium can be found as before by substituting the coefficients into the equations (13.78) and (13.79). Therefore, using the Mie scattering theory, it is possible to obtain an effective medium approach that accurately predicts the effective dispersion of sphere and coated-sphere composites in the long-wavelength limit. More general approaches for the homogenization of core–shell structures are discussed in [134].

13.2 Two-Dimensional Generalized Mie Theory

The general Foldy–Lax multiple-scattering framework allows us to naturally extend the Mie–Lorenz scattering theory to arbitrary aggregates of multiple cylinders or spheres. The resulting computational method is known as the generalized Mie theory (GMT). We introduce in this section the main ideas behind the multiple-scattering theory of arbitrary aggregates of cylinders with identical circular cross sections, or the two-dimensional (2D) GMT technique, and we discuss its three-dimensional (3D) extension in Section 13.3. The 2D GMT is a semianalytical method for solving the frequency-domain Maxwell's equations for arbitrary arrays of parallel, infinite-length, and nonoverlapping cylinders embedded in a nonabsorbing medium. The field solutions are invariant along the z-axis, making this a 2D method. A brief overview of the method is given here, though many details of the derivation can be found, for instance, in the excellent tutorial by Gagnon and Dube [720]. It should be noted that the representation used here includes interior and exterior dipole sources as originally introduced by Asatryan et al. [721]. Since we are dealing with a 2D problem, the field solutions can be represented into two distinct polarization bases: the transverse magnetic (TM), where the electric field $\mathbf{E}$ is oriented in the $\hat{z}$ direction while the magnetic field $\mathbf{H}$ is in the xy-plane, and the transverse electric (TE), in which the two fields are swapped. Regardless of polarization, all the relevant manipulations are performed on the z-oriented field, and the other components follow directly from Maxwell's equations. We simply denote the $\hat{z}$-directed field by a scalar function that will correspond to E_z for the TM polarization and with H_z for the TE polarization.

Similarly to the single-particle Mie theory, the incident, scattered, and internal electric and magnetic fields are expanded as an infinite sum of a complete set of basis functions. For the case of cylindrical aggregates, these basis functions are cylindrical Bessel functions. The exterior field $\varphi^e(\mathbf{r})$ exists only outside of the cylinders and is written as the sum of the incident and scattered fields, as follows:

$$\varphi^e(\mathbf{r}) = \varphi_0^E(\mathbf{r}) + \varphi_{sc}^E(\mathbf{r}) \tag{13.80}$$

$$\varphi_0^E(\mathbf{r}) = \sum_{l=-\infty}^{\infty} a_{nl}^{0E} J_l(k_o \rho_n) e^{jl\theta_n} \tag{13.81a}$$

$$\varphi_{sc}^E(\mathbf{r}) = \sum_{n=1}^{N} \sum_{l=-\infty}^{\infty} b_{nl} H_l^{(+)}(k_o \rho_n) e^{jl\theta_n}. \tag{13.81b}$$

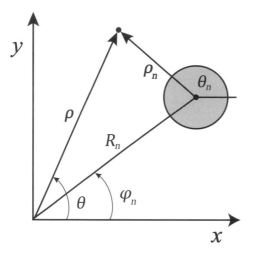

Figure 13.9 Geometric description of cylindrical aggregate coordinate systems. Polar coordinates (ρ_n, θ_n) are centered on the nth cylinder, while the polar coordinates (ρ, θ) are global.

Here $k_o = 2\pi/\lambda$ is the free-space wavenumber, and ρ_n is the the radial coordinate with respect to the nth cylinder, as indicated in Figure 13.9. The incident field $\varphi_0^E(\mathbf{r})$, which is the field produced by some specified source outside of the array, is expressed as a sum of Bessel functions of the first kind centered on the nth cylinder with expansion coefficients a_{nl}^{0E}. Exterior sources include a plane wave, a complex source beam (CSB), or a dipole radiator. Similarly, the scattered field φ_{sc}^E is written as the sum of Hankel functions of the first kind centered on each cylinder. These functions represent outward propagating fields of various angular orders l, emanating from each cylinder. We note that the phase of these functions take the form $e^{jk_o\rho}$ in the far-field, consistently with the $\exp(-j\omega t)$ time convention. The expansion coefficients b_{nl} are the polarization-dependent Mie coefficients that contain the solution of the given problem.

The field interior to the nth cylinder is similarly expanded as the sum of fields scattered from the cylinder boundary plus a contribution from any source within that cylinder:

$$\varphi_n^I(\mathbf{r}) = \varphi_{0,n}^I(\mathbf{r}) + \varphi_{sc,n}^I(\mathbf{r}) \tag{13.82}$$

$$\varphi_{0,n}^I(\mathbf{r}) = \sum_{l=-\infty}^{\infty} a_{nl}^{0I} H_l^{(+)}(k_n\rho_n)e^{jl\theta_n} \tag{13.83a}$$

$$\varphi_{sc,n}^I(\mathbf{r}) = \sum_{l=-\infty}^{\infty} c_{nl} J_l(k_o\rho_n)e^{jl\theta_n}. \tag{13.83b}$$

The expansion coefficients a_{nl}^{0I} refer to a source inside the nth cylinder.

The Mie theory is most often applied to light-scattering problems, in which case the source terms are well known, e.g., a plane-wave incident from infinity, and one would like to the find the associated interior and scattered fields. Generally, the coefficients a_{nl}^{0I} and a_{nl}^{0E} are known, and the goal is to calculate the unknown coefficients b_{nl} and c_{nl} by enforcing Maxwell's boundary conditions on the surface of each cylinder. The key to this method is writing the exterior field in terms of Bessel and Hankel functions centered on the nth cylinder only:

$$\varphi_n^E(\mathbf{r}) = \sum_l \left[a_{nl} J_l(k_o \rho_n) + b_{nl} H_l^{(+)}(k_o \rho_n) \right] e^{jl\theta_n}. \tag{13.84}$$

This is achieved by utilizing *Graf's addition theorem* [366], which enables the transformation of cylindrical Bessel functions from one frame of reference to another. This theorem is applied to rewrite the Hankel functions radiating from one cylinder in a coordinate system centered on any other specified cylinder. Once all fields are written in radial coordinates centered on any one cylinder, the boundary conditions can be applied in a straightforward manner, yielding a relationship between the known and the unknown Mie coefficients. The final result of this process consists of the linear system:

$$\mathbf{T}\mathbf{b} = \mathbf{a}_0 \tag{13.85}$$

with $\mathbf{a}_0 = \{s_{nl} a_{nl}^{0E}\}$, $\mathbf{b} = \{b_{nl}\}$, and $\mathbf{T} = \{T_{nn'}^{ll'}\}$. The transition matrix $\mathbf{T}$ and vector $\mathbf{a}_0$ contain all of the geometric, material, and source information of the scattering problem. Inverting the $\mathbf{T}$ matrix, whose explicit form can be found in [720], allows one to compute the scattering coefficients vector $\mathbf{b}$ knowing the incident ones. Additional details on the theory and implementation of the 2D GMT method can be found in our cited references [720, 722]

Despite the arbitrary geometrical arrangement of the cylinders, the scattering cross section of the aggregate can be calculated from the Mie coefficients in closed form as follows [723]:

$$\sigma_s = \frac{2}{\pi k_o} Re \sum_{\substack{nn' \\ ll'}} b_{nl} b_{n'l'}^* J_{l-l'}(k_o R_{nn'}) e^{j(l-l')\phi_{nn'}}. \tag{13.86}$$

Here $R_{nn'}$ and $\phi_{nn'}$ are the distance and angle between the n and n' cylinders. The terms $J_{l-l'}(k_o R_{nn'}) e^{j(l-l')\phi_{nn'}}$ account for the interference between fields radiated from different cylinders. If only one cylinder exists, then the preceding formula reduces to the familiar scattering cross section for a single cylinder [281]:

$$\sigma_s = \frac{2}{\pi k_o} \sum_l |b_l|^2. \tag{13.87}$$

13.2.1 Applications to Resonant Field Design

The 2D GMT method turns out to be very useful for the study the resonant field behavior of complex metal–dielectric nanostructures. Being a mesh-free spectral method,

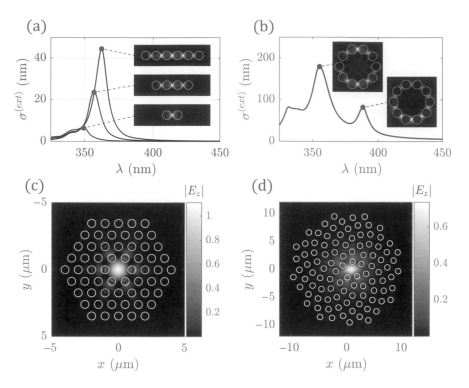

Figure 13.10 (a) TE scattering cross section of Ag nanoparticle chains with 2, 4, and 6 particles, with center-to-center separation $d_0 = 5$ nm and radii $r = 20$ nm. Insets show the electric field intensity $|E|^2$ at the peak wavelength of each spectrum. (b) TE scattering cross section of 10 Ag nanoplasmonic necklace with $d_0 = 6$ nm. Insets show the electric field intensity at the resonance peaks. The excitation for (a) and (b) is a plane wave traveling in the $+\hat{y}$ direction. (c) A localized TE mode of a triangular lattice (with hexagonal termination) with a central defect. The localized mode occurs at $k = 2.95 - j5.74 \times 10^{-5}$ (μm^{-1}). (d) Localized TE mode of a GA Vogel spiral at $k = 1.74 - j4.63 \times 10^{-5}$ (μm^{-1}). The white circles indicate the boundaries of the nanocylinders.

its accuracy is practically unmatched and its numerical efficiency can be significantly improved using advanced iterative algorithms combined with suitable T-matrix pre-conditioners [720, 724]. In what follows, we demonstrate its application to the design of "plasmonic molecules" and to resonant field localization in dielectric photonic structures. In particular, Figure 13.10a shows the computed spectra of the extinction cross section for chains of Ag nanoparticles with fixed separation $d_0 = 5$ nm and radii $r = 20$ nm and with a variable number of particles $N = 2, 4, 6$. In this case, a TE plane wave is normally incident with respect to the axis of the chain and with an electric field polarization along it, which excites the so-called *longitudinal modes* of the chain. By increasing the chain's length, its extinction cross section increases and its peak significantly red-shifts due to the resonant excitation of longitudinal plasmon modes. The inset displays the spatial distributions of the electric field intensity at the peak wavelength of each extinction spectrum, demonstrating strong nanoscale field confinement between the particles.

In recent years, near-field coupling effects in clusters of metallic nanoparticles have been the subject of intense research activities in plasmonics and nano-optics because they result in localized plasmon excitations that can "hop across" neighboring particles and transmit optical energy across devices with nanoscale footprints [713, 725–727]. More generally, the design of finite nanoparticle clusters, or "plasmonic molecules" or oligomers, provides exciting opportunities for the manipulation of optical fields at the nanoscale that are appealing to future plasmonic devices [98, 673, 728, 729]. Figure 13.10b illustrates the resonant plasmon response of a plasmonic molecule consisting of 10 Ag nanoparticles (25 nm in radius) arranged in a circular geometry with $d_0 = 6$ nm interparticle separation, i.e., a decagon "plasmonic necklace" [730]. In this case, two distinct plasmon resonances are clearly visible in the spectrum of the extinction cross section (for any given incident polarization). Generally, bright modes arise in the far-field scattering/extinction spectra of plasmonic necklaces due to hybridization of optically active (dipole) resonances that are predicted based on group-theoretic arguments. Nanoplasmonic necklaces or oligomers have been studied due to their ability to give rise to controllable Fano-type resonances [731–733], magnetic plasmon response [673, 729], as well as photonic–plasmonic coupling with polarization insensitive near-field enhancement [730]. Moreover, they can be conveniently optimized by choosing diameter of the necklace to be an integer multiple of the incident wavelength, leading to strong diffractive effects that boost near-field intensities and feed concentric nanoantennas for applications to nanoscale optical sensing and spectroscopy [734].

The 2D-GMT method can also be used to determine the resonant modes of a nanocylinder array. Modes are solutions to Maxwell's equations when no source term is present in the equation. In this case, equation (13.85) writes as follows:

$$\mathbf{T}(k)\mathbf{b} = \mathbf{0}. \tag{13.88}$$

Here the functional dependence of $\mathbf{T}$ on the complex wavevector k has been explicitly indicated. The preceding equation only has solutions when $\det[\mathbf{T}(k)] = 0$. Therefore, resonant modes are found by numerically finding complex k values that satisfy this condition. As usual, the interpretation of the complex k is that its real part is equal to the wavenumber of the mode while its imaginary part is associated with the spectral width of the mode, where modes with imaginary parts closer to zero are the most strongly localized ones (i.e., with the highest-quality factors).

A localized TE mode corresponding to a central defect state of a triangular lattice with hexagonal termination is shown in Figure 13.10c. The pillars have relative permittivity $\epsilon_r = 11$, center-to-center separation $d_0 = 1$ μm, and radii $r = 0.3$ μm. This mode has a large quality factor of $Q = \Re(k)/[2 \times \Im(k)] = 2.5 \times 10^5$. On the other hand, Figure 13.10c illustrates a localized TE mode in a GA Vogel spiral with the same permittivity, $d_0 = 1.67$ μm, and $r = 0.5$ μm. In this example, the mode is less confined and has a smaller quality factor of $Q = 3.7 \times 10^4$. Therefore, strong mode localization can be achieved in Vogel spiral geometry even in the absence of structural defects. Despite the fact that defect states in Vogel spirals are not fully understood and deserve additional studies, strong light–matter confinement can be achieved in such structures leading to the recent observations of strong-coupling and quantum electrodynamical effects [738, 739]. We will see in the next section that Vogel spirals with optimized geometries support large photonic bandgaps that open for values of the dielectric contrasts that are comparable to the ones in photonic crystals. However, Vogel spirals display stronger fluctuations in the local density of states associated to the excitation of a fractal spectrum of localized band-edge modes with unusual angular momentum properties [250, 598].

Box 13.1 What Are Photonic Bandgap Structures?

Photonic bandgap structures, or *photonic crystals*, are engineered dielectric materials with a periodic variation of the dielectric permittivity along one or more directions [26]. Photonic crystals are the optical analogues of electronic crystals in which atoms or molecules are replaced by macroscopic units with differing dielectric constants. In particular, when the corresponding refractive index contrast is sufficiently large (and neglecting materials' absorption), wave propagation can be suppressed along certain directions within a specified frequency range, giving rise to a *photonic bandgap* (PBG). These bandgap regions are generally polarization dependent but can become polarization insensitive, or complete, in certain lattice geometries such as the honeycomb lattice of dielectric rods [26, 27]. Moreover, *complete photonic bandgaps* can be more readily achieved in lattices with high rotational symmetry, which makes quasicrystals and more general isotropic structures (e.g., Vogel spirals) particularly attractive. More generally, light propagation in photonic crystals shares many of the properties of electron transport through periodic atomic potentials, which renders photonic crystals research particularly suitable to explore and exploit many fascinating phenomena of solid-state physics using light [735]. In particular, perturbing or breaking the periodicity of photonic crystal structures introduces localized or extended defect states in the photonic bandgaps where optical waves can be resonantly trapped with large quality factors, strongly enhancing light–matter interactions for a number of linear and nonlinear photonic device applications that include optical fibers, compact lasers, on-chip optical filters, and quantum optical components [27, 736, 737].

13.2.2   Applications to Photonic Bandgap Structures

We provide an application of the 2D GMT method to the study of the optical gap properties of aperiodic arrays of dielectric rods with circular cross sections, or nanopillars. In particular, we address the frequency response of finite patches of GA Vogel spiral, Eisenstein, and Gaussian arrays and identify the spectral positions of their TM photonic bandgaps by calculating the local density of states (LDOS) normalized to its free-space value (i.e., the Purcell enhancement of a line source). The periodic honeycomb lattice of dielectric rods is also investigated as a comparison since it is known to support a complete photonic bandgap region [26, 27]. In our simulations, an excitation line source is always placed in the center of the structures. However, the frequency locations and spectral widths of the excited bandgap regions do not appreciably depend on the location of the source.

As we discussed in Chapter 12, the LDOS quantifies the number of electromagnetic modes into which photons can be emitted at a given position in space. This quantity is particularly useful because it is related to experimentally observable quantities, such as transmission gaps and the spontaneous decay rate of embedded light sources inside nonhomogeneous photonic structures [92].

In the case of 2D systems, the LDOS is proportional to the total intensity radiated by the line source [721] and can be rigorously evaluated from the trace of the 2D electric Green's tensor $G^e(\boldsymbol{r}, \boldsymbol{r}')$:

$$\rho(\boldsymbol{r}; \lambda) = -\frac{4n_b^2}{c\lambda} \text{Im Tr}[G^e(\boldsymbol{r}, \boldsymbol{r}; \lambda)], \qquad (13.89)$$

where n_b is the refractive index of the embedding medium. For TE polarization, the trace in the preceding equation reduces to the following:

$$\text{Tr}[G^e(\boldsymbol{r}, \boldsymbol{r}; \lambda)] = G_{xx}^e(\boldsymbol{r}, \boldsymbol{r}; \lambda) + G_{yy}^e(\boldsymbol{r}, \boldsymbol{r}; \lambda). \qquad (13.90)$$

Calculating the LDOS requires solving the equation (13.85) with the input coefficients a_0 corresponding to a line source for both $\hat{x}$ and $\hat{y}$ orientations, and then evaluating the total fields at the position of the line source. In finite-size photonic structures, photonic bandgaps can be readily identified by the appearance of frequency regions with strongly suppressed LDOS.

The Purcell enhancement (PE) describes the modification of the radiative properties of a dipole emitter due to the coupling with the surrounding photonic environment. The Purcell enhancement $F(\boldsymbol{r}; \lambda)$ is evaluated by dividing equation (13.89) by the free-space LDOS in two dimensions, yielding the following:

$$F(\mathbf{r}; \lambda) = \frac{\rho(\boldsymbol{r}; \lambda)}{\rho_0(\boldsymbol{r}; \lambda)} = -4n_b^2 \, \text{Im} Tr[G^e(\boldsymbol{r}, \boldsymbol{r}; \lambda)] \qquad (13.91)$$

Therefore, investigating the LDOS of a localized source in a photonic structure normalized to its free-space value quantifies directly the light emission enhancement or suppression due to light-scattering effects occurring in the structure at the emitter's location. Outside of the bandgap regions, the Purcell enhancement fluctuates around unity, while its abrupt variations correspond to the excitation of optical modes with high-quality factors in the photonic structure.

In Figure 13.11, we show the LDOS maps computed considering a vertical line source located in the center of each structure. These maps are calculated by varying the emission wavelength and the radius of the pillars in order to identify the optimal parameters for the formation of spectral gaps in the investigated systems. All the maps are color-coded according to the values of $\log(LDOS)$. The honeycomb lattice is shown as a reference in Figure 13.11a and displays large spectral gaps. The fundamental gap region and two higher-order ones are clearly visible, shifting to

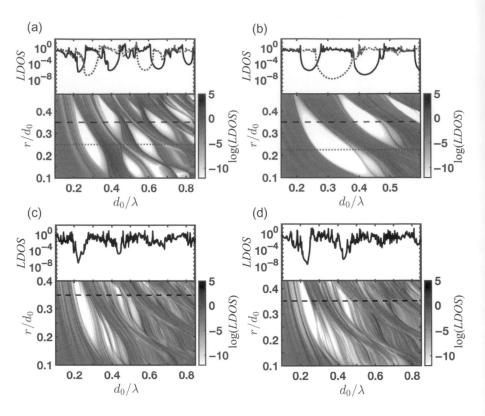

Figure 13.11 LDOS maps of dielectric nanopillars ($\epsilon = 10.5$) arranged in a honeycomb lattice (a), GA Vogel spiral (b), Gaussian prime array (c), and Eisenstein prime array (d) as a function of d_0/λ and for different values of the radius r of the nanopillars normalized with respect to their average center-to-center separation d_0. All structures have approximately 250 nanopillars except the structure in panel (b) that consists of 150 pillars. The 2D–maps are in shades of gray according to $\log(LDOS)$ and evaluated by probing with one vertically polarized (along the axes of the cylinders) dipole located at the center of each structure. The LDOS is normalized with respect to its free-space value. All the results are computed by considering $l = 5$ orders in the multipolar expansion. Insets on the top of each panel show representative 1D cuts of the LDOS plotted in semilogarithmic scale.

higher frequencies as we decrease the separation d_0 between the dielectric pillars. The Vogel spiral case, shown in Figure 13.11b, exhibits broader gap regions that are crossed by sharp resonances. Wide and isotropic TM gaps were originally discovered in GA Vogel spirals by Pollard and Parker [602] and explained, for small-enough refractive index contrast (i.e., in the weakly modulated limit), by the Bragg scattering in the characteristically isotropic Fourier space of the sunflower's spiral geometry. The fundamental structure–property relations of Vogel spiral photonic structures have been rigorously established in [250, 598]. The dashed black line and the dotted lines in Figure 13.11a,b indicate the selected r/d_0 values at which we plot the corresponding LDOS frequency spectra above each map, more clearly demonstrating the broad nature of the isotropic gaps of the Vogel spiral structure. We remark that in the strong

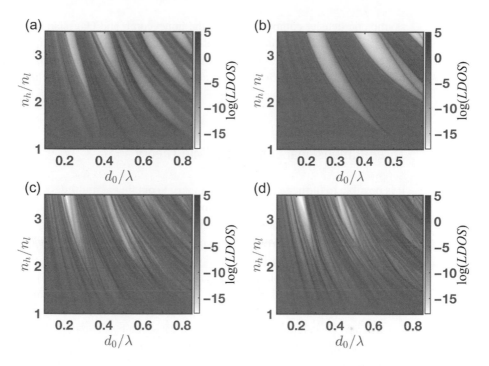

Figure 13.12 LDOS gap map of dielectric nanopillars ($\epsilon = 10.5$) arranged in a honeycomb lattice (a), GA Vogel spiral (b), Gaussian prime array (c), and Eisenstein prime array (d). All structures have approximately 250 nanopillars except the structure in (b) that consists of 150 pillars. These maps are in shades of gray according to $\log{(LDOS)}$, evaluated by probing the structures with a vertically polarized dipole located at the center. These maps are produced by considering $r/d_0 = 0.35$ as highlighted by the dashed line in Figure 13.11 and sweeping the refractive index contrast n_h/n_l. Here, d_0 refers to the average interparticle center-to-center separation. All these results are evaluated by considering $l = 5$ orders in the multipolar expansion.

modulation limit, i.e., when the dielectric permittivity contrast $\Delta \epsilon \gtrsim 12$, the field interaction with the Vogel spiral geometry becomes short range and the photonic response is dominated by the Mie resonances of the pillars, irrespective of the array's geometry. In Figure 13.11c,d, we show the results for the Gaussian and Eisenstein dielectric structures. Interestingly, we observe a characteristic fragmentation of the main gap regions into a series of subgaps separated by a complex distribution of localized modes, as also shown by the one-dimensional LDOS cuts displayed on top that correspond to the dashed line drawn at the radius-pitch ratio $r/d_0 = 0.35$. Focusing on this value, we show in Figure 13.12 the LDOS-based gap maps of the structures, computed by sweeping the refractive index contrast n_h/n_l. The results show that TM bandgaps open at reduced dielectric contrast in the isotropic Vogel spiral geometry compared to all the other investigated structures. Moreover, Gaussian and Eisenstein arrays exhibit a rich hierarchy of gap regions that can be traced back to their multifractal geometry [10]. In particular, since the Eisenstein structure has a larger asymptotic density compared to both the honeycomb and Gaussian ones, it develops TM gaps for smaller values of the refractive index contrast and is a good candidate system for the engineering of complete, i.e., polarization-insensitive, photonic bandgaps with characteristic multifractal scaling based on aperiodic deterministic geometries [740].

13.3   Three-Dimensional Generalized Mie Theory

It is possible to further extend the Mie theory to obtain the multiple-scattering solution of arbitrary clusters of identical homogeneous spheres of any size and composition. We provide in this section a self-contained account of this method following the detailed derivations provided in the papers by Mackowski [705] and Xu [706], to which we refer the reader for additional details.

For general linear problems, the scattered field from an arbitrary array of spheres can be expressed as the superposition of the individual fields scattered from each of them. However, this time the spheres in the aggregate can no longer be considered isolated and we cannot assume that they are excited by the external plane-wave incident on the aggregate. In a cluster of spheres, the electromagnetic field that excites the surface of the generic sphere j (i.e., the local exciting field) has two contributions: (i) one from the externally incident field and (ii) one from the field scattered by all the other spheres in the cluster. Therefore, each sphere is excited by unknown field coefficients that depend on all the other spheres and their positions in the aggregate.

13.3.1   Representing the Local Fields

We discuss the methodology to obtain the unknown expansion coefficients of the local exciting fields that modify the single-particle Mie–Lorenz coefficients due to the coupling among all the spheres. We refer to the geometry shown in Figure 13.13, where we consider N isotropic, homogeneous, and nonoverlapping spheres with radii a^j and known complex refractive indices N^j with $j = 1, \ldots N$. The spheres are embedded in an homogeneous and nonabsorbing medium of index N_0. Each sphere in the system is characterized by the size parameter $x^j = ka^j = 2\pi N^0 a^j / \lambda$ and relative refractive index $m^j = N^j / N^0$. The spherical coordinate system with its origin at the center of the sphere j_0 is referred to as the *primary system*. With respect to this system, the vector position of any other sphere j in the system is denoted by $\mathbf{r}_{j_0, j}$. For any pair of spheres in the aggregate, the relative position vector is defined by $\mathbf{r}_{m,k} = \mathbf{r}_{j_0,k} - \mathbf{r}_{j_0,m}$. The incident beam illuminating the aggregate of spheres is expanded in vector spherical harmonics about the centers of each sphere through known coefficients, and it may have an arbitrary profile. Similarly to the case of the single-particle Mie theory, we need to compute the scattering and internal field coefficients for each sphere in the aggregate, from which all the relevant synthetic scattering parameters (cross sections, scattering amplitude, etc.) will follow.

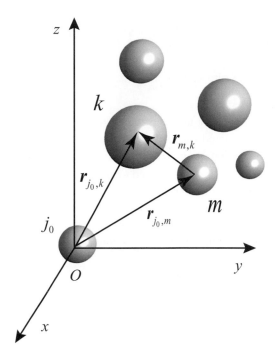

Figure 13.13 Schematics of the multiparticle scattering system used to derive the governing equations of the GMT method for multiple-sphere aggregates. Two arbitrary spheres k and m are highlighted in the aggregate. All the spheres are coupled electromagnetically by the local exciting fields. For clarity, relevant vectors used in the text for the derivation are indicated.

The first step is to expand the scattered, internal, and incident fields of an individual sphere in the cluster (say sphere j) in terms of a complete basis of vector spherical wave functions (VSWFs). Several different conventions in the definition of the field expansion coefficients can be found in the literature. Here we adopt the approach developed by Xu [706], because it recovers the well-known expressions of the single-particle Mie theory when the azimuthal index $m = 1$.

Expanding about the center of sphere j, we can write the following:

$$\mathbf{E}_s(\mathbf{r}_j) = \sum_{n=1}^{\infty} \sum_{m=-n}^{n} i\, E_{mn} \left(a_{mn}^j \mathbf{N}_{mn}^{(3)} + b_{mn}^j \mathbf{M}_{mn}^{(3)} \right) \tag{13.92}$$

$$\mathbf{H}_s(\mathbf{r}_j) = \frac{k}{\omega\mu} \sum_{n=1}^{\infty} \sum_{m=-n}^{n} E_{mn} \left(b_{mn}^j \mathbf{N}_{mn}^{(3)} + a_{mn}^j \mathbf{M}_{mn}^{(3)} \right) \tag{13.93}$$

$$\mathbf{E}_{int}(\mathbf{r}_j) = -\sum_{n=1}^{\infty} \sum_{m=-n}^{n} i\, E_{mn} \left(d_{mn}^j \mathbf{N}_{mn}^{(3)} + c_{mn}^j \mathbf{M}_{mn}^{(3)} \right) \tag{13.94}$$

$$\mathbf{H}_{int}(\mathbf{r}_j) = -\frac{k^j}{\omega\mu^j} \sum_{n=1}^{\infty} \sum_{m=-n}^{n} E_{mn} \left(c_{mn}^j \mathbf{N}_{mn}^{(3)} + d_{mn}^j \mathbf{M}_{mn}^{(3)} \right) \tag{13.95}$$

$$\mathbf{E}_0(\mathbf{r}_j) = -\sum_{n=1}^{\infty} \sum_{m=-n}^{n} i\, E_{mn} \left(p_{mn}^j \mathbf{N}_{mn}^{(1)} + q_{mn}^j \mathbf{M}_{mn}^{(1)} \right) \tag{13.96}$$

$$\mathbf{H}_0(\mathbf{r}_j) = -\frac{k}{\omega\mu} \sum_{n=1}^{\infty} \sum_{m=-n}^{n} E_{mn} \left(q_{mn}^j \mathbf{N}_{mn}^{(1)} + p_{mn}^j \mathbf{M}_{mn}^{(1)} \right). \tag{13.97}$$

where we denoted by i the imaginary unit and the vector $\mathbf{r}_j$ connects the origin of the local coordinate system centered on the jth particle with the field observation point. Note also that in the preceding formulas, differently from the single-particle Mie theory, the azimuthal expansion order m is allowed to vary in the expansions because the spherical symmetry is generally broken in the aggregate. We also introduced the following symbol [706]:

$$E_{mn} = E_0 i^n (2n+1) \frac{(n-m)!}{(n+m)!} \tag{13.98}$$

that reduces to the coefficient E_n previously defined in Section 13.1 in the case of a single particle.

Imposing the electromagnetic boundary conditions on the surface of sphere j, we can find the following expressions for the Mie–Lorenz coefficients:

$$a_{mn}^j = a_n^j p_{mn}^j \qquad\qquad b_{mn}^j = b_n^j q_{mn}^j \tag{13.99}$$

$$c_{mn}^j = c_n^j q_{mn}^j \qquad\qquad d_{mn}^j = d_n^j p_{mn}^j, \tag{13.100}$$

where $a_n^j, b_n^j, c_n^j, d_n^j$ are the expansion coefficients of the isolated jth sphere, given in Section 13.1. We should appreciate that the Mie–Lorenz coefficients of a sphere embedded in an aggregate are modified with respect to the ones of an isolated sphere by the expansion coefficients (p_{mn}^j and q_{mn}^j) of the local exciting fields.[5] In order to obtain the solution of the multisphere problem, we need to determine the expansion coefficients of the local fields that excite each sphere, as explained in the following subsection.

13.3.2 Implementing the Foldy–Lax Scheme

The local expansion coefficients are obtained using the vector addition theorem for VSWFs within the general Foldy–Lax multiple scattering framework that connects the locally exciting fields as follows:

$$\mathbf{E}_{exc}(\mathbf{r}_j) = \mathbf{E}_0(\mathbf{r}_j) + \sum_{i \neq j} \mathbf{E}^{sc}(i \to j) \tag{13.101}$$

$$\mathbf{H}_{exc}(\mathbf{r}_j) = \mathbf{H}_0(\mathbf{r}_j) + \sum_{i \neq j} \mathbf{H}^{sc}(i \to j), \tag{13.102}$$

where $\mathbf{E}^{sc}(i \to j)$ and $\mathbf{H}^{sc}(i \to j)$ denote the electric and magnetic fields that arrive at $\mathbf{r}_j$ after having been scattered by the ith particle. In order to obtain closed-form

---

[5] This is also the case for an isolated single sphere excited by a general beam that is not a plane wave.

expressions for the exciting fields on each sphere, we need to expand the externally incident field $\mathbf{E}_0$ as well as all the scattered fields about the center of the jth sphere. This step is achieved using the vector addition theorem that translates the basis set of VSWFs across different local coordinate systems centered about any other sphere. The translation theorem allows us to write the following:

$$\mathbf{M}_{mn}^{(3)}(\mathbf{r}_i) = \sum_{\nu=0}^{\infty} \sum_{\mu=-\nu}^{\nu} \left[A_{mn\mu\nu} \mathbf{M}_{\mu\nu}^{(1)}(\mathbf{r}_j) + B_{mn\mu\nu} \mathbf{N}_{\mu\nu}^{(1)}(\mathbf{r}_j) \right] \tag{13.103}$$

$$\mathbf{N}_{mn}^{(3)}(\mathbf{r}_i) = \sum_{\nu=0}^{\infty} \sum_{\mu=-\nu}^{\nu} \left[B_{mn\mu\nu} \mathbf{M}_{\mu\nu}^{(1)}(\mathbf{r}_j) + A_{mn\mu\nu} \mathbf{N}_{\mu\nu}^{(1)}(\mathbf{r}_j) \right], \tag{13.104}$$

where $\mathbf{M}_{mn}^{3}$ and $\mathbf{N}_{mn}^{3}$ are the basis VSWFs expressed with respect to an origin at the center of the ith particle, and $\mathbf{M}_{mn}^{(1)}$ and $\mathbf{N}_{mn}^{(1)}$ are the ones expressed with respect to a translated origin on the jth sphere. The translation coefficients $A_{mn\mu\nu}$ and $B_{mn\mu\nu}$ connect the local systems i and j (i.e., $i \to j$) and have known expressions [132, 706].

We are now ready to express the scattered fields that appear in the Foldy–Lax equations using an expansion in the jth local coordinate system, as follows:[6]

$$\mathbf{E}_s(\mathbf{r}_j) = \sum_{n=1}^{\infty} \sum_{m=-n}^{n} i E_{mn} \left\{ a_{mn}^i \left[\sum_{\nu=1}^{\infty} \sum_{\mu=-\nu}^{\nu} (A_{mn\mu\nu} \mathbf{N}_{\mu\nu}^{(1)} + B_{mn\mu\nu} \mathbf{M}_{\mu\nu}^{(1)}) \right] \right.$$
$$\left. + b_{mn}^l \left[\sum_{\nu=1}^{\infty} \sum_{\mu=-\nu}^{\nu} (B_{mn\mu\nu} \mathbf{N}_{mn}^{(1)} + A_{mn\mu\nu} \mathbf{M}^{(1)}) \right] \right\}, \tag{13.105}$$

and similarly for the scattered magnetic field.

Exchanging (m, n) with (μ, ν) and rearranging the expansion coefficients, we obtain the following:

$$\mathbf{E}_s(\mathbf{r}_j) = -\sum_{n=1}^{\infty} \sum_{m=-n}^{n} i E_{mn} \left[p^{i,j} \mathbf{N}_{mn}^{(1)} + q^{i,j} \mathbf{M}_{mn}^{(1)} \right] \tag{13.106}$$

$$\mathbf{H}_s(\mathbf{r}_j) = -\frac{k}{\omega\mu} \sum_{n=1}^{\infty} \sum_{m=-n}^{n} E_{mn} \left[q^{i,j} \mathbf{N}_{mn}^{(1)} + p^{i,j} \mathbf{M}_{mn}^{(1)} \right], \tag{13.107}$$

where for $i \neq j$, we defined the local coefficients [706]:

$$p_{mn}^{i,j} = -\sum_{\nu=1}^{\infty} \sum_{\mu=-\nu}^{\nu} \left[a_{\mu\nu}^i \tilde{A}_{\mu\nu mn} + b_{\mu\nu}^i \tilde{B}_{\mu\nu mn} \right] \tag{13.108}$$

$$q_{mn}^{i,j} = -\sum_{\nu=1}^{\infty} \sum_{\mu=-\nu}^{\nu} \left[a_{\mu\nu}^i \tilde{B}_{\mu\nu mn} + b_{\mu\nu}^i \tilde{A}_{\mu\nu mn} \right] \tag{13.109}$$

---

[6] In many practical applications, we do not need to include the modes with index $\nu = 0$ [706].

and introduced the new translation coefficients $\tilde{A}_{\mu\nu mn}$ and $\tilde{B}_{\mu\nu mn}$ related to the previous ones as follows [706]:

$$\tilde{A}_{\mu\nu mn} = \frac{E_{\mu\nu}}{E_{mn}} A_{mn\mu\nu} = j^{\nu-n} \frac{(2\nu+1)(n+m)!\,(\nu-\mu)!}{(2n+1)(n-m)!\,(\nu+\mu)!} A_{\mu\nu mn} \tag{13.110}$$

$$\tilde{B}_{\mu\nu mn} = \frac{E_{\mu\nu}}{E_{mn}} B_{mn\mu\nu} = j^{\nu-n} \frac{(2\nu+1)(n+m)!\,(\nu-\mu)!}{(2n+1)(n-m)!\,(\nu+\mu)!} B_{\mu\nu mn}. \tag{13.111}$$

The expansion coefficients of the local exciting fields on the jth sphere can now be expressed as follows:

$$p_{mn}^j = p_{mn}^{j,j} - \sum_{i\neq j}^{N} \sum_{\nu=1}^{\infty} \sum_{\mu=-\nu}^{\nu} \left(a_{\mu\nu}^i \tilde{A}_{\mu\nu mn} + b_{\mu\nu}^i \tilde{B}_{\mu\nu mn} \right) \tag{13.112}$$

$$q_{mn}^j = q_{mn}^{j,j} - \sum_{i\neq j}^{N} \sum_{\nu=1}^{\infty} \sum_{\mu=-\nu}^{\nu} \left(a_{\mu\nu}^i \tilde{B}_{\mu\nu mn} + b_{\mu\nu}^i \tilde{A}_{\mu\nu mn} \right), \tag{13.113}$$

where the first term on the right-hand side refers to the externally incident wave $\mathbf{E}_0$ and the other terms quantify the contribution from all the other particles to the excitation of the jth particle. Inserting the preceding results into the expressions (13.99) and (13.100) for the Mie–Lorenz coefficients, we obtain a final set of linear equations for $j = 1 \ldots N$ that provides the general solution to our problem:[7]

$$a_{mn}^j = a_n^j \left[p_{mn}^{j,j} - \sum_{i\neq j}^{N} \sum_{\nu=1}^{\infty} \sum_{\mu=-\nu}^{\nu} \left(a_{\mu\nu}^i \tilde{A}_{\mu\nu mn} + b_{\mu\nu}^i \tilde{B}_{\mu\nu mn} \right) \right] \tag{13.114}$$

$$b_{mn}^j = b_n^j \left[q_{mn}^{j,j} - \sum_{i\neq j}^{N} \sum_{\nu=1}^{\infty} \sum_{\mu=-\nu}^{\nu} \left(a_{\mu\nu}^i \tilde{B}_{\mu\nu mn} + b_{\mu\nu}^i \tilde{A}_{\mu\nu mn} \right) \right] \tag{13.115}$$

The largest order ν_{max} required for the convergence of the aggregate can be approximated by the largest value in the Mie calculations among all the constituent spheres, i.e., $\nu_{max} = max(\nu^i)$, $i = 1, 2, \ldots, N$, where we denote by ν^i the truncation multipolar expansion order for sphere i. We remark that for a single homogeneous sphere of size parameter x, a suggested convergence order is provided by the expression $x + 4\sqrt[3]{x} + 2$, which is slightly larger than x [741].

The final linear system derived in (13.114) and (13.115) contains $2N\nu_{max}(2\nu_{max} + 1)$ linear equations and can be written in matrix form. However, given the large number of unknowns, a direct solution of this system is not possible in general and asymptotic iteration methods must be used instead.

---

[7] Note that we only need to solve for the generalized Mie–Lorenz scattering coefficients a_{mn}^j and b_{mn}^j because there are simple relations connecting them with the internal field coefficients [706].

13.3.3 Scattering and Extinction of Aggregates

Once the field coefficients are obtained, the relevant cross sections of the entire aggregate can be efficiently computed using the far-field asymptotic form of the VSFWs translation theorem [742]. This allows us to refer the (angular-dependent) Mie–Lorenz scattering coefficients to an arbitrarily chosen coordinate system that may coincide, for instance, with the primary coordinate system of the j_0-sphere.

Since the effects of the aggregate can be ultimately referred to a single sphere [706, 742], it becomes possible to find explicit expressions for the elements of the amplitude scattering matrix using the following expression:

$$
\begin{bmatrix} E_{\|s} \\ E_{\perp s} \end{bmatrix} = \frac{e^{jk(r-z)}}{-jkr} \begin{bmatrix} S_2 & S_3 \\ S_4 & S_1 \end{bmatrix} \begin{bmatrix} E_{\|0} \\ E_{\perp 0} \end{bmatrix},
\tag{13.116}
$$

where $(E_{\|0}, E_{\perp 0})$ and $(E_{\|s}, E_{\perp s})$ are respectively the incident field and the scattered field components parallel and perpendicular to the scattering plane, defined by the incident z-axis and the scattered direction. With the four elements of the amplitude scattering matrix, we can construct the 4×4 *scattering matrix*, also known as the *Müller matrix*, which connects the incident and scattered Stokes parameters [281]. The scattering matrix contains all the information on the angular scattering by an aggregate of particles, and its specific form reflects general properties of the scatterers. Moreover, based on the knowledge of the amplitude matrix, it is possible to derive analytical expressions for all the other scattering parameters [742–744]. In particular, the total scattering and extinction cross sections σ_{sc} and σ_{ext} of an aggregate of spheres with individual scattering and extinction cross sections σ_{sc}^i and σ_{ext}^i are given by the following:

$$
\sigma_{sc} = \sum_{i=1}^{N} \sigma_{sc}^i = \frac{4\pi}{k^2} \sum_{i=1}^{N} \sum_{n=1}^{n_c^i} \sum_{m=-n}^{n} \Re\{a_{mn}^{i*} a_{mn}^i + b_{mn}^{i*} b_{mn}^i\}
\tag{13.117}
$$

$$
\sigma_{ext} = \sum_{i=1}^{N} \sigma_{ext}^i = \frac{4\pi}{k^2} \sum_{i=1}^{N} \sum_{n=1}^{n_c^i} \sum_{m=-n}^{n} \Re\{p_{mn}^{i*} a_{mn}^i + q_{mn}^{i*} b_{mn}^i\},
\tag{13.118}
$$

where we denote by n_c^i the truncation multipolar expansion order for sphere i. The absorption cross section σ_{abs} can be obtained by the difference between the extinction and the scattered cross sections. Dividing the preceding cross sections by the sum of the geometrical cross sections of the particles, we can obtain the corresponding efficiencies.

As an example of GMT calculations, we show in Figure 13.14 the extinction and absorption efficiency of differently arranged clusters of 5 Ag nanoparticles of radius $a = 20$ nm and 2 nm interparticle separation (edge-to-edge). The single-particle response is also provided for comparison. These results have been obtained using the freely available MATLAB toolbox CELES [745] and carefully

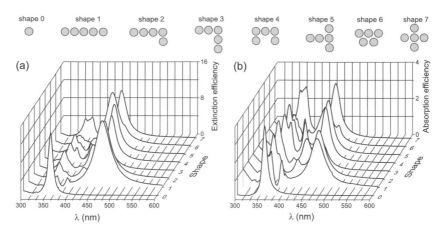

Figure 13.14 GMT calculated extinction (a) and absorption (b) efficiency for different clusters of Ag nanospheres labeled in the figure. The spheres have a radius of 20 nm and a separation of 2 nm. Up to seven multipolar orders per particles have been used in the calculation. The material dispersion data are obtained from Johnson and Christy's experimental values [112]

validated with respect to the well-established multisphere T-matrix code MSTM [746]. The computed extinction and absorption spectra feature resonant peaks that can be tuned by the geometrical arrangement of the particles for a given size and separation. These peaks originate from the excitation of longitudinal and transverse dipole and quadrupole modes in plasmon molecules. The frequency splitting between these peaks decreases for the arrays with more compact geometries (e.g., geometry 6 and 7) or by increasing the interparticle separations.

13.4   T-Matrix for Nonspherical Particles

In many practical applications of light scattering, particularly with respect to the engineering of the optical resonances of small nanoparticles [747, 748], we need to be able to quantitatively describe general aggregates of particles with nonspherical shapes. The semianalytical T-matrix approach known as the extended boundary condition method (EBCM) or the null-field method (NFM), originally proposed by P. C. Waterman in 1965 [626], enables the rigorous solution of such difficult problems. The approach is based on the vector Huygens principle [132, 749] and it exploits the linearity of Maxwell's equations and of the boundary conditions in linear optical media to establish a relationship between the incident/local and the scattered fields for particles of arbitrary shapes and for any incident field condition. This relationship requires the numerical computation of integrals over the geometrical shapes of the particles that can be very demanding for objects of arbitrary shapes. Due to its complexity, we limit our discussion to only few introductory remarks on the T-matrix calculation of an arbitrary nonspherical particle and we refer interested readers to the

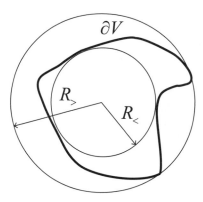

Figure 13.15 Cross-sectional view of a scattering particle bounded by a closed surface ∂V. The radius $R_>$ is the radius of the smallest circumscribed sphere and $R_<$ is the one of the largest inscribed concentric sphere.

cited literature for additional details on the theory and implementation of the powerful T-matrix method for aggregates of particles [627, 750].

13.4.1 Huygens Principle and the Null-Field Method

We introduce the T-matrix method for the calculation of the field scattered by an homogeneous object of volume V bounded by a closed surface ∂V with arbitrary shape. With reference to Figure 13.15, we denote by $R_>$ the radius of the smallest circumscribing sphere of the scattering particle centered at the origin of the coordinate system, while $R_<$ is the radius of the largest inscribed and concentric sphere.

The Huygens principle provides a convenient integral representation for the scattered field. In particular, it connects the incident field $\mathbf{E}_0$, the total field (i.e., the sum of incident plus scattered) in the region outside the particle, and the total surface field (tangential component) on the exterior face of ∂V [132] as follows:

$$\begin{cases} \mathbf{E}(\mathbf{r}') & \text{if } \mathbf{r}' \text{ is outside V} \\ 0 & \text{if } \mathbf{r}' \text{ is inside V} \end{cases}$$

$$= \mathbf{E}_0 + \int_{\partial V} \left[\mathbf{G}_0^+(\mathbf{r}',\mathbf{r}) \cdot \hat{\mathbf{n}} \times j\omega\mu\mathbf{H}(\mathbf{r}) + \nabla' \times \mathbf{G}_0^+(\mathbf{r}',\mathbf{r}) \cdot \hat{\mathbf{n}} \times \mathbf{E}(\mathbf{r}) \right] d^2\mathbf{r}'.$$

(13.119)

Note that when evaluated at the internal points (i.e., inside the volume bounded by ∂V), the preceding integral representation states the *extinction theorem* that shows how the radiation of the surface fields exactly cancels the incident wave inside the particle.

The main idea behind the T-matrix EBCM approach is to find the unknown surface field on the exterior face of ∂V by applying the Huygens principle to the points on the interior of ∂V, and then use the calculated surface fields to compute the integral in (13.119) for the points outside ∂V. This is accomplished by first expanding the incident, scattered, and internal fields in terms of a complete basis of vector spherical wave functions and then connect the surface fields on the interior and exterior faces of ∂V by applying the boundary conditions and the Huygens principle first to the interior region ($R < R_<$) and then to the exterior one ($R > R_>$). This procedure allows us to linearly interrelate the expansion coefficients of the respective fields via the T-matrix, which represents the linear relation between the scattering field coefficients and the exciting (or incident) field coefficients. The formulation can be written in the following compact matrix form:

$$\begin{pmatrix} \mathbf{p} \\ \mathbf{q} \end{pmatrix} = \begin{pmatrix} \mathbf{T}^{11} & \mathbf{T}^{12} \\ \mathbf{T}^{21} & \mathbf{T}^{22} \end{pmatrix} = \mathbf{T} \begin{pmatrix} \mathbf{a} \\ \mathbf{b} \end{pmatrix}, \tag{13.120}$$

where the expansion coefficients are arranged into column vectors, e.g., $\mathbf{a} \equiv (a_{mn})$. Similarly, the incident and the internal field coefficients as well as the internal and the scattered field coefficients can be linearly related through the matrices $\mathbf{Q}$ and $\mathbf{P}$ defined as follows [627]:

$$\begin{pmatrix} \mathbf{a} \\ \mathbf{b} \end{pmatrix} = \mathbf{Q} \begin{pmatrix} \mathbf{c} \\ \mathbf{d} \end{pmatrix} \tag{13.121}$$

$$\begin{pmatrix} \mathbf{p} \\ \mathbf{q} \end{pmatrix} = -\mathbf{P} \begin{pmatrix} \mathbf{c} \\ \mathbf{d} \end{pmatrix}. \tag{13.122}$$

The elements of $\mathbf{Q}$ and $\mathbf{P}$ are two-dimensional integrals of products of VSWFs that need to be numerically evaluated over the particle's surface and depend on the particle's size, shape, composition, and orientation with respect to the reference system. Formulas for computing the matrices $\mathbf{Q}$ and $\mathbf{P}$ for particles of any shape are listed in [128]. The T-matrix of the particle can now be obtained from the following:

$$\boxed{\mathbf{T} = -\mathbf{P}\mathbf{Q}^{-1}} \tag{13.123}$$

A fundamental property of the T-matrix method is that the elements of $\mathbf{T}$, $\mathbf{Q}$, and $\mathbf{P}$ do not depend on the incident and scattered fields but on the shape, size, composition, and orientation of the scattering particle. Therefore, the T-matrix needs to be computed only once and can then be reused in calculations involving any direction of incidence and scattering [627]. For particles with a high-degree of symmetry, e.g., rotationally symmetric particle shapes, the T-matrix is diagonal with respect to azimuthal indices m and m'. In the special case of spherical particles, the T-matrix assumes a particularly simple form:

$$\mathbf{T}^{11}_{mnm'n'} = \delta_{mm'}\delta_{nn'}b_n \tag{13.124}$$

$$\mathbf{T}^{22}_{mnm'n'} = \delta_{mm'}\delta_{nn'}a_n \tag{13.125}$$

$$\mathbf{T}^{12}_{mnm'n'} = \mathbf{T}^{21}_{mnm'n'} = 0, \tag{13.126}$$

where a_n and b_n are the Mie–Lorenz coefficients for a homogeneous sphere or their analogues for radially inhomogeneous spheres [281]. For spherical particles, all T-matrix formulas reduce to the corresponding Mie–Lorenz ones. However, for particles with nonspherical morphology and large aspect ratio (e.g., very eccentric spheroids or very large particles compared to the wavelength), the elements of $\mathbf{Q}$ may differ by orders of magnitude, thus making the computation of its inverse an ill-conditioned problem strongly subject to round-off errors. Recently, several approaches have been developed to alleviate such problems, including the use of subdomain virtual sources or the use of extended precision variables, as discussed in [301, 749]. Finally, the single-particle T-matrix method can be generalized to composite particles and arbitrary clusters of particles using the superposition T-matrix method. We refer the interested readers to the books by Dociu et al. [627] and Mishchenko et al. [301] for additional information.

14 Transport and Localization

... one has to go back in the history of physics to Faraday's lines of force if one wants to find a mnemonic device which matches Feynman's graphs in intuitive appeal.

A Guide to Feynman's Diagrams in the Many-Body Problem, Richard D. Mattuck

In this Chapter, we apply the multiple scattering theory to wave transport through continuous and discrete random media. In order to set the stage for a pedagogic discussion, we extend the diagrammatic approach to include the effects of statistical correlations, which play an important role in wave transport. In the following sections, we introduce the microscopic scattering theory and derive from it the photon diffusion equation and its fundamental solutions for the slab geometry starting from two rigorous results: (i) the Dyson's equation that governs the behavior of the average field, and (ii) the Bethe–Salpeter equation for the average field intensity (i.e., the second moment of the field). These equations underpin a number of physical phenomena that are common to both classical and quantum waves. Their application to the homogenization of metamaterials will be discussed. However, profound modifications of the photon diffusion picture arise for sufficiently strong disorder due to interference effects, giving rise to a wealth of new phenomena such as weak localization, correlated speckles, universal conductance fluctuations, and disorder-induced Anderson localization [31, 751]. A self-contained introduction to these topics is provided, while for more advanced treatments readers are referred to the cited references. Finally, we discuss random lasers with a focus on recent generalizations to photon subdiffusion regimes using the framework of fractional transport.

14.1 Scales of Radiation Transport

Let us start our discussion by introducing the relevant scales for radiation transport in a general random medium. First, we remind that the scattering mean free path ℓ_s denotes the typical distance between two successive scattering events, while the transport mean free path ℓ_{tr} measures the average distance after which light completely loses the memory of its initial direction. Moreover, the transport mean free path is connected to the scattering mean free path as follows:

$$\ell_{tr} = \frac{\ell_s}{1 - \langle \cos \theta \rangle} = \frac{1}{N \sigma_{tr}}, \tag{14.1}$$

where $\langle \cos \theta \rangle$ is the average cosine of the scattering angle for each scattering event and σ_{tr} is the cross section associated to the transport mean free path, known as the *transport scattering cross section*, which is defined as follows:

$$\sigma_{tr} = \int \int \frac{\partial \sigma}{\partial \Omega} (1 - \cos \theta) d\Omega. \tag{14.2}$$

This expression accounts for the nonisotropic nature of the differential scattering cross section of the individual particles. For isotropic scattering, we have that $\ell_s = \ell_{tr}$. Therefore, we recognize that the transport mean free path is the average distance that light travels in the sample before its propagation direction is completely randomized.

We can now understand that the regime of (weak) multiple scattering will take place when

$$\boxed{\lambda \ll \ell_s \ll L \ll L_{abs}} \tag{14.3}$$

where λ is the wavelength, L is the sample size, and $L_{abs} = 1/\alpha$ is the *absorption length* associated to the absorption coefficient α. The first inequality ensures that strong scattering and wave localization effects are negligible, the second inequality guarantees that many scattering events occur when the waves are transported through the system, and the third inequality requires small absorption in the system. Moreover, the transport of radiation in complex media can be described at roughly three different length scales [367]:

- *Microscopic scale*: this "first-principle" description is based on the solution of the full-vector Maxwell's equations provided by the Foldy–Lax multiple scattering Dyson and Bethe–Salpeter equations introduced in the previous sections.
  The diffusion and radiative transfer equations and their parameters can be derived from the microscopic theory. For example, the standard diffusion equation can be obtained from the exact Bethe–Salpeter equation under the ladder approximation. However, accurate microscopic solutions can only be obtained if the positions and shapes of all the scatterers in the system are known. The microscopic description takes into account the vector nature of light, interference effects, and the important resonance effects that occur when scatterers have dimensions comparable to the wavelength of light. These features render the microscopic scattering theory significantly more complex to work with compared to its macroscopic diffusion approximations.
- *Mesoscopic scale*: this picture describes the problem over length scales of the order of the mean free path. In this regime, the radiative transfer equation, or its integral form for a planar slab known as the Schwarzschild–Milne equation, is applicable. These transport equations follow from a detailed energy balance in the medium that results in a Boltzmann-type transport similar to the case of kinetic

theory [752]. The standard diffusion equation can be derived from the radiative transfer equation under the assumptions of small absorption, nearly isotropic scattering, and slow enough variations of the intensity.

- *Macroscopic scale*: this description is valid at scales much larger than the mean free path where the average intensity satisfies the standard diffusion equation with a diffusion constant $D = v\ell_{tr}/3$, where v is the transport speed and ℓ_{tr} the transport mean free path. This is the regime where light interference effects in multiple scattering are neglected and the transport process becomes equivalent to an uncorrelated random walk for photons.

In what follows, we will discuss light transport through random media using the first-principle (microscopic) approach applied to both continuous and discrete (i.e., multiparticle) systems.

14.2   Continuous Random Media

Examples of wave propagation through continuous random media are often found in radio-frequency (RF) remote sensing, atmospheric wave propagation, or sound waves through turbulent gases. These systems are characterized by fluctuations of the dielectric permittivity expressed as $\epsilon(\mathbf{r}) = \langle\epsilon\rangle + \epsilon_s(\mathbf{r})$, where $\epsilon_s(\mathbf{r})$ is a stochastic contribution that adds to the average permittivity $\langle\epsilon\rangle$. Due to the random nature of $\epsilon_s(\mathbf{r})$, it is not possible to solve exactly the wave equation and obtain a complete description of the system based on its exact Green function. However, it is possible to extract the information on light transport through a random system by considering configurational averages of its Green function over different realizations of the spatial disorder, i.e., by studying the so-called *statistical moments of the Green function*.

14.2.1   The Average Green Function

In Section 12.3, we have considered the dyadic Green function of a nonhomogeneous medium described by a position-dependent wavenumber. Analogously, in a random medium with spatially varying permittivity $\epsilon(\mathbf{r}) = \langle\epsilon\rangle + \epsilon_s(\mathbf{r})$, we can write the following [132]:

$$\boxed{\nabla \times \nabla \times \overleftrightarrow{\mathbf{G}}(\mathbf{r},\mathbf{r}') - \omega^2\mu\langle\epsilon\rangle \overleftrightarrow{\mathbf{G}}(\mathbf{r},\mathbf{r}') = V(\mathbf{r})\overleftrightarrow{\mathbf{G}}(\mathbf{r},\mathbf{r}') + \overleftrightarrow{\mathbf{I}}\,\delta(\mathbf{r}-\mathbf{r}')} \quad (14.4)$$

where we have defined $V(\mathbf{r}) = \omega^2\mu\epsilon_s(\mathbf{r})$. Technically, the preceding equation is a *vector stochastic differential equation* due to the fluctuating nature of the scattering potential. However, for each geometrical realization of the random medium, we can formally write its solution as a convolution-type integral equation according to the following:

$$\overleftrightarrow{\mathbf{G}}(\mathbf{r},\mathbf{r}') = \overleftrightarrow{\mathbf{G}_0}(\mathbf{r},\mathbf{r}') + \int d^3\mathbf{r}_1\,\overleftrightarrow{\mathbf{G}_0}(\mathbf{r},\mathbf{r}_1)V(\mathbf{r}_1)\overleftrightarrow{\mathbf{G}}(\mathbf{r}_1,\mathbf{r}'). \quad (14.5)$$

The dyadic Green function $\overleftrightarrow{G}_0(\mathbf{r}, \mathbf{r}')$ satisfies the wave equation of the homogeneous medium with average wavenumber $k_m = \omega\sqrt{\mu\langle\epsilon\rangle}$, namely as follows:

$$\nabla \times \nabla \times \overleftrightarrow{G}_0(\mathbf{r}, \mathbf{r}') - k_m^2 \overleftrightarrow{G}_0(\mathbf{r}, \mathbf{r}') = \overleftrightarrow{\mathbf{I}}\, \delta(\mathbf{r} - \mathbf{r}'). \tag{14.6}$$

For each realization of disorder, the integral equation (14.5) can formally be solved by the iteration method leading to the Neumann series for the dyadic total Green function:

$$\begin{aligned}
\overleftrightarrow{G}(\mathbf{r}, \mathbf{r}') &= \overleftrightarrow{G}_0(\mathbf{r}, \mathbf{r}') + \int d^3\mathbf{r}_1 \overleftrightarrow{G}_0(\mathbf{r}, \mathbf{r}_1) V(\mathbf{r}_1) \overleftrightarrow{G}_0(\mathbf{r}_1, \mathbf{r}') \\
&\quad + \int d^3\mathbf{r}_1 d^3\mathbf{r}_2 \overleftrightarrow{G}_0(\mathbf{r}, \mathbf{r}_1) V(\mathbf{r}_1) \overleftrightarrow{G}_0(\mathbf{r}_1, \mathbf{r}_2) V(\mathbf{r}_2) \overleftrightarrow{G}_0(\mathbf{r}_2, \mathbf{r}') + \cdots \\
&\quad + \int d^3\mathbf{r}_1 d^3\mathbf{r}_2 \ldots d^3\mathbf{r}_n \overleftrightarrow{G}_0(\mathbf{r}, \mathbf{r}_1) V(\mathbf{r}_1) \overleftrightarrow{G}_0(\mathbf{r}_1, \mathbf{r}_2) V(\mathbf{r}_2) \\
&\quad \times \overleftrightarrow{G}_0(\mathbf{r}_2, \mathbf{r}_3) \ldots V(\mathbf{r}_n) \overleftrightarrow{G}_0(\mathbf{r}_n, \mathbf{r}') + \cdots
\end{aligned} \tag{14.7}$$

We recognize in this expression the well-known multiple-scattering perturbation series solution of a deterministic medium. In order to introduce the contribution of randomness, we need to specify the statistical properties of the random quantity $c_s(\mathbf{r})$. For a generic random medium, this requires the specification of an infinite number of statistical moments of the random permittivity $\epsilon_s(\mathbf{r})$, practically rendering the solution impossible. Therefore, simplifying approximations on the nature of the random variability of the medium are necessary.

14.2.2 The Model of Disorder

A typical assumption is to regard $\epsilon_s(\mathbf{r})$ as a *Gaussian stationary random process*. In this case, the first and second moments of $\epsilon_s(\mathbf{r})$ are sufficient to fully describe its statistical properties, and the following is further assumed [132]:

$$\langle \epsilon_s(\mathbf{r}) \rangle = 0. \tag{14.8}$$

Moreover, for any integer n, the odd moments of $\epsilon_s(\mathbf{r})$ vanish and the *even moments can be cluster-expanded* in terms of two-point correlation functions as follows [130]:

$$\boxed{\langle \epsilon_s(\mathbf{r}_1) \ldots \epsilon_s(\mathbf{r}_{2n}) \rangle = \sum \langle \epsilon_s(\mathbf{r}_i) \epsilon_s(\mathbf{r}_j) \rangle \ldots \langle \epsilon_s(\mathbf{r}_k) \epsilon_s(\mathbf{r}_l) \rangle} \tag{14.9}$$

where the summation extends over all the possible distinct pairs of permittivities for a total of $(2n)!\,/(2^n n!)$ terms, each consisting of a product of two-point correlations. The cluster expansion of a Gaussian stationary random process only involves its two-point correlation functions due to its short correlation length ℓ_c, i.e., the values of $\epsilon_s(\mathbf{r})$ at points whose separation is larger than ℓ_c become uncorrelated. However, for more general non-Gaussian random media, the moments of $\epsilon_s(\mathbf{r})$ can be cluster-expanded, including higher-order correlations besides the two-point correlation functions. This significantly complicates the analysis to makes an effective approach based on fractional transport equations particularly attractive [22, 390]. Within the model of Gaussian disorder, we can calculate the configurationally averaged Green function, or the *mean Green function*, by averaging the Neumann's series (14.7) over different realizations of the disorder and using the cluster expansion [132]:

$$\langle \overleftrightarrow{G}(\mathbf{r},\mathbf{r}')\rangle = \overleftrightarrow{G}_0(\mathbf{r},\mathbf{r}') + \int d^3\mathbf{r}_1 d^3\mathbf{r}_2 \, \overleftrightarrow{G}_0(\mathbf{r},\mathbf{r}_1) \overleftrightarrow{G}_0(\mathbf{r}_1,\mathbf{r}_2) \overleftrightarrow{G}_0(\mathbf{r}_2,\mathbf{r}') \langle V(\mathbf{r}_1)V(\mathbf{r}_2)\rangle$$

$$+ \int d^3\mathbf{r}_1 d^3\mathbf{r}_2 d^3\mathbf{r}_3 d^3\mathbf{r}_4 \, \overleftrightarrow{G}_0(\mathbf{r}_1,\mathbf{r}_2) \overleftrightarrow{G}_0(\mathbf{r}_2,\mathbf{r}_3) \overleftrightarrow{G}_0(\mathbf{r}_3,\mathbf{r}_4) \overleftrightarrow{G}_0(\mathbf{r}_4,\mathbf{r}')$$

$$\times [\langle V(\mathbf{r}_1)V(\mathbf{r}_2)\rangle\langle V(\mathbf{r}_3)V(\mathbf{r}_4)\rangle + \langle V(\mathbf{r}_1)V(\mathbf{r}_4)\rangle\langle V(\mathbf{r}_2)V(\mathbf{r}_3)\rangle$$

$$+ \langle V(\mathbf{r}_1)V(\mathbf{r}_3)\rangle\langle V(\mathbf{r}_2)V(\mathbf{r}_4)\rangle] + \cdots \tag{14.10}$$

Note that for unbounded (continuous) random media it is important to consider permittivity fluctuations with a finite correlation length in order to guarantee the convergence of the ensemble-averaged Neumann series solution, which would otherwise diverge [130]. It is also clear that further extending the expansion series above would result in a very cumbersome expression. This motivates the introduction of formal perturbation methods based on Feynman's diagrams for the solution of general stochastic differential equations [115, 130]. Such an approach has the advantage to drastically simplify the handling of disorder-averaged perturbative expressions and to suggest a more direct physical interpretation of the underlying scattering phenomena.

14.2.3   Diagrammatics

In order to better handle the increasing complexity of expressions like the (14.10), it is convenient to improve the diagrammatic notation by introducing *single-level Feynman diagrams*, as shown in Table 14.1.

Table 14.1 Drawing conventions for the scattering diagrams of a continuous medium.

Symbol	Represented quantity	Meaning
•	$V(\mathbf{r}_i)$	Vertex over which spatial integration and contraction of dyadic indices are implied
——	$\overleftrightarrow{G}_0(\mathbf{r},\mathbf{r}')$	Dyadic Green function of the homogeneous medium
~~	$\overleftrightarrow{G}(\mathbf{r},\mathbf{r}')$	Dyadic Green function of the nonhomogeneous medium
══	$\langle \overleftrightarrow{G}(\mathbf{r},\mathbf{r}')\rangle$	Averaged (dressed) dyadic Green function of the random medium
•⌒•	$\langle V(\mathbf{r}_i)V(\mathbf{r}_j)\rangle$	Correlation links (or correlation connection)
⊛	$\overleftrightarrow{\Sigma}(\mathbf{r},\mathbf{r}')$	Dyadic self-energy or mass operator. Sum of all irreducible scattering diagrams

This vocabulary establishes a one-to-one mapping between each term in the analytical expression (14.10) and the corresponding graphical representation. As an example, let us consider the following term:

$$\underset{}{\overbrace{}} = \int d^3\mathbf{r}_1 d^3\mathbf{r}_2 \overleftrightarrow{\mathbf{G}}_0(\mathbf{r},\mathbf{r}_1)\overleftrightarrow{\mathbf{G}}_0(\mathbf{r}_1,\mathbf{r}_2)\overleftrightarrow{\mathbf{G}}_0(\mathbf{r}_2,\mathbf{r}')\langle V(\mathbf{r}_1)V(\mathbf{r}_2)\rangle, \quad (14.11)$$

which corresponds to the first integral in equation (14.10).

The meaning of the preceding diagram is as follows: it represents a convolution integral whose kernel is the algebraic product of the values of its parts, as established by Feynman's vocabulary. In this particular case, the integral has a kernel in the form of a product of three dyadic Green functions for the homogeneous medium multiplied by a correlation link between two scattering positions. At the vertex points of each diagrams, spatial integration and contraction of dyadic indices are implied. In this way, correlation effects in the two-point cluster-expansion of a Gaussian random medium appear naturally as manifestations of the underlying statistical averaging procedure.

The disorder-averaged Green function (14.10) can now be represented diagrammatically as follows:

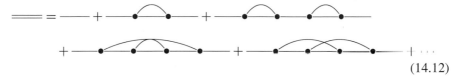

$$(14.12)$$

Due to the one-to-one nature of the introduced vocabulary, we can immediately construct the corresponding diagrams and, conversely, restore the analytical expressions from the associated diagrams. Note that this diagrammatic expansion contains only an *even number of terms*, where $\mathbf{r}_1, \ldots, \mathbf{r}_{2n}$ are combined pairwise in all possible ways. As discussed before, this produces $(2n)!\,/2^n n!$ diagrams of the nth order ($2n$-fold scattering) such that all the $2n$ vertices are connected pairwise by correlation lines in all possible ways.[1] For example, in the expansion above all possible second-order diagrams (i.e., $2n = 4$) have been shown explicitly. All these diagrams describe fourfold scattering events but are *topologically different*. Therefore, it is natural to ask what kind of different scattering process the diagrams of the same order describe. We clarify this point by considering the special case of the three second-order diagrams appearing in (14.12), which are redrawn for convenience in Figure 14.1.

In the diagrams shown in Figure 14.1a, the correlation links connect two points that belong to two distinct inhomogeneities. Therefore, this diagram describes a situation in which the wave first propagates freely from the source to the first inhomogeneity, where it is scattered twice; then it further propagates unimpeded to the second inhomogeneity, where it is again doubly scattered. The next diagram (in Figure 14.1b) also describes fourfold scattering, but this time the sequence of scattering events is different. In particular, after being scattered by the first inhomogeneity, the wave is singly scattered by the second inhomogeneity. The doubly scattered wave then propagates undisturbed to the first inhomogeneity, where it is scattered again. Finally, the triply

---

[1] Notice that the number of diagrams is also given by the double factorial $(2n - 1)!\,!$.

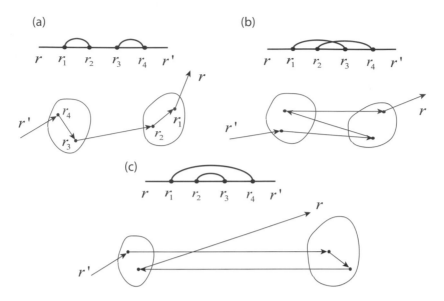

Figure 14.1 Second-order multiple scattering Feynman diagrams and corresponding scattering processes [753].

scattered wave propagates to the second inhomogeneity, where it is scattered again and finally reaches the observation point. The meaning of the diagram in Figure 14.1c follows similarly. In Figure 14.1, we illustrate the physical processes corresponding to each diagram.

In the general expansion (14.12), there appear an infinity of correlation terms, and the number of the corresponding diagrams increases more than exponentially with the order of scattering. It may therefore appear hopeless to find manageable expressions for its solutions. However, this goal can be achieved in many cases by the diagrammatic resummation technique, known as *renormalization*. In order to understand how this works, it is important to distinguish between two fundamentally different types of diagrams: *irreducible diagrams* and *reducible diagrams*. Put simply, an *irreducible diagram* is a diagram that cannot be divided into smaller diagrams without breaking a correlation link. Examples of such diagrams follow:

$$(14.13)$$

Diagrams that are not irreducible diagrams are called *reducible*. Such diagrams are obtained by connecting together irreducible diagrams. Examples of reducible diagrams follow:

$$(14.14)$$

It should be clear that reducible diagrams are automatically generated by the concatenation of irreducible diagrams. Due to the correspondence introduced by Feynman's vocabulary, if we iteratively solve an integral equation that only contains irreducible diagrams, we obtain all the reducible diagrams as solution terms. This is the key observation behind the derivation of Dyson's equation for the mean field.

14.2.4 Dyson's Equation

In order to derive an exact equation governing the average Green function, let us introduce the dyadic *self-energy (or mass) operator* $\overleftrightarrow{\Sigma}(\mathbf{r}_1, \mathbf{r}_2)$, represented symbolically by ⊛, which is the *sum of all the irreducible diagrams* in (14.12) without their end connectors:

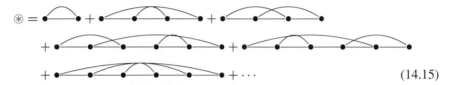

$$(14.15)$$

Therefore, all the reducible diagrams can be obtained by cascading the self-energy operator. For instance, all the reducible diagrams containing two irreducible elements are summed up by the following expression:

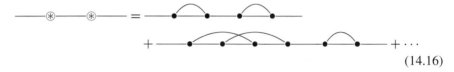

$$(14.16)$$

Note that continuing the summation process by cascading an increasing number of self-energy diagrams reproduces in the full Neumann series (14.12) for the average Green function:

$$
\mathrel{=\!\!=} = \text{———} + \text{———}\circledast\text{———} + \text{———}\circledast\text{———}\circledast\text{———}
$$
$$
+ \text{———}\circledast\text{———}\circledast\text{———}\circledast\text{———} + \cdots \tag{14.17}
$$

The reader can verify that all the diagrams, reducible and irreducible, are in fact included in the averaged Neumann series. However, the convenient diagrammatic notation allows one to simply regroup (resum) the preceding expression and obtain an alternative form of the Neumann series:

$$
\mathrel{=\!\!=} = \text{———} + \text{———}\circledast\, [\text{———} + \text{———}\circledast\text{———} + \text{———}\circledast\text{———}\circledast\text{———} + \cdots].
$$
$$\tag{14.18}$$

We recognize that within the brackets, the average Neumann series appears again. Therefore, we can write the entire expression more compactly as follows:

$$
\mathrel{=\!\!=} = \text{———} + \text{———}\circledast\mathrel{=\!\!=}. \tag{14.19}
$$

Equation (14.19) is the diagrammatic representation of *Dyson's equation* for the mean field. This equation is correct to any order of scattering (i.e., it is a rigorous result) since *it contains all the reducible and irreducible diagrams*. In its integral form, the Dyson's equation can simply be written as follows:

$$\langle \overset{\leftrightarrow}{G}(\mathbf{r},\mathbf{r}') \rangle = \overset{\leftrightarrow}{G_0}(\mathbf{r},\mathbf{r}') + \int d^3\mathbf{r}_1 d^3\mathbf{r}_2 \, \overset{\leftrightarrow}{G_0}(\mathbf{r},\mathbf{r}_1) \, \overset{\leftrightarrow}{\Sigma}(\mathbf{r}_1,\mathbf{r}_2) \langle \overset{\leftrightarrow}{G}(\mathbf{r}_2,\mathbf{r}') \rangle \qquad (14.20)$$

where $\overset{\leftrightarrow}{\Sigma}(\mathbf{r}_1,\mathbf{r}_2)$ is the dyadic self-energy operator introduced in Table 14.1. We recognize that Dyson's equation has the same mathematical structure of the Lippmann–Schwinger integral equation, of which it is in fact the generalization for the mean field. However, Dyson's equation for the *average propagator* cannot be solved without approximating the generally unknown self-energy operator. Various types of approximations originate from restricting the summation to only certain types of diagrams. For unbounded random media, several summation strategies have been developed in the context of remote sensing [132]. We will just mention here the bilocal approximation, which only considers the first diagram in the expansion of the mass operator, and the nonlinear approximation, which considers a renormalized mass operator containing all the irreducible diagrams in the Neumann series with one outer correlation [128, 130, 132].

14.2.5   Self-Energy and Nonlocal Susceptibility

Dyson's equation derived in the previous section allows us to establish a direct physical interpretation for the self-energy operator. In order to achieve this, we will use a logic that closely retraces what previously introduced in Section 12.2.5 for a system of scalar dipoles. Specifically, we first remember the defining equation of the free space dyadic Green function:

$$\nabla \times \nabla \times \overset{\leftrightarrow}{G_0}(\mathbf{r},\mathbf{r}_0) - k^2 \overset{\leftrightarrow}{G_0}(\mathbf{r},\mathbf{r}_0) = \overset{\leftrightarrow}{I}\,\delta(\mathbf{r}-\mathbf{r}_0). \qquad (14.21)$$

We then apply the operator $[\nabla \times \nabla \times -k^2]$ to Dyson's equation and obtain that the average Green function satisfies the following:

$$\nabla \times \nabla \times \langle \overset{\leftrightarrow}{G}(\mathbf{r},\mathbf{r}_0) \rangle - k^2 \langle \overset{\leftrightarrow}{G}(\mathbf{r},\mathbf{r}_0) \rangle - \int \overset{\leftrightarrow}{\Sigma}(\mathbf{r},\mathbf{r}') \langle \overset{\leftrightarrow}{G}(\mathbf{r}',\mathbf{r}_0) \rangle d^3\mathbf{r}' = \overset{\leftrightarrow}{I}\,\delta(\mathbf{r}-\mathbf{r}_0).$$
$$(14.22)$$

An equation formally identical to this result appears when solving the curl Maxwell's equation in an *isotropic medium* with *spatially nonlocal susceptibility* $\chi(\mathbf{r},\mathbf{r}')$, where the following wave equation holds [753]:

$$\nabla \times \nabla \times \mathbf{E} - k^2 \mathbf{E} = k^2 \int \chi(\mathbf{r},\mathbf{r}')\mathbf{E}(\mathbf{r}')d^3\mathbf{r}'. \qquad (14.23)$$

Comparing the last expression with equation (14.22) establishes the following correspondence:

$$\boxed{\dfrac{\Sigma(\mathbf{r},\mathbf{r}')}{k^2} = \chi(\mathbf{r},\mathbf{r}')} \qquad (14.24)$$

that is the nonlocal susceptibility of the medium. Alternatively, we can write the last expression in terms of the *nonlocal refractive index* $n(\mathbf{r}, \mathbf{r}')$:

$$n^2(\mathbf{r}, \mathbf{r}') = 1 + \frac{\Sigma(\mathbf{r}, \mathbf{r}')}{k^2} \qquad (14.25)$$

This general result shows that the self-energy operator can be regarded as the polarizability of a nonlocal medium. In particular, it shows that the origin of spatial nonlocality traces back to the presence of multiple scattering (interference) phenomena. This introduces long-range correlation effects that, within certain approximations, can be effectively modeled using fractional integral equations with power-law kernels. Moreover, such an expression holds the key to the high-frequency homogenization of metamaterials, provided their self-energy operator can be properly approximated or computed. Clearly, the spatial nonlocality of the scattering medium reflects the fact that the mean field at a given position depends on the spatial distribution of the surrounding scattering particles or inhomogeneities.

14.2.6   The Effective Medium

It is instructive to address now the case of a *statistically homogeneous random medium*, which is characterized by the following:

$$\overset{\leftrightarrow}{\Sigma}(\mathbf{r}_1, \mathbf{r}_2) = \overset{\leftrightarrow}{\Sigma}(\mathbf{r}_1 - \mathbf{r}_2) \qquad (14.26)$$

$$\langle \overset{\leftrightarrow}{\mathbf{G}}(\mathbf{r}, \mathbf{r}') \rangle = \overset{\leftrightarrow}{\mathbf{G}}_h(\mathbf{r} - \mathbf{r}'). \qquad (14.27)$$

Under these assumptions, Dyson's equation (14.19) can be transformed into a simpler algebraic equation by Fourier transforming to k-space, which allows us to express the average Green function as follows [24, 132]:

$$\overset{\leftrightarrow}{\mathbf{G}}_h(\mathbf{k}) = \left(\overset{\leftrightarrow}{\mathbf{I}} - \overset{\leftrightarrow}{\mathbf{G}}_0(\mathbf{k}) \overset{\leftrightarrow}{\Sigma}(\mathbf{k}) \right)^{-1} \overset{\leftrightarrow}{\mathbf{G}}_0(\mathbf{k}). \qquad (14.28)$$

The analogous result for *scalar waves* follows:

$$G_h(\mathbf{k}) = \frac{1}{k^2 - k_m^2 - \Sigma(\mathbf{k})}, \qquad (14.29)$$

where $k_m = \omega \sqrt{\mu \langle \epsilon \rangle}$.

The poles of the average Green function in the spectral k-domain determine the effective propagation constants of the average field $\langle \mathbf{E} \rangle$, which is referred to as the *coherent wave*. Thus it follows from equation (14.29) that the *effective propagation constant K* of the coherent field is as follows:

$$K^2 = k_m^2 + \Sigma(\mathbf{k}). \qquad (14.30)$$

It is now clear that the net effect of the disorder is to renormalize the propagation constant of the homogeneous medium to the new value provided in (14.30). In addition, we note that the self-energy $\Sigma(\mathbf{k})$ is a *nonlocal operator in space*. The nonlocality of the random medium is a consequence of the statistical correlations between different multiple scattering events. However, for weak enough scattering, the self-energy operator can be considered as independent of k (i.e., $\Sigma(\mathbf{k}) = \Sigma$) leading to a renormalized *local effective medium* with the effective propagation constant:

$$K^2 = k_m^2(\omega) + \Sigma(\omega) \qquad (14.31)$$

where here we made the frequency dependence explicit. As a result, the weakly scattering local effective medium behaves as an homogeneous medium with the Green function:

$$G(\mathbf{r} - \mathbf{r}') = \frac{e^{jK|\mathbf{r}-\mathbf{r}'|}}{4\pi|\mathbf{r} - \mathbf{r}'|}. \qquad (14.32)$$

Looking at the last equation, we would be tempted to believe that the wave propagation through the effective medium is perfectly coherent, meaning that the wave transport preserves its direction and phase properties. However, this is not the case since in general the self-energy operator is complex with a positive imaginary part [24]. As a consequence, the effective wavenumber in (14.31) is also complex with a positive imaginary part and the effective propagator (14.32) decays exponentially in space. In fact, the *coherent wave* attenuates exponentially in the sample due to the extinction introduced by the multiple-scattering losses, even in the absence of material absorption.

The fluctuating part of the field $\mathbf{E}_s(\mathbf{r})$ in a random medium is called the *incoherent field*, and the total fluctuating field in the medium $\mathbf{E}(\mathbf{r}) = \langle\mathbf{E}(\mathbf{r})\rangle + \mathbf{E}_s(\mathbf{r})$ is the sum of the coherent and incoherent parts. In a multiple scattering medium, the transport remains coherent only within a characteristic length scale, called the *scattering mean free path* ℓ_s, which is defined by the inverse imaginary part of the effective wavevector:

$$\ell_s = \frac{1}{\Im\{K\}} \qquad (14.33)$$

Therefore, in the effective medium approximation, wave transport is coherent for distances that are smaller than the scattering mean free path, and the waves propagate, maintaining their initial directions and phase relations. On the other hand, for propagation distances beyond the scattering mean free path, the coherent field has already decayed exponentially inside the medium, and transport information is carried by the average intensity, i.e., the second moment of the field. However, it is important to keep in mind that the transition between coherent and incoherent transport is difficult to exactly characterize and it is not entirely understood [280, 367, 690].

14.2.7   The Bethe–Salpeter Equation

We mentioned in the previous section that while the average propagator $\langle G\rangle$ gives us the coherent part of wave propagation, the transport beyond the mean free path can only be described by the second moment of the propagator, i.e., $\langle GG^* \rangle$. We can better appreciate this fact for scalar waves by first realizing that the (unaveraged) field intensity can be written as follows:

$$I(\mathbf{r}) \equiv |E_0(\mathbf{r})|^2 = \int \int d^3\mathbf{r}' d^3\mathbf{r}'' G(\mathbf{r}, \mathbf{r}') G^*(\mathbf{r}, \mathbf{r}'') E_0(\mathbf{r}') E_0^*(\mathbf{r}''), \qquad (14.34)$$

where $G(\mathbf{r}, \mathbf{r}')$ is the total Green function of a single realization of the random medium and E_0 the incident wave field. The propagator of the average intensity is obtained by considering the ensemble average over the disorder of the preceding expression. The situation is in fact similar to a random walk in which the diffusion dynamics is derived from the second moment of the position vector $\langle \mathbf{r}^2(t) \rangle$. Analogously, the propagation of the average optical intensity in a random medium is described by the quantity:

$$P(t, \mathbf{r}, \mathbf{r}') = \langle GG^* \rangle. \qquad (14.35)$$

When dealing with vector waves, we need to consider the ensemble average of the tensor product of the dyadic Green functions:

$$\Gamma(\mathbf{r}, \mathbf{r}', \mathbf{r}_0, \mathbf{r}_0') = \left\langle \overleftrightarrow{\mathbf{G}}(\mathbf{r}, \mathbf{r}_0) \overleftrightarrow{\mathbf{G}^*}(\mathbf{r}', \mathbf{r}_0') \right\rangle. \qquad (14.36)$$

This quantity plays a central role in the transport theory and is known as the *field correlation*. Note that each dyadic Green function that appears in $\Gamma(\mathbf{r}, \mathbf{r}', \mathbf{r}_0, \mathbf{r}_0')$ can be expanded as an infinite Neumann series (14.7), rendering the field correlation a very cumbersome expression to handle. However, we should now realize that the series expansion of the correlation contains an infinity of cross-correlation terms of the form $\langle V(\mathbf{r}_1) V^*(\mathbf{r}_1') V^*(\mathbf{r}_2') \rangle$ or $\langle V(\mathbf{r}_1) V(\mathbf{r}_2) V^*(\mathbf{r}_1') \rangle$. These terms couple the retarded and advanced Green functions together. In order to concisely represent such coupling terms using the convenient diagrammatic notation, we now need to introduce *two-level Feynman diagrams* [24, 130, 132].

Two-level diagrams are diagrammatic expressions for tensor products of advanced and retarded Green functions and are constructed such that their upper level represents the dyadic Green function and the lower level represents its complex conjugate (advanced) Green function in the dual vector space. For example:

$$\underset{\displaystyle \sim\!\!\sim\!\!\sim}{\overset{\displaystyle \sim\!\!\sim\!\!\sim}{}} \equiv \overleftrightarrow{\mathbf{G}}(\mathbf{r}, \mathbf{r}_0) \overleftrightarrow{\mathbf{G}}^*(\mathbf{r}', \mathbf{r}_0'). \qquad (14.37)$$

Remember that in the absence of disorder (i.e., for a deterministic medium), the dyadic total Green function of a nonhomogeneous medium obeys the dyadic Lippmann–Schwinger equation:

$$\sim\!\!\sim\!\!\sim \; = \; \frac{}{} + \frac{}{} \!\!\bullet\!\!\sim\!\!\sim \; . \qquad (14.38)$$

The effect of disorder is described by the second moment of the Green function, which corresponds to the two-level diagram [132]:

$$\left\langle \left(\begin{array}{c} \\ \\ \end{array} \right) \right\rangle \equiv \boxed{ \mathcal{S} } = \langle \overleftrightarrow{\mathbf{G}}(\mathbf{r}, \mathbf{r}_0) \overleftrightarrow{\mathbf{G}}^*(\mathbf{r}', \mathbf{r}_0') \rangle. \qquad (14.39)$$

In this expression, we introduced the diagrammatic representation of the intensity transport $\mathcal{S}$ operator, which is contributed by statistical correlations terms such as $\langle V(\mathbf{r}_1)V^*(\mathbf{r}_1')V^*(\mathbf{r}_2') \rangle$ or $\langle V(\mathbf{r}_1)V(\mathbf{r}_2)V^*(\mathbf{r}_1') \rangle$ that couple the advanced and retarded dyadic Green functions. Diagrammatically, the intensity transport can be represented as follows:

$$(14.40)$$

The physical interpretation of the two-level Feynman scattering diagrams is quite straightforward. To illustrate it, we discuss few selected diagrams appearing in the preceding expansion. The terms in the series describe the propagation of a wave that is *multiply scattered* from $\mathbf{r}_0$ to $\mathbf{r}$ and from $\mathbf{r}_0'$ to $\mathbf{r}'$. The scattering occurs in a way that along both paths (i.e., upper and lower levels of the Feynman diagram), the wave is scattered by different inhomogeneities of the medium. More interesting are the situations where correlations exist across the scattering paths, as illustrated in the examples shown in Figure 14.2. For instance, in Figure 14.2a, we describe a process where both the first and second wave are singly scattered by the same inhomogeneity. Further examples in Figure 14.2 illustrate a ladder diagram and a crossed diagram, which are important building blocks needed for the description of light diffusion in random media and for recurrent scattering, leading to enhanced backscattering.

Similarly to the case of standard (one-level) Feynman diagrams, it is important to distinguish two-level diagrams that have different connectivity properties (i.e., that are topologically different) and introduce reducible and irreducible ones. In particular, a

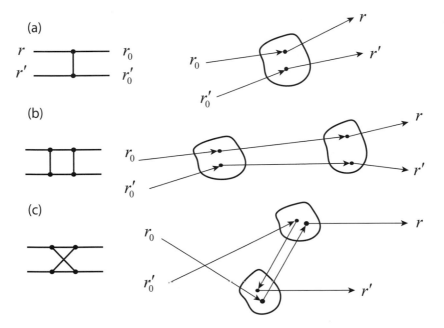

Figure 14.2 Examples of selected two- and fourfold Feynman scattering diagrams and the corresponding scattering processes [753].

reducible two-level diagram is a diagram that can be divided into smaller diagrams without breaking a correlation link. This is, for instance, the case of the following diagram:

$$\text{(14.41)}$$

since it can be cut in the middle without breaking any correlation link. All two-level diagrams that are not reducible are irreducible, and these cannot be divided without breaking a correlation link. The simplest example of an irreducible two-level diagram follows:

$$\text{(14.42)}$$

The expansion series (14.40) for the field correlation contains both reducible and irreducible two-level diagrams. However, similarly to Dyson's equation for the mean field, by concatenating irreducible diagrams we can systematically obtain all the reducible ones.

In order to derive an exact equation for the intensity transport, we first introduce the *irreducible vertex* operator $\mathcal{I}$, which consists of the *sum of all the irreducible two-level diagrams of the field correlation minus their end connectors* [132]:

$$\boxed{\mathcal{I}} = \bigg| + \bigg\times + \bigg\sqcap + \cdots \tag{14.43}$$

We recognize that the irreducible vertex is the analogue of the self-energy operator for the field correlation. Since irreducible diagrams are produced by iterating irreducible ones, we can resum *all the diagrams* that appear in the field correlation as follows:

$$\tag{14.44}$$

The term in parentheses reproduces exactly the field correlation, so that we can now write all diagrams as follows:

$$\tag{14.45}$$

Equation (14.45) is the diagrammatic representation of the *Bethe–Salpeter transport equation* for the field correlation, which is an exact result for the second moment of the field. From its diagrammatic representation, we can write down the Bethe–Salpeter equation in analytical form [132]:

$$\langle G_{ij}(\mathbf{r}, \mathbf{r}_0) G_{kl}^*(\mathbf{r}', \mathbf{r}_0') \rangle = \langle G_{ij}(\mathbf{r}, \mathbf{r}_0) \rangle \langle G_{kl}^*(\mathbf{r}', \mathbf{r}_0') \rangle$$

$$+ \sum_{m=1}^{3} \sum_{n=1}^{3} \sum_{g=1}^{3} \sum_{r=1}^{3} \int d^3\mathbf{r}_1 d^3\mathbf{r}_2 d^3\mathbf{r}_1' d^3\mathbf{r}_2' \langle G_{im}(\mathbf{r}, \mathbf{r}_1) \rangle$$

$$\times \langle G_{kg}^*(\mathbf{r}', \mathbf{r}_1') \rangle I_{mngr}(\mathbf{r}_1, \mathbf{r}_2, \mathbf{r}_1', \mathbf{r}_2') \langle G_{nj}(\mathbf{r}_2, \mathbf{r}_0) G_{rl}^*(\mathbf{r}_2', \mathbf{r}_0') \rangle$$

$$\tag{14.46}$$

where we expressed the dyadic functions in matrix notation and indicated by $I_{mngr}(\mathbf{r}_1, \mathbf{r}_2, \mathbf{r}_1', \mathbf{r}_2')$ the tensor form of the intensity operator.

In general, the intensity operator involves an infinite series of correlated scattering events, which must be approximated (i.e., truncated) in order to be able to solve the Bethe–Salpeter transport equation. Some important approximations will be discussed in the next sections.

The Bethe–Salpeter equation has its origin in quantum field theory, where it describes *bound states* of two-body quantum systems such as the positronium (e.g., electron–positron pair). Bound states have infinite lifetimes, thus their constituents interact infinitely many times and their properties are found by summing over all the possible interaction diagrams. This leads to the formal analogy with multiple light scattering where the bound states correspond to the different terms in the two coupled Green functions that make up the total light intensity transport.

In addition to the field correlation, it is convenient to introduce the *field covariance*, which corresponds to the *reducible vertex operator* diagrammatically defined as follows:

$$\boxed{\mathcal{R}} = \boxed{\mathcal{S}} - \qquad\qquad (14.47)$$

The reducible vertex operator contains the average of all the two-level diagrams in the Neumann's series for the average intensity propagator. The reducible vertex is clearly contributed by all diagrams, reducible and irreducible. However, as we will discuss later, it can be expressed in terms of all the irreducible diagrams. Therefore, using the definition of field covariance inside (14.45), we can alternatively rewrite the Bethe–Salpeter equation in terms of the field covariance as follows:

$$\boxed{\mathcal{R}} = \boxed{\mathcal{I}} + \boxed{\mathcal{I}}\ \boxed{\mathcal{R}} \qquad\qquad (14.48)$$

This equation shows that the operator $\mathcal{R}$ can be expressed in terms of the irreducible vertex operator $\mathcal{I}$:

$$\mathcal{I} = \Big| + \times + \sqcap + \sqcup + \cdots, \qquad\qquad (14.49)$$

which satisfies the operator equation:

$$\boxed{\mathcal{R}} = \boxed{\mathcal{I}} + \boxed{\mathcal{I}}\ \boxed{\mathcal{R}} \qquad\qquad (14.50)$$

This equation is an alternative formulation of the Bethe–Salpeter equation for the field covariance.

Up to this point, we have not used any simplifying assumptions in our discussion of the multiple wave scattering in a Gaussian random medium. The approach discussed was originally developed in the context of remote sensing at radio frequency and is comprehensively discussed in [128, 132]. A rigorous mathematical foundation for the diagrammatic approach can be found in [130]. In order to find explicit solutions for the Bethe–Salpeter transport equation, we need to introduce suitable approximations of the irreducible vertex operator $\mathcal{I}$. For unbounded random media, several solution strategies have been developed. We will just mention here the distorted Born approximation and the ladder approximation [128, 130, 132].

In what follows, we will focus on the case of discrete random media composed of multiple particles arranged in a random fashion, and gradually derive the photon diffusion picture starting from the general Bethe–Salpeter equation. This will be achieved within the so-called *ladder approximation*, which properly describes multiple wave scattering phenomena in the absence of interference effects (i.e, weak scattering). Before proceeding further, we need to slightly modify the diagrammatic notation in order to properly apply it to the description of multiparticle random media.

14.3   Multiparticle Random Media

The multiple-scattering formalism introduced with respect to the permittivity fluctuations in a continuous random medium can be extended naturally to discrete random media composed of multiple particles, each characterized by a T-matrix operator. The multiple scattering theory of discrete random media is known as "microscopic scattering theory," and it provides a rigorous foundation to the effective medium theories for the homogenization of the dielectric response of heterogeneous materials and metamaterials. In this section, we address the problem of discrete or multiparticle scattering systems.

In order to achieve this goal, we need to remember that in Section 12.2 we showed how to express the total T-operator of a system of many scatterers in terms of their individual T-matrix operators. In particular, we now make use of the important result in equation (12.41), which we present again here for convenience:

$$\mathbf{T} = \sum_i \mathbf{t}_i + \sum_i \sum_{j \neq i} \mathbf{t}_i \mathbf{G}_0^+ \mathbf{t}_j + \sum_i \sum_{j \neq i} \sum_{k \neq j} \mathbf{t}_i \mathbf{G}_0^+ \mathbf{t}_j \mathbf{G}_0^+ \mathbf{t}_k + \cdots \qquad (14.51)$$

where $\mathbf{T}$ is the transition operator of the system of scatterers and $\mathbf{t}_i$ are the T-operators of the individual scatterers. The terms on the right-hand side of equation (14.51) contribute to one-particle and multiparticle scattering sequences. Moreover, some multiparticle scattering terms involve the same particle more than once. This can be understood by looking, for instance, at the third contribution to the right-hand side of equation (14.51), where terms with $j = i$ and $k = j$ are excluded but terms with $k = i$ are not. Therefore, the summation in the third term can be split as follows:

$$\sum_i \sum_{j \neq i} \sum_{k \neq j} \mathbf{t}_i \mathbf{G}_0^+ \mathbf{t}_j \mathbf{G}_0^+ \mathbf{t}_k = \sum_i \sum_{j \neq i} \sum_{\substack{k \neq j \\ k \neq i}} \mathbf{t}_i \mathbf{G}_0^+ \mathbf{t}_j \mathbf{G}_0^+ \mathbf{t}_k + \sum_i \sum_{j \neq i} \mathbf{t}_i \mathbf{G}_0^+ \mathbf{t}_j \mathbf{G}_0^+ \mathbf{t}_i .$$

$$(14.52)$$

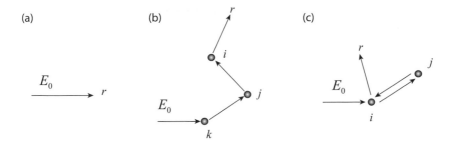

Figure 14.3 Schematic representation of the general terms contributing to the multiple scattering expansion of a discrete medium. (a) Incident field (no scattering); (b) self-avoiding three-particle scattering sequence; (c) three-particle sequence involving particle i twice. This is a recurrent scattering event.

The first contribution to equation (14.52) includes multiparticle sequences involving a given particle only once, so-called *self-avoiding sequences*, while the second one (i.e., the double summation) contains sequences involving the same particle more than once. The higher-order terms in equation (14.51) can be treated analogously. In general, the total field is composed of the incident field, and one-particle and multiparticle contributions. In turn, the multiparticle contributions can be divided into two groups: (i) self-avoiding multiparticle sequences and (ii) multiparticle sequences that involve a given particle more than once. These last terms correspond to *recurrent scattering events*. The repeated occurrence of the same particle in a multiparticle sequence is indicated by a dashed curve in the corresponding diagrammatic expression. These situations are illustrated schematically in Figure 14.3.

In Section 12.1.1, we learned that the total system's transition operator **T** is related to the Green function of the inhomogeneous medium by the following general relation:

$$\mathbf{G} = \mathbf{G}_0 + \mathbf{G}_0 \mathbf{T} \mathbf{G}_0. \tag{14.53}$$

Substituting the expansion (14.51) into the preceding equation, we readily obtain the diagrammatic expression of the system's Green function in terms of the T-operators of its individual components:

$$\mathbf{G} = -\!\!\!+ -\!\times\!- \ + -\!\times\!-\!\times\!- \ + -\!\times\!-\!\times\!-\!\times\!-$$
$$+ -\!\times\!\overset{\frown}{-}\!\times\!-\!\times\!- + -\!\times\!\overset{\frown}{-}\!\times\!-\!\times\!-\!\times\!- + \cdots \tag{14.54}$$

In the new vocabulary shown in Table 14.2, the symbol $\times$ represents the t-operator of the individual scatterers, and we also assumed for simplicity that all the particles in the system are identical.

In turn, the single-particle T-operator $t(\mathbf{r}_1, \mathbf{r}_2, \omega)$ is the sum of all the repeated scattering events from one scatterer, which is diagrammatically expressed as follows:

$$\times = \ \bullet \ + \ \bullet\!\!-\!\!\bullet \ + \ \bullet\!\!-\!\!\bullet\!\!-\!\!\bullet \ + \ \bullet\!\!-\!\!\bullet\!\!-\!\!\bullet\!\!-\!\!\bullet \ + \cdots \tag{14.55}$$

Table 14.2 Drawing conventions for the scattering diagrams of a discrete medium.

Symbol	Represented quantity	Meaning
——	$\overleftrightarrow{G_0}(\mathbf{r}, \mathbf{r}')$	Dyadic Green function of the homogeneous medium
⌢		Connection between identical scatterers (recurrent scattering)
═══	$\langle \overleftrightarrow{G}(\mathbf{r}, \mathbf{r}')\rangle$	Averaged (dressed) dyadic Green function of the random medium
×	$\mathbf{t}$	Single particle T-matrix
⊗	$\langle \mathbf{t}\rangle = \bar{\mathbf{t}}$	Ensemble averaged T-matrix

The T-matrix representation of the individual scatterer can be calculated by available analytical or numerical methods. Note that by inserting (14.55) into (14.54), we obtain again the Born–Neumann series expansion of the Green function of a discrete (deterministic) medium.

The optical properties of the discrete random medium follow by considering the disorder-averaged Green function of the multiple-scattering system, which we write as follows:

$$\langle \mathbf{G}\rangle \equiv \text{═══} = \text{—} + \text{—}\otimes\text{—} + \text{—}\otimes\text{—}\otimes\text{—} + \text{—}\otimes\text{⌢}\otimes\text{⌢}\otimes\text{—} + \cdots \quad (14.56)$$

The symbol $\otimes$ denotes the disorder-averaged T-matrix of the individual scatterers. In the present context, the terms in the preceding perturbation series that appear with dashed lines connecting identical scatterers describe *recurrent scattering events*. In this type of event, a wave that is scattered by a specific scatterer also gets scattered by at least another scatterer in the medium and then returns to the specific scatterer. For relatively weak scattering, recurrent scattering events can be neglected because the chances that a multiply scattered wave returns to a specific scatterer are very small in three-dimensional samples. This approximation is called *self-avoiding multiple-scattering (SAMS) approximation*. In many experimental situations, the scattering is weak enough so that we can neglect recurrent events. This approximation is at the origin of the Twersky multiple scattering equations that neglect all the terms corresponding to paths going through a scatterer more than once [754]. The SAMS approximation breaks down in situations of strong scattering, leading to wave localization, which approximately sets for $k\ell_s < 1$. This condition is the *Ioffe–Regel criterion* for wave localization in random media [755]. However, the effects of recurrent scattering can already be appreciated when $k\ell_s \cong 1$. Following the same reasoning of Section 14.2.4, we deduce that also for a discrete random medium the mean field obeys the Dyson equation:

$$\boxed{=\!=\!=\; = \; =\!=\!= \; + \; =\!=\!\Sigma=\!=}$$ (14.57)

where Σ denotes the *self-energy operator of the multiparticle system*. This operator can be expressed diagrammatically in terms of the disorder-averaged T-matrix [130]:

$$\Sigma = \otimes + —\!\otimes\!—\!\otimes\!—\!\otimes\!— + \otimes\!—\!\otimes\!—\!\otimes\!—\!\otimes + \cdots$$ (14.58)

Due to the multiparticle nature of the discrete medium, there are here two different groups of scattering events: (i) the ones connecting different points within the same particles and (ii) the ones connecting different particles. As a result, many more scattering diagrams will appear in the perturbative expansion of multiparticle discrete systems compared to the case of a continuous random medium. Following a widespread convention in the literature, we have used the dotted lines to link scattering events that occur *on the same scatterer*.

We can Fourier transform Dyson equation (14.57) and obtain the dressed Green function of the effective medium in a way that is formally identical to equation (14.29). The dressed Green function can be calculated exactly if we neglect all recurrent scattering events and keep only the first term in the expansion of the self-energy operator Σ. This approximation is quite accurate for dilute systems (i.e., low density of scatterers). Considering a homogeneous medium with a density of scatterers N, we have, within the validity of our approximation, that $\Sigma \approx \otimes = N\mathbf{t}$. This allows us to express the effective propagation constant in terms of the averaged T-matrix of the single scatterer, according to the following:

$$\boxed{K = \sqrt{k_m^2 + N\mathbf{t}} \approx k_m + \frac{N\mathbf{t}}{2k_m} \approx \left(k_m + \frac{N\Re\{\mathbf{t}\}}{2k_m}\right) + j\frac{N\Im\{\mathbf{t}\}}{2k_m}}$$ (14.59)

where $\Re\{\mathbf{t}\}$ and $\Im\{\mathbf{t}\}$ are the real and imaginary parts of the T-matrix. The preceding expression provides direct access to the average or effective refractive index of the scattering medium $n = K/k_0$ that is at the main goal of standard homogenization theory. We remark that the effective propagation constant, or the effective wavenumber, is determined by the complex poles of the dressed Green function, which, in many-body perturbation theory, gives the energies and lifetimes of quasiparticles. In this analogy, optical waves in the renormalized effective medium behave as quasiparticles with the real part of their effective wavenumber associated to the modification of the refractive index due to the scatterers and the imaginary part of the effective wavenumber associated to a "quasiparticle lifetime," which yields the attenuation coefficient of the field intensity due to multiple-scattering events.

Similarly to the situation of the continuous random medium, the decay constant of the field amplitude is determined by the scattering mean free path in the effective medium through the following relation:

$$\ell_s = \frac{k_m}{N \Im\{\mathbf{t}\}} = \frac{1}{N \sigma_{sc}}$$

(14.60)

where σ_{sc} is the scattering cross section of the individual scatterer and the last equality follows from the optical theorem [90, 273].

14.3.1   Homogenization of Metamaterials

Before turning our attention to the propagation of light intensity in random media, we would like to discuss in more detail an important aspect of homogenization theory known as the *coherent-potential approximation* (CPA). The interest in the homogenization of scattering media has recently been rekindled in the metamaterials community that targets the design of artificial electromagnetic media with desired electric and magnetic effective responses at ever increasing frequencies [138–140]. As discussed in Section 14.2.6, the multiple-scattering theory of random media provides a powerful framework for rigorous homogenization. However, wave scattering effects give rise to nonlocal effective medium parameters (k-dependence) that may render the description via effective parameters of limited practical interest. Establishing conditions that enable efficient parameter reduction through effective quantities in the analysis of arbitrarily scattering metamaterials is a complex issue that is beyond the scope of our discussion. In this section, we are interested to discuss the CPA approach as a strategy for the local homogenization of weakly scattering metamaterials. Since most metamaterial structures are comprised of periodic arrangements of sub-wavelength scattering units, their description often omits averaging procedures over the disorder.

The CPA provides a simple condition that determines the optimal effective wavenumber of the inhomogeneous medium by enforcing the requirement of zero overall scattering [24]. Specifically, for weak scattering where the locality condition $\Sigma(\omega, \mathbf{k}) = \Sigma(\omega)$ holds true, the CPA provides an efficient method to obtain the effective medium parameters. The main idea behind the CPA is that *there is no scattering relative to the uniform medium*. The CPA condition can be readily obtained using the result established in Section 14.2.5, which connects the self-energy operator with the effective index. For weak scattering conditions, we can rewrite equation (14.25) in local form:

$$n^2(\omega) = 1 + \frac{\Sigma(\omega)}{k^2}.$$

(14.61)

Therefore, the *no-scattering condition* becomes equivalent to the requirement:

$$\Sigma(\omega) = 0$$

(14.62)

where the disorder-averaged self-energy $\Sigma(\omega)$ is defined *relative to the effective medium*. Since the self-energy operator is directly proportional to the averaged T-matrix $\mathbf{T}$ relative to the effective medium [24], we can also express the CPA condition as follows:

$$\boxed{\mathbf{T} = 0} \qquad (14.63)$$

The relative T-matrix is related to the exact Green's matrix $\mathbf{G}$ and the effective medium Green's matrix $\mathbf{G}_e$ by the following relation:

$$\mathbf{G} = \mathbf{G}_e + \mathbf{G}_e \mathbf{T} \mathbf{G}_e. \qquad (14.64)$$

The CPA equation (14.63) expresses the fact that in the effective medium there is no scattering on overage. This condition becomes a very practical tool for metamaterials engineering when combined with the T-matrix multiple scattering expansion previously derived in Section 12.2. When the scattering is weak enough, only first-order contributions in the multiple scattering T-matrix expansion can be retained, resulting in the following:

$$\mathbf{T} \simeq \sum_i \mathbf{t}_i \qquad (14.65)$$

where $\mathbf{t}_i$ are the t-matrices of the individual scatterers in the system.

In the case of metamaterials with a periodic unit base of identical scatterers, the CPA condition simply requires the vanishing of t-matrix of such scatterer so that any refractive inhomogeneity does not scatter within the effective medium. This condition is equivalent to enforcing the vanishing of the Mie–Lorentz scattering coefficients of a plane wave that illuminates a particle from inside the effective medium. Implicit analytical expressions for the effective medium permittivity and permeability can be obtained for canonical shapes with a known t-matrix, such as core-shell spheres and cylinders. The described CPA method produces relatively simple homogenization formulas with a validity that is naturally extended beyond the quasistatic limit of the traditional Maxwell–Garnet mixing formulas. A simple implementation of the CPA procedure for metamaterials comprised of spheres of arbitrary composition and small electric size (retaining only the Mie–Lorentz dipole coefficients) is discussed in [133, 134].

14.4   Intensity Transport

In this section, we explicitly show how to use the diagrammatic approach in order to solve the general Bethe–Salpeter equation for the transport of optical waves in weakly scattering media. For simplicity, we consider only scalar waves and refer to the more specialized literature for vector generalizations [132, 655, 756].

The relevant quantity is the disorder-averaged radiation intensity, which is proportional to the modulus square of the field amplitude:

$$I(t, \mathbf{r}) \equiv \langle |\psi(t, \mathbf{r})|^2 \rangle, \qquad (14.66)$$

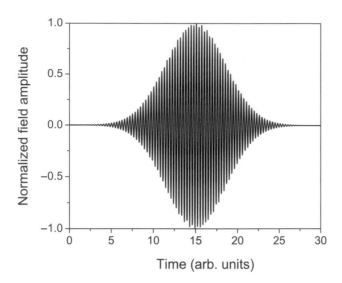

Figure 14.4 An optical wave packet in the time domain. Internal field oscillations occur at a carrier frequency ω and at wavevector $\mathbf{k}$. The envelope of the pulse, here assumed to be Gaussian, is described by the external frequency Ω in time and the wavevector $\mathbf{q}$ in space. The internal variables $(\omega, \mathbf{k})$ are much larger than the external ones $(\Omega, \mathbf{q})$, thus defining the *slowly varying envelope approximation*. The wave packet is denoted by $\Phi_{\omega, \mathbf{k}}(\Omega, \mathbf{q})$.

where ψ is the scalar field. We have already discussed how the knowledge of the total field propagator allows one to construct the *intensity propagator* by averaging the product of the advanced and retarded Green functions.

We now consider in more detail how to describe the time propagation of the intensity neglecting, for the time being, the disorder-averaging procedure. Our discussion will follow closely the derivation originally presented in [690]. First, we address the time evolution of an optical pulse that starts at a given time, $t = 0$, shown in Figure 14.4. Performing a Laplace transformation with respect to time, we can write the following:

$$I(\Omega) = \int_0^\infty dt |\psi(t, \mathbf{r})|^2 e^{j(\Omega + j\epsilon)t}, \tag{14.67}$$

where Ω, also known as the *external pulse frequency*, is the conjugate variable to the transport travel time t and represents the frequency of the envelope of the pulse. An infinitesimally small positive ϵ has been added to the integral in order to guarantee its convergence. We can now expand the pulse in its (angular) frequency components ω by taking its Fourier antitransform and write the following:

$$\psi(t) = \int_{-\infty}^{+\infty} \frac{d\omega}{2\pi} \psi(\omega) e^{j\omega t}, \tag{14.68}$$

where for convenience we dropped the explicit dependence of the pulse on the position vector **r**. In this context, the pulse frequency ω is referred to as the *internal pulse frequency*, which is simply the carrier frequency of the pulse as opposed to the generally much smaller external frequency Ω.

Inserting into (14.67) and replacing ψ with the field propagator G, we obtain an expression for the pulse intensity propagator:

$$P(\Omega) = \int_0^\infty \frac{d\omega}{2\pi} G\left(\omega + \frac{\Omega}{2} + j\epsilon\right) G^*\left(\omega - \frac{\Omega}{2} - j\epsilon\right). \tag{14.69}$$

We see that the frequency-dependent function under the sign of the integral determines completely the time dynamics of the intensity propagation.

Now let us focus on the spatial dynamics of the pulse, and also reinsert the spatial dependence and the averaging over the different realizations of the disorder. By the same logic that led to equation (14.69), this time applied with respect to the spatial Fourier transform, we obtain the *intensity propagator of the random system in the Laplace–Fourier domain*:

$$P(\Omega, \mathbf{q}) = \int_0^\infty \int_0^\infty d\omega d\mathbf{k} \left\langle G\left(\omega + \frac{\Omega}{2}, \mathbf{k} + \frac{\mathbf{q}}{2}\right) G^*\left(\omega - \frac{\Omega}{2}, \mathbf{k} - \frac{\mathbf{q}}{2}\right)\right\rangle \tag{14.70}$$

where we have used the fact that the ensemble averaging of P over an homogeneous disorder only depends on the position difference $\mathbf{r} - \mathbf{r}'$. In the preceding expression, the vector $\mathbf{k}$ is proportional to the spatial frequency of the field, and the vector $\mathbf{q}$ is proportional to the spatial frequency of the intensity. Therefore, the quantities $(\Omega, \mathbf{q})$ are considered *external* while $(\omega, \mathbf{k})$ are *internal*. In most cases, the internal pulse variables $(\omega, \mathbf{k})$ are much larger than the external variables $(\Omega, \mathbf{q})$, which is the essence of the *slowly varying envelope approximation*. It is then clear that the *relevant quantity for the intensity transport* is provided by the following [24]:

$$\Phi_{\omega, \mathbf{k}}(\Omega, \mathbf{q}) = \left\langle G\left(\omega + \frac{\Omega}{2}, \mathbf{k} + \frac{\mathbf{q}}{2}\right) G^*\left(\omega - \frac{\Omega}{2}, \mathbf{k} - \frac{\mathbf{q}}{2}\right)\right\rangle \tag{14.71}$$

In fact, once $\Phi_{\omega, \mathbf{k}}(\Omega, \mathbf{q})$ has been obtained, the intensity propagator $P(\Omega, \mathbf{q})$ can simply be recovered by integrating $\Phi_{\omega, \mathbf{k}}(\Omega, \mathbf{q})$ over the internal degrees of freedom and Laplace–Fourier transforming back into real space. Finally, we can get the desired intensity distribution $I(t, \mathbf{r})$ by integrating the known source function $S(t, \mathbf{r})$ and the pulse propagator.

Following [388, 757], we will show that the knowledge of the average t-operator of the multiparticle random medium provides crucial information on the intensity transport. In fact, making use of the diagrammatic expansion (14.56), we can readily obtain an expression for the relevant transport quantity as follows:

$$(14.72)$$

where the operator $\mathcal{R}$ introduced is the *reducible vertex* and can be expressed as a function of the irreducible vertex operator U of the discrete multi-particle system. Note that, consistent with published literature [24], we here use the symbol U for the irreducible vertex of the discrete random medium as opposed to the symbol $\mathcal{I}$ introduced in Section 14.2.7 for the continuous random medium. Moreover, the reducible operator of the discrete random medium $\mathcal{R}$ (stripped from its incoming and outgoing dressed Green functions) can be written in terms of the irreducible operator U in formal analogy with equation (14.50), which is written as follows:

$$(14.73)$$

where

$$(14.74)$$

In a translationally invariant random medium (i.e., for uniform disorder), the transport quantity $\Phi_{\omega,\mathbf{k}}(\Omega,\mathbf{q})$ obeys the Bethe–Salpeter equation (14.48), which after renaming $\mathcal{I} \rightarrow U$ can be written down explicitly:

$$\Phi_{\omega,\mathbf{k}}(\Omega,\mathbf{q}) = \langle G \rangle \langle G^* \rangle \left[1 + U \Phi_{\omega,\mathbf{k}}(\Omega,\mathbf{q}) \right] \qquad (14.75)$$

The same result can also be derived using equations (14.72) and (14.73), where we denote $G \equiv G\left(\omega + \frac{\Omega}{2}, \mathbf{k} + \frac{\mathbf{q}}{2}\right)$ and $G^* \equiv G^*\left(\omega - \frac{\Omega}{2}, \mathbf{k} - \frac{\mathbf{q}}{2}\right)$.

The Bethe–Salpeter equation (14.75) can be written in a simpler form making use of the identity [24, 388, 690]:

$$\langle G \rangle \langle G^* \rangle = \frac{\langle G \rangle - \langle G^* \rangle}{1/\langle G^* \rangle - 1/\langle G \rangle} = \frac{2j\Delta\langle G \rangle}{1/\langle G^* \rangle - 1/\langle G \rangle}, \tag{14.76}$$

where $\Delta G = (\langle G \rangle - \langle G^* \rangle)/2j$. Remembering the general expression (14.29) for the dressed Green function of the effective medium, we can directly evaluate the denominator of the expression (14.76) as follows:

$$\frac{1}{\langle G^* \rangle} - \frac{1}{\langle G \rangle} = \frac{1}{G_0^*} - \Sigma^* - \frac{1}{G_0} + \Sigma = 2\mathbf{k} \cdot \mathbf{q} + 2j\Delta\Sigma - 2\Omega\omega, \tag{14.77}$$

where $\Sigma \equiv \Sigma(\omega + \Omega/2 + j\epsilon, \mathbf{k} + \mathbf{q}/2)$ and $\Sigma^* \equiv \Sigma^*(\omega - \Omega/2 - j\epsilon, \mathbf{k} - \mathbf{q}/2)$.

In order to explicitly calculate the numerator of the expression (14.76), we need to rely on the weak-scattering (or low-density) approximation, which requires $\Delta\langle G \rangle \approx \Delta G_0$ and $\mathbf{q} \ll \mathbf{k}$. Furthermore, limiting our analysis to the stationary (i.e., CW) regime ($\Omega = 0$), we can write the following:

$$2j\Delta\langle G \rangle \approx G_0(\mathbf{k}, \omega) - G_0^*(\mathbf{k}, \omega) = (k^2 - \omega^2/c^2 + j\epsilon)^{-1} - (k^2 - \omega^2/c^2 - j\epsilon)^{-1}, \tag{14.78}$$

where we used the expression (2.134) for the k-space Green functions. Remembering that the relation (14.78) must be interpreted in the limit of $\epsilon \to 0$, we can evaluate it using the well-known formula:

$$\boxed{\lim_{\epsilon \to 0} \frac{1}{\alpha + j\beta\epsilon} = P.V. \left(\frac{1}{\alpha} \right) - j\beta\pi\delta(\alpha)} \tag{14.79}$$

(where $P.V.$ represents the Cauchy principal value) and obtain the following:

$$\Delta\langle G \rangle \approx \pi\delta(k^2 - \omega^2/c^2). \tag{14.80}$$

Our work so far reduced the Bethe–Salpeter equation (14.75) to the following simpler form:

$$(-j\mathbf{k} \cdot \mathbf{q} + \Delta\Sigma)\Phi_{\omega,\mathbf{k}}(\Omega, \mathbf{q}) = \pi\delta(k^2 - \omega^2/c^2)\left[1 + U\Phi_{\omega,\mathbf{k}}(\Omega, \mathbf{q})\right]. \tag{14.81}$$

However, even this form of the Bethe–Salpeter equation cannot be solved exactly, despite the many approximations. This is because equation (14.81) still involves the exact irreducible vertex operator U, which contains an infinity of unknown scattering diagrams. However, by considering only the contributions of special types of diagrams, it becomes possible to further approximate the irreducible vertex U and obtain a closed-form solution of the Bethe–Salpeter equation. As we will discuss in the next section, this can be accomplished assuming weak scattering of scalar waves when completely neglecting the wave nature of light.

14.4.1 The Ladder Approximation

Recurrent scattering events establish strong phase correlations among the propagating waves that are difficult to account for in multiple-scattering theory. Therefore, a decisive approximation that we should make in order to solve for the average intensity is to neglect all the *recurrent scattering diagrams*. This amounts to neglecting all the diagrams that do not share the exact same sequence of scatterers for the advanced and retarded Green functions. Physically, this is equivalent to assume *uncorrelated phases* for the waves whose average contribution to the reducible vertex is considered to vanish exactly [25]. When neglecting recurrent scattering, the advanced and retarded Green functions can only follow the same sequence of scatterers either in the same or in the opposite order. In this situation, which is described by the SAMS approximation, only two types of diagrams contribute to the reducible vertex. These are the so-called *ladder diagram* $\mathcal{L}$ and the *most crossed diagrams* $\mathcal{C}$:

$$\mathcal{L} = \quad + \quad + \quad + \quad + \cdots \tag{14.82}$$

$$\mathcal{C} = \quad + \quad + \quad + \cdots \tag{14.83}$$

Under this simplifying hypothesis, the reducible vertex can be approximated as follows:

$$\mathcal{R}(\mathbf{r}_2, \mathbf{r}_1, \Omega) = \mathcal{L}(\mathbf{r}_2, \mathbf{r}_1, \Omega) + \mathcal{C}(\mathbf{r}_2, \mathbf{r}_1, \Omega). \tag{14.84}$$

An even more severe approximation consists in neglecting the most crossed diagrams entirely, so that

$$\mathcal{R}(\mathbf{r}_2, \mathbf{r}_1, \Omega) = \mathcal{L}(\mathbf{r}_2, \mathbf{r}_1, \Omega), \tag{14.85}$$

for which:

$$U \approx \tag{14.86}$$

Therefore, in the diagrammatic expansion (14.74) for the irreducible vertex, we have neglected all the terms except the first one. This approximation is called the *ladder approximation* because under this assumption the contributions to the Bethe–Salpeter equation for the field covariance is dominated by ladder-like diagrams:

$$\mathcal{R} \approx \mathcal{L} = \quad + \quad + \quad + \quad + \cdots \tag{14.87}$$

In the terminology of quantum wave transport, this contribution to wave diffusion is called a *Diffuson* [25] and it is prevalent in the weak disorder limit $k\ell_s \gg 1$. The Diffuson can be regarded as a classical contribution to the intensity propagation that does not depend on the phases of the complex amplitudes associated to the paths from which it is constructed. Moreover, in the diffusion approximation, the Diffuson obeys the standard diffusion equation [25].

The preceding derivation clarifies exactly in what sense the ladder approximation completely neglects *all the interference effects in the multiple scattering of light*. As we will show in detail in this chapter, this transport regime results in a diffusion equation for the intensity transport that defines a standard hydrodynamic regime. This diffusion equation is equivalent to a *random walk process* described by a characteristic mean free path ℓ for the propagation of photons through the scattering medium. In this picture, all interference effects are neglected and the wave character of light is not taken into account. As a result, only the intensity and not the electric field is considered. Moreover, it is often assumed in the standard diffusion picture that *no correlations* exist among the different steps of the random walk, which makes it a Markovian process. Uncorrelated random walks are in fact primary examples of memoryless (or Markovian) stochastic process [758], discussed in detail in Appendix C. The Markovian random walk picture provides the underpinning to the successful Monte Carlo statistical description of light transport in multiply scattering media [759], such as biological tissues, which is relevant to biomedical imaging and diagnostics [760]. Spatial and temporal correlation effects in the random walk picture are discussed within the more general continuous time random walk model introduced in Chapter 8, which leads to fractional equations and anomalous diffusion phenomena. We discuss at the end of this chapter a recent application of these methods to random lasing. However, if the medium possesses *time-reversal symmetry*, correlation, and

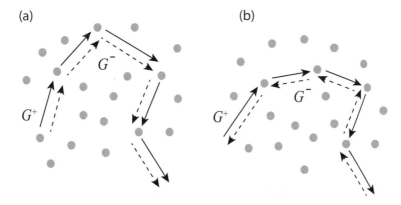

Figure 14.5 Dominant contributions to the intensity propagator $\mathcal{R}(\mathbf{r}_2, \mathbf{r}_1, \Omega)$. (a) A classical Diffuson contribution where both the advanced (G^-) and retarded (G^+) Green functions follow the same sequence of scattering events along the same direction. In this case, the contribution to the average intensity is described by the ladder approximation, which is equivalent to a standard (uncorrelated) random walk. (b) Cooperon contribution to multiple scattering. In this case, the advanced and retarded Green functions follow the sequence of scattering events in opposite directions. This situation is described by the most crossed diagrams, which are the most significant correction to the random walk model and lead to novel physical effects, such as the enhanced coherent backscattering or weak localization.

interference effects cannot be completely neglected even in the weak disorder limit. If fact, under these conditions the two Green functions follow the same sequence of scattering events *in opposite directions* resulting in constructive interference, as shown in Figure 14.5. In this situation, the coherent contribution of the most crossed diagrams leads to a significant modification of the standard diffusion picture with important physical consequences, such as coherent backscattering enhancement. In the context of quantum wave transport, the probability associated with this coherent process is called *Cooperon* due to the similarity with the diagrams investigated by L. N. Cooper in the study of superconductivity [25]. More general interference effects in diffusive systems are regarded as crossings of Diffusons and are described by specific combinations of average Green functions known as *Hikami boxes* [25]. The subtle interactions of Diffusons, Cooperons, and their crossings originate important coherent effects such as weak localization, anomalous transport, conductance fluctuations, and light intensity correlations.

14.4.2   Microscopic Derivation of Diffusion

We are now in the position to present a microscopic derivation of the standard photon diffusion equation. Looking back at the definition of the irreducible vertex in (14.74), we see that the first term only contains two t-matrices of the same scatterer that must be averaged over all possible configurations. Since the average t-matrix in an homogeneous random system does not depend on position, we can simply write $U = N\overline{tt}^*$, where N is the density of particles. Therefore, within the ladder approximation, the Bethe–Salpeter equation becomes

$$(-j\mathbf{k} \cdot \mathbf{q} + \Delta\Sigma)\Phi_{\omega,\mathbf{k}}(\Omega,\mathbf{q}) = \pi\delta(k^2 - \omega^2/c^2)\left[1 + N\overline{tt}^*\Phi_{\omega,\mathbf{k}}(\Omega,\mathbf{q})\right]. \quad (14.88)$$

Now if we approximate the self-energy operator in (14.13) by retaining only its first term, we can write the following:

$$\Delta\Sigma \equiv \frac{\Sigma - \Sigma^*}{2j} \approx N\Im\{t\}. \quad (14.89)$$

By applying the optical theorem, we can also relate the t-matrix product to the imaginary part of the t-matrix using the exact result $\Im\{t\} = \omega t t^*/4\pi c$. Now we can expand $\Phi_{\omega,\mathbf{k}}(0,\mathbf{q})$ in orders of $\mathbf{k}$ and, by retaining only the first two moments, integrate over the internal degrees of freedom $(\omega,\mathbf{k})$ to further approximate $P(\mathbf{q})$. We then obtain an equation in real space for the intensity propagator $P(\mathbf{r})$ by taking the inverse Fourier transform and realizing that each $\mathbf{k}$ is associated to a ∇ operator. Finally, by multiplying every term by the source function $S(\mathbf{r}')$ and integrating over $\mathbf{r}'$, we derive the equation that governs the steady-state (i.e., $\Omega = 0$) intensity distribution $I(\mathbf{r})$. By identifying the *transport mean free path* with $\ell_{tr} \equiv 1/N\overline{tt}^*$ and the *transport velocity* with $v_{tr} \equiv \langle\omega/k\rangle$, we can finally get the stationary diffusion equation for light:

$$\boxed{D_B\nabla^2 I(\mathbf{r}) + S(\mathbf{r}) = 0} \quad (14.90)$$

where $D_B \equiv \ell_{tr} v_{tr}/3$ is *Boltzmann's diffusion constant*. Note that v_{tr} is the average speed at which the energy is transported[2] [690].

For a point source in time and space located at the origin, the photon diffusion equation becomes

$$\boxed{\frac{\partial I(\mathbf{r},t)}{\partial t} = D_B \nabla^2 I(\mathbf{r},t) + \delta(\mathbf{r})\delta(t)} \tag{14.91}$$

Taking the Fourier transform in space $I(\mathbf{r}) \to I(\mathbf{q})$ and the Laplace transform in time $I(t) \to I(s)$ and using the initial condition $I(t=0) = 0$, we have the following:

$$sI = D_B \mathbf{q}^2 I + 1, \tag{14.92}$$

which gives

$$I = \frac{1}{s - D_B \mathbf{q}^2} \tag{14.93}$$

Transforming back in real space we obtain the well-known *Gaussian propagator* of the diffuse intensity:

$$I(\mathbf{r},t) = \frac{1}{(4\pi D_B t)^{3/2}} e^{-|\mathbf{r}|^2/4\pi D_B t}. \tag{14.94}$$

The width of the Gaussian propagator along a coordinate axis (e.g., the x-axis) determines the *mean squared displacement (MSD)* of the diffuse intensity:

$$\boxed{\langle x^2(t)\rangle = \int x^2 I(\mathbf{r},t)dx = D_B t} \tag{14.95}$$

We note that the MSD increases linearly with time, which is the hallmark of the standard diffusive transport.

14.4.3 Photonic Ohm's Law

One of the most interesting consequences of the diffusion equation (14.91) is its prediction that diffuse light transmits through a scattering medium according to the optical analogue of the Ohm's law for electronic conductors. Indeed, if we denote by T the *total diffuse transmission* (i.e., transmitted light intensity collected at all angles) through a scattering slab of thickness L, standard diffusion theory provides the following transmission scaling law:

$$T \propto \frac{z_e}{L}, \tag{14.96}$$

where $z_e \approx \ell_s$ is the so-called *extrapolation length* that identifies the plane outside the scattering medium such that diffuse intensity enters the system. The necessity of the extrapolation length in diffusion theory originates from the observation that the scattering of the incident radiation into diffuse light occurs within a "skin layer" or a

---

[2] In a random scattering medium, it is not meaningful to define a phase velocity or a group velocity since there is no well-defined wave vector and dispersion relation.

"conversion layer" with characteristic thickness of approximately one mean free path, so that we can assume $z_e \approx \ell_s$. The exact value derived from the radiative transfer theory of isotropic point scattering of scalar waves is $z_e = 0.71044609\ell_s$ [30, 761].

A heuristic derivation of the optical Ohm's law can be easily obtained by considering the diffusion equation governing the steady-state intensity transport inside a bulk scattering medium. Let us consider equation (14.91) evaluated outside the source region and in the steady state so that the time derivative vanishes. Also, we consider a slab geometry where there is no dependence on the transverse coordinates. The diffusion equation then reduces to $d^2 I(z)/dz^2 = 0$ subject to the boundary conditions $I(-z_e) = I_0$ and $I(L + z_e) = 0$. The solution is simply a linear function of z that we can write as follows:

$$I(z) = \frac{L + z_e - z}{L + 2z_e} \tag{14.97}$$

The transmission of the diffuse intensity follows from the following definition [367]:

$$T = \frac{I(z = L)}{I_0} = \frac{z_e}{L + 2z_e} \approx \frac{\ell_s}{L}, \tag{14.98}$$

where we considered $L \gg z_e$ and $z_e \approx \ell_s$.

14.4.4 Diffusion Propagator for a Slab

The photon diffusion equation can be generalized to include the effects of absorption by writing the intensity of light $I(\mathbf{r}, t)$ in the following form:

$$\frac{\partial I(\mathbf{r}, t)}{\partial t} = D_B \nabla^2 I(\mathbf{r}, t) - \frac{c}{\ell_i} I(\mathbf{r}, t) + c S(\mathbf{r}, t) \tag{14.99}$$

where c is the transport energy velocity for light inside the scattering medium, $S(\mathbf{r}, t)$ is a source term and $\ell_i = \ell_{abs} = 1/\alpha_{abs}$ is the *inelastic mean free path* corresponding to the absorption coefficient α_{abs} of the medium. It is important to realize that the diffusion equation does not hold if the intensity gradient is steep, and therefore it cannot accurately describe the diffuse intensity near the surface of the scattering medium or near the source function.

We now address photon diffusion in a slab of thickness L oriented such that the propagation axis, which is perpendicular to the interface of the slab, coincides with the z-axis. A time-independent localized source function $S(\mathbf{r}) = S_0(x, y)\delta(z - z')$ is considered. Owing to the translational invariance of the problem in the plane of the slab, we can transform the three-dimensional diffusion equation into a one-dimensional equation that involves the two-dimensional Fourier transform of the *diffuse-intensity propagator*. This propagator describes the diffuse intensity from one point in the slab to another, and it can be obtained by averaging the product of the two conjugate Green functions of the random medium.

In order to proceed, it is convenient to consider the intensity $I(\rho, z, z', t)$, where $\rho = (\mathbf{r} - \mathbf{r}')_\perp$ is the transverse vector in the plane of the slab, and the two-dimensional transform is defined as follows:

$$I(\mathbf{q}_\perp, z, z', t) = \int d^2\rho \, e^{-j\mathbf{q}_\perp \cdot \rho} I(\rho, z, z', t). \tag{14.100}$$

Using the transverse spatial decomposition in the Laplace transform, we obtain the *reduced equation* for the diffuse-intensity propagator from z' to z:

$$\frac{d^2}{dz^2} I(\mathbf{q}_\perp, z, z', \Omega) - M^2 I(\mathbf{q}_\perp, z, z', \Omega) = -\frac{\tilde{S}_0(\mathbf{q}_\perp)}{D_B} \delta(z - z'), \tag{14.101}$$

where we defined $\tilde{S}_0(\mathbf{q}_\perp) \equiv c S_0(\mathbf{q}_\perp)$ and

$$M^2 = q_\perp^2 + \frac{1}{L_a^2} + \frac{\Omega}{D_B}. \tag{14.102}$$

The quantity

$$L_a = \sqrt{\frac{D}{c\alpha}} \tag{14.103}$$

is the *absorption length* over which the flux of diffuse light has decreased by a factor e. Moreover, note that the quantity M in (14.102) corresponds to the inverse depth (i.e., decay rate) at which an incoming beam has decayed by a factor e due to the contributions of spatial dephasing, absorption, and temporal dephasing. The solutions of the reduced equation for the diffuse-intensity propagator of the slab in (14.101) are given by linear combinations of hyperbolic sine and cosine functions.

From the radiative transfer theory, it is known that the diffuse intensity at the interfaces of the slab ($z = 0$ and $z = L$) is not zero. In fact, the appropriate boundary conditions are obtained by considering zero intensity at the finite distance z_e from the plane of the slab, which depends on the refractive index contrast between the slab sample and the surrounding medium. The extrapolation length z_e can be found exactly for an isotropically scattering sample with an index-matched interface (semi-infinite medium) by solving the rigorous Milne integral equation [30, 762], yielding $z_e = 0.7104\ell_{tr}$. Solving equation (14.101) with this boundary condition the following propagator can be obtained as follows [367, 763]:

$$I(\mathbf{q}_\perp, z, z', \Omega) = \frac{\tilde{S}_0(\mathbf{q}_\perp)}{D_B} \frac{[\sinh M z_< + M z_e \cosh M z_<][B(z_>; M, L, z_e)]}{(M + M^3 z_e^2)\sinh ML + 2M^2 z_e \cosh ML} \tag{14.104}$$

where we defined $B(z_>; M, L, z_e) = \sinh M(L - z_>) + M z_e \cosh M(L - z_>)$ with $z_< = \min(z, z')$ and $z_> = \max(z, z')$. As we will show in the following section, the inversion of the Fourier and Laplace transforms can be carried out by the Cauchy residue theorem [764], and the time-dependent propagator can thus be obtained. With the knowledge of the propagator, one can calculate the diffuse intensity in a randomly disordered slab excited by any source function $S(\mathbf{r})$ using the linear superposition principle. An alternative derivation method often encountered in the engineering literature is based on the construction of "image sources" [760].

For optically thick samples with $L \gg \ell_{tr}$, equation (14.104) can be simplified [367]:

$$I(\mathbf{q}_{\perp}, z, z', \Omega) = \frac{\tilde{S}_0(\mathbf{q}_{\perp})}{D_B} \frac{[\sinh M z_<][\sinh (ML - M z_>)]}{M \sinh ML}, \tag{14.105}$$

Moreover, if we consider normal incidence, $(\mathbf{q}_{\perp} = 0)$ in the stationary limit $(\Omega = 0)$ and further neglect absorption $M = 0$ we obtain the following:

$$I(\mathbf{q}_{\perp} = 0, z, z', \Omega = 0) = \frac{\tilde{S}_0(\mathbf{q}_{\perp})}{D_B} \frac{[z_< + z_e][L - z_> + z_e]}{L + 2z_e}. \tag{14.106}$$

Assuming a delta source excitation located inside the sample at $z' = \ell_{tr}$, we can reduce the equation to the following form:

$$I(\mathbf{q}_{\perp} = 0, z, z', \Omega = 0) = \frac{\tilde{S}_0(0)}{D_B} \begin{cases} \frac{L - \ell_{tr} + z_e}{L + 2z_e}(z + z_e), & 0 \leq z \leq \ell_{tr}. \\ \frac{\ell_{tr} + z_e}{L + 2z_e}(L + z_e - z), & \ell_{tr} \leq z \leq L. \end{cases} \tag{14.107}$$

The propagator in (14.107) is a tent-shaped function that completely describes the propagation of diffuse light inside the slab (see Figure 14.6).

In the next section, we will discuss the solution of the diffusion equation in situations that are typical of many optical experiments, namely when the light source inside the sample originates from a laser beam that is incident on its surface. Expressions for the time and spatial profiles of the total reflection and transmission coefficients of diffuse light are obtained using the contour integration method [764].

14.4.5   Total Transmission and Reflection for Diffuse Light

When a laser beam with a transverse spatiotemporal profile $I_0(\mathbf{r}_{\perp}, t)$ is incident on a slab sample, the source term of the diffusion equation can be written as follows :

$$S(\mathbf{r}_{\perp}, z, t) = I_0(\mathbf{r}_{\perp}, t) e^{-z/\ell_{tr}}. \tag{14.108}$$

If the sample is thicker than the mean free path, the source term can be approximated by a point source placed at the average scattering depth:

$$S(\mathbf{r}_{\perp}, z, t) = I_0(\mathbf{r}_{\perp}, t)\delta(z - \ell_{tr}). \tag{14.109}$$

In this case, the boundary conditions for the diffusion equation can be derived by imposing no incoming energy flux at the boundary, where the flux is given by the following [763]:

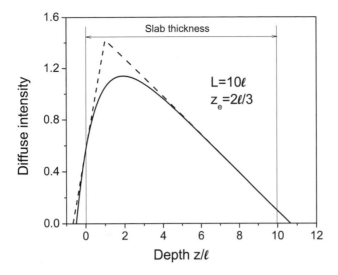

Figure 14.6 Propagators of diffuse intensity (multiplied by D_B/S_0) in a slab of thickness $L = 10\ell_{tr}$ for index-matched layers ($z_e = 2/3\ell_{tr}$). The dashed curve is the solution of the diffusion equation for a delta source positioned at $z = \ell_{tr}$. The full curve is the solution of the diffusion equation with an exponential source $S(z) = (I_0/\ell_{tr})\exp(-z/\ell_{tr})$ in the limit of the thickness L much larger than the transport mean free path ℓ_{tr}. The vertical lines identify the physical boundaries of the slab.

$$\Phi = -D_B \frac{\partial I}{\partial z}. \tag{14.110}$$

The boundary conditions are found to be the following [763]:

$$z_e \frac{\partial I}{\partial z}\Big|_{z=0} - I(z=0) = 0 \tag{14.111}$$

$$z_e - \frac{\partial I}{\partial z}\Big|_{z=L} - I(z=L) = 0. \tag{14.112}$$

As it is shown in the theory of radiative transfer, the solution can be formally extended outside the sample and the boundary conditions simplified to the following:

$$I(-z_e) = 0. \tag{14.113}$$

$$I(L+z_e) = 0. \tag{14.114}$$

The extrapolation length is given by $z_e = 2/3\ell_{tr}$ when the refractive index of the scattering material matches the one of the surrounding environment. When this is not the case, the internal reflections generated by the index mismatch need to be taken into account, which increase the value of the extrapolation length.

The Fourier–Laplace transform of the diffusion equation gives the following:

$$\frac{d^2 I}{dz^2} - M^2 I = -\frac{I_0(\mathbf{q}_\perp, \Omega)\delta(z - \ell_{tr})}{D_B}. \tag{14.115}$$

The diffuse transmission and reflection coefficients can now be properly defined:

$$R = -\Phi(z = 0) \tag{14.116}$$

$$T = \Phi(z = L). \tag{14.117}$$

Introducing the effective length of the slab $L_e = L + 2z_e$ and solving the diffusion equation subject to the boundary conditions, we can finally obtain the reflection and transmission coefficients [764–766]:

$$R = -\frac{\cosh M z_e \sinh M(\ell_{tr} + z_e - L_e)}{\sinh M L_e} I_0(\mathbf{q}_\perp, \Omega) \tag{14.118}$$

$$T = \frac{\cosh M z_e \sinh M(\ell_{tr} + z_e)}{\sinh M L_e} I_0(\mathbf{q}_\perp, \Omega). \tag{14.119}$$

Note that when the incident beam does not depend on time ($\Omega = 0$) and it is localized in the transverse direction $I_0(\mathbf{q}_\perp) = \delta(\mathbf{q}_\perp)$, we can simplify these expressions and obtain the following:

$$R = -\frac{\cosh\left(\frac{z_e}{L_a}\right) \sinh\left(\frac{\ell_{tr} + z_e - L_e}{L_a}\right)}{\sinh\left(\frac{L_e}{L_a}\right)} \tag{14.120}$$

$$T = \frac{\cosh\left(\frac{z_e}{L_a}\right) \sinh\left(\frac{\ell_{tr} + z_e}{L_a}\right)}{\sinh\left(\frac{L_e}{L_a}\right)}. \tag{14.121}$$

In the scattering regime where $L_a \gg L \gg \ell_{tr}$, these equations recover the optical Ohm's law since the total transmission becomes inversely related to the slab thickness:

$$\boxed{T \approx \frac{\ell_{tr} + z_e}{L_e}} \tag{14.122}$$

On the other hand, in the presence of significant absorption $L_a \ll L$, the total transmission always decays exponentially with the thickness:

$$T \approx e^{-L/L_a}. \tag{14.123}$$

14.4.6   Spatial Profile of Diffuse Light

We now discuss the case of a continuous-wave optical beam incident on the input surface of a scattering slab in the absence of absorption. The diffusion equation can be written as follows:

$$\frac{d^2 I}{dz^2} - q_\perp^2 I = -\frac{I_0}{D_B} \delta(z - \ell_{tr}), \tag{14.124}$$

with $I_0(\mathbf{q}_\perp, \Omega) = I_0 \delta(\Omega)$. The total reflection and transmission coefficients are given as follows [763]:

$$R = -\frac{\cosh(q_\perp z_e) \sinh[q_\perp(\ell_{tr} - z_e - L_e)]}{\sinh(q_\perp L_e)} \tag{14.125}$$

$$T = \frac{\cosh(q_\perp z_e) \sinh[q_\perp(\ell_{tr} + z_e)]}{\sinh(q_\perp L_e)}. \tag{14.126}$$

These functions have poles for $q_\perp = \frac{n\pi i}{L + 2z_e}$. To convert the expressions in direct space, we can use Cauchy's residue theorem in order to compute the inverse Fourier transform [764]. If we consider the integral path with a semicircle enclosing the half-plane with a positive imaginary part, we can write the inverse Fourier transform as follows:

$$f(x) = \frac{1}{2\pi} \int_{-\infty}^{\infty} e^{jqx} F(q) dq = j \sum_n e^{jq_n x} Res[F(q), q_n], \tag{14.127}$$

where q_n are the poles of the function $F(q)$ enclosed by the contour and Res denotes the residues of $F(q)$ at the poles.

Evaluating this expression, it is possible to obtain the following series representation for the spatial profiles of the total reflection and transmission [763, 767]:

$$R(x) = \frac{1}{L_e} \sum_{n=1}^{\infty} \sin\left[n\pi \frac{\ell_{tr} + z_e}{L_e}\right] \cos\left[n\pi \frac{z_e}{L_e}\right] e^{-n\pi|x|/L_e} \tag{14.128}$$

$$T(x) = \frac{1}{L_e} \sum_{n=1}^{\infty} (-1)^{n+1} \sin\left[n\pi \frac{\ell_{tr} + z_e}{L_e}\right] \cos\left[n\pi \frac{z_e}{L_e}\right] e^{-n\pi|x|/L_e} \tag{14.129}$$

Note that for a large transverse displacement x, the total reflectance and transmittance fall exponentially with a decay length given by L_e/π. Moreover, the central part of the transmission curve resembles a Gaussian curve with a width proportional to the thickness of the sample, as shown in Figure 14.7.

14.4.7 Time-Resolved Diffused Light

Finally, we consider an incident pulse of duration Δt much shorter than the typical time scale of transport in the medium, i.e., $\Delta t \ll L_e^2/D_B$. Under this condition, we can assume $I_0(\Omega) = 1$. Moreover, let us assume that the beam width is much larger than any length scale in the system, so that $I_0(\mathbf{q}_\perp) = 0$, and let us neglect absorption for simplicity. In this case, the relevant diffusion equation becomes

$$\Omega I = D_B \frac{d^2 I}{dz^2} + \delta(z - \ell_{tr}). \tag{14.130}$$

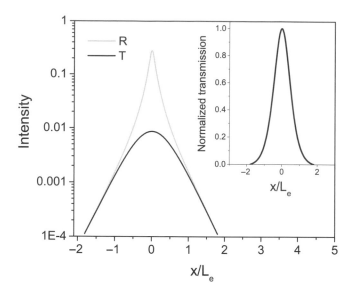

Figure 14.7 Spatial profile of diffuse transmission (the black curve) and reflection (the gray curve). The two curves decay exponentially for long displacements with a decay length L_e/π. The inset shows the normalized transmission in linear scale. The central part of the transmission is approximately a Gaussian curve with a width that is proportional to the thickness of the sample.

Solving this equation, we can compute the flux from (14.111) and hence the reflectance and reflection and transmission [764–766]:

$$R(\Omega) = - \frac{\cosh\left[\sqrt{\frac{\Omega}{D_B}}z_e\right]\sinh\left[\sqrt{\frac{\Omega}{D_B}}(\ell_{tr} + z_e - L_e)\right]}{\sinh\left[\sqrt{\frac{\Omega}{D_B}}L_e\right]} \tag{14.131}$$

$$T(\Omega) = \frac{\cosh\left[\sqrt{\frac{\Omega}{D_B}}z_e\right]\sinh\left[\sqrt{\frac{\Omega}{D_B}}(\ell_{tr} + z_e)\right]}{\sinh\left[\sqrt{\frac{\Omega}{D_R}}L_e\right]}. \tag{14.132}$$

These two expressions have poles at $\Omega = -n^2\pi^2 D_B L_e^2$. Similarly to the case discussed in the previous section, we can compute the inverse Laplace transform using Cauchy's residue theorem [764].

We recall that the inverse Laplace transform of a function $F(\Omega)$ is given as follows:

$$f(t) = \frac{1}{2\pi j}\int_{\gamma - j\infty}^{\gamma + j\infty} e^{\Omega t} F(\Omega)d\Omega, \tag{14.133}$$

where γ is greater than the real part of all the singularities of $F(\Omega)$. This integration path can be extended by adding a semicircle of infinite radius enclosing the region of the complex plane with a negative real part, such that all the poles of $F(\Omega)$ are inside the integration path C. A direct application of Cauchy's residue theorem then provides the inversion formula:

$$f(t) = \frac{1}{2\pi j} \oint_C e^{\Omega t} F(\Omega) d\Omega = \sum_n e^{\Omega_n t} Res[F(\Omega), \Omega_n].$$ (14.134)

Evaluating this expression, we can obtain the following series representations for the temporal profiles of the total reflection and transmission [767]:

$$R(t) = \frac{2D_B \pi}{L_e^2} \sum_{n=1}^{\infty} \left[n \cos \frac{n\pi z_e}{L_e} \sin \frac{n\pi(\ell_{tr} + z_e)}{L_e} \right] e^{-\frac{n^2 \pi^2 D_B}{L_e^2} t}$$ (14.135)

$$T(t) = \frac{2D_B \pi}{L_e^2} \sum_{n=1}^{\infty} \left[n(-1)^{n+1} \cos \frac{n\pi z_e}{L_e} \sin \frac{n\pi(\ell_{tr} + z_e)}{L_e} \right] e^{-\frac{n^2 \pi^2 D_B}{L_e^2} t}$$ (14.136)

It is important to note that both the transmission and the reflection decay exponentially for large times with the time constant $\tau = L_e^2 \pi^2 D_B$, as shown in Figure 14.8. Therefore, the measurement of the time-resolved reflection or transmission allows one to obtain the diffusion constant of light.

Finally, we remark that the presence of absorption in the medium moves the diffusion poles to

$$\Omega = -D_B \left[\left(\frac{n\pi}{L_e} \right)^2 + \left(\frac{1}{L_a} \right)^2 \right]$$ (14.137)

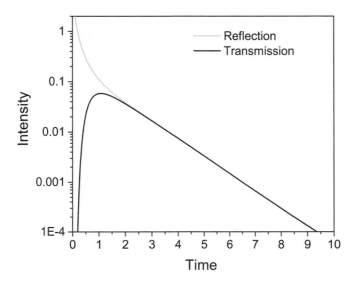

Figure 14.8 Diffuse transmission and reflection of photons in a slab. These curves decay exponentially with a time constant $L_e^2/(\pi^2 D_B)$.

without altering the general form of the series solution. However, the presence of absorption also decreases the decay constant of the time-resolved signals.

14.5   Mesoscopic Scattering Phenomena

Multiple scattering effects accumulate on length scales of many scattering mean free paths and, in the limit of weak scattering, the wave directions are completely random-ized. This gives rise to diffusive transport, which can be described in terms of the classical theory of uncorrelated random walks, i.e., Brownian motions. Therefore, the picture of photon diffusion amounts to completely neglect the wave character of light on the premise that the phase difference associated to any two transport trajectories varies randomly and is expected to vanish on average. However, deviations from this intuitive picture always arise due to interference effects that can significantly alter classical diffusion if the scattering is strong enough, which generally occurs when the wavelength is comparable to the size of the scattering particles. Moreover, genuinely novel physical effects arise due to the interplay of wave interference and strong multiple scattering. These include the coherent backscattering effect, or weak localization, which is the precursor to Anderson wave localization that occurs in the strong disorder limit in which perturbation theory and wave diffusion break down.

Coherent (interference-driven) phenomena in multiple scattering are generally very complicated to describe within the microscopic theory given their funda-mentally nonperturbative nature. Generally, they lead to a gradual modification or renormalization of the diffusion constant that acquires a dependence on the sample size near the localization threshold. In this case, the semiclassical transport theory is profoundly modified by the emergence of *correlation effects* associated to recurrent wave scattering events or cavity effects inside the random medium. These situations are described by the onset of closed loops and self-intersecting trajectories in the underlying microscopic (and random walk) picture. In particular, the formation of self-intersecting trajectories connecting two different points in a strongly scattering random medium dramatically affects its transport properties because closed-loop regions can be equivalently (i.e., with the same phase) traversed clockwise or counterclockwise, thus producing the interference effects that drive the weak localization correction of the standard diffusion (also referred as the Boltzmann limit). As a result, close to the localization threshold, wave propagation can be described by the *self-consistent theory of localization* that predicts a spatially inhomogeneous diffusion constant and a size-dependent transmission due to the enhanced probability of backscattering in the medium [24, 765, 768, 769]. Interestingly, correlation effects could also be accounted using macroscopic transport models based on fractional diffusion equations [22, 390, 391]. The onset of size-dependent effects in the physical properties of complex media due to scattering correlations is the hallmark of *mesoscopic physics* and makes random photonic media particularly suitable to explore fundamental physical effects such as universal conductance fluctuations [25].

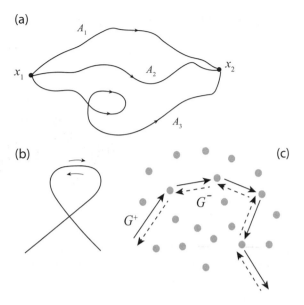

Figure 14.9 (a) Illustration of different trajectories or paths for a particle/wave to propagate from point x_1 to point x_2 in a random medium. Paths A_1 and A_2 are non- self-intersecting, and path A_3 has one self-intersection. (b) A scattering closed loop with arrows indicating time-reversed paths. (c) Advanced and retarded Green functions follow the same sequence of scattering events in opposite directions. This establishes correlations in the standard random walk model that correspond to the Cooperon correction to diffusion and lead to weak localization phenomena.

14.5.1   Weak Localization

In order to introduce the fundamental phenomenon of weak localization, which is the precursor to Anderson localization, we focus on Figure 14.9 a and consider the contribution from closed-loop trajectories to the transport of waves in a random medium. For both quantum particles and classical waves, the probability of propagation from point x_1 to point x_2 is obtained by summing over all paths connecting these two points as follows:

$$P = \sum_i |A_i|^2 + \sum_{i,j} A_i A_j^* \qquad (14.138)$$

where A_i are the complex quantum amplitudes/classical fields associated to its trajectory. As we already mentioned in the previous section, in a uniformly disordered medium no special phase relationship exists between these paths when the points x_1 and x_2 are spatially separated. In this case, the second term in equation (14.138) vanishes and no interference effects contribute to the transport (i.e., standard diffusion). However, the situation is profoundly different when the points x_1 and x_2 coincide in space, corresponding to the loops in Figure 14.9. This can be easily appreciated focusing on the loop in Figure 14.9b, which is traversed clockwise by the amplitude A_1 and counterclockwise by the *time-reversed amplitude A_2* (we assume here that no external bias, such as a magnetic field, breaks the time-reversal symmetry in the system). When inelastic scattering (absorption) is very small, the coherence of the wave amplitudes is maintained in crossing the closed loop and they interfere constructively, leading to $P = |A_1 + A_2|^2 = 4|A|^2$, where we set $A_1 = A_2 = A$ as they only differ by a unitary phase. On the other hand, the same probability for a classically diffusing particle would result in $P_d = |A_1|^2 + |A_2|^2 = 2|A|^2$ (due to the absence of path interference) that is a factor of 2 smaller. More physically, this implies that the wave interference effects increase the probability of backscattering with respect to the classical Boltzmann transport, and the contributions from closed-loop trajectories decrease the transmission (or conductivity) in the system and increase the likelihood of localization.

Therefore, we have seen that a striking example of coherence effects in multiple scattering is provided by the interference of waves that propagate in a random medium by following the same path along opposite directions, as schematized in Figure 14.9c. We note that since the waves have propagated along exactly the same path, their original phase relation is conserved regardless of the number of scattering events that they undergo (neglecting absorption), and a perfect coherence is always preserved in the direction opposite to the incident direction, i.e., the backscattering direction. Consistent with the simple argument presented previously, the probability of backscattering is enhanced by constructive interference, and the backscattered intensity in the SAMS approximation is twice as large compared to all the other directions, as illustrated in Figure 14.10. This effect is the hallmark of weak localization driven by wave interference in the multiple scattering regime, known as *coherent backscattering*

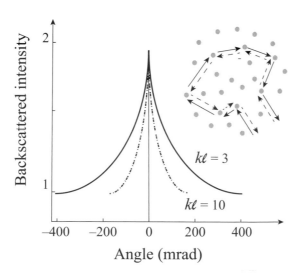

Figure 14.10 Illustration of typical coherent backscattering cone measurements on a random medium for two different values of the $k\ell$ parameter, where k is the wavenumber and ℓ_s the scattering mean free path. The inset shows the interfering paths contributing to the enhanced backscattering.

effect. The angular width W (FWHM) of the backscattering cone is inversely proportional to the scattering mean free path in the medium according to [770–772]:

$$W \approx \frac{0.7n(1-R)}{k\ell_s},$$ (14.139)

where n is the effective refractive index of the medium, $k = 2\pi n/\lambda$ is the wavenumber in the medium, and R is the diffuse reflectance coefficient. The enhancement
factor decreases below 2 once the mean free path becomes of the order of one wavelength, as a manifestation of recurrent scattering events. We point out that the coherent
backscattering effect survives after disorder averaging and can be reduced under an
external magnetic field, illustrating the importance of time-reversal symmetry for this
phenomenon to occur.

An accurate experimental determination of the enhancement factor of the coherent
backscattering cone in the recurrent scattering regime has been provided by Wiersma
et al. [645]. Enhanced backscattering of laser light from highly concentrated suspensions of polystyrene particles in water was experimentally observed by Van Albada
and Lagendijk [34] and direct experimental evidences for wave localization were
obtained by Wiersma et al. [773] using very strongly scattering semiconductor
GaAs powders and by Schuurmans et al. [772] in high-refractive macroporous
GaP networks.

The system's dimensionality plays a crucial role in determining the likelihood
of closed-loop formation leading to weak localization effects. This can be qualitatively understood within standard diffusion theory by computing the *return probability*
for the associated random walk picture [24]. This quantity measures the probability that, for times larger than an arbitrary $t_0 > 0$, the random walker returns to the
origin where it started its motion at $t = 0$. Consistent with the expression (14.94),
the fundamental solution of the diffusion equation in d spatial dimensions is given
by $P(r,t) = (4\pi Dt)^{-d/2} \exp(-r^2/4\pi Dt)$, and the return probability is obtained as
follows:

$$\boxed{\lim_{T\to\infty} \int_{t_0}^{T} P(0,t)dt = \lim_{T\to\infty} \int_{t_0}^{T} \frac{dt}{(4\pi Dt)^{d/2}}}$$ (14.140)

This integral diverges when $d = 1, 2$ independently of t_0, which implies that the random walk will certainly return to the origin for one- and two-dimensional systems.
Three-dimensional systems instead behave very differently because when $d = 3$, the
integral, which is proportional to $1/\sqrt{t_0}$, decreases toward zero for very large t_0.
This situation can be more physically understood by remembering that in a standard
diffusion process, the mean squared displacement is directly proportional to time, i.e.,
$\langle r^2 \rangle \propto t$, regardless of the space dimensionality d. Therefore, in any dimension a random walker will cover in a time t a surface proportional to $\langle r^2 \rangle$, which makes the trace
of its path appear dense when constrained to move along a line ($d = 1$) or on a plane
($d = 2$). In these cases, the random walker will visit with certainty the infinitesimal
neighborhood of any given initial point. On the other hand, when the walker is free

to move randomly in three-dimensional space, its trajectories will be very "rarefied" fractal structures [361] and the return probability will vanish because the explored surface $\langle r^2 \rangle$ cannot fill the entire volume. Since the return probability is qualitatively associated with the backscattering effect, we deduce that weak localization effects are more likely to appear in $d = 1, 2$ as opposed to $d = 3$. This qualitative picture is in fact confirmed by the scaling theory of localization [774, 775] that predicts that in one- and two-dimensional systems of infinite size, all the waves are localized regardless of the strength of disorder. On the other hand, for 3D samples, the scaling theory predicts a possible coexistence of extended and localized states in frequency regimes that are separated by an energy (or frequency) known as the "mobility edge."

The coherent backscattering phenomenon has very remarkable physical consequences since it reduces the diffusion constant from its classical Boltzmann value. Intuitively, this is so because the interference effects that increase the backscattering also tend to drag the radiation backward as if the medium were "more viscous" than it would be expected classically, i.e., neglecting interference in multiple scattering. The downward *renormalization* of the classical diffusion constant D_B is proportional to the overall scattering strength, and the effect is important only when the inelastic mean free path is long enough (low absorption). Figure 14.11 illustrates the behavior of the renormalized diffusion constant as a function of sample size as we increase the strength of disorder by reducing the scattering mean free path ℓ all the way to the localization regime. The self-consistent theory for localization [765, 768, 769] predicts the scale-dependent diffusion constant:

$$D = D_B \left(1 - \frac{\mu}{k^2 \ell^2}\right) \tag{14.141}$$

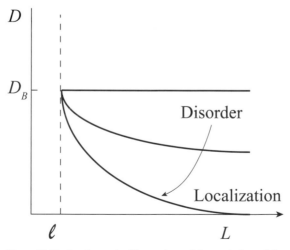

Figure 14.11 A schematic illustration of the variation of the renormalized diffusion constant with sample size L. The sample size must be larger than the mean free path ℓ in order to enter the diffusion regime. The regimes of weak scattering, strong scattering, and localization are shown following the direction of the arrow. D_B denotes the classical Boltzmann value of the diffusion constant.

where μ denotes an upper cutoff parameter that determines the location of the mobility edge, which is defined by $D = 0$ (i.e., for $\mu = 1$, the mobility edge is found at $k\ell = 1$). Van Tiggelen et al. developed a formulation of the self-consistent theory in which the diffusion constant explicitly acquires a position dependence and derived closed-form expressions for the size-dependent transmission and the lineshape of the enhanced backscattering cone [765].

This is an exceptional situation from the point of view of classical physics because it makes the renormalized diffusion constant no longer an intensive (i.e., independent of system's size) quantity such as the temperature or the electrical conductivity, etc. As shown in Figure 14.11, the sample size must be larger than the scattering mean free path in order for light diffusion to be observed. In fact, at length scales that are smaller the mean free path, the light propagation is ballistic and corresponds to wave propagation in an effective (homogeneous) medium. On the other hand, beyond the mean free path and in the weak scattering limit, the diffusion constant is independent of the sample size as expected from the classical Boltzmann transport. However, when the scattering becomes strong enough, the diffusion constant strongly decreases as a function of the sample size and it approaches an asymptotic value that can be significantly less that the classically predicted one. Finally, when the asymptotic value of the renormalized diffusion constant vanishes, the waves become localized and the interplay between strong multiple scattering and interference led to a complete breakdown of the classical diffusion picture [776]. Localized waves are nonperiodic waves that originate from self-trapping in randomly distributed scattering loops inside the sample and do not contribute to the transport of intensity. Their precise description of 3D systems is still an open problem that requires an advanced field-theoretic framework beyond our scope [777, 778].

Recent work has shown that for Hermitian (closed) systems, both Anderson and weak localization of arbitrary scalar waves originate from a universal mechanism that partitions the system into weakly coupled subregions. The boundaries of these subregions correspond to the valleys of a hidden landscape, or effective potential, that emerges from the interplay between the wave operator and the geometry of the system [779]. Since the height of the landscape along its valleys determines the coupling strength between these subregions, wave localization effects can be rigorously predicted by solving a boundary-value problem. This approach, known as the *landscape theory of localization*, has a deep mathematical foundation [780] and is actively pursued for the design of semiconductor devices [781, 782]. In what follows, we limit our discussion to introductory aspects of the localization regime, starting from the main features of Anderson wave localization.

14.6   Anderson Localization

Anderson started the field of wave localization in 1958 by showing that diffusion may be completely suppressed in a disordered lattice model under certain circumstances [31]. In particular, he proposed a tight-binding model with a static random potential and realized that sufficiently strong disorder can lead to a disruption of the metallic conduction behavior and localizes the electronic eigenstates within a finite localization length ξ. Mott argued that there exists a well-defined critical energy or *mobility edge* separating the extended from the localized states [783], eventually leading to the theory of metal-insulator transitions that emerge from the modern renormalization group approach [784].

Interference effects are a general feature of the wave nature and appear for both for optical and quantum (electron) waves. A striking demonstration of such interference effects for electron waves is provided by the Anderson localization that occurs in disordered metallic alloys [31, 751]. For electrons in strongly disordered semiconductors (or metals), the diffusion constant vanishes completely when the electron scattering mean free path becomes smaller than some critical value of the order of the wavelength. This phenomenon can be described as an interference effect that involves counterpropagating waves according to the time-reversed backscattering mechanics previously discussed in relation to weak localization. Due to the enhanced return probability of the waves that have traveled in opposite directions along the same path, the electron diffusion constant decreases dramatically (weak localization) and eventually disappears when the scattering strength is large enough. The condition

$$D(\omega) = 0 \tag{14.142}$$

generally identifies a localized wave solution, or a localized mode, at frequency ω. This localization condition also defines the *localization length* ξ beyond which diffusive transport is no longer a valid description.

In the presence of strong scattering, the renormalized diffusion constant can be expressed as follows:

$$\boxed{D(\omega) = D^{(B)}(\omega) - \delta D(\omega)} \tag{14.143}$$

where $D^{(B)}(\omega)$ is the Boltzmann classical value and $\delta D(\omega)$ is a positive correction due to interference effects. In 1D and 2D random media the correction term to the diffusion constant diverges as the system size goes to infinity ($L \rightarrow \infty$) [24]. Therefore, regardless of the amount of scattering involved, all states become localized in extended 1D and 2D scattering media. The localization lengths for 1D and 2D classical waves can be approximately obtained, respectively, as follows [24]:

$$\xi \simeq (1 + \pi)\ell_s \tag{14.144}$$

$$\xi \simeq \ell_s \exp\left(\frac{\pi}{2} K \ell_s\right), \tag{14.145}$$

where ℓ_s is the isotropic scattering mean free path and K is the wavevector of the effective medium. The preceding equations show that in 1D systems the localization length is proportional to the mean free path while in 2D systems it depends exponentially on ℓ_s and can be very large when the scattering is weak. As a result, light localization in 2D systems is much more difficult to achieve than in the 1D case even though, according to the scaling theory of localization introduced in the next section, disordered 2D systems of infinite size will eventually localize states at all frequencies. For this reason, the case $d = 2$ is referred to as the *marginal dimension* for localization to occur.

The situation for a 3D system is much more complicated and to a certain extent still controversial since when $d = 3$, the correction term to the diffusion constant does not diverge for $L \to \infty$. Therefore, in this case the possibility for $D(\omega)$ to vanish will depend on the magnitude of its coefficients, which are connected in a complicated fashion to the randomness of the model via the self-energy operator. Moreover, in this picture the local density of states and the refractive index contrast in the system play a crucial role in determining the localization transition [24]. For classical waves in 3D systems with infinite size, a convenient, though approximate, condition for light localization is provided by the so-called *Ioffe–Regel criterion* of localization:

$$\boxed{K\ell_s \leq 1} \qquad (14.146)$$

where we considered isotropic scattering particles. More generally, according to the Ioffe–Regel criterion, when the transport mean free path becomes comparable to the effective wavelength in the medium, a wave can no longer build up over one oscillation cycle of its electric field and localization occurs, as illustrated in Figure 14.12. For electromagnetic waves, the stronger scattering (i.e., the smallest values of $k\ell_s$) is obtained for wavelengths comparable to the diameter of the scattering particles, where resonance effects become important, as we can already appreciate from the Mie theory. However, there exists numerical evidence obtained within the self-consistent theory of localization that the localization of classical waves may only occur within the rather narrow frequency window: $0.5 \leq K\ell_s \leq 0.985$ [24].

The existence of a sharp range for localization can be intuitively understood because the scattering strength is becomes very small in the long wavelength limit

(a)   (b)

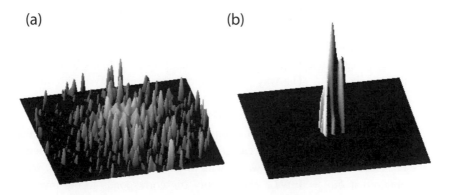

Figure 14.12 Illustration of the spatial distributions of extended and localized modes in a complex scattering medium. (a) Extended mode showing strongly fluctuating spatial patterns (i.e., speckle effect). (b) Localized mode where the radiation is confined within exponentially decaying tails [10].

while at high frequencies geometric optics describes the propagation of classical waves. However, in the geometrical limit the mean free path is expected to saturate at a scale comparable with the particle size or the interparticle separation whereas the wavenumber k (proportional to ω) will diverge. Therefore, light localization is unlikely to occur also in the high-frequency limit and it is most likely limited to an intermediate frequency range and for a specific concentration of scatterers. This, however, poses the problem to compute the effective wavenumber K since the CPA approximation may fail in the intermediate frequency regime, making quantitative predictions on localization extremely difficult [24, 690]. Electromagnetic wave localization has been at the center of intense research activities in recent years, and strong indications of localization have now been found in various disordered systems [34, 773, 776, 785–788].

The experimental demonstration of Anderson localization of photons in random media has been based on different observations that include the exponential decay of the electromagnetic waves as they propagate through a disordered medium. This approach, however, requires a delicate interpretation due to the possibility that the observed decay may be due to residual absorption losses in the samples. To avoid such an ambiguity, Chabanov et al. [786] showed that light localization can be demonstrated in quasi-one-dimensional dielectric samples by measuring the variance of the intensity fluctuations of the total transmission $T_i = \sum_j T_{ij}$. This quantity is the sum of the transmission intensities over all the output channels (i.e., directions) of the medium and is less sensitive to the presence of absorption [786]. Therefore, in such systems, localization manifests as a dramatic departure from the usual inverse exponential intensity distribution of fully developed speckle patterns.

14.6.1   Scaling Theory of Localization

Detailed calculations in the localization regime are often prohibitively complex. For this reason, a phenomenological approach, known as the *scaling theory of localization*, has been introduced [24, 774, 775]. The key observation behind the scaling theory of localization is that a localized eigenfunction should be insensitive to the system boundary conditions as the system size increases beyond its localization length ξ. On the contrary, a mobile (extended) state will be sensitive to the boundary, regardless of the sample size L. The way to characterize this sensitivity to the initial condition is by considering the frequency shift $\delta\omega$ that is the difference in the eigenvalues when the boundary conditions are changed from symmetric periodic boundary conditions to antisymmetric ones. For a homogeneous sample, this change is equivalent to adding or subtracting half of a wavelength, resulting in a wavevector change $\delta k = \pi/L$. For electromagnetic scalar waves with dispersion relation $\omega = vk$, we immediately obtain that $\delta\omega = v\pi/L$. The quantity $\delta\omega$ corresponds to the inverse of the time scale required for the wave to travel from one side of the sample to the other, thus "transmitting" the information on the switching of the boundary conditions. As a first approximation, it can be argued that in the multiple scattering regime this relevant time scale should correspond to the time it takes to diffuse across the length L of the sample. Remembering the classical expression of the diffusion length $\langle L \rangle \approx \sqrt{Dt}$ and using $\delta\omega \approx (\delta t)^{-1}$, we find a first-order scale dependence for the frequency shift [24]:

$$\delta\omega \propto \frac{1}{L^2} \qquad (14.147)$$

This quantity can be physically interpreted as the frequency width arising from the stochastic time-scale characteristic of the diffusive motion required to "communicate" to one end of the sample that the boundary conditions have been changed on the other end. This time scale reflects the obvious fact that the system's wavefunctions cannot adjust to the new boundary conditions abruptly but have to wait until the change signal has propagated diffusively through the sample length L. A quantitative measure of the sensitivity to a boundary condition change can be obtained comparing the frequency shift $\delta\omega$ with the average frequency spacing $\Delta\omega$ between two neighboring eigenvalues, introducing the dimensionless *Thouless conductance*:

$$g = \frac{\delta\omega}{\Delta\omega} \qquad (14.148)$$

where ensemble averages over the realizations of disorder are implied. For classical waves, this quantity can also be defined in terms of the transmittance, which is the sum of the transmission coefficients T_{ij} connecting all the input and output modes, i.e., $g \equiv \sum_{ij} T_{ij}$. Since localized states can feel the effects of the boundary only through their exponential tails, $\delta\omega$ must decrease exponentially with the system's size L. On the other hand, the spacing $\Delta\omega$ has the same dependence with L, i.e., $\Delta\omega \propto L^{-d}$, regardless of the localization character of the waves, implying that the Thouless conductance must decrease exponentially with the system size L. In contrast, for extended waves, $g \propto L^{d-2}$. These simple scaling arguments unveil a strong dimensional dependence of the localization behavior and suggest to investigate the scale variations of g by joining smaller samples together to form successively larger ones, which is the central idea of the scaling theory of localization.[3] For example, in the case of a one-dimensional system, the Thouless conductance would first decrease as $g \propto 1/L$ and then turn into an exponential behavior $g \propto \exp(-2L/\xi)$ in the localization regime, which implies that all the states are localized. On the other hand, when $d = 2$, the Thouless conductance remains constant in the extended regime, decreasing the chances of entering the localization regime, where it eventually decreases. Therefore, by increasing the size L of one- and two-dimensional systems, the dimensionless Thouless conductance $g(L)$ always monotonically decreases.

Let us now consider what happens in a three-dimensional system ($d = 3$). If $\delta\omega \gg \Delta\omega$, the eigenstates of a small piece of sample can easily couple to the spectrally overlapping eigenstates of neighboring samples, giving rise to a new wavefunction that is extended across the larger sample. This leads to an increase of the Thouless conductance when increasing the size L of the system according to $g \propto L$. However,

---

[3] Here we should imagine that a system of size $2L$ is made up by combining smaller blocks of size L and we ask whether or not the waves can propagate across neighboring blocks.

when $\delta\omega \ll \Delta\omega$ the eigenstates of neighboring samples cannot easily overlap through their finite width $\delta\omega$ and these states will remain localized inside the smaller samples. Since a localized mode feels its boundaries only through an evanescent tail, $\delta\omega$ will decrease exponentially with the size L of the large sample. The exponential drop of $\delta\omega$ prevails over the algebraic reduction of $\Delta\omega$, leading to a net decrease in the Thouless conductance with L, which eventually leads to localization in the infinite size limit. Therefore, for three-dimensional systems, the sign of the scaling of the conductance can flip depending on the initial values of g.

The scaling theory, originally proposed by Thouless [789], postulates that $g(L)$ is the only quantity that characterizes the localization transition by controlling the evolution of the eigenstates when we scale the linear size of the system. In fact, it follows from the previous discussion that, when $g(L)$ is initially large, it will diverge by increasing the system size and when $g(L)$ is initially small, it will vanish instead. This naturally leads to postulate a *critical conductance* g_c at which the extended regime crosses into the localized one. Summarizing the previous discussion, we can describe the extended and localized regimes in terms of the $\log g(L)$ as follows:

$$\log g(L) = \begin{cases} (d-2)\log L, & \text{for extended states,} \\ -\exp(\log L), & \text{for localized states,} \end{cases} \tag{14.149}$$

where again $g(L)$ is intended as a configurationally averaged value in the respective regimes. We also note that the derivative $d\log g(L)/(d\log L)$ will be positive for extended states but it will be negative for localized states. Therefore, if the transition from the localized to the extended regime is a continuous one, this derivative will vanish at the critical conductance g_c. This leads to introduce the universal β *scaling function* as the logarithmic derivative:

$$\boxed{\beta = \frac{d\log g(L)}{d\log L}} \tag{14.150}$$

where $g = g_c$ provides the fixed point of the scaling transformation at which the scaling of the system size produces no effects on the value of the conductance. We derive the following from the expressions in (14.149):

$$\beta \propto \begin{cases} d-2, & \text{for } g \gg 1, \\ \log g, & \text{for } g \ll 1, \\ 0, & \text{at } g = g_c. \end{cases} \tag{14.151}$$

In their seminal paper, Abrahams et al. formulated the *scaling hypothesis* in terms of the function β by requiring it to be dependent only on g, so that $\beta = f[g(L), L]$. Since the conductance g depends on the strength of disorder in the medium, the scaling hypothesis implies that any variation in the randomness of the system may be compensated by varying its size L.

Figure 14.13 illustrates the qualitative behavior of the scaling function β in the different dimensions $d = 1, 2, 3$ under the scaling hypothesis. The arrows on the curves indicate in which direction $\log g(L)$ varies when increasing L. The most intriguing

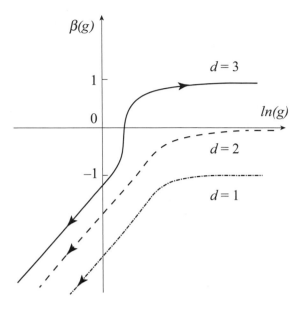

Figure 14.13 Qualitative behavior of the β scaling function versus the logarithmic conductance for different values of the space dimensionality d. Note that for $d \leq 2$, there is no zero for the β function.

feature of the three-dimensional case ($d = 3$) is the existence of the critical expo-nent g_c and a mobility edge $\log g_c$ that make light localization similar to a second-order phase transition. Moreover, in the proximity of the mobility edge, the transport becomes subdiffusive and the modes of the systems display a multifractal behavior [48, 49]. In contrast, in one- and two-dimensional systems β is always a decreasing function of the sample size L and no mobility edges exist. It follows that all the states are localized no matter how weak the randomness is.

14.7 Random Lasers

We conclude our excursion through random media by presenting their application to lasing devices. Traditional lasers require a gain or amplifying medium and a posi-tive feedback loop to turn an optical amplifier into an oscillator. Typically, positive feedback is achieved by positioning the amplifying optical medium in between two parallel and highly reflective mirrors that form an optical cavity. At each roundtrip, the confined cavity radiation gets amplified by the gain medium giving rise, above a characteristic threshold, to laser action, i.e., light amplification by stimulated emission of radiation. However, it was realized only recently that alternative feedback loop mechanisms can also result in laser action. It is now well-established that multiple light scattering in a randomly distributed amplifying medium produces long-lived photon states that experience large enough gain to overcome the lasing threshold without the need of any external mirrors or cavities. Mathematically, photon diffusion through a disordered random medium in the presence of optical gain is described by a reaction-diffusion equation that supports exponentially increasing solutions beyond a critical amplification volume, as first realized in 1968 by Letokhov in a seminal paper on the subject [790]. The resulting devices are called (nonresonant) random lasers and are now at the center of intense research activity due to their ease of fabrication and robustness combined with the small size (micron-scale) of the order of the lasing wavelength as well as their unique characteristics that include low-spatial coherence, lack of directionality, bio-compatibility, etc. An in-depth discussion on the physics and engineering applications of random lasers can be found in [791, 792].

14.7.1   The Letokhov Diffusion Model

We begin our discussion by first reviewing the main features of the Letokhov diffusion model of a nonresonant random laser, which provides the necessary background to appreciate our subsequent fractional generalizations. Mathematically, photon diffusion through in a uniform disordered random medium with optical gain is described by a reaction-diffusion equation that supports exponentially increasing solutions beyond a critical amplification volume V_{cri}, as first realized by Letokhov in his 1968 seminal paper [790]. The classical Letokhov model of a nonresonant random laser is formulated in terms of the following reaction-diffusion equation obeyed by the optical energy density:

$$\frac{\partial W(\vec{r},t)}{\partial t} = D\nabla^2 W(\vec{r},t) + \frac{v}{\ell_g} W(\vec{r},t), \tag{14.152}$$

where D is the diffusion constant of photons, v is the speed of light in the medium, and ℓ_g is its characteristic gain length. Following [790], in our work we will consider $\ell_g \gg \ell_t$, where ℓ_t is the transport mean free path. Furthermore, we have that $D = v\ell_t/2n$ [391], where n is the dimensionality of the problem. In order to more clearly understand the role of fractional operators on the critical volume for light amplification, we will focus on one-dimensional (1D) random media. Moreover, it will be convenient to work with scaled variables in order to generalize our treatment to fractional operators. Specifically, we will consider the scaling $\tau_d = \frac{l_t}{v}$ and $\tau_g = \frac{\ell_g}{v}$, which are the characteristic time for the scattering and the amplification time of a photon, respectively. In these transformed variables, equation (14.152) by τ_d can be rewritten as follows:

$$\frac{\partial W'(x,t')}{\partial t'} = D\tau_d \frac{\partial^2 W'(x,t')}{\partial x^2} + \frac{\tau_d}{\tau_g} W'(x,t'), \tag{14.153}$$

where we additionally defined the scaled time variable as $t' = \frac{t}{\tau_d}$, and $W'(x,t') = W(x,t)$. We notice that the gain coefficient can now be expressed as $\frac{\tau_d}{\tau_g} = \frac{l_t}{l_g}$, which will turn out to be a fundamental parameter in our description of the fractional random laser regimes. Equation 14.153 can be solved by the separation of variables method, which yields the separated form as follows:

$$W'(x,t') = \sum_n a_n \Psi_n(x) e^{-(B_n^2 D\tau_d - \tau_d/\tau_g)t'}, \tag{14.154}$$

where $\Psi_n(x)$ and B_n are the eigenfunctions and the eigenvalues of the following equation:

$$\frac{d^2 \Psi_n(x)}{dx^2} + B_n^2 \Psi_n(x) = 0 \tag{14.155}$$

and $W'(x,0) = \sum_n a_n \Psi_n(x)$. The constants a_n and B_n are determined by the choice of the initial-boundary conditions. This solution can be readily generalized to the three-dimensional random media by considering the equation $\nabla^2 \Psi + B_n^2 \Psi = 0$ instead. In that case, the method of separation of variables represents the solution in the form $W(x,y,z,t) = \Psi(x)\Theta(y)\Lambda(z) f(t')$, which reduces to the solution of the 1D problem $W(x,t) = \Psi(x) f(t')$ in each spatial dimension.

The boundary condition $\Psi_n(x) = 0$ is typically imposed at the extrapolation length x_e beyond the physical border of the scattering medium. Since $x_e \sim \ell_t \ll L$, we assume for simplicity $\Psi(0) = \Psi(L) = 0$. Here L is the total length of the 1D system. By imposing the boundary conditions, we obtain $B_n = \frac{n\pi}{L}$, where n is a positive integer. It is important to realize that the time-dependent part of the solution of equation (14.154) switches from an exponential decay to an exponential growth at a threshold value determined by the following:

$$- B_1^2 D\tau_d + \frac{\tau_d}{\tau_g} \geq 0 \;\Rightarrow\; -B_1^2 D + v/\ell_g \geq 0, \tag{14.156}$$

where $B_1 = \pi/L$ is the lowest-order ($n = 1$) eigenvalue. Therefore, a critical volume $V_{cri} \approx L_{cri}^3$ can be defined for the laser action according to the following:

$$V_{cri} \approx \pi^3 \left(D \frac{\ell_g}{v}\right)^{\frac{3}{2}} = \pi^3 \left(\frac{\ell_t \ell_g}{2}\right)^{\frac{3}{2}} \tag{14.157}$$

In the following sections, we generalize the Letokhov model using fractional operators and obtain analytical expressions for the critical volume in subdiffusive and superdiffusive transport regimes that can be efficiently modeled by considering fractional reaction-diffusion equations.

14.7.2 Fractional Random Lasers

Recently, the concept of a *fractional random laser* was introduced that generalizes the standard Letokhov model to fractional operators in space and time. In one-spatial dimension ($1D$), a fractional random laser obeys the following fractional reaction-diffusion model [390]:

$$D^\alpha W(x,t) = K_\alpha \frac{\partial^2 W(x,t)}{\partial x^2} + W(x,t), \tag{14.158}$$

where D^α denotes the Caputo-type time-fractional derivative of order α, K is a generalized diffusion coefficient, and $W(x,t)$ is the propagating photon flux. The preceding equation effectively describes photon diffusion in a correlated random medium.

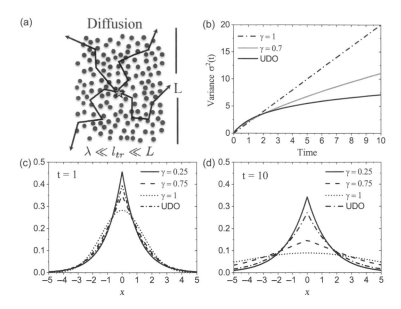

Figure 14.14 (a) Illustration of a general random walk problem in a complex scattering medium. (b) Variance of the fundamental solutions of single-order fractional diffusion equations with different diffusion exponent γ and of distributed-order (DO) fractional in time diffusion equation. (c) and (d) show corresponding probability density functions at two different times.

For one spatial dimension, a solution can be found by Fourier–Laplace inversion methods that allowed us to obtain a closed-form expression for the critical amplification volume necessary to start the laser action [390]:

$$V_{\text{cri}} = l_g^{\frac{3}{2}\alpha} \left(\frac{l_t^{2-\alpha}}{\Gamma(\alpha+1)} \right)^{\frac{3}{2}}. \tag{14.159}$$

In this expression, l_t is the transport mean free path of photons and l_g is the gain mean free path, while Γ is the Euler function. This formula clearly displays the benefits of fractional transport in complex gain media. In fact, for anomalous transport media characterized by small values of the order, the critical volume for random laser action can be significantly smaller when compared to their nonfractional counterparts (see Figure 14.14).

In their recent paper [390], Chen et al. have quantified the benefits of the subdiffusive transport in random lasing by introducing the *figure of merit* η as follows:

$$\eta_\alpha = V_{\text{cri}}/V_\alpha, \tag{14.160}$$

where V_{cri} is the critical amplification volume for the standard ($\alpha = 1$) random lasers and V_α is the volume corresponding to the single-order subdiffusive case. In all investigated cases, they proved that a smaller α value leads to a larger η value, with up to two orders of magnitude decrease in the critical amplification volume of

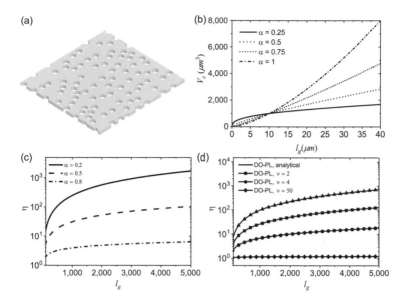

Figure 14.15 (a) Illustration of a possible implementation of a fractional random laser based on a photonic perforated membrane. (b) Critical volume with respect to the gain length for different subdiffusive time fractional orders α and for the Letokhov model. A value of transport mean free path $lt = 10\,\mu\text{m}$ was used in this example. (c) The figure of merit η obtained analytically as a function of the gain length for the time fractional single-order diffusion-reaction equation and (d) the numerical solutions obtained for the DO-U and DO-PL fractional processes with different v values. The continuous red line is the analytical solution of the DO-U case.

fractional random lasers at the largest gain length that we considered ($l = 1$ mm), as shown in Figure 14.15.

Improved efficiencies can be achieved when considering ultraslow or logarithmic-in-time photon transport mechanisms. Ultraslow photon kinetics in gain media can be effectively modeled within the distributed-order uniform model [390]:

$$\int_0^1 p\,(\alpha)D^\alpha W\,(x,t)\,d\alpha = K_p \frac{\partial^2 W\,(x,t)}{\partial x^2} + W\,(x,t), \tag{14.161}$$

where $p(\alpha)$ is the weighting function for the distributed fractional derivative and K_p the corresponding generalized diffusion coefficient. Such a system has an ultralow threshold given by [390]:

$$V_{\text{cri}} \approx 2^{\frac{3}{2}} l_t^{\,3} \log^{\frac{3}{2}} \left(\frac{l_g}{l_t} \right). \tag{14.162}$$

Distributed-order in time leads to an even stronger reduction of the lasing threshold compared to the traditional Letokhov model predictions. Recent work [390] has shown analytically that for ultraslow diffusion described by distributed-order reaction-diffusion equations the figure of merit can be as large as 100, implying a similar reduction in the active volume when considering the values for scattering and gain

Table 14.3 Figures of merit η for different fractional models with fixed $l_t = 10\ \mu\mathrm{m}$

$l_g(\mu\mathrm{m})$	$\alpha = 0.2$ η_α	$\alpha = 0.5$ η_α	$\alpha = 0.8$ η_α	DO-U η_{DO-U}	DO-PL,$\nu = 2$ η_{DO-PL}	DO-PL,$\nu = 4$ η_{DO-PL}
100	15.8	5.6	2.0	7.7	4.0	2.3
200	36.4	9.5	2.5	16.0	6.7	3.0
400	83.7	16.0	3.0	34.4	11.8	4.2
600	136.1	21.6	3.4	54.7	16.8	5.2
800	192.2	26.7	3.7	76.5	21.7	6.0
1,000	251.2	31.6	4.0	99.7	26.6	6.8

length of fabricated devices. Calculated values of η_α for different choices of ℓ_g at three fixed values of α and for $\ell_t = 10\ \mu\mathrm{m}$ are shown in Table 14.3. In all cases, a smaller α value leads to a larger η value, with up to two orders of magnitude decrease in the critical amplification volume of fractional random lasers at the largest gain length that we considered. The table also displays the corresponding values of η obtained considering ultraslow photon diffusion processes modeled by the distributed-order (DO) in time fractional transport models. This analysis shows that random lasers based on subdiffusion phenomena and ultraslow logarithmic photon transport feature significantly reduced amplification length compared to the classical Letokhov model. In particular, DO fractional lasers with uniform distribution (DO-U) are the most efficient systems among the class of DO fractional lasers and that the value of η is larger than the one possible with any single-order fractional laser case when $\ell_g/\ell_t \to \infty$.

The fractional transport analysis discussed here demonstrates the importance of photon subdiffusion phenomena for active photonic devices that could be realized in different geometries, including the photonic membrane structure shown in Figure 14.15a.

Appendix A Bound Sources

We consider the relation between the polarization vector and the induced charge distribution inside a medium. Let V be a volume bounded by the surface $S = \partial V$ containing a certain spatial distribution of polarization as illustrated in Figure A.1. We consider the case of time-independent (static) fields for simplicity, but note that identical results apply to the dynamic (time-varying fields) case. The electric potential at a point $\mathbf{r}$ outside V due to the polarization vector $\mathbf{P}(\mathbf{r})$ inside V can be obtained by superimposing the contributions of the potentials generated by the electric dipoles $\mathbf{P}(\mathbf{r}')dV'$ in the infinitesimal volume elements $dV' = d^3\mathbf{r}'$ as follows:

$$\varphi(\mathbf{r}) = \int_V \frac{\mathbf{P}(\mathbf{r}') \cdot (\mathbf{r} - \mathbf{r}')}{|\mathbf{r} - \mathbf{r}'|^3} d^3\mathbf{r}' = \int_V \mathbf{P}(\mathbf{r}') \cdot \nabla' \frac{1}{|\mathbf{r} - \mathbf{r}'|} d^3\mathbf{r}'. \tag{A.1}$$

Integrating by parts and using Gauss's theorem, we obtain the following:

$$\varphi(\mathbf{r}) = \int_{\partial V} \frac{\mathbf{P}(\mathbf{r}') \cdot \hat{\mathbf{n}}'}{|\mathbf{r} - \mathbf{r}'|} d^2\mathbf{r}' - \int_V \frac{\nabla' \cdot \mathbf{P}(\mathbf{r}')}{|\mathbf{r} - \mathbf{r}'|} d^3\mathbf{r}'. \tag{A.2}$$

The first term in equation (A.2) is the potential due to a *polarization surface charge density*:

$$\boxed{\sigma_{pol} = \mathbf{P} \cdot \hat{\mathbf{n}}} \tag{A.3}$$

while the second term in equation (A.2) is the potential due to a *polarization volume charge density*:

$$\boxed{\rho_{pol} = -\nabla \cdot \mathbf{P}} \tag{A.4}$$

The charge densities derived here are the *macroscopic bound charge contributions*, whose physical nature resides in the local accumulation of charges associated to the spatial variation of the polarization vector.

The induced polarization charges ρ_{pol} and σ_{pol} act as sources for the *depolarization field $\mathbf{E}_d$*, which arises inside a medium immersed in an external electric field $\mathbf{E}_0$. By definition, the externally applied field is the one that would exist if the material objects were not there. In standard dielectric materials, the depolarization field opposes the externally applied field. However, this is not the case for materials characterized by a negative permittivity, such as metals at optical frequencies. Interestingly, in the case of metallic nanostructures, the depolarization field at certain frequencies

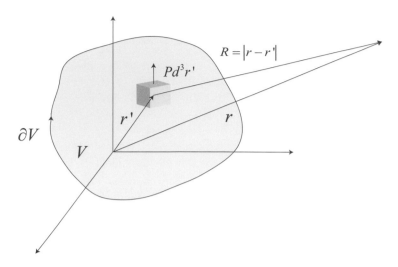

Figure A.1 Potential and field at **r** of dipole $P(\mathbf{r}')d^3\mathbf{r}'$ at $\mathbf{r}'$ inside a polarized volume V bounded by the oriented surface ∂V.

can even reinforce the externally applied field and strongly enhance the electric field values inside such particles, i.e., the *internal field*. This field, which we denote by E_{in}, is the net field inside a material:

$$E_{in} = E_0 + E_d. \tag{A.5}$$

We now turn our attention to induced current contributions. We should immediately realize that a (macroscopic) polarization current must be created by a time-varying polarization vector. This is so because from the preceding equations we have $\partial \rho_{pol}/\partial t = -\nabla \cdot \partial P/\partial t$, and by comparing this with the generally valid continuity equation $\partial \rho/\partial t = -\nabla \cdot J$ we deduce the *polarization current density* as follows:

$$\boxed{J_{pol} = \frac{\partial P}{\partial t}} \tag{A.6}$$

In magnetic materials, an additional magnetization current will be generated by the electrons circulating in atomic current loops, the so-called Amperean current,[1] as well as from the magnetic moment associated to the quantum mechanical spin. However, at the level of macroscopic classical electrodynamics, it is impossible to disentangle the electron orbital motion from the spin contribution to M, and both orbital motion and spin degrees of freedom contribute to create an effective magnetization current.

---

[1] According to recent history [793], it was Ampère's friend Augustin Fresnel who first made the conjecture that magnetism in permanent magnets such as iron is produced by currents circulating around each tiny particle of metal. Each loop of current acts effectively as minute magnets that can align to one another and produce macroscopically strong magnetic effects.

When placing a piece of magnetizable matter inside a large solenoid, we can write the following, in analogy with equation (A.5):

$$B_{in} = B_0 + B_d \qquad (A.7)$$

where B_0 is the applied magnetic flux whose sources are the free currents in the solenoid, and B_d is the field due the volume and surface induced Amperean currents J_{mag} and K_{mag}, respectively.

We now discuss how the magnetization vector gives rise to an electric current density. Let us consider a magnetized volume element V bounded by a surface $S = \partial V$. We will consider again the static case for simplicity, without loss of generality. The magnetization vector inside the volume gives rise to a vector potential:

$$A(r) = \int_V \frac{M(r') \times (r - r')}{|r - r'|^3} d^3 r' = \int_V M(r') \times \nabla' \frac{1}{|r - r'|} d^3 r', \qquad (A.8)$$

which is a sum over the elementary magnetic dipole moments $M d^3 r'$ that are present inside the volume V. Using Gauss's theorem, we obtain the following:

$$A(r) = \int_{\partial V} \frac{M(r')}{|r - r'|} \times d^2 r' + \int_V \frac{\nabla' \times M(r')}{|r - r'|} d^3 r'. \qquad (A.9)$$

We recognize in the first term of equation (A.9) the potential produced by a *magnetization surface current density*:

$$\boxed{K_{mag} = M \times \hat{n}} \qquad (A.10)$$

while the second term in equation (A.9) is the vector potential generated by a *magnetization volume current density*:

$$\boxed{J_{mag} = \nabla \times M} \qquad (A.11)$$

These equations allow us to generalize charge and current densities inside polarizable and magnetic materials by incorporating the induced contributions associated to the polarization vector $P(r, t)$ and the magnetization vector $M(r, t)$.

Appendix B Longitudinal and Transverse Vector Fields

By definition, a *longitudinal or irrotational* vector field $\mathbf{F}_{\parallel}(\mathbf{r})$ satisfies the following:

$$\nabla \times \mathbf{F}_{\parallel}(\mathbf{r}) = 0 \tag{B.1}$$

for all values of $\mathbf{r}$. The preceding definition acquires a geometric meaning when written in k-space, where we have the following:

$$j\mathbf{k} \times \mathbf{F}_{\parallel}(\mathbf{k}) = 0 \tag{B.2}$$

for all values of $\mathbf{k}$. The last expression clearly shows that a longitudinal vector field must always remain parallel to the wavevector.

A *transverse (or solenoidal)* vector field $\mathbf{F}_{\perp}(\mathbf{r})$ is characterized for all values of $\mathbf{r}$ by the following requirement:

$$\nabla \cdot \mathbf{F}_{\perp}(\mathbf{r}) = 0, \tag{B.3}$$

which in k-space becomes

$$j\mathbf{k} \cdot \mathbf{F}_{\perp}(\mathbf{k}) = 0. \tag{B.4}$$

The last expression shows that a transverse vector field remains always perpendicular to wavevector.

The k-space representation is very convenient in this context because it allows a local decomposition of any vector field into its irrotational and solenoidal components:

$$\boxed{\mathbf{F}(\mathbf{k}) = \mathbf{F}_{\parallel}(\mathbf{k}) + \mathbf{F}_{\perp}(\mathbf{k})} \tag{B.5}$$

This basic property of sufficiently regular (and bounded at infinity) vector fields follows from the *Helmholtz's decomposition theorem*, whose constructive proof is provided in the following section. However, as discussed in [85], the longitudinal-transverse field decomposition fails to be relativistically invariant because a vector field that appears to be transverse in a Lorentz frame is not necessarily transverse in another one.

B.1   Helmholtz Field Decomposition

We provide here an informal proof of the Helmholtz's decomposition theorem, which states that a sufficiently regular vector field (i.e., a smooth field that goes to zero at infinity) $\mathbf{F}(\mathbf{r})$ can be expressed as the sum of its longitudinal and transverse components:

$$\mathbf{F}(\mathbf{r}) = \mathbf{F}_{\parallel}(\mathbf{r}) + \mathbf{F}_{\perp}(\mathbf{r}). \tag{B.6}$$

The proof of this theorem provides an explicit construction of the longitudinal and transverse parts in terms of vector operations on $\mathbf{F}(\mathbf{r})$. Let us express the longitudinal and the transverse components of $\mathbf{F}(\mathbf{r})$ in terms of a scalar potential ξ and a vector potential Ψ as follows:

$$\mathbf{F}_{\parallel}(\mathbf{r}) = -\nabla \xi(\mathbf{r}) \tag{B.7}$$

$$\mathbf{F}_{\perp}(\mathbf{r}) = \nabla \times \Psi(\mathbf{r}). \tag{B.8}$$

By definition, equations (B.7) and (B.8) satisfy the requirements $\nabla \times \mathbf{F}_{\parallel}(\mathbf{r}) = 0$ and $\nabla \cdot \mathbf{F}_{\perp}(\mathbf{r}) = 0$. To prove the decomposition in equation (B.6), we then need to determine the functions ξ and Ψ everywhere in space. By considering the divergence of equation (B.6), we obtain the following:

$$\nabla \cdot \mathbf{F}(\mathbf{r}) = \nabla \cdot \mathbf{F}_{\parallel}(\mathbf{r}) = -\nabla^2 \xi(\mathbf{r}) \tag{B.9}$$

Now consider the curl of equation (B.6), which provides the following:

$$\nabla \times \mathbf{F}(\mathbf{r}) = \nabla \times \mathbf{F}_{\perp}(\mathbf{r}) = \nabla \times \nabla \times \Psi(\mathbf{r}). \tag{B.10}$$

Therefore, we have found the following:

$$\nabla^2 \xi(\mathbf{r}) = -\nabla \cdot \mathbf{F}(\mathbf{r}) \tag{B.11}$$

$$\nabla \times \nabla \times \Psi(\mathbf{r}) = \nabla \times \mathbf{F}(\mathbf{r}). \tag{B.12}$$

We now recognize that equation (B.11) has the structure of the Poisson's equation of electrostatics (i.e., $\nabla^2 \varphi = -\rho/\epsilon_0$) with a source charge equal to $\epsilon_0 \nabla \cdot \mathbf{F}$ while equation (B.12) is like Ampère's equation of magnetostatics (i.e., $\nabla \times \nabla \times \mathbf{A} = \mu_0 \mathbf{J}$, where $\mathbf{A}$ is the magnetic vector potential) with a current source term given by $\nabla \times \mathbf{F}/\mu_0$. Therefore, the solution of equations (B.11) and (B.12) can be obtained immediately using the *superposition principle* on the source terms:

$$\xi(\mathbf{r}) = \epsilon_0 \int \frac{\nabla \cdot \mathbf{F}(\mathbf{r}')}{|\mathbf{r} - \mathbf{r}'|} d^3\mathbf{r}' \tag{B.13}$$

$$\Psi(\mathbf{r}) = \frac{1}{\mu_0} \int \frac{\nabla \times \mathbf{F}(\mathbf{r}')}{|\mathbf{r} - \mathbf{r}'|} d^3\mathbf{r}'. \tag{B.14}$$

Note that equations (B.13) and (B.14) combined with equations (B.7) and (B.8) provide the desired expressions for the parallel and perpendicular components of the vector field $\mathbf{F}(\mathbf{r})$ everywhere in space (which proves Helmholtz's theorem) in terms of the divergence and the curl of $\mathbf{F}(\mathbf{r})$. Therefore, it follows as a corollary of Helmholtz's theorem that in order to determine a generic vector field $\mathbf{F}(\mathbf{r})$ *we need to specify its curl and divergence everywhere in space.*

Appendix C Essentials of Random Processes

In this appendix, we introduce essential concepts on random processes along with the basic tools that are necessary to better characterize the behavior of fields in random media. Our discussion mostly follows from the in-depth analysis in references [285, 799, 802, 804], to which we refer the interested readers.

We consider a position in space a real field variable $x(t)$ that fluctuates in time. Instead of following the behavior of the single random signal $x(t)$, we consider measuring the outcomes $x^1(t), x^2(t), x^3(t), \ldots$ of $x(t)$ in a series of experiments that constitute the so-called *ensemble of realizations* (also sample functions or sample paths) associated to the random function $x(t)$.

A *random process* consists of a large number (theoretically infinite) of realizations or paths each regarded as a separate experiment. A few realizations for a typical random process are illustrated in Figure C.1. Arguably the chief example of a real-valued random process with continuous (but nowhere differentiable!) sample paths is the Brownian motion, after Robert Brown, who discovered in 1827 the stochastic movement of pollen particles in a liquid. The Brownian motion is also known as a Wiener process after the work of Norbert Wiener, who made significant contributions to its mathematical theory.

For a typical realization $x^j(t)$ of the ensemble, we can define the corresponding time average as follows:

$$\langle x^j(t) \rangle_t = \lim_{T \to \infty} \frac{1}{2T} \int_{-T}^{T} x^j(t) dt. \tag{C.1}$$

However, it is often more convenient to introduce an additional type of average, called the *ensemble average*, which is defined as the average *over the ensemble of realizations*:

$$\langle x(t) \rangle_e = \lim_{N \to \infty} \frac{1}{N} \sum_{j=1}^{N} x^j(t). \tag{C.2}$$

Therefore, ensemble averages are measured across the ensemble. By determining the values of enough paths at a given time t, a (first-order) probability distribution for x at time t can be calculated. If another series of measurements is made at a different time t', then a second-order or joint probability distribution for x at t and for x at t' can be considered. Similarly, by performing measurements at other times, higher-order

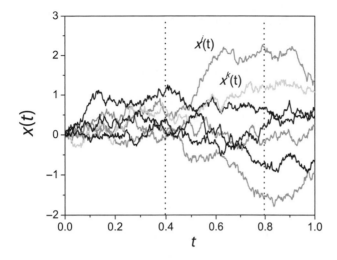

Figure C.1 Few realizations (paths) of a random process on the unit interval [0, 1]. Two representative paths are labeled in the figure. The dotted lines indicate the sampling of the process at two different times and define two random variables. By determining the values of such variables at different times over enough realizations, we can introduce the joint probability distributions as explained in the text.

probability distributions that characterize the entire ensemble can be introduced. For a *Gaussian random process*, the ensemble probability distributions are all Gaussian.

More generally, if we define the *probability density* $p(x,t)$ of an ensemble such that the quantity $p(x,t)dx$ gives the probability that the random variable $x(t)$ will take on a value in the range $(x, x + dx)$ at time t, we can write the *ensemble average* as the integral:[1]

$$\langle x(t)\rangle_e = \int xp(x,t)dx, \tag{C.3}$$

where the integration is extended over all the possible values of x. Remember that at each fixed time instant, $x(t)$ is a random variable, and the collection of random variables $x(t)$, parameterized by t, constitutes a random process. Therefore, a random process consists of an infinite collection of random variables and is specified by time-dependent probability distributions.

More general ensemble averages can be considered, for example the ensemble average of $F[x(t)]$, where $F(x)$ is any deterministic function of x (e.g., $F(x) = x^3$). Such an ensemble average can be defined by generalizing (C.3) as follows:

$$\langle F[x(t)]\rangle_e = \int F(x)p(x,t)dx. \tag{C.4}$$

It is important to realize that the knowledge of the probability density $p(x,t)$ is insufficient to completely characterize a random process. This is because $p(x,t)$

---

[1] The variable of integration x under the integration sign must be considered as time independent. In fact, only the different outcomes of $x(t)$ matter at the ensemble level, regardless of their time of occurrence.

depends only on one time argument and cannot describe statistical correlations (i.e., similarities) that may exist among random variables at different instants of times, e.g., the average of the product $x(t_1)x(t_2)$. Therefore, in order to completely characterize a random process, we need to specify the infinite sequence of higher-order (joint) probability densities:

$$p_1(x_1,t_1), p_2(x_1,x_2;t_1,t_2), p_3(x_1,x_2,x_3;t_1,t_2,t_3), \ldots, \tag{C.5}$$

where $p_1(x_1,t_1) \equiv p(x_1,t_1)$ and the quantity $p_2(x_1,x_2;t_1,t_2)dx_1dx_2$ gives the probability that the variable x takes on a value in the range (x_1,x_1+dx_1) at time t_1 and a value in the range (x_2,x_2+dx_2) at time t_2. The higher-order probability densities are similarly defined.

The knowledge of the (joint) probability densities that we have introduced enables the definition of the *autocorrelation* as the ensemble average of the product $x(t_1)x(t_2)$, which is as follows:

$$\langle x(t_1)x(t_2)\rangle_e = \int \int x_1 x_2 p_2(x_1,x_2;t_1,t_2)dx_1dx_2, \tag{C.6}$$

where we denoted $x_j \equiv x(t_j)$ with $j=1,2$.

Another useful way to characterize a random process is through the *conditional probability density* $p_n(x_n;t_n|x_1,x_2,\ldots,x_{n-1};t_1,t_2,\ldots,t_{n-1})$ that gives the probability density for the occurrence of x_n at time t_n under the condition that the process was found already with values around $x_1,x_2,\ldots,x_{n-1}$ at the previous times $t_1,t_2,\ldots,t_{n-1}$ (so that its past history is known). It follows that the conditional probability density obeys the following relation:

$$\begin{aligned} p_n(x_1,&\ldots,x_n;t_1,\ldots,t_n) \\ &= p_n(x_n;t_n|x_1,\ldots,x_{n-1};t_1,\ldots,t_{n-1})\, p_{n-1}(x_1,\ldots,x_{n-1};t_1,\ldots,t_{n-1}). \end{aligned} \tag{C.7}$$

This equation enables us to express the conditional probability in terms of joint probability densities.

The concepts introduced here can be naturally generalized to complex random functions of time, i.e., complex random processes $z(t) = x(t) + jy(t)$, where $x(t)$ and $y(t)$ are real random processes. Similarly to their real counterparts, complex random processes are characterized by the probability densities:

$$p_1(z_1,t_1), p_2(z_1,z_2;t_1,t_2), p_3(z_1,z_2,z_3;t_1,t_2,t_3), \ldots, \tag{C.8}$$

where $p_1(z_1,t_1)d^2z_1 = p(z_1,t_1)dx_1dy_1$ yields the probability that at time t_1 the random variable z takes on a value represented by a point within a small rectangular region of the complex z_1 plane of area dx_1dy_1 centered around $z_1 = x_1 + jy_1$. The second probability density is defined such that $p_2(z_1,z_2;t_1,t_2)d^2z_1d^2z_2$ gives the probability that the random variable z takes on a value represented by a point in the rectangular region $d^2z_1 = dx_1dy_1$ around $z_1 = x_1 + jy_1$ at time t_1 and a value represented by a point inside the element $d^2z_2 = dx_2dy_2$ around the point $z_2 = x_2 + jy_2$ at time t_2. Higher-order probabilities are similarly defined.

Ensemble averages involving complex random variables are defined as natural generalizations of those involving real variables. For example, the autocorrelation of a complex random process is defined as follows:

$$\langle z^*(t_1)z(t_2)\rangle_e = \int \int z_1^* z_2 \, p_2(z_1, z_2; t_1, t_2) d^2 z_1 d^2 z_2, \tag{C.9}$$

where z^* denotes the complex conjugate of z and the integration extends over all the possible values z_1 and z_2 that the complex variable z can assume.

C.1 Stationary Processes and Ergodicity

In many applications, it is often assumed that the fluctuations of optical fields reached a steady state, meaning that their statistical behavior does not change over time. This situation corresponds to a *strictly stationary random process*, whose joint probability densities $p_1, p_2, p_3, \ldots$ are invariant under translation of the origin of time. Consequently, parameters such as the mean and variance, if they are present, also do not change over time.[2]

More formally, a random process is stationary when the condition

$$p_n(z_1, \ldots, z_n; t_1 + \tau, \ldots, t_n + \tau) = p_n(z_1, \ldots, z_n; t_1, \ldots, t_n) \tag{C.10}$$

holds for all positive integer n and for all values of τ. It is easy to see that the average intensity of a stationary light field $U(t)$ (real or complex), defined by

$$\langle I(t)\rangle_e = \langle U^*(t)U(t)\rangle_e = \int |U|^2 p_1(U, t) dU, \tag{C.11}$$

does not depend on time because p_1 must remain unchanged under a translation of the time origin and therefore cannot depend on time. In addition, the stationarity property implies that that the mean $\mu = \langle x\rangle$, the mean square $\langle x^2\rangle$, and the standard deviation $\sigma^2 = \langle (x - \mu)^2\rangle$ are also not dependent on time. Simple physical examples of stationary random processes are white noise and the radiation from thermal sources.

Let us now return to consider the two previously introduced definitions for the time and ensemble averages. From its definition in (C.1), the time average depends explicitly on the particular realization $x^j(t)$ of the ensemble, and it is independent of time. On the other hand, the ensemble average in (C.2) does not depend on any particular realization but it appears to depend on time. However, when dealing with stationary random fields, the two averages often coincide. More generally, a statistically stationary random process is said to be *ergodic* if the time average of any deterministic function $F(\varphi)$ of a typical realization $\varphi \equiv x^j(t)$ of the process equals the corresponding ensemble average, or as follows:

$$\boxed{\langle F[x^j(t)]\rangle_t = \langle F[x(t)]\rangle_e} \tag{C.12}$$

---

[2] Note that the term *stationary* refers to the probability distributions and not to the samples themselves.

Therefore, a stochastic process is said to be ergodic if its statistical properties can be deduced from a single, sufficiently long, realization of the process. This definition implies that the statistical information about the ensemble of an ergodic random process $x(t)$ is already contained within a single (typical) realization $x^j(t)$ of the process. This apparently surprising result can be better understood by first subdividing a typical realization[3] $x^j(t)$ into many long segments (i.e., each of duration $2T$) and then realizing that by the ergodic hypothesis, the collection of such segments provides a valid ensemble of realizations of $x(t)$. Since such an ensemble has been constructed from the typical realization $x^j(t)$, it can be expected that statistical information derived from it or from the original realization $x^j(t)$ must be the same, which really is the meaning of the ergodic hypothesis. Even if the validity of the ergodic hypothesis cannot easily be verified in a practical context, it is often assumed in many situations of interest so that the subscripts t or e on the angular brackets are often omitted.

C.2   Autocorrelation and Cross-Correlation

The two most important expectation values that can be associated to a random process are its average, or mean, $\mu(t) \equiv \langle x(t) \rangle$ and its *autocorrelation function*:

$$\boxed{R(t_1, t_2) \equiv \langle x(t_1)x(t_2) \rangle} \qquad (C.13)$$

If the process is stationary, its mean $\mu = \langle x(t) \rangle$ and standard deviation $\sigma^2 = \langle (x - \mu)^2 \rangle$ will not depend on time and its autocorrelation function will only depend on the time difference $\tau = t_2 - t_1$. It is then possible to define a *correlation coefficient*, or *normalized autocovariance* given by the following:

$$\rho \equiv \frac{\langle [x(t) - \mu][x(t + \tau) - \mu] \rangle}{\sigma^2} = \frac{R(\tau) - \mu^2}{\sigma^2}, \qquad (C.14)$$

where

$$R(\tau) = \langle x(t)x(t + \tau) \rangle \qquad (C.15)$$

due to the stationarity of the process. Since the limiting values of ρ are ± 1, it follows that the autocorrelation must be bounded as follows:

$$-\sigma^2 + \mu^2 \leq R(\tau) \leq \sigma^2 + \mu^2. \qquad (C.16)$$

As a result, the value of the autocorrelation function can never exceed the *mean squared value* $\langle x(t)^2 \rangle = \sigma^2 + \mu^2$.

It also follows from the definition (C.15) that $R(0) = \langle x^2 \rangle$, so that the maximum value of the autocorrelation at the origin (when $\tau = 0$) equals the mean squared value for the process and $R(\tau)$ will generally decrease when τ is increasing. In particular,

---

[3] Remember that the realizations of a random process extend infinitely in time.

when $\tau \to \infty$, the values of $x(t)$ and $x(t + \tau)$ will be uncorrelated and therefore the correlation coefficient $\rho \to 0$. Therefore, the equation (C.14) implies the following:

$$R(\tau \to \infty) = \mu^2. \qquad (C.17)$$

Consistently, one of the characteristics of ergodic processes of zero average is that $R(\tau) \to 0$ when $\tau \to \infty$.

It should be clear that the autocorrelation function quantifies the statistical similarity between random variables at different times. As a result, the width of $R(\tau)$ measures the time over which there are significant statistical correlations between $x(t)$ and $x(t + \tau)$. In optics, the effective width of the autocorrelation function of a thermal source provides directly its coherence time.

The autocorrelation function satisfies a number of properties, among which the most relevant for a real random process are the following:

$$R(0) \geq 0 \qquad (C.18)$$

$$R(-\tau) = R(\tau) \qquad (C.19)$$

$$|R(\tau)| \leq R(0). \qquad (C.20)$$

In addition, it is useful to remember that the Fourier transform of the autocorrelation function is necessarily nonnegative, enabling the definition of the spectrum of a stationary random process.

The properties we have introduced can be naturally generalized to the case of a complex random process $z(t)$. In particular, for a process of zero mean, we write the autocorrelation as follows:

$$R(\tau) = \langle z^*(t)z(t + \tau) \rangle. \qquad (C.21)$$

All the previous properties still hold true, but C.20 must be complemented with the following Hermitian symmetry property:

$$R(-\tau) = R^*(\tau). \qquad (C.22)$$

A process is defined as stationary in the wide sense or *wide-sense stationary* (WSS) when its mean is independent of time and its correlation function depends only on the time difference τ. Note that a strictly stationary process is also stationary in the wide sense, but the converse is not necessarily true. For many practical engineering applications, stationarity in the wide sense is often sufficient.

The idea of *statistical similarity* quantified by the autocorrelation function can also be generalized to situations involving two different random processes, e.g., $z_1(t)$ and $z_2(t)$. These processes may represent, for instance, an optical field measured at two different spatial points. When the two random processes are *jointly stationary*,[4] their degree of statistical similarity is measured by the *cross-correlation function*:

$$\boxed{R_{12}(\tau) = \langle z_1^*(t)z_2(t + \tau) \rangle} \qquad (C.23)$$

---

[4] This happens when their joint probability density is invariant with respect to translations of the origin of time.

which satisfies the following important properties:

$$|R_{12}(\tau)| \le \sqrt{R_{11}(0)R_{22}(0)} \tag{C.24}$$

$$R_{12}(-\tau) = R_{21}^*(\tau). \tag{C.25}$$

Interested readers can find additional information on random processes and their numerous applications to physics and engineering in a number of excellent books, including the cited references [88, 363, 758, 794–796].

C.3   The Wiener–Khintchine Theorem

The Fourier decomposition of an integrable deterministic function is a well-known tool of spectral analysis. However, its generalization to stationary random processes poses some difficulties since the realizations of random processes are defined over the unbounded interval $-\infty < t < \infty$ and do not in general approach zero at infinity (i.e., they are not integrable functions). Nevertheless, with the help of generalized function theory (i.e., distributions), a rigorous theory of spectral decomposition for WSS random processes (i.e., its mean and autocovariance do not vary with respect to time) has been developed and a few important results are sketched in this section.

Let us consider a wide-sense stationary random process $z(t)$ of zero mean and formally represent a typical realization $z^j(t)$ as follows:

$$z^j(t) = \frac{1}{2\pi} \int_{-\infty}^{+\infty} \xi^j(\omega)e^{-i\omega t}d\omega, \tag{C.26}$$

where

$$\xi^j(\omega) = \int_{-\infty}^{+\infty} z^j(t)e^{i\omega t}dt. \tag{C.27}$$

If we create the product $\xi^j(\omega)\xi^{j*}(\omega')$ and take its ensemble average, after few elementary calculations we obtain the following:

$$\langle \xi^*(\omega)\xi(\omega')\rangle = \int_{-\infty}^{+\infty}\int_{-\infty}^{+\infty} R(\tau)e^{i(\omega'-\omega)t}e^{i\omega'\tau}dtd\tau, \tag{C.28}$$

where $R(\tau) = \langle z^*(t)z(t + \tau)\rangle$ is the previously defined autocorrelation function. The integration over time in (C.28) can be computed directly, yielding a Dirac delta,[5] which allows us to obtain the following:

$$\langle \xi^*(\omega)\xi(\omega')\rangle = S(\omega)\delta(\omega - \omega'), \tag{C.29}$$

where

$$S(\omega) = \int_{-\infty}^{+\infty} R(\tau)e^{i\omega\tau}d\tau. \tag{C.30}$$

---

[5] $\delta(\omega' - \omega) = \int_{-\infty}^{+\infty} e^{i(\omega'-\omega)t}dt.$

Notice that the Fourier transform that appears in this expression is well defined since the autocorrelation must decay to zero at infinity for ergodic processes. Taking the inverse Fourier transform of C.30, we have the following:

$$R(\tau) = \frac{1}{2\pi} \int_{-\infty}^{+\infty} S(\omega) e^{-i\omega\tau} d\omega. \tag{C.31}$$

The three formulas of (C.29) and (C.31) are the main results of this section. The first shows that the generalized spectral components of a wide-sense stationary random process are completely uncorrelated (i.e., delta-correlated). Moreover, the strength $S(\omega)$ of the self-correlation ($\omega = \omega'$) coincides with the Fourier transform of the autocorrelation function $R(\tau)$ of the random process. The quantity $S(\omega)$, which is formally defined by equation (C.29), is known as the *spectral density* or *power spectrum* of the stationary random process $z(t)$. The formulas (C.30) and (C.31) express the so-called *Wiener–Khintchine theorem*. This theorem expresses the fact that the spectrum of a wide-sense stationary random process $S(\omega)$ and its autocorrelation function $R(\tau)$ form a Fourier transform pair. Note that from the definition of the autocorrelation and the Wiener–Khintchine theorem, we have the following:

$$\langle x^2 \rangle = R(0) = \int_{-\infty}^{+\infty} S(\omega) d\omega. \tag{C.32}$$

Therefore, the mean squared value of a stationary random process is simply the area under the graph of its spectral density.

The Wiener–Khintchine theorem can be generalized to a pair of jointly (wide-sense) stationary random processes $z_1(t)$ and $z_2(t)$. Following a similar analysis to the one that previously led to equations (C.29) and (C.31), it is possible to establish the following:

$$\langle \xi_1^*(\omega)\xi_2(\omega') \rangle = W_{12}(\omega)\delta(\omega - \omega'), \tag{C.33}$$

where

$$W_{12}(\omega) = \int_{-\infty}^{+\infty} R_{12}(\tau) e^{i\omega\tau} d\tau. \tag{C.34}$$

Taking the inverse Fourier transform of (C.34), we have the following:

$$R_{12}(\tau) = \frac{1}{2\pi} \int_{-\infty}^{+\infty} W_{12}(\omega) e^{-i\omega\tau} d\omega. \tag{C.35}$$

In these equations, $\xi_1(\omega)$ and $\xi_2(\omega)$ are the generalized Fourier transforms of $z_1(t)$ and $z_2(t)$, the function $W_{12}(\omega)$ is called the *cross-spectral density* of $z_1(t)$ and $z_2(t)$, and the function

$$R_{12}(\tau) = \langle z_1^*(t)z_2(t + \tau) \rangle \tag{C.36}$$

is the cross-correlation function of the two random processes. The relations (C.34) and (C.35) express the *generalized Wiener–Khintchine theorem*.

C.4   Linear Systems and Random Processes

Linear systems model a variety of different situations in science and engineering and can be generally divided into two categories: (i) deterministic and (ii) random or stochastic. In the deterministic case, a given input is linearly mapped to a unique output. On the other hand, we cannot exactly predict the output of random systems with certainty, but we can compute the probability of different outputs. In this section, we discuss deterministic linear systems and their effects on both deterministic and random inputs. Additional information on this extensive and important engineering subject can be found in [363, 797].

Linear systems describe the relationship between a generic input, for instance a time signal $x_n(t)$, and the corresponding output $y_n(t)$ in terms of the action of a linear operator L on its input:

$$y_n(t) = L[x_n(t)]. \tag{C.37}$$

Linear systems satisfy the following property:

$$L\left[\sum_n a_n x_n(t)\right] = \sum_n a_n L[x_n(t)] = \sum_n a_n y_n(t). \tag{C.38}$$

Therefore, an arbitrary linear combination of inputs gives rise to the same linear combination of the corresponding outputs. This property renders linear systems extremely valuable as mathematical models for different physical situations since the response to a complex stimulus $x = \sum_n a_n x_n(t)$ can be expressed in terms of the response to individual stimuli as $y = \sum_n a_n y_n(t)$.

A particularly useful type of linear system is the one of *time-invariant (LTI) systems*, also known for spatial signals as *linear and shift-invariant (LSI) systems*. LTI and LSI linear systems have the additional property that a time translation (or a spatial shift) of the input $x_n(t - \tau)$ produces a corresponding shift of the output $y_n(t - \tau)$ or, more generally, we have the following:

$$L\left[\sum_n a_n x_n(t - \tau_n)\right] = \sum_n a_n y_n(t - \tau_n). \tag{C.39}$$

The input–output relationship of LTI systems (or LSI for space-dependent functions) is expressed in the time domain by a convolution integral:[6]

$$\boxed{y(t) = \int_{-\infty}^{+\infty} h(\tau)x(t - \tau)d\tau} \tag{C.40}$$

---

[6] This can easily be realized by using the sampling property of the Dirac delta function to represent a generic input signal $x(t) = \int x(\tau)\delta(t - \tau)d\tau$ and applying the linearity property to $y(t) = L[x(t)]$.

where $h(t; \tau) = h(t-\tau) \equiv L[\delta(t - \tau)]$ is the *impulse-response function* $h(t) \equiv h(t; 0)$ that entirely characterizes the behavior of the LTI system. An alternative description of LTI systems is obtained in the frequency domain by Fourier transforming the previous expression, which yields the following:

$$\boxed{Y(\omega) = H(\omega)X(\omega)} \tag{C.41}$$

where $X(\omega)$ and $Y(\omega)$ are the Fourier transforms of $x(t)$ and $y(t)$, respectively, and $H(\omega)$, called the *transfer function*, is the Fourier transform of the impulse-response function $h(t)$.

Let us now consider the action of an LTI system with impulse-response $h(t)$ on an input signal that is a stochastic process $X(t)$.[7] Clearly in this situation the output of the linear systems L is also a stochastic process that we denote by $Y(t)$. Then, if we consider a single realization, the action of the system is as follows:

$$Y^j(t) = \int_{-\infty}^{+\infty} h(\tau)X^j(t - \tau)d\tau. \tag{C.42}$$

If we require the input process $X(t)$ to be wide-sense stationary, we can compute the mean value of the output signal by considering the ensemble average of the previous equation, which yields the following:

$$\boxed{\langle Y(t) \rangle = \int_{-\infty}^{+\infty} h(\tau)\langle X(t - \tau) \rangle d\tau = \langle X(t) \rangle \int_{-\infty}^{+\infty} h(\tau)d\tau = \langle X(t) \rangle H(0)} \tag{C.43}$$

Therefore, the mean value of the output process equals the mean value of the input process multiplied by the value of the transfer function at the origin (i.e., the d.c. component), which is the area under the curve of the impulse response function $h(t)$.

The autocorrelation function of the output process can be easily obtained by evaluating the following ensemble average:

$$R_{YY}(t, t + \tau) = \langle Y(t)Y(t + \tau) \rangle. \tag{C.44}$$

Using the input–output convolution relation (C.42) inside the previous expression and after rearranging the order of integration, we obtain the following:

$$\langle Y(t)Y(t + \tau) \rangle = \int_{-\infty}^{+\infty} \int_{-\infty}^{+\infty} \langle X(t - t')X(t + \tau - t'') \rangle h(t')h(t'')dt'dt''. \tag{C.45}$$

Assuming wide-sense stationarity for the process $X(t)$, the previous equation reduces to the following:

$$R_{YY}(\tau) = \int_{-\infty}^{+\infty} \int_{-\infty}^{+\infty} R_{XX}(\tau + t' - t'')h(t')h(t'')dt'dt''. \tag{C.46}$$

---

[7] Analogous conclusions will hold for the case of an LSI system.

We can rewrite the last result in simpler form as follows:

$$R_{YY}(\tau) = R_{XX}(\tau) * h(\tau) * h(-\tau) \tag{C.47}$$

where the symbol $*$ represents the convolution operation.

The mean square value of the output process is immediately obtained as follows:

$$\langle Y^2(t) \rangle = R_{YY}(0) = \int_{-\infty}^{+\infty} \int_{-\infty}^{+\infty} R_{XX}(t' - t'')h(t')h(t'')dt'dt''. \tag{C.48}$$

Note that the output process $Y(t)$ is also wide-sense stationary when the input process $X(t)$ is wide-sense stationary since in the last equations there is no explicit time dependence.

An analysis similar to that which is presented here leads us to compute the cross-correlation of the input and output processes as follows:

$$R_{XY}(t, t + \tau) \equiv \langle X(t)Y(t + \tau) \rangle = \int_{-\infty}^{+\infty} \langle X(t)X(t + \tau - t') \rangle h(t')dt'. \tag{C.49}$$

When $X(t)$ is wide-sense stationary, the previous expression reduces to the following:

$$R_{XY}(\tau) = \int_{-\infty}^{+\infty} R_{XX}(\tau - t')h(t')dt' = R_{XX}(\tau) * h(\tau) \tag{C.50}$$

In a similar way, we can also obtain the following:

$$R_{YX}(\tau) = \int_{-\infty}^{+\infty} R_{XX}(\tau + t')h(t')dt' = R_{XX}(\tau) * h(-\tau) \tag{C.51}$$

In addition, it is easy to prove the following relationships based on equations (C.46), (C.51), and (C.50):

$$R_{YY}(\tau) = R_{XY}(\tau) * h(-\tau) \tag{C.52}$$

$$R_{YY}(\tau) = R_{YX}(\tau) * h(\tau). \tag{C.53}$$

We now derive the spectral density of the output process $Y(t)$, always assuming a wide-sense input process $X(t)$, by Fourier transforming equation (C.46). When the impulse response is a real function,[8] we immediately obtain the following important result:

$$S_{YY}(\omega) = S_{XX}(\omega)|H(\omega)|^2 \tag{C.54}$$

This expression allows us to compute the mean squared value of the output $Y(t)$, or the *mean power of the output process*, in the following form:

$$P_Y \equiv R_{YY}(0) = \int_{-\infty}^{+\infty} S_{YY}(\omega)d\omega = \int_{-\infty}^{+\infty} S_{XX}(\omega)|H(\omega)|^2 d\omega, \tag{C.55}$$

where we used the Wiener–Khintchine theorem.

---

[8] Therefore, the Fourier transform of $h(-t)$ equals $H^*(\omega)$.

C.5 Karhunen–Loéve Expansion of Random Processes

This important result allows us to express a random process, not necessarily Gaussian, as a linear combination, with uncorrelated random coefficients, of the eigenfunctions of its autocorrelation function. The orthogonal basis functions used in this representation are determined by the covariance function of the process. In many applications, particularly in relation to the theory of partial optical coherence of waves discussed in Section 6.4, it is useful to represent a random process in terms of a *complete orthonormal system* (CONS) of functions $\{\phi_n(t)\}$ (with $n = 0, 1, \ldots$) in $\mathbf{L}^2([-T/2, +T/2])$ that satisfy the following:

$$\int_{-T/2}^{T/2} \phi_m(t)\phi_n^*(t)dt = \begin{cases} 1 & \text{if } n = m. \\ 0 & \text{if } n \neq (m). \end{cases} \tag{C.56}$$

Since $\{\phi_n(t)\}$ are a complete set of basis functions in the interval $(-T/2, T/2)$, we can expand any (suitably regular) function, including the sample functions $x^j(t)$ of the random process $x(t)$, as follows:

$$x^j(t) = \sum_{n=0}^{\infty} b_n \phi_n(t), \tag{C.57}$$

where the expansion coefficient is given by the following:

$$b_n = \int_{-T/2}^{T/2} x^j(t)\phi_n^*(t)dt. \tag{C.58}$$

If we now consider a random process with a given autocorrelation function $R(t_1, t_2)$, we can ask how to find the particular set of basis functions $\{\phi_n(t)\}$ such that the expansion coefficients b_n are uncorrelated, i.e.:

$$\langle b_n b_m^* \rangle = \begin{cases} \lambda_m & \text{if } m = n \\ 0 & \text{if } m \neq n \end{cases} \tag{C.59}$$

Assuming for simplicity a zero mean wide-sense stationary random process (so that the correlation coincides with the variance), the condition that must be satisfied by the orthonormal set $\{\phi_n(t)\}$ in order to achieve uncorrelated coefficients can be readily obtained by substituting the expression of the expansion coefficients (C.58) into (C.59), yielding the following:

$$\langle b_n b_m^* \rangle = \int_{-T/2}^{T/2} \left[\int_{-T/2}^{T/2} R(t_1, t_2)\phi_m(t_1)dt_1 \right] \phi_n^*(t_2)dt_2. \tag{C.60}$$

Therefore, the coefficients b_n will be uncorrelated when the set of basis functions $\{\phi_n(t)\}$ is chosen to satisfy the following integral equation:

$$\boxed{\int_{-T/2}^{T/2} R(t_1, t_2)\phi_m(t_1)dt_1 = \lambda_m \phi_m(t_2)} \tag{C.61}$$

This equation shows that the required orthonormal basis functions are the eigenfunctions of a *Fredholm integral equation* of the second kind with a kernel given by the correlation function of the process under consideration. We can now write the correlation of the expansion coefficients as follows:

$$\langle b_n b_m^* \rangle = \int_{-T/2}^{T/2} \lambda_m \phi_m(t_2) \phi_n^*(t_2) dt_2 = \begin{cases} \lambda_m & \text{if } m = n \\ 0 & \text{if } m \neq n, \end{cases} \quad \text{(C.62)}$$

which clearly shows the uncorrelated nature of the expansion coefficients.

Therefore, we have shown that a zero mean, wide-sense stationary random process $x(t)$ can be represented using an orthonormal set of functions as follows:

$$\boxed{x(t) = \sum_{n=0}^{\infty} b_n \phi_n(t)} \quad \text{(C.63)}$$

where the expansion coefficients satisfy the following:

$$\langle b_n b_m^* \rangle = \lambda_n \delta_{nm} \quad \text{(C.64)}$$

(δ_{nm} is the Kronecker symbol), and the eigenfunctions $\{\phi_n(t)\}$ are the solutions of the integral equation (C.61). This important representation of a random process is known as *Karhunen–Loéve expansion*.

In general, adding more terms in the Karhunen–Loéve expansion corresponds to adding oscillations of increasing frequency and decreasing amplitude. Note that due to the uncorrelated nature of the expansion coefficients, the Karhunen–Loéve expansion *naturally decorrelates the random process*. Moreover, if the process is Gaussian, the expansion coefficients are the independent standard Gaussian random variables $N(0, 1)$, i.e., with zero average and variance equal to one.

Finally, the Karhunen–Loéve expansion allows one to express the correlation function $R(t_1, t_2)$ in an infinite series of "modes." This can be shown by substituting the series expansion (C.57) into the definition of the correlation function and using the delta-correlation property of the expansion coefficients. Simple algebra leads to the following result:

$$R(t_1, t_2) = \sum_{n=0}^{\infty} \lambda_n \phi_n(t_2) \phi_n^*(t_1), \quad \text{(C.65)}$$

where each term in the series is a *mode of the correlation function*. The expansion series for a number of important processes will be addressed in the following section.

C.6   Gaussian Processes and Their Representations

Gaussian random processes are widespread in science and engineering, where they model a large number of complex physical situations. Important examples include colored noise, Brownian motion, and the Ornstein–Uhlenbeck process describing, for instance, random drift-diffusion motion and Johnson noise [758, 798].

C.6.1 Gaussian Vectors and Random Processes

The random variables in the collection $\{X_i\}$ with $i = 1 \ldots n$ are said to be *jointly Gaussian* if the linear combination:

$$\sum_{i=1}^{n} a_i X_i \tag{C.66}$$

is a Gaussian random variable for any real number a_i. If that is the case, the column vector $\mathbf{X} = [X_1, X_1, \ldots, X_n]^T$ (where $[\cdot]^T$ denotes the transpose) is called a *Gaussian random vector*. An important property of jointly Gaussian random variables is that their joint probability density is completely determined by their mean and covariance matrices. More specifically, a Gaussian random vector $\mathbf{X}$ with mean $\boldsymbol{\mu}$ and covariance matrix $\boldsymbol{\Sigma}$ is fully characterized by the n-variate probability density function:

$$p(\mathbf{x}) = \frac{1}{\sqrt{(2\pi)^n}\sqrt{|\Sigma|}} \exp\left[-\frac{1}{2}(\mathbf{x} - \boldsymbol{\mu})^T \Sigma^{-1}(\mathbf{x} - \boldsymbol{\mu})\right], \tag{C.67}$$

where $|\Sigma|$ denotes the determinant of the covariance matrix.

We can now precisely define a Gaussian random process. We say that a stochastic process indexed by a continuous parameter t (space or time) is a *Gaussian process* if and only if for any finite set of indices $t_1, t_2, \ldots, t_n$ the random variables $X(t_1), X(t_2), \ldots, X(t_n)$ form a Gaussian random vector. Equivalently, every finite collection of those random variables admits a multivariate normal distribution. Gaussian processes are called second-order processes because their mean and covariance functions completely define their behavior. In particular, a Gaussian stochastic process $X(t)$ is uniquely characterized by two functions: its mean[9] $\mu_t = \langle X(t) \rangle$ and its covariance $C(t, s) \equiv \langle [X(t) - \mu_t][X(s) - \mu_t] \rangle$.

In the next sections, we introduce a few notable examples of Gaussian and non-Gaussian random processes and discuss their connection with stochastic differential equations.

C.6.2 Brownian Motion and Colored Noise

The one-dimensional *Brownian motion* is a fundamental building block the theory of stochastic processes [796]. The standard Brownian motion, or standard *Wiener process* $W(t)$ defined over the interval $[0, T]$, is the continuous-time stochastic process that satisfies the following conditions:

- $W(0) = 0$ (with probability one).
- $W(t)$ is continuous in the time variable t.

---

[9] Note that for a generally non-stationary Gaussian process the mean depends on time.

- $W(t)$ has statistically independent increments, i.e., the increments $W(t) - W(s)$ and $W(v) - W(u)$ are independent for $0 \leq s < t < u < v \leq T$.
- For $0 \leq s < t \leq T$, the increment, or Brownian step $W(t) - W(s)$, obeys a normal distribution with zero mean and variance equal to $t - s$. Equivalently, $W(t) - W(s) \sim \sqrt{t - s} N(0, 1)$.

The Brownian motion is the continuum limit of a discrete random walk of increasingly more frequent and increasingly smaller steps. The computer simulation of the Brownian motion is based directly on its definition: after discretization of the interval $[0, T]$ in small steps $dt = T/N$, where N is the desired number of samples, a vector of independent Gaussian increments $d\mathbf{W} = \sqrt{dt} N(0, 1)$ is created. Discretized realizations (or paths) of the Brownian motion are then simply obtained by computing the *cumulative sum* on the vector $d\mathbf{W}$. Three representative paths of the Brownian motion computed as described previously are shown in Figure C.2.

General Gaussian processes such as the Brownian motion can be represented according to their Karhunen–Loéve expansion as follows:

$$G(t) = \sum_{k=1}^{\infty} \xi_k \sqrt{\lambda_k} \phi_k(t), \tag{C.68}$$

where ξ_k are mutually independent standard Gaussian variables, and λ_k and $\phi_k(t)$ are, respectively, the eigenvalues and eigenfunctions of the integral equation obeyed by the covariance $C(t, s)$:

$$\int_0^T C(t, s) \phi_k(t) dt = \lambda_k \phi_k(t), \tag{C.69}$$

for $k \geq 1$ and where the eigenfunctions of the covariance are a CONS with uniform convergence in $\mathbf{L}^2([0, T])$.

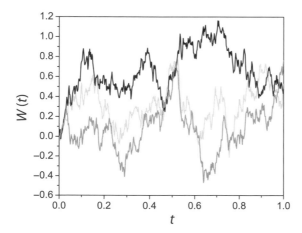

Figure C.2 Three different realizations (paths) of discretized Brownian motion on the unit interval $[0, 1]$. $N = 500$ samples were used in the simulations.

In the special case of the Brownian motion, the covariance function is $C(t,s) = \min(t,s)$. For $t,s \in [0,1]$, the eigenfunctions of this covariance are as follows:

$$\phi_k(t) = \sqrt{2} \sin\left[\left(k - \frac{1}{2}\right)\pi t\right], \tag{C.70}$$

and the corresponding eigenvalues are as follows:

$$\lambda_k = \frac{1}{\left(k - \frac{1}{2}\right)^2 \pi^2}. \tag{C.71}$$

Therefore, the Karhunen–Loéve expansion of the standard Brownian motion is equal to the following:

$$W(t) = \sqrt{2} \sum_{k=1}^{\infty} \xi_k \frac{\sin\left[\left(k - \frac{1}{2}\right)\pi t\right]}{\left(k - \frac{1}{2}\right)\pi} \tag{C.72}$$

Gaussian white noise is characterized by a delta Dirac covariance that can be expressed as $C(t,s) = \sigma^2 \delta(t - s)$, and therefore has infinite variance. This property reflects the fact that a white noise process assumes independent values at each instant of time (i.e., complete lack of correlations). This property cannot be observed in any natural phenomena since it implies a diverging room mean squared energy, However, Gaussian white noise is often used to model random processes with extremely small correlation. The standard one-dimensional Brownian motion can be generalized to any number of dimensions. In particular, an n-dimensional Brownian motion is a continuous stochastic process defined by the vector $W_t = [W_1(t), \ldots, W_n(t)]^T$, where the components $W_i(t)$ are mutually independent standard one-dimensional Brownian motions. The Brownian motion has many important regularity and scaling properties, including nondifferentiability and self-similarity. Rigorous discussions can be found in [799].

Gaussian processes with finite correlation, such as the Brownian motion that we have introduced, form a general class of processes often referred to as *colored noise*. An alternative way of representing *correlated Gaussian noise*, or colored noise, is through their Fourier series expansion[799]:

$$G_C(t) = \sum_{k=-\infty}^{\infty} e^{-jkt} a_k \xi_k \tag{C.73}$$

where ξ_k are i.i.d. standard Gaussian random variables $N(0,1)$, $j = \sqrt{-1}$, and a_k are expansion coefficients. When the coefficients a_k are all equal to the same constant, the preceding Fourier expansion represents Gaussian white noise. On the other hand, when the expansion coefficients depend on the summation index k, the underlying process is called colored in analogy to light dispersion. In particular, when $|a_k|^2 \propto 1/k^\alpha$ the corresponding noise process is called $1/f^\alpha$ noise. Brownian noise corresponds to

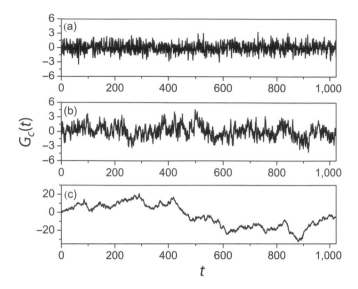

Figure C.3 Three representative samples of discretized colored noise processes showing the effect of progressively increasing correlations. (a) $\alpha = 0$ (white noise), (b) $\alpha = 1$ (pink noise), and (c) $\alpha = 2$ (brown noise).

the special case $\alpha = 2$, referred to as *red noise* in this context. The process with $\alpha = 1$ is known as *pink noise*.

When the value of α is progressively decreased from 2 (Brownian motion) to 0 (white noise), the corresponding Gaussian processes become less and less correlated, and in the limit of $\alpha = 0$ we obtain uncorrelated white noise, as shown in Figure C.3. When the basis functions in (C.73) are chosen as the eigenfunctions of the covariance function, the series representation of the noise reduces to its Karhunen–Loéve expansion. The concept of correlated noise can be extended to processes with arbitrary value of the exponent α, including negative numbers. It is interesting to notice that Gaussian $1/f^{\alpha}$ processes with $0 < \alpha \leq 2$ have been experimentally discovered in many physical relaxation processes that occur in complex dynamical systems. An interesting review of this topic can be found in [800].

C.7   Additional Examples of Random Processes

We present here a few more examples of notable stochastic processes. In particular, we focus on a number of additional Gaussian processes that can be derived from the standard Brownian motion. We introduce first the *Brownian bridge* W_B, which is a Brownian motion with fixed values at both the extremes of the interval $[0, T]$, i.e., with the additional constraint $W_B(T) = 0$. A Brownian bridge $W_B(t)$ in $[0, 1]$ can be derived from the corresponding standard Brownian motion $W(t)$ by setting the following:

$$W_B(t) = W(t) - t W(1) \tag{C.74}$$

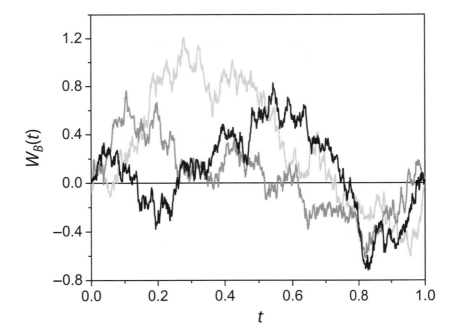

Figure C.4 Three different realizations (paths) of discretized Brownian bridge on the unit interval $[0, 1]$. $N = 1,000$ samples have been used in the simulations.

with $0 \leq t \leq 1$. This process admits the following Karhunen–Loéve expansion:

$$W_B(t) = \sum_{k=1}^{\infty} \xi_k \frac{\sqrt{2}\sin(\pi k t)}{\pi k} \qquad (C.75)$$

where again ξ_k are mutually independent standard Gaussian random variables. A bridge over $[0, T]$ has zero average and variance $t(T - 1)/T$, implying that the most uncertainty is found in the middle of the bridge. Importantly, in contrast to the standard Brownian motion, the increments in a Brownian bridge are not independent because the values at both its extremes are pinned. Three different realizations (paths) of a discretized Brownian bridge on the unit interval are displayed in Figure C.4.

Another important example of a Gaussian process often encountered in the modeling of complex and fractal systems is the *fractional Brownian motion* (fBm). Unlike the standard Brownian motion, the fractional Brownian motion W_H is a one-parameter family of Gaussian processes with correlated increments and covariance given by the following:

$$C(t, s) = \frac{1}{2}(|s|^{2H} + |t|^{2H} - |t - s|^{2H}), \qquad (C.76)$$

where $H \in (0, 1)$ is called the *Hurst parameter*. Similar to the Brownian motion, the fBm also starts at zero and is defined over the interval $[0, T]$. The Hurst parameter controls the degree of correlation between the increments of the fBm as well as the regularity of its trajectories (sample paths). The fBm provides a convenient approach

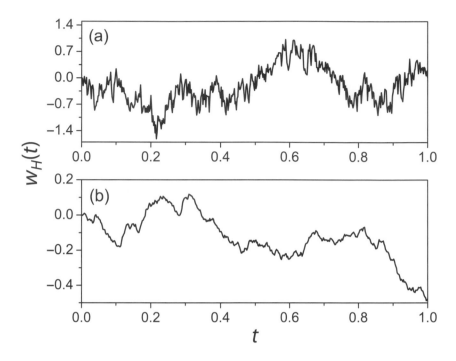

Figure C.5 Discretized sample paths of fractional Brownian motion on the unit interval $[0, 1]$ with $H = 1/4$ (a) and with $H = 3/4$ (b). $N = 500$ samples have been used in the simulations.

to generate correlated disorder with the specified spectral density $S(k) \propto 1/k^\alpha$. The spectral exponent α is directly related to the Hurst exponent H by the simple relation $\alpha = 2H + 1$. In general, the larger the value of H, the smoother the paths become. We note that when $H = 1/2$, then the fBm reduces to the standard Brownian motion. However, when $H > 1/2$ the increments $X_H(t) = W_H(t + 1) - W_H(t)$ of the fBm process are positively correlated, i.e., trends are likely to continue over long periods, while when $H < 1/2$ the increments are negatively correlated and strong variations in the process are expected. Therefore, the Hurst exponent captures the persistent character of the fBm increments. This characteristic behavior is displayed in Figure C.5.

The fBm has several remarkable properties, including the stationarity of its increments $W_H(t) - W_H(s) \sim W_H(t - s)$ and self-similarity due to the homogeneous nature of its covariance. In particular, for any real constant a we can write the following:

$$W_H(at) \sim |a|^H W_H(t). \tag{C.77}$$

Moreover, the fBm is the only self-similar Gaussian process. More information on the properties of the fBm can be found in [801].

We now introduce the *Ornstein–Uhlenbeck process* (O-U) Gaussian process, which, differently from the Brownian motion, it is a stationary process. Interestingly, the O-U process can be defined through a variable transformation on the standard

Brownian motion and it is used to describe the velocity of a Brownian particle in the presence of friction. Due to its relevance in physics and engineering applications, the O-U process can be considered the prototype of any noisy relaxation process [758]. For example, the dynamics of an elastic spring strongly damped by friction in the presence of random thermal fluctuations is described by the O-U process.

The O-U process $U(t)$ can be defined as the solution of a relaxation differential equation driven by white noise, first introduced by Langevin in 1908 as follows [758]:

$$\frac{dU(t)}{dt} = -\alpha U(t) + \beta \eta(t) \tag{C.78}$$

where $\eta(t)$ is the fast varying Gaussian white noise,[10] α is equal to the dissipation (friction) coefficients experienced by the particle, and $\beta = \sqrt{2D}$ is a real constant proportional to the diffusion coefficients D of the randomly transported particle.

Since the white noise process is the time derivative of the Wiener process $\eta(t) = dW(t)/dt$, we can rewrite equation C.78 as follows:

$$dU(t) = -\alpha U(t)dt + \beta dW(t) \tag{C.79}$$

The equation in this form is the stochastic Langevin equation, which is a special case of the *Itô drift-diffusion* stochastic differential equation (SDE) for the more general process $X(t)$:

$$dX(t) = \alpha(X(t),t)dt + \beta(X(t),t)dW(t), \tag{C.80}$$

where $\alpha(X(t),t)$ and $\beta(X(t),t)$ are functions of $X(t)$ and possibly of time as well, called the generalized drift and diffusion coefficients, respectively. The solutions of the stochastic equation (C.80) are known as the *Itô diffusions*. Stochastic differential equations, which are solved by continuous time stochastic processes, have become standard tools in the modeling of diffusive processes in the physical and biological sciences as well as in economics and finance. However, differently from usual ordinary differential equations, SDEs are always specified in differential form because many interesting stochastic processes, such as the Brownian motion, are continuous but nowhere differentiable. As a result, the stochastic differential $dW(t)$ that appears in equation (C.80) should always be thought of as the random process $N(0,1)\sqrt{dt}$ (where $N(0,1)$ is a standard Gaussian random variable with zero average and unit variance) and not as a traditional differential of a continuous function. It follows that the Itô SDE in the form (C.80) cannot simply be divided by dt because dW is not a standard differential. The existence and uniqueness theorems for general SDEs (i.e., Lipschitz conditions on the drift and diffusion terms) typically impose more stringent constraints than their deterministic counterparts [802]. However, the solutions

---

[10] When written as in (C.78), the Langevin equation is not justified mathematically since the delta-correlated white noise term is not a regular continuous function. The rigorous framework of stochastic calculus is required in order to correctly interpret the Langevin and similar stochastic differential equations.

of drift-diffusion SDEs in the form (C.80) can be readily obtained numerically using explicit time-stepping methods such the Euler–Maruyama (EM) method and the more accurate Milstein's method [802, 803]. More sophisticated numerical schemes are discussed in [799].

The formal solution of the general SDE equation C.80 can be expressed in terms of the *Itô stochastic integral*:

$$X(t) = X(0) + \int_0^t \alpha(X(s), s)ds + \int_0^t \beta(X(s), s)dW(s) \qquad (C.81)$$

Stochastic integrals and stochastic differential equations are the primary tools of *stochastic calculus*, which is a sophisticated field of mathematics that has found numerous applications in the modeling of complex physical systems [799, 802, 804]. Stochastic calculus can be loosely defined as the infinitesimal calculus of nondifferentiable functions. One of its central ideas is the one of integrating a given stochastic process with respect to another stochastic process, usually a Wiener process, leading to the notion of *stochastic integration*.

Applying the general tool sets of stochastic calculus to the Langevin equation (C.79), we can reduce its solution to the stochastic integral form:[11]

$$U(t) = U(0)e^{-\alpha t} + \beta \int_0^t e^{-\alpha(t-s)}dW(s). \qquad (C.82)$$

Using elementary averaging properties of stochastic integrals [758, 804], it is possible to obtain analytical expressions for the average value and for the variance of the O-U process, as follows:

$$\langle U(t) \rangle = U(0)e^{-\alpha t} \qquad (C.83)$$

$$\sigma_U^2(t) = \langle (U(t) - \langle U(t) \rangle)^2 \rangle = \frac{D}{\alpha}(1 - e^{-2\alpha t}). \qquad (C.84)$$

We can appreciate from these solutions that the O-U process converges to the solutions of the standard Brownian motion in the limit of $\alpha \to 0$. In particular, by expanding for small α values the expression within the parenthesis in equation (C.84), we obtain the well-known variance solution $\sigma_U^2 = \langle X^2(t) \rangle \to 2Dt$ that characterizes Brownian motion, and it is the hallmark of standard diffusion.

A useful representation of the Ornstein–Uhlenbeck process $U(t)$ connects it directly with the Brownian motion $W(t)$ as follows [799]:

$$U(t) = e^{-t/\sigma}W(e^{\frac{2t}{\sigma}}), \qquad (C.85)$$

where $\sigma = \alpha^{-1}$ and $t \geq 0$. Moreover, the O-U process is characterized by the exponential covariance function:

$$C(t, s) \propto \exp(-|t - s|/\sigma). \qquad (C.86)$$

---

[11] This requires the following change of variable $Y(t) = X(t)\exp(\alpha t)$ before taking the integration.

An explicit Karhunen–Loéve expansion for the O-U process can be obtained based on its covariance function. However, this requires the accurate numerical solution of a transcendental equation [805]. For stochastic processes with more general covariance functions $C(t, s)$, it may be impossible to find explicitly its eigenvalues and eigenfunctions. In such cases, the Karhunen–Loéve expansion must be obtained by solving the covariance integral equation (C.69) numerically.

We finally consider the *geometric Brownian motion* (GBM) $X_G(t)$, which is an important example of a non-Gaussian process with applications to econometrics and quantitative finance due to the role it plays in option pricing and assets dynamics, where it solves the Black–Scholes stochastic differential equation [382]. The motivation behind this process is related to the fact that the Brownian motion, as a Gaussian process, may assume both positive and negative values, which is not a desirable property in the financial modeling of prices. This fact leads us to consider the exponential of a Brownian motion with drift. Following this approach, the GBM can be defined as the continuous time stochastic process:

$$X_G(t) = X_G(0) \exp\left[\left(\mu - \frac{\sigma^2}{2}\right) t + \sigma W(t)\right], \tag{C.87}$$

where $W(t)$ is the standard Brownian motion, μ is the drift constant, and σ is a diffusion (volatility) constant. Note that the stochastic process within the square bracket of the preceding equation is a Brownian motion with drift parameter $\mu - \sigma^2/2$ and diffusion parameter σ, and the geometric Brownian motion is simply the exponential of this process. Three representative paths of the OU and GBM processes are displayed in Figure C.6. We finally remark that the GBM is the solution of the following stochastic differential equation:

$$dX_G(t) = X_G(t)(\mu dt + \sigma dW(t)). \tag{C.88}$$

A direct application of the useful *Itô formula* [799] to the function $Y(t) = \log(X_G(t))$ shows that the logarithm of the GBM satisfies the stochastic equation:

$$dY(t) = \left[\left(\mu - \frac{\sigma^2}{2}\right) dt + \sigma dW(t)\right]. \tag{C.89}$$

This is the Itô drift-diffusion equation obeyed by the logarithm of the GBM process. As a result, for any value of t, the random variable $X_G(t)$ follows a non-Gaussian *log-normal distribution* with expected value and variance given by the following:

$$\langle X_G(t) \rangle = X_0 e^{(\mu + \sigma^2/2)t} \tag{C.90}$$

$$\sigma_G^2(t) = X_0^2 e^{(2\mu + \sigma^2)t} \left(e^{\sigma^2 t} - 1\right), \tag{C.91}$$

where $X_0 = X_G(0)$. A log-normal distribution is a continuous probability distribution of a random variable whose logarithm is normally distributed. A random variable that is log-normally distributed takes only positive real values. The log-normal distribution is a strongly asymmetric probability density function describing the product of a large number of independent and identically distributed random variables. Its analytical

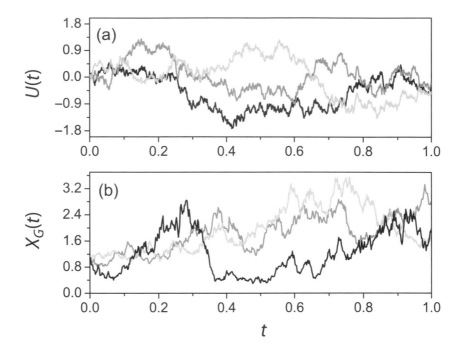

Figure C.6 (a) Different realizations (paths) of the Ornstein–Uhlenbeck process on the unit interval generated according to the analytical formula (C.85) with $\sigma = 0.5$. (b) Different realizations (paths) of the geometric Brownian motion on the unit interval. In these examples, we considered $\mu = 1$ and $\sigma = 2$, with initial condition $X_G(0) = 1$ and $dt = 0.002$.

expression can be found immediately by considering the logarithm of the product and by applying the CLT to the resulting sum of logarithms (i.e., considering the CLT in the logarithmic domain). Log-normal distributions typically describes *multiplicative processes* that occur in many different fields of science and engineering ranging from geology, physics, chemistry, biology, neuroscience, social sciences, and finance. An in-depth discussion of the many surprising properties and applications of the log-normal distribution can be found in [806, 807].

C.7.1   The Itô Formula and Its Derivation

The Itô formula is a very useful result that connects the process $X(t)$ obeying the equation (C.80) and the process $Y(t) = f(X(t),t)$, where $f(x,t)$ is a continuous function with continuous first and second derivatives. Specifically, when $X(t)$ obeys the SDE (C.80), the Itô formula provides the *stochastic differential* of $Y(t)$:

$$dY(t) = df = \left(\frac{\partial f}{\partial t} + a \frac{\partial f}{\partial x} + \frac{b^2}{2} \frac{\partial^2 f}{\partial x^2} \right) dt + b \frac{\partial f}{\partial x} dW(t). \qquad (C.92)$$

The Itô formula is the stochastic analogue of the chain rule of differential calculus, to which it reduces when $b = 0$. Notice that a second-order derivative appears in the linear expansion term due to the fact that a second-order change in the variable $W(t)$ is proportional to a first-order change in the variable t (always remember that $dW(t) = N(0, 1)\sqrt{dt}$).

The derivation of the Itô formula proceeds from quite simply from the Taylor series expansion of the function $f(x, t)$ (up to second order in X) and retaining only first-order stochastic terms using both equation (C.80) and the heuristic rule $dW^2 = dt$:

$$df = \frac{\partial f}{\partial t} dt + \frac{\partial f}{\partial x} dx + \frac{1}{2} \frac{\partial^2 f}{\partial x^2} dx^2 + \cdots \qquad (C.93)$$

The formula (C.92) follows when we substitute in the preceding expression $x \to X(t)$ and $dx \to a dt + b dW(t)$ and collect only first-order terms in dt and dW (all quadratic differentials can be neglected in the limit $dt \to 0$).

As a simple application example of the Itô formula, we derive the analytical solution of the SDE:

$$dX_G(t) = X_G(t)(\mu dt + \sigma dW(t)), \qquad (C.94)$$

where σ and μ are constants. As we commented previously, the preceding equation is solved by the geometric Brownian motion that $X_G(t)$. We first consider the composite process $Y(t) = f(X_G(t)) = \log(X_G(t))$ and write the following:

$$df = \left(\frac{\partial f}{\partial X_G} dX_G + \frac{1}{2} \frac{\partial^2 f}{\partial X_G^2} \right) (dX_G)^2. \qquad (C.95)$$

Substituting from the SDE (C.94) and retaining only first-order terms, we obtain the following:

$$df = \frac{1}{X_G} (\mu X_G dt + \sigma X_G dW) - \frac{1}{2} \sigma^2 dt = \left(\mu - \frac{\sigma^2}{2} \right) dt + \sigma dW, \qquad (C.96)$$

which is a new drift-diffusion process for the variable $Y(t) = \log(X_G(t))$. Integrating the preceding equation, it follows:

$$\log(X_G(t)) = \log(X_G(0)) \sigma W(t) + \left(\mu - \frac{\sigma^2}{2} \right) t. \qquad (C.97)$$

Taking the exponentiation, we finally obtain the desired analytical solution:

$$X_G(t) = X_G(0) \exp \left[\left(\mu - \frac{\sigma^2}{2} \right) t + \sigma W(t) \right]. \qquad (C.98)$$

Appendix D Derivation of the Foldy–Lax Equations

We show how to rigorously establish the Foldy–Lax multiple-scattering equations following the derivation in [132] (see Figure D.1). Let us first considering the simple case of a scattering system that contains only two arbitrary particles. The particles are described by (single-particle) transition operators $\mathbf{t}_i$ and $\mathbf{t}_j$, which in principle include near-field and far-field effects and are excited by an incident wave $|\psi_0\rangle$. We have seen that each particle will give rise to a scattered field $\mathbf{G}_0^+ \mathbf{t}_n |\psi_0\rangle$, where $n = i, j$, so that the total field originated by the "direct" scattering from each particle can be written as follows:

$$|\psi\rangle = |\psi_0\rangle + \mathbf{G}_0^+ \mathbf{t}_i |\psi_0\rangle + \mathbf{G}_0^+ \mathbf{t}_j |\psi_0\rangle + \cdots \qquad (D.1)$$

However, higher-order scattering effects must be included in order to properly describe multiple scattering. As we will now see, this phenomenon complicates the definition of the exciting field on each particles, which will no longer coincide with the external (i.e., the incident) field $|\psi_0\rangle$. In particular, notice that the term $\mathbf{G}_0^+ \mathbf{t}_j \mathbf{G}_0^+ \mathbf{t}_i |\psi_0\rangle$ describes a second-order term where the field scattered from particle i is further scattered from particle j. Thus to second-order scattering terms, we can write the following:

$$|\psi\rangle = |\psi_0\rangle + \mathbf{G}_0^+ \mathbf{t}_i |\psi_0\rangle + \mathbf{G}_0^+ \mathbf{t}_j |\psi_0\rangle + \mathbf{G}_0^+ \mathbf{t}_j \mathbf{G}_0^+ \mathbf{t}_i |\psi_0\rangle + \mathbf{G}_0^+ \mathbf{t}_i \mathbf{G}_0^+ \mathbf{t}_j |\psi_0\rangle . \quad (D.2)$$

We can continue this process to higher orders in scattering and write a total field expansion that contains an *infinite number of terms*:

$$\begin{aligned}
|\psi\rangle = &|\psi_0\rangle + \mathbf{G}_0^+ \mathbf{t}_j |\psi_0\rangle + \mathbf{G}_0^+ \mathbf{t}_i |\psi_0\rangle \\
&+ \mathbf{G}_0^+ \mathbf{t}_j \mathbf{G}_0^+ \mathbf{t}_i |\psi_0\rangle + + \mathbf{G}_0^+ \mathbf{t}_i \mathbf{G}_0^+ \mathbf{t}_j |\psi_0\rangle \\
&+ \mathbf{G}_0^+ \mathbf{t}_j \mathbf{G}_0^+ \mathbf{t}_i \mathbf{G}_0^+ \mathbf{t}_j |\psi_0\rangle + \mathbf{G}_0^+ \mathbf{t}_i \mathbf{G}_0^+ \mathbf{t}_j \mathbf{G}_0^+ \mathbf{t}_i |\psi_0\rangle \\
&+ \mathbf{G}_0^+ \mathbf{t}_j \mathbf{G}_0^+ \mathbf{t}_i \mathbf{G}_0^+ \mathbf{t}_j \mathbf{G}_0^+ \mathbf{t}_i |\psi_0\rangle + \mathbf{G}_0^+ \mathbf{t}_i \mathbf{G}_0^+ \mathbf{t}_j \mathbf{G}_0^+ \mathbf{t}_i \mathbf{G}_0^+ \mathbf{t}_j |\psi_0\rangle + \cdots
\end{aligned} \qquad (D.3)$$

We can rearrange the infinite series (D.3) as follows:

$$\begin{aligned}
|\psi\rangle = &|\psi_0\rangle + \mathbf{G}_0^+ \mathbf{t}_j (|\psi_0\rangle + \mathbf{G}_0^+ \mathbf{t}_i |\psi_0\rangle + \mathbf{G}_0^+ \mathbf{t}_i \mathbf{G}_0^+ \mathbf{t}_j |\psi_0\rangle \\
&+ \mathbf{G}_0^+ \mathbf{t}_i \mathbf{G}_0^+ \mathbf{t}_j \mathbf{G}_0^+ \mathbf{t}_i |\psi_0\rangle + \cdots) + \mathbf{G}_0^+ \mathbf{t}_i (|\psi_0\rangle + \mathbf{G}_0^+ \mathbf{t}_j |\psi_0\rangle \\
&+ \mathbf{G}_0^+ \mathbf{t}_j \mathbf{G}_0^+ \mathbf{t}_i |\psi_0\rangle + \mathbf{G}_0^+ \mathbf{t}_j \mathbf{G}_0^+ \mathbf{t}_i \mathbf{G}_0^+ \mathbf{t}_j |\psi_0\rangle + \cdots).
\end{aligned} \qquad (D.4)$$

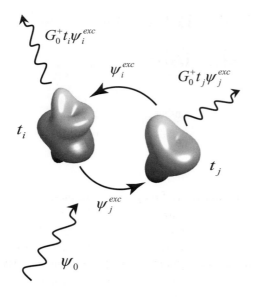

Figure D.1 Schematic representation of two-particle coupling within the self-consistent Foldy–Lax scheme of multiple scattering.

The first sum in equation (D.4) is the field that excites particle j as the last particle, and the second sum represents the field that interacts with particle i as the last particle. We therefore rigorously define the locally exciting fields for the two particles $|\psi_n^{exc}\rangle$ with $n = i, j$ as follows:

$$
\begin{aligned}
|\psi_i^{exc}\rangle = |\psi_0\rangle + \mathbf{G}_0^+ \mathbf{t}_j |\psi_0\rangle + \mathbf{G}_0^+ \mathbf{t}_j \mathbf{G}_0^+ \mathbf{t}_i |\psi_0\rangle \\
+ \mathbf{G}_0^+ \mathbf{t}_j \mathbf{G}_0^+ \mathbf{t}_i \mathbf{G}_0^+ \mathbf{t}_j |\psi_0\rangle + \mathbf{G}_0^+ \mathbf{t}_j \mathbf{G}_0^+ \mathbf{t}_i \mathbf{G}_0^+ \mathbf{t}_j \mathbf{G}_0^+ \mathbf{t}_i |\psi_0\rangle + \cdots
\end{aligned}
\tag{D.5}
$$

and

$$
\begin{aligned}
|\psi_j^{exc}\rangle = |\psi_0\rangle + \mathbf{G}_0^+ \mathbf{t}_i |\psi_0\rangle + \mathbf{G}_0^+ \mathbf{t}_i \mathbf{G}_0^+ \mathbf{t}_j |\psi_0\rangle \\
+ \mathbf{G}_0^+ \mathbf{t}_i \mathbf{G}_0^+ \mathbf{t}_j \mathbf{G}_0^+ \mathbf{t}_i |\psi_0\rangle + \mathbf{G}_0^+ \mathbf{t}_i \mathbf{G}_0^+ \mathbf{t}_j \mathbf{G}_0^+ \mathbf{t}_i \mathbf{G}_0^+ \mathbf{t}_j |\psi_0\rangle + \cdots
\end{aligned}
\tag{D.6}
$$

These defined quantities represent the fields that excite the two particles at the end of the infinite sequence of multiple scattering events that occur between the two. These final exciting fields are the self-consistent excitation fields that include all orders of multiple scattering in the system. As a result, the two particles give rise to "final scattered fields" $|\psi_i^s\rangle$ and $|\psi_j^s\rangle$ that can be written as follows:

$$
|\psi_i^s\rangle = \mathbf{G}_0^+ \mathbf{t}_i |\psi_i^{exc}\rangle
\tag{D.7a}
$$

$$
|\psi_j^s\rangle = \mathbf{G}_0^+ \mathbf{t}_j |\psi_j^{exc}\rangle .
\tag{D.7b}
$$

The total field of the multiple scattering problem is then given by the following:

$$
|\psi\rangle = |\psi_0\rangle + |\psi_i^s\rangle + |\psi_j^s\rangle = |\psi_0\rangle + |\psi^s\rangle ,
\tag{D.8}
$$

where $|\psi^s\rangle = |\psi_i^s\rangle + |\psi_j^s\rangle$. Now we observe that we can rewrite the series in equation (D.5) as follows:

$$\begin{aligned}|\psi_i^{exc}\rangle = |\psi_0\rangle + \mathbf{G}_0^+ \mathbf{t}_j(|\psi_0\rangle + \mathbf{G}_0^+ \mathbf{t}_i |\psi_0\rangle + \mathbf{G}_0^+ \mathbf{t}_i \mathbf{G}_0^+ \mathbf{t}_j |\psi_0\rangle \\ + \mathbf{G}_0^+ \mathbf{t}_i \mathbf{G}_0^+ \mathbf{t}_j \mathbf{G}_0^+ \mathbf{t}_i |\psi_0\rangle + \cdots).\end{aligned} \qquad (D.9)$$

We now recognize that the term in parentheses in this expression coincides exactly with equation (D.6), namely it is equal to $|\psi_j^{exc}\rangle$! Therefore, we can finally write the following system:

$$|\psi_i^{exc}\rangle = |\psi_0\rangle + \mathbf{G}_0^+ \mathbf{t}_j |\psi_j^{exc}\rangle \qquad (D.10a)$$

$$|\psi_j^{exc}\rangle = |\psi_0\rangle + \mathbf{G}_0^+ \mathbf{t}_i |\psi_i^{exc}\rangle. \qquad (D.10b)$$

These equations are the self-consistent Foldy–Lax multiple-scattering equations for the rigorously defined final exciting fields of the two particles. Once the exciting fields are solved, the total scattered field can be obtained using equations (D.7a), (D.7b), and (D.8). We notice finally that the Foldy–Lax equations are exact, since they are derived from Maxwell's equations *without any approximation* and only require the knowledge of the single-particle transition matrices. Moreover, their generalization to a system of N particles is immediate:[1]

$$|\psi_i^{exc}\rangle = |\psi_0\rangle + \sum_{j \neq i}^{N} \mathbf{G}_0^+ \mathbf{t}_j |\psi_j^{exc}\rangle, \qquad (D.11)$$

where $i = 1, \cdots, N$. The total scattered field is given by the following:

$$|\psi^s\rangle = \sum_{i=1}^{N} |\psi_i^s\rangle \qquad (D.12a)$$

$$|\psi_i^s\rangle = \mathbf{G}_0^+ \mathbf{t}_i |\psi_i^{exc}\rangle \qquad (D.12b)$$

and following is the total field:

$$|\psi\rangle = |\psi_0\rangle + |\psi^s\rangle. \qquad (D.13)$$

The Foldy–Lax multiple-scattering equations (D.11)–(D.13) consist of N equations in the N unknowns $|\psi_i^{exc}\rangle$ for $i = 1, \cdots, N$. In principle, they can be solved numerically for a system of N arbitrary scattering particles of which the single-particle T-matrices are known or can be found. Their solution is very complex in general, but it assumes a simple form in the case of N point scatterers appendix. Besides being pedagogically very instructive, the solution for point scatterers also provides a rigorous foundation for the theory of radiative transfer [132].

---

[1] Note that the term $\mathbf{G}_0^+ \mathbf{t}_i |\psi_i^{exc}\rangle$ is absent because the exciting field of particle i does not act on itself.

Appendix E The Multifractal Formalism

The multifractal spectrum was originally introduced by Frisch and Parisi [597] in the investigation of the energy dissipation of turbulent fluids. The multifractal spectrum is a powerful tool for the characterization of complex signals because it can efficiently resolve their local fluctuations. Examples of multifractal structures/phenomena are commonly encountered in dynamical systems theory (e.g., strange attractors of nonlinear maps), physics (e.g., diffusion-limited aggregates, turbulence), engineering (e.g., random resistive networks, image analysis), geophysics (e.g., rock shapes, creeks), and even in finance (e.g., dynamics of stock markets). Multifractal measures involve singularities of different strengths, and their $D(\alpha)$ spectrum displays generally a single humped shape (i.e., downward concavity) that extends over a compact interval $[\alpha_{min}, \alpha_{max}]$, where α_{min} (respectively α_{max}) correspond to the strongest (respectively the weakest) singularities.[1] The maximum value of $D(\alpha)$ corresponds to the (average) box-counting dimension D_f of the multifractal object, while the difference $\Delta\alpha = \alpha_{max} - \alpha_{min}$ can be used as a parameter reflecting the randomness of the intensity measure [808].

The complex geometry of multifractal sets can alternatively be described by a parameter called the generalized dimensions D_q that was originally introduced to characterize strange attractors in the phase space of dissipative dynamical systems that exhibit chaotic behavior [809]. The generalized dimension is related to the scaling of the qth moments of a distribution and it is defined as follows:

$$D_q = \frac{1}{q-1} \lim_{\ell \to 0} \left[\frac{\log \mu(q, \ell)}{\log \ell} \right] \tag{E.1}$$

where $\mu(q, \ell)$ is equal to

$$\mu(q, \ell) = \sum_i P_i(\ell)^q \sim \ell^{-\tau(q)}. \tag{E.2}$$

Here, $P_i(\ell)$ is a probability calculated as the integral of the quantity under analysis (the number of boxes in our case) over the ith box. Equation (E.2) can be interpreted

---

[1] On the other hand, the spectra of homogeneous or monofractal patterns are not humped but converge to a certain value.

in the following way: if the qth moments of $P_i(\ell)$ are proportional to some power $\tau(q)$ of the box size ℓ, then $\mu(q,\ell)$ distinguishes intertwined regions that scale in different ways according to the exponent $\tau(q)$, also named the mass exponent [48]. Specifically, $\tau(q) = (q-1)D_q$ shows how the moments of a given distribution scale with the box size ℓ. If $\tau(q)$ is a nonlinear function of q, we call the distribution of measures multifractal [48, 810].

The two exponents $D(\alpha)$ and $\tau(q)$ are related to each other by means of a *Legendre transformation* [48, 601, 810, 811]:

$$\tau(q) = \alpha(q)q - D[\alpha(q)] \quad \text{where} \ \ q = \frac{dD(\alpha)}{d\alpha}$$

$$D(\alpha) = q(\alpha) - \tau[q(\alpha)] \quad \text{where} \ \ \alpha = \frac{d\tau(q)}{dq}. \tag{E.3}$$

These equations unveil a deep connection with the thermodynamic formalism of statistical mechanics, where $\tau(q)$ and q are conjugate thermodynamic variables to $D(\alpha)$ and α [601]. In this context, the function $\mu(q,\ell)$ is formally analogous to the partition function $Z(\beta)$, while $\tau(q)$ can be interpreted as the free energy. Its Legendre transform $D(\alpha)$ is thus the analogue of entropy, while α corresponds to the energy E. Notably, the characteristic parabolic shape of $D(\alpha)$ resembles the functional dependence of the entropy from E.

In order to implement the multifractal formalism, we could derive $\tau(q)$ from the log-log plot of ℓ versus $\mu(q,\ell)$ by using a least-fit square method. Then, D_q and $D(\alpha)$ could be obtained from equations (E.1) and (E.3), respectively. However, the derivation of $D(\alpha)$ through the Legendre transformation is known to suffer from numerical inaccuracies [48, 601, 810]. In order to overcome this obstacle, the Chhabra–Jensen method can be implemented. The corresponding algorithm is particularly useful for the multifractal analysis of point patterns and bitmap images, and it is implemented in the routine FracLac (ver. 2.5) [812] developed for the National Institutes of Health (NIH)-distributed Image-J software package [813].

The Chhabra–Jensen method computes $D(\alpha)$ directly from the data [601]. By defining the one-parameter family $\hat{\mu}_i(q,\ell)$ through the following relation:

$$\hat{\mu}_i(q,\ell) = \frac{P_i(\ell)^q}{\sum_j P_j(\ell)^q}, \tag{E.4}$$

the singularity strength α and the multifractal spectrum $D(\alpha)$ are given by the following:

$$\alpha = \lim_{\ell \to 0} \frac{\sum_i \hat{\mu}_i(q,\ell) \log[P_i(\ell)]}{\log \ell} \tag{E.5}$$

$$f[\alpha(q)] = \lim_{\ell \to 0} \frac{\sum_i \hat{\mu}_i(q,\ell) \log[\hat{\mu}_i(q,\ell)]}{\log \ell}. \tag{E.6}$$

Specifically, the numerators of equations (E.5) and (E.6) are evaluated for each value of q for decreasing box sizes. $f[\alpha(q)]$ and $\alpha(q)$ are then extrapolated from the slopes of the plots of $\sum_i \hat{\mu}_i(q,\ell) \log[\hat{\mu}_i(q,\ell)]$ and $\sum_i \hat{\mu}_i(q,\ell) \log[P_i(\ell)]$ versus $\log \ell$, respectively. Therefore, α and $D(\alpha)$ are calculated directly from the data by performing a general box-counting procedure. More details on the explicit derivation of equations (E.5) and (E.6) can be found in [48, 601, 810].

Appendix F Aperiodic Arrays and Spectral Graph Theory

The primary objects of graph theory are *graphs* or *networks*, which are mathematical structures used to model pairwise relations between objects. Graphs, initially conceived as convenient tools for the study of molecular structures in chemistry, are now among the primary concepts of discrete mathematics and found numerous applications to physics (e.g., tight-binding models), neuroscience, linguistics, and computer science, where they model networks flows and communication streams, data structures, etc. [814, 815]. In what follows, we apply graph theory to better analyze the distinctive geometry of aperiodic patterns. This is achieved by studying the graph structure of the Delaunay triangulation associated to each point pattern. In addition, we will investigate the spectral properties of Laplacian operator defined on graphs. Before we proceed with our discussion, let us introduce some basic notions of graph theory.

A graph is composed of vertices (nodes, or points) that are connected by line segments called edges. More precisely, a simple[1] graph is an ordered pair $G = (V, E)$ comprising a set V of vertices together with a set E of edges such that each edge is associated with two vertices. A weighted graph $G = (V, E, w)$ is a graph in which a numerical weight (a real number w) has been assigned to every edge. The *degree* of a vertex in an unweighted graph is the number of edges in which it is involved. Simple graphs are named regular if every vertex has the same degree d. When dealing with weighted graphs, we should consider the *weighted degree* of a vertex, which is the sum of the weights of the attached edges.

The key idea of spectral graph theory is to study graphs through the spectral properties (i.e., eigenvalues, and eigenvectors) of symmetric matrices associated to them, such as the *adjacency matrix* or the *graph Laplacian matrix*. In short, graph spectral theory addresses the relationship between structural and spectral properties of graphs. Since we can associate a graph to any aperiodic point pattern via its Delaunay triangulation, we propose here to apply spectral graph theory methods in order to investigate the structure–property relationships of the aperiodic graphs associated to deterministic aperiodic arrays.

---

[1] In this context, the adjective "simple" means that the graph is unweighted (each edge has the same unit length) and undirected, and it does not contain loops or multiple edges connecting the same pair of vertices.

The first operator that is typically associated to a graph G is represented by the *adjacency matrix* A_G defined by the following:

$$A_G(a,b) = \begin{cases} w(a,b) & \text{if } (a,b) \in E \\ 0 & \text{otherwise} \end{cases} \tag{F.1}$$

For a graph with n vertices, A_G is an $n \times n$ symmetric matrix[2] that associates to each pair of connected vertices in the graph the weight of their connecting edge. If the graph is unweighted, then the weight $w(a,b) = 1$ for connected vertices and $w(a,b) = 0$ otherwise. Clearly the elements of the matrix indicate whether pairs of vertices are adjacent or not in the graph. In our study, the weights of the vertices of the graphs are simply the lengths of edges of the Delaunay triangulation corresponding to the aperiodic point patterns. The adjacency matrix of undirected simple graphs is symmetric and therefore it has a complete set of real eigenvalues and an orthogonal eigenvector basis. The set of eigenvalues of a graph is the *spectrum of the graph*. Many important structural properties of a graph are "encoded" in its spectrum.

We then consider the the *degree matrix* D_G that is a diagonal matrix that stores along its diagonal all the values of the degrees of each vertice in the graph:

$$D_G(a,b) = \begin{cases} \mathbf{d}(a) & \text{if } a = b, \\ 0 & \text{otherwise,} \end{cases} \tag{F.2}$$

where $\mathbf{d}(a)$ is a vector containing the (weighted) degrees of all the nodes.

Based on the adjacency and degree matrices, we can define the $n \times n$ *graph Laplacian* matrix as:

$$L_G = D_G - A_G \tag{F.3}$$

and the *random walk Laplacian* W_G (or 'diffusion matrix') as follows:

$$W_G = D_G^{-1} L_G = I - D_G^{-1} A_G. \tag{F.4}$$

The Laplacian matrices of weighted graphs arise in many different applications. For example, they appear when applying the finite difference discretization schemes to solve Laplace equations with Neumann boundary conditions. They also arise in the modeling of diffusion through networks, where a quantity ϕ diffuses over time through a graph G. Given the initial values for the diffusing quantity ϕ, as time progresses they diffuse smoothly to their neighbors via the edge connections, and eventually the whole system settles to the same value at equilibrium. This process is called *graph diffusion*.[3] In the case of a simple path,[3] the eigenvalues and eigenvectors of the graph Laplacian corresponds to the oscillation frequencies and vibrational

---

[2] The adjacency matrix is always symmetric because we are considering only simple graphs, and simple graphs are undirected.

[3] A path in a graph is a finite sequence of edges that connect a sequence of vertices that are all distinct from one another.

modes of the corresponding discretized string. For simple graphs, the Laplacian matrix has nonnegative eigenvalues and stores the degree of each node of the graph along its diagonal. We will discover later in this section that the eigenvectors of the graph Laplacian associated to the edge-weighted Delaunay triangulation of aperiodic arrays allows us to identify distinctive *geometrical resonances* in aperiodic arrays.

The smallest nonzero eigenvalue of L_G is called the *spectral gap*. The second smallest eigenvalue of L_G is the *algebraic connectivity* AC of the graph G. This eigenvalue is greater than 0 if and only if G is a connected graph. The magnitude of AC reflects how well connected the graph G is overall. The *connectivity* of a graph is defined as the minimum number of elements (nodes or edges) that need to be removed in order to disconnect the remaining nodes. The algebraic connectivity is bounded above by the vertex connectivity of the graph. Finally, we mention an important variant of the graph Laplacian that is known as the *random walk Laplacian*. The random walk Laplacian of G encodes the dynamics of a random walk on the graph. The random walk on a weighted graph moves from a vertex a to its neighbor b with probability proportional to $w(a,b)$. The walk matrix is used to study the evolution of the probability distribution of a random walk. As the eigenvalues and eigenvectors of W_G provide information about the behavior of a random walk on G, they also provide information about the graph itself.

In Table F.1, we summarize the important graph-based topological parameters associated to the unweighted Delaunay triangulations of each array, all considering $S\lambda^2 = 4.19$. We can observe that all the arrays have comparable average interarticle separation (scaled by the wavelength) $\langle d \rangle / \lambda$ but vary in their minimum interparticle distance d_{min}/λ significantly. Due to its homogeneously disordered nature, the random array features the smallest d_{min}/λ. In order to deepen our understanding of geometrical structures, we also computed their average node degree $\langle D \rangle$, which is the average number of links per node. $\langle D \rangle$ can readily be obtained from the diagonal entries of the graph Laplacian matrix. We notice that the average node degree does not vary significantly across the different structures due to the homogeneous nature of their Delaunay graph. However, significant differences can be observed by considering the Pearson correlation coefficient R, also known in graph theory as the

Table F.1 Geometrical and graph-based parameters of the structures.

Structure	d_{min}/λ	$\langle d \rangle / \lambda$	$\langle D \rangle$	R	C_N	C_{WS}	AC
Square	0.50436	8.4195	5.752	−0.15535	0.38633	0.44005	0.017206
Penrose	0.34091	8.4625	5.917	−0.29777	0.39162	0.41858	0.018345
Random	0.011587	8.042	5.926	−0.15252	0.38313	0.43558	0.022469
μ-spiral	0.29777	7.9442	5.942	−0.17738	0.40156	0.41145	0.01878
π-spiral	0.17491	8.0401	5.838	0.17011	0.39641	0.41163	0.01591
τ-spiral	0.40869	7.9687	5.936	−0.15421	0.39988	0.40776	0.021046

assortativity coefficient. This parameter measures the correlations between nodes of similar degree and it can be computed as follows [815]:

$$R = \frac{1}{\sigma_q^2} \sum_{k_i, k_j} [e(k_i, k_j) - q(k_i)q(k_j)], \tag{F.5}$$

where k_j is the node degree of the jth node, $q(k_j) = (k_j + 1)p(k_j + 1)/\sum_i k_i p(k_i)$, and σ_q^2 is the standard deviation of the node distribution $q(k_j)$. In general, R lies between -1 and 1. Positive values of R indicate a correlation between nodes of similar degree, while negative values indicate relationships between nodes of different degree (nonassortativity). Typical social networks are assortative as opposed to biological networks that, due to their tendency toward a disassortative equilibrium state, are often found to be disassortative [815]. Interestingly, the graph topology of all our structures is disassortative except the τ-spiral, which is assortative. This property reflects the more regular and homogeneous geometry of this structure.

Next we calculated the *Newman global clustering coefficient C* of the graphs, which measures to what extent nodes in a graph tend to cluster together. This metric is based on cluster triplets, which are three connected nodes. The global clustering coefficient is defined as follows [814, 815]:

$$C_N = \frac{3N_T}{P_2}, \tag{F.6}$$

where N_T is the number of triangles[4] and P_2 is the number of paths of length 2 in the network that can be obtained by the formula $P_2 = \sum_{i=1}^{N} k_i(k_i - 1)/2$. The number of triangles can be obtained directly by using the spectral property of the adjacency matrix as $N_T = Tr(A^3)/6$. Only small differences can be detected in the global clustering behavior of the graphs. Furthermore, it is of interest to consider the local clustering properties of the graphs, which are captured by their Watts–Strogatz index $C_{WS} = \sum_{i=1}^{N} 2t_i/N[k_i(k_i - 1)]$, where N is the total number of nodes.

We will now study more in detail the spectral properties of the graph Laplacian L_G. The algebraic connectivity AC of each graph, which indicates how well connected a given graph G is overall, is listed in Table F.1.

We then turn our attention to the eigenvectors of the graph Laplacian, which form an expansion base for the characterization of the graph geometry of the array. We consider here the weighted graph Laplacian with weights equal to the edge lengths of the Delaunay triangulation. In Figure F.1, we show some representative eigenvectors of the weighted graph Laplacian. Their spatial structures correspond to the *geometrical resonances* of deterministic aperiodic arrays. We notice that the spatial distributions of the geometrical resonances with the smallest eigenvalues are similar to the (scalar) resonant modes of a cylindrical cavity described by a wavefunction $\phi_{n,m}(\rho) = J_m(k_\rho \rho) \exp(im\theta)$ characterized by radial and azimuthal mode indices. However, due to the nonhomogeneous nature of the underlying aperiodic graph, these resonances are not perfectly circularly symmetric, as it can be easily appreciated in the

---

[4] A triangle is formed by three closed triplets, one centered at each node.

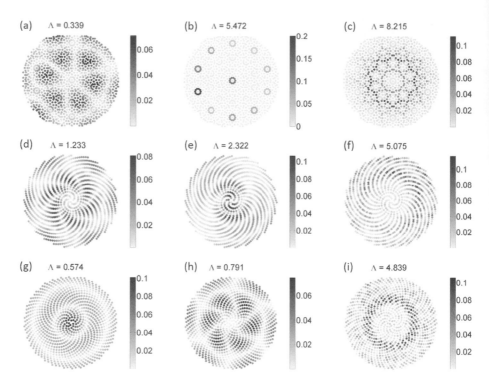

Figure F.1 Representative eigenvectors of the edge-length weighted graph Laplacian matrix for different aperiodic structures: (a)–(c) for an N = 1,108 Penrose array with (eigenvalues equal to 0.3398, 5.4721, and 8.2152, respectively); (d)–(f) for a μ-spiral array (eigenvalues equal to 1.2338, 20.0862, and 21.6709, respectively); (g)–(i) for a τ-spiral array (eigenvalues equal to 0.7915, 4.8394, and 20.0284).

case of the Vogel spirals (Figure F.1d,g). By investigating the geometrical resonances with larger eigenvalues, we discovered that they are similar to the most localized electromagnetic field resonances of the vector point patterns, discussed in Section 12.4.3 (compare with Figure 12.10). This resemblance is remarkable since the modes of the graph Laplacian originate from purely geometrical/interconnectivity properties of the aperiodic graphs. Therefore, graph theory reveals, at minimal computational cost, important features of the optical resonances driven by the aperiodic geometry of the arrays.

Appendix G Essentials of Fractional Calculus

On September 30, 1695, Guillaume de l'Hôpital wrote to Leibniz inquiring about a particular notation he had used to denote the nth derivative of the linear function $f(x) = x$, which is as follows:

$$\frac{d^n f(x)}{dx^n}. \tag{G.1}$$

Specifically, de l'Hôpital's question to Leibniz was: "What would the result be if $n = 1/2$?" Leibniz's response followed: "An apparent paradox, from which one day useful consequences will be drawn." The mathematical approach of fractional calculus, born out of that simple yet far-seeing answer, later grew into a rich subject that has been studied by the best minds in mathematics, including Fourier, Euler, Laplace, Liouville, and Riemann, to name a few. In what follows, we review the basic notions of fractional calculus mostly based on references [816–818], to which we refer the interested readers.

The starting point toward the development of fractional operators is the *Cauchy formula* that reduces the calculation of the $n$fold primitive (i.e., repeated integration formula) of a (locally) absolutely integrable and causal function[1] $f(t)$, with $t \in \mathbb{R}^+$, to the single integral of convolution type:

$$J^n f(t) = \frac{1}{(n-1)!} \int_0^t (t - \tau)^{n-1} f(\tau) d\tau, \tag{G.2}$$

where J^n indicates the integral operator of integer order n. The Cauchy formula motivates the Riemann–Liouville (RL) definition of the *fractional integral* of order α, which is valid for $\alpha \in \mathbb{R}^+$:

$$\boxed{J^\alpha f(t) = \frac{1}{\Gamma(\alpha)} \int_0^t (t - \tau)^{\alpha-1} f(\tau) d\tau} \tag{G.3}$$

where we introduced the Euler's Γ function. The preceding RL definition of the fractional integer operator J^α naturally extends its integer-order counterpart J^n and reduces to the Cauchy formula when n is an integer, since $\Gamma(n) = (n-1)!$. We set by convention $J^0 f(t) = f(t)$, so that $J^0 = I$, the identity operator.

---

[1] Remember that a causal function $f(t)$ satisfies the condition $f(t) = 0$ if $t < 0$.

The fractional integral operator J^α satisfies the semigroup property:

$$J^\alpha J^\beta = J^{\alpha+\beta} = J^\beta J^\alpha \quad \alpha, \beta > 0, \tag{G.4}$$

which also implies the commutative property of fractional integrals. The nth-order integer derivative operator, denoted by D^n, also satisfies the same semigroup and commutative properties. However, the operators J^n and D^n do not commute with each other in general, i.e., $J^n D^n \neq D^n J^n$. Indeed, $D^n J^n = I$ but $J^n D^n f(t) = f(t) - \sum_{k=0}^{n-1} f^k(0) \frac{t^k}{k!}$, where f^k denotes the kth derivative. This asymmetry creates the need for two different definitions of the fractional derivative, which are the Riemann–Liouville derivative and the Caputo derivative, which we introduce in this appendix.

We can now introduce the two main definitions of fractional derivatives. Setting $m - 1 < \alpha \leq m$ with $m \in \mathbb{N}$, the RL fractional derivative of order α, denoted by D^α_{RL}, is defined as the *left inverse* of the operator J^α, i.e., $D^\alpha f(t) = D^m J^{m-\alpha} f(t)$, corresponding to the following:

$$D^\alpha_{RL} f(t) = \begin{cases} \frac{d^m}{dt^m} \left[\frac{1}{\Gamma(m-\alpha)} \int_0^t (t - \tau)^{m-1-\alpha} f(\tau) d\tau \right] & m - 1 < \alpha < m \\ \frac{d^m f(t)}{dt^m} & \alpha = m. \end{cases} \tag{G.5}$$

On the other hand, the Caputo fractional derivative, denoted by D^α, is defined as the right inverse of the operator J^n according to the following:

$$D^\alpha f(t) = \begin{cases} \frac{1}{\Gamma(m-\alpha)} \int_0^t (t - \tau)^{m-1-\alpha} \frac{d^m f(\tau)}{dt^m} d\tau & m - 1 < \alpha < m \\ \frac{d^m f(t)}{dt^m} & \alpha = m. \end{cases} \tag{G.6}$$

The two definitions of the fractional derivative are related by the following [392]:

$$D^\alpha_{RL} f(t) = D^\alpha f(t) + \sum_{k=0}^{m-1} \frac{t^{k-\alpha}}{\Gamma(k - \alpha + 1)}. \tag{G.7}$$

This definition of the Caputo derivative is more restrictive than the Riemann–Liouville one because it requires the absolute integrability of the derivative of order m. However, the Caputo fractional derivative has the very desirable feature that it vanishes when applied to a constant, which is not necessarily the case for D_{RL} when α is not an integer. Moreover, the Caputo derivative is more suitable for the solution of fractional differential equation models that can be treated by the Laplace transform technique. In fact, its Laplace transform requires the knowledge of the bounded initial values of the function $f(t)$ and of its integer derivatives of order $k = 1, 2, \ldots, m - 1$ in analogy with the case when $\alpha = m$:

$$\mathcal{L}(D^\alpha f(t), s) = s^\alpha \hat{f}(s) - \sum_{k=0}^{m-1} s^{\alpha-1-k} f^k(0^+), \tag{G.8}$$

where $\mathcal{L}$ denotes the Laplace transform operator and $m - 1 < \alpha \leq m$. Therefore, when solving fractional differential equations, the Caputo derivative allows one to use integer-order initial conditions, which have a clear physical meaning. This will not happen if using the RL (left-hand) derivative, which requires one to specify fractional-order initial conditions (of unclear meaning). For all these reasons, the Caputo derivative is often preferred in many physics and engineering applications.

Fractional derivatives are often computed using the Grunwald–Letnikov approach, which makes use of a repeated differentiation formula analogous to the Cauchy formula for iterated integration. When n is a positive integer, the repeated differentiation formula reads as follows:

$$d^n f(x) = \lim_{h \to 0} \frac{(-1)^n}{h^n} \sum_{m=0}^{n} (-1)^m \binom{n}{m} f(x + mh). \tag{G.9}$$

This expression can be generalized to noninteger order α if the binomial coefficient is understood using its Γ function expression as follows:

$$\boxed{D_{GL}^{\alpha} f(x) = \lim_{h \to 0} \frac{(-1)^{\alpha}}{h^{\alpha}} \sum_{m=0}^{\infty} (-1)^m \binom{\alpha}{m} f(x + mh)} \tag{G.10}$$

which is the Grunwald–Letnikov (GL) fractional derivative. Using the following identity:

$$\binom{-n}{m} = (-1)^m \frac{(n + m - 1)!}{(n - 1)! \, m!}, \tag{G.11}$$

we can extend the preceding property to real numbers α using the Γ function:

$$\binom{-\alpha}{m} = (-1)^m \frac{\Gamma(\alpha + m)}{\Gamma(\alpha) m!}. \tag{G.12}$$

Therefore, we can define the GL fractional integral of order α as follows:

$$\boxed{D_{GL}^{-\alpha} f(x) = \lim_{h \to 0} h^{\alpha} \sum_{m=0}^{\infty} \frac{\Gamma(\alpha + m)}{\Gamma(\alpha) m!} f(x - mh)} \tag{G.13}$$

The two expressions (G.10) and (G.13) are known as GL *differintegrals*[2] and are computed in Figure G.1 for the function $y = x$ for different values of α. While the RL definition of fractional integrals and derivatives is more suitable for analytical studies, the RL definitions are preferred for numerical implementations. Finally, it can be shown that the GL and RL formulations of fractional calculus are in fact entirely equivalent. See, for instance, Gorenflo et al. [816].

---

[2] Note that α can now assume any real value. Therefore, "fractional calculus" is a misnomer because the order α is not limited to rational numbers.

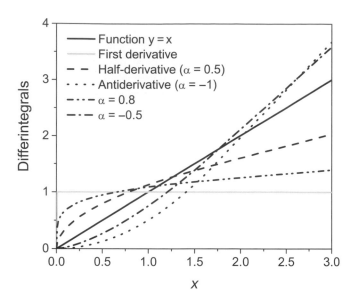

Figure G.1 GL differintegrals of the function $y = x$ for different values of the order α.

G.1   Fractional Differential Equations

Let us consider a linear fractional differential equation of real order α, with $0 < \alpha \leq 1$ for simplicity:

$$D^\alpha f(t) = \lambda f(t), \qquad \lambda \in \mathbb{R}. \tag{G.14}$$

Performing the Laplace transform, we obtain the following:

$$\hat{f}(s) = f(0)\frac{s^{\alpha-1}}{s^\alpha - \lambda}. \tag{G.15}$$

The function $(s^{\alpha-1})/(s^\alpha - \lambda)$ is the Laplace transform of the Mittag–Leffler function $E_\alpha(\lambda t^\alpha)$, which is defined as follows:

$$E_\alpha(z) := \sum_{n=0}^{\infty} \frac{z^n}{\Gamma(\alpha n + 1)}, \qquad \alpha > 0, \qquad z \in \mathbb{C} \tag{G.16}$$

The Mittag–Leffler function plays a special role in fractional calculus that is analogous to the one of exponential functions in the solution of ordinary differential equations [816]. In fact, the solution of linear fractional differential equations of order α is usually expressed in terms of a linear combination of derivatives and integrals of the Mittag–Leffler function. In Figure G.2, we show the fundamental solutions of the fractional relaxation-oscillation equation:

$$D^\alpha x(t) + \omega_0^2 x(t) = \delta(t), \tag{G.17}$$

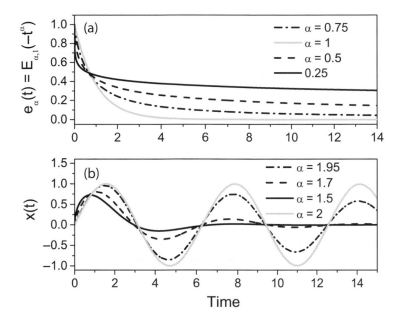

Figure G.2 (a) Fundamental solutions of the fractional relaxation equation for different values of the order α of the derivative. (b) Fundamental solutions of the fractional oscillation model for different values of the order α. Note the significant damping that occurs in the solutions when decreasing the order α. Here we considered $\omega_0 = 1$.

where D^α is the Caputo fractional derivative. The preceding model describes a *fractional oscillator*. However, differently from the standard, i.e., the integer-order harmonic oscillator, the solutions of its fractional counterpart correspond to a relaxation behavior for $0 < \alpha \leq 1$ and to an oscillatory for $1 < \alpha \leq 2$, as reflected in Figure G.2a,b, respectively. These solutions can be conveniently obtained using the Laplace transform method, which yields the following:

$$x(t) = \sum_{k=0}^{m-1} J^k E_\alpha(-t^\alpha) x^k(0) - \frac{d}{dt}\left[E_\alpha(-t^\alpha)\right]. \tag{G.18}$$

Another important class of functions that appears in fractional calculus in relation to general fractional diffusion problems is the Fox H functions, which are defined via the Mellin–Barnes integral representation:

$$H_{p,q}^{m,n}(x) = \frac{1}{2\pi i} \int_{\mathcal{L}} \mathcal{H}_{p,q}^{m,n} x^s ds, \tag{G.19}$$

where $\mathcal{L}$ is a suitable path in the complex plane and $\mathcal{H}_{p,q}^{m,n}$ is defined as follows:

$$\mathcal{H}_{p,q}^{m,n}(s) = \frac{A(s)B(s)}{C(s)D(s)}, \tag{G.20}$$

where

$$A(s) = \prod_{j=1}^{m} \Gamma(b_j - s\beta_j), \qquad B(s) = \prod_{j=1}^{n} \Gamma(1 - a_j - \alpha_j s)$$

and

$$C(s) = \prod_{j=m+1}^{q} \Gamma(1 - b_j - \beta_j s), \qquad D(s) = \prod_{j=n+1}^{p} \Gamma(a_j - \alpha_j s).$$

Here $a_j, b_j \in \mathbb{C}$ and $\alpha_j, \beta_j \in \mathbb{R}^+$, $0 \le n \le p$ and $1 \le m \le q$. The H functions are also denoted as follows:

$$H_{p,q}^{m,n}(x) = H_{p,q}^{m,n}\left[x \left| \begin{matrix} (a_j, \alpha_j)_{j=1,\dots p} \\ (b_j, \beta_j)_{j=1,\dots q} \end{matrix} \right. \right]. \tag{G.21}$$

For additional details, see [396, 819].

The Fox H functions are the solutions of the fractional diffusion equation with both spatial and time fractional operators. For example, in the case of fractional time diffusion, Mainardi et al. [397] found an explicit expression of the fundamental solution $W(x,t)$ (i.e., the Green' function) of the problem in the following form:

$$W(x,t) = \frac{1}{\sqrt{4K_\alpha t^\alpha}} H_{1,1}^{1,0}\left[\frac{|x|}{\sqrt{K_\alpha t^\alpha}} \left| \begin{matrix} (1 - \frac{\alpha}{2}, \frac{\alpha}{2}) \\ (0,1) \end{matrix} \right. \right], \tag{G.22}$$

where the relevant H function is the following:

$$H_{1,1}^{1,0}(x) = H_{1,1}^{1,0}\left[x \left| \begin{matrix} (1 - \frac{\alpha}{2}, \frac{\alpha}{2}) \\ (0,1) \end{matrix} \right. \right]. \tag{G.23}$$

G.2   Fundamental Solutions of Fractional Diffusion

In this section, we provide more detailed information on the analytical and numerical methods utilized to obtain the results presented in Section 14.7.2. In particular, we discuss the solutions shown in Figure G.3 that correspond to the Green functions or fundamental solutions of subdiffusive, distributed-order (DO-U and DO-PL), and superdiffusive transport models. The fundamental solutions are obtained by solving the initial-value Cauchy problem for different fractional diffusion equations.

The single-order subdiffusion Cauchy problem is defined by the following [400]:

$$\begin{cases} D^\alpha W(x,t) = K_\alpha \frac{\partial^2 W(x,t)}{\partial x^2}. \\ W(x,0) = \delta(x). \end{cases} \tag{G.24}$$

The equation solution is discussed in detail in [400]. The Laplace domain expression of the fundamental solution $\tilde{W}(x,s)$, where s is the Laplace domain variable, can be written as follows:

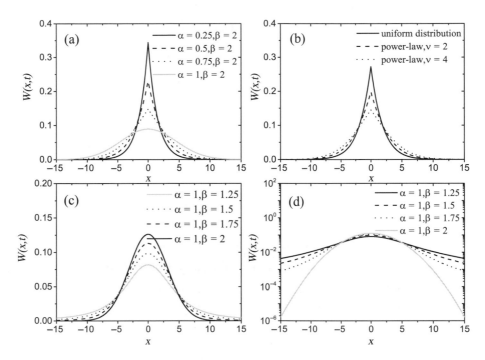

Figure G.3 (a) The fundamental solution $W(x, t = 10)$ of single-order subdiffusion for different fractional time-derivative order α, (b) The fundamental solution $W(x, t = 10)$ of DO diffusion follows power-law distribution function for $v = 1$(DO-U), $v = 2$, and $v = 4$, (c) The fundamental solution $W(x, t = 10)$ of superdiffusion for different fractional spatial derivative β in linear scale. (d) The fundamental solution $W(x, t = 10)$ of superdiffusion as shown in (c) in the logarithmic scale

$$\tilde{W}(x, s) = \frac{1}{2\sqrt{K_\alpha}} s^{\frac{\alpha}{2}-1} \exp\left(-\frac{|x|}{\sqrt{K_\alpha}} s^{\frac{\alpha}{2}}\right). \tag{G.25}$$

Following the approach in [820], we perform a numerical inverse Laplace transform and plot in Figure G.3a the single-order subdiffusion fundamental solutions $W(x, t)$ for different α at time $t = 10$ and for a generalized diffusion coefficient $K_\alpha = 1$. We can observe that for $\alpha = 1$, the result recovers to the well-known Gaussian fundamental solution [400]. However, when decreasing α, the $W(x, t)$ solution increases in amplitude at the center, where it develops a characteristic kink and shows heavy tails for large values of the space coordinate x. This shows that a smaller order α of the generalized time derivative leads to a slower diffusion process. The fundamental solution of the DO-fractional diffusion is obtained by solving the equation [400]:

$$\begin{cases} \int_0^1 p(\alpha)[D^\alpha W(x, t)]d\alpha = \tilde{K} \frac{\partial^2 W(x, t)}{\partial x^2} d\alpha. \\ W(x, 0) = \delta(x). \end{cases} \tag{G.26}$$

The fundamental solution $W(x,t)$ in the Laplace domain is expressed as follows [400]:

$$\tilde{W}(x,s) = \frac{I(s)^{\frac{1}{2}}}{2s\sqrt{\tilde{K}}\exp\left[-|x|\left(I(s)/\sqrt{\tilde{K}}\right)^{\frac{1}{2}}\right]}. \tag{G.27}$$

In Figure G.3b, we show the numerically obtained fundamental solution for the DO-U model at $t = 10$. For the DO-PL case, the expression of $I(s)$ is as follows:

$$I(s) = \int_0^1 s^\alpha v\alpha^{v-1}d\alpha = v\left[\gamma(v, -\log(s))\right](-\log(s))^{-v}, \tag{G.28}$$

where the $\gamma(v, -\log(s))$ is the lower incomplete gamma function defined as $\gamma(s,x) = \int_0^x t^{s-1}e^{-t}dt$. Implementing directly the inverse Laplace transform on equation (G.28) is difficult because s is contained in the lower limit of the integral that defines γ. Therefore, we expanded this integral using the holomorphic extension of $\gamma(v, -\log(s))$ following [821], and we obtain the following:

$$\gamma(v, -\log(s)) = \left[-\log(s)\right]^v \Gamma(v)s \sum_{k=0}^\infty \frac{(-\log(s))^k}{\Gamma(v+k+1)}. \tag{G.29}$$

Combining equations (G.27)–(G.28), we can now perform the numerical inverse Laplace transform and obtain the fundamental solution for different DO-PL cases. We plot in the Figure G.2b the fundamental solutions of DO-PL with $v = 2$ and $v = 4$ at $t = 10$. We can observe that by increasing v from 1 to 4, the fundamental solution is decreasing in its value at the center and spreads spatially with longer tails indicating a faster diffusion. Therefore, the slower diffusion for the DO-PL case with $v = 1$, which coincides with the DO-U model, leads to larger η values as shown in Table 14.3.

Interestingly there is an anlytical expression for the fundamental solution of the DO-PL model for $t \gg 1$, which is given by the following [822]:

$$W(x,t) \approx \frac{1}{\sqrt{4\tilde{K}}}\left[\frac{\Gamma(v+1)}{\log(t)^v}\right]^{1/2}\exp\left\{-\left(\frac{\Gamma(v+1)}{\tilde{K}}\right)^{1/2}\frac{|x|}{\log(t)^{v/2}}\right\} \tag{G.30}$$

We find an excellent agreement between our numerical results obtained with the holomorphic extension method and the preceding asymptotic analytical expression when $t > 500$, which further validates our approach.

Finally, the fundamental solution for the superdiffusive case can be obtained by solving the following equation:

$$\begin{cases} \frac{\partial W(x,t)}{\partial t} = K_\beta D^\beta W(x,t). \\ W(x,0) = \delta(x). \end{cases} \tag{G.31}$$

Using Fourier transformation in space, we can derive the fundamental solution $W(k,t)$ in the following form [391]:

$$\tilde{W}(k,t) = \exp(-K_\beta t \,|k|), \tag{G.32}$$

where k is the Fourier domain argument. In Figure G.3c,d, we show the corresponding fundamental solutions $W(x,t)$ at $t = 10$ obtained by performing numerical Fourier inversion. We can observe that by decreasing β, the fundamental solution will develop longer (heavy) tails in the spatial domain, which imply the divergence of its finite moments. This is more evident in Figure G.3d, where we plot the solutions using semilogarithmic scale. In Figure G.3d, we also plot in green the fundamental solution of the standard diffusion model ($\alpha = 1, \beta = 2$), which clearly shows slower diffusion fronts compared to the superdiffusive cases.

Appendix H Mie–Lorenz Field Components

In this appendix, we list the explicit expressions for the field components derived from the Mie–Lorentz theory. The meanings of the different symbols appearing in this list are discussed in Chapter 13.

$$E_{i,r} = \frac{\cos\phi \sin\theta}{\rho^2} \sum_{n=1}^{\infty} [-jn(n+1)]\pi_n \psi_n \tag{H.1}$$

$$E_{i,\theta} = \frac{\cos\phi}{\rho} \sum_{n=1}^{\infty} E_n(\pi_n \psi_n - j\tau_n \psi_n') \tag{H.2}$$

$$E_{i,\phi} = \frac{\sin\phi}{\rho} \sum_{n=1}^{\infty} E_n(j\pi_n \psi_n' - \tau_n \psi_n) \tag{H.3}$$

$$E_{s,r} = \frac{\cos\phi \sin\theta}{\rho^2} \sum_{n=1}^{\infty} E_n[ja_n n(n+1)\pi_n \xi_n] \tag{H.4}$$

$$E_{s,\theta} = \frac{\cos\phi}{\rho} \sum_{n=1}^{\infty} E_n(ja_n \tau_n \xi_n' - b_n \pi_n \xi_n) \tag{H.5}$$

$$E_{s,\phi} = \frac{\sin\phi}{\rho} \sum_{n-1}^{\infty} E_n(b_n \tau_n \xi_n - ja_n \pi_n \xi_n') \tag{H.6}$$

$$E_{int,r} = \frac{\cos\phi \sin\theta}{\rho^2} \sum_{n=1}^{\infty} E_n[-jd_n n(n+1)\pi_n \psi_n] \tag{H.7}$$

$$E_{int,\theta} = \frac{\cos\phi}{\rho} \sum_{n=1}^{\infty} E_n(c_n \pi_n \psi_n - jd_n \tau_n \psi_n') \tag{H.8}$$

$$E_{int,\phi} = \frac{\sin\phi}{\rho} \sum_{n=1}^{\infty} E_n(jd_n \pi_n \psi_n' - c_n \tau_n \psi_n) \tag{H.9}$$

$$H_{i,r} = -\frac{k}{\omega\mu}\frac{\sin\phi\sin\theta}{\rho^2}\sum_{n=1}^{\infty}E_n[jn(n+1)]\pi_n\psi_n \tag{H.10}$$

$$H_{i,\theta} = -\frac{k}{\omega\mu}\frac{\sin\phi}{\rho}\sum_{n=1}^{\infty}E_n(j\tau_n\psi_n' - \pi_n\psi_n) \tag{H.11}$$

$$H_{i,\phi} = -\frac{k}{\omega\mu}\frac{\cos\phi}{\rho}\sum_{n=1}^{\infty}E_n(j\pi_n\psi_n' - \tau_n\psi_n) \tag{H.12}$$

$$H_{s,r} = \frac{k}{\omega\mu}\frac{\sin\phi\sin\theta}{\rho^2}\sum_{n=1}^{\infty}E_n[jb_n n(n+1)\pi_n\xi_n] \tag{H.13}$$

$$H_{s,\theta} = \frac{k}{\omega\mu}\frac{\sin\phi}{\rho}\sum_{n=1}^{\infty}E_n(jb_n\tau_n\xi_n' - a_n\pi_n\xi_n) \tag{H.14}$$

$$H_{s,\phi} = \frac{k}{\omega\mu}\frac{\cos\phi}{\rho}\sum_{n=1}^{\infty}E_n(jb_n\pi_n\xi_n' - a_n\tau_n\xi_n) \tag{H.15}$$

$$H_{int,r} = -\frac{k_{int}}{\omega\mu_{int}}\frac{\sin\phi\sin\theta}{\rho^2}\sum_{n=1}^{\infty}E_n[jc_c n(n+1)\pi_n\psi_n] \tag{H.16}$$

$$H_{int,\theta} = -\frac{k_{int}}{\omega\mu_{int}}\frac{\sin\phi}{\rho}\sum_{n=1}^{\infty}E_n(jc_n\tau_n\psi_n' - d_n\pi_n\psi_n) \tag{H.17}$$

$$H_{int,\phi} = -\frac{k_{int}}{\omega\mu_{int}}\frac{\cos\phi}{\rho}\sum_{n=1}^{\infty}E_n(jc_n\pi_n\psi_n' - d_n\tau_n\psi_n) \tag{H.18}$$

References

[1] A. N. Whitehead, *Introduction to Mathematics*. Henry Holt and Company, New York, 1911.

[2] R. Falk and C. Konold, "Making sense of randomness: implicit encoding as a basis for judgment," *Psychological Review*, vol. 104, no. 2, pp. 301–318, 1997.

[3] S. Dehaene, *The Number Sense: How the Mind Creates Mathematics*. Oxford University Press, New York, 2011.

[4] H. Poincaré, "Sur le probléme des trois corps et les équations de la dynamique," *Acta Mathematica*, vol. 13, pp. A3–A270, 1890.

[5] P. J. Steinhardt, *The Second Kind of Impossible: The Extraordinary Quest for a New Form of Matter*. Simon and Schuster, New York, 2018.

[6] S. Torquato, "Hyperuniform states of matter," *Physics Reports*, vol. 745, pp. 1–95, 2018.

[7] J. P. Allouche and J. O. Shallit, *Automatic Sequences: Theory, Applications, Generalizations*. Cambridge University Press, New York, 2009.

[8] M. Schroeder, *Number Theory in Science and Communication: With Applications in Cryptography, Physics, Digital Information, Computing, and Self-Similarity*. Springer-Verlag, Berlin, 2009.

[9] L. Dal Negro, Y. Chen, and F. Sgrignuoli, "Aperiodic photonics of elliptic curves," *Crystals*, vol. 9, pp. 482–509, 2019.

[10] F. Sgrignuoli, S. Gorsky, W. A. Britton, R. Zhang, F. Riboli, and L. Dal Negro, "Multifractality of light in photonic arrays based on algebraic number theory," *Communications Physics*, vol. 3, pp. 1–9, 2020.

[11] C. Janot, *Quasicrystals: A Primer*. Oxford University Press, Oxford, 1994.

[12] C. de Lange and T. Janssen, "Incommensurability and recursivity: lattice dynamics of modulated crystals," *Journal of Physics C: Solid State Physics*, vol. 14, no. 34, pp. 5269–5292, December 1981. [Online]. Available: https://doi.org/10.1088%2F0022-3719%2F14%2F34%2F009

[13] D. R. Hofstadter, "Energy levels and wave functions of Bloch electrons in rational and irrational magnetic fields," *Physical Review B*, vol. 14, pp. 2239–2249, September 1976. [Online]. Available: https://link.aps.org/doi/10.1103/PhysRevB.14.2239

[14] P. G. Harper, "The general motion of conduction electrons in a uniform magnetic field, with application to the diamagnetism of metals," *Proceedings of the Physical Society. Section A*, vol. 68, no. 10, pp. 879–892, October 1955. [Online]. Available: https://doi.org/10.1088%2F0370-1298%2F68%2F10%2F305

[15] S. N. Evangelou and J.-L. Pichard, "Critical quantum chaos and the one-dimensional harper model," *Physical Review Letters*, vol. 84, pp. 1643–1646, February 2000. [Online]. Available: https://link.aps.org/doi/10.1103/PhysRevLett.84.1643

[16] R. Wang, M. Röntgen, C. V. Morfonios, F. A. Pinheiro, P. Schmelcher, and L. Dal Negro, "Edge modes of scattering chains with aperiodic order," *Optics Letters*, vol. 43, no. 9, pp. 1986–1989, May 2018. [Online]. Available: http://ol.osa.org/abstract.cfm?URI=ol-43-9-1986

[17] A. Dareau, E. Levy, M. B. Aguilera, et al., "Revealing the topology of quasicrystals with a diffraction experiment," *Physical Review Letters*, vol. 119, p. 215304, November 2017. [Online]. Available: https://link.aps.org/doi/10.1103/PhysRevLett.119.215304

[18] F. Baboux, E. Levy, A. Lemaître, et al., "Measuring topological invariants from generalized edge states in polaritonic quasicrystals," *Physics Review B*, vol. 95, p. 161114, April 2017. [Online]. Available: https://link.aps.org/doi/10.1103/PhysRevB .95.161114

[19] S. Abe and H. Hiramoto, "Fractal dynamics of electron wave packets in one-dimensional quasiperiodic systems," *Physics Review A*, vol. 36, pp. 5349–5352, 1987.

[20] R. Ketzmerick, K. Kruse, S. Kraut, and T. Geisel, "What determines the spreading of a wave packet?" *Physical Review Letters*, vol. 79, pp. 1959–1963, 1997.

[21] R. Ketzmerick, G. Petschel, and T. Geisel, "Slow decay of temporal correlations in quantum systems with cantor spectra," *Physical Review Letters*, vol. 69, pp. 695–698, August 1992. [Online]. Available: https://link.aps.org/doi/10.1103/PhysRevLett.69.695

[22] L. Dal Negro and S. Inampudi, "Fractional transport of photons in deterministic aperiodic structures," *Scientific Reports*, vol. 7, no. 1, p. 2259, 2017.

[23] M. Gardner, *Penrose Tiles to Trapdoor Ciphers*. W. H. Freeman, New York, 1989.

[24] P. Sheng, *Introduction to Wave Scattering, Localization and Mesoscopic Phenomena*, 2nd ed. Springer, Berlin, 2006.

[25] E. Akkermans and G. Montambaux, *Mesoscopic Physics of Electrons and Photons*. Cambridge University Press, New York, 2007.

[26] J. D. Joannopoulos, S. G. Johnson, J. N. Winn, and R. D. Meade, *Photonic Crystals: Molding the Flow of Light*, 2nd ed. Princeton University Press, Princeton, 2008.

[27] K. Sakoda, *Optical Properties of Photonic Crystals*, 2nd ed. Springer, Berlin, 2005.

[28] L. Brillouin, *Wave Propagation in Periodic Structures*. McGraw-Hill, New York, 1946.

[29] M. I. Mishchenko, "125 years of radiative transfer: enduring triumphs and persisting misconceptions," *AIP Conference Proceedings*, vol. 1531, no. 1, pp. 11–18, 2013. [Online]. Available at: https://aip.scitation.org/doi/abs/10.1063/1.4804696

[30] S. Chandrasekhar, *Radiative Transfer*. Dover, New York, 1960.

[31] P. W. Anderson, "Absence of diffusion in certain random lattices," *Physical Review*, vol. 109, pp. 1492–1505, 1958.

[32] P. W. Anderson, "The question of classical localization: a theory of white paint?" *Phylosophical Magazine B*, vol. 52, no. 3, pp. 505–509, 1985.

[33] Y. Kuga and A. Ishimaru, "Retroreflectance from a dense distribution of spherical particles," *Journal of the Optical Society of America A*, vol. 1, p. 831, 1984.

[34] M. P. van Albada and A. Lagendijk, "Observation of weak localization of light in a random medium," *Physical Review Letters*, vol. 55, p. 2692, 1985.

[35] P. Wolf and G. Maret, "Weak localization and coherent backscattering of photons in disordered media," *Physical Review Letters*, vol. 55, p. 2696, 1985.

[36] S. John, "Electromagnetic absorption in a disordered medium near a photon mobility edge," *Physical Review Letters*, vol. 53, p. 2169, 1984.

[37] D. Shechtman, I. Blech, D. Gratias, and J. W. Cahn, "Metallic phase with long-range orientational order and no tranlsational symmetry," *Physical Review Letters*, vol. 53, pp. 1951–1953, 1984.

[38] D. Levine and P. J. Steinhardt, "Quasicrystals: a new class of ordered structures," *Physical Review Letters*, vol. 26, pp. 2477–2480, 1984.

[39] M. Senechal, *Quasicrystals and Geometry*. Cambridge University Press, Cambridge, 1995.

[40] T. Janssen, G. Chapuis, and M. de Boissieu, *Aperiodic Crystals: From Modulated Phases to Quasicrystals*. Oxford University Press, Oxford, 2007.

[41] R. Merlin, K. Bajema, R. Clarke, F. Y. Juang, and P. K. Bhattacharya, "Quasiperiodic GaAs-AIAs heterostructures," *Physical Review Letters*, vol. 55, pp. 1768–1770, 1985.

[42] B. Kohmoto, H. Sutherland, and K. Iguchi, "Localization of optics: quasiperiodic media," *Physical Review Letters*, vol. 58, p. 2436, 1987.

[43] M. Born and E. Wolf, *Principles of Optics*, 7th ed. Cambridge University Press, Cambridge, 1999.

[44] L. Dal Negro, C. J. Oton, Z. Gaburro, et al., "Light transport through the band-edge states of fibonacci quasicrystals," *Physical Review Letters*, vol. 90, p. 055501, February 2003. [Online]. Available: https://link.aps.org/doi/10.1103/PhysRevLett.90.055501

[45] M. Kohmoto, L. P. Kadanoff, and C. Tang, "Localization problem in one dimension: mapping and escape," *Physical Review Letters*, vol. 50, pp. 1870–1872, 1983.

[46] M. Kolá and M. K. Ali, "Attractors of some volume-nonpreserving fibonacci trace maps," *Physical Review A*, vol. 39, pp. 6538–6544, June 1989. [Online]. Available: https://link.aps.org/doi/10.1103/PhysRevA.39.6538

[47] W. Gellerman, M. Kohmoto, B. Sutherland, and P. C. Taylor, "Localization of light waves in fibonacci dielectric multilayers," *Physical Review Letters*, vol. 72, pp. 633–636, 1994.

[48] M. Schreiber and H. Grussbach, "Multifractal wave functions at the Anderson transition," *Physical Review Letters*, vol. 67, no. 5, pp. 607–610, 1991.

[49] S. Faez, A. Strybulevych, J. H. Page, A. Lagendijk, and B. A. van Tiggelen, "Observation of multifractality in Anderson localization of ultrasound," *Physical Review Letters*, vol. 103, p. 155703, October 2009. [Online]. Available: https://link.aps.org/doi/10.1103/PhysRevLett.103.155703

[50] T. Fujiwara, M. Kohmoto, and T. Tokihiro, "Multifractal wave functions on a fibonacci lattice," *Physical Review B*, vol. 40, pp. 7413–7416, October 1989. [Online]. Available: https://link.aps.org/doi/10.1103/PhysRevB.40.7413

[51] M. Kolar, M. K. Ali, and F. Nori, "Generalized Thue–Morse chains and their physical properties," *Physical Review B*, vol. 43, pp. 1034–1047, 1991.

[52] Z. Cheng, R. Savit, and R. Merlin, "Structure and electronic properties of Thue–Morse lattices," *Physical Review B*, vol. 37, pp. 4375–4382, March 1988. [Online]. Available: https://link.aps.org/doi/10.1103/PhysRevB.37.4375

[53] N.-h. Liu, "Propagation of light waves in Thue–Morse dielectric multilayers," *Physical Review B*, vol. 55, pp. 3543–3547, February 1997. [Online]. Available: https://link.aps.org/doi/10.1103/PhysRevB.55.3543

[54] L. Dal Negro, M. Stolfi, Y. Yi, et al., "Photon band gap properties and omnidirectional reflectance in Si/SiO Thue–Morse quasicrystals," *Applied Physics Letters*, vol. 84, no. 25, pp. 5186–5188, 2004.

[55] S.-F. Cheng and G.-J. Jin, "Trace map and eigenstates of a Thue–Morse chain in a general model," *Physical Review B*, vol. 65, p. 134206, March 2002. [Online]. Available: https://link.aps.org/doi/10.1103/PhysRevB.65.134206

[56] X. Jiang, Y. Zhang, S. Feng, K. C. Huang, Y. Yi, and J. D. Joannopoulos, "Photonic band gaps and localization in the Thue–Morse structures," *Applied Physics Letters*, vol. 86, no. 20, p. 201110, 2005. [Online]. Available: https://doi.org/10.1063/1.1928317

[57] M. Kohmoto, "Localization problem and mapping of one-dimensional wave equations in random and quasiperiodic media," *Physical Review B*, vol. 34, pp. 5043–5047, October 1986. [Online]. Available: https://link.aps.org/doi/10.1103/PhysRevB.34.5043

[58] C. Tang and M. Kohmoto, "Global scaling properties of the spectrum for a quasiperiodic Schrödinger equation," *Physical Review B*, vol. 34, pp. 2041–2044, August 1986. [Online]. Available: https://link.aps.org/doi/10.1103/PhysRevB.34.2041

[59] M. Kohmoto, B. Sutherland, and C. Tang, "Critical wave functions and a cantor-set spectrum of a one-dimensional quasicrystal model," *Physical Review B*, vol. 35, pp. 1020–1033, 1987.

[60] M. Baake, U. Grimm, and D. Joseph, "Trace maps, invariants, and some of their applications," *International Journal of Modern Physics B*, vol. 07, no. 06n07, pp. 1527–1550, 1993. [Online]. Available: https://doi.org/10.1142/S021797929300247X

[61] M. Kohmoto and Y. Oono, "Cantor spectrum for an almost periodic Schrodinger equation and a dynamical map," *Physics Letters A*, vol. 102, no. 4, pp. 145–148, 1984. [Online]. Available: www.sciencedirect.com/science/article/pii/0375960184909289

[62] K. Esaki, M. Sato, and M. Kohmoto, "Wave propagation through Cantor-set media: chaos, scaling, and fractal structures," *Physical Review E*, vol. 79, p. 056226, 2009.

[63] M. Kolář and M. K. Ali, "One-dimensional generalized fibonacci tilings," *Physical Review B*, vol. 41, pp. 7108–7112, April 1990. [Online]. Available: https://link.aps.org/doi/10.1103/PhysRevB.41.7108

[64] M. Kolář and F. Nori, "Trace maps of general substitutional sequences," *Physical Review B*, vol. 42, pp. 1062–1065, July 1990. [Online]. Available: https://link.aps.org/doi/10.1103/PhysRevB.42.1062

[65] E. Maciá and F. Domínguez-Adame, "Physical nature of critical wave functions in Fibonacci systems," *Physical Review Letters*, vol. 76, pp. 2957–2960, 1997.

[66] J.-P. Desideri, L. Macon, and D. Sornette, "Observation of critical modes in quasiperiodic systems," *Physical Review Letters*, vol. 63, pp. 390–393, July 1989. [Online]. Available: https://link.aps.org/doi/10.1103/PhysRevLett.63.390

[67] N. Ferralis, A. W. Szmodis, and R. D. Diehl, "Diffraction from one- and two-dimensional quasicrystalline gratings," *American Journal of Physics*, vol. 72, no. 9, pp. 1241–1246, 2004. [Online]. Available: https://doi.org/10.1119/1.1758221

[68] T. Hattori, N. Tsurumachi, S. Kawato, and H. Nakatsuka, "Photonic dispersion relation in a one-dimensional quasicrystal," *Physical Review Letters, B*, vol. 50, pp. 4220–4223, August 1994. [Online]. Available: https://link.aps.org/doi/10.1103/PhysRevB.50.4220

[69] H. Noh, J. Yang, S. V. Boriskina, et al., "Lasing in Thue–Morse structures with optimized aperiodicity," *Applied Physics Letters*, vol. 98, p. 201109, 2011.

[70] L. Dal Negro, *Optics of Aperiodic Structures: Fundamentals and Device Applications*. Pan Stanford Publishing, Singapore, 2014.

[71] L. Dal Negro and S. V. Boriskina, "Deterministic aperiodic nanostructures for photonics and plasmonics applications," *Laser Photonics Review*, vol. 6, pp. 1–41, 2011.

[72] Z. V. Vardeny, A. Nahata, and A. Agrawal, "Optics of photonic quasicrystals," *Nature Photonics*, vol. 7, pp. 177–187, 2013.

[73] W. Steurer and D. Sutter-Widmer, "Photonic and phononic quasicrystals," *Journal of Physics D: Applied Physics*, vol. 40, pp. R229–R247, 2007.

[74] R. Lifshitz, A. Arie, and A. Bahabad, "Photonic quasicrystals for nonlinear optical frequency conversion," *Physical Review Letters*, vol. 95, p. 133901, 2005.

[75] D. S. Wiersma, "Disordered photonics," *Nature Photonics*, vol. 7, pp. 188–196, May 2013.

[76] W. Pauli, *Theory of Relativity*. Dover Publications, New York, 1958.

[77] A. Messiah, *Quantum Mechanics*. Dover Publications, New York, 1999.

[78] B. E. A. Saleh and M. C. Teich, *Fundamentals of Photonics*, 2nd ed. John Wiley, Hoboken, 2007.

[79] A. Yariv and Y. Pochi, *Photonics: Optical Electronics in Modern Communications*, 6th ed. Oxford University Press, New York, 2007.

[80] J. C. Maxwell, "On physical lines of force," *Philosophical Magazine*, vol. 21 (parts I and II) and 23 (parts III and IV), pp. 1–48, 1861–1862.

[81] M. Faraday, *Experimental Researches in Electricity*. Dover Publications, New York, 1965.

[82] J. D. Jackson, *Classical Electrodynamics*, 3rd ed. John Wiley, 1998.

[83] A. Garg, *Classical Electromagnetism in a Nutshell*. Princeton University Press, Princeton, 2012.

[84] J. D. Jackson, "From Lorenz to Coulomb and other explicit gauge transformations," *American Journal of Physics*, vol. 70, p. 917, 2002.

[85] C. Cohen-Tannoudji, J. Dupont-Roc, and G. Grynberg, *Photons and Atoms. Introduction to Quantum Electrodynamics*. Wiley-VCH, Strauss GmbH, Morlenbach 2004.

[86] O. L. Brill and B. Goodman, "Causality in the Coulomb Gauge," *American Journal of Physics*, vol. 35, pp. 832–837, 1967.

[87] M. O. Scully and M. S. Zubairy, *Quantum Optics*. Cambridge University Press, Cambridge, 2002.

[88] L. Mandel and E. Wolf, *Optical Coherence and Quantum Optics*. Cambridge University Press, New York, 1995.

[89] R. Courant and D. Hilbert, *Methods of Mathematical Physics*, 6th ed., vols. 1–2. Interscience Publishers, New York, 1966.

[90] R. G. Newton, *Scattering Theory of Waves and Particles*. Dover Publications, New York, 2002.

[91] N. Zettili, *Quantum Mechanics: Concepts and Applications*, 2nd ed. John Wiley, UK, 2009.

[92] L. Novotny and B. Hecht, *Principles of Nano-Optics*, 2nd ed. Cambridge University Press, New York, 2012.

[93] F. N. H. Robinson, *Macroscopic Electromagnetism*, 3rd ed. Pergamon Press, Oxford, 1973.

[94] E. M. Landau, E. M. Lifshitz, and L. P. Pitaevskii, *Electrodynamics of Continuous Media*. Elsevier, Amsterdam, 1984.

[95] Y. R. Shen, *The Principles of Nonlinear Optics*. John Wiley, Menlo Park, 1984.

[96] R. P. Feynman, R. B. Leighton, and M. Sands, *The Feynman Lectures on Physics*, vol. 2. Addison-Wesley, Palo Alto, 1964.

[97] S. A. Maier, *Plasmonics: Fundamentals and Applications*. Springer, New York, 2007.

[98] T. V. Shahbazyan and M. I. Stockman, *Plasmonics: Theory and Applications*. Springer, Dordrecht, 2013.

[99] M. L. Brongersma and P. G. Kik, *Surface Plasmon Nanophotonics*. Springer, 2007.

[100] C. T. Tai, *Dyadic Green's Functions in Electromagnetic Theory*, 2nd ed. IEEE Press, New York, 1993.

[101] W. C. Chew, *Waves and Fields in Inhomogeneous Media*. IEEE Press, New York, 1995.

[102] J. Bladel, "Some remarks on green's dyadic for infinite space," *IEEE Transactions on Antennas and Propagation*, vol. AP-9, pp. 563–566, 1961.

[103] R. F. Harrington, *Field Computation by Moment Methods*. Macmillan, New York, 1968.

[104] D. Livesay and K. Chen, "Electromangetic fields induced inside arbitrarily shaped biological bodies," *IEEE Transactions on Microwave Theory and Techniques*, vol. MTT-22, no. 12, pp. 1273–1280, 1974.

[105] J. van Bladel, *Electromagnetic Fields*, 2nd ed. IEEE Press and John Wiley, Hoboken, 2007.

[106] J. van Bladel, *Singular Electromagnetic Fields and Sources*. IEEE Press, Piscataway, 1991.

[107] G. Rubinacci and A. Tamburrino, "A broadband volume integral formulation based on edge-elements for full-wave analysis of lossy interconnects," *IEEE Transactions on Antennas and Propagation*, vol. 54, no. 10, pp. 2977–2989, 2006.

[108] G. Miano, G. Rubinacci, and A. Tamburrino, "Numerical modelling of the interaction of nanoparticles with electromagnetic waves," *Compel*, vol. 26, no. 3, pp. 586–599, 2007.

[109] G. Miano, G. Rubinacci, and A. Tamburrino, "Numerical modeling for the analysis of plasmon oscillations in metallic nanoparticles," *IEEE Transactions on Antennas and Propagation*, vol. 58, no. 9, pp. 2920–2933, 2010.

[110] L. Dal Negro, G. Miano, G. Rubinacci, A. Tamburrino, and S. Ventre, "A fast computation method for the analysis of an array of metallic nanoparticles," *IEEE Transactions on Magnetics*, vol. 45, no. 3, pp. 1618–1621, 2009.

[111] H. A. Lorentz, *The Theory of Electrons and Its Applications to the Phenomena of Light and Radiant Heat*, 2nd ed. Stechert and Co, New York, 1916.

[112] P. B. Johnson and R. W. Christy, "Optical constants of the noble metals," *Physical Review B*, vol. 6, no. 12, pp. 4370–4379, 1972.

[113] M. E. Peskin and D. V. Schroeder, *An Introduction to Quantum Field Theory*. Westview Press, Boulder, 1995.

[114] L. D. Landau, "Über die bewegung der elektronen in kristallgitter," *Zeitschrift fur Physik Sowjetunion*, vol. 3, pp. 644–645, 1933.

[115] R. D. Mattuck, *A Guide to Feynman Diagrams in the Many-Body Problem*, 2nd ed. Dover Publications, New York, 1976.

[116] G. Grosso and G. Pastori Parravicini, *Solid State Physics*, 2nd ed. Academic Press, Oxford, 2014.

[117] H. Haug and S. W. Koch, *Quantum Theory of the Optical and Electronic Properties of Semiconductors*, 4th ed. World Scientific Publishing, Singapore, 2004.

[118] V. M. Agranovich and V. L. Ginzburg, *Crystal Optics with Spatial Dispersion and Excitons*. Springer Verlag, Berlin, 1984.

[119] J. Homola, S. S. Yee, and G. Gauglitz, "Surface plasmon resonance sensors: review," *Sensors and Actuators B*, vol. 54, pp. 3–15, 1999.

[120] H. Raether, *Surface Plasmons on Smooth and Rough Surfaces and on Gratings*. Springer-Verlag, Berlin, 1988.

[121] E. D. Palik, *Handbook of Optical Constants*. Academic Press, Orlando, 1985.

[122] K. M. McPeak, S. V. Jayanti, J. P. Kress, et al., "Plasmonic films can easily be better: rules and recipes," *ACS Photonics*, vol. 2, no. 12, pp. 326–333, 2015.

[123] A. Sihvola, *Electromagnetic Mixing Formulas and Applications*. Institution on Engineering and Technology, London, 2008.

[124] J. C. Maxwell-Garnett, "Colours in metal glasses and in metallic films," *Philosophical Transactions of the Royal Society Series A*, vol. 203, pp. 385–420, 1904.

[125] D. A. G. Bruggeman, "Berechnung verschiedener physikalischer konstanten von heterogenen substanzen. i. dielektrizitätskonstanten und leitfähigkeiten der mischkörper aus isotropen substanzen," *Annals of Physics*, vol. 416, pp. 636–664, 1935.

[126] P. Mallet, C. A. Guérin, and A. Sentenac, "Maxwell-Garnett mixing rule in the presence of multiple scattering: derivation and accuracy," *Physical Review B*, vol. 72, pp. 014 205–1–014 205–9, 2005.

[127] B. T. Draine, "The discrete-dipole approximation and its application to interstellar graphite grains," *Astrophysical Journal*, vol. 333, pp. 848–872, 1988.

[128] L. Tsang, J. A. Kong, and R. T. Shin, *Theory of Microwave Remote Sensing*. John Wiley and Sons, New York, 1985.

[129] L. Tsang and J. A. Kong, "Multiple scattering of electromagnetic waves by random distributions of discrete scatterers with coherent potential and quantum mechanical formalism," *Journal of Applied Physics*, vol. 51, pp. 3465–3485, 1980.

[130] V. Frisch, *Wave Propagation in Random Medium, in Probabilistic Methods in Applied Mathematics*, vol. 1, Bharuch-Reid Ed. Academic Press, New York, 1968.

[131] C. A. Guérin, P. Mallet, and A. Sentenac, "Effective-medium theory for finite-size aggregates," *Journal of the Optical Society of America A*, vol. 23, pp. 349–358, 2006.

[132] L. Tsang, J. A. Kong, and K. Ding, *Scattering of Electromangetic Waves*, vol. III. John Wiley, New York, 2000.

[133] Y. Wu, J. Li, Z. Zhang, and C. T. Chan, "Effective medium theory for magnetodielectric composites: beyond the long-wavelength limit," *Physical Review B*, vol. 74, no. 085111, pp. 1–9, 2006.

[134] B. A. Slovick, Z. G. Yu, and S. Krishnamurthy, "Generalized effective-medium theory for metamaterials," *Physical Review B*, vol. 89, no. 155118, pp. 1–5, 2014.

[135] S. Torquato and J. Kim, *Nonlocal Effective Electromagnetic Wave Characteristics of Composite Media: Beyond the Quasistatic Regime, Phys. Rev. X 11, 021002–2021*.

[136] M. C. Rechtsman and S. Torquato, "Effective dielectric tensor for electromagnetic wave propagation in random media," *Journal of Applied Physics*, vol. 103, no. 8, p. 084901, 2008. [Online]. Available: https://doi.org/10.1063/1.2906135

[137] Y. Chen, L. Lu, G. E. Karniadakis, and L. Dal Negro, "Physics-informed neural networks for inverse problems in nano-optics and metamaterials," *Optics Express*, vol. 28, no. 8, pp. 11618–11633, 2020.

[138] L. Solymar and E. Shamonina, *Waves in Metamaterials*. Oxford University Press, 2009.

[139] V. M. Shalaev and A. K. Sarychev, *Electrodynamics of Metamaterials*. World Scientific, 2007.

[140] N. Engheta and R. W. Ziolkowski, *Metamaterials. Physics and Engineering Explorations*. John Wiley and IEEE Press, Canada, 2006.

[141] M. Silveirinha and N. Engheta, "Tunneling of electromagnetic energy through subwavelength channels and bends using ϵ-near-zero materials," *Physical Review Letters*, vol. 97, no. 157403, 2006.

[142] N. Engheta, "Pursuing near-zero response," *Science*, vol. 340, pp. 286–287, 2013.

[143] R. A. Shelby, D. R. Smith, and S. Schultz, "Experimental verification of a negative index of refraction," *Science*, vol. 292, pp. 77–79, 2001.

[144] J. B. Pendry, "Negative refraction makes a perfect lens," *Physical Review Letters*, vol. 85, pp. 3966–3969, 2000.

[145] D. R. Smith, W. J. Padilla, D. C. Vier, S. C. Nemat-Nasser, and S. Schultz, "Composite medium with simultaneously negative permeability and permittivity," *Physical Review Letters*, vol. 84, pp. 4184–4187, 2000.

[146] V. G. Veselago, "The electrodynamics of substances with simultaneously negative values of ϵ and μ," *Soviet Physics Uspekhi,* vol. 10, pp. 509–514, 1967.

[147] J. C. Bose, "On the rotation of plane of polarization of electric waves by a twisted structure," *Proceedings of the Royal Society*, vol. 63, pp. 146–152, 1898.

[148] C. Forestiere, A. J. Pasquale, A. Capretti, et al., "Genetically optimized plasmonic nanoarrays," *Nano Letters*, vol. 12, no. 4, pp. 2037–2044, 2012.

[149] Y. Dong and S. Liu, "Topology optimization of patch-typed left-handed metamaterial configurations for transmission performance within the radio frequency band based on the genetic algorithm," *Journal of Optics*, vol. 14, no. 105101, pp. 1–9, 2012.

[150] C. Della Giavampaola and N. Engheta, "Digital metamaterials," *Nature Materials*, vol. 13, pp. 1115–1121, 2014.

[151] T. J. Cui, M. Q. Qi, J. Zhao, and Q. Cheng, "Coding metamaterials, digital metamaterials and programmable metamaterials," *Light Science and Applications*, vol. 3, no. 218, pp. 1–9, 2014.

[152] G. B. Whitham, *Linear and Nonlinear Waves*. John Wiley and Sons, Hoboken, 1999.

[153] L. A. Ostrovsky and A. I. Potapov, *Modulated Waves. Theory and Applications*. Johns Hopkins University Press, Baltimore, 1999.

[154] P. K. Murphy, *Machine Learning. A Probabilistic Perspective*. MIT Press, Cambridge, 2012.

[155] I. Goodfellow, Y. Bengio, and A. Courville, *Deep Learning*. MIT Press, Cambridge, 2016.

[156] K. Funahashi, "On the approximate realization of continuous mappings by neural networks," *Neural Networks*, vol. 2, pp. 183–192, 1989.

[157] K. Hornik, M. Stinchcombe, and H. White, "Multilayer feedforward networks are universal approximators," *Neural Networks*, vol. 2, pp. 359–366, 1989.

[158] S. Haykin, *Neural Networks and Learning Machines*, 3rd ed. Pearson, New York, 2009.

[159] Y. Sun, Z. Xia, and U. S. Kamilov, "Efficient and accurate inversion of multiple scattering with deep learning," *Optics Express*, vol. 26, no. 11, pp. 14 678–14 688, 2018.

[160] Y. Sanghvi, Y. Kalepu, and U. K. Khankhoje, "Embedding deep learning in inverse scattering problems," *IEEE Transactions on Computational Imaging*, vol. 6, pp. 46–56, 2019.

[161] M. Raissi, P. Perdikaris, and G. E. Karniadakis, "Physics-informed neural networks: a deep learning framework for solving forward and inverse problems involving nonlinear partial differential equations," *Journal of Computational Physics*, vol. 378, pp. 686–707, 2019.

[162] L. Lu, X. Meng, Z. Mao, and G. E. Karniadakis, "Deepxde: a deep learning library for solving differential equations," *eprint arXiv:1907.04502*, 2019.

[163] A. Sommerfeld, *Optics*, vol. 4, *Lectures on Theoretical Physics*. Academic Press, New York, 1954.

[164] A. Wang and A. J. Prata, "Lenslet analysis by rigorous vector diffraction theory," *Journal of the Optical Society of America A*, vol. 12, no. 5, pp. 1161–1169, 1995.

[165] A. S. Marathay and J. F. McCalmont, "Vector diffraction theory for electromagnetic waves," *Journal of the Optical Society of America A*, vol. 18, no. 10, pp. 2585–2593, 2001.

[166] J. Braat and P. Török, *Imaging Optics*. Cambridge University Press, Cambridge, 2019.

[167] P. Török, P. R. T. Munro, and E. E. Kriezis, "Rigorous near- to far-field transformation for vectorial diffraction calculations and its numerical implementation," *Journal of the Optical Society of America A*, vol. 23, no. 3, pp. 713–722, 2006.

[168] W. Hsu and R. Barakat, "Stratton-chu vectorial diffraction of electromangetic fields by apertures with application to small-fresnel-number systems," *Journal of the Optical Society of America A*, vol. 11, no. 2, pp. 623–629, 1994.

[169] J. Kim, Y. Wang, and X. Zhang, "Calculation of vectorial diffraction in optical systems," *Journal of the Optical Society of America A*, vol. 35, no. 4, pp. 526–535, 2018.

[170] H. Bethe, "Theory of diffraction by small holes," *Physical Review*, vol. 66, pp. 163–182, 1944.

[171] G. Kirchhoff, "Zur theorie der lichtstrahlen," *Weidemann Ann.*, vol. 2, no. 18, pp. 663–695, 1883.

[172] N. Mukunda, "Consistency of Rayleigh's diffraction formulas with Kirchhoff's boundary conditions," *Journal of the Optical Society of America*, vol. 52, no. 3, pp. 336–337, 1962.

[173] J. W. Goodman, *Introduction to Fourier Optics*, 4th ed. W. H. Freeman, New York, 2017.

[174] H. Stark, *Applications of Optical Fourier Transforms*. Academic Press, New York, 1982.

[175] C. A. Balanis, *Antenna Theory*, 4th ed. John Wiley, Hoboken 2016.

[176] M. C. Rechtsman, J. M. Zeuner, Y. Plotnik, et al., "Photonic floquet topological insulators," *Nature*, vol. 496, pp. 196–200, 2013.

[177] S. Stützer, Y. Plotnik, Y. Lumer, et al. "Photonic topological Anderson insulators," *Nature*, vol. 560, no. 7719, pp. 461–465, 2018.

[178] B. H. Kolner, "Space-time duality and the theory of temporal imaging," *IEEE Journal of Quantum Electronics*, vol. 30, no. 8, pp. 1951–1963, 1994.

[179] T. Poon and T. Kim, *Engineering Optics with Matlab*, 2nd ed. World Scientific Publishing, Singapore, 2018.

[180] H. M. Ozaktas, Z. Zalevsky, and M. Alper Kutay, *The Fractional Fourier Transform with Applications in Optics and Signal Processing*. John Wiley, New York, 2001.

[181] H. Mendlovic and H. M. Ozaktas, "Fractional fourier transforms and their optical implementation: I," *Journal of the Optical Society of America A*, vol. 10, pp. 1875–1881, 1993.

[182] H. Mendlovic and H. M. Ozaktas, "Fractional fourier transforms and their optical implementation: Ii," *Journal of the Optical Society of America A*, vol. 10, pp. 2522–2531, 1993.

[183] B. J. West, M. Bologna, and P. Grigolini, *Physics of Fractal Operators*. Springer, New York, 2003.

[184] F. Intonti, N. Caselli, N. Lawrence, J. Trevino, D. S. Wiersma, and L. Dal Negro, "Near-field distribution and propagation of scattering resonances in vogel spiral arrays of dielectric nanopillars," *New Journal of Physics*, vol. 15, no. 8, p. 085023, 2013.

[185] J. A. Stratton and L. J. Chu, "Diffraction theory of electromagnetic waves," *Physical Review*, vol. 56, pp. 99–107, 1939.

[186] J. A. Stratton, *Electromagnetic Theory*. McGraw-Hill, New York, 1941.

[187] A. S. B. Holland, *Introduction to the Theory of Entire Functions*. Academic Press, New York and London, 1973.

[188] R. P. Boas, *Entire Functions*. Academic Press, New York, 1954.

[189] J. Lindberg, "Mathematical concepts of optical superresolution," *Journal of Optics*, vol. 14, no. 083001, pp. 1–23, 2012.

[190] A. J. den Dekker and A. van den Bos, "Resolution: a survey," *Journal of the Optical Society of America A*, vol. 14, pp. 547–557, 1997.

[191] A. Vijayakumar and S. Bhattacharya, *Design and Fabrication of Diffractive Optical Elements with MATLAB*. SPIE Press, Bellingham, 2017.

[192] Y. Chen, W. Britton, and L. Dal Negro, "Phase-modulated axilenses for infrared multiband spectroscopy," *Optics Letters*, vol. 45, no. 8, pp. 2371–2374, 2020.

[193] W. A. Britton, Y. Chen, F. Sgrignuoli, and L. Dal Negro. "Phase-modulated axilenses as ultracompact spectroscopic tools," *ACS Photonics*, vol. 7, no. 10, 2731–2738, 2020.

[194] Y. Chen, W. Britton, and L. Dal Negro, "Design of infrared microspectrometer based on phase-modulated axilenses," *Applied Optics*, vol. 59, pp. 5532–5538, 2020.

[195] J. E. Harvey and L. Forgham, "The spot of arago: new relevance for an old phenomenon," *American Journal of Physics*, vol. 52, pp. 243–247, 1984.

[196] K. Iizuka, *Engineering Optics*, 3rd ed. Springer, New York, 2008.

[197] M. V. Berry and M. R. Dennis, "Natural superoscillations in monochromatic waves in d dimensions," *Journal of Physics A: Mathematical and Theoretical*, vol. 42, no. 2, p. 022003, 2009.

[198] F. M. Huang, Y. Chen, F. Javier Garcia de Abajo, and N. I. Zheludev, "Optical super-resolution through super-oscillations," *Journal of Optics A: Pure and Applied Optics*, vol. 9, pp. S285–S288, 2007.

[199] E. T. F. Rogers, J. Lindbergn, T. Roy, et al., "A super-oscillatory lens optical microscope for subwavelength imaging," *Nature Materials*, vol. 11, pp. 432–435, 2012.

[200] F. M. Huang and N. I. Zheludev, "Super-resolution without evanescent waves," *Nano Letters*, vol. 9, no. 3, pp. 1249–1254, 2009.

[201] M. V. Berry and S. Popescu, "Evolution of quantum superoscillations and optical superresolution without evanescent waves," *Journal of Physics A: Mathematical and General*, vol. 39, pp. 6965–6977, 2006.

[202] A. Kempf, "Black holes, bandwidths and Beethoven," *Journal of Mathematical Physics*, vol. 41, pp. 2360–2374, 2000.

[203] P. J. S. G. Ferraira and A. Kempf, "Superoscillations: faster than the Nyquist rate," *IEEE Transactions on Signal Processing*, vol. 54, no. 10, pp. 3732–3740, 2006.

[204] E. T. F. Rogers and N. I. Zheludev, "Optical super-oscillations: sub-wavelength light focusing and super-resolution imaging," *Journal of Optics*, vol. 15, no. 094008, pp. 1–23, 2013.

[205] R. Wang, F. A. Pinheiro, and L. Dal Negro, "Spectral statistics and scattering resonances of complex primes arrays," *Physical Review B*, vol. 97, no. 024202, pp. 1–11, 2018.

[206] A. Vasara, J. Turunen, and A. T. Friberg, "Realization of general nondiffracting beams with computer-generated holograms," *Journal of the Optical Society of America A*, vol. 6, no. 11, pp. 1748–1754, 1989.

[207] A. Vijayakumar, P. Parthasarathi, S. S. Iyengar, et al., "Conical Fresnel zone lens for optical trapping," *International Conference on Optics and Photonics 2015*, vol. 9654, 2015.

[208] W. T. Chen, A. Y. Zhu, V. Sanjeev, M. Khorasaninejad, Z. Shi, E. Lee, and F. Capasso, "A broadband achromatic metalens for focusing and imaging in the visible," *Nature Nanotechnology*, vol. 13, no. 3, p. 220, 2018.

[209] S. Zhang, A. Soibel, S. A. Keo, et al., "Solid-immersion metalenses for infrared focal plane arrays," *Applied Physics Letters*, vol. 113, no. 11, p. 111104, 2018.

[210] W. A. Britton, Y. Chen, F. Sgrignuoli, and L. Dal Negro, "Phase-modulated axilenses as ultracompact spectroscopic tools," *ACS Photonics*, vol. 7, no. 10, pp. 2731–2738, 2020. [Online]. Available: https://doi.org/10.1021/acsphotonics.0c00762

[211] J. F. Nye, *Natural Focusing and Fine Structure of Light*. Institute of Physics Publishing, 1999.

[212] R. Salem, M. A. Foster, A. C. Turner, D. F. Geraghty, M. Lipson, and A. L. Gaeta, "Optical time lens based on four-wave mixing on a silicon chip," *Optics Letters*, vol. 15, no. 33, pp. 1047–1049, 2008.

[213] A. Klein, T. Yaron, E. Preter, H. Duadi, and M. Fridman, "Temporal depth imaging," *Optica*, vol. 4, no. 5, pp. 502–506, 2017.

[214] J. T. Mendonça and P. K. Shukla, "Time refraction and time reflection: two basic concepts," *Physica Scripta*, vol. 65, no. 2, pp. 160–163, 2002.

[215] Y. Xiao, D. N. Maywar, and G. P. Agrawal, "Reflection and transmission of electromagnetic waves at a temporal boundary," *Optics Letters*, vol. 39, no. 3, pp. 574–577, 2014.

[216] A. M. Shaltout, K. G. Lagoudakis, J. van de Groep, et al., "Spatiotemporal light control with frequency-gradient metasurfaces," *Science*, vol. 365, pp. 374–377, 2019.

[217] Y. Zhou, M. Z. Alam, M. Karimi, et al., "What is the temporal analog of reflection and refraction of optical beams?" *Nature Communications*, vol. 11, no. 2180, 2020.

[218] B. W. Plansinis, W. R. Donaldson, and G. P. Agrawal, "What is the temporal analog of reflection and refraction of optical beams?" *Physical Review Letters*, vol. 115, no. 183901, 2015.

[219] J. Ye and S. T. e. Cundiff, *Femtosecond Optical Frequency Comb: Principle, Operation and Applications*. Springer Science + Business Media, Boston, 2005.

[220] N. Picqué and T. W. Hänsch, "Frequency comb spectroscopy," *Nature Photonics*, vol. 13, pp. 146–157, 2019.

[221] V. Lakshminarayanan, A. K. Ghatak, and K. Thyagarajan, *Lagrangian Optics*. Springer Science, New York, 2002.

[222] M. V. Berry and C. Upstill, "Catastrophe optics: morphologies of caustics and their diffraction patterns," *Progress in Optics*, vol. 18, pp. 257–346, 1980.

[223] S. Dupré, "Optics, pictures and evidence: Leonardo's drawings of mirrors and machinery," *Early Science and Medicine*, vol. 10, no. 2, p. 211–236, 2005.

[224] V. I. Arnold, *Catastrophe Theory*, 2nd ed. Springer, Berlin, 1986.

[225] T. Poston and I. Stewart, *Catastrophe Theory and Its Applications*. Pitman, London, 1978.

[226] R. Gilmore, *Catastrophe Theory for Scientists and Engineers*. John Wiley, New York, 1981.

[227] R. Thom, *Structural Stability and Morphogenesis. An Outline of a General Theory of Models*. W. A. Benjamin Inc., Reading, 1975.

[228] G. A. Siviloglou and D. N. Christodoulides, "Accelerating finite energy airy beams," *Optics Letters*, vol. 32, no. 8, p. 979–981, 2007.

[229] M. V. Berry and N. Balazs, "Nonspreading wave packets," *American Journal of Physics*, vol. 47, pp. 264–267, 1979.

[230] G. A. Siviloglou, J. Broky, A. Dogariu, and D. N. Christodoulides, "Observation of accelerating airy beams," *Physical Review Letters*, vol. 99, no. 213901, 2007.

[231] J. Baumgartl, M. Mazilu, and K. Dholakia, "Optically mediated particle clearing using airy wavepackets," *Nature Photonics*, vol. 2, pp. 675–678, 2008.

[232] J. F. Nye, "The motion and structure of dislocations in wavefronts," *Proceedings of the Royal Society of London. Series A*, vol. 378, pp. 219–239, 1981.

[233] J. F. Nye and M. Berry, "Dislocations in wave trains," *Proceedings of the Royal Society of London. Series A*, vol. 336, pp. 165–190, 1974.

[234] M. R. Dennis, K. O'Holleran, and M. J. Padgett, "Singular optics: optical vortices and polarization singularities," *Progress in Optics*, vol. 53, pp. 293–363, 2009.

[235] M. Berry, J. F. Nye, and F. Wright, "The elliptic umbilic diffraction catastrophe," *Philosophical Transactions of the Royal Society*, vol. 291, pp. 453–484, 1979.

[236] C. Kharif and E. Pelinovsky, "Physical mechanisms of the rogue wave phenomenon," *European Journal of Mechanics – B/Fluids*, vol. 22, pp. 603–63 437, 2003.

[237] A. R. Osborne, *Nonlinear Ocean Waves and the Inverse Scattering Transform*. Academic Press, New York, 2010.

[238] D. R. Solli, C. Ropers, P. Koonath, and B. Jalali, "Optical rogue waves," *Nature*, vol. 450, pp. 1054–1057, 2007.

[239] J. M. Dudley, F. Dias, M. Erkintalo, and G. Genty, "Instabilities, breathers and rogue waves in optics," *Nature Photonics*, vol. 8, pp. 755–764, 2014.

[240] J. J. Metzger, R. Fleischmann, and T. Geisel, "Statistics of extreme waves in random media," *Physical Review Letters*, vol. 112, p. 203903, May 2014. [Online]. Available: https://link.aps.org/doi/10.1103/PhysRevLett.112.203903

[241] A. Mathis, L. Froehly, S. Toenger, F. Dias, G. Genty, and J. M. Dudley, "Caustics and rogue waves in an optical sea," *Scientific Reports*, vol. 5, no. 12822, 2015.

[242] A. Safari, R. Fickler, M. J. Padgett, and R. W. Boyd, "Generation of caustics and rogue waves from nonlinear instability," *Physical Review Letters*, vol. 119, no. 203901, 2017.

[243] S. Coles, *An Introduction to Statistical Modeling of Extreme Values*. Springer-Verlag, London, 2001.

[244] F. Sgrignuoli, Y. Chen, S. Gorsky, W. A. Britton, and L. Dal Negro, "Optical rogue waves in multifractal photonic arrays," *Physical Review B*, vol. 103, no. 19, 2021.

[245] P. Coullet, "Optical vortices," *Optics Communications*, vol. 73, pp. 403–408, 1989.

[246] A. M. Yao and M. J. Padgett, "Orbital angular momentum: origins, behavior and applications," *Advances in Optics and Photonics*, vol. 3, pp. 161–204, 2011.

[247] J. P. Torres and L. Torner, Eds., *Twisted Photons. Applications of Light with Orbital Angular Momentum*. Wiley-VCH, Weinheim, 2011.

[248] L. Allen, M. W. Beijersbergen, R. J. C. Spreeuw, and J. P. Woerdman, "Orbital angular-momentum of light and the transformation of Laguerre–Gaussian laser modes," *Physical Review A*, vol. 45, pp. 8185–8189, 1992.

[249] S. Chavez-Cerda, M. Padgett, I. Allison, et al., "Holographic generation and orbital angular momentum of high-order mathieu beams," *Journal of Optics B: Quantum Semiclassical Optics*, vol. 4, pp. S52–S57, 2002.

[250] S. F. Liew, H. Noh, J. Trevino, L. Dal Negro, and H. Cao, "Localized photonic band edge modes and orbital angular momenta of light in a golden-angle spiral," *Optics Express*, vol. 19, pp. 23631–23642, 2011.

[251] N. Lawrence, J. Trevino, and L. Dal Negro, "Control of optical orbital angular momentum by vogel spiral arrays of metallic nanoparticles," *Optics Letters*, vol. 37, pp. 5076–5078, 2012.

[252] G. Molina-Terriza, J. P. Torres, and L. Torner, "Twisted photons," *Nature Physics*, vol. 3, pp. 3015–310, 2007.

[253] D. Grier, "A revolution in optical manipulation," *Nature Physics*, vol. 424, pp. 810–816, 2003.

[254] A. Mair, A. Vaziri, G. Weihs, and A. Zeilinger, "Entanglement of the orbital angular momentum states of photons," *Nature*, vol. 412, pp. 313–316, 2001.

[255] A. Baev, P. N. Prasad, H. Ren, M. Samoc, and M. Wegener, "Metaphotonics: an emerging field with opportunities and challenges," *Physics Reports*, vol. 594, pp. 1–60, 2015. [Online]. Available: www.sciencedirect.com/science/article/pii/S0370157315003361

[256] F. Capasso, "The future and promise of flat optics: a personal perspective," *Nanophotonics*, vol. 7, no. 6, pp. 953–957, 2018.

[257] N. Yu and F. Capasso, "Flat optics with designer metasurfaces," *Nature Materials*, vol. 13, pp. 139–150, 2014.

[258] N. Yu, P. Genevet, M. A. Kats, et al., "Light propagation with phase discontinuities: generalized laws of reflection and refraction," *Science*, vol. 334, pp. 333–337, 2011.

[259] X. Ni, N. K. Emani, A. V. Kildishev, A. Boltasseva, and V. M. Shalaev, "Broadband light bending with plasmonic nanoantennas," *Science*, vol. 335, p. 427, 2012.

[260] D. Lin, P. Fan, E. Hasman, and M. L. Brongersma, "Dielectric gradient metasurface optical elements," *Science*, vol. 345, pp. 298–301, 2014.

[261] C. Wu, N. Arju, G. Kelp, et al., "Spectrally selective chiral silicon metasurfaces based on infrared fano resonances," *Nature Communications*, vol. 5, no. 3892, pp. 1–9, 2014.

[262] M. Khorasaninejad and F. Capasso, "Metalenses: versatile multifunctional photonic components," *Science*, vol. 358, no. 1146, pp. 1–8, 2017.

[263] P. Genevet and F. Capasso, "Holographic optical metasurfaces: a review of current progress," *Reports on Progress in Physics*, vol. 78, no. 024401, pp. 1–19, 2015.

[264] F. Ding, A. Pors, and S. I. Bozhevolnyi, "Gradient metasurfaces: a review of fundamentals and applications," *Reports on Progress in Physics*, vol. 81, no. 2, p. 026401, 2017.

[265] Y. Zhao, X. Liu, and A. Alù, "Recent advances on optical metasurfaces," *Journal of Optics*, vol. 16, no. 123001, pp. 1–14, 2014.

[266] G. D. Mahan, *Many-Particle Physics*. Plenum Press, New York, 1990.

[267] A. J. Devaney, *Mathematical Foundation of Imaging, Tomography and Wavefield Inversion*. Cambridge University Press, Cambridge, 2012.

[268] H. Liu, D. Liu, H. Mansour, P. T. Boufounos, L. Waller, and U. S. Kamilov, "Seagle: sparsity-driven image reconstruction under multiple scattering," *IEEE Transactions on Computational Imaging*, vol. 4, no. 1, pp. 73–86, 2018.

[269] D. Colton and R. Kress, "Looking back on inverse scattering theory," *SIAM Review*, vol. 60, no. 4, pp. 779–807, 2018.

[270] J.-M. Jin, *Theory and Computation of Electromangetic Fields*, 2nd ed. John Wiley & Sons, Hoboken, 2010.

[271] M. Kaku, *Quantum Field Theory. A Modern Introduction*. Oxford University Press, New York, 1993.

[272] P. J. Sementilli, B. R. Hunt, and M. S. Nadar, "Analysis of the limit to superresolution in incoherent imaging," *Journal of the Optical Society of America A*, vol. 10, pp. 2265–2276, 1993.

[273] R. G. Newton, "Optical theorem and beyond," *American Journal of Physics*, vol. 44, no. 7, p. 639, 1976.

[274] E. Feenberg, "The scattering of slow electrons by neutral atoms," *Physical Review*, vol. 40, p. 40, 1932.

[275] H. Levine and J. Schwinger, "On the theory of diffraction by an aperture in an infinite plane screen I," *Physical Review*, vol. 74, p. 958, 1948.

[276] H. C. van de Hulst, "On the attenuation of plane waves by obstacles of arbitrary size and form," *Physica*, vol. 15, p. 740, 1949.

[277] J. A. Lock, J. T. Hodges, and G. Gouesbet, "Failure of the optical theorem for Gaussian-beam scattering by a spherical particle," *Journal of the Optical Society of America A*, vol. 12, pp. 2708–2715, 1995.

[278] M. J. Berg, C. M. Sorensen, and A. Chakrabarti, "Extinction and the optical theorem. Part I. Single particles," *Journal of the Optical Society of America A*, vol. 25, no. 7, pp. 1504–1513, 2008.

[279] M. J. Berg, C. M. Sorensen, and A. Chakrabarti, "Extinction and the optical theorem. Part II. Multiple particles," *Journal of the Optical Society of America A*, vol. 25, no. 7, pp. 1514–1520, 2008.

[280] M. I. Mishchenko and L. D. Travis, *Multiple Scattering of Light by Particles*. Cambridge Univesrity Press, New York, 2006.

[281] C. F. Bohren and D. R. Huffman, *Absorption and Scattering of Light by Small Particles*. Wiley-VCH, Weinheim, 2004.

[282] G. C. Reali, "Reflection from dielectric materials," *American Journal of Physics*, vol. 50, no. 12, pp. 1133–1136, 1982.

[283] V. C. Ballenegger, "The Ewald–Oseen extinction theorem and extinction lengths," *American Journal of Physics*, vol. 67, no. 7, pp. 599–605, 1999.

[284] J. W. Goodman, *Statistical Optics*, 2nd ed. John Wiley and Sons., Greenwood Village, 2007.

[285] E. Wolf, *Introduction to the Theory of Coherence and Polarization of Light*. Cambridge University Press, 2007.

[286] E. Wolf, "A macroscopic theory of interference and diffraction of light from finite sources ii. fields with a spectral range of arbitrary width." *Proceedings of the Royal Society of London*, vol. 230, p. 246–265, 1955.

[287] E. Wolf, "Unified theory of coherence and polarization of random electromagnetic beams," *Physical Letters A*, vol. 312, pp. 263–267, 2003.

[288] E. Wolf, "Correlation-induced changes in the degree of polarization, the degree of coherence and the spectrum of random electromagnetic beams on propagation," *Optics Letters*, vol. 28, pp. 1078–1080, 2003.

[289] J. Tervo, T. Setälä, and A. T. Friberg, "Theory of partially coherent electromagnetic fields in the space–frequency domain," *Journal of the Optical Society of America A*, vol. 21, pp. 2205–2215, 2005.

[290] G. Gbur and T. D. Visser, "The structure of partially coherent fields, in progress in optics, Emil Wolf ed." *Progess in Optics*, vol. 55, pp. 285–341, 2010.

[291] A. Labeyrie, S. G. Lipson, and P. Nisenson, *An Introduction to Optical Stellar Interferometry*. Cambridge University Press, Cambridge, 2006.

[292] A. Dogariu and E. Wolf, "Spectral changes produced by static scattering on a system of particles," *Optics Letters*, vol. 23, pp. 1340–1342, 1998.

[293] G. Gbur and E. Wolf, "Determination of density correlation functions from scattering of polychromatic light," *Optics Communications*, vol. 168, pp. 39–45, 1999.

[294] A. C. Schell, "Multiple plate antenna," Ph.D. Thesis, Massachusetts Institute of Technology, 1961.

[295] K. A. Nugent, "A generalization of Schell's theorem," *Optics Communications*, vol. 79, pp. 267–269, 1990.

[296] P. P. Ewald, "Introduction to the dynamical theory of X-ray diffraction," *Acta Crystallographica*, vol. A25, pp. 103–108, 1969.

[297] D. L. Goodstein, *States of Matter*. Dover Publications, Mineola, 1985.

[298] S. Torquato and F. H. Stillinger, "Local density fluctuations, hyperuniformity, and order metrics," *Physical Review E*, vol. 68, p. 041113, 2003.

[299] S. Torquato, G. Zhang, and F. H. Stillinger, "Ensemble theory for stealthy hyperuniform disordered ground states," *Physical Review X*, vol. 5, p. 021010, 2015.

[300] C. M. Sorensen, "Light scattering by fractal aggregates: a review," *Aerosol Science and Technology*, vol. 35, pp. 648–687, 2001.

[301] M. I. Mishchenko, L. D. Travis, and A. A. Lacis, *Scattering, Absorption and Emission of Light by Small Particles*. Cambridge Univesrity Press, Edinburgh, 2002.

[302] G. M. Conley, M. Burresi, F. Pratesi, K. Vynck, and D. S. Wiersma, "Light transport and localization in two-dimensional correlated disorder," *Physical Review Letters*, vol. 112, p. 143901, 2014.

[303] D. S. Sivia, *Elementry Scattering Theory*. Oxford University Press, New York, 2011.

[304] J. Hansen and I. R. McDonald, *Theory of Simple Liquids with Applications to Soft Matter*, 4th ed. Academic Press, San Diego, 2013.

[305] K. Khare, *Fourier Optics and Computational Imaging*. John Wiley and Ane Books Pvt. Ltd., 2016.

[306] J. M. Cowley, *Diffraction Physics*. Elsevier, 1995.

[307] T. Inui, Y. Tanabe, and Y. Onodera, *Group Theory and Its Applications in Physics*. Sprringer-Verlag, Berlin and Heidelberg, 1990.

[308] H. N. Kritikos, "Radiation symmetries of antenna arrays," *Journal of the Franklin Institute*, vol. 295, no. 4, pp. 283–292, 1973.

[309] S. Y. K. Lee, J. J. Amsden, et al., "Spatial and spectral detection of protein mono-layers with deterministic aperiodic arrays of metal nanoparticles," *PNAS*, vol. 107, pp. 12 086–12 090, 2010.

[310] S. V. Boriskina, S. Y. K. Lee, J. J. Amseden, F. Omenetto, and L. Dal Negro, "Formation of colorimetric fingerprints on nano-patterned deterministic aperiodic surfaces," *Optics Express*, vol. 18, no. 14, pp. 14 568–14 576, 2010.

[311] J. Trevino, C. Forestiere, G. Di Martino, S. Yerci, F. Priolo, and L. Dal Negro, "Plasmonic-photonic arrays with aperiodic spiral order for ultra-thin film solar cells," *Optics Express*, vol. 20, pp. A418–A430, 2012.

[312] V. Pierro, V. Galdi, G. Castaldi, I. M. Pinto, and L. B. Felsen, "Radiation properties of planar antenna arrays based on certain categories of aperiodic tilings," *IEEE Transactions on Antennas and Propagation*, vol. 53, no. 2, pp. 635–644, 2005.

[313] S. Lang, *Algebraic Number Theory*. Addison-Wesley, Reading, 1970.

[314] H. Cohen, *A Course in Computational Algebraic Number Theory*. Springer-Verlag, Berlin, 1993.

[315] T. J. Dekker, "Prime numbers in quadratic fields," *CWI Quarterly*, vol. 7, pp. 367–394, 1994.

[316] M. Baake and U. Grimm, *Aperiodic Order*, vol. 1 *A Mathematical Invitation*. Cambridge University Press, New York, 2013.

[317] L. Dal Negro, D. T. Henderson, F. Sgrignuoli, "Wave transport and localization in prime number landscapes", Frontiers in Physics, 9, 490 (2021).

[318] S. Berthier, *Iridescences: The Physical Colors of Insects*. Springer, New York, 2007.

[319] S. Kinoshita, *Structural Colors in the Realm of Nature*. World Scientific, 2008.

[320] S. Gorsky, R. Zhang, A. Gok, et al., "Directonal light emission enhancement from led-phosphor converters using dielectric vogel spiral arrays," *Applied Physics Letters Photonics*, vol. 3, pp. 126 103–126 114, 2018.

[321] S. Gorsky, W. A. Britton, Y. Chen, et al., "Engineered hyperuniformity for directional light extraction," *Applied Physics Letters Photonics*, no. 4, p. 110801, 2019.

[322] D. Levine and P. J. Steinhardt, "Quasicrystals. I. Definition and structure," *Physical Review B*, vol. 34, pp. 596–616, 1986.

[323] E. Maciá-Barber, *Quasicrystals: Fundamentals and Applications*. CRC Press, Boca Raton, 2021.

[324] M. Queffélec, *Substitution Dynamical Systems – Spectral Analysis*. Springer-Verlag, Berlin, 2010.

[325] M. Schroeder, *Fractals, Chaos, Power Laws*. W. H. Freeman, New York, 1991.

[326] J. M. Luck, "Cantor spectra and scaling of gap widths in deterministic aperiodic systems," *Physical Review B*, vol. 39, pp. 5834–5849, 1989.

[327] E. Maciá-Barber, *Aperiodic Structures in Condensed Matter: Fundamentals and Applications*. CRC Press Taylor and Francis, Boca Raton, 2009.

[328] E. Maciá, "The role of aperiodic order in science and technology," *Reports on Progress in Physics*, vol. 69, pp. 397–441, 2006.

[329] M. Dulea, M. Johansson, and R. Riklund, "Localization of electrons and electromagnetic waves in a deterministic aperiodic system," *Physical Review B*, vol. 45, pp. 105–114, 1992.

[330] M. Dulea, M. Johansson, and R. Riklund, "Unusual scaling of the spectrum in a deterministic aperiodic tight-binding model," *Physical Review B*, vol. 47, pp. 8547–8551, 1993.

[331] L. Dal Negro, N. N. Feng, and A. Gopinath, "Electromagnetic coupling and plasmon localization in deterministic aperiodic arrays," *Journal of Optics A: Pure and Applied Optics*, vol. 10, p. 064013, 2008.

[332] L. Kroon, E. Lennholm, and R. Riklund, "Localization-delocalization in aperiodic systems," *Physical Review B*, vol. 66, p. 094204, 2002.

[333] L. Kroon and R. Riklund, "Absence of localization in a model with correlation measure as a random lattice," *Physical Review B*, vol. 69, p. 094204, 2004.

[334] F. J. García de Abajo, R. Gómez-Medina, and J. J. Sáenz, "Full transmission through perfect-conductor subwavelength hole arrays," *Physical Review E*, vol. 72, no. 016608, pp. 1–4, 2005.

[335] F. J. García de Abajo, J. J. Sáenz, I. Campillo, and J. S. Dolado, "Site and lattice resonances in metallic hole arrays," *Optics Express*, vol. 14, no. 1, pp. 7–18, 2006.

[336] S. Zou and G. C. Schatz, "Narrow plasmonic/photonic extinction and scattering line shapes for one and two dimensional silver nanoparticles arrays," *Journal of Chemical Physics*, vol. 121, pp. 12 606–12 612, 2004.

[337] S. Zou, N. Janel, and G. C. Schatz, "Silver nanoparticle array structures that produce remarkable narrow plasmon lineshapes," *Journal of Chemical Physics*, vol. 120, pp. 10 871–10 875, 2004.

[338] T. W. Ebbesen, H. J. Lezec, H. F. Ghaemi, T. Thio, and P. A. Wolff, "Extraordinary optical transmission through sub-wavelength hole arrays," *Nature*, vol. 391, pp. 667–669, 1998.

[339] F. J. García de Abajo, "Colloquium: light scattering by particles and hole arrays," *Reviews of Modern Physics*, vol. 79, pp. 1267–1290, 2007.

[340] A. Authier, *Dynamical Theory of X-Ray Diffraction*. Oxford University Press, Oxford, 2001.

[341] R. W. Wood, "Anomalous diffraction gratings," *Physical Review*, vol. 48, pp. 928–936, 1935.

[342] U. Fano, "The theory of anomalous diffraction gratings and of quasi-stationary waves on metallic surfaces (Sommerfeld's waves)," *Journal of the Optical Society of America*, vol. 31, pp. 213–222, 1941.

[343] U. Fano, "Effects of configuration interaction on intensities and phase shifts," *Physical Review*, vol. 124, pp. 1866–1878, 1961.

[344] T. Matsui, A. Agrawal, A. Nahata, and Z. V. Vardeny, "Transmission resonances through aperiodic arrays of subwavelength apertures," *Nature*, vol. 446, pp. 517–521, 2007.

[345] F. Przybilla, C. Genet, and T. W. Ebbesen, "Enhanced transmission through Penrose subwavelength hole arrays," *Applied Physics Letters*, vol. 89, no. 121115, 2006.

[346] J. V. Bellissard, A. Bovier, and J. Ghez, "Gap labelling theorems for one-dimensional discrete Schrödinger operators," *Reviews in Mathematical Physics*, vol. 4, pp. 1–37, 1992.

[347] L. Dal Negro and N. N. Feng, "Spectral gaps and mode localization in Fibonacci chains of metal nanoparticles," *Optics Express*, vol. 15, pp. 14 396–14 403, 2007.

[348] F. M. Huang, N. I. Zheludev, Y. Chen, and F. J. García de Abajo, "Focusing of light by a nano-hole array," *Applied Physics Letters*, vol. 90, no. 091119, 2007.

[349] F. Sgrignuoli, R. Wang, F. Pinheiro, and L. Dal Negro, "Localization of scattering resonances in aperiodic Vogel spirals," *Physical Review B*, vol. 99, p. 104202, 2019.

[350] Y. Chen, F. Sgrignuoli, and L. Dal Negro, "Optical superoscillations of prime number arrays," in preparation.

[351] J. Illian, A. Penttinen, H. Stoyan, and D. Stoyan, *Statistical Analysis and Modelling of Spatial Point Patterns*. John Wiley, Chichester, 2008.

[352] S. Torquato, A. Scardicchio, and C. E. Zachary, "Point processes in arbitrary dimension from fermionic gases, random matrix theory, and number theory," *Journal of Statistical Mechanics*, 2008.

[353] A. Gabrielli, B. Jancovici, M. Joyce, J. L. Lebowitz, L. Pietronero, and F. Sylos Labini, "Generation of primordial cosmological perturbations from statistical mechanical models," *Physical Review D*, vol. 67, p. 043506, 2003.

[354] S. Torquato, *Random Heterogeneous Materials: Microstructure and Macroscopic Properties*. Springer-Verlag, New York, 2012.

[355] K. Huang, *Statistical Mechanics*, 2nd ed. John Wiley and Sons, California, 1987.

[356] A. Isihara, *Condensed Matter Physics*. Oxford University Press, New York and Oxford, 1991.

[357] L. S. Ornstein and F. Zernike, "Accidental deviations of density and opalescence at the critical point of a single substance," *Royal Netherlands Academy of Arts and Sciences (KNAW). Proceedings*, vol. 17, p. 793–806, 1914.

[358] J. K. Percus and G. J. Yevick, "Analysis of classical statistical mechanics by means of collective coordinates," *Physical Review*, vol. 110, p. 1–13, 1958.

[359] L. Tsang and K. Kong, J. A. Ding, *Scattering of Electromangetic Waves: Numerical Simulations*, vol. 3. John Wiley, New York, 2001.

[360] M. S. Wertheim, "Exact solution of the Percus–Yevick integral equation for hard spheres," *Physical Review Letters*, vol. 20, p. 321–323, 1963.

[361] D. ben Avraham and S. Havlin, *Diffusion and Reactions in Fractals and Disordered Systems*. Cambridge University Press, Cambridge, 2000.

[362] J. W. Goodman, *Speckle Phenomena in Optics: Theory and Applications*. Ben Roberts and Company, US, 2015.

[363] A. Papoulis, *Probability, Random Variables, and Stochastic Processes*, 3rd ed. McGraw-Hill, New York, 1991.

[364] E. Jakeman and K. D. Ridley, *Modeling Fluctuations in Scattered Waves*. Taylor and Francis, New York, 2006.

[365] O. Kallenberg, *Foundations of Modern Probability*, 2nd ed. Springer-Verlag, New York, 2002.

[366] M. Abramowitz and I. A. Stegun, *Handbook of Mathematical Functions*. Dover, New York, 1965.

[367] M. C. W. van Rossum and T. M. Nieuwenhuizen, "Multiple scattering of classical waves: microscopy, mesoscopy and diffusion," *Reviews of Modern Physics*, vol. 71, pp. 313–371, 1999.

[368] F. Scheffold and G. Maret, "Universal conductance fluctuations of light," *Physical Review Letters*, vol. 81, pp. 5800–5803, December 1998. [Online]. Available: https://link.aps.org/doi/10.1103/PhysRevLett.81.5800

[369] B. Shapiro, "Large intensity fluctuations for wave propagation in random media," *Physical Review Letters*, vol. 57, pp. 2168–2171, October 1986. [Online]. Available: https://link.aps.org/doi/10.1103/PhysRevLett.57.2168

[370] S. Feng, C. Kane, P. A. Lee, and A. D. Stone, "Correlations and fluctuations of coherent wave transmission through disordered media," *Physical Review Letters*, vol. 61, pp. 834–837, August 1988. [Online]. Available: https://link.aps.org/doi/10.1103/PhysRevLett.61.834

[371] S. Feng and P. A. Lee, "Mesoscopic conductors and correlations in laser speckle patterns," *Science*, vol. 251, no. 4994, pp. 633–639, 1991. [Online]. Available: https://science.sciencemag.org/content/251/4994/633

[372] J. Bertolotti, E. G. van Putten, C. Blum, A. Lagendijk, W. L. Vos, and A. P. Mosk, "Non-invasive imaging through opaque scattering layers," *Nature*, vol. 491, no. 7423, pp. 232–234, November 2012. [Online]. Available: https://doi.org/10.1038/nature11578

[373] I. M. Vellekoop, A. Lagendijk, and A. P. Mosk, "Exploiting disorder for perfect focusing," *Nature Photonics*, vol. 4, no. 5, pp. 320–322, May 2010. [Online]. Available: https://doi.org/10.1038/nphoton.2010.3

[374] A. P. Mosk, A. Lagendijk, G. Lerosey, and M. Fink, "Controlling waves in space and time for imaging and focusing in complex media," *Nature Photonics*, vol. 6, no. 5, pp. 283–292, May 2012. [Online]. Available: https://doi.org/10.1038/nphoton.2012.88

[375] H. Yilmaz, E. G. van Putten, J. Bertolotti, A. Lagendijk, W. L. Vos, and A. P. Mosk, "Speckle correlation resolution enhancement of wide-field fluorescence imaging," *Optica*, vol. 2, no. 5, pp. 424–429, May 2015. [Online]. Available: www.osapublishing.org/optica/abstract.cfm?URI=optica-2-5-424

[376] S. K. Sahoo, D. Tang, and C. Dang, "Single-shot multispectral imaging with a monochromatic camera," *Optica*, vol. 4, no. 10, pp. 1209–1213, 2017. [Online]. Available: www.osapublishing.org/optica/abstract.cfm?URI=optica-4-10-1209

[377] R. Klages, G. Radons, and Igor M. Sokolov (eds.), *Anomalous Transport: Foundations and Applications*. Wiley-VCH Verlag GmbH and Co. KGaA, Weinheim, 2008.

[378] P. de Gennes, "On a relation between percolation theory and the elasticity of gels," *Journal de Physique Lettres*, vol. 37, pp. 1–2, 1976.

[379] J. Gouyet, *Physics and Fractal Structures*. Masson and Springer, Paris, 1996.

[380] P. Lévy, *Calcul des probabilités*. Gauthier-Villars, Paris, 1925.

[381] B. Mandelbrot, "Stable paretian random functions and the multiplicative variation of income," *Econometrica*, vol. 29, no. 4, p. 517–543, 1961.

[382] J. Bouchaud and J. Potters, *Theory of Financial Risk and Derivative Pricing*, 2nd ed. Cambridge University Press, Cambridge 2003.

[383] B. V. Gnedenko and A. N. Kolmogorov, *Limit Distributions for Sums of Independent Random Variables*. Addison-Wesley, Cambridge, 1954.

[384] E. W. Montroll and G. H. Weiss, "Random walks on lattices," *Journal of Mathematical Physics*, vol. 6, pp. 167–181, 1965.

[385] H. Scher and E. W. Montroll, "Anomalous transit-time dispersion in amorphous solids," *Physical Review B*, vol. 12, pp. 2455–2477, 1975.

[386] J. Klafter and I. M. Sokolov, *First Steps in Random Walks*. Oxford University Press, Oxford, 2011.

[387] P. Barthelemy, J. Bertolotti, and D. S. Wiersma, "A lévy flight for light," *Nature*, vol. 453, pp. 495–498, 2008.

[388] J. Bertolotti, *Light Transport beyond Diffusion*. Ph.D. Thesis, University of Florence, 2007.

[389] F. Sgrignuoli and L. Dal Negro, "Subdiffusive light transport in three-dimensional subrandom arrays," *Physical Review B*, vol. 101, no. 214204, 2020.

[390] Y. Chen, A. Fiorentino, and L. Dal Negro, "A fractional diffusion random laser," *Scientific Reports*, vol. 9, no. 1, pp. 1–14, 2019.

[391] R. Metzler and J. Klafter, "The random walk's guide to anomalous diffusion: a fractional dynamics approach," *Physics Reports*, vol. 339, no. 1, pp. 1–77, 2000. [Online]. Available: www.sciencedirect.com/science/article/pii/S0370157300000703

[392] R. Gorenflo and F. Mainardi, "Fractional calculus: integral and differential equations of fractional order," *arXiv preprint arXiv:0805.3823*, 2008.

[393] M. Kwaśnicki, "Ten equivalent definitions of the fractional Laplace operator," *Fractional Calculus and Applied Analysis*, vol. 20, no. 1, pp. 7–51, September 2017. [Online]. Available: http://arxiv.org/abs/1507.07356v2;http://arxiv.org/pdf/1507.07356v2

[394] A. Lischke, G. Pang, M. Gulian, et al., "What is the fractional Laplacian?" November 2018. [Online]. Available: http://arxiv.org/abs/1801.09767v2;http://arxiv.org/pdf/1801.09767v2

[395] F. Mainardi, A. Mura, G. Pagnini, and R. Gorenflo, "Sub-diffusion equations of fractional order and their fundamental solutions," in *Mathematical Methods in Engineering*, K. Taḍ, J. A. Tenreiro Machado, and D. Baleanu, Eds. Springer, 2007, pp. 23–55.

[396] F. Mainardi and G. Pagnini, "The Wright functions as solutions of the time-fractional diffusion equation," *Applied Mathematics and Computation*, vol. 141, no. 1, pp. 51–62, August 2003. [Online]. Available: www.sciencedirect.com/science/article/pii/S009630030200320X

[397] F. Mainardi, Y. Luchko, and G. Pagnini, "The fundamental solution of the space-time fractional diffusion equation," *arXiv preprint cond-mat/0702419*, 2007.

[398] A. Blumen, G. Zumofen, and J. Klafter, "Transport aspects in anomalous diffusion: Lévy walks," *Physical Review A*, vol. 40, pp. 3964–3973, October 1989. [Online]. Available: https://link.aps.org/doi/10.1103/PhysRevA.40.3964

[399] A. Chaves, "A fractional diffusion equation to describe Lévy flights," *Physics Letters A*, vol. 239, no. 1, pp. 13 – 16, 1998. [Online]. Available: www.sciencedirect.com/science/article/pii/S037596019700947X

[400] F. Mainardi, A. Mura, G. Pagnini, and R. Gorenflo, "Time-fractional diffusion of distributed order," *Journal of Vibration and Control*, vol. 14, no. 9–10, pp. 1267–1290, 2008. [Online]. Available: https://doi.org/10.1177/1077546307087452

[401] A. V. Chechkin, R. Gorenflo, and I. M. Sokolov, "Retarding subdiffusion and accelerating superdiffusion governed by distributed-order fractional diffusion equations," *Physical Review E*, vol. 66, p. 046129, October 2002. [Online]. Available: https://link.aps.org/doi/10.1103/PhysRevE.66.046129

[402] M. Florescu, S. Torquato, and P. J. Steinhardt, "Designer disordered materials with large, complete photonic band gaps," *PNAS*, vol. 106, no. 49, pp. 20 658–20 663, 2009.

[403] R. D. Batten, F. H. Stillinger, and S. Torquato, "Classical disordered ground states: super-ideal gasses and stealth and equi-luminous materials," *Journal of Applied Physics*, vol. 104, p. 033504, 2008.

[404] C. E. Zachary and S. Torquato, "Hyperuniformity in point patterns and two-phase random heterogeneous media," *Journal of Statistical Mechanics*, p. P12015, 2009.

[405] Y. Jiao, T. Lau, H. Hatzikirou, M. Meyer-Hermann, J. C. Corbo, and S. Torquato, "Avian photoreceptor patterns represent a disordered hyperuniform solution to a multiscale packing problem," *Physical Review E*, vol. 89, p. 022721, February 2014. [Online]. Available: https://link.aps.org/doi/10.1103/PhysRevE.89.022721

[406] S. Torquato, G. Zhang, and M. de Courcy-Ireland, "Uncovering multiscale order in the prime numbers via scattering," *Journal of Statistical Mechanics*, no. 093401, pp. 1–15, 2018.

[407] S. Torquato, G. Zhang, and M. D. Courcy-Ireland, "Hidden multiscale order in the primes," *Journal of Physics A: Mathematical and Theoretical*, vol. 52, no. 13, p. 135002, March 2019. [Online]. Available: https://doi.org/10.1088%2F1751-8121%2Fab0588

[408] Z. Ma and S. Torquato, "Random scalar fields and hyperuniformity," *Journal of Applied Physics*, vol. 121, no. 244904, pp. 1–15, 2017.

[409] G. H. Hardy and E. M. Wright, *An Introduction to the Theory of Numbers*. Oxford University Press, New York, 2008.

[410] O. Leseur, R. Pierrat, and R. Carminati, "High-density hyperuniform materials can be transparent," *Optica*, vol. 3, no. 7, pp. 763–767, 2016.

[411] D. G. Kendall, "On the number of lattice points inside a random oval," *Quarterly Journal of Mathematics*, vol. 19, pp. 1–26, 1948.

[412] D. G. Kendall and R. A. Rankin, "On the number of points of a given lattice in a random hypersphere," *Quarterly Journal of Mathematics*, vol. 4, pp. 178–189, 1953.

[413] G. H. Hardy, "On the expression of a number as the sum of two squares," *Quarterly Journal of Mathematics*, vol. 46, pp. 263–283, 1915.

[414] J. Beck, "Irregularities of distribution i," *Acta Mathematica*, vol. 159, pp. 1–49, 1987.

[415] A. Gabrielli, M. Joyce, and S. Torquato, "Tilings of space and superhomogeneous point processes," *Physical Review E*, vol. 77, p. 031125, 2008.

[416] J. Kim and S. Torquato, "Methodology to construct large realizations of perfectly hyperuniform disordered packings," *Physical Review E*, vol. 99, p. 052141, May 2019. [Online]. Available: https://link.aps.org/doi/10.1103/PhysRevE.99.052141

[417] L. S. Froufe-Pérez, M. Engel, J. J. Saénz, and F. Scheffold, "Band gap formation and anderson localization in disordered photonic materials with structural correlations," *PNAS*, vol. 114, no. 36, pp. 9570–9574, 2017.

[418] G. J. Aubry, L. S. Froufe-Pérez, U. Kuhl, O. Legrand, F. Scheffold, and F. Mortes-sagne, "Experimental evidence for transparency, band gaps and anderson localization in two-dimensional hyperuniform disordered photonic materials," *arXiv preprint arXiv:2003.00913*, 2020.

[419] F. Sgrignuoli, S. Torquato and L. Dal Negro, "Localization in three-dimensional stealthy hyperuniform disordered systems," *arXiv:2109.03894*, 2021.

[420] E. C. Oğuz, J. E. S. Socolar, P. J. Steinhardt, and S. Torquato, "Hyperuniformity of quasicrystals," *Physical Review B*, vol. 95, no. 054119, pp. 1–10, 2017.

[421] E. C. Oğuz, J. E. S. Socolar, P. J. Steinhardt, and S. Torquato, "Hyperuniformity and anti-heperuniformity in one-dimensional substitution tilings," *Acta Crystallographica*, vol. A75, pp. 3–13, 2019.

[422] N. M. Korobov, *Exponential Sums and Their Applications*. Springer, Dordrecht, 1992.

[423] C. Lemieux, *Monte Carlo and Quasi-Monte Carlo Sampling*. Springer, New York, 2009.

[424] H. Weyl, "Über die gleichverteilung von zahlen mod. eins." *Mathematische Annalen*, vol. 77, pp. 313–352, 1916.

[425] L. Kuipers and H. Niederreiter, *Uniform Distribution of Sequences*. John Wiley, New York, 1974.

[426] S. J. Miller and R. Takloo-Bighash, *An Invitation to Modern Number Theory*. Princeton University Press, Princeton, 2006.

[427] H. Niederreiter, "Low-discrepancy and low-dispersion sequences," *Journal of Number Theory*, vol. 30, pp. 51–70, 1988.

[428] M. Mckay, R. Beckman, and W. Conover, "A comparison of three methods for selecting values of input variables in the analysis of output from a computer code," *Technometrics*, vol. 21, pp. 239–245, 1979.

[429] O. Bohigas, R. U. Haq, and A. Pandey, "Higher-order correlations in spectra of complex systems," *Physical Review Letters*, vol. 54, no. 15, p. 1645, 1985.

[430] M. L. Mehta, *Random Matrices*. Elesvier, Amsterdam, 2004.

[431] M. Gardner, "Mathematical games: the remarkable lore of the prime number," *Scientific American*, vol. 210, pp. 120–128, 1964.

[432] G. H. Hardy and J. E. Littlewood, "Some problems of partitions numerorum III: on the expression of a number as a sum of primes," *Acta Mathematica*, vol. 44, no. 1, pp. 1–70, 1923.

[433] C. Radin, *Miles of Tiles*. AMS, Providence, 1999.

[434] C. Radin, "The pinwheel tilings of the plane," *Annals of Mathematics*, vol. 139, pp. 661–702, 1994.

[435] F. Sgrignuoli and L. Dal Negro, "Hyperuniformity and wave localization in pinwheel scattering," *Physics Review B*, vol. 103, no. 22, 224202, 2021.

[436] W. Schwarz and J. Spilker, *Arithmetical Functions. An Introduction to Elementary and Anlytic Properties of Arithmetic Functions and to Some of Their Almost-Periodic Properties*. Cambridge University Press, Cambridge, 1994.

[437] J. Sander, J. Steuding, and R. Steuding, *From Arithmetic to Zeta-Functions*. Springer International Publishing AG, Switzerland, 2016.

[438] S. Ramanujan, "On certain trigonometrical sums and their applications in the theory of numbers," *Transactions of the Cambridge Philosophical Society*, vol. 22, no. 13, pp. 259–276, 1918.

[439] T. M. Apostol, *Introduction to Analytic Number Theory*. Springer-Verlag, New York, 1976.

[440] H. Wendt, P. Abry, and S. Jaffard, "Bootstrap for empirical multifractal analysis," *IEEE Signal Processing Magazine*, vol. 1053, no. July, pp. 38–48, 2007.

[441] H. Wendt and P. Abry, "Multifractality tests using bootstrapped wavelet leaders," *IEEE Transactions on Signal Processing*, vol. 55, no. 10, pp. 4811–4820, 2007.

[442] S. Mallat, *A Wavelet Tour of Signal Processing*, 3rd ed. Elsevier, 2009.

[443] K. Falconer, *Fractal Geometry: Mathematical Foundations and Applications*, 3rd ed. John Wiley and Sons Ltd, Chichester, 2014.

[444] A. Arneodo, E. Bacry, and J. F. Muzy, "The thermodynamics of fractals revisited with wavelets," *Physica A: Statistical Mechanics and Its Applications*, vol. 213, pp. 232–275, 1995.

[445] A. Arneodo, G. Grasseau, and M. Holschneider, "Wavelet transform of multifractals," *Physical Review Letters*, vol. 61, pp. 2281–2284, 1988.

[446] S. Jaffard, "Wavelet techniques in multifractal analysis," *Proceedings of Symposia Pure Mathematics, Americal Mathematical Society*, vol. 72, no. 2, pp. 91–152, 2004.

[447] S. Jaffard, B. Lashermes, and P. Abry, "Wavelet leaders in multifractal analysis," in *Wavelet Analysis and Applications*, T. Quian, M. I. Vai, X. Yuesheng, Eds. Birkhäuser, pp. 219–264, 2006.

[448] S. Jaffard, "Multifractal formalism for functions. part 2: Self-similar functions," *SIAM Journal on Mathematical Analysis*, vol. 28, no. 4, pp. 971–998, 1997.

[449] G. Tenenbaum, *Introduction to Analytic and Probabilistic Number Theory*, vol 46, *Cambridge Studies in Advanced Mathematics*. Cambridge University Press, 1995.

[450] B. Mazur and W. Stein, *Prime Numbers and the Riemann Hypothesis*. Cambridge University Press, New York, 2016.

[451] E. Schmidt, "Über die anzahl der primzahlen unter gegebener grenze," *Mathematische Annalen*, vol. 57, pp. 195–204, 1903.

[452] G. H. Hardy and J. E. Littlewood, "Contributions to the theory of the riemann zeta-function and the theory of the distribution of primes," *Acta Mathematica*, vol. 41, p. 119–196, 1916.

[453] L. Schoenfeld, "Sharper bounds for the Chebyshev functions $\theta(x)$ and $\psi(x)$. ii," *Mathematics of Computation*, vol. 30, no. 134, pp. 337–360, 1976.

[454] D. Schumayer and D. A. W. Hutchinson, "Colloquium: physics of the Riemann hypothesis," *Reviews of Modern Physics*, vol. 83, no. 2, pp. 307–330, 2011.

[455] J. B. Conrey, "The Riemann hypothesis," *Notices of the American Mathematical Society*, vol. 50, no. 3, pp. 341–353, 2003.

[456] B. Riemann, "Ueber die anzahl der primzahlen unter einer gegebenen grösse," *Monatsberichte der Berliner Akademie*, pp. 1–9, 1859.

[457] G. Tenenbaum and M. M. France, *The Prime Numbers and Their Distribution*, vol. 6. American Mathematical Society, 2000.

[458] J. Steuding, *An Introduction to the Theory of L-Functions*. A course given at the Autonoma University, Madrid, 2005.

[459] H. M. Edwards, *Riemann's Zeta Function*. Academic Press, New York and London, 1974.

[460] A. Ivic, *The Riemann Zeta Function: The Theory of the Riemann Zeta Function with Applications*. John Wiley and Sons, New York, 1985.

[461] H. Wu and D. W. L. Sprung, "Riemann zeros and a fractal potential," *Physical Review E*, vol. 48, no. 4, pp. 2595–2598, 1993.

[462] D. Schumayer, B. P. van Zyl, and D. A. W. Hutchinson, "Quantum mechanical potentials related to the prime numbers and Riemann zeros," *Physical Review E*, vol. 78, no. 5, p. 056215, 2008.

[463] H. Bohr, "Über eine quasi-periodische eigenshaft dirichletscher reihen mit anwendung auf dirichletschen l-functione," *Mathematische Annalen*, vol. 85, pp. 115–122, 1922.

[464] R. Crandall and C. Pomerance, *Prime Numbers. A Computational Perspective*, 2nd ed. Springer, New York, 2005.

[465] H. Riesel and G. Gohl, "Some calculations related to Riemann's prime number formula," *Mathematics of Computations*, vol. 24, no. 112, pp. 969–983, 1970.

[466] E. Bombieri, *Problems of the Millennium: The Riemann Hypothesis*. Clay Mathematics Institute, 2008.

[467] C. S. R. B. W. A. Borwein, P., "Localization of waves," in *The Riemann Hypothesis. A Resource for the Afficionado and Virtuoso Alike*, C. S. R. B. W. A. Borwein, P., Ed. Springer, New York, 2008, pp. 3–7.

[468] A. M. Odlyzko, "On the distribution of spacings between zeros of the zeta function," *Mathematics of Computation*, vol. 48, no. 177, pp. 273–308, 1987.

[469] A. Goetschy and S. E. Skipetrov, "Euclidean random matrices and their applications in physics," *arXiv:1303.2880*, 2013.

[470] A. Goetschy and S. E. Skipetrov, "Non-Hermitian Euclidean random matrix theory," *Physical Review E*, vol. 84, pp. 011 150–1–011 150–10, 2011.

[471] G. E. Mitchell, A. Richter, and H. A. Weidenmüller, "Random matrices and chaos in nuclear physics: nuclear reactions," *Reviews of Modern Physics*, vol. 82, no. 4, pp. 2845–2901, 2010.

[472] P. Bourgade and J. P. Keating, "Quantum chaos, random matrix theory, and the Riemann ζ-function," *Séminaire Poincaré XIV*, pp. 115–153, 2010.

[473] C. W. J. Beenakker, "Random-matrix theory of quantum transport," *Reviews of Modern Physics*, vol. 69, no. 3, pp. 731–808, 1997.

[474] T. Guhr, A. Müller-Groeling, and H. A. Weidenmüller, "Random-matrix theories in quantum physics: common concepts," *Physics Reports*, vol. 299, pp. 189–425, 1998.

[475] A. D. Mirlin, "Statistics of energy levels and eigenfunctions in disordered systems," *Physics Reports*, vol. 326, pp. 259–382, 2000.

[476] S. E. Skipetrov and M. E. Sokolov, "Absence of Anderson localization of light in a random ensemble of point scatterers," *Physical Review Letters*, vol. 112, pp. 023 905–1–023 905–5, 2013.

[477] S. E. Skipetrov and A. Goetschy, "Eigenvalue distributions of large Euclidean random matrices for waves in random media," *Journal of Physics A: Mathematical and Theoretical*, vol. 44, pp. 065 102–065 127, 2011.

[478] A. Goetschy and S. E. Skipetrov, "Euclidean matrix theory of random lasing in a cloud of cold atoms," *European Physics Letters*, vol. 96, pp. 34 005–p1–34 005–p6, 2011.

[479] T. K. Timberlake and J. M. Tucker, "Is there quantum chaos in the prime numbers?" *arXiv:quant-ph/0708.2567*, 2007.

[480] M. V. Berry and M. Robnik, "Semiclassical level spacings when regular and chaotic orbits coexist," *Journal of Physics A*, vol. 17, no. 12, p. 2413, 1984.

[481] J. Steuding, *Value-Distribution of L-Functions*. Springer-Verlag, Berlin and Heidelberg, 2007.

[482] K. Ireland and M. Rosen, *A Classical Introduction to Modern Number Theory*, 2nd ed. Springer-Verlag, New York, 1990.

[483] G. Everest and T. Ward, *An Introduction to Number Theory*. Springer-Verlag, London 2005.

[484] J. E. Littlewood, "Distribution des nombres premiers," *Comptes rendus de l'Académie des Sciences*, vol. 158, pp. 1869–1872, 1914.

[485] A. Granville and G. Martin, "Prime number races," *American Mathematical Monthly*, vol. 113, no. 1, pp. 1–33, 2006.

[486] M. Rubinstein and P. Sarnak, "Chebyshev's bias," *Experimental Mathematics*, vol. 3, no. 3, pp. 173–197, 1994.

[487] R. J. Lemke Oliver and K. Soundararajan, "Unexpected biases in the distribution of consecutive primes," *PNAS*, vol. 113, no. 31, pp. E4446–E4454, 2016.

[488] T. Tao, "The dichotomy between structure and randomness, arithmetic progressions, and the primes," *arXiv:math/0512114v2*, 2005.

[489] E. Szemerédi, "On sets of integers containing no k elements in arithmetic progression," *Acta Arithmetica*, vol. 27, pp. 299–345, 1975.

[490] T. Tao, "The Gaussian primes contain arbitrarily shaped constellations," *Journal d'Analyse Mathématique*, vol. 99, pp. 109–176, 2006.

[491] Y. Zhang, "Bounded gaps between primes," *Annals of Mathematics*, vol. 179, no. 3, pp. 1121–1174, 2014.

[492] G. Zhang, F. Martelli, and S. Torquato, "The structure factor of primes," *Journal of Physics A: Mathematical and Theoretical*, vol. 51, no. 11, p. 115001, February 2018. [Online]. Available: https://doi.org/10.1088%2F1751-8121%2Faaa52a

[493] P. X. Gallagher, "On the distribution of primes in short intervals," *Mathematika*, vol. 23, no. 1, p. 4–9, 1976.

[494] S. Torquato, A. Scardicchio, and C. E. Zachary, "Point processes in arbitrary dimension from fermionic gases, random matrix theory, and number theory," *Journal of Statistical Mechanics*, vol. 2008, no. 11, p. P11019, 2008.

[495] M. Wolf, "Multifractality of prime numbers," *Physica A: Statistical Mechanics and Its Applications*, vol. 160, no. 1, pp. 24–42, 1989. [Online]. Available: www.sciencedirect.com/science/article/pii/0378437189904615

[496] M. Wolf, "1/f noise in the distribution of prime numbers," *Physica A: Statistical Mechanics and Its Applications*, vol. 241, no. 3, pp. 493–499, 1997. [Online]. Available: www.sciencedirect.com/science/article/pii/S0378437197002513

[497] P. Bak, C. Tang, and K. Wiesenfeld, "Self-organized criticality: an explanation of the 1/f noise," *Physical Review Letters*, vol. 59, pp. 381–384, July 1987. [Online]. Available: https://link.aps.org/doi/10.1103/PhysRevLett.59.381

[498] S. Ares and M. Castro, "Hidden structure in the randomness of the prime number sequence?" *Physica A: Statistical Mechanics and Its Applications*, vol. 360, no. 2, pp. 285–296, 2006. [Online]. Available: www.sciencedirect.com/science/article/pii/S0378437105006473

[499] J. H. Silverman, *A Friendly Introduction to Number Theory*, 4th ed. Pearson Education Inc., 2013.

[500] L. J. Lange, "An elegant continued fraction for π," *American Mathematical Monthly*, vol. 106, no. 5, pp. 456–458, 1999.

[501] T. W. Cusick and M. E. Flahive, *The Markov and Lagrange Spectra*. Mathematical Surveys and Monographs no. 30, American Mathematical Society, Providence, 1994.

[502] C. G. Moreira, "Geometric properties of the Markov and Lagrange spectra," *Annals of Mathematics*, vol. 188, no. 1, pp. 145–170, 2018.

[503] H. Bohr, *Almost Periodic Functions*. Julius Springer, Berlin, 1933.

[504] H. Bohr, "Zur theorie der fastperiodischen funktionen i," *Acta Mathematica*, vol. 45, p. 29127, 1925.

[505] A. S. Besicovitch, *Almost Periodic Functions*. Dover Publications, New York, 1954.

[506] R. L. Cooke, "Almost-periodic functions," *American Mathematical Monthly*, vol. 88, no. 7, pp. 515–526, 1981.

[507] C. Corduneanu, *Almost Periodic Oscillations and Waves*. Springer, Berlin and Heidelberg, 2008.

[508] M. V. Berry and M. Tabor, "Level clustering in the regular spectrum," *Proceedings of the Royal Society of London. Series A*, vol. A356, pp. 375–394, 1977.

[509] Z. Rudnick, P. Sarnak, and A. Zaharescu, "The distribution of spacings between the fractional parts of $n^2\alpha$," *Inventiones Mathematicae*, vol. 145, pp. 37–57, 2001.

[510] Z. Rudnick and P. Sarnak, "The pair correlation function of fractional parts of polynomials," *Communications in Mathematical Physics*, vol. 194, pp. 61–70, 1998.

[511] E. Beltrami, *What Is Random?* Copernicus, Springer-Verlag, New York, 1999.

[512] M. S. Bartlett, "Chance or chaos?" *Journal of the Royal Statistical Society*, vol. 153, no. 3, pp. 321–347, 1990.

[513] Z. Rudnick and A. Zaharescu, "The distribution of spacings between fractional parts of lacunary sequences," *Forum Mathematicum*, vol. 14, no. 5, pp. 691–712, 2002.

[514] Z. Rudnick and A. Zaharescu, "A metric result on the pair correlation of fractional parts of sequences," *Acta Arithmetica*, vol. 89, no. 3, pp. 283–293, 1999.

[515] H. Cohen, *A Classical Invitation to Algebraic Numbers and Class Fields*. Springer-Verlag, New York, 1978.

[516] J. R. Goldman, *The Queen of Mathematics. A Historically Motivated Guide to Number Theory*. A. K. Peters Ltd., Natick, 2004.

[517] I. Stewart and D. Tall, *Algebraic Number Theory and Fermat's Last Theorem*, 3rd ed. A. K. Peters Ltd., Canada, 2002.

[518] E. Gethner, S. Wagon, and B. Wick, "A stroll through the Gaussian primes," *American Mathematical Monthly*, vol. 105, pp. 327–333, 1998.

[519] N. Tsuchimura, "Computational results for Gaussian moat problem," *IEICE Transactions*, vol. 88-A, pp. 1267–1273, 2005.

[520] I. Vardi, "Prime percolation," *Experimental Mathematics*, vol. 7, no. 3, pp. 275–289, 1998.

[521] P. P. West and B. D. Sittinger, "A further stroll into the Eisenstein primes," *American Mathematical Monthly*, vol. 124, no. 7, pp. 609–620, 2017.

[522] D. Bressoud and S. Wagon, *A Course in Computational Number Theory*. John Wiley and Sons, Hoboken, 2000.

[523] J. Renze, S. Wagon, and B. Wick, "The Gaussian zoo," *Experimental Mathematics*, vol. 10, no. 2, pp. 161–173, 2001.

[524] A. Weil, "On some exponential sums," *PNAS*, vol. 34, no. 5, pp. 204–207, 1948.

[525] H. Niederreiter and J. Rivat, "On the correlation of pseudorandom numbers generated by inversive methods," *Monatshefte für Mathematik*, vol. 153, pp. 251–264, 2008.

[526] W. Trappe and L. C. Washington, *Introduction to Cryptography with Coding Theory*, 2nd ed. Pearson Prentice Hall, Upper Saddle River, 2006.

[527] A. R. Meijer, *Algebra for Cryptologists*. Springer International Publishing, Switzerland, 2016.

[528] R. S. Calinger, *Leonhard Euler: Mathematical Genius in the Enlightenment*. Princeton and Oxford University Press, Princeton, 2016.

[529] C. Cobeli and A. Zaharescu, "On the distribution of primitive roots mod p," *Acta Arithmetica*, vol. 83, no. 2, pp. 143–153, 1998.

[530] Z. Rudnick and A. Zaharescu, "The distribution of spacings between small powers of a primitive root," *Israel Journal of Mathematics*, vol. 120, pp. 271–287, 2000.

[531] J. Hoffstein, J. Pipher, and J. H. Silverman, *An Introduction to Mathematical Cryptography*. Springer, New York, 2008.

[532] S. Y. Lee, G. F. Walsh, and L. Dal Negro, "Microfluidics integration of aperiodic plasmonic arrays for spatial-spectral optical detection," *Optics Express*, vol. 21, no. 4, pp. 4945–4957, 2013.

[533] J. H. Silverman, *The Arithmetic of Elliptic Curves*. Springer Science & Business Media Berlin and New York, 2009.

[534] L. C. Washington, *Elliptic Curves Number Theory and Cryptography*. Chapman and Hall/CRC, Boca Raton, 2008.

[535] Clay Mathematics, *Millennium Problems*. Available online: www.claymath.org/millennium-problems.

[536] R. Taylor, "Automorphy for some ℓ-adic lifts of automorphic mod ℓ galois representations. ii." *Publications Mathematiques de l'Institut des Hautes Etudes Scientifiques*, vol. 108, pp. 183–239, 2008.

[537] M. Blum and S. Micali, "How to generate cryptographically strong sequences of pseudo-random bits," *SIAM Journal on Computing*, vol. 13, no. 4, pp. 850–864, 1984.

[538] L. Blum, M. Blum, and M. Shub, "A simple unpredictable pseudo-random number generator," *SIAM Journal on Computing*, vol. 15, no. 2, pp. 364–383, 1986.

[539] G. Marsaglia, "Random numbers fall mainly in the planes," *PNAS*, vol. 61, no. 1, pp. 25–28, 1968.

[540] E. E. Fenimore and T. M. Cannon, "Coded aperture imaging with uniformly redundant arrays," *Applied Optics*, vol. 17, no. 3, pp. 337–347, February 1978. [Online]. Available: http://ao.osa.org/abstract.cfm?URI=ao-17-3-337

[541] S. R. Gottesman and E. E. Fenimore, "New family of binary arrays for coded aperture imaging," *Applied Optics*, vol. 28, no. 20, pp. 4344–4352, October 1989. [Online]. Available: http://ao.osa.org/abstract.cfm?URI=ao-28-20-4344

[542] M. J. Cieślak, K. A. A. Gamage, and R. Glover, "Coded-aperture imaging systems: past, present and future development past, present and future development – a review." *Radiation Measurements*, vol. 92, pp. 59–71, 2016.

[543] G. J. Chaitin, *Algorithmic Information Theory*. Cambridge University Press, 1987.

[544] G. J. Chaitin, *Thinking about Gödel and Turing. Essays on Complexity, 1970–2007*. World Scientific, Singapore, 2007.

[545] G. J. Chaitin, *The Limits of Mathematics*. Springer-Verlag, Singapore, 1998.

[546] G. J. Chaitin, "Information-theoretic limitations of formal systems," *Journal of the ACM*, vol. 21, no. 3, pp. 403–434, 1974.

[547] E. Nagel and N. J. R., *Gödel's Proof. Revised Edition*. New York University Press, New York, 2001.

[548] R. M. Smullyan, *A Beginner's Guide to Mathematical Logic*. Dover Publications, Mineola, 2014.

[549] C. E. Shannon, "A mathematical theory of communication," *Bell System Technical Journal*, vol. 27, no. 3, pp. 379–423, 1948.

[550] S. M. Pincus, "Approximate entropy as a measure of system complexity," *PNAS*, vol. 88, pp. 2297–2301, 1991.

[551] M. Costa, A. L. Goldberger, and C. K. Peng, "Multiscale entropy analysis of complex physiologic time series," *Physical Review Letters*, vol. 89, no. 6, p. 068102, 2002.

[552] M. Costa, A. L. Goldberger, and C. K. Peng, "Multiscale entropy analysis of biological signals," *Physical Review E*, vol. 71, no. 021906, 2005.

[553] E. S. Keeping, *Introduction to Statistical Inference*. Dover Publications Inc., New York, 1995.

[554] F. P. Ramsey, "On a problem of formal logic," *Proceedings of the London Mathematical Society*, vol. s2-30, pp. 264–286, 1930.

[555] D. Micciancio and S. Goldwasser, *Complexity of Lattice Problems: A Cryptographic Perspective*. Kluwer Academic Publishers, Dordrecht, 2002.

[556] R. Zamir, *Lattice Coding for Signals and Networks*. Cambridge Univerity Press, Cambridge, 20014.

[557] C. A. Pickover, *The Math Book: From Pythagoras to the 57th Dimension, 250 Milestones in the History of Mathematics*. Sterling, New York, 2009.

[558] K. M. Dunbabin, *Mosaics of the Greek and Roman World*. Cambridge Univeristy Press, Cambridge, 1999.

[559] E. Broug, *Islamic Geometric Patterns*. Thames and Hudson, London, 2008.

[560] J. Kepler, *Harmonice Mundi, Book II*. Lincii, 1619.

[561] E. S. Fyodorov, "Simmetrija na ploskosti [symmetry in the plane]," *Zapiski Imperatorskogo Sant-Petersburgskogo Mineralogicheskogo Obshchestva [Proceedings of the Imperial St. Petersburg Mineralogical Society]*, vol. 28, pp. 245–291, 1891.

[562] B. Grünbaum and G. C. Shephard, *Tilings and Patterns*, 2nd ed. Dover, New York, 2016.

[563] H. S. M. Coxeter, *Regular Polytopes*. Dover Publications Inc., New York, 1973.

[564] H. Minkowski, *Diophantische Approximationen*. Druck und Verlag Von B. G. Teubner, Leipzig, 1907.

[565] E. S. Fedorov, *Das Kristallreich: Tabellen zur Kristallochemischen Analyse*. Academy of Sciences, St. Petersburg, 1920.

[566] A. Schoenflies, *Kristallsystem und Kristallstruktur*. Teubner, 1891.

[567] W. Barlow, "Über die geometrischen eigenschaften homogener starrer strukturen und ihre anwendung auf krystalle, [on the geometrical properties of homogeneous rigid structures and their application to crystals]," *Zeitschrift für Krystallographie und Mineralogie*, vol. 23, pp. 1–63, 1894.

[568] H. Hiller, "The crystallographic restriction in higher dimensions," *Acta Crystallographica*, vol. A41, pp. 541–544, 1985.

[569] N. G. de Bruijn, "Algebraic theory of Penrose's non-periodic tilings of the plane," *Indagationes Mathematicae, Proceedings of the Koninklijke Nederlandse Akademie van Wetenshappen Series*, vol. A84, no. 1, pp. 38–66, 1981.

[570] R. V. Moody and J. Patera, "Quasicrystals and icosians," *Journal of Physics A: Mathematical and General*, vol. 26, pp. 2829–2853, 1993.

[571] S. van Smaalen, *Incommensurate Crystallography*. Oxford University Press, Oxford, 2007.

[572] R. Berger, "The undecidability of the domino problem," *Memoirs American Mathematical Society*, vol. 66, pp. 1–72, 1966.

[573] E. Jeandel and M. Rao, "An aperiodic set of 11 Wang tiles," *eprint arXiv:1506.06492*, pp. 1–40, 2015.

[574] R. Penrose, "Pentaplexity," *Bulletin of the Institute for Mathematics and Applications*, vol. 10, pp. 266–271, 1974.

[575] P. M. de Wolff and W. van Aalst, "The four-dimensional space group of $\gamma - na_2co_3$," *Acta Crystallographica A*, vol. 28, p. 111, 1972.

[576] P. M. de Wolff, "The pseudo-symmetry of modulated crystal structures," *Acta Crystallographica A*, vol. 30, pp. 777–785, 1974.

[577] L. Bieberbach, "Über die bewegungsgruppen der n-dimensional en euklidischen räume mit einem endlichen fundamental bereich," *Matematische Annallen*, vol. 72, pp. 400–412, 1912.

[578] E. Asher and A. Janner, "Algebraic aspects of crystallography I: space groups as extensions," *Helvetica Physica Acta*, vol. 38, pp. 551–572, 1965.

[579] E. Asher and A. Janner, "Algebraic aspects of crystallography II: non-primitive translations in space groups," *Communications in Mathematical Physics*, vol. 11, pp. 138–167, 1968.

[580] G. Fast and T. Janssen, "Determination of n-dimensional space groups by means of an electronic computer," *Journal of Computational Physics*, vol. 7, pp. 1–11, 1971.

[581] "Icru report of the executive committee," *Acta Crystallographica A*, vol. 48, p. 922, 1992.

[582] D. A. W. Thompson, *On Growth and Form*. Dover, New York, 1992.

[583] P. Prusinkiewicz and A. Lindenmayer, *The Algorithmic Beauty of Plants*. Springer-Verlag, New York, 1990.

[584] P. Ball, *Nature's Patterns*. Oxford University Press, New York, 2009.

[585] G. J. Mitchison, "Phyllotaxis and the Fibonacci series," *Science*, vol. 196, pp. 270–275, 1977.

[586] J. A. Adam, *A Mathematical Nature Walk*. Princeton University Press, Princeton, 2009.

[587] R. V. Jean, *Phyllotaxis*. Cambridge University Press, New York, 1995.

[588] C. Bonnet, *Recherches sur l'usage des feuilles dans les plantes*. E. Luzac, fils., Göttingen and Leyden, 1754.

[589] I. Adler, D. Barabe, and R. V. Jean, "A history of the study of phillotaxis," *Annals of Bothany*, vol. 80, pp. 231–244, 1997.

[590] A. M. Turing, "The chemical basis of morphogenesis," *Philosophical Transactions of the Royal Society London*, vol. 237B, pp. 37–52, 1952.

[591] I. Adler, "A model of contact pressure in phyllotaxis," *Journal of Theoretical Biology*, vol. 45, pp. 1–79, 1974.

[592] M. Naylor, "Golden, $\sqrt{2}$, and π flowers: a spiral story," *Mathematics Magazine*, vol. 75, pp. 163–172, 2002.

[593] L. Dal Negro, N. Lawrence, and J. Trevino, "Analytical light scattering and orbital angular momentum spectra of arbitrary Vogel spirals," *Optics Express*, vol. 20, pp. 18 209–18 223, 2012.

[594] D. S. Simon, N. Lawrence, J. Trevino, L. Dal Negro, and A. V. Sergienko, "High Capacity quantum Fibonacci coding for key distribution," *Physical Review A*, vol. 87, p. 032312, 2013.

[595] H. E. Stanley, "Multifractal phenomena in physics and chemistry (review)," *Nature*, vol. 335, pp. 405–409, 1988.

[596] B. B. Mandelbrot, "An Introduction to multifractal distribution functions," in *Fluctuations and Pattern Formation*, H. E. Stanley and N. Ostrowsky, Ed. Kluwer, Dordrecht and Boston, 1988, 345–360.

[597] U. Frisch and G. Parisi, "Fully developed turbulence and intermittency." *New York Academy of Sciences, Annals*, vol. 357, 359–367, 1980.

[598] J. Trevino, S. F. Liew, H. Noh, H. Cao, and L. Dal Negro, "Geometrical structure, multifractal spectra and localized optical modes of aperiodic Vogel spirals," *Optics Express*, vol. 20, pp. 3015–3033, 2012.

[599] L. Dal Negro, R. Wang, and F. A. Pinheiro, "Structural and spectral properties of deterministic aperiodic optical structures," *Crystals*, vol. 6, pp. 161–195, 2016.

[600] T. Halsey, M. H. Jensen, L. P. Kadanoff, I. Procaccia, and B. I. Shraiman, "Fractal measures and their singularities: the characterization of strange sets," *Physical Review A.*, vol. 33, pp. 1141–1151, 1986.

[601] A. Chhabra and R. V. Jensen, "Direct determination of the $f(\alpha)$ singularity spectrum," *Physical Review Letters*, vol. 62, pp. 1327–1330, 1989.

[602] M. E. Pollard and G. J. Parker, "Low-contrast bandgaps of a planar parabolic spiral lattice," *Optics Letters*, vol. 34, pp. 2805–2807, 2009.

[603] A. Hof, "On diffraction by aperiodic structures," *Communications in Mathematical Physics*, vol. 169, pp. 25–43, 1995.

[604] M. Baake and U. Grimm, "Mathematical diffraction of aperiodic structures," *Chemical Society Reviews*, vol. 41, pp. 6821–6843, 2012.

[605] M. Baake and U. Grimm, "Kinematic diffraction from a mathematical viewpoint," *Zeitschrift für Kristallographie*, vol. 226, pp. 711–725, 2011.

[606] A. Janner and T. Janssen, "Symmetry of periodically distorted crystals," *Physical Review B*, vol. 15, pp. 643–658, 1977.

[607] E. Bombieri and J. E. Taylor, "Quasicrystals, tilings, and algebraic number theory: some preliminary connections," *Contemporary Mathematics*, vol. 64, pp. 241–264, 1987.

[608] R. V. Moody, "Model sets: a survey". In *From Quasicrystals to More Complex Systems. Centre de Physique des Houches*, vol 13, F. Axel, F. Dénoyer, J. P. Gazeau, Ed. Springer, Berlin, Heidelberg, 2000.

[609] M. Baake and R. V. Moody, "Weighted Dirac combs with pure point diffraction," *Journal für die reine und angewandte Mathematik (Crelle)*, vol. 573, pp. 61–94, 2004.

[610] M. Baake and U. Grimm, "Diffraction of limit periodic point sets," *Philosophical Magazine*, vol. 91, pp. 2661–2670, 2011.

[611] P. Collet and J. P. Eckmann, *Iterated Maps on the Interval as Dynamical Systems.* Birkhäuser, Boston, 1980.

[612] M. Reed and B. Simon, *Methods of Modern Mathematical Physics, vol. 1. Functional Analysis.* Academic Press, San Diego, 1980.

[613] C. Godrèche and J. M. Luck, "Multifractal analysis in reciprocal space and the nature of the fourier transform of self-similar structures," *Journal of Physics A*, vol. 23, pp. 3769–3797, 1990.

[614] S. Aubry, C. Godrèche, and J. M. Luck, "Scaling properties of a structure intermediate between quasiperiodic and random," *Journal of Statistical Physics*, vol. 51, pp. 1033–1075, 1988.

[615] A. Hof, "On scaling in relation to singular spectra," *Journal of Statistical Physics*, vol. 184, pp. 567–577, 1997.

[616] M. Höffe and M. Baake, "Surprises in diffuse scattering," *Zeitschrift für Kristallographie*, vol. 215, pp. 441–444, 2000.

[617] M. Baake, F. Gähler, and U. Grimm, "Spectral and topological properties of a family of generalised Thue–Morse sequences," *Journal of Mathematical Physics*, vol. 53, no. 032701, pp. 1–24, 2012.

[618] M. Baake and M. Höffe, "Diffraction of random tilings: some rigorous results," *Journal of Statistical Physics*, vol. 99, pp. 219–261, 2000.

[619] E. Bombieri and J. E. Taylor, "Which distributions of matter diffract? An initial investigation," *Journal de Physique*, vol. 47, no. 7, pp. 3–19, 1986.

[620] D. Lenz, "Aperiodic order and pure point diffraction," *Philosophical Magazine*, vol. 88, no. 13, pp. 2059–2071, 2008.

[621] M. Baake, D. Frettlöh, and U. Grimm, "Pinwheel patterns and powder diffraction," *Philosophical Magazine*, vol. 87, no. 18, pp. 2831–2838, 2007.

[622] G. H. Hardy and M. Riesz, *The General Theory of Dirichlet's Series*. Cambridge University Press, Cambridge, 1915.

[623] D. Ghisa, "Fundamental domains and analytic continuation of general Dirichlet series," *British Journal of Mathematics and Computer Science*, vol. 25, pp. 100–116, 2015.

[624] D. Ghisa and A. Horvat-Marc, "Geometric aspects of denseness theorems for Dirichlet functions," *Journal of Advances in Mathematics and Computer Science*, vol. 25, pp. 1–11, 2017.

[625] D. Ghisa and A. Horvat-Marc, "Geometric aspects of quasi-periodic property of Dirichlet functions," *Advances in Pure Mathematics*, vol. 8, pp. 699–710, 2018.

[626] P. C. Waterman, "Matrix formulation of electromagnetic scattering," *Proceedings of IEEE*, vol. 53, pp. 805–811, 1965.

[627] A. Doicu, T. Wriedt, and Y. Eremin, *Light Scattering by Systems of Particles*. Springer, Berlin, 2006.

[628] M. Born, "Quantenmechanik der stoßvorgänge," *Physikalische Zeitschrift*, vol. 38, pp. 803–827, 1926.

[629] H. M. Nussenzveig, *Causality and Dispersion Relations*. Academic Press, New York, 1972.

[630] M. Reed and B. Simon, *Methods of Modern Mathematical Physics*, vol. 3: *Scattering Theory*. Academic Press, San Diego, 1979.

[631] D. A. Bykov and L. L. Doskolovich, "Numerical methods for calculating poles of the scattering matrix with applications in grating theory," *Journal of Lightwave Technology*, vol. 31, no. 5, p. 793, 2013.

[632] D. Felbacq, "Numerical computation of resonance poles in scattering theory," *Physical Review E*, vol. 64, p. 047702, 2001.

[633] N. A. Gippius and S. G. Tikhodeev, "Application of the scattering matrix method for calculating the optical properties of metamaterials," *Physics – Uspekhi*, vol. 52, no. 9, p. 967, 2009.

[634] M. Nevière, E. Popov, and R. Reinisch, "Electromagnetic resonances in linear and nonlinear optics: phenomenological study of grating behavior through the poles and zeros of the scattering operator," *Journal of the Optical Society of America A*, vol. 12, no. 3, p. 513, 1995.

[635] E. Anemogiannis, E. N. Glytsis, and T. K. Gaylord, "Efficient solution of eigenvalue equations of optical waveguiding structures," *Journal of Lightwave Technology*, vol. 12, no. 12, p. 2080, 1994.

[636] J. von Neumann and E. Wigner, "Über merkwürdige diskrete eigenwerte," *Physikalische Zeitschrift*, vol. 30, no. 50, pp. 465–467, 1929.

[637] Y. Plotnik, O. Peleg, F. Dreisow, M. Heinrich, S. Nolte, A. Szameit, and M. Segev, "Experimental observation of optical bound states in the continuum," *Physical Review Letters*, vol. 107, p. 183901, October 2011. [Online]. Available: https://link.aps.org/doi/10.1103/PhysRevLett.107.183901

[638] L. H. Guessi, R. S. Machado, Y. Marques, et al., "Catching the bound states in the continuum of a phantom atom in graphene," *Physical Review B*, vol. 92, p. 045409, July 2015. [Online]. Available: https://link.aps.org/doi/10.1103/PhysRevB.92.045409

[639] C. W. Hsu, B. Zhen, A. D. Stone, J. D. Joannopoulos, and M. Soljačić, "Bound states in the continuum," *Nature Reviews Materials*, vol. 1, no. 16048, 2016.

[640] F. Sgrignuoli, M. Röntgen, C. Morfonios, P. Schmelcher, and L. Dal Negro, "Compact localized states of open scattering media: a graph decomposition approach for an ab initio design." *Optics Letters*, vol. 44, pp. 375–378, 2019.

[641] D. C. Marinica, A. G. Borisov, and S. V. Shabanov, "Bound states in the continuum in photonics," *Physical Review Letters*, vol. 100, no. 183902, 2008.

[642] A. Kodigala, T. Lepetit, Q. Gu, B. Bahari, Y. Fainman, and B. Kanté, "Lasing action from photonic bound states in continuum," *Nature*, vol. 541, pp. 196–199, 2017.

[643] V. Mocella and S. Romano, "Giant field enhancement in photonic resonant lattices," *Physical Review B*, vol. 92, p. 155117, October 2015. [Online]. Available: https://link.aps.org/doi/10.1103/PhysRevB.92.155117

[644] S. Romano, G. Zito, S. Torino, et al., "Label-free sensing of ultralow-weight molecules with all-dielectric metasurfaces supporting bound states in the continuum," *Photonics Research*, vol. 6, no. 7, pp. 726–733, July 2018. [Online]. Available: http://www.osapublishing.org/prj/abstract.cfm?URI=prj-6-7-726

[645] D. S. Wiersma, M. P. van Albada, B. A. van Tiggelen, and A. Lagendijk, "Experimental evidence for recurrent multiple scattering events of light in disordered media," *Physical Review Letters*, vol. 74, pp. 4193–4196, May 1995. [Online]. Available: https://link.aps.org/doi/10.1103/PhysRevLett.74.4193

[646] L. Foldy, "The multiple scattering of waves. I. General theory of isotropic scattering by randomly distributed scatterers," *Physical Review*, vol. 67, pp. 107–119, 1945.

[647] M. Lax, "Multiple scattering of waves," *Reviews of Modern Physics*, vol. 23, pp. 287–310, 1951.

[648] L. D. Landau and E. M. Lifshitz, *Quantum Mechanics: Non-Relativistic Theory*, 3rd ed. Butterworth-Heinemann, Oxford 2003.

[649] M. Rusek, J. Mostowski, and A. Orlowski, "Random green matrices: from proximity resonances to Anderson localization," *Physical Review A*, vol. 61, pp. 022 704–1–022 704–6, 2000.

[650] F. A. Pinheiro, M. Rusek, A. Orlowski, and B. A. van Tiggelen, "Probing Anderson localization of light via decay rate statistics," *Physical Review E*, vol. 69, pp. 02 605–1–026 605–4, 2004.

[651] M. Rusek, A. Orlowski, and J. Mostowski, "Localization of light in three-dimensional random dielectric media," *Physical Review E*, vol. 53, pp. 4122–4130, 1996.

[652] M. Rusek and A. Orlowski, "Analytical approach to localization of electromagnetic waves in two-dimensional random media," *Physical Review E*, vol. 51, pp. R2763–R2766, 1995.

[653] A. Christofi, F. A. Pinheiro, and L. Dal Negro, "Probing scattering resonances of Vogel's spirals with the Green's matrix spectral method," *Optics Letters*, vol. 41, pp. 1933–1936, 2016.

[654] F. A. Pinheiro, "Statistics of quality factors in three-dimensional disordered magneto-optical systems and its applications to random lasers," *Physical Review A*, vol. 78, pp. 023 812-1-023 812-8, 2008.

[655] M. I. Mishchenko, *Electromagnetic Scattering by Particles and Particle Groups.* Cambridge University Press, New York, 2014.

[656] P. de Vries, D. V. van Coevorden, and A. Lagendijk, "Point scatterers for classical waves," *Reviews of Modern Physics*, vol. 70, pp. 447–466, 1998.

[657] B. T. Draine and P. J. Flatau, "Discrete-dipole approximation for scattering calculations," *Journal of the Optical Society of America A*, vol. 11, pp. 1491–1499, 1994.

[658] M. A. Chaumet and A. G. Hoekstra, "The discrete dipole approximation: an overview and recent developments," *Journal of Quantitative Spectroscopy and Radiative Transfer*, vol. 106, pp. 558–589, 2007.

[659] J. J. Goodman, B. T. Draine, and P. J. Flatau, "Application of fast-Fourier-transform techniques to the discrete-dipole approximation," *Optics Letters*, vol. 16, pp. 1198–1200, 1991.

[660] T. Feng, Y. Xu, W. Zhang, and A. E. Miroshnichenko, "Ideal magnetic dipole scattering," *Physical Review Letters*, vol. 118, no. 173901, pp. 1–6, 2017.

[661] A. I. Kuznetsov, A. E. Miroshnichenko, Y. H. Fu, J. Zhang, and B. Luk'yanchuk, "Magnetic light," *Scientific Reports*, vol. 2, p. 492, 2012.

[662] A. García-Etxarri, R. Gómez-Medina, L. S. Froufe-Pérez, et al., "Strong magnetic response of submicron silicon particles in the infrared," *Optics Express*, vol. 19, pp. 4815–4825, 2011.

[663] S. Kruk and Y. Kivshar, "Functional Meta-optics and nanophotonics governed by Mie resonances," *ACS Photonics*, vol. 4, no. 11, pp. 2638–2649, 2017. [Online]. Available: https://doi.org/10.1021/acsphotonics.7b01038

[664] I. Sersic, C. Tuambilangana, T. Kampfrath, and A. F. Koenderink, "Magnetoelectric point scattering theory for metamaterial scatterers," *Physical Review B*, vol. 83, p. 245102, June 2011. [Online]. Available: https://link.aps.org/doi/10.1103/PhysRevB.83.245102

[665] S. D. Jenkins and J. Ruostekoski, "Theoretical formalism for collective electromagnetic response of discrete metamaterial systems," *Physical Review B*, vol. 86, p. 085116, August 2012. [Online]. Available: https://link.aps.org/doi/10.1103/PhysRevB.86.085116

[666] S. D. Jenkins, J. Ruostekoski, Nust Papasimakis, S. Savo, and N. I. Zheludev, "Many-body subradiant excitations in metamaterial arrays: experiment and theory," *Physical Review Letters*, vol. 119, p. 053901, Aug 2017. [Online]. Available: https://link.aps.org/doi/10.1103/PhysRevLett.119.053901

[667] P. C. Chaumet and A. Rahmani, "Coupled-dipole method for magnetic and negative refraction materials," *Journal of Quantitative Spectroscopy and Radiative Transfer*, vol. 110, pp. 22–29, 2009.

[668] G. W. Mulholland, C. F. Bohren, and K. A. Fuller, "Light scattering by agglomerates: coupled electric and magnetic dipole method," *LANGMUIR*, vol. 10, pp. 2533–2546, 1994.

[669] M. A. Green, C. F. Bohren, and K. A. Fuller, "Self-consistent optical parameters of intrinsic silicon at 300 K including temperature coefficients," *Solar Energy Materials and Solar Cells*, vol. 92, pp. 1305–1310, 2008.

[670] B. García-Camara, F. Moreno, F. González, and O. J. F. Martin, "Light scattering by an array of electric and mangetic nanoparticles," *Optics Express*, vol. 18, pp. 10001–10015, 2010.

[671] B. García-Camara, R. Alcaraz de la Osa, J. Saiz, F. González, and F. Moreno, "Directionality in scattering by nanoparticles: Kerker's null-scattering conditions revisited," *Optics Letters*, vol. 36, pp. 728–730, 2011.

[672] S. Person, M. Jain, Z. Lapin, J. J. Sáenz, G. Wicks, and L. Novotny, "Demonstration of zero optical backscattering from single nanoparticles," *Nano Letters*, vol. 13, pp. 1806–1809, 2013.

[673] N. Liu, S. Mukherjee, K. Bao, et al., "Magnetic plasmon formation and propagation in artificial aromatic molecules," *Nano Letters*, vol. 12, pp. 364–369, 2012.

[674] J. M. Wylie and J. E. Sipe, "Quantum electrodynamics near an interface," *Physical Review A*, vol. 30, pp. 1185–1193, 1984.

[675] R. Pierrat and R. Carminati, "Spontaneous decay rate of a dipole emitter in a strongly scattering disordered environment," *Physical Review A*, vol. 81, p. 063802, 2010.

[676] C. Chamon and C. Mudry, "Density of states for dirty d-wave superconductors: a unified and dual approach for different types of disorder," *Physical Review B*, vol. 63, p. 100503(R), 2001.

[677] T. S. Grigera, V. Martin-Mayor, G. Parisi, and P. Verrocchio, "Phonon interpretation of the boson peak in supercooled liquids," *Nature*, vol. 422, p. 289, 2003.

[678] A. Amir, Y. Oreg, and Y. Imry, "Localization, anomalous diffusion, and slow relaxations: a random distance matrix approach," *Physical Review Letters*, vol. 105, p. 070601, 2010.

[679] C. E. Maximo, N. Piovella, P. W. Courteille, R. Kaiser, and R. Bachelard, "Spatial and temporal localization of light in two dimensions," *Physical Review A*, vol. 92, p. 062702, 2015.

[680] S. E. Skipetrov, "Finite-size scaling analysis of localization transition for scalar waves in a three-dimensional ensemble of resonant point scatterers," *Physical Review B*, vol. 94, p. 064202, 2016.

[681] P. Massignan and Y. Castin, "Three-dimensional strong localization of matter waves by scattering from atoms in a lattice with a confinement-induced resonance," *Physical Review A*, vol. 74, p. 013616, 2006.

[682] F. A. Pinheiro and L. C. Sampaio, "Lasing threshold of diffusive random lasers in three dimensions," *Physical Review A*, vol. 73, p. 013826, 2006.

[683] A. Goetschy and S. E. Skipetrov, "Euclidean matrix theory of random lasing in a cloud of cold atoms," *Europhysics Letters*, vol. 96, p. 34005, 2011.

[684] B. Gremaud and T. Wellens, "Speckle instability: coherent effects in nonlinear disordered media," *Physical Review Letters*, vol. 104, p. 133901, 2010.

[685] E. Akkermans, A. Gero, and R. Kaiser, "Photon localization and Dicke superradiance in atomic gases," *Physical Review Letters*, vol. 101, p. 103602, 2008.

[686] A. A. Svidzinsky, J. T. Chang, and M. O. Scully, "Cooperative spontaneous emission of n atoms: many-body eigenstates, the effect of virtual Lamb shift processes, and analogy with radiation of n classical oscillators," *Physical Review A*, vol. 81, p. 053821, 2010.

[687] L. Bellando, A. Gero, E. Akkermans, and R. Kaiser, "Cooperative effects and disorder: a scaling analysis of the spectrum of the effective atomic hamiltonian," *Physical Review A*, vol. 90, p. 063822, 2014.

[688] S. E. Skipetrov and I. M. Sokolov, "Magnetic-field-driven localization of light in a cold-atom gas," *Physical Review Letters*, vol. 114, p. 053902, 2015.

[689] B. A. van Tiggelen, R. Maynard, and T. M. Nieuwenhuizen, "Theory for multiple light scattering from Rayleigh scatterers in magnetic fields," *Physical Review E*, vol. 53, p. 2881, 1996.

[690] A. Lagendijk and B. A. van Tiggelen, "Resonant multiple scattering of light," *Physics Reports*, vol. 270, pp. 143–215, 1996.

[691] S. E. Skipetrov, "Finite-size scaling of the density of states inside band gaps of ideal and disordered photonic crystals," *European Physical Journal B*, vol. 93, no. 70, pp. 1–6, 2020.

[692] V. Efimov, "Energy levels arising from resonant two-body forces in a three-body system," *Physical Letters B*, vol. 33, p. 563, 1970.

[693] M. E. Zoorob, M. D. B. Carlton, G. J. Parker, J. J. Baumberg, and Netti, "Complete photonic bandgaps in 12-fold symmetric quasicrystals," *Nature*, vol. 404, pp. 740–743, 2000.

[694] F. Haake, *Quantum Signatures of Chaos*. Springer-Verlag, 2001.

[695] J. T. Chalker, I. V. Lerner, and R. A. Smith, "Random walks through the ensemble: linking spectral statistics with wave-function correlations in disordered metals," *Physical Review Letters*, vol. 77, pp. 554–557, 1996.

[696] S. S. Evangelou and J. L. Pichard, "Critical quantum chaos and the one-dimensional Harper model," *Physical Review Letters*, vol. 84, pp. 1643–1646, 2000.

[697] E. Maciá, "On the nature of electronic wave functions in one-dimensional self-similar and quasiperiodic systems," *ISRN Condensed Matter Physics*, vol. ID 165943, pp. 1–35, 2014.

[698] G. Labeyrie, F. de Tomasi, J.-C. Bernard, C. A. Müller, C. Miniatura, and R. Kaiser, "Coherent backscattering of light by cold atoms," *Physical Review Letters*, vol. 83, pp. 5266–5269, December 1999. [Online]. Available: https://link.aps.org/doi/10.1103/PhysRevLett.83.5266

[699] J. Billy, V. Josse, Z. Zuo, et al., "Direct observation of Anderson localization of matter waves in a controlled disorder," *Nature*, vol. 453, pp. 891–894, 2008.

[700] G. Mie, "Beiträge zur optik trüber medien, speziell kolloidaler metallösungen," *Annalen der Physik*, vol. 330, pp. 377–445, 1908.

[701] A. Clebsch, "Ueber die reflexion an einer kughelfläche," *Journal für Mathematik*, vol. 61, pp. 195–262, 1863.

[702] L. Lorenz, "Lysbevaegelsen i og unustden for en af plane lysbolger belyst kugle," *Det Kongelige Danske Videnskabernes Selskabs Skrifter*, vol. 6, pp. 1–62, 1890.

[703] P. Debye, "Der lichtdruck auf kugeln von beliebigem material," *Annalen der Physik*, vol. 335, pp. 57–136, 1909.

[704] W. Hergert and T. Wriedt, *The Mie Theory: Basics and Applications*. Springer-Verlag, Berlin, 2012.

[705] D. W. Mackowski, "Calculation of total cross sections of multiple-sphere clusters," *Journal of the Optical Society of America A*, vol. 11, pp. 2851–2861, 1991.

[706] Y.-l. Xu, "Electromagnetic scatteiring by an aggregate of spheres," *Applied Optics*, vol. 34, pp. 4573–4588, 1995.

[707] A. E. Krasnok, A. E. Miroshnichenko, P. A. Belov, and Y. S. Kivshar, "Huygens optical elements and Yagi-Uda nanoantennas based on dielectric nanoparticles," *JETP Letters*, vol. 94, no. 8, pp. 593–598, 2011.

[708] C. Pfeiffer and A. Grbic, "Metamaterial huygens' surfaces: tailoring wave fronts with reflectionless sheets," *Physical Review Letters*, vol. 110, no. 197401, pp. 1–5, 2013.

[709] Y. Chen and L. Dal Negro, "Pole-zero analysis of scattering resonances of multilayered nanospheres," *Physical Review B*, vol. 98, no. 235413, 2018.

[710] P. Grahn, A. Shevchenko, and M. Kaivola, "Electromagnetic multipole theory for optical metamaterials," *New Journal of Physics*, vol. 14, no. 093033, pp. 1–11, 2012.

[711] N. A. Butakov and J. A. Schuller, "Designing multipolar resonances in dielectric metamaterials," *Scientific Reports*, vol. 6, no. 38487, pp. 1–8, 2016.

[712] V. M. Shalaev and S. Kawata, *Nanophotonics with Surface Plasmons*. Elsevier, Amsterdam, 2007.

[713] U. Kreibig and M. Vollmer, *Optical Properties of Metal Clusters*. Springer-Verlag, Berlin, 1995.

[714] M. Kerker, D. S. Wang, and C. L. Giles, "Electromagnetic scattering by magnetic spheres," *Journal of the Optical Society of America*, vol. 73, pp. 765–767, 1983.

[715] J. Olmos-Trigo, C. Sanz-Fernández, D. R. Abujetas, et al., "Kerker conditions upon lossless, absorption, and optical gain regimes," *Physical Review Letters*, vol. 125, p. 073205, Aug 2020. [Online]. Available: https://link.aps.org/doi/10.1103/PhysRevLett .125.073205

[716] W. Liu and Y. S. Kivshar, "Generalized Kerker effects in nanophotonics and meta-optics [invited]," *Optics Express*, vol. 26, no. 10, pp. 13 085–13 105, May 2018. [Online]. Available: www.opticsexpress.org/abstract.cfm?URI=oe-26-10-13085

[717] M. S. Wheeler, J. S. Aitchison, and M. Mokahedi, "Coated nonmagnetic spheres with a negative index of refraction at infrared frequencies," *Physical Review B*, vol. 73, p. 045105, 2006.

[718] W. T. Doyle, "Optical properties of a suspension of metal spheres," *Physical Review B*, vol. 39, pp. 9852–9858, 1989.

[719] M. S. Wheeler, J. S. Aitchison, and M. Mokahedi, "Coated nonmagnetic spheres with a negative index of refraction at infrared frequencies," *Physical Review B*, vol. 72, p. 193103, 2005.

[720] D. Gagnon and L. J. Dube, "Lorenz–Mie theory for 2D scattering and resonance calculations," *Journal of Optics*, vol. 17, p. 103501, 2015.

[721] A. A. Asatryan, K. Busch, R. C. McPhedran, L. C. Botten, C. M. de Sterke, and N. A. Nicorovici, "Two-dimensional green tensor and local density of states in finite-sized two-dimensional photonic crystals," *Waves in Random Media*, vol. 3, pp. 9–25, 2013.

[722] P. A. Martin, *Multiple Scattering. Interaction of Time-Harmonic Waves with N Obstacles*. Cambridge University Press, New York, 2006.

[723] S. Gorsky and L. Dal Negro, "2D generalized Lorentz–Mie theory with a MATLAB accompaniment," *internal report*, 2020.

[724] A. Cerjan, y. D. Chong, and A. D. Stone, "Steady-state ab initio laser theory for complex gain media," *Optics Express*, vol. 23, no. 5, pp. 6455–6477, 2015.

[725] M. Quinten, A. Leitner, J. R. Krenn, and F. R. Aussnegg, "Electromagnetic energy transfer via linear chains of silver nanoparticles," *Optics Letters*, vol. 23, pp. 1331–1333, 1998.

[726] S. A. Maier, P. G. Kik, H. A. Atwater, S. Meltzer, and A. A. G. Requicha, "Local detection of electromagnetic energy transport below the diffraction limit in metal nanoparticle plasmon waveguides," *Nature Materials*, vol. 2, pp. 229–232, 2003.

[727] J. A. Fan, C. Wu, B. K., J. Bao, et al., "Self-assembled plasmonic nanoparticle clusters," *Science*, vol. 328, pp. 1135–1138, 2010.

[728] P. Nordlander, C. Oubre, E. Prodan, K. Li, and N. I. Stockman, "Plasmon hybridization in nanoparticle dimers," *Nano Letters*, vol. 4, pp. 899–903, 2004.

[729] H. Wang, D. W. Brandl, P. Nordlander, and N. J. Halas, "Plasmonic nanostructures: artificial molecules," *Accounts of Chemical Research*, vol. 40, pp. 53–62, 2007.

[730] A. Pasquale, B. M. Reinhard, and L. Dal Negro, "Engineering photonic-plasmonic coupling in metal nanoparticle necklaces," *ACS Nano*, vol. 5, pp. 6578–6585, 2011.

[731] J. Ye, F. Wen, H. Sobhani, et al., "Plasmonic nanoclusters: near field properties of the Fano resonance interrogated with sers," *Nano Letters*, vol. 12, pp. 1660–1667, 2012.

[732] M. Hentschel, S. M., R. Vogelgesang, H. Giessen, A. P. Alivisatos, and N. Liu, "Transition from isolated to collective modes in plasmonic oligomers," *Nano Letters*, vol. 10, pp. 2721–2726, 2010.

[733] N. A. Mirin, B. K., and P. Nordlander, "Fano resonances in plasmonic nanoparticle aggregates," *Journal of Physical Chemistry A*, vol. 113, pp. 4028–4034, 2009.

[734] A. Pasquale, B. M. Reinhard, and L. Dal Negro, "Concentric necklace nanolenses for optical near-field focusing and enhancement," *ACS Nano*, vol. 6, pp. 4341–4348, 2012.

[735] P. Marcoš and C. M. Soukoulis, *Wave Propagation. From Electrons to Photonic Crystals and Left-Handed Materials*. Princeton University Press, Princeton, 2008.

[736] I. A. Sukhoivanov and I. V. Guryev, *Photonic Crystals. Physics and Practical Modeling*. Springer, Berlin, 2009.

[737] R. M. de al Rue and C. Seassal, "Photonic crystal devices: some basics and selected topics," *Laser and Photonics Reviews*, vol. 6, pp. 564–587, 2012.

[738] O. Trojak, S. Gorsky, F. Sgrignuoli, et al., "Cavity quantum electro-dynamics with solid-state emitters in aperiodic nano-photonic spiral devices," *Applied Physics Letters*, vol. 117, no. 12, 124006, 2020.

[739] O. J. Trojak, S. Gorsky, C. Murray, et al., "Cavity-enhanced light-matter interaction in Vogel-spiral devices as a platform for quantum photonics." *Applied Physics Letters*, vol. 118, no. 1, 011103, 2021.

[740] L. Dal Negro, Y. Chen, S. Gorsky, and F. Sgrignuoli, "Aperiodic bandgap structures for enhanced quantum two-photon sources," *Journal of the Optical Society of America B*, 38, C94–C104 (2021).

[741] W. J. Wiscombe, "Improved Mie scattering algorithms," *Applied Optics*, vol. 19, pp. 1505–1509, 1980.

[742] Y.-l. Xu, "Electromagnetic scattering by an aggregate of spheres: far-field," *Applied Optics*, vol. 36, pp. 9496–9508, 1997.

[743] Y.-l. Xu, "Electromagnetic scattering by an aggregate of spheres: asymmetry parameter," *Physical Letters A*, vol. 249, pp. 30–36, 1998.

[744] Y.-l. Xu, "Electromagnetic scattering by an aggregate of spheres: theoretical and experimental study of the amplitude scattering matrix," *Physical Review E*, vol. 58, pp. 3931–3948, 1998.

[745] A. Egel, L. Pattelli, G. Mazzamuto, D. S. Wiersma, and U. Lemmer, "Celes: Cuda-accelerated simulation of electromagnetic scattering by large ensembles of spheres," *Journal of Quantitative Spectroscopy and Radiative Transfer*, vol. 199, pp. 103–110, 2017. [Online]. Available: www.sciencedirect.com/science/article/pii/S0022407317301772

[746] D. Mackowski and M. Mishchenko, "A multiple sphere T-matrix Fortran code for use on parallel computer clusters," *Journal of Quantitative Spectroscopy and*

Radiative Transfer, vol. 112, no. 13, pp. 2182–2192, 2011, [Online]. Available: www .sciencedirect.com/science/article/pii/S0022407311001129

[747] C. Forestiere, G. Miano, S. V. Boriskina, and L. Dal Negro, "The role of nanoparticle shapes and deterministic aperiodicity for the design of nanoplasmonic arrays," *Optics Express*, vol. 17, pp. 9648–9661, 2009.

[748] G. Iadarola, C. Forestiere, L. Dal Negro, F. Villone, and G. Miano, "GPU-accelerated T-matrix algorithm for light-scattering simulations," *Journal of Computational Physics*, vol. 231, pp. 5640–5652, 2012.

[749] M. I. Mishchenko, L. D. Travis, and D. W. Mackowski, "T-matrix computations of light scattering by nonspherical particles: a review," *Journal of Quantitative Spectroscopy and Radiative Transfer*, vol. 55, pp. 535–575, 1996.

[750] D. W. Mackowski and M. I. Mishchenko, "Calculation of the T matrix and the scattering matrix for ensembles of spheres," *Journal of the Optical Society of America A*, vol. 13, pp. 2266–2278, 1996.

[751] E. Abrahams, Ed., *50 Years of Anderson Localization*. World Scientific Publishing, Singapore, 2010.

[752] S. Harris, *An Introduction to the Theory of the Boltzmann Equation*. Holt, Rinehart, and Winston, New York, 1971.

[753] S. M. Rytov, Y. A. Kravstov, and V. I. Tatarskii, *Principles of Statistical Radiophysics 4. Wave Propagation Through Random Media*. Springer-Verlag, Berlin, 1989.

[754] V. Twersky, "On propagation in random media of discrete scatterers," *Proceedings of Symposia in Applied Mathematics*, vol. 16, pp. 84–116, 1964.

[755] A. F. Ioffe and A. R. Regel, "Non-crystalline, amorphous, and liquid electronic semiconductors," *Progress in Semiconductors*, vol. 4, pp. 237–291, 1960.

[756] F. Borghese, P. Denti, and R. Saija, *Scattering from Model Nonspherical Particles: Theory and Applications to Environmental Physics*, 2nd ed. Springer-Verlag, Berlin, 2007.

[757] B. A. van Tiggelen, *Multiple Scattering and Localization of Light*. Ph.D. Thesis, University of Amsterdam, 1992.

[758] D. S. Lemons, *An Introduction to Stochastic Processes in Physics*. The Johns Hopkins University Press, Baltimore, 2002.

[759] I. Lux and L. Koblinger, *Monte Carlo Particle Transport Methods: Neutron and Photon Calculations*. CRC Press, Boston, 1991.

[760] L. V. Wang and H.-I. Wu, *Biomedical Optics. Principles and Imaging*. John Wiley and Sons, Hoboken, 2007.

[761] G. Placzek and W. Seidel, "Milne's problem in transyort theory," *Physical Review*, vol. 72, pp. 550–555, 1947.

[762] H. C. van de Hulst, *Multiple Light Scattering*. Academic Press, New York, 1980.

[763] J. X. Zhu, D. J. Pine, and D. A. Weitz, "Internal reflection of diffusive light in random media," *Physical Review A*, vol. 44, no. 6, pp. 3948–3959, 1991.

[764] G. B. Arfken, *Mathematical Methods for Physicists*, 5th ed. Academic Press, San Diego, 2001.

[765] B. A. van Tiggelen, A. Lagendijk, and D. S. Wiersma, "Reflection and transmission of waves near the localization threshold," *Physical Review Letters*, vol. 84, pp. 4333–4336, May 2000. [Online]. Available: https://link.aps.org/doi/10.1103/PhysRevLett.84.4333

[766] P. Barthelemy, *Light transport beyond diffusion*. PhD Thesis, 2009.

[767] A. Z. Genack and J. M. Drake, "Relationship between optical intensity, fluctuations and pulse propagation in random media," *Europhysics Letters*, vol. 11, no. 4, pp. 331–336, 1990.

[768] D. Vollhardt and P. Wölfle, "Diagrammatic, self-consistent treatment of the Anderson localization problem in $d \leq 2$ dimensions," *Physical Review B*, vol. 22, pp. 4666–4679, 1980. [Online]. Available: https://link.aps.org/doi/10.1103/PhysRevB.22.4666

[769] W. Gotze, "A theory for the conductivity of a fermion gas moving in a strong three-dimensional random potential," *Journal of Physics C: Solid State Physics*, vol. 12, no. 7, pp. 1279–1296, 1979. [Online]. Available: https://doi.org/10.1088%2F0022-3719%2F12%2F7%2F018

[770] E. Akkermans, P. E. Wolf, and R. Maynard, "Coherent backscattering of light by disordered media: analysis of the peak line shape," *Physical Review Letters*, vol. 56, pp. 1471–1474, April 1986. [Online]. Available: https://link.aps.org/doi/10.1103/PhysRevLett.56.1471

[771] M. B. van der Mark, M. P. van Albada, and A. Lagendijk, "Light scattering in strongly scattering media: multiple scattering and weak localization," *Physical Review B*, vol. 37, pp. 3575–3592, March 1988. [Online]. Available: https://link.aps.org/doi/10.1103/PhysRevB.37.3575

[772] F. J. P. Schuurmans, M. Megens, D. Vanmaekelbergh, and A. Lagendijk, "Light scattering near the localization transition in macroporous gap networks," *Physical Review Letters*, vol. 83, pp. 2183–2186, September 1999. [Online]. Available: https://link.aps.org/doi/10.1103/PhysRevLett.83.2183

[773] D. S. Wiersma, P. Bertolini, A. Lagendijk, and R. Righini, "Localization of light in a disordered medium," *Nature*, vol. 390, pp. 671–673, 1997.

[774] E. Abrahams and P. A. Lee, "Scaling description of the dielectric function near the mobility edge," *Physical Review B*, vol. 33, pp. 683–689, January 1986. [Online]. Available: https://link.aps.org/doi/10.1103/PhysRevB.33.683

[775] E. Abrahams, P. W. Anderson, D. C. Licciardello, and T. V. Ramakrishnan, "Scaling theory of localization: absence of quantum diffusion in two dimensions," *Physical Review Letters*, vol. 42, pp. 673–676, March 1979. [Online]. Available: https://link.aps.org/doi/10.1103/PhysRevLett.42.673

[776] A. Lagendijk, B. A. van Tiggelen, and D. S. Wiersma, "Fifty years of Anderson localization," *Physics Today*, vol. 62, no. 8, pp. 24–29, 2009.

[777] S. Hikami, "Anderson localization in a nonlinear-σ-model representation," *Physical Review B*, vol. 24, pp. 2671–2679, September 1981. [Online]. Available: https://link.aps.org/doi/10.1103/PhysRevB.24.2671

[778] C. Tian, "Supersymmetric field theory of local light diffusion in semi-infinite media," *Physical Review B*, vol. 77, p. 064205, February 2008. [Online]. Available: https://link.aps.org/doi/10.1103/PhysRevB.77.064205

[779] M. Filoche and S. Mayboroda, "Universal mechanism for Anderson and weak localization," *PNAS*, vol. 109, no. 37, pp. 14761–14766, 2012. [Online]. Available: https://www.pnas.org/content/109/37/14761

[780] D. N. Arnold, G. David, M. Filoche, D. Jerison, and S. Mayboroda, "Computing spectra without solving eigenvalue problems," *SIAM Journal on Scientific Computing*, vol. 41, no. 1, pp. B69–B92, 2019. [Online]. Available: https://doi.org/10.1137/17M1156721

[781] D. N. Arnold, G. David, D. Jerison, S. Mayboroda, and M. Filoche, "Effective confining potential of quantum states in disordered media," *Physical Review Letters*,

vol. 116, p. 056602, February 2016. [Online]. Available: https://link.aps.org/doi/10.1103/PhysRevLett.116.056602

[782] M. Filoche, M. Piccardo, Y.-R. Wu, C.-K. Li, C. Weisbuch, and S. Mayboroda, "Localization landscape theory of disorder in semiconductors. I. Theory and modeling," *Physical Review B*, vol. 95, p. 144204, Apr 2017. [Online]. Available: https://link.aps.org/doi/10.1103/PhysRevB.95.144204

[783] N. F. Mott, "Electrons in disordered structures," *Advanced Physics*, vol. 16, no. 61, pp. 49–144, 1967.

[784] C. Di Castro and R. Raimondi, *Statistical Mechanics and Applications in Condensed Matter*. Cambridge University Press, 2015.

[785] A. Z. Genack and N. Garcia, "Observation of photon localization in a three-dimensional disordered system," *Physical Review Letters*, vol. 66, pp. 2064–2067, April 1991. [Online]. Available: https://link.aps.org/doi/10.1103/PhysRevLett.66.2064

[786] A. A. Chabanov, M. Stoytchev, and A. Z. Genack, "Observation of photon localization in a three-dimensional disordered system," *Nature*, vol. 404, pp. 850–853, 2000.

[787] J. Wang and A. Z. Genack, "Transport through modes in random media," *Nature*, vol. 471, pp. 345–348, 2011. [Online]. Available: https://doi.org/10.1038/nature09824

[788] M. Davy and A. Z. Genack, "Selectively exciting quasi-normal modes in open disordered systems," *Nature Communications*, vol. 9, 2018. [Online]. Available: https://doi.org/10.1038/s41467-018-07180-3

[789] D. J. Thouless, "Electrons in disordered systems and the theory of localization," *Physics Reports*, vol. 13, pp. 93–142, 1974.

[790] V. S. Letokhov, "Generation of light by a scattering medium with negative resonance absorption," *Soviet Journal of Experimental and Theoretical Physics*, vol. 26, p. 835, 1968.

[791] H. Cao, "Lasing in random media," *Waves in Random Media*, vol. 13, no. 3, pp. R1–R39, 2003.

[792] H. Cao, "Review on latest developments in random lasers with coherent feedback," *Journal of Physics A: Mathematical and General*, vol. 38, pp. 10 497–10 535, 2005.

[793] N. Forbes and B. Mahon, *Faraday, Maxwell, and the Electromangetic Field*. Prometheous Books, Amherst, 2014.

[794] S. Ross Sheldon, *Stochastic Processes*, 2nd ed. John Wiley and Sons, Inc., New York, 1996.

[795] W. Paul and J. Baschnagel, *Stochastic Processes*, 2nd ed. Springer-Verlag, New York, 2013.

[796] Z. Schuss, *Theory and Applications of Stochastic Processes*. Springer, New York, 2010.

[797] W. C. van Etten, *Introduction to Random Signals and Noise*. John Wiley and Sons, 2005.

[798] D. T. Gillespie, "The mathematics of Brownian motion and Johnson noise," *American Journal of Physics*, vol. 64, no. 3, pp. 225–240, 1996.

[799] Z. Zhang and G. E. Karniadakis, *Numerical Methods for Stochastic Partial Differential Equations with White Noise*. Springer International Publishing AG, Gewerbestrasse, 2017.

[800] P. Häunggi and P. Jung, "Colored noise in nynamical systems," in *Advances in Chemical Physics*, I Prigogine and S. A. Rice, Ed. Wiley Online Library, 1995. [Online] Available: https://onlinelibrary.wiley.com/doi/abs/10.1002/9780470141489.ch4

[801] B. Mandelbrot and J. W. van Ness, "Fractional Brownian motions, fractional noises and applications," *SIAM Review*, vol. 10, no. 4, p. 422–437, 1968.

[802] G. J. Lord, C. E. Powell, and T. Shardlow, *An Introduction to Computational Stochastic PDEs*. Cambridge University Press, New York, 2014.

[803] D. J. Higham, "An algorithmic introduction to numerical simulation of stochastic differential equations," *SIAM Review*, vol. 43, no. 3, pp. 525–546, 2001.

[804] N. G. Van Kampen, *Stochastic Processes in Physics and Chemistry*, 3rd ed. Elsevier, Amsterdam, 2007.

[805] M. Jardak, C. H. Su, and G. E. Karniadakis, "Spectral polynomial chaos solutions of the stochastic advection equation," *Journal of Scientific Computing*, vol. 17, p. 319–338, 2002.

[806] E. Limpert, W. Stahel, and M. Abbt, "Lognormal distributions across the sciences: keys and clues," *BioScience*, vol. 51, no. 5, p. 341–352, 2001.

[807] E. L. Crow and K. Shimizu, *Lognormal Distributions: Theory and Applications*. Dekker, New York, 1988.

[808] E. L. Albuquerque and M. G. Cottam, "Theory of elementary excitations in quasiperiodic structures," *Physics Reports*, vol. 379, pp. 225–337, 2003.

[809] P. Grassberger and I. Procaccia, "Characterization of strange attractors," *Physical Review Letters*, vol. 50, no. 5, pp. 346–349, 1983.

[810] T. Nakayama and K. Yakubo, *Fractal Concepts in Condensed Matter Physics*, vol. 140. Springer Science & Business Media, New York, 2013.

[811] T. C. Halsey, P. Meakin, and I. Procaccia, "Scaling structure of the surface layer of diffusion-limited aggregates," *Physical Review Letters*, vol. 56, no. 8, pp. 854–857, 1986.

[812] A. Karperien, "Fraclac for ImageJ, Version 2.5." 1999–2007.

[813] W. S. Rasband, *ImageJ*, http://imagej.nih.gov/ij/. U. S. National Institutes of Health, Bethesda, 1997–2011.

[814] M. E. J. Newman, *Networks. An Introduction*. Oxford University Press, New York, 2010.

[815] E. Estrada and P. Knight, *A First Course in Network Theory*. Oxford University Press, New York, 2015.

[816] R. Gorenflo, A. A. Kilbas, F. Mainardi, and S. V. Rogosin, *Mittag–Leffler Functions, Related Topics and Applications*, vol. 2. Springer-Verlag, Berlin and Heidelberg, 2014.

[817] F. Mainardi, *Fractional calculus and waves in linear viscoelasticity: an introduction to mathematical models*. World Scientific, 2010.

[818] D. Baleanu, K. Diethelm, E. Scalas, and J. J. Trujillo, *Fractional calculus: models and numerical methods*. World Scientific, 2012, vol. 3.

[819] F. Mainardi, G. Pagnini, and R. Saxena, "Fox h functions in fractional diffusion," *Journal of Computational and Applied Mathematics*, vol. 178, no. 1, pp. 321–331, 2005, Proceedings of the Seventh International Symposium on Orthogonal Polynomials, Special Functions and Applications. [Online]. Available: www.sciencedirect.com/science/article/pii/S0377042704003826

[820] J. Abate and W. Whitt, "A unified framework for numerically inverting laplace transforms," *INFORMS Journal on Computing*, vol. 18, no. 4, pp. 408–421, 2006. [Online]. Available: https://doi.org/10.1287/ijoc.1050.0137

[821] J. D. Buckholtz, "Concerning an approximation of Copson," *Proceedings of the American Mathematical Society*, vol. 14, no. 4, pp. 564–568, 1963.

[822] A. V. Chechkin, J. Klafter, and I. M. Sokolov, "Fractional Fokker–Planck equation for ultraslow kinetics," *Europhysics Letters*, vol. 63, no. 3, pp. 326–332, August 2003. [Online]. Available: https://doi.org/10.1209%2Fepl%2Fi2003-00539-0

Index